WITHDRAWN FROM
TSC LIBRARY

REFERENCE
DO NOT REMOVE
FROM LIBRARY

Encyclopedia of
Antarctica and the Southern Oceans

Encyclopedia of
Antarctica and the Southern Oceans

Edited by

B. Stonehouse
Scott Polar Research Institute

JOHN WILEY & SONS, LTD

Copyright © 2002   John Wiley & Sons Ltd, The Atrium, Southern Gate, Chichester,
West Sussex PO19 8SQ, England

Telephone (+44) 1243 779777

Email (for orders and customer service enquiries): cs-books@wiley.co.uk
Visit our Home Page on www.wileyeurope.com or www.wiley.com

All Rights Reserved. No part of this publication may be reproduced, stored in a retrieval system or transmitted in any form or by any means, electronic, mechanical, photocopying, recording, scanning or otherwise, except under the terms of the Copyright, Designs and Patents Act 1988 or under the terms of a licence issued by the Copyright Licensing Agency Ltd, 90 Tottenham Court Road, London W1T 4LP, UK, without the permission in writing of the Publisher. Requests to the Publisher should be addressed to the Permissions Department, John Wiley & Sons Ltd, The Atrium, Southern Gate, Chichester, West Sussex PO19 8SQ, England, or emailed to permreq@wiley.co.uk, or faxed to (+44) 1243 770571.

This publication is designed to provide accurate and authoritative information in regard to the subject matter covered. It is sold on the understanding that the Publisher is not engaged in rendering professional services. If professional advice or other expert assistance is required, the services of a competent professional should be sought.

Other Wiley Editorial Offices

John Wiley & Sons Inc., 111 River Street, Hoboken, NJ 07030, USA

Jossey-Bass, 989 Market Street, San Francisco, CA 94103-1741, USA

Wiley-VCH Verlag GmbH, Boschstr. 12, D-69469 Weinheim, Germany

John Wiley & Sons Australia Ltd, 33 Park Road, Milton, Queensland 4064, Australia

John Wiley & Sons (Asia) Pte Ltd, 2 Clementi Loop #02-01, Jin Xing Distripark, Singapore 129809

John Wiley & Sons Canada Ltd, 22 Worcester Road, Etobicoke, Ontario, Canada M9W 1L1

British Library Cataloguing in Publication Data

A catalogue record for this book is available from the British Library

ISBN 0-471-98665-8

Typeset in 10/12pt Times by Laserwords Private Limited, Chennai, India
Printed and bound in Great Britain by Antony Rowe Ltd, Chippenham, Wiltshire
This book is printed on acid-free paper responsibly manufactured from sustainable forestry in which at least two trees are planted for each one used for paper production.

Contents

List of Advisory Editors	vii
List of Contributors	ix
Introduction	xi
A–Z Entries	1–297

Appendix A: Agreed Measures for the Conservation of Antarctic Fauna and Flora	298	Study Guide: Climate and Life	345
		Study Guide: Exploration	348
Appendix B: Convention for the Conservation of Antarctic Seals	302	Study Guide: Geography	354
		Study Guide: Geology and Glaciology	358
Appendix C: Convention on the Conservation of Antarctic Marine Living Resources	307	Study Guide: Information Sources	360
		Study Guide: National Interests in Antarctica	364
Appendix D: Protocol on Environmental Protection to the Antarctic Treaty	315	Study Guide: Protected Areas under the Antarctic Treaty	369
Appendix E: Text of the Antarctic Treaty	336	Study Guide: Southern Oceans and Islands	374
Further reading	340	A–Z Listing of Encyclopedia Entries	378

List of Advisory Editors

Dr Peter D. Clarkson   Earth Sciences
Dr Klaus Dodds   Politics and Political Geography
Dr David J. Drewry   Glaciology
Robert K. Headland   Historical Geography
Dr John C. King   Climate and Atmosphere
Dr David W. H. Walton   Life Sciences

List of Contributors

Adie, Dr Raymond J. (RJA) Formerly: British Antarctic Survey, Cambridge UK

Bertram, Esther K. R. (EKRB) Dept. of Geography, Royal Holloway, Egham, Surrey, TW20 0CX UK.

Bonner, W. Nigel (WNB) Formerly: British Antarctic Survey. Deceased

Brigham, Dr Lawson W. (LWB) Scott Polar Research Institute, University of Cambridge

Clarke, Prof. Andrew C. (ACC) British Antarctic Survey: Natural Environment Research Council, High Cross, Madingley Road, Cambridge CB5 0ET UK

Clarkson, Dr Peter D. (PDC) Scientific Committee on Antarctic Research: Scott Polar Research Institute, Lensfield Road, Cambridge CB2 1ER UK

Crame, Dr James A. (JAC) British Antarctic Survey: Natural Environment Research Council, High Cross, Madingley Road, Cambridge CB5 0ET UK

Croxall, Dr John P. (JPC) British Antarctic Survey: Natural Environment Research Council, High Cross, Madingley Road, Cambridge CB5 0ET UK

Doake, Dr Christopher S. M. (CSMD) British Antarctic Survey: Natural Environment Research Council, High Cross, Madingley Road, Cambridge CB5 0ET UK

Dodds, Dr Klaus (KD) Dept. of Geography, Royal Holloway Egham, Surrey, TW20 0CX UK

Drewry, Dr David J. (DJD) Office of the Vice-Chancellor, University of Hull, Cottingham Road, Hull. HU6 7RX UK

Francis, Dr Jane E. (JEF) Dept. of Earth Science, University of Leeds, Leeds UK

Garrett, Dr S. W. (SWG) Chevron Canada Resources: 500, Fifth Ave. SW, Calgary, Alberta, Canada T2P 0LT

Hambrey, Prof. M. J. (MJH) Dept. of Earth Sciences, Moores University, Byrom St., Liverpool L3 3AF, UK

Headland, Robert K. (RKH) Scott Polar Research Institute: University of Cambridge, Lensfield Road, Cambridge CB2 1ER UK

Hermichen, Claudia (CH) Bennewitz, Germany. Formerly: Scott Polar Research Institute

Houston, Brent (BH) Bozeman, Montana, USA

Jones, Dr Anna (AJ) British Antarctic Survey: Natural Environment Research Council, High Cross, Madingley Road, Cambridge CB5 0ET UK

King, Dr John C. (JCK) British Antarctic Survey: Natural Environment Research Council, High Cross, Madingley Road, Cambridge CB5 0ET UK

Lewis Smith, Dr Ronald. I. (RILS) British Antarctic Survey: Natural Environment Research Council, High Cross, Madingley Road, Cambridge CB5 0ET UK

Lorius, Dr Claude (CL) Laboratoire de Glaciologie et de Géophysique de l'Environment, BP 96 38402 St Martin d'Heres, France

McIntyre, Dr Neil (NmcI) Mullard Space Science Laboratory: University College, London, Holmbury St. Mary, Dorking, Surrey RH5 6NT, UK

Mills, William J. (WJM) Scott Polar Research Institute: University of Cambridge, Lensfield Road, Cambridge CB2 1ER UK

Rees, Dr W. Gareth (WGR) Scott Polar Research Institute: University of Cambridge, Lensfield Road, Cambridge CB2 1ER UK

Stonehouse, Dr Bernard (BS) Scott Polar Research Institute: University of Cambridge, Lensfield Road, Cambridge CB2 1ER UK

Swithinbank, Dr Charles W. (CWS) Scott Polar Research Institute: University of Cambridge, Lensfield Road, Cambridge CB2 1ER UK

Walton, Dr D. W. H. (DWHW) British Antarctic Survey: Natural Environment Research Council, High Cross, Madingley Road, Cambridge CB5 0ET UK

West, Dr Janet (JW) Scott Polar Research Institute: University of Cambridge, Lensfield Road, Cambridge CB2 1ER UK

Introduction

A continent larger than Australia in an ocean broader than the Atlantic: between them Antarctica and the Southern Ocean cover more than one eleventh of Earth's surface. Antarctica, at the southern end of the world, differs in many ways from its six sister continents. It is the highest continent, by far the coldest and most remote, and unique in lacking indigenous human population. 150 years ago geographers knew it only as a tentative outline at the bottom of the world – an incomplete ring on polar projections, a squiggle on Mercator's charts. The interior was first penetrated less than a century ago. Much of the coast and practically all of the ice-bound interior were first mapped, from sledging traverses and aerial photographs, by explorers who are alive today.

The oceans and islands surrounding Antarctica have longer histories of human contact. The southern extremities of the Pacific, Atlantic and Indian oceans, and the great Southern Ocean upon which all three converge, were explored mainly during the seventeenth to nineteenth centuries. Many of their islands were discovered and charted by early nineteenth century sealers. Stripped of their seals, few of the islands offered opportunities for further exploitation or colonization: most were forgotten within a generation. Only two groups, the temperate Falkland Islands and Tristan da Cunha, came to support permanent habitation. The Auckland Islands, Campbell Island and Iles Kerguelen supported shorter-term farming or fishing enterprises.

Dearth of land resources has been balanced by wealth in the surrounding oceans. First to be exploited were fur seals and elephant seals, providing respectively skins and oil during the early nineteenth century. From 1904 South Georgia became an important centre of Southern Ocean whaling, and seven decades of both land-based and pelagic whaling yielded an almost continuous output of oil, protein, bone-meal, and latterly pharmaceutical products. When competition from vegetable oils hastened the collapse of whaling, attention shifted to the southern oceans' huge stocks of marketable fish, squid and crustaceans. The potential resource of krill awaits only the discovery of a profitable market.

Several of the southern islands became the sites of long-term research stations. Of these the oldest, Orcadas on the South Orkney Islands, will achieve its first centenary in 2003. From 1944 Antarctica itself began to support permanent stations. Currently the southern continent, islands and oceans are of great interest to scientists, an interest that has stimulated the development of a unique political regime for the region. Antarctica and its neighbouring islands south of 60°S are claimed by seven nations, but currently administered under the Antarctic Treaty by an international consortium involving over 40 nations. Islands and oceans north of 60°S remain outside the Treaty, though some of their living resources come under the influence of Treaty instruments and management regulations.

Currently, too, the area is growing in importance as an attraction for tourists. Antarctica is no longer the remote land at the southern end of the world. Both continent and oceans lie within a few hours' travel of populated areas. Tourism has discovered their accessibility during the summer months, their scenic beauty, and the fascination of their wildlife, geology and history of exploration. Numbers of passenger cruises and flights to the area are as yet small, but tourism is growing and diversifying – a stimulating challenge to the Antarctic Treaty System that aspires to manage the continent and its environs.

These are the geographical areas and some of the issues that we cover in 'Encyclopedia of Antarctica and the Southern Oceans'. The encyclopedia was conceived some 15 years ago in a different format and context, and has evolved into its present form during the past five years. The magnitude of the region covered, and the range of scientific disciplines and philosophical issues germane to it, provided a predictable initial challenge, matched by the speed at which new knowledge and understanding of the region have been accumulating. The function of an encyclopedia is to reflect lasting values within current understanding, rather than the ephemera of latest findings. The business of the compiling editor and advisory editors has been to sort one from the other – a difficult task when new information on the region accumulates daily. A further complication has been the fact that almost all the contributors and editors involved are researchers, spending several months of each year in the Antarctic or southern ocean regions. Their contemporary knowledge is reflected in the content of the encyclopedia, but assembling it required patience and forbearance on their part as well as mine. This has been matched by the patience of John Wiley's editorial and production team, who coped gently but firmly with our itinerant habit and published the encyclopedia on time.

An encyclopedia reports on, and gains substantially from, the published works of many who are not listed as direct contributors. The authenticity of geographical locations, place-names and their history within this work depends almost entirely on the outstanding scholarship of Geoffrey Hattersley-Smith and F. G. Alberts, whose key publications on Antarctic place-names (see Further Reading) are essential tools for every Antarctic researcher. Similarly the painstaking work of many historians (notably A. G. E. Jones on the early nineteenth century sealers and navigators, and R. K. Headland on the whole gamut of southern exploration), and of biologists, geographers, political scientists and others past and present who have contributed to Antarctic knowledge, is also recorded throughout this work. Inevitably it has not been possible to accord the chapter-and-verse recognition that would appear in research papers: authors instead have the satisfaction of knowing that much of what they so painstakingly discovered has now become common lore. In several entries, particular key works are listed as further reading. As this encyclopedia is intended for general readership as much as for scholars and researchers, I have restricted references mainly to books that will be accessible to the general public. Scholars who use the encyclopedia as a first reference will have little difficulty in finding the research papers on which their continuing studies depend: they and others may find the **Study Guide: Information Sources**, compiled by William Mills and Esther Bertram, particularly helpful.

Among those listed as contributors I thank especially Ray Adie, David Walton and Andrew Clarke, who have been associated with this venture from its inception. I thank also Peter Clarkson, a constant source of good advice on many topics, who also checked much of the typescript. Philip Stickler provided the map section; Esther Bertram was an invaluable assistant who produced the in-text maps and other contributions. Claudia Hermichen, while a library student at the Scott Polar Research Institute, brought order to the biographical section. R. K. Headland's encyclopedic knowledge of Antarctic history and current affairs has contributed extensively, and I thank him especially for checking most of the final typescript. Brent Houston at short notice strengthened and enhanced the ornithological entries. These among others have added their knowledge of Antarctica and helped to improve accuracy. Entries written by authors other than myself bear their names: those without names are my own, as are any remaining errors or omissions: I should welcome help from readers in amending them.

I have been fortunate throughout in being based at the Scott Polar Research Institute, University of Cambridge, with excellent library facilities and a circle of helpful colleagues. I thank Prof. K. Richards, former Director of the Institute, for allowing the publication of photographs from its archives. Royalties from this work will benefit the Institute's library and information services. I thank also many members of British Antarctic Survey, our neighbours across the city of Cambridge, some as contributors, others as advisors who have helped with quick responses to queries.

I thank the editorial staff at John Wiley & Sons, Ltd and Helen Heyes (proof-reader) and Susan Dunsmore (copy-editor). Lastly I thank my wife Sally, without whose constant support and forebearance this work could never have been completed.

Bernard Stonehouse
Editor

Aagaard, Bjarne. (1873–1956). Norwegian shipping manager and polar historian. Born in 1873, he went to sea at the age of fifteen, later managing shipping offices in Scotland, Germany and Hong Kong. On retiring in 1925 he devoted himself to Norwegian interests in Antarctica, publishing books and articles, and influencing both popular and political opinion toward Norway's annexation of **Bouvetøya**, **Peter I Øy** and **Dronning Maud Land**. His most scholarly work, *Fangst og forskning I Sydishavet* (*Hunting and Exploration in the Southern Ocean*), a four-volume study of whaling and exploration, was published between 1930 and 1950. Aagaard died on 29 Sept. 1956.
(CH/BS)

Abbot Ice Shelf. 72°45′S, 96°00′W. Massive ice shelf off Eights and Bryan coast of Ellsworth Land, West Antarctica, linking MacNamara, Dustin and Thurston Islands with the mainland. The shelf was named for Rear Admiral J. Lloyd Abbot Jr. USN.

Abbott, George P. (d.1923). British seaman and explorer. A petty officer in the Royal Navy, Abbott joined the **British Antarctic (*Terra Nova*) Expedition 1910–13**, serving with the Northern Party at Cape Adare and wintering on Inexpressible Island. He served in the navy throughout World War I.

Ablation. Processes by which mass, in the form of snow and ice, is lost from glaciers, floating ice and snow surfaces. The term is generally used to cover melting, sublimation, evaporation, and wind erosion, but does not exclude avalanching and calving.

Ablation Point. 70°48′S, 68°21′W. Oasis area on the east coast of Alexander Island, facing George VI Sound. An area of approximately 186 km² (73 sq miles), extending from 70°45′S to 70°55′S, and from 68°40′W to the coast, has been designated SSSI No. 29 (Ablation Point – Ganymede Heights). It shows a range of geomorphological features including raised beaches, moraines and patterned ground, with freshwater ponds supporting a diverse fauna.

Ablation Valley. 70°48′S, 68°30′W. A relatively ice-free valley on the east coast of Alexander Island, descending into George VI Sound.

Aboa Station. 73°03′S, 13°25′W. Finnish summer-only research station in the Kraulfjella, inland from Kronprinsesse Märtha Kyst of Dronning Maud Land.

Accumulation of snow and ice. Processes by which snow and ice are added to glaciers, floating ice and snow surfaces, including precipitation, air-borne drifting, condensation and physical transport by avalanching, and accretion of freezing water.

Adare, Cape. 71°17′S, 170°15′E. Prominent cape forming the eastern entrance to Robertson Bay, Pennell Coast, Victoria Land. It was the site of the first recorded landing in East Antarctica, by the **Norwegian (Tønsberg) Whaling Expedition 1893–95**, and of the first over-wintering on land, by the **British Antarctic (*Southern Cross*) Expedition 1898–1900**. The northern party of the **British Antarctic (*Terra Nova*) Expedition 1910–13** wintered there in 1911. Remains of the expedition huts are designated HSM No. 22, and an area of about 0.5 km² (0.2 sq miles) has been designated SPA No. 29 (Cape Adare), to allow the application of a management plan for the site.

Adelaide Island. 67°15′S, 68°30′W. Large island forming the northeastern side of Marguerite Bay, Antarctic Peninsula, first sighted in 1832 by British sealing captain John Biscoe and named by him for the contemporary British queen.

Adelaide (Station T). 67°46′S, 68°55′W. British Antarctic Survey research station and runway on the southwestern corner of Adelaide Island, Loubet Coast,

Graham Land. The station (Stephenson House) was built in February 1961 and further buildings were added in the following two seasons. It was occupied year-round as a centre for survey, geology, meteorology and flying operations, using a ski runway on the nearby ice piedmont, until March 1977, when this function was diverted to Rothera Station. In 1984 ownership was transferred to the government of Chile, which now operates it as Teniente Carvajal Station.

Adélie Coast. 67°00′S, 139°00′E. Coast of Terre Adélie, East Antarctica. Facing the Dumont d'Urville Sea and bounded by Point Alden in the west and Pourquoi Pas Point in the east, it consists mainly of ice cliffs with emergent archipelagos of small islands. The coast was discovered in 1840 by Capt. J. Dumont d'Urville, leader of the French Naval Expedition 1837–40, who named it for his wife. It is currently the site of the French permanent research station, **Dumont d'Urville Station**.

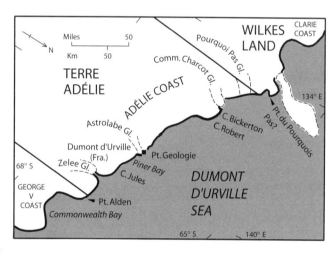

Adélie Coast

Adélie Land. See **Terre Adélie**.

Adélie penguin. *Pygoscelis adeliae*. Second-largest of the three species of pygoscelid (brush-tailed) penguins, adélies stand 70 cm (28 in) tall and weigh about 5 kg (11 lb). Distinguished by their feathered bill and white eye ring, they live mainly on rocky coasts of continental Antarctica and the near-Antarctic islands, feeding on krill and surface-living fish. They breed in colonies of several hundreds to many thousands of pairs, building nests of pebbles, bones and feathers during November and December. The two eggs are chalky-white, the chicks silver-grey, moulting to grey-brown. Breeding success depends largely on availability of food during a short period from December to February, which is intimately linked with fluctuations in the distribution of inshore sea ice. Adélies winter mainly on the pack ice far from land. World population is estimated at 2,600,000 pairs.

Admiralty Bay. 62°08′S, 58°27′W. Large bay on south side of King George Island, South Shetland Islands. Discovered about 1820, it was charted by the British sealer Capt. George Powell and named for the British Board of Admiralty. In the early twentieth century it provided moorings for whaling factory ships. It currently accommodates Henryk Arctowski and Commandante Ferraz, respectively Polish and Brazilian year-round research stations, Machu Picchu and Pieter J. Lenie, respectively Peruvian and United States summer-only stations, and an Ecuadorian refuge. An area of approximately 17.5 km² (6.8 sq. miles) along the western shore, immediately south of Henryk Arctowski Station, is designated SSSI No. 8 (Western shore of Admiralty Bay) to facilitate studies of birds and seals.

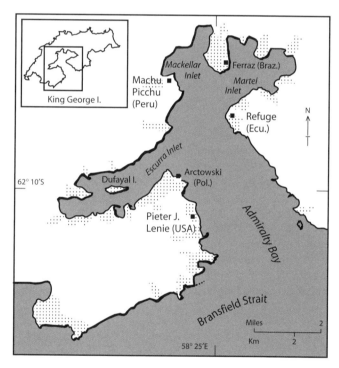

Admiralty Bay

Admiralty Bay (Base G). 62°05′S, 58°25′W. Former British research station at Martel Inlet, Keller Peninsula, King George Island, South Shetland Islands. The station was established by the Falkland Islands. Dependencies Survey in January 1947 and occupied for two months: it was then occupied continuously from the following January to January 1961, when it was closed. The huts were removed in 1995–96, leaving concrete foundations which have since accumulated an interesting collection of lichens.

Admiralty Mountains. 71°45′S, 168°30′E. Major mountain range of northeastern Victoria Land, with peaks rising above 4000 m (13 120 ft). It was named in 1841 by its discoverer, Capt. J. C. Ross RN, for the British Board of Admiralty.

Agreed Measures for the Conservation of Antarctic Fauna and Flora. Measures for Antarctic conservation drawn up in 1964 at the Third Consultative Meeting of the Antarctic Treaty. Generally known as the 'Agreed Measures', these were enhanced by additions arising from subsequent consultative meetings. The Treaty had originally identified 'the preservation and conservation of living resources in the Antarctic' as an issue of common interest, and the First Consultative Meeting in July 1961 provided, in Recommendation I–VIII, general rules of conduct for the preservation and conservation of living resources. The Agreed Measures provided more definitive statements of conservation objectives.

The text of the Agreed Measures appears in **Appendix A**. In summary: following general statements of policy in Articles I–V, Article VI of the Measures prohibited the killing, wounding or capturing of any native mammal or bird (except for whales, which were covered by the 1946 International Whaling Convention). Participating governments were authorized to permit killing or capture to provide (a) indispensable food for man or dogs; (b) specimens for scientific study or information; or (c) specimens for museums or zoological gardens. Permits needed to be specific, to ensure that no more animals were taken than could normally be replaced the following breeding season, and to ensure that the variety of species and balance of the particular ecosystem were maintained. Permits for taking fur seals of the genus *Arctocephalus* and Ross seals could be issued only for compelling scientific purposes.

Article VII required governments to take measures that would minimize harmful interference with normal living conditions of native mammals and birds, and to alleviate pollution of coastal waters. While it was recognized that the presence of a scientific station might cause inevitable disturbance, such activities as allowing dogs, helicopters, explosives or firearms to disturb nearby bird colonies were to be avoided. Article VIII provided for the designation of Specially Protected Areas (SPAs) where natural ecological systems, or areas of outstanding scientific interest could be protected. Entry to such areas except by permit was prohibited. Article IX forbade the introduction of non-indigenous species of animals or plants, though permits could be issued to allow importation of sledge dogs, domestic or laboratory animals and plants, the latter including viruses, bacteria, yeasts and fungi. Introductions had to be subject to careful control, and to removal or destruction once their purpose was served. Reasonable precautions had to be taken to avoid accidental introductions of alien species.

Later Consultative Meetings added further measures to enhance conservation objectives. The fourth meeting reviewed provisions for specially protected species and areas. The fifth meeting invited governments to draw up and circulate lists of historic monuments deemed worthy of preservation and protection. The sixth provided further protection for SPAs, and the seventh spelled out more detailed prescriptions for this important category of protected site. The Eighth Consultative Meeting of 1989 provided for the designation of Sites of Special Scientific Interest (SSSIs), to include areas where scientific investigations were being carried out, or where exceptional scientific interest required long-term protection. SSSIs were designated for limited periods and a management plan was later required for each site. The XVth Consultative Meeting required effective management plans for SPAs. The Agreed Measures and their subsequent additions are now mainly of historical value: the **Protocol on Environmental Protection to the Antarctic Treaty** of 1991 overtook and replaced them.

Ahlmannryggen. 71°50′S, 2°25′W. (Ahlmann Ridge). Mountain ridge, mainly ice-covered, extending north from Borgmassivet toward Fimbulisen, western Dronning Maud Land. It was named for Swedish glaciologist H. W. Ahlmann.

Aiken, Alexander. British seaman who served as boatswain in *Terra Nova* during the voyage to relieve *Discovery*, **British National Antarctic Expedition 1901–4**.

Aitcho Islands. 62°24′S, 59°47′W. Group of small, mainly low-lying islands straddling English Strait, South Shetland Islands. The islands were charted by Discovery Investigations in 1935: the name is derived from the initials of the British Hydrographic Office.

Akademik Vernadsky Station. 65°15′S, 64°16′W. Ukrainian permanent research station operating year-round at Marina Point, Galindez Island, Argentine Islands, Biscoe Archipelago. Until February 1996 it was the British Antarctic Survey research station Faraday (Base F), which was first occupied in May 1954. Faraday replaced an earlier British station built in 1953, which in turn replaced an earlier station built in 1947: see **Faraday Station (Base F)**. Faraday and Akademik Vernadsky jointly are the longest continuously running station in the Antarctic Peninsula area.

Albatross Island. 54°01′S, 37°20′W. Island in Bay of Isles, on the north coast of South Georgia, named for the wandering albatrosses that breed on its tussock-covered slopes.

Wandering Albatross nesting in tussock grass, Albatross Island. Photo: BS

Albatrosses, southern oceanic. Large flying birds of the order Procellariiformes, family Diomedidae, with long narrow wings on which they characteristically soar and glide. The vernacular name includes the three largest species, but is often extended to include smaller forms alternatively named **mollymawks**. Wandering albatrosses *Diomedea exulans*, royal albatrosses *D. epomophora* and Amsterdam Island albatrosses *D. amsterdamensis* the largest living seabirds, have a bill-to-tail length of about 1.2 m (4 ft), wing span 3.5 m (11.5 ft). Distinguished from lesser albatrosses by their all-white underwings and mostly-white bodies, they differ only slightly from each other. Wandering albatrosses are darker as juveniles, becoming progressively paler with age. Royal albatrosses have more prominent nostril tubes, a black cutting edge along the bill, and all-white tails throughout life. Amsterdam Island albatrosses are similar but generally darker. The albatrosses hunt by soaring and gliding in the zone of strong westerly winds, and feed by short plunging dives from the surface, catching mainly fish and squid. Wandering albatrosses breed on islands throughout the zone, royal albatrosses only on New Zealand's South Island, Campbell Island, and the Auckland and Chatham Islands. Amsterdam Island albatrosses breed only on Ile Amsterdam. All select wind-swept moorland or cliff-top sites to facilitate landing and take-off. Their chicks take over a year to reach independence: young birds begin courtship from about their fifth year onward.

Albedo. Reflectance: a measure of the degree to which a part of Earth's surface reflects solar energy. Of the incident high-frequency solar energy that impinges on the outer atmosphere, about one quarter is absorbed in passing through the atmosphere, and one quarter is reflected back by clouds or back-scattered by particles in the atmosphere. Of the remaining half that reaches Earth's surface, some is reflected back into space, the amount reflected depending on the albedo or reflectance of the surface. Snow and ice particularly reflect much energy away, dark rocks and vegetation reflect relatively little (see Table 1). The proportion not reflected is absorbed, warming the surface and contributing to snow-melting and evaporation of water. Thus high albedo results in low surface warming: ice fields in consequence tend to be self-sustaining.

Table 1 Typical values of albedo in polar regions, i.e. the percentage of incoming short-wave radiation reflected by the surface. For derivation of percentages see **Solar radiation**. Data from Lockwood (1974) and other sources

Surface	Albedo (%)
Calm water	2–5
Moist soil	8–10
Dry rock or soil	12–18
Green vegetation	10–25
Ice covered with water layer	25–30
Rough sea ice, open pack ice	30–40
Cobbles, sandy beaches	30–40
Snow-covered soils	30–40
Thawing snow, wet sea ice	40–75
White-capped seas	75–85
Dry clean snow	85–95

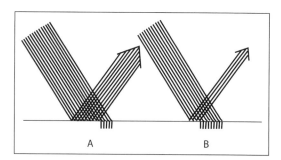

With high albedo (A), much of the incident radiation is reflected away, and only a little is absorbed to melt the snow or warm the ground. With low albedo (B), less is reflected and more absorbed. A lower total incidence can result in higher ground temperatures

Alexander Island. 71°00′S, 70°00′W. Large mountainous island off the western flank of Antarctic Peninsula, separated from the mainland English Coast by George VI Sound. It was discovered by the Russian explorer Fabian von Bellingshausen, and named by him for the contemporary Russian tsar.

Alexandra, Cape. 67°45′S, 68°35′W. Prominent dark cape forming the southeastern corner of Adelaide Island. Discovered in 1909 by J.-B. Charcot, it was named by him for the contemporary British queen.

Alexandra Mountains. 77°25′S, 153°30′W. Group of mountains forming the northern flank of Edward VII Peninsula, Marie Byrd Land, West Antarctica. Discovered in 1902 by the **British National Antarctic Expedition 1901–4**, they were named for the contemporary British queen.

Alfred-Faure Station. 46°26′S, 51°52′E. French permanent research station operating year-round on Ile de la Possession, Iles Crozet. The station has operated continuously since 1963.

Allardyce Range. 54°25′S, 36°33′W. The main mountain range of central South Georgia, rising to 2935 m (9627 ft) in Mt. Paget, the island's highest point. The range was named for Sir W. L. Allardyce, who was Governor of the Falkland Islands and Dependencies 1904–14 during the early days of whaling.

Almirante Brown Station. 64°53′S, 62°53′W. Argentine research station on Coughtrey Peninsula, Paradise Bay, Danco Coast. Opened as a naval station in 1951, it was named for Admiral Guillermo Brown (1777–1857), first commandant of the Buenos Aires Navy. The station was almost completely burnt down in 1981. Partly rebuilt, it was transferred to Instituto Antártico Argentino, which currently uses it from time to time for summer biological and oceanographic research.

Amery Ice Shelf. 74°15′S, 125°00′W. Large triangular ice shelf in Prydz Bay, between Lars Christensen and Ingrid Christensen coasts, fed mainly by the Lambert Glacier. The shelf was named by Sir Douglas Mawson in 1931 for British diplomat W. B. Amery, who was prominent in promoting British claims in Antarctica.

Ames Range. 75°42′S, 132°20′W. Range of ice-covered mountains forming a northern offshoot of Flood Range, inland from the Hobbs Coast, Marie Byrd Land. The range was named for Joseph Ames, father-in-law of US explorer Richard Byrd.

Amphibolite Point. 60°41′S, 45°21′W. Prominent point on south coast of Coronation Island, South Orkney Islands, named for an exposure of dark metamorphic rock.

Amsterdam, Ile. 37°50′S, 77°30′E. A warm-temperate island of the southern Indian Ocean, 4500 km (2812 miles) from south-eastern Africa and 3500 km (2187 miles) from south-western Australia. An oval volcanic island, roughly 10 km (6.2 miles) long and 8 km (5 miles) wide, it rises to a central plateau at 600 m (1968 ft) and a complex of cones up to 880 m (2886 ft) high. Its nearest neighbour, Ile St Paul 80 km (50 miles) away, and next nearest Iles Kerguelen 1400 km (875 miles) away, share with Ile Amsterdam a longitudinal submarine ridge. Of volcanic origin, Ile Amsterdam is the eastern remnant of a much larger island, composed of interbedded lavas and tuffs, most of which has disappeared under the sea. It lies in relatively warm water close to the subtropical convergence. The climate is mild and windy, with mean annual temperature 13.5 °C, and little seasonal variation: the range of monthly means is 5.8 °C. A little over 100 cm of rain falls per year: winds are westerly and persistent. Originally forested close to sea level, with tussock grassland and bog above, the island has been devastated repeatedly by fire and also by the grazing of introduced cattle, pigs, sheep, goats and rabbits. Currently its forest is confined to small remnant stands, and its grasslands, contaminated by thistles and other alien species, are grazed by feral cattle. Norway rats, house mice and feral cats are abundant. The island supports breeding rockhopper penguins and a small population of an endemic species of great albatross: there are also breeding stocks of yellow-nosed and sooty albatrosses, two smaller petrels, terns and skuas. Landbirds include pintails and waxbills. Amsterdam Island fur seals breed on the coasts.

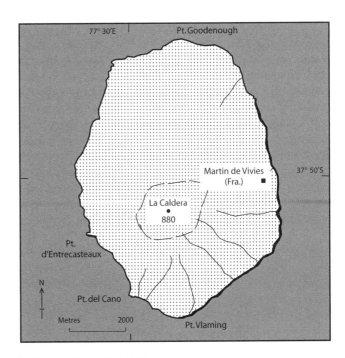

Ile Amsterdam. Height in m

Sighted in March 1522 by Juan Sebastian de Elcano, Ile Amsterdam was first charted and named by the Dutch navigator Willem de Vlamingh in 1696–97, and subsequently visited by sealers from the late eighteenth century onward. France claimed them in 1843, and French entrepreneurs released cattle in 1871. The island has supported a permanent scientific station, La Roche Godon, since 1949. It shares with Ile St Paul the protected status of a French national park.

Amundsen, Roald Engelbregt Gravning. (1872–1928).

Norwegian polar explorer who led expeditions through the Northwest Passage and to the South Pole. Born near Christiania (Oslo), into a prosperous ship-building family, Amundsen was inspired by his countryman Fridtjof Nansen to become a polar explorer. Training as a ship's officer, he gained sea time in sealing ships, and sailed as second mate of *Belgica* on the **Belgian Antarctic Expedition 1897–99**. He trained also in geomagnetism under the German physicist Georg von Neumayer. In 1903–6 with six companions he sailed the small fishing boat *Gjøa* from Baffin Bay westward through the Arctic archipelago of northern Canada, living in close contact with local Inuits. During the first winter he sledged with dog teams to the North Magnetic Pole. In the second summer *Gjøa* emerged into the Beaufort Sea, becoming the first ship to achieve the Northwest Passage. In the second winter, again with Inuits, he sledged 450 km (250 miles) inland from King Point to the Alaskan settlement of Eagle to telegraph news of his progress. *Gjøa* arrived in San Francisco in October 1906.

Amundsen's success earned him Nansen's support in borrowing the research vessel *Fram* for a bid to reach the North Pole. Frederick Cook's claim to have reached the pole, announced in September 1909, reduced Amundsen's enthusiasm for a northern journey. Sailing from Norway in June 1910, ostensibly for the Arctic, he switched objectives and headed instead for the South Pole (see **Norwegian South Polar Expedition 1910–12**). By the end of January 1911 he was established at 'Framheim' on the Ross Ice Shelf, and by midwinter had laid the first depots southward. Planting the Norwegian flag at the Pole on 14 December was his reward for meticulous planning and professional approach to exploration.

Amundsen made a fortune from shipping in World War I. He returned to the Arctic in 1918 with his own ship *Maud*, in which he sailed and drifted along the Siberian coast. Interested in flight as a means of exploration, in May 1926 he crossed the Arctic basin from Svalbard to Alaska in the airship *Norge*. In June 1928 he was lost with two companions while flying over the Arctic Ocean in search of a missing Italian expedition.

Further reading: Amundsen, (1912); Huntford (1979).

Roald Amundsen (1872–1928). Photo: Scott Polar Research Institute

Amundsen Coast. 85°30′S, 162°00′W. Part of the coast of Victoria Land, East Antarctica, facing the Ross Ice Shelf and bounded by Liv Glacier and Scott Glacier. It consists of spectacular ice-laden mountains, alternating with glaciers that feed into the ice shelf. The coast was identified in 1961 and named for the Norwegian explorer Roald Amundsen.

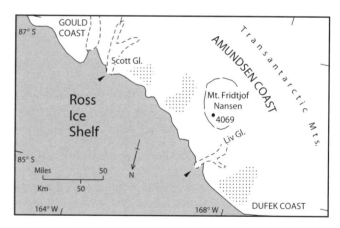

Amundsen Coast

Amundsen–Scott Station. 90°S. United States research station operating year-round at the South Geographical Pole. Opened on 24 January 1957, it stands at an elevation of 2835 m (9298 ft), some 653 m (2034 ft) lower than Vostok Station on the high plateau, and is correspondingly warmer throughout the year. Minimum temperatures around −80° occur in August, maxima around −13°C in January or February. The station, consisting of insulated buildings, many of them under a geodesic dome, is maintained entirely by air from McMurdo Station on the coast.

Amundsen Sea. 73°S, 112°W. Coastal sea off West Antarctica between Thurston Island and Siple Island, bordered by the Walgreen and Bakutis coasts of Marie Byrd Land. There is no definitive northern boundary. This is an area of heavy snowfall: coasts are almost entirely icebound, generating large tabular bergs. Inshore waters are ice-filled and virtually unnavigable throughout the year except by icebreaker. The sea was named for the Norwegian explorer Roald Amundsen.

Anare Mountains. 70°55′S, 166°00′E. Range extending north–south inland from Oates Coast, Victoria Land, named for the Australian National Antarctic Research Expedition.

Anchor ice. Submerged ice that has grown on the sea bed or lake bed by nucleation around organic and inorganic materials from supercooled water.

Andersen Harbour. 64°19′S, 62°56′W. Harbour between Eta Island and Omega Island, Melchior Islands, Palmer Archipelago. It was used by early twentieth-century whalers and probably named for the Norwegian whaler Capt. O. Andersen.

Anderson, William Ellery. (1919–92). British soldier and explorer. Born in Cheltenham, he joined the army in 1939, serving through World War II (Military Cross 1943) and the Korean War (bar to MC 1951). Retiring from the Army in 1954 he joined the Falkland Islands Dependencies Survey, spending a year as leader of Base D, when he led several dog-sledging surveys. Later he took part in mountaineering in the Himalayas and ocean cruising, and became a prison governor.

Anderson Massif. 79°10′S, 84°45′W. High outlying massif on the northeastern flank of the Heritage Range, Ellsworth Mountains, Ellsworth Land.

Andersson, Johan Gunnar. (1874–1960). Swedish geologist, archaeologist and explorer. After training as a geologist he participated in two expeditions to the Arctic. Appointed second in command of the **Swedish South Polar Expedition 1901–4** during its second year, he landed with two companions at Hope Bay in 1902, hoping to sledge to the expedition base on Snow Hill Island. This proving impossible, the party wintered in a makeshift stone hut, joining the Snow Hill group in the following spring. In 1906 Andersson became Professor of Geography at Uppsala University and head of the Geological Survey of Sweden. He died on 29 Oct. 1960.
(CH)

Andvord Bay. 64°50′S, 62°37′W. Glacier-lined bay on the southwest side of Arctowski Peninsula, Danco Coast, Graham Land, charted by A. de Gerlache in 1898 and named for Norwegian R. Andvord, the Belgian Consul in Christiania (Oslo).

Annenkov Island. 54°29′S, 37°05′W. A tall, tussock-covered island off the southern coast of South Georgia, rising to Olstad Peak (650 m, 2133 ft). Capt. James Cook in 1775 named it 'Pickersgill Island' for one of his lieutenants, but Capt. Fabian von Bellingshausen, recharting in 1820, mistakenly ascribed Pickersgill to a group further southeast, and named this larger island for Lt. M. Annenkov, one of his officers.

Antarctic. Descriptive name of the southern polar region. Adjectivally the term 'Antarctic' implies 'opposite to the Arctic', i.e to the northern polar region underlying Arctos or Ursa Major, the constellation of the great she-bear. Astronomers recognized that Arctos appeared to rotate about Polaris, the northern pole star. Thus the constellation came to identify the northernmost regions of both celestial and terrestrial globes, and 'Antarctic' typified southern regions. Fourteenth-century geographers described as 'Antarctic' a range of southern lands, from the southern shore of the Mediterranean Sea to virtually the whole area south of the equator. As explorers penetrated the southern continents and oceans, the term was progressively restricted to remaining unknown regions, though in seventeenth-century literature 'antarctic' could be applied to anything that was distant, remote or opposite to the norm, including opinions and philosophical concepts. Ultimately geographers recaptured the term, using it more precisely to describe the southern polar region, finally applying it to the southern continent itself. See **Antarctic region** and **Antarctica**.

Antarctic and Southern Ocean Coalition (ASOC). A coalition of conservation organizations and pressure groups that, in its own words, 'leads the national and international campaigns to protect the biological diversity and pristine wilderness of Antarctica, including its oceans and marine life .... We work closely with the key users of Antarctica, including scientists, tourists and governments to ensure that activities have minimal environmental impacts. We conduct legal and policy research and analysis, and produce educational materials.' The Antarctica Project, ASOC's secretariat based in Washington DC, is a non-profit organization that welcomes membership. A website is available: see **Study Guide: Information Sources**.

Antarctic Circle. Parallel of latitude drawn conventionally in 66°32′S, forming the northern boundary of an area in which, for at least one day each year, the sun fails to rise at the winter solstice (21 June) and to set at the summer solstice (22 December). Length of continuous night or day increases from 24 hours at the Circle to six months at the South Pole. The Circle runs close to the shore of East Antarctica, but well to the north of the bulk of West Antarctica, cutting through the base of Antarctic Peninsula north of Adelaide Island. It has nothing to show for itself on the ground, and no direct ecological or political significance. Its position varies slightly from year to year as the tilt of Earth increases or decreases.

Antarctic fringe islands. Islands standing on the shelf of continental Antarctica, geographically close enough to the mainland to come directly under continental influences (cf. **Antarctic islands**). Examples are Alexander Island, Palmer Archipelago and the Windmill Islands. For details of these and other fringe islands, see individual entries.

Antarctic islands. Islands of the Southern Ocean that stand between Antarctica and the mean northern limit of **pack ice**. They include the South Shetland Islands, South Orkney Islands, South Sandwich Islands, Peter I Øy, Balleny Islands and Scott Island. This category excludes those islands that stand on the continental shelf close to Antarctica (cf. **Antarctic fringe islands**). Climatically Antarctic islands share many characteristics of the continental coast. Ice or semi-permanent snowfields cover much of their land area. Liable throughout the year to snowstorms, they have short, cool summers with mean monthly air temperatures close to 0 °C. In winter, sea ice inhibits the warming effects of the sea: mean monthly temperatures fall below −10 °C on the coast, lower among the mountains. Often dry, with less than 100 mm precipitation throughout the year, the islands appear bare of vegetation, though green patches of moss and grass occur in sheltered, damp areas.

Antarctic Peninsula. 65°00′S, 62°00′W. The large peninsula, mountainous and ice-mantled, that extends between the Bellingshausen Sea and the Weddell Sea. From mainland West Antarctica it extends for about 1400 km (875 miles) toward South America, little more than 100 km (63 miles) across along Graham Land, the northern half, but widening to 200 km (125 miles) or more across Palmer Land in the south. The summit, a peneplaned plateau, is ice-covered and dissected by glaciers. On the eastern flank the glaciers merge into the extensive Larsen Ice Shelf bordering the Weddell Sea: on the warmer western flank they merge into narrow piedmont ice shelves or flow directly into the sea along a deeply dissected coastline. Geologically the peninsula forms a southern continuation of the Andes, diverted eastward by plate tectonics into the Scotia Arc.

Together with its offlying islands, it forms a geographical and climatic province, the Maritime Antarctic, milder and biologically richer than any other sector of Antarctica. The first part of the Antarctic continent to be seen and charted, it has since 1819 been the most frequently visited sector. Its readily-accessible west coast is currently the most often visited, with several permanent and currently disused scientific stations: since 1958 it has been a popular venue for summer visits by cruise ships. Politically it is claimed by Britain as part of British Antarctic Territory. To Argentines it is Tierra de San Martin, to Chileans Tierra de O'Higgins, both names honouring national heroes of the early nineteenth century wars of independence.

Mountains and glaciers of Antarctic Peninsula. Photo: BS

Antarctic region. The southern polar region, including continental Antarctica, much or all of the surrounding Southern Ocean, and some of the island groups within that ocean. There is no general agreement on how far north the region extends. At least four northern boundaries are in common use.

A *geographical* Antarctic region is circumscribed by the Antarctic Circle (latitude 66°32′S) to include much but not all of the continent and some of its neighbouring seas. This definition has little ecological or climatic validity. Its main value lies in allowing direct geographic comparisons with the area of identical size (21.81 million km², 8.52 million sq miles) enclosed by the Arctic Circle. Within the Antarctic Circle lies a high ice-covered continent, almost entirely desert, populated only by a few hundred scientists and support staff. Within the Arctic Circle lies an ice-covered sea ringed by extensive lands, some desert, some forested, some even cultivated, with cities, towns and smaller settlements and a human population of over two million.

A *political* Antarctic region, defined by the **Antarctic Treaty System**, is circumscribed by latitude 60°S, to include all of continental Antarctica and Antarctic Peninsula and the southern islands of the Scotia Arc. This definition excludes the South Sandwich Islands, South Georgia and other small islands and island groups of the Southern Ocean. The area so defined is approximately 34.42 million km^2 (13.44 million sq miles), more than half as large again as the area enclosed by the Antarctic Circle.

An *oceanographic* Antarctic region, recognized mainly by marine biologists, is circumscribed by the Antarctic **Convergence** or Polar Front, a line or narrow zone in southern oceans marking the limit of northward spread of polar surface waters: this is also the northern limit of the Southern Ocean. Its mean position remains stable over decades and possibly centuries, shifting slightly from summer to winter. One instrument of the Treaty System, the **Convention on the Conservation of Antarctic Marine Living Resources**, defines the Convergence for its own legal purposes in terms of ten sets of coordinates that approximate to its mean position. The area thus defined is approximately 46.16 million km^2 (18.03 million sq miles), more than twice that enclosed by the Antarctic Circle.

A *climatic* Antarctic region, of more interest to terrestrial ecologists, is circumscribed by the 10 °C summer isotherm, a line connecting points at which the mean temperature of the warmest month (normally January or February) does not fall below 10 °C. This line ranges mainly over the oceans, barely touching land except in southwestern Chile and Tierra del Fuego. For most of its length it runs parallel with and slightly north of the Antarctic Convergence, enclosing a similar or marginally greater area.

Antarctic Sound. 63°26′S, 56°39′W. Channel between d'Urville Island and Trinity Peninsula. Though discovered and navigated by earlier explorers, it was charted in 1902 by the **Swedish South Polar Expedition 1901–4** and named for the expedition ship.

Antarctic Treaty System. System of international governance currently providing for the government and management of Antarctica. The term includes the **Antarctic Treaty** (1959) and recommendations and instruments arising from it (see Appendices A–E). Specific instruments are **the Convention for the Conservation of Antarctic Seals** (1972), the **Convention on the Conservation of Antarctic Marine Living Resources** (1980), and the **Protocol on Environmental Protection to the Antarctic Treaty** (1991). A **Convention on the Regulation of Antarctic Mineral Resource Activities** (1988) was prepared and opened for signature but never adopted.

The Treaty was negotiated in Washington in 1959 between the 12 nations that participated in Antarctic research during the International Geophysical Year 1957–58. The text of the treaty appears in Appendix E. It provided for later accession by other interested states. Any state that is a member of the United Nations may accede as a non-consultative party, and non-member states may join with the consent of the consultative parties. States that undertake substantial Antarctic scientific research (usually but not always signified by their establishing a permanent research station) may be accepted into an inner circle of consultative parties, which participate fully in the decision-making process of the treaty meetings. Signatories to the Treaty currently (2002) include 27 consultative and 18 non-consultative parties (see below).

Antarctic Treaty Consultative Meetings were at first held in alternate years. As the delegates began to concern themselves more and more with management, the amount of business generated required more frequent opportunities for contact and discussion, and from 1991 consultative meetings became annual events. Held in member countries in rotation by English alphabetical order, the meetings occupy two to three weeks, in which delegates and retinues of advisors exchange information, discuss matters of common interest, and seek conclusions. Until 1995 delegates agreed on 'measures in furtherance of the principles and objectives of the Treaty', to be promulgated as recommendations on which consensus has been reached. In 1995 the terminology of agreements changed to 'measures', 'resolutions' and 'decisions' (see below).

Non-consultative parties and some other interested bodies (for example, the **International Association of Antarctica Tour Operators**) are invited to attend the meetings as experts. Special Consultative Meetings are from time to time convened to accept new consultative parties, and to discuss issues that are too involved to be covered in annual meetings, for example, the elaboration of conventions. Special Meetings of Experts are occasionally called to discuss such technical matters as telecommunications, air safety and environmental monitoring.

There is as yet no permanent secretariat. Despite agreement by many states that one was needed for the Treaty Organization to fulfil its self-appointed stewardship functions, delegates were for long unable to reach the necessary consensus on funding, location and other issues. Buenos Aires, Argentina, was eventually selected as a location in July 2001.

Legislation arising from the Treaty

Most Treaty states provide national legislation to cover the activities of their own citizens in Antarctica. Mandatory Recommendations of Treaty meetings (after 1995 called Measures: see below), when ratified by individual governments, become obligations upon those governments, to be embodied when necessary in their legislation. Thus citizens of each individual state, when visiting the Treaty area, are bound not by law arising directly from the Treaty, but by the laws and regulations of their own country, drawn up in accordance with Treaty deliberations. Since the inception of the Treaty over 200 Recommendations have emerged on a wide range of topics from the issue of commemorative postage stamps to conservation. However, the Treaty's most far-reaching deliberations are contained in its three major instruments – the conventions on conservation of seals and marine living resources, and the protocol on environmental protection.

A survey presented at the XIXth Antarctic Treaty Consultative Meeting in May 1995 showed that, of all Recommendations made at consultative meetings since 1983, none had in fact entered into force. In every case at least one party, sometimes several, had omitted to ratify – a comment, perhaps, on the priorities given to Antarctic matters in the national legislatures of the consultative parties. Parties agreed to act as though the Recommendations were in force until ratification was completed. At the same meeting Recommendations were replaced with three new categories of enactment: Measures, Resolutions and Decisions. *Measures* are the equivalent of the former Mandatory Recommendations, that needed to be taken into national legislation and ratified. *Resolutions* are the equivalent of former Hortatory Recommendations, that governments were encouraged to follow but did not need to ratify or take into national legislation. *Decisions* relate to procedural matters within the Treaty System. The distinction was made particularly to ensure faster enactment of Resolutions and Decisions, that do not require action by governments but become operative at the time of adoption, normally in the final report of the meeting.

Treaty objectives

Arising as it did at the height of the Cold War and during a period of tensions over sovereignty between Argentina, Britain and Chile, the Treaty aimed to continue the use of Antarctica for peaceful purposes (Article 1), specifically for the promotion of free international scientific cooperation, including exchanges of scientists (Articles 2 and 3). In deference alike to claimant nations, non-claimants and those that reserved the right to claim in the future, the Treaty dissociated itself from sovereignty issues (Article 4). To defend the continent from contemporary proposals that it be used for nuclear experimentation and dumping, the Treaty prohibited nuclear explosions and the disposal of radioactive wastes within its area (Article 5), defined as south of 60°S, specifically including ice shelves but excluding the high seas (Article 6).

The Treaty provided for mutual inspections of stations, installations and operations, requiring parties to exchange advance notice of their operations (Article 7) and specifying that observers and scientists working with nations other than their own would be subject only to their own national jurisdiction (Article 8). It provided for consultative meetings, specifying some of the issues to be discussed (Article 9), and requiring all contracting parties to ensure that the principles and purposes of the Treaty were observed (Article 10). It provided means of resolving disputes between parties (Article 11), mechanisms for modification and amendment, and an opportunity for major review of the Treaty after 30 years (Article 12), and means of accession by other states (Article 13). The final article required four language versions of the Treaty (English, French, Russian and Spanish) to be deposited in the archives of the Government of the United States, and for copies to be passed to governments of all signatory and acceding states.

Accession to the Treaty

Soon after its publication the Treaty was acceded to by Poland, and before the end of the 1960s by Czechoslovakia, Denmark and the Netherlands. During the 1970s Romania, the German Democratic Republic, Brazil, Bulgaria and the German Federal Republic acceded, and Poland was admitted to consultative status. During the 1980s no fewer than 18 countries acceded, particularly during the period 1983–88 when the Convention on the Regulation of Antarctic Mineral Resource Activities (**CRAMRA**) was under discussion. During the same decade 11 acceded countries set up 'research operations' (some little more than temporarily occupied huts) that were deemed to qualify them for consultative status. After the collapse of CRAMRA and the establishment in its place of a long-term moratorium on minerals development, interest among newcomers declined. However, the wave of accession had recruited two of the world's most populous states, China and India. Delegates could (and did) claim that the Treaty was supported by over 70 per cent of the world's population.

Review of the Treaty

On the thirtieth anniversary of the Treaty in 1991, delegates to the XVIth Antarctic Treaty Consultative Meeting reaffirmed its objectives and values, declaring that the Treaty had 'united countries active in Antarctica in a uniquely successful agreement for the peaceful use of a continent'. Research and cooperation between the Treaty parties had 'signalled to the world that nations can work together for their mutual benefit and for the benefit of international peace and cooperation ... to preserve a major part of this planet, for the benefit of all mankind, as a zone of peace, where the environment is protected and science is pre-eminent'. The delegates drew attention to a record of conservation starting in 1964 with the **Agreed Measures for the Conservation of Antarctic Fauna and Flora**, and culminating in the 1991 **Protocol on Environmental Protection** that had established 'a comprehensive legally binding regime for ensuring that activities that parties undertake in Antarctica are consistent with protection of the Antarctic environment and its dependent and associated ecosystems'.

Outlining Antarctica's fundamental role in understanding global environmental processes, the declaration termed it 'a pristine laboratory, of world-wide significance', enabling research on the detection and monitoring of such phenomena as the depletion of atmospheric ozone, global warming and sea level changes. Meteorological research had provided data essential to southern hemisphere forecasting, glaciological research provided information on the heat exchange budget and Antarctica's influence on

Table 2 Membership of the Antarctic Treaty, with dates of joining

State	Associate membership	Full membership
Argentina		23 June 1961
Australia		23 June 1961
Austria		25 August 1987
Belgium		26 July 1960
Brazil	16 May 1975	12 September 1983
Bulgaria	11 September 1978	25 May 1998
Canada	4 May 1988	
Chile		23 June 1961
China, People's Republic	8 June 1983	7 October 1985
Colombia	31 October 1989	
Cuba	16 August 1984	
Czech Republic[1]	14 June 1962	
Denmark	20 May 1965	
Ecuador	15 September 1987	19 November 1990
Estonia	15 June 1992	
Finland	15 May 1984	9 October 1989
France		16 September 1960
Germany[2]	5 February 1979	3 March 1981
Greece	8 January 1987	
Guatemala	31 July 1991	
Hungary	27 January 1984	
India	19 August 1983	12 September 1983
Italy	18 March 1981	5 October 1987
Japan		4 August 1960
Korea, North		21 January 1987
Korea South	28 November 1986	9 October 1989
Netherlands	30 March 1967	19 November 1990
New Zealand		1 November 1960
Norway		24 August 1960
Papua New Guinea[3]		16 March 1981
Peru	10 April 1981	9 October 1989
Poland	8 June 1961	29 July 1977
Romania	15 September 1971	
Russia[4]		2 November 1960
Slovakia[1]	14 June 1962	
South Africa		21 June 1960
Spain	31 March 1982	21 September 1988
Sweden	24 April 1984	21 September 1988
Switzerland	15 November 1990	
Turkey	24 January 1996	
United Kingdom		31 May 1960
United States of America		18 August 1960
Ukraine	28 October 1992	
Uruguay	11 January 1980	7 October 1985
Venezuela	24 March 1990	

Notes: [1]The Czech Republic and Slovakia succeeded to the Treaty on the partitioning of Czechoslovakia, 1 January 1993.
[2]The German Democratic Republic became an associate 19 November 1974 and a consultative member 5 October 1987: Germany was unified 3 October 1990.
[3]Papua New Guinea succeeded to the Treaty after gaining independence from Australia.
[4]Russia inherited membership from the former Soviet Union in December 1991.
Source: R. K. Headland, unpublished.

weather and climate. Geological and geophysical research gave insights into global geological history and the formation of continents.

Antarctica, added the delegates, was particularly well suited to the study of solar–terrestrial interactions and cosmic rays, and its extreme environment provided unique opportunities to study environmental adaptations of organisms. Biological research yielded data essential to informed decision-making about marine living resources. Human biology and medicine gave information on the physiological adaptation of man to extreme climates and isolation. The Treaty parties had ensured that the results of these important research efforts were freely available to all mankind.

A record of success?

The severest critics of the Treaty System might concede that much of this self-assessment is true. Starting in a period of international hostility and suspicion, the Treaty has provided a stable framework for discussion between interested nations, maintaining the Antarctic region free from hostilities, and encouraging governments to support expensive scientific research that they would not otherwise have considered.

Early criticism of the Treaty System centred mainly on (i) its establishment as a consortium (some said an 'exclusive club') of a few powerful states with money to spend on Antarctic research, and (ii) its failure for many years to share its deliberations with those outside. Later criticisms concern the instruments through which the organization has sought to manage and govern the affairs of a continent. Even before the inception of **CCAMLR** in the mid-1970s, environmentalists were starting to show interest in the last continent. Their concerns included the damage already done during former exploitation by sealing and whaling, the damage currently being inflicted by ill-managed and insensitive scientific operations, and the possibilities of further damage from new exploitation of Antarctica's remaining resources notably minerals and potential for tourism. The secrecy with which a small group of interested countries conducted negotiations on Antarctica attracted the hostility both of conservation groups and of the United Nations General Assembly. The former were concerned with environmental protection: the latter considered that a large area of the world had been taken over by an 'exclusive club' of powerful states, and sought a more equitable sharing of whatever resources Antarctica might offer. Since the XVth Consultative Meeting, and especially since the failure of CRAMRA, delegates have found it expedient to take the public more into their confidence, to the extent of inviting the participation of non-governmental organizations and agencies in their deliberations.

Like any other working international agreement, the Treaty has found bases for practical consensus in fields of conflicting interests. When some critics, for example, find its insistence on scientific pre-eminence undemocratic, exclusive and short-sighted and others favour stronger regulatory protection against non-scientific activities, the delegates point to the even-handed approach of the 1991 Protocol on Environmental Protection, which they claim treats scientific and non-scientific activities alike. Perhaps the main difficulties faced by the Treaty parties lie in their efforts to manage a continent in the absence of the ability to generate simple, relevant legislation. The cumbersome machinery of the Treaty and its instruments generates voids in understanding between those who work in Antarctica and the multinational diplomats who seek to regulate their activities.

Remote and with no indigenous human population, Antarctica provides a forum in which may be explored important issues in areas that are more politically loaded, from the deep sea bed to the moon. If the Treaty illustrates 'the inevitable outcome of any discussions between nations with varying interests, that is, the lowest common denominator of agreement [doing] little more than to preserve the *status quo* of the late 1950s' (Beck 1986), it has shown also 'qualities of foresight in dealing with issues before they become insoluble, restraint in the pursuit of national interest and recognition of the national interests of others' (Heap 1994), and has clearly provided much food for thought and discussion in international law and politics. *Further reading*: Chaturvedi (1996); Jørgensen-Dahl and Østreng (1991).

Antarctica. Southernmost of Earth's seven continents, and fifth in order of area, Antarctica as it appears on maps occupies slightly less than one tenth of Earth's land surface. It is the highest, coldest and most isolated continent, unique in being almost completely ice-covered. Roughly comma-shaped, some 4500 km (2800 miles) across at its widest, the continent surrounds the South Geographical Pole eccentrically, as though pulled some 650 km (400 miles) out of symmetry in the direction of India. Beneath the ice of the comma body lie several smaller land masses, parts of which appear as scattered outcrops and more continuous extents of exposed rock. The comma tail, **Antarctic Peninsula**, 1200 km (750 miles) long and pointing toward South America, shows more exposed rock than the rest of the continent.

The ice cap occupying 97.6 per cent of the continental area has a mean surface elevation of over 2000 m (6560 ft), making this by far the highest continent. Asia, the next highest, is less than half as high with a mean elevation of 960 m (3150 ft). Exposed rock outcrops include the extensive **Transantarctic Mountains** of East Antarctica and the **Sentinel Range** of West Antarctica, the latter including **Vinson Massif**, the continent's highest peak at 4901 m (16 075 ft). Of the coast, only about 2 per cent appears as cliffs or beaches, much of it along Antarctic Peninsula. The rest is made up of ice cliffs, backed by long ice slopes that over-ride and hide the true shoreline.

The area of the continent is calculated in different ways for different purposes:

- Minimally, including Alexander Island but excluding ice shelves (that extend to unknown distances beyond the hidden coast), ice rises (representing ice-covered islands within the shelves) and exposed islands on the continental shelf: 12.39 million km² (4.84 million sq miles).
- Including the larger ice rises and islands over 500 m high (Alexander, Bear, Berkner, Rooseveldt, Ross and Thurston Islands) but excluding ice shelves and all smaller ice rises and islands: 12.48 million km² (4.88 million sq miles).
- Including islands and ice rises joined to the continent by ice shelves, but excluding the ice shelves themselves and other islands on the continental shelf: 12.51 million km² (4.89 million sq miles)
- Including ice shelves, ice rises and islands, but excluding other islands on the continental shelf (the area most directly comparable with areas of other continents): 13.97 million km² (5.46 million sq miles).

By any of these calculations, Antarctica is the fifth-largest continent.

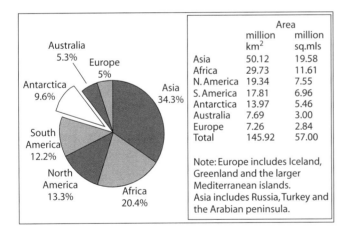

Antarctica in the family of continents, showing percentages of the world's total land area

The Antarctic ice cap, with mean thickness of 2160 m (7087 ft) and estimated volume 30.11 million km³ (5.15 million cubic miles), contains roughly 90 per cent of the world's ice. There are few visible mountain ranges. A relatively featureless surface buries an underlying topography of mountains, valleys and lakes to a mean estimated depth of 1880 m (6166 ft), in several places exceeding 4000 m (13120 ft). The greatest depth of ice so far recorded, near 70°S, 135°E in East Antarctica, is over 4700 m (15400 ft). About 11 per cent of the sheet consists of floating ice, notably in the extensive Ross and Ronne-Filchner ice shelves.

The high mean elevation makes Antarctica by far the coldest continent. Winter and summer alike, the coldest place on earth is close to the highest point of the Antarctic ice cap (see **Vostok Station**). The weight of ice holds the underlying continent down, depressing the continental shelf some 300–400 m lower than that of any other continent. Without its ice cap Antarctica would be a relatively smaller continent, rating second-highest among the continents with a mean bed-rock elevation of about 860 m (2820 ft).

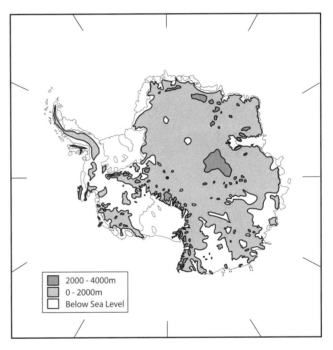

Bedrock of continental Antarctica, shown as it would appear after the land surface had risen isostatically following removal of the overlying ice. The Gamburtsev Subglacial Mountains and Vostok Subglacial Highlands occupy the central area of East Antarctica: other mountain ranges, e.g. the Ellsworth and Transantarctic mountains, and uplands of Dronning Maud Land and Antarctic Peninsula can be identified from current maps. Map based on Drewry (1983) and other sources

Geographically the continent is made up of two distinct provinces, separated by a wide channel running between the bights of the Ross and Weddell seas. Identified originally as the eastern and western provinces, they are currently called **East Antarctica** and **West Antarctica**. Some geographers argued that these names became confusing and meaningless in the far south, and introduced Greater and Lesser Antarctica to replace them. However, a general agreement among cartographers to represent Antarctica with the Greenwich meridian central clarified the issue. East Antarctica lies predominantly in the eastern hemisphere, West Antarctica in the western hemisphere, and East and West Antarctica are now more widely used. Geologists occasionally refer to them respectively as Gondwana and Andean provinces, in recognition of their different origins.

Antarctica was the last continent to be discovered, and remains the only continent without a permanent human population. Seven nations claim parts of the continent and its offlying islands. Other nations accept, dispute or reject their claims. The continent is governed by a consortium of nations that have acceded to the Antarctic Treaty (see **Antarctic Treaty System**).

Antarctica, East. See **East Antarctica**.

Antarctica, Greater. See **East Antarctica**.

Antarctica, Lesser. See **West Antarctica**.

Antarctica Project. See **Antarctic and Southern Ocean Coalition (ASOC)**.

Antarctica, West. See **West Antarctica**.

Antártida Argentina. Triangular sector of West Antarctica and adjacent islands bounded by 25°W and 74°W, and extending from 60°S to the South Pole. The east–west limits were defined by a Resolution of the Comisión Nacional del Antártico of March 1947. The territory is regarded by the Argentine government as part of Argentina. It overlaps geographically with British Antarctic Territory and Territorio Antártico Chileno.

Antipodes Islands. 49°41′S, 178°46′E. A group of cool-temperate islands of the South Pacific Ocean, 800 km (500 miles) east-southeast of New Zealand's South Island. Antipodes Island and the offlying Windward Islands, Bollons Islands and Leeward Island, form a small grass-covered archipelago arising from the eastern edge of the Cambell submarine plateau. Formed volcanically within the last million years, they consist of layers of lava and scoria that are eroding rapidly to form precipitous cliffs and stacks. The main island, some 6–7 km (3.7–4.4 miles) across, rises to a central plateau 300 m (984 ft) above sea level: the highest peak, Mt Galloway, rises to 366 m (1201 ft). There are no long-term meteorological records, but the climate is cool and windy, with mean annual temperature about 8 °C and estimated rainfall 150 cm per year. Over 60 species of flowering plants and ferns have been recorded, forming rich tussock meadows with patches of shrubs and upland bog. Erect-crested and rockhopper penguins nest on the cliffs and in gullies; other breeding seabirds include three species of albatross, giant petrels, several species of smaller petrel and brown skuas. Endemic landbirds include snipe, pipits and two species of parakeet: redpolls, dunnocks and starlings also breed there, presumably having crossed from New Zealand.

The islands were discovered in 1800 by Capt Henry Waterhouse RN, on a voyage from Australia to England. Though often visited by sealers, and the site of several

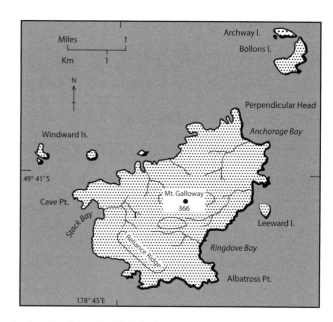

Antipodes Islands. Height in m

wrecks, they appear never to have suffered occupation and remain relatively pristine. The only introduced mammals are house mice. Now protected as a New Zealand National Nature Reserve, they may be visited only by authorized scientific parties.
Further reading: Fraser (1986).

Anvers Island. 64°33′S, 63°35′W. Largest and southernmost island in Palmer Archipelago. Discovered in 1832 by sealing captain John Biscoe, who may have landed at Biscoe Bay to take possession of the coast, it was re-surveyed in 1898 by A. de Gerlache, of the **Belgian Antarctic Expedition 1895–98** and named by him for the Belgian province. Mountainous and almost completely ice covered, its highest peak, Mt. Français, rises to 2825 m (9181 ft). Arthur Harbour, on the southern coast, was the site of former British Base N, and is currently the site of Palmer Station, a permanent United States research station.

Anvers Island (Base N). 64°46′S, 64°05′W. Former British research station in Arthur Harbour, Anvers Island, Palmer Archipelago. The station was built in February 1955 by the Falkland Islands Dependencies Survey to survey and conduct geological research on the island, and was occupied until November 1958. In 1963 the hut was loaned to the United States Government for use by scientists at the nearby Palmer Station. It was used again from 1969 as an air facility base, for a ski runway on the neighbouring ice shelf. The hut was destroyed by fire in December 1971.

Aramis Range. 70°37′S, 67°00′E. Southernmost of three northern ranges of Prince Charles Mountains, Mac.Robertson Land, named for one of Alexander Dumas's three musketeers.

Archer, Colin. (1823–1921). Norwegian ship designer, of Scottish descent, who designed and built the expedition ship *Fram*, launching it from his Larvik shipyard in 1892. The ship was built for F. Nansen for a voyage across the Arctic polar basin, later used by O. Sverdrup in the Arctic and by R. Amundsen on his expedition to the South Pole.

Archer, Walter William. (d. 1944). British seaman and master cook. He served for almost two years with the **British Antarctic (*Terra Nova*) Expedition** as shipboard cook, then transferred to the shore party for the final year. He left the navy after World War I and ran a catering business in London. He died on 28 January 1944.

Arctowski, Henryk. (1871–1958). Polish meteorologist and polar explorer. Born in Poland, educated in Belgium, Paris, London and Zürich, Arctowski served with the **Belgian Antarctic Expedition 1897–99**. In charge of the geological, oceanographical and meteorological programme, he recorded weather during the long drift of *Belgica* in the Bellingshausen Sea. Until 1909 he was connected with the Belgian Weather Service, later joining the New York Public Library Service, where he became Chief of the Science Division. In 1920 he returned to Poland as Professor of Geophysics and Meteorology at the University of Lwow. Lecturing in the USA at the outbreak of war in 1939, he remained there with an appointment in the Smithsonian Institution. He died in Washington on 20 February 1958. The permanent Polish research station in Admiralty Bay, King George Island, South Shetland Islands, is named for him. (CH/BS)

Arctowski Peninsula. 64°45′S, 62°25′W. A mountainous peninsula between Andvord Bay and Wilhelmina Bay, forming the northern side of Errera Channel, on the Danco Coast of Graham Land. Discovered by the **Belgian Antarctic Expedition 1897–99**, it was named for the Polish scientist Henryk Arctowski, a member of the expedition.

Arctowski Station. See **Henryk Arctowski Station**.

Ardery Island. 66°22′S, 110°28′E. A small rocky island of the Windmill Islands, Vincennes Bay, Budd Coast, noted for its rich avifauna, including nesting Antarctic petrels and Antarctic fulmars. Together with neighbouring Odbert Island, which is similarly endowed, it has been designated Specially Protected Area No. 3 (Ardery Island and Odbert Island). The total area is approximately 1.9 km^2 (0.75 sq miles).

Ardley, Richard Arthur Blyth. (1906–42). British seaman and hydrographer. A merchant navy and Royal Naval Reserve officer, in 1929 he was appointed Third Officer of RRS *Discovery II*. He made five cruises, taking particular interest in hydrographic survey and birds: he wrote Discovery Reports on the birds of the South Orkney Islands, and the qualities of RRS *Discovery II* as an oceanic research ship. Retiring as Chief Officer in 1937, he became assistant harbour master at Haifa, and was killed on active service in Tobruk on 12 September 1942.

Ardley Island. 62°13′S, 58°56′W. Low-lying island on the west side of Maxwell Bay, King George Island, South Shetland Islands, 0.5 km (0.3 miles) east of Fildes Peninsula and close to the Russian station Bellingshausen, the Chilean station Teniente Marsh, and the Chinese station Great Wall. It was named for the Discovery Investigations hydrographer Lt R. A. B. Ardley, RNR, who surveyed Maxwell Bay in 1935, and accommodates the Argentine refuge Ballve and other refuges and summer stations. Noted for an unusually rich flora and avifauna, the island is designated Site of Special Scientific Interest No. 33 (Ardley Island).

Arena Valley. 77°50′S, 160°59′E. An ice-free valley opening into the southern side of Taylor Glacier: part of the Victoria Land dry-valley system.

Argentine Islands. 65°15′S, 64°16′W. Group of islands in the Wilhelm Archipelago, Graham Coast, discovered by J.-B. Charcot and named by him for the Argentine Republic. The islands accommodated the southern base of the **British Graham Land Expedition** during their first winter, and Base F (Faraday) of the Falkland Islands Dependencies Survey, later British Antarctic Survey, from January 1947 to February 1996. The station, on Galindez Island, is now Akademik Vernadsky, a Ukrainian research facility.

Argus, Dome. 81°00′S, 77°00′E. Ice dome over 4000 m (13 120 ft) above sea level on the high plateau of Princess Elizabeth Land. Also called Dome A, this is the highest-known point on the Antarctica ice sheet.

Armitage, Albert Borlase. (1864–1943). British seaman and explorer. Born in Scotland on July 2 1864, he was brought up in Yorkshire. In 1878 he joined the merchant navy, and was later commissioned in the Royal Naval Reserve. In 1894–97 he took part in the Jackson-Harmsworth Expedition to Franz Josef Land, appointed second in command with responsibilities for meteorology, magnetometry, astronomy and survey. Returning to sea as chief officer, he was appointed mate and navigating officer of SY *Discovery*, the newly-built ship of Scott's **British National Antarctic Expedition 1901–4.** Scott quickly made him second-in-command of the expedition. Given responsibility for surveying among the Victoria Land mountains west of McMurdo Sound, Armitage led the sledging parties that explored Ferrar Glacier, ascending to about 2750 m (9000 ft) and pioneering the route that Scott later used to reach the polar plateau. On return to England he resumed his career in the merchant navy, becoming master of several passenger ships and serving throughout World War I. He retired in 1924 as commodore of the Pacific and Orient Line, and a captain RNR. During World War II, in his late seventies, he served as an air raid warden. Armitage wrote several books on his Antarctic experiences, including *Cadet to Commodore* (1925), a biography. He died on 31 October 1943.
Further reading: Armitage (1905).

Armytage, Bertram. (1869–1943). Australian soldier and explorer. Born in New South Wales, he was educated at Melbourne and Jesus College, Cambridge, then served in the Australian army, seeing action in the South African War. He joined the **British Antarctic (*Nimrod*) Expedition 1907–9**, in charge of ponies, and was involved in several sledging journeys.

Arrival Heights. 77°49′S, 166°39′E. Heights overlooking Hut Point, Ross Island, from which members of the **British National Antarctic Expedition 1901–4** watched for the arrival of relief ships. An area of approximately 1.1 km^2 (0.4 sq miles) has been designated SSSI No. 2 (Arrival Heights), on the grounds that it is electromagnetically quiet, offering ideal conditions for recording minute signals in upper atmosphere research.

Arrowsmith Peninsula. 67°15′S, 67°15′W. A large mountainous peninsula 80 km (50 miles) long on Loubet Coast, Graham Land, named for Edwin Arrowsmith, a former governor of the Falkland Islands.

Arthur Harbour. 64°46′S, 64°04′W. Harbour between Bonaparte Pt. and Norsel Pt., on the southwest coast of Anvers Island, Palmer Archipelago, named in 1955 for the then governor of the Falkland Islands, Sir Oswald Arthur. From 1955–58 it was the site of Base N of the Falkland Islands Dependencies Survey. In 1963 the station hut was loaned to the US Government during the building of Palmer Station. The original hut was destroyed by fire in 1971: the buildings of the US station now dominate the harbour.

Artigas Station. 62°11′S, 58°54′W. Uruguayan research station near Profound Lake, Fildes Peninsula, King George Island, South Shetland Islands. Opened in December 1984, it operated from time to time as a summer-only station, and currently operates year-round.

Asgard Range. 77°37′S, 161°30′E. Range separating Wright and Taylor valleys, in the Dry Valleys oasis area of Victoria Land. It was named for the mountain home of the gods in Norse mythology.

ASOC. See **Antarctic and Southern Ocean Coalition (ASOC).**

Astrolabe Island. 63°17′S, 58°40′W. Isolated island in Bransfield Strait, rising to 560 m (1837 ft). It was discovered by Capt. J. Dumont d'Urville during the **French Naval Expedition 1837–40**, and named for one of his ships.

Asuka Station. 71°31′S, 24°8′E. Japanese research station in the Sør Rondane, 120 km (75 miles) inland from Breidvika, Prinsesse Ragnhild Kyst, Dronning Maud Land. Established in December 1984, it was used intermittently for summer parties, and for a wintering party in 1988.

Athos Range. 70°13′S, 64°50′E. Northernmost of three ranges making up the northern end of Prince Charles Mountains, Mac.Robertson Land, named for one of Alexander Dumas's three musketeers.

Atkinson, Edward Leicester. (1882–1929). British naval surgeon and explorer. Born in Trinidad, he read medicine at St. Thomas's Hospital London, qualifying in 1906. In 1908 he joined the Royal Navy as a surgeon, developing special interests in parasitology. He served with the fleet, and in 1910 volunteered to join Scott's **British Antarctic (*Terra Nova*) Expedition 1910–13**, as surgeon to the main party and parasitologist. On the bid for the South Pole he sledged with the supporting parties as far as the upper depot on the Beardmore Glacier. After the loss of the polar party, he remained at Cape Evans for the second winter, as senior officer taking charge, and as medical officer taking care of several patients through a difficult winter. In 1913 he led the sledging party that discovered Scott's final camp on the Ross Ice Shelf. He served with distinction in World War I, being twice decorated and several times wounded. He continued in the navy after the war, reaching the rank of commander, and died at sea on 20 February 1929.

Atmosphere. The gaseous envelope some 1000 km thick surrounding Earth, forming both a shield from high-energy radiation and a heat trap that provides tolerable levels of temperature and humidity at the surface. Close to Earth's surface the atmosphere is made up of nitrogen (78 per cent), oxygen (21 per cent), traces of other gases including argon, carbon dioxide, water vapour, methane, ammonia, oxides of nitrogen, sulphur, ozone and hydrogen, and contains also particles of rock dust, plant and animal material, soot and other products of earthly origin. Held by gravity close to Earth's surface, about 99 per cent of the mass of atmosphere lies packed within the lowest 30 kms. Above, it thins rapidly with height to an indeterminate outer border.

Five concentric layers of atmosphere are distinguished by their temperature characteristics:

- *Troposphere*, closest to Earth, lying some 6–8 km (23 000 ft) deep over the poles and more than twice as deep (17 km, 56 000 ft) over the equator. Within it, temperatures decrease with height. Its outer boundary is the tropopause.
- *Stratosphere*, rising to about 50 km (165 000 ft), in which temperatures increase with height.
- *Mesosphere*, rising to about 85 km (280 000 ft), in which temperatures again fall with altitude.
- *Thermosphere*, in which temperatures again rise.
- *Exosphere*, where the gases are so rarified and molecules so far apart that temperature measurement becomes meaningless.

Atmospheric physicists distinguish two zones of ionization that cut across the above scheme:

- *Ionosphere*, a layer from about 20 km (66 000 ft) upward where atoms and molecules become photo-ionized, i.e. charged electrically by solar energy, particularly from about 80 km in the mesosphere and thermosphere.
- *Magnetosphere*, layer of varying height above Earth's surface in which all particles are charged and their movements controlled by Earth's magnetic field. Its height is least toward the sun, greatest away from it.

The ionosphere is particularly important in reflecting radio waves. Changes in it due to changes in solar output may

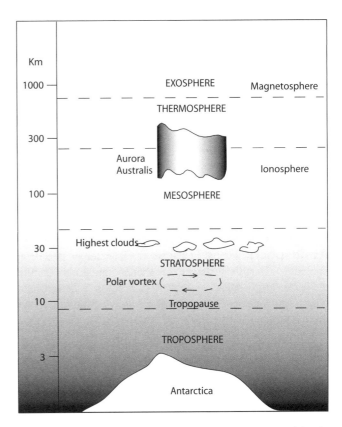

Layers in the atmosphere. The vertical scale is logarithmic: 30 km = 18.75 miles; 100 km = 62.5 miles

interfere seriously with radio communications. Over the Antarctic region the stratosphere includes the polar vortex, in which occurs the seasonal **ozone hole**, and the ionosphere accommodates the often-spectacular southern lights or **aurora australis**.

Atmospheric circulation. Pressure systems that most often affect the Antarctic region are (a) a persistent anticyclone centred over the continent, producing still conditions inland and easterly winds along the coasts, and (b) a constant succession of depressions or cyclones circling the continent from west to east, bringing prevailing westerlies to the Southern Ocean, the sub-Antarctic islands, Scotia Arc and Antarctic Peninsula. These systems generate either systemic winds, caused by pressure differences due to differential atmospheric heating, or **katabatic winds** and **barrier winds** caused by topography.

The central anticyclone over Antarctica produces clear skies or high dispersed cloud and light winds, with very little precipitation: much of the high plateau is a dry desert. Despite the constant cold, many who work there find it a pleasant, stable environment, free from the melting snow, gusting winds and generally unpredictable weather of the coastal **'banana belt'**. Weather in coastal regions, including Antarctic Peninsula and the islands of the Scotia Arc, is dominated by a succession of depressions, which circle the continent eastward at rates of 600 to 1000 km per day. Each depression brings a recognizable weather sequence of lowering cloud, warm winds, rain or snow, often the heavy snowfall that feeds and perpetuates coastal glaciers and ice shelves. Their departure is marked by a brief spell of colder air and clear skies. Depressions driving inland from the coast bring systemic winds and heavy snowfall, often to the lower plateau of West Antarctica, less frequently to East Antarctica. To the warmer Southern Ocean islands successive depressions bring year-round overcast, dismal climates and buffeting winds, with rain and sleet in summer, sleet and heavy snowfall in winter.

Atmospheric Heat Transport. Transfer of heat from one part of Earth's surface to another by movements within the lower atmosphere, an important process in determining polar temperatures. The atmosphere above Antarctica loses more heat to space by infra-red radiation than it gains from incoming **solar radiation**. Satellite measurements show an annual net cooling of 90 Wm^{-2} at the top of the atmosphere for the region south of 70°S. To balance this cooling, the atmosphere transports heat from mid-latitudes into the polar regions by three main pathways:

- *Mean atmospheric circulation*: the outflow of cold katabatic winds at low levels and compensating inflow of warmer air aloft represent a net transport of heat into the Antarctic atmosphere which balances about 60 per cent of the net radiative cooling.
- *Depressions*: cyclones circling around Antarctica and spiralling in towards the continent transport much heat from lower latitudes, providing about 35 per cent of the warming.
- *Quasi-stationary planetary waves*: atmospheric waves with lengths of several thousand km, forced by high mountains and rough terrain. Since, except in Antarctica itself, there is little large-scale orography in mid- to-high latitudes of the southern hemisphere, these contribute only about 5 per cent of the total heat transport to Antarctica.

(JCK)

Auckland Islands. 50°40'S, 165°52' and 166°10'E. An archipelago of warm temperate islands in the southern South Pacific Ocean. They stand 400 km (250 miles) south of New Zealand on the Campbell submarine plateau, a close cluster of two large islands and several smaller islets and stacks. Auckland Island, the largest, is 30 km (18.7 miles) long, up to 20 km (12.5 miles) across, and rises to over 600 m (1968 ft), with precipitous western cliffs and a deeply indented eastern seabord. Adams Island immediately to the south is 22 km (13.7 miles) long and up to 8 km (5 miles)

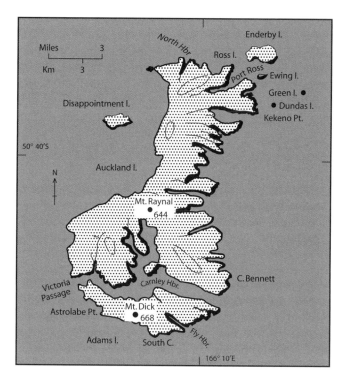

Continually eroded by heavy seas, the Auckland Islands have high western cliffs, gentler eastern slopes. There is no permanent land ice. Heights in m

wide, rising to 667 m (2188 ft) in Mt Dick, the archipelago's highest peak. Enderby, Rose, Ewing and Dundas Islands to the north and Disappointment Island to the west are much smaller. All are volcanic islands, the products of two main centres of activity that produced extensive flows of lava some 15–20 million years ago. This was followed by prolonged marine erosion, particularly along the western seaboard, and glaciation. Basement granites appear on Musgrave Peninsula in the south of Auckland Island. There is no current volcanic activity or permanent ice. The climate is mild and damp, with mean annual temperature 8.2 °C, range of monthly means 5.9 °C, rainfall of 150–200 cm and constant westerly winds.

The islands support a rich flora including over 200 species of flowering plant and fern. Sheltered eastern flanks of the main island feature a forest of flowering southern rata trees: Ewing Island is forested with *Olearia*. Uplands carry scrub, tussock meadows, moors and bogs. There is a rich seabird fauna including rockhopper, erect-crested and yellow-eyed penguins, wandering, royal, shy and light-mantled sooty albatrosses, many species of petrel, tern, skua and shag. Endemic pipits, teal, rail, banded dotterel, tomtits and snipe breed on the larger islands. Fur seals, elephant seals and sea lions are abundant.

The group was discovered in 1806 by the British sealer Abraham Bristow, and long became a haunt of sealers and whalers. Moriories and Maoris from New Zealand formed

small settlements during the 1840s, and from 1849 to 1852 **Enderby Settlement**, a whaling and farming settlement with population of up to 300, was established at Port Ross, Enderby Island. There were later attempts at sheep farming. Many ships were wrecked on the islands. Cape Expeditions, a New Zealand operation, placed coast watchers on the island 1941–45. Rabbits, house mice, domestic cats, cattle, sheep, pigs and goats have been introduced at different times with varying levels of success and environmental damage. The islands are currently administered by New Zealand as a National Nature Reserve and may be visited only by permission.
Further reading: Fraser (1986).

Aurora australis. Brightly coloured curtains or streamers of light that appear from time to time in the night sky over parts of the Antarctic region, caused by a stream of electrons travelling from the sun and impinging on the atmosphere. Beyond the outer limits of atmosphere the electrons become entrapped in Earth's magnetic field, and are channelled along lines of magnetic force that converge toward the geomagnetic poles. On striking the atmosphere, primary electrons produce showers of secondary and tertiary electrons which, some 300 km above Earth's surface, begin to collide with gaseous atoms and molecules. Exchanges of energy occur, resulting in emissions of radiation at characteristic frequencies, some of which are visible as light. Atomic oxygen high in the upper atmosphere emits a blood-red colour which appears to fill the sky: lower down it produces arcs, rays and curtains of green. Nitrogen struck by high-energy particles that have penetrated deep into the atmosphere emits a distinctive rust-red, often appearing as a fringe below green curtains. At higher levels nitrogen molecules produce rays of vivid blue-violet.

Auroras appear in an oval ring of the mesosphere at altitudes between 80 and 300 km, occurring most often along a broad ring, the southern auroral zone, some 2000 km in diameter, centred about the geomagnetic pole. Though present by day and night, they become visible only against a dark sky, producing shifting displays that typically extend over several hundred km^2. They are most spectacular during periods of intense solar activity, when the stream of incoming electrons is strongest. Each night brings a unique performance, that shifts from evening to morning with the rotation of Earth, and from minute to minute with gusts of the solar wind. Close to the geomagnetic pole, where lines of force impinge vertically, the curtains and rays of light appear to hang overhead. Away from the pole they appear inclined, converging on a distant point. Because the geomagnetic pole is located in East Antarctica, auroras occur most frequently and brilliantly over the Pacific sector, where observers see them every second or third night. They are much rarer over Antarctic Peninsula and the Weddell Sea, seldom showing more than a distant glow in the southern sky.

Australasian Antarctic Expedition 1911–14. The first Australian scientific expedition to Macquarie Island and Antarctica. Douglas **Mawson**, a Yorkshire-born Australian physicist and geologist, who had served in Victoria Land with Shackleton's **British Antarctic (*Nimrod*) Expedition 1907–9**, in 1910 planned a predominantly scientific and survey expedition to the coast of East Antarctica immediately south of Australia. He bought *Aurora*, a 35-year-old wooden sealer of 386 tons with auxiliary steam power, refitted it for scientific work, and recruited a team of 31, mainly scientists and technologists, to undertake a substantial research programme. Expedition equipment included 49 dogs, radio communications and a Vickers REP monoplane, a light aircraft with detachable sledge-runner undercarriage. Antarctica's first heavier-than-air machine was damaged before it reached Antarctica and never flew, but became a propellor-driven tractor. Mawson planned originally for three bases on the Antarctic mainland (subsequently settling for two), and one on Macquarie Island that would provide a radio link between the expedition and homeland. *Aurora* (Capt. John King Davis) left Hobart, Tasmania, on 2 December 1911, carrying the Antarctic party, accompanied by *Toroa*, a transport that landed a team of five on Macquarie Island. On 4 January 1912 the southern party sighted the ice cliffs of the mainland. After brief exploration along the coast Mawson found an ice-free area, which he called Cape Denison, and set up his main base of 18 men. On 19 January Davis in *Aurora* headed westward along the coast, establishing a smaller Western Base at the west end of Shackleton Ice Shelf. Leaving a party of eight under the leadership of Frank **Wild**, *Aurora* returned to Hobart, arriving on 12 March 1912.

Cape Denison turned out to be one of the world's windiest locations, subject to frequent katabatic gales and hurricanes pouring from the inland ice cap. During the first year winds averaged almost 80 km (50 miles) per hour (see **home of the blizzard**). Despite this Mawson and his companions maintained a comprehensive programme of magnetic, auroral, meteorological and biological observations, and extensive coastal and inland sledging journeys. Constant offshore winds destabilized local sea ice, so most of the sledging was done on inland ice crossed by dangerously crevassed ice streams. In August Mawson, B. E. S. Ninnis and C. T. Madigan established an ice-cave depot, 'Aladdin's cave', 8 km (5 miles) inland, from which their sledging journeys started. In September three sledging parties reconnoitred east, south and west, constantly hampered by cold and violent winds, establishing routes for longer journeys later in the season. Not until November did the weather improve enough for the five main parties, each of three men, to explore more thoroughly.

An Eastern Coastal man-hauling party led by meteorologist C. T. Madigan left on 8 November 1912, exploring 270 miles first over shelf ice, crossing the Mertz Glacier Tongue and descending to the fast ice, crossing the Ninnis Glacier Tongues and continuing eastward over the fast ice to within sight of Cape Freshfield. They returned via Horn Bluff and Penguin Point, to reach Aladdin's Cave on 15 January 1913. A Near-East party led by geologist F. L. Stillwell meanwhile man-hauled and mapped in the eastern area between Cape Denison and Mertz Glacier. A Southern party led by magnetician R. Bage left on 10 November 1912, heading inland up the continental slope toward the South Magnetic Pole. On 22 December, 70°36'S, at a distance of 480 km (301 miles) from base and an elevation of almost 1800 m (5900 ft), they estimated that the Magnetic Pole still lay 80 km (50 miles) ahead of them, but had to turn back. They returned to Cape Denison on 10 January 1913. A Western party led by engineer F. H. Bickerton left on 20 November and took the tractor sledge on a shorter journey among the coastal mountains. The sledge, converted from the damaged aircraft, with a dangerously whirling propellor, proved temperamental and at times unmanageable. On 20 November, when the party left with stores for Aladdin's cave, it successfully pulled a heavy load one mile uphill in three minutes. On 3 December the engine ran roughly and on the following day gave out altogether. Thereafter the party man-hauled, on a journey that took them over 270 km (170 miles) westward over relatively featureless ice slopes to 138°E, the longitude of Cape Robert. They returned on 17 January 1913.

Mawson himself led the longest journey with companions Ninnis, a British army officer, and X. Mertz, an experienced Swiss mountaineer and skier, taking three sledges with dog teams. Leaving base with Madigan's party on 8 November, they parted company four days later to head southeastward over the icecap. Crossing the heavily-crevassed Mertz and Ninnis ice streams, they continued eastward to a point beyond Cape Freshfield. Preparing to return, they concentrated the remaining loads onto two sledges. On 14 December, some 500 km east of Cape Denison and almost 730 m (2400 ft) above sea level, Ninnis, his sledge and team of six dogs disappeared down a crevasse. With them went much of the food, including rations for the remaining six dogs and essential camping equipment. Mawson and Mertz turned for home, making slow progress and killing the starving dogs one by one to eke out the rations. On 1 January 1913, still 160 km (100 miles) from base, Mertz complained of abdominal pains. Weakening rapidly, he died on 7 January, leaving Mawson to struggle home alone. He returned to Cape Denison in the afternoon of 8 February, only to learn that, after a long wait, *Aurora* had sailed that morning to relieve the Western Base. Bad weather and advancing winter made it impossible for Davis to return that season, so Mawson and six companions remained to continue their scientific observations for a second winter.

At the Western Base weather conditions were only marginally better throughout the year, heavy snowfall combining with strong winds to create frequent blizzard conditions and huge snow drifts. The base was built 585 m (640 yards) from the ice cliffs of Shackleton Ice Shelf, which at this point extended some 27 km (17 miles) from the ice-clad coast. In early March Wild made a reconnaissance and depot-laying journey south, taking 25 days to reach a point some 40 km (25 miles) inland, 800 m (2600 ft) up on the icecap. On 22 August he led a depot-laying journey eastward over Shackleton Ice Shelf to Hippo Nunataks, Cape Charcot, returning on 15 September after experiencing winds that tore the tent to pieces and a near-miss by an avalanche. On 26 September Dr S. E. Jones led a journey to the southwest over old fast ice to Helen Glacier, returning by an overland route on which a blizzard delayed them for 17 days. The main eastern summer journey, led by Wild, started at the end of October. The party of four, with three dogs, rounded Hippo Nunataks and Cape Gerlache, and headed south to Mt. Barr Smith, a nunatak over 1200 m (4000 ft) high, which they reached on 22 December. They returned to base on 5 January 1913. Meanwhile Jones led a man-hauling party of three westward to Kaiser Wilhelm Land. On the outward journey along the ice shelf they diverted to the sea ice at a group of islands (later to be called the Haswell Islands), where they discovered a large colony of emperor penguins, then only the second-known, and the first-known colony of Antarctic petrels. On 22 November they reached and climbed **Gaussberg**, linking with the discoveries of the **German South Polar Expedition 1901–3**. The party returned to base on 21 January 1913.

The station on Macquarie Island, a prefabricated hut on the shore of North-East Bay, on the northern end of the island, operated as part of the expedition from 22 December 1911 to 18 November 1914. During this period the five members of the party, led by meteorologist G. F. Ainsworth, completed the first topographical and biological surveys of the island, maintaining tidal and meteorological observations, and operating irregular radio schedules with Australia and Cape Denison from a transmitter on Wireless Hill.

Capt. Davis had arrived at Cape Denison in *Aurora* on 13 January 1913, expecting to clear the base. Madigan's party returned on 16 January and Bickerton's two days later. When Mawson's party had failed to return, Davis departed early on 8 February on the 2400 km (1500 mile) cruise to relieve the Western Base, leaving the relief party of six with a year's supplies of food and fuel. Later in the day Davis received a radio message to say that Mawson had arrived, but strong winds prevented him from returning

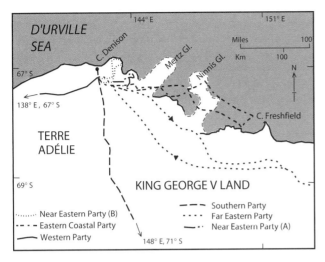

Major sledging journeys from the main expedition base, Cape Denison

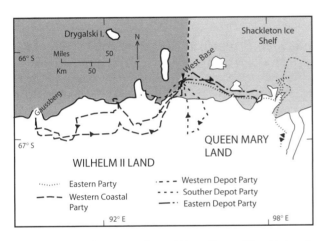

Major sledging journeys from West Base, Shackleton Ice Shelf

to pick the party up. Reaching the Western Base on 23 February, he embarked the men and returned as directly as possible to Hobart. In the following season he left Hobart on 19 November, relieved the Macquarie Island base, and ran south to reach Cape Denison on 14 December. *Aurora* made a short voyage of biological and oceanographic exploration along the Antarctic coast, returning to Australia on 26 February 1914.
Further reading: Mawson (1915).

Australian Antarctic Territory. Triangular sector of East Antarctica bounded by 45°E (bordering Dronning Maud Land) and 160°E (bordering Ross Dependency), and extending from 60°S to the South Pole, but excluding **Terre Adélie**. The territory was defined by a British Order in Council of February 1933, provision being made for its administration by the Government of Australia. Responsibility was accepted by Australia in 1936. The territory includes Enderby, Kemp, Mac.Robertson, Princess Elizabeth, Wilhelm II, Queen Mary, Wilkes and George V lands.

Automatic weather stations. See **Weather stations**.

Avalanche. Mass of snow and ice, with or without included rock, sliding rapidly downhill. An avalanche occurs when a mass of material, held in place on a slope by inertia and friction, is loosened and able to accelerate downhill. Loosening may result from melting, releasing water that acts as a lubricant: small earthquakes and other vibrations are contributory causes. Snow avalanches may result from changes in crystalline structure deep within the snowpack, providing surfaces that slide readily over each other. Dry snow in motion gathers air within its mass, and is likely to travel further and faster than wet snow. Water-sodden snow is ponderous and destructive, and may solidify on coming to rest. Avalanches are a constant feature in mountainous regions of Antarctica, particularly in regions of high snowfall, and in summer when diurnal melting is most likely.

Avery Plateau. 66°50′S, 65°30′W. Ice plateau rising to 1800 m between the Loubet and Foyn coasts of Graham Land, Antarctic Peninsula, named for sealing Capt. G. Avery of the cutter *Lively*, who in 1832 accompanied Capt. J. Biscoe in exploring the Loubet Coast.

Avian Island. 67°46′S, 68°54′W. Low island off the southwestern tip of Adelaide Island, named for its extensive colonies of breeding birds and seals. Originally designated SSSI No. 30, the island with its surrounding littoral zone has been redesignated SPA No. 21 (Avian Island).

B

Bach Ice Shelf. 72°00′S, 72°00′W. Semi-circular ice shelf occupying a large bay in the southern end of Alexander Island, opening onto Ronne Entrance. It was named for the German composer J. S. Bach, one of a suite of names in the area commemorating musicians.

Bacharach, Alfred Louis. (1891–1966). British food chemist and musicologist. He developed research interests in medicine and pharmacy, advising British and polar expeditions on nutrition. Bacharach Nunatak, Loubet Coast, Antarctica was named for him. He died on 17 July 1966. (CH)

Bage, Robert. (c. 1888–1915). Australian engineer and explorer. A graduate in engineering of Melbourne University and officer in the Royal Australian Engineers. Joining the **Australasian Antarctic Expedition 1911–14,** he was assigned to the Main Base Party, with responsibility for chronometers, astronomical observations, tidal records and magnetometry. Bage led the southern sledging party to the South Magnetic Pole. He served with the Australian army in World War I and was killed in action in the Dardanelles Campaign.

Bagshawe, Thomas Wyatt. (1901–76). British polar explorer. Born on 18 April 1901, he entered Cambridge University to study geology. He abandoned his studies when invited to join the **British Imperial Antarctic Expedition 1920–21**, and sailed south to Deception Island in a whaling ship with fellow-expeditioner Maxime **Lester**. The leader John L. Cope and second-in-command George Hubert Wilkins arrived later and the party transferred to Andvord Bay, on the Danco Coast of Graham Land where they intended to set up a base. However, shortage of funds curtailed the expedition, and Cope and Wilkins left. Bagshawe and Lester decided to remain, setting up a base in a stranded water boat on a point between Paradise Harbour and Andvord Bay. Though ill-equipped and with few comforts, they lived on Waterboat Point from January 1921 to January 1922, completing a year-round programme of scientific observations, including a study of the breeding biology of the neighbouring gentoo penguins. Bagshawe later joined the family engineering business in Dunstable, Bedfordshire, and became involved in local affairs, supporting and endowing Luton Museum, of which he became honorary director. In 1949 he was appointed High Sheriff of his country. He died at Worthing on 28 January 1976.

Bailey Ice Stream. 79°00′S, 30°00′W. Ice stream flowing west-southwest from the high plateau of Coats Land to Filchner Ice Shelf. The stream was named for British glaciologist J. T. Bailey, who had studied it, and was killed in a crevasse accident nearby.

Bailey Peninsula. 66°17′S, 110°32′E. An ice-free area between Newcomb and O'Brian bays, close to Casey Station, Budd Coast. An area of approximately 0.5 km^2 (0.2 sq miles), designated SSSI No. 16 (Northeast Bailey Peninsula), supports very rich communities of lichens, mosses and liverworts.

Bakewell, William L. (1888–1969). American seaman and explorer. Born in Illinois, USA, he became a seaman, and was a dockside worker in Montevideo when Shackleton's British **Imperial Transantarctic Expedition 1914–17** passed through. He stowed away in *Endurance*: on discovery he joined the crew, and underwent all the privations of wintering on sea ice and at Point Wild. On return to South America he became a sheep farmer, later farming in the USA. He died at Marquette, Michigan on 21 May 1969.

Bakutis Coast. 74°45′S, 120°00′W. Part of the coast of Marie Byrd Land, West Antarctica, facing the Amundsen Sea. Bounded by Dean Island in the east and Cape Herlacher in the west, the coast consists of large ice covered islands embedded in the Getz Ice Shelf. It was delineated mainly from aerial photographs by USN Operation Highjump, 1946–47, named for Rear Adml. Fred E. Bakutis

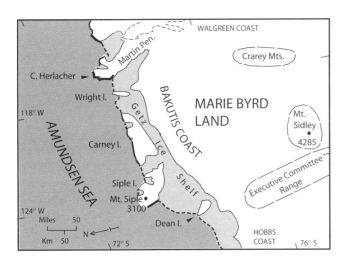

Bakutis Coast

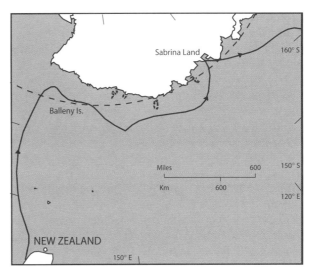

John Balleny's voyages

USN, who commanded US Naval Support Force Antarctica 1965–67.

Balleny, John. (d. 1839). Early nineteenth-century British navigator and explorer. Born in the mid-to-late eighteenth century, Balleny is on record in 1798 as part-owner of a cargo ship, *Blenheim*, and from 1814 onward as master of several ships trading along the east coast of England. In July 1838 he was given command of a small schooner, *Eliza Scott*, 138 tons, owned by the London-based whaling and sealing firm of Messrs **Enderby**, and commissioned to explore for new sealing lands south of New Zealand. Accompanied by the cutter *Sabrina*, 47 tons (Capt. Thomas Freeman), he reached New Zealand in early December and Campbell Island on 16 January 1839. From there they sailed south-southeast. On 1 February, in 69°S, 174°E, they encountered the edge of the pack ice. Easing westward in search of a passage through the ice, on 9 February Balleny saw the cluster of ice-covered islands in 66°S that now bear his name. They approached one of the larger islands, where Capt Freeman jumped ashore to collect rocks from a narrow beach. On 2 March, further west and in 65°S, both Balleny and Freeman sighted land to the south–possibly part of what is now the coast of Victoria Land. Unable to approach closer, the two ships headed northward. On the night of 24 March, in foul weather, *Sabrina* was lost with all hands. *Eliza Scott* returned safely to Britain, with little to show for the venture but a new group of islands and possible new stretch of ice-girt continental coast. The main islands of the **Balleny Islands** group, Young, Borradaile, Buckle and Sturge, bear the names of unfortunate shareholders in the enterprise: the nearby coast is named for the cutter *Sabrina* that did not return.

Further reading: Jones (1992); Mills (1905).

Balleny Islands. 66°55′S, 163°20′E. Chain of three large, ice-covered volcanic islands (Sturge, Buckle and Young) and three small islands (Sabrina, Borradaile, Row), oriented northeast–southwest, 250 km (156 miles) from Cape Adare, Victoria Land. Sturge Island, the southernmost and largest, rises to Brown Peak (1705 m, 5592 ft), the islands' highest point. The group was discovered in 1839 by British sealing captain John Balleny. Geologically it is young, representing activity not more than 10 000 years ago. The islands are ringed by steep cliffs, testifying to rapid erosion. Intermittent fumarolic activity was reported

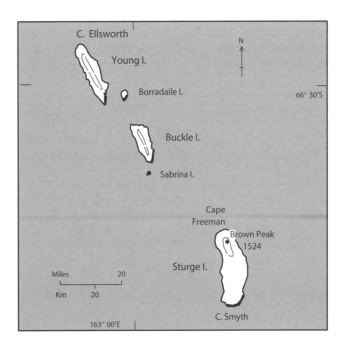

Balleny Islands. Steep-sided and heavily glaciated, the islands are usually surrounded by pack ice and difficult to access. Heights in m

on Buckle Island between 1839 and the end of the century, but has not appeared since. Sabrina Island is designated Specially Protected Area No. 4. The islands are administered by New Zealand as part of the Ross Dependency.

Banana belt. Derisory term applied to coastal regions of Antarctica by scientists and support staff who work in colder inland areas.

Banks, Joseph. (1743–1820). British expedition botanist. Born 15 February 1743, son of a wealthy land-owner, Joseph became interested in botany while at Christ Church College, Oxford. Personal wealth allowed him to travel as a scientist and collector, first in HMS *Niger* to Labrador and Newfoundland (April 1766–January 1767), later in the bark HMS *Endeavour* on Capt. James **Cook**'s first expedition to the Southern Ocean (August 1768–July 1771), and in the brig *Sir Lawrence* to Iceland (July–November 1772). Subsequently he worked on his collections, publishing extensive and valuable accounts of his travels. Appointed President of the Royal Society in 1778, he was awarded a baronetcy in 1781. Banks died on 19 June 1820.
Further reading: Lysaght (1971); O'Brian (1987).

Banzare Coast. 67°00′S, 126°00′E. Part of the coast of Wilkes Land, East Antarctica. Bounded by Cape Southard in the east and Cape Morse and Porpoise Bay in the west, the coast consists almost entirely of ice cliffs and extensive ice shelves. It was explored and photographed from the air by the **British, Australian and New Zealand Antarctic Research Expedition 1930–31**, the initials of which form its name.

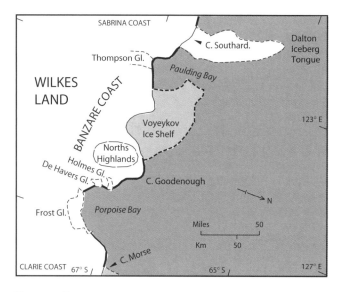

Banzare Coast

Barff Peninsula. 54°19′S, 36°18′W. A long, tussock-covered peninsula terminating in Barff Point, and forming a sheltering eastern arm for Cumberland Bay, South Georgia. Point and peninsula were named for Lt. A. D. Barff RN, who surveyed the bay in 1906.

Barlas, William. (1888–1941). British colonial magistrate. Born in Scotland in 1888, he went to the Falkland Islands as a teacher in 1908. In 1920 he was appointed deputy magistrate on South Georgia, and magistrate in 1928. Over the years he gained much practical experience of the island and the whaling industry. He was killed by an avalanche close to Grytviken on 2 September 1941.

Barne, Michael. (1877–1961). British seaman and explorer. A Royal Navy officer, he volunteered for Scott's **British National Antarctic Expedition 1901–4,** in which he took responsibility for magnetometry and soundings, and was involved in several sledge journeys. He served with distinction in World War I, retiring from the navy with the rank of captain in 1919. Cape Barne, McMurdo Sound, is named for him.

Barrier. Antiquated name for an ice cliff terminating a major ice shelf, extended to include the ice shelf itself. The term was widely used by early twentieth-century expeditions to describe the Ross Ice Shelf and its terminal cliff: see also **ice front**.

Barrier winds. Winds that blow parallel to mountain ranges and other natural barriers. In parts of Antarctica, cold air flowing at low levels is diverted by a high mountain barrier, generating a pressure gradient that increases the flow to a strong wind. Such barrier winds are observed, for example, along the Transantarctic Mountains fringing the Ross Ice Shelf, and along the eastern (Weddell Sea) side of Antarctic Peninsula. In this latter region the southerly barrier wind helps to drive sea ice northwards out of the Weddell Sea. By bringing cold continental air northwards, it keeps the climate to the east of the Peninsula much colder than along the relatively milder maritime west coast, and brings anomalously cold conditions to the South Orkney Islands.
(JCK)

Barwick Valley. 77°21′S, 161°10′E. An ice-free valley opening into Victoria Valley, part of the Victoria Land dry-valley system. Named for New Zealand biologist R. E. Barwick, it is regarded as one of the least disturbed and contaminated of the dry valleys, with a unique polar desert ecosystem. An area of approximately 279 km² has been designated SSSI No. 3 (Barwick Valley).

Bastin, François. (1920–69). Belgian polar meteorologist. Born in Angleur on 26 November 1920, he escaped from occupied Belgium to Britain in 1942 and joined the Belgian section of the Royal Air Force, specializing in meteorology. Continuing in the peacetime Belgian meteorological service, in 1957 he organized a meteorological programme for the Belgian International Geophysical Year expedition to Antarctica, and in 1959 led the second expedition at Base Roi-Baudouin. On return he became chief of operations of Centre National de Récherches Polaires de Belgique. He died on 6 October 1969 following a motor accident.

Beacon Dome. 86°08′S, 146°25′W. High dome-like mountain rising to 3010 m. (9873 ft), forming part of Watson Escarpment, La Gorce Mts., at the southern end of the Transantarctic Mountains. The dome is named for its Beacon series of sedimentary rocks.

Beaked whale, southern. *Berardius arnuxii*. A small, rarely seen whale of the cool southern oceans, this is the southern counterpart of the better-known Baird's beaked whale of the north Pacific Ocean. Known mainly from beached specimens and skeletal remains, it grows to 10 m long and weighs up to 8 tonnes. Specimens are occasionally seen in Antarctic waters, though easily confused with the smaller and possibly more common southern **bottlenose whale**. They appear to feed in deep water, mainly on fish and squid.
Further reading: Carwardine (1995); Watson (1981).

Bear Peninsula. 74°36′S, 110°50′W. A mountainous, ice-covered peninsula of Walgreen Coast, Marie Byrd Land, discovered during United States Navy Operation Highjump 1946–47, and named for USS *Bear*, the flagship of the **United States Antarctic Service Expedition 1939–41**, aircraft from which had earlier surveyed the coast.

Beaufort Island. 76°56′S, 166°56′E. A steep-sided, semi-circular island of volcanic origin standing 21 km (13 miles) north of Ross Islands in the Ross Sea. Ice-capped on its western, weather side, with steep eastern cliffs, it forms part of the Ross Archipelago. It was charted by Capt. J. C. Ross RN in 1841 and named for the British Hydrographer to the Navy, Capt. F. Beaufort RN. The island accommodates an emperor penguin colony on nearshore sea ice at the north end, and a large colony of Adélie penguins on a raised beach at the southwestern end. Approximately 18.4 km² (7.19 sq miles) in area, it is designated SPA No. 5 (Beaufort Island), to protect its substantial and varied avifauna and preserve the natural ecological system as a reference area.

Béchervaise, John Mayston. (1910–98). Australian geographer and expeditioner. He served with the Australian National Antarctic Research Expeditions as officer-in-charge at Heard Island 1952–53, and at Mawson 1955–56 and 1959–60. An art historian, he worked also as a school teacher and lecturer, and led several geographical expeditions.

Beckmann Fjord. 54°03′S, 37°12′W. Fjord off the east side of Bay of Isles, South Georgia, named in 1912 by American ornithologist R. C. Murphy for Capt. Beckmann, gunner of the whale-catcher *Don Ernesto*, who was killed in an accident nearby.

Bedrock surface of Antarctica. Echo-sounding and other techniques have provided data on the bedrock topography and allowed speculation on the bedrock geology. The illustration in the **Antarctica** entry shows Antarctica as it might appear several million years after the ice is removed when, freed of the weight of the ice, it has risen to a level of isostatic equilibrium with the surrounding Earth's crust. The area of bedrock underlying East Antarctica appears much smaller than its former ice-mantled counterpart, and is crossed by several ranges of high mountains. The South Pole itself lies on an extensive lowland plain. A deep, narrow sea separates East Antarctica from the neighbouring archipelago of West Antarctica with huge, shallow embayments occupying the former sites of the Ross and Filchner-Ronne ice shelves. Deep fjords and basins separate the West Antarctic islands, which are mountainous and now similar in character to the offlying South Shetland Islands.

Belgian Antarctic Expedition 1897–99. The first of several expeditions encouraged by deliberations of the 1895 sixth **International Geographical Congress**. The Belgian Antarctic Expedition was initiated and led by Baron Adrien **de Gerlache de Gomery**, a lieutenant in the Royal Belgian Navy, with the support of the Belgian government, the Brussels Geographical Society, and public and private donations. In 1896 de Gerlache bought a three-masted, steam powered Norwegian sealing vessel *Patria* (250 tons) renaming it *Belgica*. In the following year he assembled an international team, including Belgian magnetic observer Emile Danco, two Polish physical scientists Henryk Arctowski and Antoine Dobrowolski, and a Romanian botanist, Emile Racovitza. Officers and crew included both Belgians and Norwegians. The Norwegian second mate, Roald **Amundsen**, and American ship's surgeon Dr Fredrick **Cook**, were later to make their individual marks in polar exploration. Leaving Ostend on 24 August 1897, de Gerlache headed first for Rio de Janiero, where he picked up Cook, then for Tierra del Fuego, where he spent some weeks in survey work, eventually reaching the South Shetland Islands on 20 January. From there *Belgica* sailed south into a wide channel (now Gerlache Strait) between Antarctic mainland and an archipelago of large islands, for the first time distinguishing Antarctic Peninsula from the mountainous islands now collectively called Palmer Archipelago – which Gerlache named Liège, Brabant and Anvers for Belgian provinces. On 22 January 1898, during a severe storm, Carl Wiencke, a young seaman, fell overboard and was swept away.

After a painstaking survey of the strait, *Belgica* passed out through the islands on 12 February into the open sea, heading south toward the Antarctic Circle. Despite the late season, de Gerlache hoped to make a landfall in the unknown region between Biscoe's Adelaide Island and Bellingshausen's Alexander Land. On 3 March, in 71°30′S, well to the west of Alexander Land, the ship became trapped in pack ice, during the next 13 months drifting over 1000 km (625 miles) with the pack ice. The expedition became the first on record to winter south of the Antarctic Circle, indeed mostly south of 70°S. Captive within the pack ice, *Belgica* zig-zagged in several directions, but overall south, mainly over a continental shelf some 450 m (1500 ft) deep. On starry nights the navigating officer, George Lecointe, tracked the ship's wanderings by astronomical observations. The sun disappeared from mid-May for over two months, plunging the expedition into a state of despondency that de Gerlache, never an inspiring leader, was unable to combat. Lt. Danco died in early June, and several others of the ship's company became melancholic. However, oceanographic and marine biological observations continued. The survival and ultimate success of the expedition, especially during the long, trying winter, were due mainly to the efforts of Amundsen and Cook, both of whom had previous experience of working through Arctic winters.

Belgica remained fast in the ice for the winter, spring, and much of the second summer. After strenuous efforts to free her by blasting and sawing, release came finally on 14 March 1899, and the ship returned to Punta Arenas and home. Maps, charts and impressive volumes of scientific results were published by the Belgian government, setting a useful precedent for the publication of results from later voyages and expeditions.

Further reading: Cook (1900).

Belgicafjella. 72°35′S, 31°15′E. (Belgica Mountains). Isolated group of low mountains and nunataks in the Thorshavenheiane, eastern Dronning Maud Land. Discovered by a Belgian survey party, they were named for the ship of the **Belgian Antarctic Expedition 1897–99**.

Belgrano II Station. See **General Belgrano stations.**

Bellingshausen, Fabian Gottlieb Benjamin von. (1778–1852). (Thaddeus Thaddevich Bellingshausen). German naval officer in the Imperial Russian Navy, explorer of the Southern Ocean. Born 18 August 1778 in Osel, Estonia, Bellingshausen joined the Imperial Russian Navy and served as a midshipman on Admiral Adam von Kreusenstern's round-the-world expedition of 1803–6. An able and determined navigator, in 1819 he was given command of the **Russian Naval Expedition 1819–21,** including two ships, *Vostok* and *Mirnyy*, the latter under the command of Capt. M. P. Lazarev. The expedition circumnavigated the world, making important contributions to exploration in Antarctica and the Southern Ocean, and also among islands in the southern Pacific Ocean. Bellingshausen ended a distinguished naval career as admiral commanding the Baltic port of Kronstadt.

Fabian Gottlieb von Bellingshausen (1778–1852). Photo: SPRI

Bellingshausen Island. 59°25′S, 27°03′W. Easternmost island of the Southern Thule group, at the southern end of the South Sandwich Islands: a roughly triangular island, consisting of a volcanic cone and crater, partly ice-capped, rising to 253 m (830 ft) and resting on a broad field of consolidated lava. The island was charted in 1819 by the Russian explorer F. von Bellingshausen and later named for him. Geologically one of the youngest islands in the group, it has not been seen to be volcanically active, but there are several points of fumarolic activity around the cone and on the lava platform.

Bellingshausen Sea. 71°S, 85°W. Marginal sea off West Antarctica between Alexander Island and Thurston Island, bordered by the English and Bryant coasts of Ellsworth Land. There is no definitive northern boundary: Peter I Øy, 500 km (300 miles) from the coastal ice shelf, is generally deemed to lie in this sea. The coasts are ice-bound throughout the year: even in summer coastal waters are heavily invested with pack ice and icebergs. The sea was named for Fabian von Bellingshausen, leader of the **Russian Naval Expedition (1819–21)**, who was probably the first to enter its waters.

Bellingshausen Station. 62°11′S, 58°58′W. Russian permanent year-round station on the shore of Ardley Cove, King George Island, South Shetland Islands. It has operated continuously since its establishment in February 1968.

Belt of pack ice. Extensive, elongate area of pack ice, shaped by winds and tidal movements, often a hindrance to ship navigation.

Bennett, Messrs. Daniel, and Son. A London-based firm of oil merchants, prominent in sealing and whaling during the late eighteenth and early nineteenth centuries. Entering the trade in the 1780s, Daniel and his son William, based at Wapping on the lower River Thames, expanded their business rapidly, sending many ships to the Southern Ocean. Like their rivals in the trade, Messrs **Enderby**, the Bennetts gave their masters opportunities to discover new sealing grounds. While in their employ **George Powell** discovered the South Orkney Islands, and **Peter Kemp** discovered a sector of the coast of Antarctica.

Berg. See **Iceberg**

Berg, Thomas Erik. (1933–69). American polar pedologist. Born in La Crosse, Wisconsin, he was educated at La Crosse State College and the University of Wisconsin. From the late 1950s Berg worked closely with Prof. R. F. Black in studying permafrost and patterned ground, particularly on Ross Island and in the Victoria Land Dry Valleys, but also on the Antarctic Peninsula. In 1965 he joined the staff of the Research Council of Alberta, extending his studies to permafrost and periglacial features of southern Canada. He was killed in a helicopter accident in Victoria Land on 19 November 1969.

Berg Ice Stream. 73°42′S, 78°20′W. Ice stream flowing northward into Carroll Inlet, English Coast, Ellsworth Land. The stream was identified from aerial photographs and named for Capt. H. Berg USN, who commanded the icebreaker USS *Eltanin* 1964–65.

Bergschrund. Crevasse at the head of a glacier, between a containing rock wall and the main body of ice.

Bergy bit. Small **iceberg**, usually of glacier ice, of diameter less than 10 m, with less than 5 m showing above sea level.

Berkner, Lloyd Viel. (1905–67). American polar physicist and administrator. Born in Milwaukee, Wisc. on 1 February 1905, he trained himself in radio and became a ship's radio officer. In 1927 he completed a degree course in electrical engineering at the University of Minnesota. After postgraduate studies at George Washington University, he served successively with the US Bureau of Lighthouses and the Bureau of Standards, Department of Commerce. Joining **Byrd's First Antarctic Expedition 1928–30** as radio engineer, he became familiar with problems of long-distance radio propagation, and on returning to the Bureau of Standards undertook research in this field. As a US Navy officer in World War II he worked on proximity fuses and airborne radar. After the war he held a succession of research and administrative posts in government, universities and international science, helping to set up the International Geophysical Year 1957–58 and chairing the International Council of Scientific Unions and the national Academy of Science's Space Science Board. Berkner Island, a subglacial island between the Filchner and Ronne ice shelves, is named for him. He died on 4 June 1967.

Berkner Island. 79°30′S, 47°30′W. Totally ice-covered island some 320 km (200 miles) long, forming a dome approximately 975 m (3200 ft) high between the Ronne and Filchner ice shelves. Discovered in 1947, it was named for US radio engineer and physicist Lloyd V. Berkner.

Bernacci, Louis Charles. (1876–1942). British scientist and explorer. Born in Belgium of Italian and Belgian parents, he was brought up in England and Tasmania, training at the Melbourne Observatory. He joined the **British Antarctic (*Southern Cross*) Expedition 1898–1900** as meteorologist and magnetometry observer, and later the **British National Antarctic Expedition 1901–4** in a similar capacity. He served in the Royal Navy during both World Wars, but was invalided out in 1942 and died shortly after.

Bertram, George Colin Lawder. (1911–2001). British polar explorer and biologist. Born on 27 November 1911, Bertram read zoology at St John's College, Cambridge. While still an undergraduate he took part in expeditions to Bjornøya, Svalbard and Scoresby Sound, East Greenland, the latter in company with French explorer Dr J.-B. Charcot. After studying coral reefs in the Red Sea, Bertram joined the **British Graham Land Expedition 1934–37**. He studied seals that were killed for food, writing his conclusions in a scientific report and a PhD thesis. On returning to the UK in 1939 he wrote *Arctic Antarctic: The Technique of Polar Travel*. Early in World War II he was involved in research on Arctic clothing, and in 1940 joined the Colonial Office as a fisheries officer and adviser. After the war he returned to Cambridge, becoming a fellow and tutor of St. John's College, and part-time Director of the Scott Polar Research Institute 1949–56. Concerned with eugenics, in 1959 he wrote *Adam's Brood: Hopes and Fears of a Biologist*. In later life he studied manatees (*In Search of Mermaids: the Manatees of Guyana*, 1963), Pribilof Island fur seals and other ecological issues, and published two books of memoirs (*Antarctica, Cambridge, Conservation and Population: A Biologist's Story*, 1987, and *Memories and Musings of an Octogenarian Biologist*, 1992). He died in Graffham, Sussex, on 11 January 2001.

Bertrand, Kenneth J. (d. 1978). Geomorphologist and Antarctic historian. He was a member of the US Advisory Committee on Antarctic Names 1947–73, serving as chairman from 1962. He published *Americans in Antarctica, 1775–1948* (American Geographical Society, 1971) which is a comprehensive history of American Antarctic involvement. Bertrand Ice Piedmont is named for him. He died on 17 December 1978.
(CH)

Bickerton, Francis H. (1889–1954). Australian engineer and explorer. Born in Oxford, England in 1889, he trained in engineering, joining the **Australasian Antarctic Expedition 1911–14** as electrical engineer and motor expert. At Cape Denison he took charge of the air tractor sledge and wireless installations, and led the western sledging party across Terre Adélie to Cape Robert. He died on 21 August 1954.

Bigourdan Fjord. 67°33′S, 67°23′W. Fjord between Pourquoi Pas Island and Arrowsmith Peninsula, Falliéres Coast of Graham Land, opening onto northern Marguerite Bay. The fjord was named by J.-B. Charcot for the French astronomer Guillaume Bigourdan.

Bingham, Edward William. (1901–93). British expedition medical officer. Born in Dungannon, Co. Tyrone, he graduated in medicine in 1926 and was commissioned in the Royal Navy. He served with the British Arctic Air Route Expedition 1930–31, and in 1932 served in HMS *Challenger*, a hydrographic ship working off Labrador, where he over-wintered, gaining further familiarity with cold and dog driving. As medical officer on the **British Graham Land Expedition 1934–37** he took part in the final long sledging journey across Graham Land. In 1945–47 he led the newly-formed **Falkland Islands Dependencies Survey**, in which he passed on sledging and camping techniques to a new generation of British explorers. Following more mundane service appointments, he retired from the navy with the rank of commander in 1957. He died after a long and active retirement on 1 September 1993.

Bird Island. 54°00′S, 38°03′W. Steep, tussock-covered island off the northwestern end of South Georgia, from which it is separated by Bird Sound. The island was named by Capt. James **Cook** in January 1775 for the thousands of birds that he saw on and around the island. The ecology of the island's seabirds and seals has been studied intermittently since 1957, continuously since 1982, at a biological station of British Antarctic Survey in Jordan Cove.

Bird Island Station (Base BI). 54°00′S, 38°03′W. British research station at Freshwater Inlet, Jordan Cove, Bird Island, South Georgia. Established especially for biological research, the site was first occupied in 1957: the station was built in the following year and occupied by British and United States summer parties to 1964. Extra living quarters were added in 1963 by the United States Antarctic Research Program. Year-round studies began in 1983. The station has since been occupied continuously, except for a period of 25 weeks, April to mid-September 1982, when it was evacuated during and following the Argentine armed invasion of South Georgia.

Birds, southern oceanic. The most prominent animals of the southern regions, birds, breed on the shores and inland nunataks of Antarctica and on all the island groups of the southern oceans. Nearly all are seabirds. Some

Breeding seabirds. x = breeding locality

Locality	Antarctic continent	Antarctic Peninsula	South Shetland Islands	South Orkney Islands	South Sandwich Islands	Peter I Øy	Balleny Islands	South Georgia	Bouvetøya	Heard and MacDonald Islands	Iles Kerguelen	Iles Crozet	Macquarie Island	Marion and Prince Edward Islands	Falkland Islands	Campbell Island	Auckland Islands	Iles Amsterdam and St Paul	Tristan da Cunha & Gough Island
PENGUINS																			
Emperor penguin	x	x																	
King penguin								x		x	x	x	x	x	x				
Adélie penguin	x	x	x	x	x	x	x		x										
Chinstrap penguin		x	x	x	x	x		x	x										
Gentoo penguin		x	x	x	x			x			x	x	x	x	x	x			
Macaroni penguin		x	x	x	x			x	x	x	x	x		x	x	x			
Rockhopper penguin										x	x	x	x	x	x	x	x	x	x
Royal penguin													x						
Magellanic penguin															x				
ALBATROSSES AND MOLLYMAWKS																			
Wandering albatross								x			x	x		x		x	x	?	x
Royal albatross																x	x		
Amsterdam Island albatross																		x	
Black-browed albatross								x			x	x		x		x	x		
Grey-headed albatross								x			x	x	x	x		x			
Yellow-nosed albatross												x		x				x	x
Shy albatross																	x		
Sooty albatross											x	x		x				x	x
Light-mantled sooty albatross								x			x	x	x	x		x	x		
FULMARS																			
Antarctic giant petrel	x	x	x	x	x			x	x	x	x	x	x	x					
Hall's giant petrel								x			x	x	x	x	x	x			x
Southern fulmar	x	x	x	x	x	x		?	x										
Antarctic petrel	x																		
Cape petrel	x	x	x	x	x	x	x	x	x	x	x	x							
Snow petrel	x	x	x	x	x	x	x	x	x										

(continued overleaf)

BIRDS, SOUTHERN OCEANIC

Breeding seabirds. x = breeding locality (*continued*)

Locality	Antarctic continent	Antarctic Peninsula	South Shetland Islands	South Orkney Islands	South Sandwich Islands	Peter I Øy	Balleny Islands	South Georgia	Bouvetøya	Heard and MacDonald Islands	Iles Kerguelen	Iles Crozet	Macquarie Island	Marion and Prince Edward Islands	Falkland Islands	Campbell Island	Auckland Islands	Iles Amsterdam and St Paul	Tristan da Cunha & Gough Island
PRIONS, ETC																			
Antarctic prion	x	x	x	x	x			x		x	x	x	x	?	?		x		
Broad-billed prion												x		x				x	x
Fulmar prion										x	?								
Fairy prion								x		?		?	x	x	x		x		
Slender-billed prion											x	x	?	?	x				
Great-winged petrel											x	x		x				?	x
White-headed petrel											x	x	x				x		
Atlantic petrel																			x
Kerguelen petrel											x	x		x					x
Soft-plumaged petrel								x				x		x					x
Blue petrel										x	x	x	x	x	?				
Grey petrel											x	x	x	x		x		?	x
White-chinned petrel								x			x	x	?	x	x	x	x	?	x
Greater shearwater															x				x
Sooty shearwater													x		x	x	x		
Little shearwater																		?	
Flesh-footed shearwater																		x	
STORM PETRELS																			
Wilson's storm-petrel	x	x	x	x	x	?	?	x		x	x	x		x					
Black-bellied storm-petrel	?	x	x	x		?	?	x			x	x		?	x		x	?	x
Grey-backed storm-petrel								x			x	x	x	?	x		x		?
White-faced storm-petrel																	x	?	?
DIVING PETRELS																			
South Georgia diving-petrel								x		x	x	x	?	x		x			
Common diving-petrel								x		x	x	x	x	x	x	x	x	?	x

BIRDS, SOUTHERN OCEANIC

Breeding seabirds. x = breeding locality

Locality	Antarctic continent	Antarctic Peninsula	South Shetland Islands	South Orkney Islands	South Sandwich Islands	Peter I Øy	Balleny Islands	South Georgia	Bouvetøya	Heard and MacDonald Islands	Iles Kerguelen	Iles Crozet	Macquarie Island	Marion and Prince Edward Islands	Falkland Islands	Campbell Island	Auckland Islands	Iles Amsterdam and St Paul	Tristan da Cunha & Gough Island
CORMORANTS, SHAGS																			
Imperial Shag		x	x	x	x			x			x	x	x	x	x				
Heard Shag										x									
Rock shag																			
Auckland Islands shag																	x		
Campbell Island shag																x			
GULLS, SKUAS, TERNS																			
Kelp gull		x	x	x	x			x			x	x	x	x	x	x	x		
Dolphin gull															x				
Brown skua		x	x	x	x	?		x	x	x	x	x	x	x	x	x	x	x	x
South polar skua	x	x	x	x	?	?													
Antarctic tern		x	x	x	x			x	?	x	x	x	x	x	x	x	x	x	
Kerguelen tern										?	x	x		x					x
SHEATHBILLS																			
Snowy sheathbill		x	x	x	x			x											
Black-faced sheathbill										x	x	x		x					

Compiled from various sources by B. Houston and B. Stonehouse

of the world's largest seabird flocks feed in the oceans, and the islands are home to some of the world's densest concentrations of breeding seabirds. In both species and individuals, these far outnumber birds that breed or feed on land (see below). Nearly all southern breeding species leave their breeding areas in autumn to winter at sea (see **migration, seasonal**), some on the pack ice, some in temperate or tropical southern latitudes, a few north of the equator. Very few species of northern hemisphere breeding bird fly south to feed in southern waters: Arctic terns and sooty shearwaters are conspicuous exceptions. There is no massive annual incursion of northern birds to breed on southern islands, corresponding to the northward migration of waterfowl and seabirds to Arctic areas.

The table lists nine species of **penguin**, 38 species of **albatross** and **petrel**, five species of **cormorant**, and eight species of **gull**, **skua**, **tern** and **sheathbill** that breed on Antarctic, sub-Antarctic and temperate islands. Though

Chinstrap penguins parading to and from the very large breeding colony at Baily Head, Deception Island, South Shetland Islands

penguins are often thought to typify the region, petrels far outnumber other kinds of bird in species diversity, numbers of individuals and biomass. For further details see entries under group names. Non-breeding vagrant birds appear from time to time on southern islands, mostly from South America, New Zealand and some islands have attracted either native or introduced **land birds** from the nearest land masses, for example parrots and chaffinches from Australasia.

Biscoe, John. (*c.* 1794–1848). British mariner and explorer. Born in Enfield, near London, Biscoe joined the Royal Navy in 1812, where he quickly achieved promotion to midshipman: on his discharge in 1815 he was rated Acting-Master. His subsequent career is unrecorded until, in 1830, he was given command of the brig *Tula* (150 tons), a sealing vessel owned by the London firm of Messrs. **Enderby**, whalers and sealers. On 10 July, in company with the smaller cutter *Lively* (Capt. Smith, later Capt. G. Avery), Biscoe left London for his first voyage of sealing and exploration to the Southern Ocean. He reached the Falkland Islands on 9 November, where he took on water and provisions. Leaving on 27 November, Biscoe and Avery sailed eastward to the South Sandwich Islands. Between 19 and 31 December they visited Montagu Island, where they attempted unsuccessfully to land and take seals, and explored southward into fields of icebergs and heavy pack ice. Bearing southeastward and east, in latitudes below 66°S, Biscoe reported several possible sightings of ice-covered land. From 24 February 1831, following a more certain sighting in 51°E, he discovered a narrow sector of mountainous coast that, despite bad weather and contrary winds, he attempted to chart. Naming it Enderby Land, he

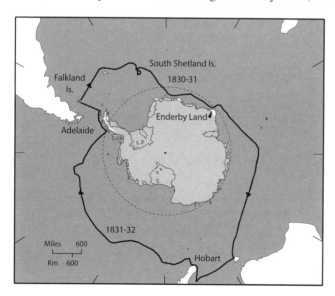

Biscoe and Avery's voyages from the Falkland Islands to Tasmania via Enderby Land, and from Tasmania back to the Falkland Islands via Adelaide Island and Graham Land

identified Cape Ann (named for his mother) and four prominent peaks that he named Mts Charles, George and Henry for his employers the Enderby brothers, and Mt Gordon for their brother-in-law. This was the first sighting of bare, relatively ice-free mountains in East Antarctica. Plagued by bad weather and scurvy among the crew, Biscoe retired northeastward, reaching Hobart, Van Dieman's Land (Tasmania) on 10 May. *Lively*, which had parted company in early March, rejoined him in August.

Biscoe and Avery left Hobart on 8 October 1831, visiting the Chatham and Bounty Islands off southern New Zealand, and setting course for the South Shetland Islands. After a voyage of four months, mostly in high latitudes and latterly among fleets of icebergs, on 15 February 1832 he sighted land with 'a most imposing and beautiful appearance.' that he named Adelaide Island for the British queen. During the following two weeks he continued northeastward, marking onto his chart the archipelago of low, ice-covered islets now called the Biscoe Islands, backed by a high, mountainous coast which he made several attempts to reach. On 21 February he 'pulled into a large inlet', now impossible to identify accurately, but probably on the coast of Anvers Island. There he took possession of the land for King William IV. The coast was later named Graham Land for Sir James Graham, then First Lord of the Admiralty. Biscoe and Avery continued on to the South Shetland Islands, where *Tula* was badly damaged. They returned to the Falkland Islands in late April, where *Lively* foundered during a storm. The crew were saved, but Biscoe in *Tula* returned to Britain alone, reaching London on 8 February 1833. His journals and notes testify to meticulous seamanship and record-keeping, and a lively and well-informed interest in natural phenomena, including weather, the Aurora Australis, formation of sea ice, seabirds, whales and seals. For his Antarctic discoveries he received gold medals from the Royal Geographical Society and the Paris Geographical Society.

Three months after his return Biscoe was again at sea on a voyage south, commanding the brig *Hopeful* with Lt Henry Rea, RN on board as hydrographer, and the brig *Rose* in attendance. By 23 October, when the expedition reached the Falkland Islands, command had passed to Rea: Biscoe's health may not have recovered from the hardships of his first expedition. *Hopeful* and *Rose* sailed for the South Shetland Islands and Adelaide Island, where *Hopeful* was crushed in the pack ice. Biscoe completed a second circumnavigation of Antarctica in *Rose*, returning to London in 1835. He continued in command of Enderby ships for several years, mostly in New Zealand and Australian waters. A voyage in 1838–39 in *Lady Emma* took him from Sydney, Australia to Campbell Island and a third circumnavigation of Antarctica. The circumstances of his death are uncertain: he is believed to have died at sea on passage from Hobart to London in 1848, leaving a wife and family in poverty. *Further reading*: Mill (1905); Jones (1992).

Biscoe Islands. 66°00′S, 66°30′W. Chain of low-lying, snow-covered islands 30 km (19 miles) long, off the Graham and Loubet coasts of Graham Land. Extending from the Pitt Islands and Reynaud Island in the northeast to the Barcroft Islands, bordering Matha Strait, in the southwest, they were discovered by British sealing captain J. Biscoe in 1832 and later named for him.

Biscoe Point. 64°49′S, 63°49′W. Rocky promontory on the southeast side of Biscoe Bay, southern Anvers Island. The ground supports stands of Antarctic flowering plants growing from well-developed loam, with a rich soil biota. An area of approximately 2.7 km² (6.9 sq miles), including foreshore, an islet and coastal waters, has been designated SSSI No. 20 (Biscoe Point).

Bismarck Strait. 64°51′S, 64°00′W. Strait between southern Anvers Island and Wiencke Island, Palmer Archipelago, and the Wauwermans Islands of Wilhelm Archipelago. Charted by German whaler Capt. Eduard Dallmann in 1874, it was named for the German statesman and contemporary Chancellor, Otto von Bismarck.

Bjåland, Olav. (1872–1961). Norwegian explorer. Born in Morgedal, Telemark, he became a carpenter and accomplished skier. Joining the **Norwegian South Polar Expedition 1910–12**, he was one of the five members of the party that reached the South Pole. After the expedition he returned to Norway and to his former pursuits, living quietly in Morgedal until his death in May 1961.
(CH)

Black Coast. 71°45′S, 62°00′W. Part of the east coast of Palmer Land, West Antarctica, bounded by Cape Boggs in the north and Cape Mackintosh in the south. Facing the Weddell Sea, it includes a range of mountains through which descend glaciers that merge into a narrow ice shelf, pierced by ice-covered islands. The coast was identified in 1940 by the United States Antarctic Service Expedition 1939–41, and named for Cdr Richard B. Black USN, commander of the expedition's East Base.

Black Island. 78°12′S, 166°25′E. Mountainous island with prominent central peak of 1040 m (3411 ft), embedded in the western Ross Ice Shelf, forming part of Ross Archipelago: so named for its black volcanic rock.

Blaiklock Island. 67°33′S, 67°04′W. Low, ice-free island separating Bigourdan and Bourgeois fjords at the boundary of Loubet and Fallières coasts of Antarctic Peninsula, eastern Marguerite Bay. Charted by and named for British surveyor K. V. Blaiklock, the island accommodates a refuge hut built in March 1957, for use by survey parties from Horseshoe Island (Base Y) and other stations in the area. The hut is included with Base Y in the designation of an Historic Monument.

Blissett, Harry Arthur. (*c*.1878–1955). British seaman and expedition member. As a private in the Royal Marines he joined the **British National Antarctic Expedition 1901–4**, taking part in several sledging journeys, including a visit to the emperor penguin colony at Cape Crozier. In later life he moved to New Zealand, working in the prison service. He died in Christchurch on 14 August 1955.
(CH)

Blowing snow. Clouds of snow particles raised by wind to heights of several metres above ground, with corresponding loss of visibility.

Blue whale. *Balaenoptera musculus*. Measuring up to 33 m (108 ft) and weighing up to 120 tonnes, blue whales are the largest of all the whales and possibly the largest-ever living animals. Dark blue-grey, with pale mottled patches, they are found in offshore waters of all the world's oceans, including both polar regions. There are separate northern and southern stocks: those of the southern oceans are marginally bigger than those of the north. Both stocks winter in warm temperate and tropical waters, where females give birth to their calves. In spring they migrate poleward, singly or in groups of two or three, to take advantage of rich feeding grounds along the ice edge. **Krill** and other euphausiids appear to be their main foods. Calves that are born some 7 m (23 ft) long and weighing 2.5 tonnes grow in a single season to 16 m (52.5 ft) and 25 tonnes, mainly on their mother's milk. Adults also fatten rapidly, laying down

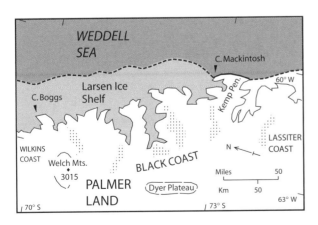

Black Coast

thick blubber (subdermal fat) that provides a food store for their lean months in warmer waters. During the period of Southern Ocean whaling (1905–87) southern blue whales were the prime target. In the 1930–31 season alone over 30 000 were killed in Antarctic waters. Up to the mid-1930s they represented over 50 per cent of all whales killed annually: by the 1950s this had dropped to 10 per cent, and by the 1960s to less than 6 per cent. Remnant stocks appear now to be recovering slowly: there is a current estimated world population of 10 000–15 000.
Further reading: Small (1971); Watson (1981).

Blue Whale Harbour. 54°04′S, 37°01′W. Harbour on the north coast of South Georgia, named for the largest species of whale that was hunted off the island.

Bonner, William Nigel. (1928–94). British seal biologist and administrator. Born in London, he read zoology at University College, London. He spent 18 months at a two-man base on South Georgia 1953–55, working on penguins and seals, and again wintered in 1959 as a sealing inspector. Later he joined British Antarctic Survey as a biologist, retiring in May 1988 as Head of Biology. He died on 27 August 1994.

Booth Island. 65°05′S, 64°00′W. Rugged island off the Graham Coast of Antarctic Peninsula, forming part of Wilhelm Archipelago. It was discovered by E. Dallmann during the **German Whaling and Sealing Expedition 1873–74** and named for Oskar or Stanley Booth of Hamburg, who supported the expedition. Lt. A. de Gerlache later named the island Wandel Island: the name has been transferred to the main peak (980 m, 3214 ft). In 1903 J.-B. Charcot wintered his ship *Le Français* in a small bay, Port Charcot, on the northern coast of the island.

Borchgrevink, Carsten Egeberg. (1864–1934). Norwegian-Australian explorer who led the first expedition to winter on Antarctica. Born in Christiania (Oslo) to a Norwegian father and English mother, Borchgrevink trained in forestry. In 1888 he emigrated to Australia, becoming a school teacher in New South Wales. In 1894 he volunteered as seaman and naturalist aboard the sealer *Antarctic* (Capt. L. Kristensen), which was fitting out in Melbourne for a whaling expedition to Antarctica (**Norwegian (Tønsberg) Whaling Expedition 1893–95**). The expedition saw few whales and caught none, but penetrated the ice of the Ross Sea, landing at Possession Island and Cape Adare. This experience led Borchgrevink to visit England to raise funds for an Antarctic expedition of his own. Though a bumptious and self-aggrandizing approach made him unpopular in official circles, he received encouragement and financial backing from British newspaper proprietor George Newnes, enough to launch the small but enterprising **British Antarctic (*Southern Cross*) Expedition 1898–1900**. Despite the limited but real successes of this expedition, Borchgrevink failed to gain acceptance as a *bona fide* explorer. He returned to Norway, where he lived uneventfully and died in comparative poverty. Belatedly, the Royal Geographical Society awarded him its Patrons Medal in 1930.
Further reading: Borchgrevink (1901); Jones (1992).

Carsten Egeberg Borchgrevink (1864–1934). Photo: SPRI

Borchgrevink Coast. 73°00′S, 169°30′E. Part of the coast of Victoria Land, East Antarctica, facing the Ross Sea. Bounded by Cape Adare in the north and Cape Washington in the south, it consists mainly of steep ice-mantled mountains and glaciers, with fringing ice shelves and glacier tongues. The coast was named for Australian explorer Carsten E. Borchgrevink, who discovered it and explored the immediate area during two expeditions of 1894–95 and 1898–1900.

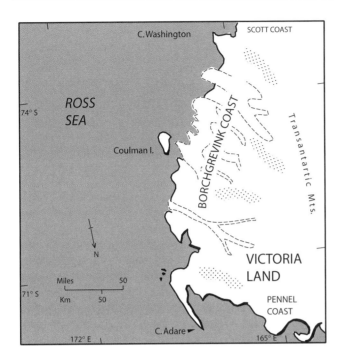

Borchgrevink Coast

Borga Station. 72°58′S, 3°48′W. A small South African research station inland among **Borgmassivet**, Dronning Maud Land, established in May 1969 for summer traverse parties.

Borgmassivet. 72°45′S, 3°30′W. (Castle Massif). Mountain massif 50 km (31 miles) long forming the southern end of Ahlmannryggen, western Dronning Maud Land. Clustered around Borga, a high peak of 2546 m (8351 ft), are several other peaks of similar height. The area was photographed by the **German Antarctic (*Schwabenland*) Expedition 1938–39**, and explored from the ground by the **Norwegian-British-Swedish Expedition 1949–52**.

Borradaile Island. 66°35′S, 162°45′E. Ice-covered island between Buckle Island and Young Island of the Balleny Islands, off the Oates Coast of Victoria Land. The group was discovered in 1839 by sealing captain J. Balleny, who named the island for W. Borradaile, a subscriber to the expedition.

Bottlenose whale, southern. *Hyperoodon planifrons*. Measuring up to 7.5 m and weighing up to 4 tonnes, in life these are readily confused with slightly larger southern **beaked whales**, which they closely resemble. Similarities include the general body shape, prominent beak and bulbous forehead. The main anatomical differences lie in the shorter and more triangular beak, larger and more prominent paired front teeth (which erupt only in males, and are seldom discernible in live specimens). Found in cool and cold waters of the southern oceans, they are known mainly from beached specimens and skeletal remains on southern islands and mainland coasts. However, southern bottlenose whales have been reported as present all the year round in the coldest waters of the Southern Ocean, for example in coastal polynyas of the Weddell Sea. Variously reported as predominantly blue or brown, with paler underparts, they appear in groups of three, four or more, usually over deep water, and are believed to hunt at great depths for fish and squid. Little else is known of their biology.
Further reading: Carwardine (1995); Watson (1981).

Boucot Plateau. 82°25′S, 155°40′E. Ice plateau in the Geologists Range of Victoria Land, identified from aerial photographs and named for US geologist A. J. Boucot.

Bounty Islands. 47°45′S, 179°02′E. Archipelago of warm-temperate islands and stacks 700 km (437 miles) east-southeast of New Zealand's South Island. Clustering in three groups on the eastern flank of the Campbell submarine plateau, they are the eroded, sea-washed remnants of a once-larger granitic landmass. The highest, Funnel Island in the central group, rises to 73 m (239 ft). There is little soil, and vegetation is limited to salt-tolerant algae and lichens. There are no climatic records, but mean annual temperature is likely to be about 10 °C. Breeding seabirds include erect-crested penguins, Salvin's mollymawk, endemic prions and shags, and cape petrels, terns and southern black-backed gulls. The islands are the main breeding ground of New

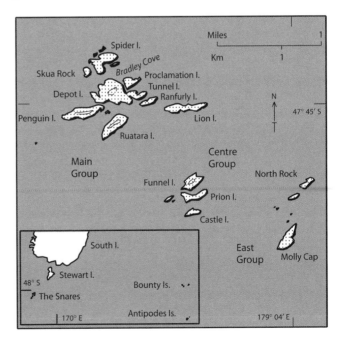

Bounty Islands

Zealand fur seals. Discovered in 1788 by Capt William Bligh RN during his ill-fated voyage in HMS *Bounty*, they were frequently visited by sealers during the nineteenth century. A refuge hut was established in the 1880s, but no castaways are known to have used it. The islands are awash during winter storms, and lack of sheltered flat ground makes it unlikely that human settlement will ever be attempted. They have the protected status of a New Zealand National Nature Reserve, with limited access for visitors.
Further reading: Fraser (1986).

Bourgeois Fjord. 67°40′S, 67°05′W. Fjord between Pourquoi Pas Island and the main Fallières Coast of Graham Land, opening into northern Marguerite Bay. The fjord was named by J.-B. Charcot for the French military cartographer Col. J. E. Bourgeois.

Bouvet de Lozier, Jean-Baptiste Charles. (1705–86). French navigator and explorer. Indentured in youth to a seafaring career, Bouvet de Lozier reached the rank of lieutenant in the French Compagnie des Indes. He proposed to his employers an expedition to explore the southern Atlantic Ocean in search of a benign southern continent, supposedly discovered in the early sixteenth century by Paulmyer de Gonneville. In 1738 he was assigned a fleet of two ships, the 280-ton transport *Aigle* under his own command, and the 200-ton transport *Marie* commanded by Capt. Nicolas-Pierre Guyot. The ships left Lorient on 19 July, and in early November headed southeastward from Santa Catarina, Brazil, following what was believed to be de Gonville's course. Meeting icebergs and brash ice in 48°S (approximately the latitude of Paris in the north), he pressed southward through cold air and fog to approximately 54°S. On 1 January 1739 he discovered a high, snow-covered headland. For almost two weeks fog, pack ice and contrary winds kept him from sailing closer. Naming his discovery Cape Circumcision, he departed eastward, hoping somewhat optimistically that it would prove to be part of the great southern continent. Continuing along the edge of the pack ice, he achieved a passage of approximately 7000 km (4400 miles) in latitudes below 52°S, noting and describing tabular icebergs, penguins and seals, but finding no further indications of land. Lost for many years and subsequently rediscovered, the headland proved only to be part of a tiny remote island, perhaps the world's remotest island, now called Bouvetøya.

Bouvetøya. 54°25′S, 3°21′E. Uninhabited sub-Antarctic island of the Southern Ocean, 9 km (5.6 miles) long, 7 km (4.4 miles) wide and rising to 780 m (2558 ft). Probably the world's most isolated island, it stands 4800 km (3000 miles) east of Cape Horn and 3000 km (1875 miles) southwest of Cape Town, marking the southern end of the submarine mid-Atlantic ridge. Over 96 per cent of its rugged surface is ice-capped: the ice fills and spills from a central caldera, topping the strata of lava and scoria that make up the bulk of the island. Steam rises from the caldera rim, a reminder of the island's relatively recent volcanic origins. Steep cliffs, constantly eroded by the sea, flank it on all sides. The island is made up mainly of alkaline basalts. There is a meagre flora of algae, lichens and mosses, and a fauna mainly of cliff-nesting sea birds, Adélie, chinstrap and macaroni penguins, and Antarctic fur seals.

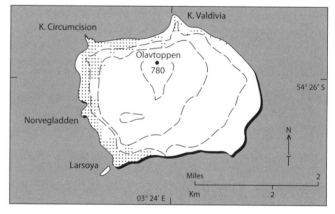

Bouvetøya. Height in m

Discovered in January 1739 by the French navigator Jean-Baptiste-Charles de Lozier Bouvet, Bouvetøya was frequently lost in mist and pack ice and rediscovered several times. Its position was finally fixed by the Norwegian expeditions of 1927–29 that mapped it and studied its flora. Lacking anchorages, harbours, or sufficient flat ground to support even a small research station, Bouvetøya was left in isolation. Later visitors during the 1950s discovered a beach and raised platform in the northwest corner. At first thought to represent a lava flow, this was later identified as a landslide, and in 1977 Norwegian scientists installed an

Steep cliffs and tidewater glaciers of ice-capped Bouvetøya. Photo: BS

automatic weather station. Fur seals too found this feature soon after its appearance and, despite its rapid erosion by the sea, continue to use it as a breeding ground. The island was annexed by Norway in 1930 and is administered as a nature reserve. Landing is prohibited except under permit.

Bowers, Henry Robertson (Birdie). (1883–1912). British seaman and explorer. Born at Greenock, he entered the Royal Navy as a cadet. In 1905 he was commissioned in the Royal Indian Marine. He joined Scott's **British Antarctic (*Terra Nova*) Expedition 1910–13**, initially as stores officer in *Terra Nova*, but eventually became a popular and invaluable member of the shore party. He took part in several sledging journeys, including, with Wilson and Cherry-Garrard, the winter journey to Cape Crozier to collect early embryos of emperor penguins. Bowers was one of four in the support group that accompanied the polar party (Scott, Evans, Oates and Wilson) up the Beardmore Glacier and across the polar plateau. In a last-minute change of plan on the plateau, Scott decided to include him as a fifth member of the polar party. They reached the South Pole on 17 January 1912. All died some eight weeks later of cold and malnutrition on the return journey.

Bowers Mountains. 71°10′S, 163°15′E. Range of mountains rising to 2600 m (8528 ft), forming the eastern flank of Rennick Glacier, inland from Oates Coast, Victoria Land, They were named for Lt. Henry R. Bowers of the **British Antarctic (*Terra Nova*) Expedition**.

Bowling Green Plateau. 79°42′S, 158°36′E. Ice plateau abutting the Cook Mountains, Victoria Land, named for Bowling Green State University, Ohio, USA.

Bowman Coast. 68°10′S, 65°00′W. Part of the east coast of southern Graham Land, Antarctic Peninsula, facing the Weddell Sea. Bounded by Cape Northrop in the north and Cape Agassiz in the south, the coast consists of mountains rising to a high interior plateau, with glaciers descending to the broad Larsen Ice Shelf. It was identified and photographed by Sir Hubert Wilkins on his flight of 20 December 1928, and named for US geographer and academic Isaiah Bowman.

Boyd Strait. 62°50′S, 62°00′W. Strait separating Snow Island and Smith Island at the western end of the South Shetland Islands. It was surveyed and named by James Weddell in 1823 for Capt. David Boyd, RN.

Boyd, Vernon Davis. (d. 1965). American expedition machinist. Boyd took part in **Byrd's Second Antarctic Expedition 1933–35**, and the **United States Antarctic Service Expedition 1939–41**, servicing the tractors and generators. Commissioned in the US Marine Corps, he served after World War II in Operation Highjump 1946–47 and Operation Windmill 1947–48, and took part in planning for Operation Deep-Freeze 1955–56. He died on 29 May 1965, aged 57.

Brabant Island. 64°15′S, 62°20′W. Ice-covered, mountainous island, second largest of Palmer Archipelago, off Danco Coast, Antarctic Peninsula. It was charted in 1898 by Lt. A. de Gerlache and named for the Belgian province. Its highest peak is Mt Parry (2520 m, 5904 ft). The island was surveyed by the British Joint Services Expedition 1983–85 led by Cdr J. R. Furse RN.

Bransfield, Edward. (1783–1852). British naval hydrographer who charted the **South Shetland Islands**. Born in southern Ireland, Bransfield first went to sea with the merchant fleet, but in June 1803 was impressed into the Royal Navy as an ordinary seaman. In August 1805 he was advanced to able seaman. During a varied career on the lower deck he became a competent navigator, and in 1811 was promoted to midshipman, in 1813 to second master. In 1817 he was appointed to HMS *Andromache* for service as master in South America. When Capt. William Smith, of the brig *Williams*, reported his discovery of 'New South Britain' (the South Shetland Islands) to the British naval authority in Valparaiso, the commanding officer, Capt W. H. Shirreff RN, chartered the brig for a hydrographic expedition to the islands. He placed Bransfield, whose abilities he had already noted, in charge of the survey, with three midshipmen to help. Bransfield completed a survey of the newly-discovered group, circumnavigating them and almost certainly sighting the mainland of Antarctic Peninsula. His name is commemorated in a peak and an island close to the tip of the Peninsula, and in Bransfield Strait to the south of the South Shetland Islands. Bransfield left the navy in

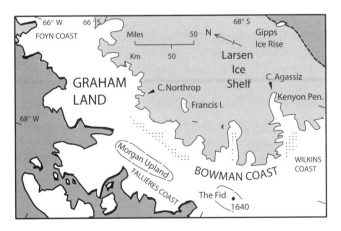

Bowman Coast

1821 for the merchant service, commanding several ships in general trade. He retired to Brighton, where he died on 31 October 1852.
Further reading: Jones (1992).

Bransfield Island. 63°11′S, 56°36′W. Ice-capped island off the northeast tip of Antarctic Peninsula, named for E. **Bransfield**, the Royal Navy hydrographer who in 1820 surveyed the South Shetland Islands and neighbouring waters.

Bransfield Strait. Strait approximately 100 km (62.5 miles) wide between the South Shetland Islands and Antarctic Peninsula, charted by and named for British naval hydrographer Edward **Bransfield**. An area of 1100 km^2 (430 sq miles) off Low Island, South Shetland Islands, between 63°20′S and 63°35′S; 61°45′W and 62°30′W, has been designated SSSI No. 35 (Western Bransfield Strait). Including inshore, shelf and deeper waters, it is one of two known sites in the region that are suitable for bottom trawling for fish and other benthic organisms.

Brash ice. Clusters of floating ice, with constituent fragments no more than 2 m (6.5 ft) across, derived from the disintegration of floes or bergy bits.

Breakbones Plateau. 57°04′S, 26°41′W. Rough, rocky plateau formed of overlapping lava fields on Candlemas Island, South Sandwich Islands.

Bridgeman Island. 62°04′S, 56°44′W. A small volcanic island of the eastern South Shetland Islands, standing in isolation in Bransfield Strait and rising to a single pyramidic peak of 240 m. (787 ft). Discovered in 1820, a remnant of a once-larger volcano, the island was reported to be in eruption in 1821, and passing ships reported fumaroles up to 1890. It has since been quiescent.

Brisbane, Matthew. (*c.* 1787–1833). British mariner and explorer, who in February 1823 accompanied James **Weddell** to a record latitude of 74°15′S in the Weddell Sea. Little is known of his early career. Born in Perth, Scotland, possibly into a seafaring family, Brisbane appears in the shipping record in 1822 as master of the cutter *Beaufoy*; this may have been his first command on a sealing venture. Successfully completing the historic voyage with Weddell, he returned to London in June 1824. Continuing as a sealer in South America and the southern islands, he was shipwrecked three times – twice in Tierra del Fuego (1827, 1830) and also on South Georgia (1829). In February 1833 he was present at Port Louis, in the Falkland Islands, when local settlers mutinied against the Buenos Aires government representative. Along with others, he was forcibly expatriated to Montevideo by the visiting US naval corvette *Lexington*. Returning to the settlement later in the same year, possibly as an acting governor under a British regime, he was murdered by the mutineers on 26 August.
Further reading: Jones (1992).

Bristol Island. 59°02′S, 26°31′W. One of the largest of the South Sandwich Islands, toward the southern end of the chain. Mainly ice-covered, the island rises to the near-symmetrical cone of Mt. Darnley (1100 m, 3608 ft). Freezland Rock, a pinnacle reef 300 m (975 ft) high, stands among lesser reefs on the western side. Discovered and charted by the British explorer Capt. J. Cook, the island was named for the Earl of Bristol. It was reported to be in violent eruption in December 1935, and again in 1956.

Bristow, Abraham. British sealing captain who in 1806 discovered the Auckland Islands. Details of his birth and death are unknown. One of several ships' masters of the same surname, Abraham Bristow first appears in shipping records in 1797 as master of *Speedy*, a whaling ship of 313 tons owned by the London firm of Messrs Enderby. From 1800 he commanded *Ocean*, a cargo ship under the same house flag. In August 1806, on passage from Tasmania to Cape Horn, Bristow sighted a large group of islands, which appeared to be 'of moderate height', with a good harbour at the north end, and likely to be well stocked with seals. On a return voyage in the following year, commanding the whaler *Sarah*, he entered the harbour that he called 'Sarah's Bosom' (later renamed Port Ross), where he landed and took formal possession of the group. He named them for Lord Auckland, a family friend, and released pigs ashore for the sustenance of any seamen who might be stranded there. Bristow continued as a master for several years, disappearing from the records after about 1820.
Further reading: Jones (1992).

Britannia Range. 80°05′S, 158°00′E. Mountain range rising to several peaks of over 2400 m (7872 ft) inland from Shackleton Coast, Victoria Land. The range was named for the training ship HMS *Britannia*, through which many polar officers entered the Royal Navy.

British Antarctic (*Nimrod*) Expedition 1907–9. Following his premature return from the **British National Antarctic Expedition 1902–4,** Ernest Shackleton resolved to take his own expedition to Antarctica, with the triple objectives of reaching the South Geographic and South Magnetic Poles, and exploring King Edward VII Land, flanking the eastern edge of the Ross Ice Shelf.

With difficulty he raised financial support from private and government sources, and secured reluctant backing from the Royal Geographical Society. The expedition sailed from London on 30 July 1907 in the wooden sealing vessel *Nimrod* (Capt. Rupert England), reaching Lyttelton, New Zealand on 23 November. Heavily laden with stores and equipment, including a substantial living hut, an Arrol-Johnston motor car, nine dogs and ten ponies, and with a complement of 39 crew and staff, the ship left port on 1 January 1908. To save coal, the steamship *Koonya* towed *Nimrod* south as far as the edge of the pack ice, which was reached on 14 January. Thereafter *Nimrod* pressed on alone through the ice, sighting the Ross Ice Shelf on 23 January. Shackleton had planned to establish his base in a deep embayment of the ice shelf, discovered in 1903. However, the bay had disappeared, and pack ice was pressing in: he decided instead to run for the more familiar McMurdo Sound and western shore of Ross Island. Landing began at Cape Royds on 3 February. On 22 February *Nimrod* departed for New Zealand, leaving a shore party of 15.

Interior of Shackleton's hut, Cape Royds. Photo: BS

Exploration began with a party of six under William Edgeworth David, which between 5 and 11 March 1908 made a first ascent of Mt. Erebus, immediately behind the base. However, Shackleton's main effort was directed toward achieving the geographic pole. Depots were laid southward from September onward, and the main party of four (Shackleton, Jameson Adams, Eric Marshall and Frank Wild), accompanied by four ponies, started south on 28 October 1908. Crossing the ice shelf along the route pioneered during the **British National Antarctic Expedition 1902–4** by Scott, Wilson and Shackleton, they discovered a route south up the **Beardmore Glacier**. On the polar plateau they continued southward, man-hauling on severely-reduced rations. On 9 January 1909 they reached a point in 88°23′S, 162°E, some 97 statute miles from the pole, where their physical condition forced them to turn back. Retracing their steps, they returned to McMurdo Sound, where on 1 March they finally reached the waiting *Nimrod*.

On 5 October 1908 a second party, consisting of geologists W. E. David and Douglas Mawson, and surgeon Alister Mackay, set out westward across McMurdo Sound to seek the magnetic pole. Though their first few miles across the sea ice made use of the motor car, the rest of the journey involved strenuous man-hauling. From Terra Nova Bay they ascended Larsen Glacier to the polar plateau, continuing southwestward under the guidance of the magnetic dip needle. On 15 January 1909 they had reached a point in 72°42′S where the angle of dip was only 15′ from the vertical. There, they calculated, within 24 hours the shifting pole would come to them. On the following day they hoisted the Union Flag and claimed the area for the British Empire. The party retraced its steps to the coast and was picked up by *Nimrod* on 3 February. From 9 December 1908 a third party, consisting of geologist Raymond Priestley and assistants Bertram Armytage and Sir Philip Brocklehurst, explored the mountains and glaciers on the western flank of McMurdo Sound, including the Ferrar Glacier and the dry (glacier-free) valleys of Victoria Land. They were picked up by *Nimrod* on 26 January 1909.

With all hands safely aboard, *Nimrod* returned to Lyttleton on 25 March. Despite failure to reach the geographic pole, Shackleton was knighted for his achievements, and established as a leading polar explorer. His two-volume account of the expedition (see *Further reading*) includes summaries of the results by his team of scientists, followed by a series of scientific reports.

Further reading: Shackleton 1909.

British Antarctic (*Southern Cross*) Expedition 1898–1900. The Norwegian-born, Australian-based adventurer C. E. **Borchgrevink**, who in 1895 had landed with his fellow-countrymen Henrik Bull and Leonard Kristensen at Cape Adare, Victoria Land (**Norwegian (Tønsberg) Whaling Expedition 1893–95**), determined to return with an expedition to explore inland from the Cape. Neither a scientist nor a professional seaman, Borchgrevink gained little support from geographical sources in Australia, Britain or Norway, but finally obtained encouragement and financial backing from the British newspaper proprietor George Newnes. In the wooden steam whaler *Southern Cross* (Capt. Bernhard Jensen), the expedition sailed from London on 22 August 1898 and from Hobart, Tasmania on 19 December. From meagre financial resources Borchgrevink had equipped the party with sledges, sledge-dogs, two Lapp dog handlers, and kayaks, and assembled a promising team of young scientists including physicist L. Bernacchi, hydrographer W. Colbeck and assistant zoologist H. B. Evans, all of whom distinguished themselves on later polar expeditions.

Southern Cross encountered pack ice off the Balleny Islands and for almost seven weeks tried to force a way southward. Returning north to open water, Kristensen re-entered a much lighter field of pack ice further east, quickly breaking through and making good progress toward Cape Adare. There Borchgrevink landed on 17 February 1899, establishing two prefabricated huts (Camp Ridley) on a beach below the cape. The ship left for New Zealand, and the party of ten prepared to over-winter, becoming the second expedition to winter south of the Antarctic Circle, the first to do so on the continent itself. Borchgrevink quickly found that the base was badly sited for achieving his main objective – to sledge inland with dog teams. Parties crossed the sea ice of Robertson Bay to Duke of York Island, penetrating a few miles into the interior, but ice cliffs and steep glaciers made further progress impossible. The sun disappeared on 15 May and was absent for ten weeks until 29 July. Winter passed uneventfully and all but one of the party retained their health. The Norwegian biologist, Nicolai Hanson, died of an enteric disease on 14 October 1899, achieving the sad distinction of being the first to be buried on Antarctica.

Southern Cross returned on 28 January 1900, and the expedition left Camp Ridley on 2 February, moving southward along the Victoria Land coast to Possession Island, Wood Bay, Franklin Island, Beaufort Island and Ross Island. Heading eastward along the Ross Ice Shelf, Colbeck recorded that the ice shelf had receded some 48 km (30 miles) since the visit of James Clark Ross some 60 years earlier. This enabled the ship on 11 February to achieve a record furthest-south of 78°21'S. On 19 February Borchgrevink, Colbeck and one of the Lapp dog-handlers stepped onto the ice shelf at a low point and made a brief sledging excursion southward, achieving a further record in 78°50'S. The expedition returned to New Zealand on 30 March, and to England in June 1900 with a creditable collection of scientific specimens, charts and data.

Further reading: Bernacchi (1901); Borchgrevink (1901).

British Antarctic (*Terra Nova*) Expedition 1910–13. Expedition in which Capt. R. F. Scott achieved the South Pole, but died on the return journey. In 1909 Scott, a serving officer in the Royal Navy, planned a second expedition that would secure the South Pole for Britain. Supported by both government and private funds, he sailed in *Terra Nova*, a converted Scottish whaler, barque-rigged with auxiliary engine, under the command of Cdr H. L. L. Pennell. *Terra Nova* sailed from London on 1 June 1910, and from Lyttleton, New Zealand on 26 November. Scott's party included several of his former colleagues from the **British National Antarctic Expedition 1901–4**: his equipment included 19 ponies, 30 dogs and three tractors. Hopes of establishing a base at Cape Crozier were defeated by heavy surf: he landed instead at Cape Evans, McMurdo Sound, where by mid-January his hut was built and all the stores were ashore.

First Western Party: *Terra Nova* cruise

With unloading finished, on 27 January 1911 *Terra Nova* took a four-man geological party – Edgar Evans, Frank Debenham and Charles Wright, led by Griffith Taylor – across McMurdo Sound, landing them at Butter Point to explore inland among the western mountains. They remained in the field for three months, mapping first the adjacent valleys of the Ferrar and Taylor glaciers (the latter a 'dry' valley devoid of snow or ice), then the more complex western flank of the Koettlitz Glacier. The party returned to Hut Point in March, and to Cape Evans on 13 April.

After landing the Western Party, *Terra Nova* left for a cruise along the Barrier (the front of the Ross Ice Shelf), with the intention of establishing a second, smaller base toward the eastern end, close to Edward VII Land. The most likely site was an indentation in the Barrier that Shackleton had discovered and named Bay of Whales. Entering the bay, they found it already occupied by *Fram*, the ship of Amundsen's **Norwegian South Polar Expedition 1910–12**. Amundsen had established his base Framheim three miles to the south, from which he intended a bid for the South Pole. Victor Campbell, charged with setting up the second base, had no wish to establish it so close to another expedition. He returned with *Terra Nova* to Cape Evans, where he reported the discovery, then moved north to Ridley Beach, Cape Adare, where Borchgrevink had wintered with the **British Antarctic (*Southern Cross*) Expedition 1898–1900**. There the expedition's Northern Party (see below) landed and established its base on 18 February 1911. *Terra Nova* headed north for a brief cruise along the northern coast of Victoria Land (now the Oates Coast), then returned to winter in New Zealand.

The autumn sledging party (less E. R. G. R. Evans) preparing to leave Cape Evans on 26 January 1911. Left to right: Crean, Wilson, Keohane, Bowers, Gran, Scott, Forde, Meares, Cherry-Garrard, Oates, Atkinson. Photo: Scott Polar Research Institute

Autumn, winter and spring journeys

When *Terra Nova* left Cape Evans on 25 January, Scott immediately took advantage of good weather to lay depots of food and fuel along the polar route south. Twelve men, 8 ponies and 26 dogs headed southward across the fast ice to Hut Point and beyond, establishing depots as far as 'One Ton Depot' in 79°28′S. The party returned to Cape Evans on 13 April. On 27 June a three-man party (Edward Wilson, Apsley Cherry-Garrard and 'Birdie' Bowers) left Cape Evans to visit the emperor penguin colony at Cape Crozier, hoping to arrive early enough in the season to secure embryos from newly-laid eggs. To the difficulties of a mid-winter journey they added the hazards of experimental rations, returning exhausted on 2 August with three embryos. In October, at the first signs of spring, Scott led further depot-laying journeys south, in preparation for the expedition's main polar journey.

The polar journey

On 3 November the polar party and support groups left from Hut Point, a cavalcade including 16 men, 2 tractors, 9 ponies and a dog team. Setting off in good weather, they made excellent progress, though within three days both tractors broke down, reducing their drivers to man-hauling, and increasing the loads of the rest of the team. Ponies were shot at intervals to feed both dogs and men, gradually diminishing the cavalcade. Two of the party, Bernard Day and Frank J. Hooper, turned back on 24 November. At the gateway to the Beardmore Glacier, reached on 9 December, the remaining ponies were shot. Cecil Meares, Demetri Gerof and the dogs started up the glacier, but according to plan turned back on 11 December, leaving only three man-hauling teams of four in the field. On 21 December, more than halfway up the glacier, Edward Atkinson, Apsley Cherry-Garrard, Patrick Keohane and Charles Wright turned back. On 4 January, now at the rim of the polar plateau, Henry ('Birdie') Bowers, Tom Crean, Teddy Evans and William Lashley were scheduled to return, but Scott made a last-minute decision to keep Bowers in the polar party. So Bowers, Edgar Evans, Lawrence ('Titus') Oates, Scott and Edward Wilson formed the party of five that continued across the plateau. Reaching the South Pole on 18 January 1912, they found to their acute disappointment a black tent with messages indicating that Amundsen had already reached the Pole almost five weeks earlier. Scott's party trudged slowly back across the plateau, weakening gradually on short, inadequate rations. All managed to descend the Beardmore Glacier, but on the lower slopes Edgar Evans fell and suffered concussion, dying on 18 February. The remaining four, by now suffering severely from cold and malnutrition, continued in bad weather across the ice shelf. Oates walked out of the tent to his death on 16 or 17 March: Bowers, Scott and Wilson survived only a few days longer.

The final support party (Teddy Evans, Lashley and Crean) had also encountered difficulties on returning across the ice shelf: in the final stage Evans had collapsed, and Crean walked the last 60 km (35 miles) to Hut Point alone to seek help. Atkinson and Demitri were there with the dog teams, awaiting a break in the weather. They brought back Lashley and Evans: later a group led by Cherry-Garrard sledged south to One Ton Depot in an unavailing search for the polar party.

Second Western Party

On 18 November 1911, after the polar party had left Cape Evans, Griffith Taylor, Frank Debenham, Trygve Gran and Robert Forde made a second expedition to the western mountains, essentially continuing northward the geological work of Taylor's earlier group. Man-hauling across the Sound to Butter Point, they visited Cape Bernacci, Dunlop Island and Cape Roberts, reaching Granite Harbour on 30 November. From there they explored inland up the deeply incised valley of the Mackay Glacier to Mt Suess. On 15 January 1912 the party moved back to Cape Roberts to await a pre-arranged pick-up by *Terra Nova*. As the ship was delayed, in early February they hauled back to Butter Point and started across the fast ice toward Cape Evans. They were picked up by *Terra Nova* on 14 February.

The second season

Terra Nova left Lyttelton on 15 December 1911, relieved the northern base at Cape Adare on 4 January 1912, and dropped the Northern Party off at Terra Nova Bay three days later. Held in the pack ice for three weeks, she eventually reached Cape Evans on 6 February. Pennell brought 7 mules, a present from the government of India, 14 new dogs, and stores for a second year. The polar party had not returned and search parties were out seeking it. The ship recovered the Western Party from the sea ice on 14 February, and made several attempts to recover Campbell's party from Terra Nova Bay. However, Pennell was running short of fuel: obliged at all costs to leave before the Sound froze over, he left on 3 March. At Cape Evans Atkinson took command, with a reduced complement of 12. Before winter set in he continued to send out search parties along the trail as far as One Ton Depot. By 1 May all hands were back at the base, where they slipped into a winter routine of scientific observations and preparation for spring journeys. Depot laying began in September, and on 30 October a major sledging party, with seven mules and two dog teams, headed south from Hut Point on the final search for the missing polar team. On 12 November, 11 miles beyond One Ton Depot, they found the tent with the bodies of Bowers, Scott and Wilson: Evans and Oates had died earlier. Taking diaries and scientific specimens, they collapsed the tent onto the bodies, built a snow cairn over it, and returned to Cape Evans. In early December Raymond Priestley led a party of six up Mount Erebus, a climb he had previously made as a member of Shackleton's British Antarctic (*Nimrod*) Expedition 1907–9.

Terra Nova left Lyttelton on 14 December 1912 and relieved the Cape Evans base on 18 January. On 20 January Atkinson led a party up Observation Hill, behind Hut Point, to erect a cross commemorating the five who had died on their return journey from the South Pole.

The Northern Party

After landing and establishing a base close to Borchgrevink's old hut at Cape Adare, Campbell and his five companions (see above) settled to a programme of biological, meteorological and geological study that, together with limited sledging excursions, provided a busy first year. Relieved by *Terra Nova* on 6 January 1912, Campbell asked to be moved to Evans Cove, Terra Nova Bay, for a six-week programme of mainly geological exploration. Dropped on 8 January with rations for six weeks, they expected to be picked up in mid-February. When the time came, ice floes packed the approaches to Terra Nova Bay: the ship could not pick them up. The party spent a difficult winter in an igloo and cave on Inexpressible Island, living mainly on seal and penguin meat. On 30 September they set out to walk along the coast to Cape Evans, surviving on depots left by Shackleton's *Nimrod* expedition of 1907–9. They reached Hut Point on 6 November and Cape Evans on the following day.

This expedition is remembered mainly for the tragic failure of Scott's polar party: it could as well be remembered for the solid scientific achievements of the other, less spectacular and more fortunate parties in the western mountains. *Further reading*: Huxley (1914); Priestley (1914).

British Antarctic Survey (BAS). The scientific and administrative body responsible for managing most British operations in Antarctica, particularly in **British Antarctic Territory**, but with a wider mandate to promote and encourage international cooperation in both Arctic and Antarctic regions. Arising in 1961 from the Falkland Islands Dependencies Survey, on the advent of the Antarctic Treaty, it continued under management of a government department until 1967, when BAS was transferred to the Natural Environment Research Council. From headquarters outside Cambridge, it currently (2002) operates four wintering stations, a summer station, two supply and research ships

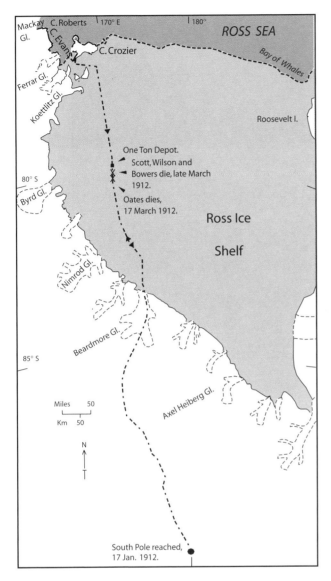

British Antarctic (*Terra Nova*) Expedition 1910–13: route of the South Polar Party

and five aircraft (see **Study Guide: National interests in Antarctica**).

British Antarctic Territory. Territory in the South American sector of Antarctica designated by a British Government Order in Council of 26 February 1962, lying between longitudes 20°E and 80°E, and south of latitude 60°S, and including Graham Land, Coats Land, Alexander Island and the South Shetland and South Orkney Islands. Formerly part of the **Falkland Islands Dependencies**, the territory was separated administratively from islands north of 60°S following Britain's accession to the **Antarctic Treaty**. It is administered by a High Commissioner. British Antarctic Territory overlaps geographically with Antártida Argentina and Territorio Antártico Chileno.

British, Australian and New Zealand Antarctic Research Expedition 1929–31. Popularly termed BANZARE, this shipborne expedition explored the coasts of what is now Australian Antarctic Territory by sea and air. Concerned that the French government had in 1924 filed claim to Terre Adélie, within the sector of East Antarctica immediately south of Australia, Sir Douglas **Mawson** and a group of influential fellow-Australians sought to prod their reluctant government into filing a similar claim to the rest of the sector. Discussions at the London Imperial Conference of 1926 precipitated action: the Australian, British, and New Zealand governments sponsored a scientific and survey cruise that would establish Australian interests in this sector of Antarctica. Given command of the expedition, Mawson secured the use of RRS *Discovery* under Capt. J. K. **Davis**. The ship was equipped with a Gipsy Moth floatplane, that was carried on deck and lowered over the side by crane. With considerable help from private sponsors, notably industrialist Sir MacPherson Robertson, the expedition sailed from Cape Town in October 1929.

Discovery visited Iles Crozet, Iles Kerguelen and Heard Island, then headed south into the pack ice. Reaching 65°41′S, the ship worked westward to 66°57′S, 71°57′E, where on 26 December they were within soundings and in sight of an extensive ice-covered coast. Continuing westward, on 31 December the ship was again stopped by ice in 65°10′E. A reconnaissance flight by pilots S. A. C. Campbell and E. Douglas revealed a continuing coastline and group of offshore islands, named the Douglas Islands. From there *Discovery* charted many miles of new coastline – mainly ice cliffs dotted with islands and steeply-rising rocky capes, which Mawson named for MacPherson Robertson. On 12 January 1930 they sighted Kemp Land, and landed on a small island, Possession Island, where Mawson raised the Union Flag and claimed land and islands between 73°E and 47°E as British territory. On 14 January *Discovery* met *Norvegia*, a whaling ship commanded by Capt. H. **Riiser-Larsen**, that had approached from the west, engaged on a similar survey expedition on behalf of Norwegian whaling magnate Lars Christensen. The two commanders agreed to regard longitude 45°E as the mutual boundary for their work. *Discovery* spent more time charting Enderby Land before heading north on 26 January.

The ship wintered in Australia. On 22 November 1930 the expedition again headed south, first to Macquarie Island, then to Cape Denison, east of Terre Adélie, where Mawson had led the **Australasian Antarctic Expedition 1911–14**. Under Capt. K. N. McKenzie, *Discovery*'s officers re-charted the coast of Terre Adélie and, as the voyage continued westward during January and February 1931, they charted and surveyed, both from the air and from the ship, the coasts of Wilkes Land and Queen Mary, Wilhelm II and Princess Elizabeth lands.

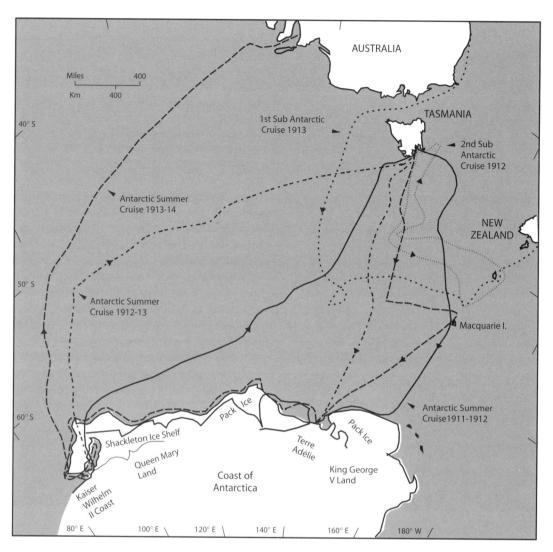

British, Australian and New Zealand Antarctic Research Expedition 1929–31: routes of the expedition ship RRS *Discovery*

Finally on 19 February, having returned to Mac.Robertson Land, RRS *Discovery* turned north, reaching Hobart one month later. In both seasons Mawson took opportunities to claim territory on behalf of Britain. His claims were subsequently affirmed by a British Order in Council dated February 1933 that established **Australian Antarctic Territory**.

Further reading: Mawson (1932).

British Graham Land Expedition 1934–37. A small British expedition, shoestring-funded by private subscriptions, the Royal Geographical Society and the British Government, the BGLE explored and mapped the southern Antarctic Peninsula area. The leader, John Rymill, a young Australian surveyor, drew together a team of volunteers including scientists, surveyors, seamen, an aircraft pilot and radio operator. Most were associated with Cambridge University: many had already served on Arctic expeditions. Their ship, an elderly wooden Brittany topsail schooner of 130 tons, with twin diesel engines, was reinforced for work in ice and re-named *Penola*. The expedition was equipped with a small De Havilland Fox Moth reconnaissance aircraft, operating from floats or skis, and a motor launch. Of 65 sledge dogs bought from Greenland, many died of distemper *en route* and had to be replaced by others from Labrador. The main purpose of the expedition was to explore and map in the area east and south of Marguerite Bay, which Sir Hubert **Wilkins** and Lincoln **Ellsworth** had described and photographed in their earlier flights.

Penola sailed from London on 10 September 1934 and from the Falkland Islands on 7 January 1935, reaching Port

Lockroy, Wiencke Island, on 24 January. There they picked up stores, the aircraft and the sledge dogs that RRS *Discovery II*, survey ship of the **Discovery Investigations**, had brought south for them. Engine problems made it impossible for *Penola* to head much further south without extensive repairs. After reconnaissance by aircraft and motor boat, the expedition sailed only 30 miles south to the Argentine Islands, where they set up a two-storey living hut with attached hangar, and allowed the ship to freeze in for the winter.

Breeding and rearing more huskies, training the new teams, refitting *Penola* and exploring over the sea ice, surveying and making scientific collections as far south as Cape Evensen kept the 16-man team busy during winter and spring. In December they cut *Penola* free from the fast ice, and on 3 January 1936 the ship returned to Deception Island to pick up timber and mail. During her absence the remaining team packed up in preparation for moving to an advanced base that would allow them to explore extensively further south. *Penola* returned on 27 January, and on 16 February, following a long reconnaissance by the aircraft, sailed for Marguerite Bay.

There the expedition established a southern base on a group of inshore islands that they named for Frank **Debenham**, Professor of Geography at Cambridge University. On 12 March *Penola* sailed north to the Falkland Islands and South Georgia, leaving behind a wintering party of nine men and 93 dogs. Following extensive reconnaissance flights both north and south of the base, members of the group made winter sledging journeys south to lay depots for a later summer journey, and north to explore the fjord system of northern Marguerite Bay. During an August reconnaissance flight Rymill discovered George VI Sound, an ice-filled rift between Alexander Island and the mainland. The main sledging journeys of spring and summer travelled far down the Sound and eastward from it to the high plateau of Graham Land, rationalizing and placing in context the aerial observations of Wilkins and Ellsworth, in particular disposing of their reported channels to the Weddell Sea, and demonstrating structural continuity between Antarctic Peninsula and Western Antarctica. *Penola* returned in late February, clearing the southern base on 12 March, and returning to England on 4 August 1937.

For its size and budget the BGLE was a remarkably successful venture, providing a foundation for the further exploration of southern Antarctic Peninsula in years to come.
Further reading: Rymill (1938).

British Imperial Antarctic Expedition 1920–22.

Planned as a major expedition involving 12 aircraft in a polar flight, this expedition consisted finally of two men wintering and undertaking scientific observations in a converted waterboat on the west Graham Land coast. Dr John Cope, former member of the *Aurora* party of the Imperial Transantarctic Expedition 1914–17, and Hubert George **Wilkins**, Australian pilot and adventurer, planned originally to fly from King Edward VII Land to the South Pole, in relays using 12 redundant bombers of the Royal Air Force. Receiving little support, they modified the plan to explore southward along the central plateau of Graham Land, Antarctic Peninsula. To this end in January 1921 Cope and Wilkins, accompanied by Thomas Wyatt **Bagshawe**, a 19-year-old graduate in geology, and Maxime Charles **Lester**, a ship's officer only a few years older, made a reconnaissance south along the peninsula in a whale catcher provided by Norwegian whaler Lars Christensen. Seeking a base from which to climb to the plateau, they were landed in Andvord Bay, on the Danco Coast, together with a 28-ft lifeboat, seven sledge dogs and stores for two years. After several attempts to reach the plateau, they gave up, Cope and Wilkins deciding to abandon the expedition and leave with the whalers. Bagshawe and Lester, having travelled so far, determined to stay. Transferring to Paradise Harbour, they found and modified an abandoned water boat – a small wooden boat used by the whalers to transport fresh water. In this they overwintered, ill-equipped and poorly rationed, with little fuel and few comforts. The two made tidal observations and an excellent year-round series of meteorological and biological studies, completing a first-ever breeding study of the gentoo penguins that thronged their island in spring. Whalers relieved the expedition in January 1922, astonished to find Bagshawe and Lester cheerful, in good health, and in no hurry to leave after their winter ordeal.
Further reading: Bagshawe (1939).

British National Antarctic Expedition 1901–4.

Inspired by the Sixth International Geographical Congress, and sponsored jointly by the British government, the Royal Society, the Royal Geographical Society and private donations, this major expedition explored the Ross Sea region of Antarctica under the leadership of Lt. R. F. **Scott** RN. RRS *Discovery*, a sailing ship with auxiliary steam engine, was built especially for the expedition. Manned by both Royal Navy and merchant service personnel, with a group of civilian scientists, the ship left England in August 1901 for New Zealand, calling at Madeira, South Trinidad and Cape Town. They first encountered the edge of the pack ice in the southern Indian Ocean, landed briefly on Macquarie Island, and reached Lyttelton on 29 November. After replacing stores and equipment that had been damaged *en route*, and stocking up with new provisions, the expedition sailed on 21 December.

Crossing the Antarctic Circle on 3 January 1902, they made a brief landing at Cape Adare, then followed the route of James Clarke Ross along the coast of Victoria Land to

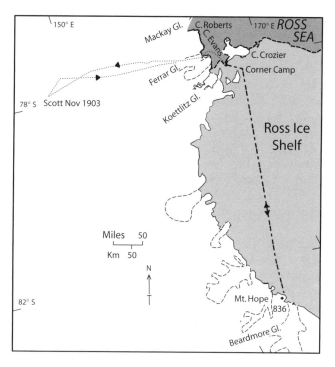

British National Antarctic Expedition 1901–4: major sledging journeys

McMurdo Sound. Landing briefly to leave a message at Cape Crozier, they headed eastward along the Barrier or front of the Ross Ice Shelf, discovering at the eastern end an ice-bound coast and mountains – new land that Scott named for his monarch, King Edward VII. On 4 February, stopping briefly in the Bay of Whales, an indentation in the Ross Ice Shelf, Scott put up a captive hydrogen balloon to 244 m (800 ft). Returning to McMurdo Sound, the party erected a store hut and magnetic observatories ashore and allowed *Discovery* to be frozen into the sea ice for the winter.

Few of the team had previous experience of polar travel or living. In the cramped quarters afforded by the ship, the small scientific team pursued their studies in geology, marine biology and ornithology, making short forays to collect specimens. Though the expedition had brought sledge dogs, it lacked skilled handlers or drivers. Scott used the dogs only reluctantly, preferring to rely on man-hauling, as tried and tested by earlier naval expeditions. The first major sledging expedition, involving Scott, Edward Wilson and Ernest Shackleton with 19 ill-trained and malnourished dogs, left the ship on 2 November 1902, exploring inland across the Ross Ice Shelf toward the South Pole. By 20 December only 14 of the dogs remained alive, and the men were suffering severely from snow-blindness and incipient scurvy. On 30 December they reached 82°16′S, their furthest south, close to the foothills of the Queen Alexandra Range of the Transantarctic Mountains. During the return journey the rest of the dogs died or were killed, and Shackleton collapsed, possibly with scurvy, and was unable to haul. They returned to the ship on 3 February 1903, completely exhausted. Meanwhile a second party led by Albert Armitage had left *Discovery* on 29 November and explored westward among the Victoria Land mountains, finding in a huge glacier (named for expedition geologist Hartley Ferrar) a route to an extensive plateau behind the mountains. They returned on 19 January.

Five days later the relief ship *Morning* (Capt. W. Colbeck) arrived off the McMurdo Sound ice edge, ready to help *Discovery* to break free and return home. Some 13 km (8 miles) of solid ice still separated the two ships. By late February it became apparent that the ice would not break out. *Morning* transferred stores and on 3 March 1903 sailed north, taking some of the overwintering team including a reluctant but still unfit Shackleton. The reduced complement aboard *Discovery* settled for a second winter. Sledging began again on 7 September, when a party of six, led by Charles Royds and including Wilson, man-hauled to the emperor penguin colony at Cape Crozier, hoping to find fertile eggs. They were too late: all the penguins had half-grown chicks, confirming that this species incubated through the coldest months of winter. On 26 October Scott led a nine-man party westward, exploring further the routes that Armitage had pioneered in the western mountains. With petty officers Evans and Lashley, Scott climbed a crumpled staircase of glaciers to stand for the first time on the polar plateau. They marched many miles across the bleak, featureless plain, turning back on 30 November, and returning to *Discovery* just in time to celebrate Christmas 1903. From then on all hands were employed in trying to free the ship. On 5 January 1904 two relief ships, *Morning* (Capt. William Colbeck) and *Terra Nova* (Capt. Harry MacKay), appeared at the ice edge, still some 32 km (20 miles) to the north. They brought instructions that Scott should if necessary abandon *Discovery*. Fortunately during early February the ice began to break back, and on 16 February the three ships cleared McMurdo Sound together.

Scott's first expedition produced many volumes of scientific results, particularly in the field of glaciology, geology and both terrestrial and marine biology. Local surveys mapped the McMurdo Sound region to high standards of accuracy. The major sledging journeys indicated that the way to the South Pole almost certainly lay across the Ross Ice Shelf, through the mountains of southern Victoria Land, and probably across a high, central polar plateau.

Further reading: Scott (1905); *Reader's Digest* (1985); Savours (1966).

British Naval Expedition 1839–43. A world-wide expedition of hydrographic and magnetic surveys, particularly to record magnetic variation in the shipping lanes of

HMSs *Erebus* and *Terror* in calm waters off Ross Island, McMurdo Sound. Source: Ross 1847

the southern oceans. One of three major national expeditions that explored the southern oceans during this period, this British operation was led by Captain James Clark **Ross**, a Royal Navy officer of scientific ability with considerable experience of Arctic sea ice. Already discoverer of the North Magnetic Pole, Ross had a personal ambition to discover its southern counterpart. Equipped with two stoutly built monitors (heavy gunships) HMSs *Erebus* and *Terror*, the latter commanded by Cdr Francis Crozier, Ross left England in September 1839. On the route south he set up observatories and determined the magnetic characteristics of St Helena, Cape Town and Iles Kerguelen. Reaching Hobart, Tasmania in August 1840, he stayed for several weeks, taking further observations and preparing his ships for work in the Antarctic. Apart from magnetometric research, Ross constantly encouraged the ornithological research of one of his surgeons, Robert McCormick, and the botanical research of his assistant surgeon, Joseph Dalton **Hooker**.

Erebus and *Terror* sailed south on 12 November, visiting the Auckland Islands and Campbell Island, then continuing south along longitude 169°E to the edge of the pack ice. New Year's Day 1841 found the two ships pressing confidently into the ice. To his surprise Ross discovered a relatively easy way through into a polynya, or anomalous stretch of open water, within which he continued southward for several days. To the west there gradually emerged an ice-bound coast, backed by extensive ranges of mountains rising to 2400 m (8000 ft) and more. The coast he named Victoria Land for the reigning British queen: the mountains became Admiralty Range. Ross's enthusiasm for his discovery was muted: magnetometer measurements showed that the magnetic pole he coveted lay 800 km (500 miles) beyond the mountains, and far beyond his reach. On 12 January he and Cdr Crozier landed on a small islet, Possession Island, where they toasted the Queen and claimed the territory for Britain. Further south he discovered a larger island with an active volcano alongside an extensive bay. The volcano he called Mount Erebus for his ship, the bay **McMurdo Sound** for his First Officer, and other features were named for other ship's officers. The island was later named for Ross himself. To the east beyond Ross Island stretched an ice wall over 250 miles long with cliffs 150 ft high, fronting a huge plain of ice that extended southward as far as he could see. Later explorers came to know this as the Ross Barrier or **Ross Ice Shelf**. Close to the ice cliffs Ross reached the unprecedented latitude of 78°4′S.

Ross returned to Hobart on 6 April 1841, and left for a second southern cruise on 23 November. The ships returned to what is now the **Ross Sea**, encountering heavy pack ice and bad weather. They traced the great ice shelf further east, but made no further geographical discoveries. In early March 1842 they headed north, then turned east toward the Falkland Islands, which they reached in five weeks. Ross wintered in the sheltered harbour of Port Louis, setting up a magnetic observatory and enjoying a few weeks' rest and recreation ashore. Asked to adjudicate on whether Port Louis or Port William, further south, would provide a better setting for a future seat of government, Ross unhesitatingly chose Port William, paving the way for the founding of Stanley, the modern capital.

The third southern voyage began on 17 December 1842. Ross hoped to explore around d'Urville's recently discovered 'Louis Philippe's Land' and perhaps follow Weddell's track into the Weddell Sea. Heavy pack ice prevented both operations: he made relatively minor landfalls off the tip of Antarctic Peninsula, and on 5 March 1843 reached the edge of the pack ice in 71°30′S, 14°51′W. It was too late in the season to do more: *Erebus* and *Terror* turned north for home. They reached Britain on 4 September 1843.

Ross's expedition was lauded for his location of the South Magnetic Pole and magnetometric measurements across the Southern Ocean. His discovery of the sea that bears his name, his penetration of the polynya (now known to form almost every year along the Victoria Land coast) and delineation of the Ross Ice Shelf, provided an important highway to the interior of Antarctica, and ultimately to the South Pole.

Further reading: Ross (1847).

Brocklehurst, Sir Philip Lee, Bart. (1887–1975).

British polar explorer. Born on 7 March 1887 at Swythamley Park, Staffordshire, he inherited the baronetcy in 1903. While an undergraduate at Trinity Hall, Cambridge, he joined the **British Antarctic (*Nimrod*) Expedition 1907–9** as a scientific assistant. The youngest member of the group, he sustained frostbite on the ascent of Mt Erebus and subsequently lost a toe, but took part in several other journeys, exploring the Taylor Valley and Ferrar Glacier. Commissioned in a yeomanry regiment, Brocklehurst served in World War I with the Life Guards, and in 1918–20 with the Egyptian Army. During World War II he commanded a brigade of the Arab Legion. In peace time he tended his estate. He died on 28 January 1975, the last survivor of the *Nimrod* expedition.

Bruce, Wilfred Montague. (1874–1953). British polar seaman. Born in Scotland on 26 October 1874, he became an officer in the merchant navy and RNR. The brother-in-law of Robert Falcon Scott, he served with the **British Antarctic (*Terra Nova*) Expedition 1910–13**, first by helping to collect and transport sledge dogs and ponies from Vladivostok, later as second officer in *Terra Nova* on both the southern voyages. He served in the navy in World War I, retiring as a Captain RNR shortly after the war to take up pig farming. He died on 21 September 1953.

Bruce, William Spiers. (1867–1921). Scottish physician, oceanographer and polar explorer. Born in London of Scottish parents, Bruce trained in natural sciences and medicine at the universities of Edinburgh and Aberdeen. In 1892–93 he joined the whaling ship *Balaena* (see **Dundee Whaling Expedition 1892–93**), as ship's surgeon and naturalist, on a voyage to the South American sector of the Southern Ocean. In 1895–96 he was director of the meteorological observatory on Ben Nevis. In 1897 he joined the Jackson-Harmsworth Expedition to Franz Josef Land, and later worked with other private expeditions on Novaya Zemlya and Svalbard. Despite considerable financial problems he organized and led the **Scottish National Antarctic Expedition 1902–4**, which surveyed the South Orkney Islands, and established a meteorological station which has since, under Argentine management, provided the longest continuous record of observations in the Antarctic region. He also explored south into the Weddell Sea, discovering Coats Land. On return he became director of the Scottish Oceanographical Laboratory, a post that enabled him to complete and publish the reports of the expedition. Bruce returned to Svalbard in 1906, and again in 1912, 1914 and 1919, on a variety of scientific projects, and received many scientific awards for his polar research. He died in Edinburgh in October 1921.

Further reading: Bruce (1896; 1911); Speak (1992).

Bruce Plateau. 66°00′S, 64°00′W. Ice plateau at about 1830 m., extending across Graham Land from the Gould and Erskine glaciers to Flandres Bay. It was named for the Scottish explorer W. S. Bruce.

Brunt Ice Shelf. 75°40′S, 25°00′W. Broad ice shelf bordering the Caird Coast of Coats Land, between Dawson-Lambton Ice Stream and Stancomb-Wills Glacier Tongue. It is maintained by a constant flow of ice from the Brunt Icefalls. Named for British physicist D. Brunt, since January 1956 it has been occupied by Halley (Base Z), a succession of British research stations concerned especially with geophysics and upper atmosphere studies.

Bryan Coast. 73°35′S, 84°00′W. Part of the coast of Ellsworth Land, West Antarctica, facing the Bellingshausen Sea. Bounded by Pfrogner Point in the west and Rydberg Peninsula in the east, the coast consists of steep ice cliffs fronted in part by ice shelves, and backed by a featureless rising plain of ice. It was delineated from aerial photographs and named for Rear Admiral George S. Bryan USN, hydrographer of the US Navy.

William Spiers Bruce (1867–1921). Photo: Scott Polar Research Institute

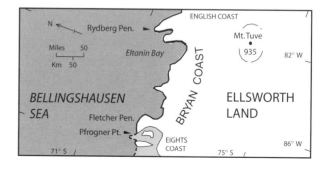

Bryan Coast

Buckle Island. 66°50′S, 163°12′E. Central island of the Balleny Islands group; a long, narrow volcanic island, largely ice-covered and ringed by steep cliffs. The island was named for J. W. Buckle, a subscriber to John Balleny's sealing expedition of 1839, on which it was discovered. Intermittent fumarolic activity was reported on the island between its discovery and the end of the century, but has not appeared since.

Bucknell, Ernest Selwyn. (1926–2001). New Zealand expedition cook. Born in Upper Hutt on 28 October 1926, he qualified as a fitter and turner with New Zealand Railways, and for several years became a licensed deer, possum and wallaby culler with the Wildlife Division. In 1956 he became cook to the Ross Sea party of the Commonwealth Transantarctic Expedition, serving for two years at Scott Base. Returning to the Wildlife Division, he resumed work on control and management in New Zealand. He retired in 1991, and died on 14 March 2001.

Budd Coast. 66°30′S, 112°00′E. Part of the coast of Wilkes Land, East Antarctica, facing the Davis Sea, and bounded by the Hatch Islands in the west and Cape Waldron in the east. Discovered during the US Exploring Expedition (1838–42), the coast was named for Thomas. A. Budd USN, acting Master of USS *Peacock*, one of the expedition ships.

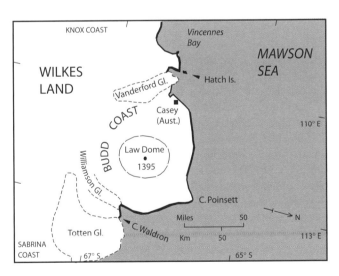

Budd Coast

Bunger Hills. 67°17′S, 100°47′E. Group of low, ice-free hills, alternating with morainic shingle flats and shallow lakes. Between the edge of the continental ice plain and the western half of Shackleton Ice Shelf, at the eastern end of Knox Coast, it is an oasis or anomalously ice-free area of about 1000 km² (390 sq miles) of land, islands and lakes. The area was discovered in February 1947 during a photo-reconnaissance flight of Operation Highjump, and named for the pilot, Lt. Cdr David E. Bunger USN, who landed his seaplane on one of the larger lakes. From October 1956 Bunger Hills accommodated the Soviet research station Oazis, which in January 1959 briefly became the Polish station Antoni Dobrowolski.

Burdwood Bank. 54°00′S, 60°00′W. Submarine bank south of the Falkland Islands, extending east-to-west between Tierra del Fuego and South Georgia, and rising to about 100 m (328 ft) below sea level. It forms part of the northern arm of the **Scotia Arc**.

Bursey, Jacob. (d. 1980). US Coast Guard officer and explorer. Born in Newfoundland, he served as a dog driver with **Byrd's First Antarctic Expedition 1928–30** and the **United States Antarctic Service Expedition 1939–41**, wintering at Little America III. Later he acted as technical advisor to the US Navy Antarctic Expedition (Operation Deep-Freeze I) 1955–56. He died on 23 March 1980. (CH)

Byers Peninsula. 62°38′S, 61°05′W. A long peninsula forming the western end of Livingston Island, South Shetland Islands, named for an American shipowner, James Byers, whose sealing ships visited the area in the 1820s. The area is rich in freshwater lakes, outcrops of fossil-bearing Jurassic and Cretaceous strata and early nineteenth century sealing camps. Approximately 65.7 km² (26 sq miles) of ice-free land and littoral, originally designated SPA No. 10, are currently redesignated SSSI No. 6 (Byers Peninsula), to protect the area for research.

Byrd, Richard Evelyn, Jr. (1888–1957). American aviator and explorer of both polar regions. Born on 25 October 1888 into a patrician Virginian family, Byrd was educated at the University of Virginia and the US Naval Academy, Annapolis, graduating in 1912. When an accident in 1916 left him with a damaged leg, he was rated unfit for sea duties and relegated to deskwork in the naval reserve. However, he trained as a pilot in the newly-formed naval air service, and in 1918 was involved in planning the first-ever long-distance flights of US Navy seaplanes across the Atlantic Ocean. In 1925, when the Navy considered sending one of its airships on a trans-Arctic flight from Alaska to Svalbard, Byrd was one of the few officers with expertise in logistics. The flight never took place, but Byrd saw the possibility of using large aircraft to explore unknown areas of the Arctic basin.

As a reserve officer, he was free to plan an expedition of his own to the Arctic, seeking help both from the Navy and from private sponsors. The Navy had already agreed to lend aircraft to a veteran explorer, Donald Macmillan, for work in Greenland. Byrd was seconded to the expedition as Macmillan's planning and liaison officer, and gained his first Arctic flying experience over west Greenland and Ellesmere Island. His experience of polar operations was put to good use in 1926, the following year, when Byrd and Floyd Bennett flew a ski-equipped, tri-engined Fokker aircraft from Svalbard to the North Pole, a return journey of 1300 miles over featureless sea ice. From this expedition he learned useful lessons about long-distance polar flying and navigation, and also how to organize and manage low-budget expeditions, making full use of media publicity. In late June 1927 he crossed the Atlantic by air from Newfoundland to France, and in the following year took his first expedition to Antarctica.

Byrd's First Antarctic Expedition 1928–30 was primarily a bid to reach the South Pole by air. This was the spectacular objective for which he gained both private funding and public support. However, Byrd used his knowledge of aviation logistics to explore and photograph wide areas of King Edward VII Land, and gave full support to a scientific team, under Laurence **Gould**, who in one of the longest-ever recorded dog-sledging journeys surveyed the newly-discovered Rockefeller Mountains.

Promoted to Rear Admiral in the US Naval Reserve, Byrd made full use of publicity, gaining for himself a heroic image which he exploited in promoting further exploration. **Byrd's Second Antarctic Expedition 1933–35** continued scientific and geographical exploration in what was becoming the 'American' sector of Antarctica. Again Byrd combined strong sledging programmes, using dog teams and tractors, with wide-ranging aerial photographic survey. On returning to the USA he sought to establish claims to huge tracts of Antarctica on behalf of his country, a move that the US government neither ratified nor supported. His success did, however, persuade the government that Antarctic exploration was no longer a matter for private enterprise. In 1939 he was appointed head of a newly-formed Antarctic agency and given command of the **United States Antarctic Service Expedition 1939–41**, an expedition almost entirely government-funded with two bases, one at Little America and the other on Stonington Island, Marguerite Bay, Antarctic Peninsula. Planned to last for several years with changing personnel, the USAS expedition ended under threat of war in 1941.

In World War II Byrd advised on cold-weather clothing and equipment, and in planning long-range air routes for war in the Pacific Ocean. He took part in the **United States Navy Antarctic Development Project 1946–47** (Operation Highjump), revisiting the continent and again flying to the South Pole. In 1955 he initiated the new wave of US Antarctic research preparatory to the International Geophysical Year Programs. He died aged 68 on 11 March 1957.
Further reading: Byrd (1931), (1935)

Byrd, Richard E. III. (1920–88). American naval officer. The son of Rear Admiral Richard Byrd, he studied at Harvard before joining the Navy in 1942. After service in World War II he took part in Operation Highjump 1946–47 and Operation Deep Freeze I 1955–56, in which he was ADC to his father. Mount Byrd, Marie Byrd Land, was named for him. He died on 3 October 1988.
(CH)

Byrd's First Antarctic Expedition 1928–30. An expedition to the Ross Sea sector of West Antarctica that included the first flight toward the South Pole. Richard Evelyn **Byrd**, US aviator and naval reserve officer, experienced in logistic planning, in 1926 had flown over Greenland and Arctic Canada, and made a well-publicized flight from Spitsbergen towards the North Pole. In 1928 he made use of contacts in society, government, the US Navy and news media to mount an Antarctic expedition to fly to the South Pole.

Two ships, *City of New York* (515 tons) and *Eleanor Bolling* (800 tons, named for Byrd's mother), left New York respectively in August and September 1928 for New Zealand, sailing south for Antarctica on 2 December. *Eleanor Bolling* towed the smaller ship for eight days to the edge of the pack ice, where she transferred stores and returned to New Zealand. For a further 13 days *City of New York* was towed through the pack ice by *C. A. Larsen*, a large Norwegian whaling factory ship, casting off in open water on 23 December and reaching Bay of Whales, at the eastern end of Ross Ice Shelf, on 28 December. Here, 11 km (7 miles) inland from the ice edge, Byrd established 'Little America', an overwintering station and camp for 42 men and 80 sledge dogs. The equipment included 650 tons of stores, with three monoplane aircraft (a tri-motored Ford, a Fokker Universal and a Fairchild), plus a number of small tractors. On 27 January 1929 Byrd, with pilot Bernt Balchen and radioman Harold June, flew eastward over Edward VII Land, discovering the Rockefeller Mts. There were further reconnaissance flights during February and early March, and on 7 March dog teams left Little America to lay depots to the south, for a longer journey in the following spring.

Operations resumed in October, when more depots were laid in preparation for the major scientific thrust – a sledging expedition, led by geologist Dr Laurence M. Gould, to explore the Queen Maud Range some 800 km (500 miles) south across the Ross Ice Shelf. Gould's party of six sledgers and five dog teams set out on 4 November. Meanwhile preparations were made for the expedition's more widely heralded feat, the flight to the South Pole, a distance of over 1280 km (800 miles) involving a climb to over 3660 m (12 000 ft). At a time when few knew anything of planning for long-distance flights, Byrd was examining the special problems of flying at the limits of aircraft endurance and navigation in polar regions. Hazards included extreme cold, requiring lubricants that would not solidify, warming engines before flights and keeping them warm during refuelling stops, and keeping condensation moisture out of the fuel. Predicting weather, particularly upper winds which drastically affected fuel consumption, range and course-setting, was difficult in an area where there were no weather stations. Finding direction was a major difficulty where magnetic compasses were unreliable: Byrd used a sun compass, like a sun-dial in reverse, that gave direction far more reliably throughout the 24 hours of daylight.

The flight began at 3.29 a.m. on 28 November. Byrd, photographer Ashley McKinley, radioman Harold June and pilot Bernt Balchen followed closely the route from Bay of Whales taken by Amundsen some 18 years earlier. By 9.00 they were climbing a steep glacier close to the Axel Heiberg. To reach the polar plateau they had to jettison emergency food bags, clearing the glacier ice by a few hundred feet. Then a steady flight took them to the Pole, which they reached at 1.15 p.m. By late afternoon they had returned to the foot of the glacier, landed and refuelled from a depot. Six hours later they landed in Little America, after a flight totalling almost 16 hours. Through December and January Byrd's aircraft overflew and photographed many square miles of territory, bringing back excellent photographs of mountains, glaciers and ice coasts. Meanwhile Gould's sledging party made a detailed survey of the Queen Maud Range, collecting both scientific data and the accurate astronomical fixes that consolidated the aerial photography. During nine weeks in the field they covered over 2100 km (1300 miles), returning to Little America on 19 January 1930.

The expedition returned to New Zealand in March 1930. Through radio, cine-films and the printed word Byrd reached a far wider public than any explorer before him, and the US public especially responded with adulation.
Further reading: Byrd (1931).

Byrd's Second Antarctic Expedition 1933–35.

A second and more thorough exploration by dog sledge and aircraft of the Ross Sea sector of West Antarctica. Fired by the success of his first expedition of 1928–30, Richard Byrd immediately began planning further exploration along similar lines. Despite acute economic depression in the USA, he secured considerable support from industries and the public. His two expedition ships, the 59-year-old barquentine *Bear of Oakland* (703 tons) and freighter *Jacob Ruppert* (8500 tons), left the USA respectively in September and October. Sailing from Wellington on 12 December, *Jacob Ruppert* reached the edge of the pack ice off the coast of King Edward VII Land on 20 December. There Byrd made several unsuccessful attempts to reconnoitre south in the Curtiss-Wright Condor float plane, *William Horlick*. Reaching the Bay of Whales on 17 January 1934, the party reoccupied and extended Little America, providing for 56 men, 150 sledge dogs, 4 aircraft (including an autogyro), several tractors, and 3 Guernsey cows and a calf, the gift of a milk company. *Bear of Oakland* arrived on 30 January. After unloading, Byrd used the ship in a further attempt to explore eastward along the King Edward VII Land coast, making little progress against heavy pack ice. *Jacob Ruppert* left Bay of Whales on 5 February, *Bear of Oakland* three weeks later.

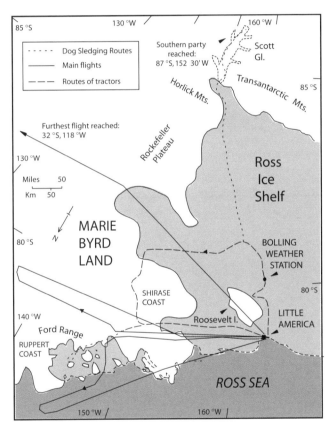

Byrd's Second Antarctic Expedition 1933–35

In the autumn twilight, sledging parties with tractors and dogs laid depots far into the interior, on 16 March setting up the Bolling Advance Weather Base, a small station in 80°08′S, 163°57′W on the Ross Ice Shelf. Here Byrd elected to overwinter alone. On 22 March he transferred command of the expedition to physicist Thomas Poulter, and was flown out to the shack. For several weeks Byrd maintained radio contact, transmitting weather observations, and for a time seemed content in his solitude.

However, his condition deteriorated: comrades who risked darkness and cold to visit in August found him unkempt, possibly poisoned by fumes from a stove, and brought him back to base.

From late September sledging parties set out to east and south, exploring in areas that had been discovered during the previous expedition. On 27 September a tractor train led by Harold June left base to lay depots and explore in the Edsel Ford Mountains, and on 14 October a sledging party led by biologist Paul Siple headed eastward to explore Marie Byrd Land. A party led by geologist Q. A. Blackburn explored south in the Queen Maud Range. All brought back excellent data, including geological and biological specimens, and astronomical fixes that supported the aerial photography. From mid-November Byrd organized a series of long-range flights which established that mountains and high plateaux lay in every direction behind the Ross Ice Shelf. Meanwhile physicists working closer to Little America established that the ice shelf under the base was 90 m (300 ft) thick, afloat in water over 600 m (2000 ft) deep. Antarctica was clearly a single ice-covered continent.

All parties returned to Little America in late January, and the two ships cleared the base on 6 February.

Again Byrd returned to the USA in triumph. His two expeditions, meticulously planned in the Amundsen tradition, had filled a huge gap in the map of Antarctica, produced a generation of polar scientists and technicians, and given the United States a leading role in the exploration of the southern continent. Though official US policy could not support Byrd's claims to the newly-discovered territories, the government was satisfied that his well-publicized achievements would strengthen possible future claims, to match those currently being lodged by Britain, New Zealand, Norway, France and other nations.
Further reading: Byrd (1935).

Byrd Station. 80°01'S, 119°32'W. American inland research station established on Rockefeller Plateau, Marie Byrd Land, in February 1957. It was used both as an overwintering station and for summer traverses across West Antarctica, particularly during the late 1950s and early 1960s.

C

Caird Coast. 76°00′S, 24°00′W. Part of the coast of Coats Land, East Antarctica, facing the Weddell Sea, and bounded by Stancomb-Wills Ice Stream in the northeast, Hayes Glacier in the southwest. The coast consists almost entirely of ice cliffs, was named by Sir Ernest Shackleton for Sir James Caird, a patron of the **Imperial Trans-Antarctic Expedition 1914–1917**.

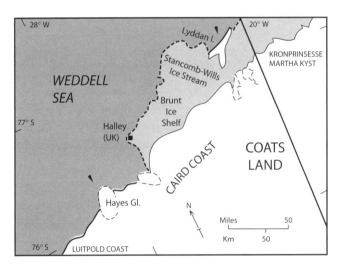

Caird Coast

California Plateau. 86°04′S, 145°10′W. Extensive ice plateau rising to 3270 m by Scott Glacier, Queen Maud Mountains, named for the University of California.

Calving. Massive breaking-away of ice from the face of a glacier or ice shelf, or fragmenting of an iceberg.

Campbell, Victor Lindsey Arbuthnot. (1875–1956). British seaman and explorer. Born on 20 August 1875 in Brighton, Sussex, he served briefly in the merchant navy before receiving a commission in the Royal Navy in 1895. In 1901 he resigned his commission to live as a country gentleman, but in 1910 joined Capt. Robert F. Scott's British Antarctic (*Terra Nova*) Expedition 1910–13 as chief mate. Given command of the Northern Party, he and five companions spent a year at Cape Adare, a second year exploring and sledging along the Victoria Land coast. The party wintered in a snow cave on Inexpressible Island, eventually sledging back to the main base at Cape Evans. In World War I he saw active service at Gallipoli, Jutland and in northern Russia, and was decorated three times. Retiring from the navy in 1922 he moved to Newfoundland, where he farmed and fished. He served briefly in World War II in Trinidad and Canada. He died on 19 November 1956.

Campbell Island. 52°34′S, 169°10′E. A cool-temperate island of the southern Pacific Ocean, Campbell Island stands on an extensive submarine platform 620 km (387 miles) south of New Zealand's South Island. Raised by volcanic activity six to eight million years ago, it has since been eroded by sea and glaciers into a rough, deeply indented dome up to 18 km (11.2 miles) across: the highest point, Mt Honey, rises to 569 m (1866 ft). There are several small offlying islets. The island consists mainly of alkaline to transitional basalts: no recent volcanic activity has been reported. The climate is cool and damp, with mean annual temperature 6.7 °C, range of mean monthly temperatures 4.7 °C, 140 cm of rain per year, and persistent westerly winds. The flora includes over 200 species of flowering plants and ferns, including many alien species from New Zealand. Woody shrubs and ferns dominate coastal areas, forming miniature forests in gullies sheltered from the winds. Uplands support tussock meadows, bogs and peat-based moors. Yellow-eyed, rockhopper and erect-crested penguins breed among the coastal vegetation: royal, black-browed and grey-headed albatrosses, giant petrels, many smaller petrels, shags, skuas and terns also breed in abundance, especially on offshore islands. Pipits, dunnocks, redpolls and an endemic teal breed among the vegetation. Elephant seals, fur seals and sea lions are seasonally abundant on the beaches.

Discovered by sealing captain Frederick Hasselburg in January 1910, the island was visited irregularly by sealers throughout the nineteenth century, and by whalers hunting its inshore waters up to 1916. From 1895 to 1931 it was farmed for sheep, and a wartime meteorological station was sited at the head of Perseverance Harbour. Bird populations on the main island have been severely affected by introduced Norway rats and cats, and feral sheep have altered the tussock meadows; however, many of the small offlying islands remain relatively pristine. Campbell Island is a New Zealand National Nature Reserve from which the public are excluded, except under permit.

Further reading: Bailey and Sorensen (1962); Fraser (1986).

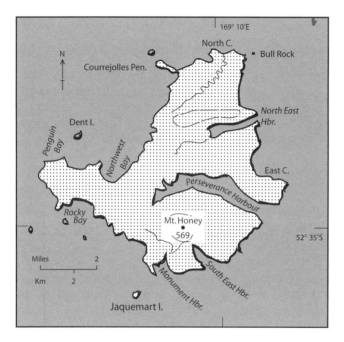

Campbell Island. Height in m

Campbell Island Station. 52°33′S, 169°09′E. New Zealand research station at the head of Perseverance Harbour, Campbell Island. Established in 1941 as a wartime meteorological and coast-watching station, it operated continuously until it was closed in 1995.

Canada Glacier. 77°37′S, 163°05′E. A glacier feeding into Taylor Valley, part of the dry valley system of Victoria Land. An area of approximately $1.2\,km^2$ (0.5 sq miles) between the glacier tongue and the shore of Lake Fryxell has been designated SSSI No. 12 (Canada Glacier–Lake Fryxell), protecting research on moraine deposits, ancient lake levels, and an unusually rich assembly of algae and mosses.

Canadian sealing expedition 1901–2. The sealing ships *Beatrice L. Corkrum* (Capt. R. Balcom) and *Edward Roy* (Capt. F. W. Gilbert) from Halifax, Nova Scotia, made exploratory sealing voyages around the Falkland Islands.

Capt Balcom also visited the South Shetland Islands. These voyages proved profitable enough to encourage further sealing visits by ships from Nova Scotia during the following decade.

Candlemas Islands. 57°03′S, 26°43′W. A group of two small volcanic islands, Candlemas Island and Vindication Island, together with several outlying rocks, toward the northern end of the South Sandwich Islands. The group was discovered by Capt. James Cook on 2nd February (Candlemas Day) 1775.

Cape Geddes (Base C). 60°41′S, 44°34′W. British research station at Ferguslike Peninsula, Laurie Island, South Orkney Islands. Built in January 1946, it was named Cardinall House and occupied for meteorological and survey observations for one year only. In March 1947 operations moved to a new station on Signy Island (Base H). The hut remains, currently in good order, for use as a refuge.

Capitán Arturo Prat Station. 62°30′S, 59°41′W. Chilean naval station, established in February 1947 on Guesalaga Peninsula, Greenwich Island, South Shetland Islands. Chile's first permanent year-round research station, originally 'Soberania', it was renamed for a Chilean naval hero. The station has operated continuously since its establishment.

Carvajal Station. See **Teniente Carvajal Station**.

Casey Station. 66°17′S, 110°32′E. Australian permanent year-round research station on Bailey Peninsula, Windmill Islands, Budd Coast of East Antarctica. The original station was built in 1969, as a replacement for nearby **Wilkes Station**, a United States International Geophysical Year base dating from 1957, which was taken over by Australia in 1959 and occupied for ten years. The present station, some 500 m (547 yards) from 'Old Casey', was opened in 1988–89.

Caughley Beach. 77°10′S, 166°40′E. A biologically important beach and hinterland 1 km (0.6 miles) north of Cape Bird, Ross Island, situated between two breeding groups of Adélie penguins. An area of $0.3\,km^2$ between the beach and the icecap of Mt. Bird above, scheduled SSSI No. 10, contains extensive stands of moss, algae, and lichens, which are the subject of long-term research. SPA 20 lies within the area of SSSI 10. The beach was named for New Zealand biologist Graham Caughley.

Cavendish, Thomas. (1560–92). English navigator, privateer and explorer. Born in Suffolk, England, he matriculated at Cambridge University in 1576. His first long voyage in 1585 was with Sir Richard Grenville to Virginia. In July 1586 he fitted out his own expedition of three ships to emulate Sir Francis Drake's recent voyage to the west coast of South America. In southern Patagonia he discovered Puerto Deseado, naming it for his ship *Desire*, then passed through the Strait of Magellan in January and February 1587. He looted settlements and shipping along the coast of South America, and captured a Spanish treasure ship off California. Completing the third world circumnavigation, he returned to Britain in November 1588. In August 1591 he sailed again, this time in *Leicester*, accompanied by Capt. **John Davis** in *Desire*. After attempting unsuccessfully to traverse the Strait of Magellan, he returned to the southern Atlantic Ocean, parting company with Capt. Davis off Patagonia. Davis independently discovered the Falkland Islands: Cavendish died at sea *en route* to St Helena in June 1592.

CCAMLR. See **Convention on the Conservation of Antarctic Marine Living Resources**.

CCAS. See **Convention for the Conservation of Antarctic Seals**.

***Challenger* Expedition 1872–76.** A worldwide oceanographic expedition by HMS *Challenger*, a corvette of the Royal Navy, that included visits to the Southern Ocean and some of the southern islands. The purpose was to explore the sea bed and its living creatures, particularly to determine routes for submarine telegraph cables, and to examine and trace the origins of deep-sea water masses. The ship, a spar-decked corvette of 2306 tonnes with auxiliary steam, equipped with steam winches for deep-sea trawling, dredging and sounding, was commanded successively by Capt. George Strong Nares RN and Capt. Frank Tourle Thomson, RN. Accommodation included laboratory and work-space for six civilian scientists directed by C. Wyville Thompson.

In a voyage mainly to temperate and tropical oceans, HMS *Challenger* visited Tristan da Cunha and Nightingale Island in September 1873, Marion Island on 26 December, Iles Crozet and Kerguelen and Heard Island in January and early February 1874. From Iles Kerguelen the ship ran 550 miles (880 km) southward into the Southern Ocean, toward an appearance of land reported some 30 years earlier by **Wilkes**. The first tabular iceberg was seen on 11 February: pack ice was encountered on 15 February, and on the following day HMS *Challenger* became the first steam-assisted ship to cross the Antarctic Circle. Skirting eastward along the edge of the pack ice, the ship reached a point some 15 miles (24 km) from the reported position of Wilkes's 'Termination Land', in deep water with no land in sight. From there HMS *Challenger* returned to warmer waters, shaping course for Melbourne, Australia.

After a long Pacific Ocean cruise the ship returned to the Atlantic Ocean via the Strait of Magellan (December 1875), and the Falkland Islands (January 1876). Results of the expedition, contained in 50 volumes, were published intermittently over the following 20 years. Those relating to the southern oceans included the first detailed charts and systematic accounts of many of the southern islands, and important monographs on the anatomy of penguins and petrels.

Further reading: Campbell, Lord George (1876). Linklater (1972); Thomson (1877).

***Chanticleer* Expedition 1828–29.** In response to a request by the British scientist Sir Edward Sabine for data on forces of gravity and terrestrial magnetism in the southern hemisphere, the Admiralty in 1828 despatched the sloop HMS *Chanticleer* to the South Shetland Islands, then the most southerly known and charted land. In command was Capt. Henry Foster, an able navigator and scientist. Foster sailed from Staten Island on 21 December 1828, reaching the South Shetland Islands on 3 January 1829 and passing through the chain to a small island in 63°44′S (now called Chanticleer Island), in Palmer Archipelago. There he landed, taking possession of the island for Britain. Returning northward, he entered the crater of Deception Island, anchoring off a small inlet at the northeastern end of the harbour where he set up his observatory. The harbour is now called Port Foster, the inlet Pendulum Cove, from the instrument used in determining gravity. Foster completed his observations successfully on 8 March and returned to South America. A minimum mercury-in-glass thermometer that he left on the island was recovered by a sealer, W. H. Smiley in 1842, showing a minimum temperature reading of $-5°F$ ($-21°C$).

Further reading: Webster (1834).

Charcot, Jean-Baptiste Etienne Auguste. (1867–1936). French explorer and physician. Born in Neuilly-sur-Seine to a wealthy medical family, Charcot trained in medicine at the Institut Pasteur and practised for a time as a cancer specialist. However, his main interest was polar exploration. In August 1903, accompanied by a small team of scientists, he sailed in his own ship, *Le Français*, to Antarctic Peninsula. After joining briefly in the search for the missing **Swedish South Polar Expedition**, he sailed south into Gerlache Strait and wintered the ship in fast ice off Booth (Wandel) Island. During the second summer he charted his way southward to Alexander Island and named many features of the western Antarctic Peninsula coast. Prevented by heavy pack ice from exploring further south, he resolved to return to Antarctica with a stronger and better ship.

Jean-Baptiste Charcot (1867–1936). Photo: Scott Polar Research Institute

Returning to France, Charcot commissioned a new, specially built expedition ship, which he called *Pourquoi-pas?* Sailing south to the Peninsula in August 1908, with a party of scientists and hydrographers, he explored the South Shetland Islands and the northern Peninsula, overwintering off Petermann Island, slightly further south than before. In the following summer he charted more accurately his discoveries of the previous expedition, then explored the unknown coast of the peninsula south of Adelaide Island, penetrating far into Marguerite Bay. South again in the Bellingshausen Sea he sailed beyond Alexander Land to within sight of Peter I Øy, discovering a further ice-bound coast now called Charcot Island.

The scientific results of Charcot's two expeditions, and his achievements as a polar explorer, received wide acclaim. He became director of the maritime research laboratory of École Pratique de Hautes Études, to which he dedicated his ship. During World War I he served in both the Royal Navy and the French Navy. During the 1920s and 1930s he took *Pourquoi-pas?* north on ten summer expeditions, providing opportunities for young scientists of several nations to experience polar research. In the final voyage of 1936, the scientist known widely as 'the polar gentleman' died along with most of his crew when the ship foundered off Iceland.
Further reading: Charcot (1906; 1911); Malaurie (1989).

Charcot Island. 69°57′S, 75°25′W. Ice-covered island in the Bellingshausen Sea, discovered by the French Antarctic (*Pourquoi Pas?*) Expedition in 1910 and named by J.-B. Charcot for his father, neurologist Dr J. M. Charcot.

Charcot, Port. 65°04′S, 64°00′W. Harbour in the north side of Booth Island, Wilhelm Archipelago, where the French Antarctic (*Français*) Expedition 1903–5 overwintered in their ship. The harbour was named by the leader, Dr J.-B. Charcot, for his father, neurologist Dr J. M. Charcot.

Charlie, Dome. 75°00′S, 125°00′E. Ice dome 3200 m. (10 496 ft.) above sea level in eastern Wilkes Land. Also called Dome C and Dome Circe, this was an important site of ice-core drilling during the 1970s, and is currently the site of **Concordia**, a joint French/Italian research facility.

Cheeseman, Al.. (d. 1943). Canadian polar pilot. In 1929 he was a member of Sir Hubert **Wilkins** 's expedition to Graham Land: later he flew with Wilkins in the search for Soviet aviators missing in the Arctic. An officer in the Royal Canadian Air Force, he died on active service in January 1943.

Cheetham, Alfred. (d. 1918). British polar seaman. Born in Liverpool, a merchant navy boatswain working out of Hull, and Royal Naval Reservist, he first visited Antarctica as boatswain in *Morning*, the relief ship sent to McMurdo Sound in 1902–3 and 1903–4 to relieve the **British National Antarctic Expedition 1901–4**. Shackleton, who returned with the ship, invited him to join the ship's company of the **British Antarctic (*Nimrod*) Expedition 1907–9**, as boatswain and third officer. He served also as boatswain in the **British Antarctic (*Terra Nova*) Expedition 1910–13**, and again as third officer with Shackleton on the **Imperial Trans-Antarctic Expedition 1914–17**. Returning to the sea in World War I, he was lost when his ship was torpedoed in 1918.

Cherry-Garrard, Apsley. (1886–1959). British explorer. Educated at Christ Church, Oxford, he joined the **British Antarctic (*Terra Nova*) Expedition 1910–13** as assistant zoologist, spending two years ashore at Cape Evans. Though hampered by poor eyesight, he took part in all the major sledging activities, including the winter journey to Cape Crozier to collect emperor penguin embryos, depot-laying journeys, and search parties to seek, and eventually to find, the ill-fated polar party. He is best remembered for his classic account of the expedition, *The Worst Journey in the World* (1922).

Apsley Cherry-Garrard on his return from Cape Crozier. Photo: Scott Polar Research Institute

Chinstrap adult and chick, Antarctic Peninsula. Photo: BS

Chinstrap penguin. *Pygoscelis antarctica.* Smallest and noisiest of the pygoscelid (brush-tailed) penguins, chinstraps stand 75 cm (30 in) tall and weigh about 4.5 kg (10 lb). Distinguished by their white cheeks and narrow band of black feathers under the chin, they live almost entirely in the maritime Antarctic. Though vagrants have been seen on many other southern islands, breeding is restricted to islands of the Scotia Arc, Antarctic Peninsula and Bouvetøya: they are particularly well represented on the South Sandwich Islands. Chinstraps breed in colonies of several hundreds to many thousands of pairs, building nests of pebbles, bones and feathers, often close to colonies of Adélies or gentoos, though seldom intermixing. The two eggs are chalky-white, the chicks silver-grey. They feed mainly on krill and surface-living fish. World population is estimated at 7 500 000 pairs.

Christensen, Lars. (1884–1965). Norwegian whaling entrepreneur and explorer. Born near Sandefjord, son of Chr. Christensen and heir to a wealthy ship-owning and whaling business, he served in the firm's shipping offices. In 1892–94 his father Christen financed the Norwegian (Sandefjord) Whaling Expeditions to explore possibilities for whaling in Antarctic waters, and in 1905 sent his factory ship *Admiralen* to operate off the South Shetland Islands. From 1907 Lars developed his own shipping operations, starting whaling in Chile and other southern waters. His company A/S Condor made profit processing already-flensed whale carcasses in Grytviken, South Georgia, and A/S Hvalen operated the factory ship *Hvalen* off the South Shetland Islands. In 1922, on the death of his father, he took over the family business, in which southern whaling figured significantly. Between 1926 and 1937 he organized a series of exploring expeditions, the **Norwegian (*Christensen*) Whaling Expeditions 1926–37**, that firmly established his country's interests in the Southern Ocean and Antarctica. Concerned at first to explore for new lands from which whaling could be conducted without payment of foreign dues, he used his catchers, factory ships and transports to explore Bouvetøya, Peter I Øy and the Southern Ocean. Subsequently his navigating officers and pilots photographed and charted most of the coastal sector of East Antarctica that, as a result of his enterprise, in 1939 became Dronning Maud Land. His interest extended also into neighbouring areas of Enderby, Mac.Robertson and Princess Elizabeth lands, where coasts are named both for himself and for his wife Ingrid. Christensen financed scientific research and survey and, in honour of his father, founded the Kommandør Chr. Christensen Hvalfangstmuseum in his home town. He died on 10 December 1965. *Further reading*: Christensen (1935).

Churchill Mountains. 81°30'S, 158°30'E. Block of mountains forming part of Shackleton Coast, on the western flank of Ross Ice Shelf, between Byrd and Nimrod glaciers. Individual mountains were identified and named by the **British National Antarctic Expedition 1901–4**, but the range as a whole was identified from later aerial surveys and named for British statesman Sir Winston Churchill.

Cierva Point. 64°10'S, 60°57'W. Peninsula of Danco Coast, the site of the Argentine research station Primavera. An area of approximately 51.8 km² (20 sq miles), including the peninsula and several islands, but excluding the station site, was designated SSSI No. 15 (Cierva Point), protecting long-term avian and botanical studies.

Circumcision, Port. 65°11'S, 64°10'W. Inlet on the southeast side of Petermann Island, Wilhelm Archipelago, where the **French Antarctic (*Pourquoi-pas?*) Expedition 1908–10** overwintered in their ship. The harbour was named for the Christian festival celebrated on 1 January, on which it was discovered.

Cirque Fjord. 67°18'S, 58°39'E. Glacier-filled fjord separating Law Promontory from the Kemp Coast, opening into Stefansson Bay. It was discovered and photographed during flights by the Norwegian (Christensen) Whaling Expedition of 1936–37.

Cirque glacier. Small glacier in a semi-circular or horseshoe-shaped embayment that it has carved on a mountain flank.

Clarence Island. 61°13'S, 54°06'W. Easternmost island of the South Shetland Islands, discovered by Edward Bransfield in 1820 and named for the English prince (later King William IV).

Clarie Coast. 66°30'S, 133°00'E. Part of the coast of Wilkes Land, East Antarctica, bounded by Cape Morse in the west and Pointe du *Pourquoi Pas?* in the east. The coast consists mainly of ice cliffs with emergent islands, backed by a rising ice-covered plain. It was named in 1840 for Clarie Jacquinot, wife of the captain of *Zelée*, the ship that accompanied J. Dumont d'Urville on the **French Naval Expedition 1837–40**. The coast is named Wilkes Coast by some authorities.

Clark, Robert Selbie. (1882–1950). British polar marine biologist. Born and educated in Scotland, he became a zoologist in marine research stations in Edinburgh (where

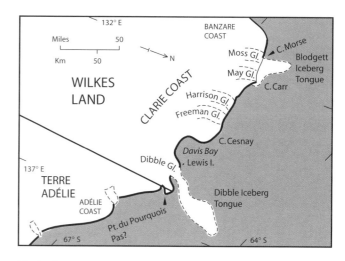

Clarie Coast

he worked with W. S. Bruce on biological material from the Scottish National Antarctic Expedition) and Plymouth. As a biologist he joined the **Imperial Trans-Antarctic Expedition 1914–1917**, amassing a collection of specimens and notes from South Georgia and the Weddell Sea, all of which were lost with the loss of *Endurance*. On return from the expedition he served with the Royal Navy in World War I, then returned to marine biological and fisheries research, retiring as director of the Torry Marine Research Laboratory, Aberdeen. He died on 29 September 1950.

Clark Peninsula. 66°15'S, 110°36'E. Peninsula on the north side of Newcomb Bay, close to Casey Station. An area of approximately 12.1 km² (4.7 sq miles), extending from the peninsula to Löken Moraines, was designated SSSI No. 17 to protect long-term research on moss and lichen communities.

Clerke Rocks. 55°01'S, 34°42'W. An isolated group of rocks some 64 km (40 miles) east-southeast of Cooper Island, South Georgia. A visible fragment of the mainly submerged northern arm of the Scotia Arc, they spread like broken teeth over 11 km (7 miles) of sea, the highest rising to 331 m (1085 ft). Shags roost on the highest rocks: the seas surrounding are popular feeding grounds for sea birds and fur seals.

Climate change. Antarctic studies of climate change and variability are limited because **climatic records** are short and geographically limited. Few climatological stations operated before the International Geophysical Year 1957–58 (IGY). Most stations are coastal: only two, Vostok and Amundsen-Scott, provide records longer than 20 years for the East Antarctica plateau. Apart from the west coast of Antarctic Peninsula, West Antarctica is almost devoid of long-term observations. A limited number of observations are available from early expeditions. On a longer timescale, increasing information is available from studies of **ice cores**, providing information of great relevance to long-term changes.

Over much of Antarctica little change in year-to-year mean temperatures has been recorded. At the south polar station Amundsen-Scott, annual mean temperatures have fallen slightly since the IGY. Around the East Antarctica coast during the same period, annual means have risen slightly, but the trend is insignificant among normal year-to-year variations. The west coast of Antarctic Peninsula between Anvers Island and Marguerite Bay shows more significant warming: annual means have risen by 2.5 °C since the late 1940s – one of the largest warming trends seen anywhere on Earth in recent years. Warming has caused noticeable changes in the western Peninsula environment. Snow and ice cover have receded from small low-lying islands, allowing greatly increased colonization by native plant species. The **Wordie Ice Shelf** in Marguerite Bay declined in area from 2000 km² in 1966 to 700 km² in 1989 and has since virtually disappeared. Similar losses are currently occurring from further north along the peninsula. Causes of warming are as yet uncertain but complex interactions of atmosphere, ocean and sea ice make local climate particularly sensitive to changes. In particular it is not clear if the warming is purely local in cause or whether it represents an amplification of the much smaller trends seen in regional and global temperatures.

Further reading: King and Turner (1997).
(JCK)

Climatic data: sources. Meteorological and climatological data collected at Antarctic stations are generally archived by the national polar institute or meteorological service of the country that operates the station. As yet there is no single data centre where such observations can be conveniently accessed. This situation is currently (2001) being addressed by the READER (REference Antarctic Data for Environmental Research) project being undertaken by the Physics and Chemistry of the Atmosphere (PACA) group of SCAR. The aim of this project is to make a comprehensive dataset of climatological observations for Antarctica available from a single source. Until this dataset becomes available, the Internet provides a convenient means of accessing data collected by some national operators. Useful web-sites include:

Alfred Wegener Institute for Polar and Marine Research, Bremerhaven, Germany: http://www.awi-bremerhaven.de; Antarctic Meteorological Research Center, University of Wisconsin: http://uwamrc.ssec.wisc.edu/; Australian Antarctic Division: http://www.antdiv.gov.au; British Antarctic Survey: http://www.antarctica.ac.uk.
(JCK)

Climatic records. The earliest southern hemisphere weather records – the raw materials from which climatic records accumulate – are found in ships' logs of seventeenth- and eighteenth-century exploring expeditions. The earliest reliable observations from the Antarctic region are probably those of James Cook, recorded during his voyage of 1776: limitations are lack of standardization of instruments and methods of recording. The scientific expeditions of the mid-nineteenth century recorded with a higher level of accuracy. While most were itinerant, a few spent days or weeks at moorings, yielding the first continuous records from fixed stations. Late nineteenth and early twentieth century geographical expeditions provided the first year-round records from fixed observatories and the first analyses of records in modern terms. The first long-term Antarctic land station, established by the Scottish National Expedition 1903–4, began observations on the South Orkney Islands in April 1903. Under Argentine management, their station **Orcadas** has operated ever since, providing the longest continuous record for an Antarctic station. A second run of observations, begun in January 1905 at **Grytviken** whaling station, South Georgia, was continued at the British government station at King Edward Point until the Argentine invasion of South Georgia in April 1982. This provided the second-longest continuous record for the region. The longest continuous records from continental stations date only from the late 1940s and early 1950s.

Whalers and explorers provided intermittent records through the 1920s and 1930s. In the early 1940s began a succession of long-term government-run stations, at first in the South American sector, later spreading to other regions of the Antarctic coast. The **International Geophysical Year (1957–58)** brought in many more stations, filling gaps along the coast and occupying strategic points on the inland plateau including the South Pole and the Pole of Inaccessibility. For the first time there existed a continent-wide network of observatories, maintaining radio schedules that added Antarctic data instantaneously to the world network. A few of these early stations still operate: several closed after thirty or more years of continuous observations. Currently about 25 stations are manned for regular three-hourly or six-hourly synoptic observations – by world standards a very low coverage for so large an area. Data gathered at three-hourly or six-hourly intervals are passed by radio or satellite links to collective centres for both local and world-wide use. Meteorologists in South America, South Africa and Australasia have come to rely on data from the far south for both daily and seasonal weather forecasting. Records accumulated over many years, enhanced by climatic data from ice cores and other sources, are being used to detect longer-term climatic trends.

Further reading: King and Turner (1997).

Climatic zones. As might be expected for so wide an area, there is no single 'Antarctic climate'. Though cold dominates the whole region, there are marked differences between the intense cold experienced on the high polar plateau and the comparative mildness of the coast, and between the continental coast and that of the peripheral islands. Though 'climate' includes temperature, wind, cloudiness and other factors (see **Weather, Antarctic**), temperature alone is singled out in defining climatic zones. Continental Antarctica is divisible into five climatic zones. Islands of the Southern Ocean fall into two zones, Antarctic and sub-Antarctic, and the islands north of the Antarctic Convergence fall into two further zones, cool temperate and warm temperate (see **Southern islands**). Table 1 shows the annual march of mean monthly temperatures for representative stations in each of these zones, a selection of which are illustrated in the diagrams.

Continental high plateau

Characterized by extreme cold throughout the year, often with clear skies of feathery cirriform clouds, light winds and constant extreme cold. Precipitation is mostly very light, in the form of fine ice spicules, amounting to a few cm per year. Snowfall is rare, mostly from depressions that occasionally penetrate from the coasts, bringing flurries of snow and stronger winds. There is a wide annual temperature range, consistent with a continental climate. Extreme chilling at the snow surface causes persistent **temperature inversions**. Temperatures vary mainly according to altitude. Mean monthly temperatures fall rapidly between February and April, then remain level or descend gently for the next three or four months (**kernlos effect**), usually with a minimum in August. Vostok Station, high on the East Antarctica plateau, currently records Earth's lowest winter temperatures annually, in July or August. Plateau Station, which operated only for three years, was probably slightly colder, and lower mean temperatures may be expected at higher points on the plateau, or among the peaks of Ellsworth Mountains. Amundsen-Scott Station, at the South Pole, stands lower in altitude and is markedly warmer throughout the year.

Continental low plateau

Characterized by higher temperatures throughout the year (a consequence of lower elevation), and a smaller annual temperature range. Though dominated by persistent anticyclones, bringing clear skies, calm air and little precipitation, atmospheric pressure and weather are more variable due to frequent incursions of depressions, bringing spells of overcast skies, strong winds and heavy snowfall. There are no permanent research stations in this area: Byrd Station (1530 m), and Siple Station (1050 m), operated intermittently by US researchers, yielded the only available recent climatic records.

Continental high latitude coasts

Characterized by cold winters and short, only slightly warmer summers. Depressions bring frequent changes in weather, including persistent cloud and year-round snow.

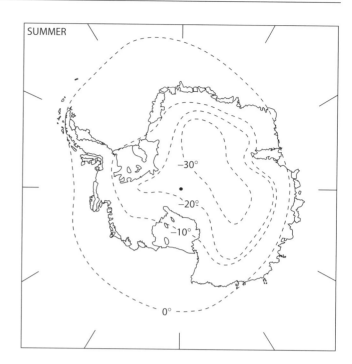
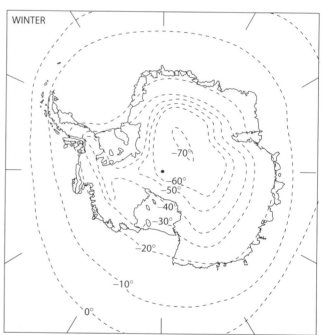

Mean midsummer and midwinter surface temperatures (°C) for continental Antarctica and the Southern Ocean

Table 1 Mean monthly and annual temperatures, and temperature ranges (°C), at selected Antarctic and southern oceanic island stations. Data from World Meteorological Organization (1971), Kuhn and others (1975), Schwerdtfeger (1984), Jones and Limbert (1987) and other sources. All data are from records gathered continuously for over ten years, except for Siple and Russkaya (each 5 years), Plateau (3 years) and Oazis (2 years)

Station	Position	Ht. (m)	Jan	Feb	Mar	Apr	May	Jun	Jul	Aug	Sep	Oct	Nov	Dec	Mean	Range
Continental high plateau stations																
Amundsen-Scott	90°00′S	2835	−27.9	−40.2	−54.3	−57.3	−57.3	−58.2	−59.9	−59.7	−58.4	−50.7	−38.4	−27.7	−49.3	32.2
Plateau	79°15′S, 40°30′E	3625	−33.9	−44.4	−57.2	−65.8	−66.4	−69.0	−68.0	−71.4	−65.0	−59.5	−44.4	−32.3	−56.4	39.1
Vostok	78°27′S, 106°52′E	3488	−32.3	−44.3	−58.0	−64.9	−65.9	−65.1	−67.0	−68.3	−66.3	−57.1	−43.4	−32.3	−55.4	36.0
Continental low plateau stations																
Byrd	80°01′S, 120°00′W	1530	−14.7	−19.8	−27.7	−29.7	−33.0	−34.1	−35.6	−36.7	−36.6	−30.2	−21.4	−14.4	−27.9	22.3
Siple	75°55′S, 83°55′W	1054	−11.9	−18.7	−24.9	−29.0	−27.1	−30.9	−30.5	−36.7	−30.3	−26.6	−17.7	−13.1	−24.8	24.8
Continental high-latitude coastal stations																
Belgrano	78°00′S, 38°48′W	50	−6.3	−13.2	−22.0	−26.2	−29.6	−32.0	−33.5	−32.9	−31.0	−21.7	−12.7	−6.1	−22.2	27.4
McMurdo	77°50′S, 166°30′E	24	−3.1	−8.8	−17.6	−21.1	−23.3	−23.5	−25.8	−26.9	−25.0	−19.5	−9.9	−3.1	−17.4	23.8
Halley	75°31′S, 26°30′W	35	−4.8	−9.8	−16.6	−20.0	−24.4	−26.6	−28.9	−28.5	−26.2	−19.2	−11.6	−5.3	−18.5	24.1
Russkaya	74°42′S, 136°51′W	100	−2.5	−6.7	−8.9	−13.1	−13.8	−18.4	−20.0	−20.6	−20.3	−14.1	−8.6	−5.0	−12.7	18.1
SANAE	70°19′S, 2°21′W	52	−4.1	−8.9	−14.5	−19.3	−21.2	−23.4	−27.1	−27.2	−25.2	−18.8	−10.8	−5.2	−17.1	23.1
Continental low-latitude coastal stations																
Davis	68°35′S, 77°59′E	13	0.6	−2.6	−9.3	−13.1	−15.7	−15.6	−17.3	−17.1	−16.5	−12.5	−5.1	0.0	−10.3	17.9
Mawson	67°36′S, 62°55′E	8	0.1	−4.4	−10.3	−14.5	−16.1	−16.8	−17.8	−18.8	−17.7	−13.2	−5.4	−0.3	−11.3	18.9
Dumont d'Urville	66°40′S, 140°01′E	40	−0.7	−4.2	−8.7	−12.7	−14.7	−16.0	−16.2	−16.8	−16.1	−13.2	−7.0	−1.7	−10.7	16.1
Mirnyy	66°33′S, 93°01′E	30	−1.6	−5.3	−10.1	−13.9	−15.3	−15.6	−16.6	−17.4	−17.2	−13.5	−7.0	−2.4	−11.3	15.8
Oazis	66°18′S, 100°34′E	28	1.9	−2.3	−5.2	−7.5	−11.3	−20.5	−17.3	−16.4	−16.5	−11.2	−3.5	1.7	−9.1	22.4
Antarctic Peninsula stations																
San Martin	68°08′S, 67°07′W	4	0.4	−1.3	−3.1	−6.2	−9.4	−13.1	−14.4	−13.2	−10.4	−7.3	−3.3	0.1	−6.8	14.8
Rothera	67°34′S, 68°08′W	15	1.0	0.1	−1.6	−3.7	−6.8	−8.8	−12.6	−11.8	−9.4	−7.2	−3.3	0.2	−5.3	13.6
Faraday/Vernadsky	65°15′S, 64°16′W	9	0.3	0.1	−1.0	−3.4	−5.8	−8.1	−10.7	−11.0	−8.3	−5.1	−2.6	−0.4	−4.7	11.3
Marambio	64°14′S, 56°43′W	198	−1.9	−3.3	−6.8	−12.3	−12.9	−14.9	−16.3	−16.2	−12.7	−6.4	−4.3	−1.9	−9.2	14.4
Esperanza	63°24′S, 56°59′W	8	0.2	−1.3	−3.7	−7.5	−9.3	−11.4	−11.8	−10.6	−7.4	−4.1	−2.1	−0.2	−5.8	12.0
Antarctic island stations																
Deception Island	62°59′S, 60°43′W	8	1.4	1.1	0.1	−2.1	−4.3	−6.3	−8.0	−7.7	−4.8	−2.4	−1.0	0.5	−2.8	9.4
Arturo Prat	62°29′S, 59°40′W	5	0.8	1.4	−0.1	−1.5	−3.2	−5.4	−7.7	−8.2	−5.3	−2.2	−1.3	0.4	−2.7	9.6
Bellingshausen	62°12′S, 58°56′W	16	1.1	1.2	0.1	−1.8	−4.0	−6.0	−7.1	−7.3	−4.6	−2.6	−1.2	0.4	−2.6	8.4
Admiralty Bay	62°05′S, 58°25′W	10	1.3	1.3	0.1	−2.4	−4.9	−6.5	−8.5	−7.4	−4.4	−1.4	−0.8	0.7	−2.7	9.8
Orcadas	60°44′S, 44°44′W	6	0.3	0.5	−0.6	−3.0	−6.7	−9.8	−10.5	−9.8	−6.4	−3.4	−2.1	−0.5	−4.3	11
Sub-Antarctic island stations																
South Georgia	54°18′S, 36°30′W	3	4.7	5.4	4.6	2.5	0.2	−1.5	−1.5	−1.5	0.1	1.7	3.0	3.8	1.8	6.9
Heard Island	53°06′S, 72°31′E	4	3.3	3.5	3.0	2.4	1.3	−0.4	−0.6	−0.8	−1.2	−0.2	0.5	2.1	1.3	4.7
Cool temperate island stations																
Macquarie Island	54°30′S, 158°54′E	30	6.8	6.7	6.2	5.1	4.2	3.3	3.2	3.2	3.4	3.8	4.5	5.9	4.7	3.6
Iles Kerguelen	49°12′S, 70°12′E	20	7.2	7.7	7.1	5.7	3.7	2.3	1.9	2.0	2.2	3.3	4.7	6.2	4.5	5.8
Marion Island	46°53′S, 37°52′E	25	6.8	7.3	7.2	5.9	4.7	4.1	3.6	3.3	3.2	4.4	5.1	5.7	5.1	4.1
Falkland Islands	51°42′S, 57°52′W	50	8.7	9.0	8.2	5.8	3.9	2.4	2.2	2.6	3.4	5.2	7.0	7.7	5.3	6.8
Warm temperate island stations																
Tristan da Cunha	37°03′S, 12°19′W	10	17.4	18.2	17.0	15.7	14.3	13.2	12.0	11.8	11.8	12.8	14.2	16.3	14.6	6.4
Gough Island	40°21′S, 09°53′W	10	14.3	15.0	13.6	12.6	11.5	10.7	9.6	9.6	9.7	10.3	11.7	14.0	11.9	5.4

Coasts in latitudes higher than about 65°S are likely to be invested year-round with fast ice, which reduces possibilities of summer warming from the open sea. Several high-latitude 'coastal' stations are sited several miles inland on ice shelves, and are under strong continental influences throughout the year. Belgrano, McMurdo, Halley, SANAE and Russkaya stations have provided some of the longest continuous climatic records for this zone.

Continental low latitude coasts

Antarctica's milder coastal climates are warmer, even in winter, than many winter areas of central Europe, Asia, Scandinavia and the American mid-west. The lower slopes of the continent, the coast and the offlying islands are relatively warm, even in winter. Only the continuing presence of snow and ice keeps them from rising far above freezing point in summer. Low cloud cover is often persistent, precipitation often heavy, brought in by circling depressions. Winds tend to be strong, mainly **katabatic** from the icy inland slopes. Despite their wide longitudinal range, stations in this zone show almost identical patterns of climates, typified by Mawson, Dumont d'Urville and Mirnyy. Oazis and Davis, sited in ice-free oasis areas, are relatively free of snow, and become anomalously warm in summer.

Antarctic Peninsula

Markedly maritime climates with cold winters and warmer summers. Stations along the western flank of the peninsula are consistently warmer than those in similar latitudes along the icebound eastern flank, or along the low latitude continental coasts. Weather patterns are dominated by depressions circulating mainly from the west, bringing predictable successions of low clouds, precipitation and winds. Rain is occasionally recorded in summer. Representative stations are Rothera, San Martin, Faraday/Vernadsky, Esperanza and Marambio.

Antarctic islands

Maritime climates similar to those of western Antarctic Peninsula, though milder throughout the year, with winter temperatures depressed by the presence of sea ice. The South Orkney Islands are consistently cooler than the South Shetlands, due to chilling influences from the Weddell Sea. Persistent low clouds, rain and sleet are frequent in summer, heavy snow in winter. These are typified by Decepcion, Arturo Prat, Bellingshausen, Henryk Arctowski and Orcadas stations. Orcadas, in Scotia Bay, South Orkney Islands, established in March 1903, has the longest continuous meteorological record of any Antarctic station.

Sub-Antarctic islands

Southern Ocean islands north of the northern limit of sea ice have markedly oceanic climates, with cool summers and only slightly cooler winters. Successive depressions bring cloudy skies, frequent rain in summer, snow in

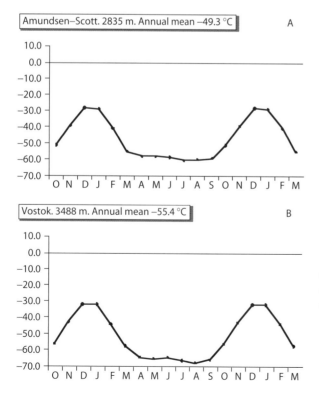

A. *High plateau station.* At Amundsen–Scott, rapid spring warming from September brings peak means in December and January, followed by rapid cooling February to March, with a very slow decline to a level minimum (kernlos) in July to September: annual range of means 32.2 °C.

B. *High plateau station.* At Vostok, in a lower latitude but higher elevation on the ice cap, the pattern is similar to that at Amundsen–Scott, but uniformly colder. Annual range of means 36.0 °C.

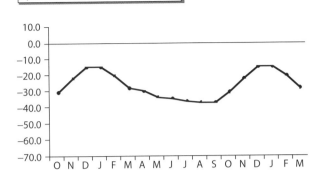

C. *Low plateau station.* Byrd has similar summer peaks at higher mean temperatures: rapid autumn cooling is followed by a continuous decline through winter to a September minimum: annual range 22.3 °C.

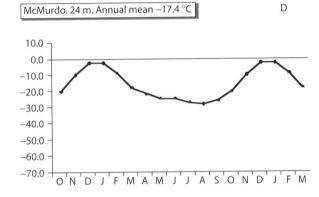

D. *High latitude coastal station.* At McMurdo, the pattern is similar to B but warmer throughout the year: persistent fast ice prevents summer warming from the sea. Annual range 23.8 °C.

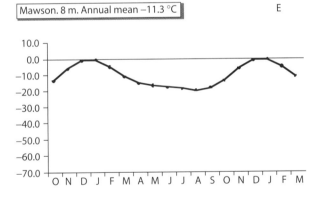

E. *Low latitude coastal station.* At Mawson, winters remain cold but summers are warmer: thinner inshore ice or nearby open sea allow one monthly mean above freezing point. Annual range 18.9 °C.

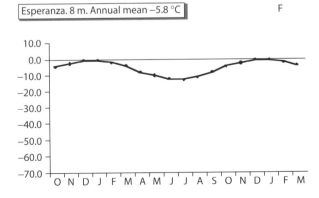

F. *Antarctic Peninsula station.* At Esperanza, on the northern tip of Antarctic Peninsula, winters are warmer than at Mawson, summers slightly warmer, with three monthly means above freezing point. Annual range 13.6 °C.

64 CLIMATIC ZONES

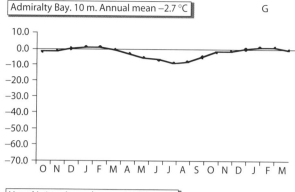

G. *Southern Scotia Arc station.* At Admiralty Bay, South Shetland Islands, fast ice persists for over half the year and winters remain cold, but there are four summer months with means above freezing point and a near-maritime annual range of 9.8 °C.

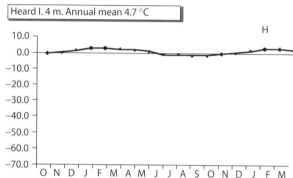

H. *Sub-Antarctic island station.* At Heard Island there is no winter sea ice, so winters are distinctly warmer. Seven months have means above freezing point: the annual mean too is above freezing point, and the annual range of monthly means is only 4.7 °C.

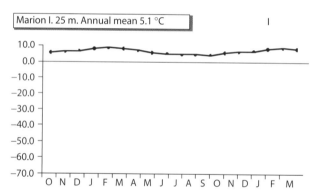

I. *Cool temperate island.* Marion Island, north of the Antarctic Convergence, has an equitable climate with temperatures controlled almost entirely by the surrounding sea. Frosts are rare, and no monthly mean falls below 0 °C.

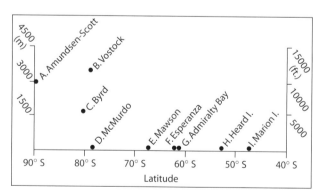

The relative latitudes and heights above sea level of stations A–I

winter and year-round strong winds, often intensified by local topography. These are typified by South Georgia (Grytviken,) and Heard Island.

Cool temperate islands

Islands north of the Antarctic Convergence, in what oceanographers call 'Subtropical' surface waters, have year-round temperate climates. Those standing close to the Convergence have markedly cooler, cloudier and wetter climates than those further north, with strong, persistent west winds. A convenient boundary is the 10 °C isotherm for the warmest month, which in either hemisphere approximates to the tree line. Typical cool-temperate islands are the Falkland Islands (Stanley) in the south Atlantic Ocean, and Macquarie Island, Iles Kerguelen (Port au Français) and Marion Island in the south Indian Ocean.

Warm temperate islands

Islands standing north of the 10 °C isotherm for the warmest month have warm, relatively benign climates, drier and sunnier than their southern counterparts, with intermittent westerly winds which occasionally blow very strongly. Representative stations are Tristan da Cunha and Gough Island, in the southern Atlantic Ocean.

Clissold, Thomas Charles. (d. 1963). British polar seaman. A Royal Navy artificer, he joined the **British Antarctic (*Terra Nova*) Expedition 1910–13** as a cook, but served ashore also as a dog-handler and sledger on depot-laying journeys. During World War I he served in the Middlesex Regiment and the Royal Flying Corps. On demobilization he emigrated to New Zealand, joining the Transport Department, from which he retired in 1953. He died on 20 October 1963, aged 77.

Clo unit. Unit of insulation used in studies of clothing required by humans in polar regions. One clo is equivalent to the amount of clothing needed by a human at rest in an ambient temperature of 20 °C. See also **Cold, human responses** and **Wind chill**.

Clothier Harbour. 62°22′S, 59°40′W. Harbour on the northwest side of Robert Island, South Shetland Islands, named by US sealers for a sealing vessel that sank there in 1820.

Clowes, Archibald John. (*c*.1901–60). British marine chemist. A graduate of the Royal College of Science, he joined Discovery Investigations in 1924. He spent two years at the Marine Biology Station, South Georgia, and made many cruises in RRS *William Scoresby* and RRS *Discovery*, concerned particularly with the chemistry and hydrology of the Southern Ocean. After service with the Royal Navy in World War II, he returned briefly to Discovery Investigations, but left in 1947 for Cape Town, where he worked with the South African Fisheries Laboratory.

Coal Harbour. 54°02′S, 37°57′W. Harbour on the southwest coast of South Georgia, used as a bunkering depot by early twentieth-century whalers.

Coasts, Antarctic. For ease of reference, cartographers have divided most of the coastline of continental Antarctica into 55 separate lengths of coast. Named mainly for explorers or expeditions, these provide useful addresses for coastal features. The coasts, with their hinterlands, are as follows: for further details and maps see individual entries.

Coasts of East Antarctica

Adélie Coast (Terre Adélie)
Amundsen Coast (Ross Dependency)
Banzare Coast (Wilkes Land)
Borchgrevink Coast (Victoria Land)
Budd Coast (Wilkes Land)
Caird Coast (Coats Land)
Clarie Coast (Wilkes Land)
Dufek Coast (Ross Dependency)
George V Coast (George V Land)
Hillary Coast (Victoria Land)
Ingrid Christensen Coast (Princess Elizabeth Land)
Kemp Coast (Kemp Land)
Knox Coast (Wilkes Land)
Kronprinsesse Märtha Kyst (Dronning Maud Land)
Kronprins Olav Kyst (Dronning Maud Land)
Lars Christensen Coast (Mac.Robertson Land)
Leopold and Astrid Coast (Princess Elizabeth Land)
Luitpold Coast (Coats Land)
Mawson Coast (Mac.Robertson Land)
Oates Coast (Oates Land)
Pennell Coast (Victoria Land)
Prins Harald Kyst (Dronning Maud Land)
Prinsesse Astrid Kyst (Dronning Maud Land)
Prinsesse Ragnhild Kyst (Dronning Maud Land)
Queen Mary Coast (Queen Mary Land)
Sabrina Coast (Wilkes Land)
Scott Coast (Victoria Land)
Shackleton Coast (Ross Dependency)
Wilhelm II Coast (Wilkes Land)

Coasts of West Antarctica

Bakutis Coast (Marie Byrd Land)
Black Coast (Palmer Land)
Bowman Coast (Graham Land)
Bryan Coast (Ellsworth Land)
Danco Coast (Graham Land)
Davis Coast (Graham Land)
Eights Coast (Ellsworth Land)
English Coast (Palmer Land)
Fallières Coast (Graham Land)
Foyn Coast (Graham Land)
Gould Coast (Marie Byrd Land)
Graham Coast (Graham Land)
Hobbs Coast (Marie Byrd Land)
Lassiter Coast (Palmer Land)
Loubet Coast (Graham Land)
Nordenskjöld Coast (Graham Land)
Orville Coast (Ellsworth Land)
Oscar II Coast (Graham Land)
Ruppert Coast (Marie Byrd Land)
Rymill Coast (Palmer Land)
Saunders Coast (Marie Byrd Land)
Shirase Coast (Marie Byrd Land)
Siple Coast (Marie Byrd Land)
Walgreen Coast (Marie Byrd Land)
Wilkins Coast (Palmer Land)
Zumberge Coast (Ellsworth Land)

Coats Land. Area of East Antarctica between the Weddell Sea and 20°W, (adjacent to Dronning Maud Land), extending south to 82°S. It consists mainly of ice plains and slopes bounded by the steep ice cliffs of Luitpold Coast and Caird Coast, that together form the eastern shore of the Weddell Sea. Mountains include the Theron Mtns, the Shackleton Range and several nunataks. The northeast coast was discovered by the **Scottish National Antarctic Expedition 1902–4**, and named for Scottish industrialists James and Andrew Coats, patrons of the expedition. Southern coastal areas were charted by the **German Antarctic Expedition 1911–12** and British **Imperial Trans-Antarctic Expedition 1914–16**. Coats Land forms part of British Antarctic Territory, and since 1956 has been the site of Halley, a succession of British research stations. A small sector is claimed by the Argentine government as part of Antártida Argentina.

Colbeck, William. (1871–1930). British seaman and navigator. Born and educated in Hull, Yorkshire, he was apprenticed to the merchant navy at the age of 15, and later commissioned in the Royal Naval Reserve. He joined the **British Antarctic (*Southern Cross*) Expedition 1898–1900**, serving as navigating officer in the ship, and wintering as cartographer, magnetician and meteorologist with the shore party. He took part in the sledging programme from Camp Ridley: Colbeck Bay, Victoria Land, is named for him. In 1903 he was appointed master of *Morning*, one of the relief ships of the **British National Antarctic Expedition 1901–4**; Scott named Cape Colbeck, Edward VII Peninsula, in his honour. After further service at sea he became marine surveyor and bailiff in Liverpool. His son **William Robinson Colbeck** also served in Antarctica.

Colbeck, William Robinson. (*c.*1906–1986). British seaman and navigator. The son of Capt. **William Colbeck**, he was apprenticed to the merchant navy. As a young officer he was commissioned in the Royal Naval Reserve, and at the age of 23 joined the **British, Australian and New Zealand Antarctic Research Expedition 1929–31** as second officer and navigator in *Discovery*. Much of the charting on both voyages became his responsibility. Colbeck Archipelago, off Mawson Coast, is named for him. At the end of the second cruise he was promoted to first officer for the voyage back to Britain. He obtained his master's certificate and, after several more years at sea, became Marine Surveyor to the Liverpool Mersey Docks and Harbour Board, and a Fellow of the Institute of Navigation.

Colbeck Archipelago. 67°26′S, 60°58′E. Group of rocky islands east of Taylor Glacier on the Mawson Coast of Kemp Land, East Antarctica, charted during the **British, Australian and New Zealand Antarctic Research Expedition 1929–31** and named for W. R. Colbeck, second officer of the expedition ship RRS *Discovery*.

Cold climate survival strategies. Behavioural and physiological activities by which plants and animals overcome environmental challenges. Polar and alpine organisms are subject to sub-zero temperatures continuously in winter and intermittently in summer. While in some habitats deep winter snow provides insulation from the lowest temperatures, almost all such organisms are from time to time at risk from temperatures low enough to cause serious tissue damage. Some habitats have the additional stress of aridity, or high salt concentrations. Solutions are varied, but usually a combination of physiological and ecological strategies.

Land plants

Lichens photosynthesize at temperatures well below 0 °C: carbon dioxide exchange has been recorded even at −18.5 °C. Antarctic lichens, among the most cold-resistant plants known, have returned to normal metabolism after being plunged into liquid nitrogen at −196 °C. Equally remarkable is their ability to survive frequent wetting and drying cycles without serious damage to their cell membranes, though some of the cell contents leak out. These attributes allow lichens to take advantage of short growing seasons, low summer temperatures and other conditions that would destroy most plants: they have colonized even the most remote nunataks on the continent.

Mosses too are tough, though less able than lichens to cope with Antarctic extremes. They are in consequence less widely distributed than lichens, favouring the warmer and wetter coastal areas. Moss species that form tight cushions are among the hardiest, sometimes growing on rock exposures alongside crustose lichens. Moss photosynthesis has been recorded down to −10 °C and the hardiest species can apparently withstand the freeze–thaw and wet–dry cycling characteristic of exposed continental habitats. However, mosses are at their best and show their greatest diversity in the relatively mild conditions of Antarctic Peninsula and nearby islands. Some species form deep **moss peat** banks, or in wetter areas moss carpets.

Algae and fungi, less advanced plants than lichens and mosses, might be expected to have less chance of survival than the more organized cryptogams yet both have been found in soils in the most barren parts of Antarctica. In the dry valleys of Victoria Land endolithic communities live within the fabric of the rocks. Inside the coarse-grained Beacon Sandstone live a lichen, a fungus and an alga, all as separate layers growing through the spaces between the rock crystals. Both green and blue-green algae have been found growing within other rocks, as well as in cracks and in the spaces underneath loose boulders. Blue-green cyanobacteria are among the hardiest plants known. As fixers of atmospheric nitrogen, they may be crucial in providing nearly all the nitrogen for simple terrestrial and aquatic communities away from the influence of birds.

Yeasts, bacteria, fungi and micro-algae, the commonest Antarctic microbes, seem particularly tolerant of extreme climates and low nutrient levels. Yeasts have been found at 2800 m on Mount Howe, the world's most southerly mountain, in 87°S. It is not clear how yeasts cope so well with extreme conditions. One possible explanation is that their cells remain quiescent in the most stressful situations, becoming active only when water and nutrients arrive simultaneously. This opportunist lifestyle is seen also in yeast populations in the more favourable maritime Antarctic, which increase rapidly to peak numbers at snow melt when available nutrients are at their highest, then decline steadily throughout the rest of the summer.

The well-studied strategies adopted by temperate flowering plants are widespread among a range of plants in polar environments. As growth ceases in autumn, plants accumulate high concentrations of sugars or sugar alcohols in their cells. These depress the freezing point of the cellular contents and stop the formation of lethal ice crystals, except at very low temperatures. Lichens, inhabiting the most exposed habitats and thus exposed to the lowest winter temperatures (below −55 °C), may survive by losing water from their cells, in the very low humidities characteristic of Antarctic winters, effectively freeze-drying.

Land animals

Terrestrial arthropods live at temperatures that immobilize their temperate counterparts, and their slow rates of development are counterbalanced by longevity. Two strategies are used to survive extreme cold: to avoid freezing or to tolerate freezing. As examples of avoidance, arthropods are able to supercool their body fluids, i.e. maintain them in a liquid state at very low temperatures, completely avoiding the formation of ice crystals which would disrupt the cells and tissues and cause damage or death. They thereby avoid intracellular freezing, often to temperatures below −25 °C, but at lower temperatures eventually freeze and die. This group are described as freezing-susceptible. Curiously, this type of cold hardiness is not limited to species inhabiting polar or low temperature environments. Desert centipedes, a mite of temperate latitudes and even tropical arthropods have been found to have a significant level of cold resistance by supercooling. A second, less common strategy is adopted by animals that allow ice crystals to form within the tissues but outside the cells. Called freezing-tolerant, these animals possess nucleating agents to initiate crystal formation in extra-cellular liquids, but protect their cell contents accumulating glycerol as an antifreeze. The larvae of wingless midges *Belgica antarctica* are apparently the only freezing-tolerant invertebrates in the Antarctic, though adult midges are freezing-susceptible. Especially tolerant of ice formation, drought and salt stress are Antarctic nematodes, which have been found in some of the driest and most saline areas of Victoria Land.

Glycerol and a range of sugar alcohols (e.g. sorbitol, mannitol, trehalose) commonly found in plants are all involved in invertebrate cold hardiness. As yet no general theory has been proposed to explain why some species are freezing-susceptible, some freezing-tolerant and some apparently possess both adaptations at different stages in their life cycles.
Further reading: Vincent (1988).
(DWHW)

Cold, human responses. Humans from tropical or temperate regions are poorly equipped by nature for polar life: those moving to work for a few months in Antarctica have little opportunity to acclimatize. Some outdoor workers quickly develop tolerance to cold in exposed hands and faces, and depend on plentiful food and activity to maintain high metabolic output and keep themselves warm. All who work in Antarctica rely for comfort on developing favourable microclimates, particularly in the forms of housing and clothing,. Those who winter in the '**banana belt**' (warmer coastal areas) seldom experience temperatures below −30 °C. Those on the colder coasts and in the interior may need to cope with temperatures much lower even in summer. Modern polar clothing favours lightweight, durable, windproof fabrics, padded for insulation and designed with vents that allow sweat to escape freely. Taking as standard the **clo unit** (equivalent to the amount of clothing needed by a human at rest in an ambient temperature of 20 °C), men and women working actively at temperatures down to −45 °C usually find a thickness of 3 clo sufficient to keep warm. Sedentary workers at similar temperatures need an extra unit. For temperatures below −50 °C, additional clo make for garments that are cumbersome, hamper activity and increase effort, so on balance become less efficient. Electrically-warmed clothing may be considered as an alternative, but requires reliable sources of power. In these environments people work outside for a time, then take shelter to warm up. So long as outside workers can move and exercise, they can usually maintain comfort in short spells down to −80 °C. At Antarctica's coldest station, **Plateau**, heated gloves were available, but shunned by observers who maintained instruments outdoors at temperatures around −80 °C.

Difficulties of maintaining aircraft and other mechanical vehicles increase considerably below −45 °C. Fuels and oils become viscous, metal tools become brittle and tend to fracture, and working parts seize up due to differential contraction. Aircraft landing on the high plateau, for example at Amundsen-Scott Station, keep their engines running to avoid the difficulties that arise if they are switched off and allowed to cool down.

Colonization of Antarctic environments. Recruitment of plants and animals to Antarctic ecosystems. Colonizing propagules, including spores, seeds, plant fragments and invertebrates have been recovered by various air sampling techniques. Traps on the continent and southern islands have collected airborne pollen and spores from South American plants: sampling from ships has netted fragments of plants and invertebrates. We do not know how many propagules travel this way, nor their viability. One species of moss new to the Antarctic has recently become established, possibly after such a journey, on freshly-deposited volcanic ash at Deception Island, South Shetland Islands. Genetic analysis of mosses suggests that there have been multiple introductions to the continent from outside, rather than redistribution from an initial introduction.

Propagules are transported also in packing materials, foods, construction materials and unsterilized soil. Microbiological examination of materials used for drilling in the **Victoria Land Dry Valleys** showed the presence of a wide range of alien bacteria, despite advance planning to minimize introductions. On Signy Island, South Orkney Islands, imported soil used in plant growth studies now supports an introduced population of a wingless fly, *Enteromoptera murphyi*. Annual meadow grass *Poa annua* grew and produced seeds on Deception Island; *Poa pratensis* has grown and flowered at a site on the Danco Coast, and over 70 species of flowering plant have been accidentally introduced to the southern islands.
(DWHW)

Coman, Francis Dana. (c.1896–1952). American expedition doctor. After graduating in medicine he served in Labrador with the International Grenfell Association. While on the staff of Johns Hopkins University, he became a medical officer on the Byrd First Antarctic Expedition 1928–30, and in 1934–35 was the medical officer in *Wyatt Earp* during Lincoln Ellsworth's second expedition. He died on 28 January 1952.

Commandante Ferraz Station. 62°05′S, 58°23′W. Brazilian permanent year-round research station. It was established on Martel Inlet, Keller Peninsula, Admiralty Bay, King George Island in February 1984, close to the site of the former Base G, Falkland Islands Dependencies Survey, now removed. The station gives ready access to the island's ice cap.

Commerson's dolphin. *Cephalorhynchus commersonii*. Small, piebald dolphins up to 1.4 m (4.5 ft) long, this species is restricted to cool southern waters. They are particularly prominent around Tierra del Fuego and the Falkland Islands, but have been reported also off Tristan da Cunha, Gough Island and Iles Kerguelen, and at widely dispersed points close to and even south of the Antarctic Convergence. Distinguishing characteristics are their blunt, rounded dorsal fin, white chin, and broad white stripe extending on either side of the body, from behind the eyes almost to the base of the dorsal fin. They move in groups of five or six, sometimes gathering in much larger assemblies of up to 100. Small groups often accompany small boats, and take a friendly interest in divers and swimmers. Little is known of their biology.

Commonwealth Meteorological Expedition 1913–15. A two-year meteorological operation to Macquarie Island. After the **Australasian Antarctic Expedition 1911–14** had operated a meteorological station successfully for two years on Macquarie Island, the Australian Commonwealth government decided to maintain the station for a further period. The relief ship *Endeavour* was lost with all hands after effecting the first relief in December 1914, and wartime difficulties caused the station to be abandoned one year later.

Commonwealth Range. 84°15′S, 172°20′E. A range flanking the eastern side of Beardmore Glacier, Dufek Coast, Victoria Land. It was named by Shackleton for the Commonwealth of Australia, which strongly supported his **British Antarctic (*Nimrod*) Expedition 1907–9**.

Commonwealth Trans-Antarctic Expedition 1955–58. The first expedition to cross Antarctica from the Weddell Sea to the Ross Sea. Instigated and led by Dr V. E. Fuchs, who had gained experience as leader of the Falkland Islands Dependencies Survey, the expedition was supported by government funding from Britain, South Africa, Australia and New Zealand. Plans called for stations to be set up as far south as possible on either side of the continent, for depots to be laid inland on either side, and for a party with dog teams, tractors and air support to make the crossing of 3200 km (2000 miles), via the South Pole. Fuchs planned and led the Weddell Sea operations and the crossing. Operations in and from the Ross Sea were planned and led by the New Zealand explorer Sir Edmund Hillary, who in 1953 had been one of the first two men to climb Mount Everest.

MV *Theron*, a Canadian sealing vessel, left London on 14 November 1955, laden with equipment for the Weddell Sea base. After the customary tussle with pack ice, aided by

Snocat tractor negotiating a crevasse. Photo: Trans Antarctic Expedition and Scott Polar Research Institute

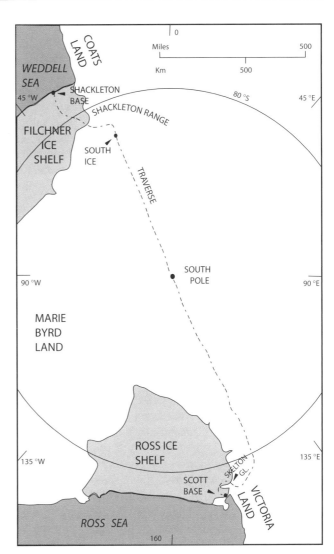

Commonwealth Trans-Antarctic Expedition 1955–58

reconnaissance from a small floatplane, she docked alongside the fast ice in Vahsel Bay and began to unload. Bad weather hampered operations, and *Theron* was compelled to depart in early February, leaving much of the stores and equipment on the fast ice. Though the advance party of eight, led by Ken V. Blaiklock, worked hard to move everything inshore and onto the ice shelf, bad weather continued. Some vital material, including a tractor and many tons of fuel, were lost. Camping in a packing case, the eight men spent the first bitter winter building a living hut, Shackleton.

There was time for little else before the relief ship MV *Magga Dan* appeared on 13 January 1957, bringing the main party with more equipment and replacement fuel. Now the first task was aerial reconnaissance, to find a site for a depot and base along a likely route south of the Shackleton Mountains. A site was found some 450 km (281 miles) inland, and a large depot and small base hut were flown in. The main party under Fuchs wintered at Shackleton: a three-man party, again led by Blaiklock, wintered in the advance base. Meanwhile, on the New Zealand side, a party under Hillary had arrived in McMurdo Sound in early January 1957, set up Scott Base at Pram Point, Ross Island, and had begun to seek routes to the polar plateau suitable for dog teams and tractors. In this they eventually succeeded, finding a difficult but possible route up the Skelton Glacier.

Both parties were confined to their bases during the winter months, preparing for the major event in spring. On 8 October 1957 Fuchs led a reconnaissance party of tractors to mark out an overland route to South Ice. At about the same time the first field parties were leaving Scott Base to begin their programme of depot-laying for the crossing party. After considerable difficulties with crevasses and hard, wind-packed sastrugi, Fuchs and his tractor train reached South Ice on 13 November. Fuchs was flown back to Shackleton to bring the main party forward, leaving on 24 November and reaching South Ice again on 21 December. From there the crossing party started out on 24 December. Two dog teams led the way, reconnoitring and marking out a route for four Sno-cats (large tractors with four independently suspended caterpillar tracks), three weasels and a Muskeg tractor (all double-tracked vehicles). The procession reached Amundsen-Scott, the south polar station, on 19 January. There Fuchs found Hillary waiting for him. The New Zealander's allotted task was completed with the establishment of a final depot 800 km (500 miles) from the Pole, but Hillary had seen no reason to stop there. He and two companions had completed the journey in their Ferguson tractors, arriving on 4 January 1958. By a short head, three barely modified farm tractors became the first vehicles to be driven to the South Pole. The combined parties left Amundsen-Scott on 24 January, arriving at Scott Base on 2 March. The journey of 3453 km (2158 miles) had been completed in 99 days. HMNZS *Endeavour*, the re-supply ship of the Ross Sea contingent, left Scott Base for Wellington three days later.

During the crossing, gravity readings, seismic soundings, and glaciological and meteorological observations were gathered systematically, and geological and other scientific data were collected at both bases.
Further reading: Fuchs and Hillary (1958).

Community development. Development of communities of plants and animals in habitats, following colonization. Recent retreat of ice in the Antarctic region has exposed extensive areas of bare rock and mineral soils. The earliest colonizers are usually microbial: bare surfaces, whether newly-exposed rock faces or mineral particles resulting from **weathering**, are colonized by bacteria, fungi, algae and cyanobacteria. Crustose lichens may establish themselves directly onto rock or small patches of organic matter, either dead or living. In time more lichens and mosses become established, taking advantage of cracks and depressions to gain a foothold. As a surface gradually accumulates lichens and mosses, organic matter gathers between and below the plants, providing a substrate for more microbial life. Crucial at each stage of community development is the provision of nutrients. Coastal sites that receive wind-blown spray from the sea in summer, and sites near bird colonies, usually have more abundant nutrients, including key compounds of nitrogen and phosphorus, than remote inland nunataks. High nutrient concentrations can be toxic: only a few species of plant can grow directly under bird perches or in rookeries, but areas a few metres downwind from a colony may receive just the right inputs in dust and aerosols for community development. Green algae and cyanobacteria bind soil together in the form of microscopic rafts, with particles trapped between the algal strands and covered in bacteria, providing islands for the growth of macro-algae, mosses and epiphytic lichens. During this development the primary species are succeeded by secondary species requiring an organic base for establishment. In favourable habitats an almost closed community of lichens and mosses develops. Development is slow because of slow growth and short cold summers. Species diversity is low, and sexual reproduction is limited to a few species; most plants rely on asexual reproduction by dispersal of vegetative fragments. The two species of Antarctic flowering plant ripen viable seed only in summers when conditions are exceptionally favourable. The rapid recent spread of both species shows that such summers are becoming more common.

Among the plant community there is a complementary development of an animal community. To protozoa, rotifers, nematodes and tardigrades, living in the water films around mineral particles, are gradually added a greater diversity of invertebrates including mites and insects, as new microhabitats become available. Detailed life cycles have been described for some common mites and springtails but little is known about diet, reproductive success or the importance of their role in the community.
Further reading: Vincent (1988).
(DWHW)

COMNAP. See **Council of Managers of National Antarctic Programmes**.

Compañia Argentina de Pesca. Whaling company founded by Norwegian whaler Capt. C. A. Larsen, that for over 60 years operated a land station on South Georgia. Having failed to interest Norwegian or British entrepreneurs in possibilities for whaling from South Georgia (see **Norwegian (Sandefjord) Whaling Expeditions 1892–94**), Larsen returned south as master of *Antarctic*, expedition ship of the **Swedish South Polar Expedition 1901–4**. Following the loss of his ship and rescue by the Argentine navy, Larsen was delivered to Buenos Aires, where he took the opportunity again to raise the question of whaling on South Georgia. Three foreign residents, Norwegian P. Christophersen, German-American H. H. Schlieper and Swede E. Tornquist combined to form the Compañia Argentina de Pesca Sociedad Anónima, registered in Buenos Aires on 29 February 1904. Larsen returned to Sandefjord, Norway to buy the equipment for the whaling station, shipping it out in the transport *Louise* and steam whale-catcher *Fortuna*. The ships arrived at Grytviken, a harbour in Cumberland Bay that had earlier been used by sealers, on 16 November 1904. Work began immediately on the prefabricated station: the first whale, a humpback, was harpooned close by on 22 December, and processed two days later. Taking mainly humpback whales, the station flourished and returned substantial profits from its first season onward.

Neither Larsen nor the Compañia had informed the Government of the Falkland Islands of their intentions. Not knowing of their operations, the Government in 1905 leased the island to a Chilean-based company, the South Georgia Exploration Co., which intended to farm cattle and sheep, explore for minerals and take seals for oil. Representatives of the company arrived in Grytviken in August 1905 to find the Compañia Argentina de Pesca already in possession.

The 'plan' or flensing platform at the whaling station, Grytviken, South Georgia. Photo: Scott Polar Research Institute

Matters were settled amicably: the Argentine Compañia applied for and received a government lease that, while regulating its position, made it liable to taxation and controls. The Chilean company, never more than speculative, faded from the scene: its rights were eventually bought by another whaling enterprise. Several other companies opened whaling stations or moored factory ships in South Georgia's sheltered harbours, introducing strong competition for generous but finite local stocks of whales. Catchers were soon travelling to more distant feeding grounds to bring in a wider variety of species, including right whales and rorquals. Employing a mixture of Norwegian, British and Argentine workers, with fortunes that fluctuated from season to season, the Compañia Argentina de Pesca operated its station continuously through two world wars and several cycles of industrial depression. It finally succumbed in 1965 when Grytviken, the first station to be established on South Georgia, became the last to close.
Further reading: Hart (2001); Headland (1984).

Concordia Station. 75°00′S, 125°00′E. A joint French and Italian permanent research station at Dome Charlie (Dome C), on the central ice cap of East Antarctica, about 3200 m (10 496 ft) above sea level (see **Charlie, Dome**). Established in 1999–2000, the station is re-supplied mainly by air from Dumont d'Urville and Terra Nova Bay. It provides facilities for studies of glaciology, meteorology and upper atmosphere phenomena, micrometeorites, geomagnetism, atmospheric chemistry and human adaptation.

Convention for the Conservation of Antarctic Seals. Instrument of the **Antarctic Treaty System** regulating the hunting of six species of seal in the Antarctic Treaty area. During the late 1960s, when commercial sealers were investigating possibilities of exploiting Antarctic seals on pack ice, it became apparent that conservation measures agreed under the Antarctic Treaty could not limit the right of any nation to fish or hunt on the high seas south of 60°S. Several species of Antarctic seal that congregate on the sea ice were therefore open to uncontrolled exploitation. To bring such species under protection, the Treaty nations in 1972 agreed a Convention for the Conservation of Antarctic Seals (CCAS), which came into force in 1978.

For the full text of the convention see Appendix B. Article 1 defines the convention as applying to seas south of 60°S. It affirms provisions of Article IV of the Treaty, concerning sovereignty issues, and lists the six species to which the convention may be applicable (southern elephant seal *Mirounga leonina*, leopard seal *Hydrurga leptonyx*, Weddell seal *Leptonychotes weddelli*, crabeater seal *Lobodon carcinophagus*, Ross seal *Ommatophoca rossi*, and the southern fur seals *Arctocephalus* spp.). Article 2 concerns implementation: contracting parties agree that the seals shall be killed only in accordance with the provisions of the convention, each party to provide appropriate laws, regulations and other measures. Article 3 introduces an annex specifying measures which the parties will adopt, relating to permissible catches, protected and unprotected species, open and closed seasons, open and closed areas, designation of reserves, limits relating to sex, size, or age for each species, and other practical matters by which hunting may be regulated. Under Article 4 seals may be killed or captured under permit (a) to provide indispensable food for men and dogs; (b) to provide for scientific research; and (c) to provide specimens for museums, educational or cultural institutions. Article 6 provides for a meeting of parties to be convened 'at any time after commercial sealing has begun', with a view to establishing both a system of control and inspection and a commission to perform the necessary functions. This article provides also for 'considering other proposals' including the provision of further regulatory measures and moratoria. Article 7 provides for a meeting of contracting parties within five years of entry into force of the Convention, and at least every five years thereafter, to review the operation of the Convention.

The annex of the Convention prohibits entirely the hunting of fur seals and elephant seals (the two species that were most heavily exploited in the past) and also of Ross seals, the biology of which is little known. It permits the annual capture of up to 175 000 crabeater seals, 12 000 leopard seals and 5000 Weddell seals. Weddell seals may not be taken between 1 September and 31 January, and the period 1 March to 31 August is a closed season for all species. The annex provides for six regional sealing zones to be closed in numerical sequence, and for three reserves around the South Orkney Islands, the southwestern Ross Sea and Edisto Inlet, on the Victoria Land coast, where killing or capture are prohibited (**Sealing zones and reserves**). Hunters are required to provide annual details of their operations, numbers killed and other information.

In the absence of any reported commercial sealing operations, no review meeting of the Convention was deemed necessary during 1983, the fifth anniversary year. The first review meeting was held in September 1988. Substantial hunting by two Soviet ships in 1986–87, unannounced officially but considered by conservation groups to mark the start of a commercial enterprise, raised the question of how, in the absence of notice by a contracting party, the start of commercial sealing could be defined for the purposes of Article 6. It became apparent during the meeting that the Soviet activity did not mark the start of commercial sealing. Contracting parties reported themselves satisfied with the way the Convention had operated, recommending only minor amendments to the annex. Environmental groups disagreed, dismissing the Convention as ineffectual and putting an alternative case for a complete moratorium on Antarctic sealing.

Further reading: Chaturvedi (1996).

Convention on the Conservation of Antarctic Marine Living Resources (CCAMLR).

An instrument of the **Antarctic Treaty System** regulating the conservation of marine living organisms south of the Antarctic Convergence. Commercial fishing that developed in the Southern Ocean during the 1960s involved several kinds of fin fish, squid and krill, all of which form part of the food webs that involve also sea birds, seals and whales. Forecasts suggested that exploiting the Southern Ocean might double the world's annual fishing catch. Biologists feared that severe fishing could not fail to affect the ecology of the Southern Ocean as a whole. From their concern arose CCAMLR, an agreement that seeks to cover the conservation (including rational use) of all marine living resources found south of the Antarctic Convergence. Conceived during the late 1970s, CCAMLR was signed in 1980 and came into effect in April 1982. The purpose was to regulate catches of Southern Ocean krill and fish through an approach based on ecosystem management, by ensuring stable recruitment of stocks, maintaining the balance of ecological relationships between harvested, dependent and related populations, and prohibiting changes in population that were not likely to be reversible within a few decades.

The Convention is administered through a Commission, with a secretariat based in Hobart, Tasmania. The 23 members of the Commission have regulatory powers to set catch limits, define fishing areas, name protected species, designate open and closed seasons and areas, regulate effort and method of harvesting, appoint observers to monitor shipboard activities, etc. It is advised by a scientific committee, which receives reports and recommendations from working groups on fisheries monitoring, fisheries interactions and ecosystem monitoring.

Summary of the Convention

The full text of the Convention's 33 articles and annex appear in **Appendix C**. Article I declares the Convention to apply to Antarctic marine living resources of the area south of 60°S latitude and the area between that latitude and the Antarctic Convergence, and defines living resources as the populations of fin fish, molluscs, crustaceans and all other species of living organism, including birds, found south of the Convergence. The Convergence is redefined for purposes of the Convention in terms of coordinates of latitude and longitude (see **Convergence, Antarctic**). Article II outlines the objective of the Convention, which is the conservation of marine living resources (further defining conservation to include rational use), and summarizes the principles to be applied in harvesting and associated activities. Article III requires Contracting Parties, whether or not they are party to the Antarctic Treaty, not to engage in activities contrary to the principles and purposes of the Treaty. Articles IV and V require such parties to be bound respectively by Articles IV and VI of the Treaty, and by the Agreed Measures for the Conservation of Antarctic Fauna and Flora. Article VI notes that nothing in the Convention shall derogate from rights and obligations of Contracting Parties under the International Convention for the Regulation of Whaling, and the Convention for the Conservation of Antarctic Seals. Article VII establishes the Commission for the Conservation of Marine Living Resources and defines its membership: Articles VIII to XIII cover the legal personality, privileges and immunities, functions, conservation measures, implementation and other properties of the Commission, including the establishment of its headquarters in Hobart, Tasmania. Articles XIV to XX deal with the establishment, functions and management of the Scientific Committee for the Conservation of Marine Living Resources, including its languages, budget, and provisions for information gathering. Articles XXI to XXXIII cover compliance, inspection, disputes and withdrawal from the Convention.

Conserving fin fish

Attempts to exploit stocks of fin fish date back to the early twentieth century, when whalers on South Georgia sent barrels of salted fish to South America. Abundance of whales during the early years made the fish resource insignificant. As whaling profits declined, the companies looked again at fishing, using purse seines and small trawls during the 1930s, 1950s and 1960s. At least one effort was judged impracticable because the very high catches burst the nets. The USSR developed the first successful Antarctic fishing operation off South Georgia, taking marbled notothenia *Notothenia rossii*. Nearly all the catch was filleted and frozen for human consumption. Annual catches rose quickly, reaching a peak of 403 000 tonnes in 1969–70, and thereafter crashing. Attention was transferred to the waters around Iles Kerguelen, with similar results. Extending the fishing to other species, mainly icefish *Champsocephalus gunnari,* and to areas south of 60°S, provided further harvests. Other nations joined the fishery, including Poland, the German Democratic Republic, Japan and Bulgaria. Thus by the time that CCAMLR became effective, harvesting had reduced many stocks of Antarctic fish to levels far below those calculated to produce the maximum sustainable yield.

CCAMLR introduced regulatory measures for South Georgia, and these are used by the government to manage the licensed fishery within the 200 mile EEZ created in 1993. France also created a 200-mile EEZ around each of its sub-Antarctic island groups to manage the fishing there. For the Southern Ocean as a whole, a Working Group on Fish Stock Assessment continues to review and assess the status of fish stocks and develop effective management strategies. Its efforts have been extended to include exploratory fisheries for deepwater crabs and squid around South Georgia.

Conserving krill

Krill, the shrimp-like crustacean *Euphausia superba* that forms the common food of baleen whales and of many seals and sea birds, is a resource of vast magnitude. The perception that reduction in numbers of whales due to commercial hunting had left a large surplus of their main food that would be available for a fishery, stimulated much interest in krill stocks. Because it swarms in summer, krill is relatively easy to locate with conventional fish-finding echo-sounders, and to capture using midwater trawls fitted with fine liners. Krill fishing began in tandem with conventional fin-fishing. The USSR dominated the field, taking the first catches in the 1961–62 season. Japan followed in 1972–73, and later many other nations, notably Chile, Poland, the German Democratic Republic, South Korea and Taiwan joined in. Annual catches first exceeded 100 000 tonnes in 1976–77, and 500 000 tonnes in 1981–82. Thereafter catches fluctuated between about 200 000 and 450 000 tonnes during the late 1980s and early 1990s, falling to 87 000 tonnes in 1993.

Though it has proved difficult to estimate total stocks of krill in the Southern and South Atlantic oceans, even the heaviest annual catches so far remain a tiny fraction of the tens of millions of tonnes that form the most conservative estimates. The Commission has been cautious in its approach to the krill fishery, applying its first conservation measures to local stocks in 1991. Though krill is so plentiful that a general reduction due to fishing seems unlikely, it is prudent to recall that the same was believed true of whales. Furthermore, even local reductions may have severe consequences for predators with limited foraging ranges from shore, for example, lactating fur seals or penguins rearing chicks. Krill can be processed for human consumption, either as peeled tail-meats or as a paste prepared by squeezing the flesh through perforated plates. Neither has proved popular: currently most of the catch is used to make krill meat, a product rich in oil and protein, but of low value for use in animal feed. Should a larger market develop for krill as human food, higher prices and keener fishing might be expected. Conservationists are particularly concerned to maintain krill stocks, as any radical reduction in this key species would be likely to have far-reaching effects on many stocks of Antarctic birds and mammals.

Conserving squid

Squid are important components in the diet of Antarctic predators, notably seals and penguins. Though clearly abundant, they move too fast to be caught in bulk by trawling, and are hunted commercially by 'jigging' lines from specially equipped ships. Major squid fisheries developed during the 1980s near New Zealand and the Falkland Islands. Squid have not so far been harvested on a commercial basis in Antarctic waters, though there have been some unsuccessful pilot expeditions.

Is CCAMLR succeeding?

CCAMLR provides for Antarctic waters a single regulatory body, providing standard methods of stock survey, precautionary regulations based on ecosystem considerations, and an inspection system. Critics draw attention to certain phrases in the text of the Convention that have proved difficult to interpret. There are practical biological problems, for example in applying the concept that populations may be maintained at levels that ensure the greatest net annual increment, when harvesting takes place at different levels in the food chain. Statistical bases have proved difficult to establish, and difficulties have been experienced in detecting changes of abundance in the various stocks that may be harvested, or the stocks dependent on them. Perhaps the greatest difficulty facing the Commission is lack of resources to combat unauthorized fishing (especially of the high value Patagonian toothfish) which can clearly undermine all the management effort. New initiatives such as the Catch Documentation Scheme are addressing 'pirate' fishing problems. Supporters argue that, despite a slow start and the severe criticisms of conservationists, CCAMLR has provided a basis for managing Southern Ocean ecosystems, and its recommendations grow steadily in practicality and effectiveness. Though it cannot yet control fishing activities to the extent that they are controlled in other oceans, it is 'an acceptable political compromise', capable of laying the foundations on which the Southern Ocean may successfully be managed.

Further reading: Everson (2000); www.ccamlr.org
(WNB/BS)

Convention on the Regulation of Antarctic Mineral Resource Activities (CRAMRA). A proposed instrument of the **Antarctic Treaty System** that attempted to regulate the exploitation of minerals within the Treaty area. Discussions lasting over seven years preceded the formal presentation of the Convention in 1988. The Convention never came into force.

Coal and iron minerals discovered by early explorers, later discoveries of other, rarer minerals, and the drilling in 1973 of traces of natural gas from the continental shelf, showed clearly that Antarctica's rock formations, no less than those of any other continent, include minerals of economic interest. Though mineral exploitation on a commercial scale had never seriously been contemplated, the Scientific Committee on Antarctic Research in 1977 reported to Antarctic Treaty consultative parties on environmental problems that might arise should mining or drilling begin. The Zumberg Report, published in 1979, summarized contemporary knowledge of mineral occurrences in Antarctica, and alerted the world to environmental impacts that might be expected from resource development in the Antarctic region. Environmentalists alerted by the report intensified their campaign for Antarctica to be protected completely from mining and other forms of potentially harmful exploitation. To the Treaty parties, a more immediate concern was the likely impact of mining on the Antarctic Treaty itself, especially on the issue of sovereignty embodied in Article 4. In the view of one international lawyer, 'the exploitation of minerals is perhaps the strongest manifestation of exclusive territorial sovereignty' (Triggs 1987). Claimant states that had surrendered relatively trivial rights over living resources would be unlikely to forego more important, and possibly more lucrative, rights arising from mineral exploitation. States that disregarded claims would have no reason to support claimants in attempts to regulate exploitation within claimed territories.

Formal negotiations toward a minerals regime took place within the framework of the Fourth Special Antarctic Treaty Consultative Meeting, which began in Wellington in June 1982. Subsequent meetings were held in eight other capitals, finally returning to Wellington in January 1988. An atmosphere of diplomatic secrecy fuelled the deepest suspicions of antagonists. Prominent in opposition were conservationists who wanted Antarctica to remain undeveloped, and non-Treaty members of the United Nations who wanted Antarctic resources to be shared by all, rather than by the self-elected coterie of Treaty states. Those who sought a workable regime faced many problems arising from weaknesses inherent in the Treaty itself, not least the requirement for consensus on every issue between all consultative parties, and uncertainty as to the legal bases on which jurisdiction for management could be exercised.

Presented at the final meeting, the convention's 67 articles provided for a Commission to monitor information relevant to mineral exploitation, designate prohibited and potentially exploitable areas, and perform a range of regulatory functions covering prospecting, exploration and development of mining operations, with strong emphasis on environmental protection. From the outset CRAMRA was savaged by conservation organizations. Though it prohibited mineral exploitation activities that did not pass strict environmental tests, subject to consensus decisions, it was opposed broadly on the grounds that mineral exploitation was unnecessary in Antarctica and should be banned rather than regulated. More temperate critics, including many scientists experienced in Antarctic research programmes, regarded the convention as flawed, over-complicated, unworkable, or inadequate to provide the environmental protection that by this time was agreed essential. Two claimant nations, France and Australia, declined to ratify the Convention. Other nations demurred, and CRAMRA foundered.

From the wording of the Convention were salvaged environmental protection concepts that, in a series of special consultative meetings in 1990 and 1991, became the **Protocol on Environmental Protection to the Antarctic Treaty** (Appendix D). Under Article 7 of the Protocol, any activity relating to mineral resources, other than scientific research, is specifically prohibited. Thus Treaty members are effectively barred from mining or drilling for oil and gas within the Treaty area for a period of 50 years from 1991, when the Protocol received general acceptance.

Further reading: Barnes (1982); Triggs (1987); Zumberge (1979).

Convergence, Antarctic. A maritime boundary defining the northern limit of the Southern Ocean, also known as the Polar Front. It forms where cool Antarctic surface waters, flowing outward from the shores of the continent, disappear below warmer, less dense waters of subtropical origin (see map: **Southern islands**). Winding sinuously between latitudes 45°S and 60°S, its position varies slightly from day to day and season to season. It can sometimes be seen as a line or narrow zone of turbulence in the ocean: more often its presence is detected by a relatively sudden shift in sea temperature, e.g. 3°C over a horizontal distance of 3–4 km, accompanied by mist or fog. The boundary has great significance both for marine plants and animals and, through its effects on climate, for the terrestrial flora and fauna of Antarctic and sub-Antarctic islands. The Convergence is also generally accepted as a northern ecological boundary to the Antarctic. Under Article I (4) of the **Convention on the Conservation of Antarctic Marine Living Resources**, the Convergence is deemed to be a line joining ten sets of coordinates that approximate to its mean position.

Cook, James. (1728–78). British navigator and explorer. Born 27 October 1728 at Marton, near Middlesborough, England, Cook was the son of an agricultural foreman. The family moved to Aireyholme and he attended the village school at Great Ayton. Aged 16, he served briefly in a shop in the fishing village of Staithes, then was apprenticed to a Quaker ship owner in Whitby, where he learnt seamanship and navigation in coal-carrying ships. Achieving rapid promotion, he was rated first mate in 1752 and master three years later. In 1755 he joined the Royal Navy, quickly rising from able seaman to master during the Anglo-French wars (1756–63), and seeing action off France and eastern Canada. His charting of parts of the St Lawrence River was an important preliminary to General Wolfe's capture of Quebec. After the war he continued hydrographic work mainly in Newfoundland and Labrador. Cook's meticulous chartwork, interests in astronomy and other branches of science, and ability to command small vessels, singled him out for advancement. In 1768 he was promoted to lieutenant and commissioned to lead a scientific expedition to observe the transit of Venus across the sun. For the expedition ship he chose a Whitby collier, commissioned as *Endeavour* (see below). Returning successfully in 1771 with enhanced reputation, he was promoted to captain and commissioned to lead a second expedition to explore the southern oceans. Between 1772 and 1775 he circumnavigated the ocean in high southern latitudes; crossing the Antarctic Circle on 17 January 1773, and exploring South Georgia and southern South Sandwich Islands. Leaving for a third expedition in 1776, he confirmed the discovery of Prince Edward Islands and Iles Crozet and Kerguelen in the southern Indian Ocean. He was killed during a fracas in Hawaii on 14 February 1778.

Cook was noted for his humane interest in the men serving under him. He found no difficulty in crewing his vessels, even for long voyages, sometimes in appalling conditions. At a time when poor diet regularly decimated ship's crews, Cook made every effort to provide variety, particularly of antiscorbutics that would help to prevent scurvy.

Cook's voyages

Cook made three major voyages of discovery in the southern oceans. His first, in the naval bark *Endeavour,* was primarily a scientific expedition to observe a transit of Venus – a passage of the planet across the face of the sun that was of particular interest to astronomers. The ship left Plymouth on 26 August 1768, sailing for Rio de Janeiro and Cape Horn. On board was a scientific party including botanists Joseph **Banks** and Dr Daniel **Solander**. *Endeavour* reached Tahiti in April 1769. After observing the transit successfully, Cook headed through the Society Islands and south and southwestward across the Pacific Ocean. On 7 October he sighted New Zealand, and during the following five months circumnavigated and surveyed the coasts of both main islands. Taking departure from Cape Farewell on 1 April 1770, he sailed northward along the east coast of Australia, passing with difficulty and some damage through the Great Barrier Reef and making a brief landing in New Guinea. From there *Endeavour* proceeded to Cape Town, returning to England in July 1771.

Cook spent a year ashore. Promoted to captain, he prepared for a second expedition, this time to explore the southern oceans in search of the unknown continent that was believed to lie within them. With two ships, *Resolution* and *Adventure*, both converted colliers, with the latter under command of Capt. Tobias Furneaux, he sailed from Plymouth on 16 July 1772, heading south for Cape Town. Continuing southward in December, he reached the edge of the pack ice. Skirting eastward, on 17 January 1773 he made the first crossing of the Antarctic Circle in 39°E, some 200 miles from the coast of what is now western Dronning Maud Land. Turning eastward he crossed the Indian Ocean in a higher latitude than any navigator before him, passing well south of Kerguelen-Trémarec's landfall (**Iles Kerguelen**) and southern Australia. Finding no indications of land anywhere along his course, he reached southern

James Cook 1728–78

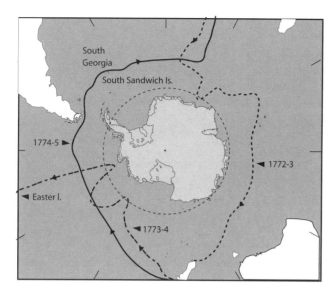

Cook's voyages in southern waters. The first leg (1772–73) of Cook's second voyage took him from Cape Town, south across the Antarctic Circle and on to Dusky Sound, New Zealand. The second leg (1773–74) took him from New Zealand southeastward into the southern Pacific Ocean pack ice: after crossing the Circle twice he headed north to Easter Island and Polynesia. After returning to New Zealand, his third leg (1774–75) took him again across the southern Pacific, but in a lower latitude, to Tierra del Fuego and the discovery of South Georgia and the South Sandwich Islands

New Zealand on 25 March, and wintered in warmer reaches of the Pacific Ocean.

In spring he headed south again from New Zealand, crossing the Pacific Ocean close to the edge of the pack ice. On 30 January 1774, surrounded by loose pack ice and large tabular ice bergs, Cook reached his furthest south, a point in 71°10'S, 106°54'W, some 40 miles from the coast of West Antarctica. Cook made no claim to have seen land, but judged correctly that so great a mass of ice could have accumulated only on land nearby. Extricating his ship from the ice fields with some difficulty he continued eastward, charting Easter Island before turning north and exploring more of the tropical and temperate Pacific Ocean. In December 1774 he returned to Tierra del Fuego. Early in the following year he continued eastward, surveying the north coast of South Georgia and discovering the South Sandwich Islands, both of which he claimed for his King. Cook's search still left room for a southern polar continent, but greatly reduced both the area in which it could lie, and the economic value to be expected from it. He returned to Britain in mid-1775, justly pleased with the success of his second expedition.

Accounts of his voyages, published as books, attracted considerable public attention: the simplicity and directness of his observations still appeal. Noting that South Georgia lay in a latitude similar to that of his native Yorkshire, and describing it as 'savage and horrible', he wrote: 'The wild rocks raised their lofty summits till they were lost in the clouds, and the valleys lay covered with everlasting snow. Not a tree was to be seen, nor a shrub even big enough to make a toothpick ... Who would have thought that an island of no greater extent than this, situated between the latitude of 54° and 55°, should, in the very height of summer, be in a manner wholly covered with many fathoms deep with frozen snow, but more especially the southwest coast? The very sides and craggy summits of the lofty mountains were cased with snow and ice; but the quantity which lay in the valleys is incredible; and at the bottom of the bays, the coast was terminated by a wall of ice of considerable height.' His discovery that South Georgia was an island, and not the continent he was seeking, afforded him no great disappointment, 'for to judge of the bulk by the sample, it would not be worth the discovery'.

The presence of so much ice in temperate latitudes, and the massive tabular bergs he had encountered further south, convinced Cook that there remained a continent, as yet unseen, in the far Southern Ocean. 'Countries condemned to everlasting rigidity by Nature, never to yield to the warmth of the sun, for whose wild and desolate aspect I find no words; such are the countries we have discovered; what then may those resemble which lie still further to the south?' Of anyone possessing the resolution and fortitude to push further south, he added, 'I shall not envy him the fame of his discovery, but I make bold to declare that the world will derive no benefit from it.'

Now firmly established as the nation's leading maritime explorer, Cook led a third expedition (1776–80). Again in *Resolution*, he was accompanied by Capt. James Clerke in *Adventure*, this time mainly to examine the north Pacific Ocean. In passage through the Indian Ocean he visited and named the Prince Edward Islands and confirmed the recent French discoveries of Iles Crozet and Kerguelen. After a successful foray into north Pacific waters, Cook returned to Hawaii. He was killed in a skirmish with natives on 14 February 1778.

Further reading: Stamp and Stamp (1978); Beaglehole (1955, 1961, 1967).

Cook Ice Shelf. 68°40'S, 152°30'E. Wide ice shelf along the George V Coast, East Antarctica, occupying a bay between Cape Freshfield and Cape Hudson. It was named by the Australasian Antarctic Expedition 1911–14 for J. Cook, then Australian prime minister.

Cook Island. 59°27'S, 27°10'W. Central island of Southern Thule, the three southernmost islands of the South Sandwich Islands group. Approximately 6 km (3.75 miles) long and 4 km (2.5 miles) across, the island is of volcanic origin, currently non-active and almost completely ice-capped, with precipitous ice or rock cliffs. Mt Harmer, the highest point, rises to 1075 m (3526 ft). It was charted by Capt. James Cook in 1775.

Cool temperate islands. Southern Oceanic islands that lie between the Antarctic Convergence and the 10 °C summer isotherm. They include the Falkland Islands in the southern Atlantic Ocean, Marion Island, the Prince Edward Islands, Iles Crozet, Iles Kerguelen, Macquarie Island, Campbell Island and the Antipodes Islands. For a map and table see **Study Guide: Southern Oceans and Islands**. For more detailed accounts see entries for individual islands and groups.

Cooper Sound. 54°48′S, 35°47′W. Channel between Cooper Island and the eastern end of the main island of South Georgia.

Cooperation Sea. 68°S, 70°E. Coastal sea off East Antarctica, between Cape Boothby and the western end of West Ice Shelf, bordered by the Mawson, Lars Christensen and Ingrid Christensen coasts. There is no definitive northern boundary. Fast ice and pack ice extend off the coast in winter: in summer pack ice and bergs move westward, driven by easterly currents and offshore winds. The name appears mainly on Russian maps.

Coppermine Peninsula. 62°23′S, 59°42′W. Ice-free peninsula on the west side of Robert Island, South Shetland Islands, named for traces of copper minerals in some of the rocks. The coastal flats are unusually rich in vegetation and in bird and seal breeding colonies. The Chilean refuge **Risopatron** is occasionally used by summer visiting parties. An area of approximately 0.9 km^2 (0.3 sq miles) has been designated SPA No. 16 (Coppermine Peninsula).

Cormorants, southern oceanic. Flying birds of the order Pelecaniformes, family Phalacrocoracidae, cormorants or shags are divers and foot-swimmers with long necks, narrow bills, webbed feet and stubby, shooting-stick tails. Strictly inshore feeders, they are seldom seen more than a few miles from their breeding grounds. During dives lasting one to several minutes they catch fish and bottom-living invertebrates. Imperial shags *Phalacrocorax atriceps*, with body length 60 cm (2 ft) and wing span 1.2 m (4 ft), are commonest in the maritime Antarctic, nesting throughout the Scotia Arc and along the western flank of Antarctic Peninsula south to Marguerite Bay. A small population nests also on Heard Island. Iles Kerguelen support a local

Imperial shags. Photo: BS

Kerguelen shag *P. verrucosus*. Rock shags *P. magellanicus* breed on the Falkland Islands and southern coasts of Argentina and Chile. Their nests, accumulating over several years, are piles of seaweed and feathers held together by guano, accommodating two to four bluish-white eggs. Cormorants are year-round residents on the warmer islands, summer migrants to areas further south that are invested by sea ice in winter.

Cornice. Ice and snow forming an overhang, usually along the lee of a ridge or cliff face.

Coronation Island. 60°38′S, 45°35′W. Largest of the South Orkney Islands, a mountainous, glaciated island with steep rocky cliffs and tide-water glaciers. An area of approximately 88.5 km^2 (35 sq miles), between Conception Point, Wave Peak and Foul Point on the north side of the island has been designated SPA No. 18. Including both ice-free and glaciated terrain, it supports large seabird colonies, lichen-dominated cliffs, and sublittoral ecosystems typical of the maritime Antarctic environment.

Cosgrove Ice Shelf. 73°32′S, 100°45′W. Circular ice shelf occupying a deep bay between King Peninsula and Canisteo Peninsula of the Walgreen Coast, Ellsworth Land, West Antarctica. The shelf was named for Lt J. R. Cosgrove USNR, a staff officer of United States Naval Support Force during Operation Deep-Freeze 1967–68.

Cosmonaut Sea. 68°S, 40°E. Coastal sea off East Antarctica, between Riiser-Larsen Halvøya and Tange Promontory, bordered by Prins Harald Kyst and Kronprins Olav Kyst. There is no definitive northern boundary. Fast ice and pack ice extend off the coast in winter: in summer pack ice and bergs move westward, driven by easterly currents and offshore winds, sometimes providing difficulties of access to the Japanese station Syowa. The name, which appears mainly on Russian maps, commemorates cosmonauts of the Soviet Union.

Cotton Plateau. 82°54′S, 159°40′E. Ice plateau of the Queen Elizabeth Range, Victoria Land, named for New Zealand geomorphologist Sir Charles Cotton.

Coughtrey Peninsula. 64°54′S, 62°53′W. A small peninsula in Paradise Harbour, on the Danco Coast of Graham Land, mapped in 1912–13 by a visiting geologist: the origin of the name is obscure. Since 1950 the peninsula has been the site of an Argentine research station, Almirante Brown.

Council of Managers of National Antarctic Programmes (COMNAP). A committee federated to the Scientific Committee on Antarctic Research concerned with matters of practical management and logistics within Antarctic national research programmes. Managers (i.e. officials responsible for implementing the national programmes) review operational matters, seek solutions to common operational problems, discuss national responses to common issues and responses to SCAR on logistical matters, and review management aspects of projected scientific programmes. COMNAP and its subcommittee SCALOP arose in 1988 to provide management bodies in which the requirements of scientists could be reconciled with the practicalities of running national programmes. The council produces an annual report.

Couzens, Thomas. (d. 1959). New Zealand expeditioner. A Lieutenant in the Royal New Zealand Army Corps, and member of the New Zealand Geological and Survey Expedition, he died when the tractor he was driving fell into a crevasse in Victoria Land.

Covadonga Harbour. 63°19′S, 57°55′W. Inlet in the northeast corner of Huon Bay, south of Cape Legoupil, Trinity Peninsula, named for a Chilean naval frigate.

Crabeater seal. *Lobodon carcinophagus*. These seals are most plentiful in open pack ice, especially at the fringes of the winter pack ice. Large adults of either sex measure 2.8 to 3 m (9 to 10 ft) long and weigh up to 300 kg (660 lb). The fur is fawn to slate grey, usually unspotted. Monogamous, they form pairs in September on the ice, males defending a small area around the females. Single pups are born in early-to-mid-October. Many crabeater seals carry conspicuous scars and open cuts, thought to be inflicted by leopard seals: newly weaned crabeaters are especially vulnerable. Females attain sexual maturity at 3–5 years and probably live another 15 years. The teeth are cusped, providing a strainer that retains krill, their main food. Estimated to number over 15 million, they are probably the world's most populous seal.
Further reading: Laws (1993).
(JPC)

CRAMRA. See **Convention on the Regulation of Antarctic Mineral Resource Activities**.

Crean, Thomas. (1877–1938). British seaman and explorer. Born in Annascaul, Ireland, he enlisted in the Royal Navy in 1893. As an AB he volunteered for service with the **British National Antarctic Expedition 1901–4**, playing a full role in activities ashore including several sledging journeys. Rated petty officer on return to the navy, he was invited by Scott to join the **British Antarctic (*Terra Nova*) Expedition 1910–13**, as an expert sledger and pony handler. As a member of the final support party, he accompanied the polar party far up the Beardmore Glacier. On the return journey he walked the last 56 km (35 miles) alone to secure help for a stricken colleague, for which he was awarded the Albert Medal. His third and last expedition was with Shackleton's **Imperial Trans-Antarctic Expedition 1914–17**, in which he was one of six who crossed from Elephant Island to South Georgia in the lifeboat *James Caird*, and one of three who crossed South Georgia to seek help from the Stromness whaling station. He served in World War I, was commissioned, and in 1920 retired from the navy, returning to southern Ireland to run the 'South Polar Inn' in his home village. He died on 27 July 1938.

Crevasse. Near-vertical rift in an ice sheet, formed where the sheet is under sufficient tension to fracture. Found in most extensive ice sheets, crevasses are notorious for their dangers to travellers. Open ones are easy to see, but those bridged by drifted snow (see **Snow bridge**) are a hazard to sledgers, vehicles and to landing aircraft. Because the weight of an ice sheet restricts tensile stresses to near-surface layers, crevasses do not usually penetrate down to bedrock. Their depths often depend on ice temperature. In alpine glaciers they are seldom deeper than 30 m: in the colder and more brittle ice sheets of polar regions they can be considerably deeper. On valley glaciers the orientation of crevasses gives an indication of the active stress fields. A general feature is that side crevasses always point upstream towards the centre of the glacier. More complicated crevasse fields are often found on ice streams, for example, a pattern of diamond-shaped blocks is common.
(CSMD)

Cross, Jacob. (1876–1946). British seaman and explorer. Born at Little Clacton, he entered the Royal Navy in 1891. As a seaman petty officer first class he volunteered for the **British National Antarctic Expedition 1901–4**, taking part in the sledging and other shoreside activities, and serving as assistant to naturalist Edward Wilson, becoming particularly adept at bird skinning. He served with the navy throughout World War I, retiring into civilian government employment in Kent. In retirement he became a noted breeder of Sealyham terriers. He died on 8 July 1946.

Crosson Ice Shelf. 75°05′S, 109°25′W. Triangular shelf in a bay formed by the arm of Bear Peninsula, western Walgreen Coast, Marie Byrd Land. The shelf is named for Cdr W. E. Crosson, USN, a construction officer during Operation Deep-Freeze 1973.

Crozet, Iles. 46°15′S, 50°15′E. A scattering of three small groups of cool-temperate volcanic islands and rocks, extending over 160 km (100 miles) of the southwestern Indian Ocean, with a scattering of outliers. Ile de la Possession, the central island, is 18 km (11.2 miles) long and 11 km (6.9 miles) wide, rising to a peak of 934 m (3064 ft). Ile de l'Est, a neighbour 15 km (9.4 miles) away, is of similar area with a taller central ridge rising to 1090 m (3575 ft). About 100 km (62 miles) to the west lies Ile aux Cochons, a round island 9 km (5.6 miles) in diameter with a central cone 775 m (2542 ft) high. Outlying Ile des Pingouins and Iles des Apôtres are relatively tiny. The group shares a submarine platform with the Prince Edward Islands. Though the islands are clearly of volcanic origin, with scoria cones and reefs and ridges representing thin flows of basaltic lava, none has been active in historic times. Damp, cloudy and windy, with mean annual temperature about 4.9 °C and range of monthly temperatures 5.3 °C, they are currently without glaciers, though small patches of ice remain year-round on the peaks of the taller eastern islands. Predominantly green at sea level, they have a rich flora of grasses, shrubs and mosses, including tussock meadows close to sea level and fellfield above 100 m (328 ft). Sea birds breed at all levels, fertilizing the vegetation with their droppings. The islands support huge colonies of king, macaroni and rockhopper penguins, and several species of albatross and lesser petrel. Breeding Antarctic fur seals and southern elephant seals are again plentiful after nineteenth-century hunting.

Discovered in 1772 by Marc Macé Marion du Fresne, the islands were of interest particularly to nineteenth-century sealers, who stripped them of fur seals and elephant seals. Formally claimed by France, since 1938 they have held the protected status of a national park. A scientific station, Alfred-Faure, was established on Ile de la Possession in 1962. Since then the group has been thoroughly studied by French scientists. Biologically and geologically they are perhaps the best known of all the southern islands.

Crozier, Cape. 77°31′S, 169°24′E. Large, ice-covered cape forming the eastern end of Ross Island. Discovered during the **British Naval Expedition 1839–43**, it was named for Capt. R. M. Crozier, commander of HMS *Terror*. An area of approximately 19 km^2 (7.5 sq miles), including an extensive Adélie penguin colony, overlooking the site of an emperor penguin colony in the nearby Ross Ice Shelf, has been designated SSSI No. 4 (Cape Crozier).

Emperor penguin colony on the sea ice off Cape Crozier, Ross Island. Photo: BS

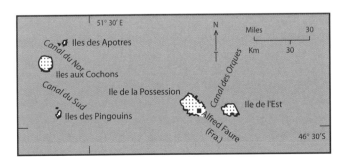

Iles Crozet

Crust. Surface layer of hard snow, usually formed by wind, or by warming in the sun and re-freezing. To travellers this is often a welcome development, making a snow surface easier to walk, ski or sledge over.

Cruzon, Richard L. (d. 1970). US Naval commander. Graduating from the US Naval Academy in 1919, he became second-in-command to Richard Byrd in USS *Bear*, cargo ship of the United States Antarctic Service Expedition 1939–41. After service in the Pacific theatre during World

War II, he commanded the fleet of 13 ships involved in Operation Highjump, in Antarctic waters in 1946–47. He retired as a vice admiral in 1954 and died on 15 April 1970.

Cryptogam Ridge. 74°21′S, 164°42′E. A prominent ridge near the summit of **Mt Melbourne**, Victoria Land. The mountain is noted for its mild geothermal activity, which supports a unique community of mosses, algae, and soil microfauna. An area of approximately 0.7 km² (0.3 sq miles), including most of the ridge, has been designated SPA No. 22. The protected area lies within the broader boundaries of SSSI 24 (summit of Mt Melbourne).

Crystal Sound. 66°23′S, 66°30′W. Broad waterway between the southern Biscoe Islands and the Graham Coast, Graham Land, extending south to Liard Island.

Currents, convergences and divergences. Strong ocean surface currents are generated mainly by winds, which move surface waters before them at speeds proportional to the square of wind speeds. Wind-induced motion penetrates to a depth of about 100 m, but within this surface layer (called the Ekman layer, for the Swedish oceanographer who first identified and explored it) other forces come into play. Friction between water particles reduces the speed of movement, and Coriolis force, induced by the earth's rotation, redirects the motion, in the southern hemisphere bringing a northern component to a westerly flow, a southern component to an easterly flow. In the **Southern Ocean**, the dominant surface current immediately surrounding Antarctica is the East Wind Drift: further north the dominant current is the stronger and more persistent West Wind Drift.

Where surface winds from contrary directions converge, the Ekman effect tends to force the resulting contra-flowing water masses either apart or together. Where water masses are parted, replacement water wells up from below: a zone where this is happening is called a *divergence*, and the resulting water movement is an upwelling. A marked divergence occurs between the East Wind and West Wind drifts. Where the masses are driven together, the less-dense mass over-rides the denser: the zone is a *convergence*, and the water movement is a downwelling. The Antarctic Convergence or Polar Front, the northern limit of the Southern Ocean, marks the junction where warm, less-dense waters from the north over-ride cold dense waters from the south. Upwelling, occurring for example at the Antarctic Divergence, may bring nutrient-rich waters up from below and stimulate plankton production. Mixing of waters due to downwelling may also cause local enrichment. Both convergences and divergences tend therefore to provide rich feeding grounds for sea birds.

Czegka, Victor H. (1880–1973). American explorer. Born on 21 May 1880 in Landsdron, Bohemia, he moved to the USA and in 1905 enlisted in the US Marine Corps. He served in both of Byrd's Antarctic expeditions, in the first as a machinist, in the second as manager and supply officer. He assisted also in preparations for the United States Antarctic Service Expedition 1939–41. Czegka died on 2 February 1973.

D

Dakshin Gangotri. 70°05′S, 12°00′E. Former Indian research station on Prinsesse Astrid Kyst, East Antarctica. India's first station in the Antarctic area, it was established in 1984 as a wintering station, and replaced in 1989 by **Maitri Station**.

Dallmann Bay. 64°20′S, 62°55′W. Bay between Anvers and Brabant Islands, Palmer Archipelago, named for its discoverer Capt. E. Dallmann. An area of approximately 710 km^2 (277 sq miles) has been designated SSSI No. 36 (East Dallman Bay), as a site close to Palmer Station suitable for bottom trawling for fish and other benthic organisms.

Dallmann Laboratory. 62°14′S, 58°40′W. German research station in Potter Cove, King George Island, established in January 1994, and operated intermittently in conjunction with Argentine Jubany Station.

Damoy Point. 64°49′S, 63°31′W. Headland on Wiencke Island, Danco Coast, that accommodates both Argentine and British refuge huts. The British Antarctic Survey hut, built in 1973, was used intermittently to house personnel and stores in transit from ships to a local airstrip on Doumer Island, a function that ceased after the opening of the Rothera Station airstrip in 1992.

Dana, James Dwight. (1813–95). American naturalist. While studying natural sciences at Yale University he joined the **United States Exploring Expedition 1838–42** as geologist and mineralogist, later taking on additional responsibilities as marine zoologist. In these fields he contributed substantially to the final expedition reports. Returning to Yale, he became Professor of Natural History and a leading naturalist of his day, publishing textbooks on geology and mineralogy.

Danco Coast. 64°42′S, 62°00′W. Part of the west coast of Graham Land, Antarctic Peninsula, facing Gerlache Strait, and extending from Cape Sterneck in the northeast to Cape Renard in the southwest. Deeply indented with bays and scattered with islands, the coast has steep mountains rising to the central Graham Land plateau, and glaciers descending to narrow piedmont ice shelves. It was explored in February 1821 by the American sealer Capt. John Davis, who may have landed in Hughes Bay, later charted by the Belgian Antarctic Expedition 1895–98, and named for the geophysicist, Emile Danco, who died during the expedition.

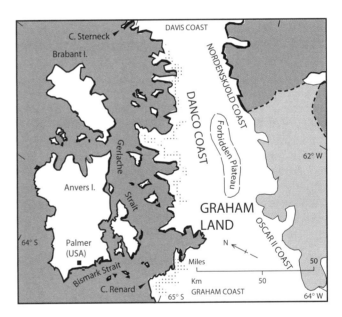

Danco Coast

Danco Island (Base O). 64°44′S, 62°36′W. British research station at Danco Island, Errera Channel, Danco Coast. Built in February 1956 for local survey and geological research, the station was occupied year-round until February 1959. Since then it has been used intermittently by small survey parties.

Darwin Mountains. 79°51′S, 156°15′E. A range emerging from the polar ice cap behind the Hillary Coast of Victoria Land, flanked by the Hatherton and Darwin Glaciers.

It was named for Leonard Darwin, a former Honorary Secretary of the Royal Geographical Society.

Dater, Henry Murray. (1909–74). American polar historian. Born in Brooklyn, New York on 28 February 1909, he read history at Yale University and in France. In 1936–43 he taught history at Kent State University, then joined the US Navy as a historian, with special interests in aviation history. In 1946 he became a civilian historian in the Office of the Chief of Naval Operations. Ten years later he became staff historian to Rear Admiral Richard Byrd. From then he concentrated on American polar history, collecting and archiving data and records from ongoing operations, writing extensively on current history topics, and alerting naval personnel to their historical heritage. In 1960 he became founder editor of *Bulletin of the US Antarctic Projects Officer*. He made six visits to Antarctica, and served for many years on the US Advisory Committee on Antarctic Names. He died in Washington, DC, on 26 June 1974.

David, Tannat William Edgeworth. (1858–1934). Australian geologist and explorer. Born in Wales, he studied geology at New College, Oxford and the Royal College of Science, London. In 1882 he was appointed assistant geological surveyor to the Government of New South Wales, and in 1891 became Professor of Geology in the University of Sydney. He joined the **British Antarctic (*Nimrod*) Expedition 1907–9** as geologist, leading the first ascent of Mt Erebus and the first party to reach the South Magnetic Pole. He served with distinction in the Australian Army during World War I, and was knighted in 1920. Returning to academic life, he retired in 1924 to complete his major work on the geology of Australia.

Davis, John. (*c.* 1550–1605). (alternatively Davys). English navigator and explorer who discovered the Falkland Islands. Born at Sandridge, Devon, England, Davis took to seafaring early in life, becoming both a skilled seaman and navigator. Between 1585 and 1587, convinced of the existence of a Northwest Passage to the East Indies, he established with like-minded neighbours a North-West Company to finance voyages of exploration. He led three Arctic expeditions toward eastern Canada, discovering Davis Strait and Cumberland Sound, and explored the west Greenland coast. In 1588 he commanded the ship *Black Dog* in the fleet that opposed the Spanish Armada. In the following year, in command of *Desire*, Davis accompanied Capt. **Thomas Cavendish** in *Leicester* on a further expedition toward the Northwest Passage, hoping to enter it from the Pacific Ocean end. In the South Atlantic Ocean, contrary winds prevented the two ships from navigating the Strait of Magellan. In ships severely disabled by storms, Cavendish and Davis parted company off Patagonia. Drifting eastward, on 14 August 1592 Davis entered a sound between two large, low-lying islands, which he named Falkland Sound for the current First Lord of the Admiralty. This was the first recorded visit to the islands now known as the Falkland Islands. He returned to England in the following year. Later voyages included visits to the Azores Islands, the Indian Ocean and the East Indies. Davis invented a navigational quadrant, a forerunner of the sextant, and wrote of his northern travels in *Worldes Hydrographical Description* (1595). He published also a mariner's handbook, *The Seaman's Secrets* (1599). He was killed in a fight with Japanese pirates off the East Indies.

Davis, John King. (1884–1967). British seaman and navigator. Born in London, he was apprenticed in the merchant navy in 1900, gaining his first mate's ticket in 1906. He joined Shackleton's **British Antarctic (*Nimrod*) Expedition 1907–9** as chief officer of the ship: having qualified as extra master in 1908, he was able to take command for the final voyage back to London. In 1911 Douglas Mawson invited him to command *Aurora* in the **Australasian Antarctic Expedition 1910–14**, in which he made three summer voyages of relief and exploration, as well as oceanographic cruises between Tasmania and East Antarctica. He declined an offer to join Shackleton's **Imperial Trans-Antarctic Expedition 1914–17**, and became involved in World War I. However, in 1916–17 he took command of *Aurora* for the voyage that relieved the Ross Sea party. In 1920 he became Australian Commonwealth Director of Navigation. He returned to Antarctica in 1929–30 in command of *Discovery* during the first season of the **British, Australian and New Zealand Antarctic Research Expedition 1929–31**. He died in Melbourne on 7 May 1967.

Davis Coast. 64°00′S, 60°00′W. Part of the west coast of Graham Land, Antarctic Peninsula, facing Gerlache Strait, and extending from Cape Kjellman in the east to Cape Sterneck in the west. The coast was sighted by Edward Bransfield in 1820 and explored in February 1821 by the New England sealer Capt. John Davis, for whom it is named. Some authorities call it Palmer Coast, for the American sealer and explorer Capt. Nathaniel B. Palmer.

Davis Sea. 66°S, 92°E. Coastal sea off East Antarctica between the eastern end of West Ice Shelf and Shackleton Ice Shelf, bordered by Queen Mary Coast and Wilhelm II Coast. There is no definitive northern boundary. Fast ice

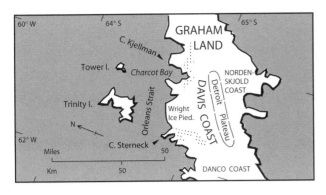

Davis Coast

and pack ice extend off the coast in winter: in summer pack ice and bergs move westward, driven by easterly currents and offshore winds. The name was given originally by Mawson's Australasian Antarctic Expedition 1911–14 to a stretch of sea between Drygalski Island and Termination Ice Tongue, but was later extended to a wider area: it honours Capt. J. K. Davis, commander of the expedition ship *Aurora* and second-in-command of the expedition.

Davis Station. 68°35′S, 77°59′E. Permanent year-round Australian research station on Ingrid Christensen Coast, Princess Elizabeth Land, bordering the Vestefold Hills oasis area. Established in 1957 as an Australian International Geophysical Year station, it operated almost continuously to December 1964, when it was closed. The station reopened in January 1969 and has operated continuously since then.

Davis Valley Ponds. 82°27′S, 51°21′W. See **Forlidas Pond**.

Davys, John. See **Davis, John**.

Day length. Proportion of the day in which the sun appears above the horizon, usually expressed as number of hours of sunshine per 24-hour period. At the spring and autumn equinoxes day length is 12 hours all over the world. In equatorial and tropical regions it varies only slightly from a mean of 12 hours between summer and winter. In temperate latitudes there is a wider variation, yielding long (e.g. 18–20 hour) summer days and correspondingly long winter nights. At the polar circles day length extends to 24 hours on midsummer day, and decreases to zero on midwinter day. Beyond the circles the sun in summer remains above the horizon, and in winter disappears below it, for days or weeks on end. At the geographical poles, days and seasons merge: the sun remains above the horizon for the six summer months, then remains below it for the six months of winter.

In calculating actual day-to-day variations in possible hours of sunshine, it is necessary to take into account effects of refraction, and the diameter of the sun's disc. Taking a mean horizontal refraction of 37′ of arc, and adding 16′ for the semi-diameter of the sun, part at least of the sun's disc may be visible for part of every day of the year up to 60 miles inside the polar circle, and stays visible on at least one day per year as far as 60 miles outside it. In particular atmospheric circumstances local refraction may be twice as high, on some evenings keeping the sun even longer above

Table 1 Hours of daylight on the 21st of each month, and dates of start and end of 24-hour periods of light and darkness, in latitudes 50° to 90° north and south. These times and dates are correct for 1989, and vary slightly from year to year. Data calculated by Dr Gareth Rees

		Latitude south					Latitude north				
		90°	80°	70°	60°	50°	50°	60°	70°	80°	90°
January		24.0	24.0	24.0	17.5	15.6	8.8	7.1	2.5	0.0	0.0
February		24.0	24.0	16.5	14.7	13.9	10.5	9.8	8.3	2.3	0.0
March		24.0	12.4	12.2	12.1	12.1	12.2	12.3	12.5	13.6	24.0
April		0.0	0.0	7.7	9.4	10.2	14.1	15.1	17.1	24.0	24.0
May		0.0	0.0	2.0	7.0	8.7	15.7	17.6	24.0	24.0	24.0
June		0.0	0.0	0.0	5.9	8.0	16.4	18.9	24.0	24.0	24.0
July		0.0	0.0	1.6	7.0	8.7	15.7	17.7	24.0	24.0	24.0
August		0.0	0.0	7.6	9.4	10.2	14.1	15.1	17.2	24.0	24.0
September		24.0	12.2	12.1	12.1	12.1	12.3	12.4	12.6	13.1	24.0
October		24.0	24.0	16.6	14.8	13.9	10.4	9.7	8.2	0.8	0.0
November		24.0	24.0	24.0	17.5	15.6	8.8	7.1	2.4	0.0	0.0
December		24.0	24.0	24.0	18.9	16.4	8.1	5.9	0.0	0.0	0.0
24-h day	Starts	21/9	17/10	18/11	–	–	–	–	16/5	13/4	19/3
	Ends	22/3	24/2	21/1	–	–	–	–	27/7	29/8	24/9
24-h night	Starts	23/3	18/4	25/5	–	–	–	–	26/11	22/10	25/9
	Ends	20/9	26/8	18/7	–	–	–	–	16/1	19/2	18/3

the horizon. Absence of the sun above the horizon does not of course indicate total darkness. For example, in latitude 70° north or south, there is twilight for three or four hours daily throughout winter, usually enough for travelling and other outdoor activities.

Deacon, George Edward Raven. (1906–84). British polar oceanographer. After completing a degree in chemistry at King's College, London, he joined Discovery Investigations to examine the chemistry of sea water. He served both at the South Georgia marine laboratory and at sea, identifying water masses of the Southern Ocean from their temperature and chemical composition, and tracking their movements and dynamics. During World War II he was involved in research on submarine detection and wave patterns. In 1949 he became first director of the National Institute of Oceanography. Deacon was knighted on retirement in 1971. He died on 16 November 1984, shortly after the publication of his book summarizing a lifetime study of the Southern Ocean.
Further reading: Deacon (1984).

Debenham, Frank. (1883–1965). Australian expedition geologist and geographer. Born on 26 December in Bowral, New South Wales, he was educated at the University of Sydney (BA 1904, BSc 1910), latterly studying geology with Prof. David Edgeworth. In 1910 he joined the **British Antarctic (*Terra Nova*) Expedition 1910–13**, in which he made extensive geological and topographic surveys of the mountains and glaciers west of McMurdo Sound. In Cambridge to write up the results, he joined the army early in World War I, and saw action as an infantry officer in Salonika. On demobilization he returned to Cambridge, where he was appointed successively lecturer in surveying, reader in geography, and (in 1931) the university's first Professor of Geography. Instrumental in founding the Scott Polar Research Institute, he was also its first director. Retiring in 1949, he worked in Africa and wrote extensively on polar and geographical matters. He died in Cambridge on 23 November 1965.

De Bougainville, Comte Louis-Antoine. (1729–1811). French navigator, scientist and explorer. Born in Paris, Louis-Antoine was educated in law and mathematics, then joined the army (1753–63), seeing service in Arcadia (French Canada). He transferred to the navy and, with the support of St Malo merchants, led three expeditions to the 'Iles Malouines' (Falkland Islands), establishing Port Louis, a settlement for French colonists displaced from Arcadia. In 1766, when the islands were sold to Spain, de Bougainville led a French naval expedition to explore and establish further colonies. First visiting the Falkland Islands to negotiate their transfer, he spent several weeks surveying Tierra del Fuego and the Strait of Magellan, then explored the southern Pacific Ocean in a vain search for Terra Australis Incognita. His team of naturalists, led by J.-P. Commerson, made extensive collections from Pacific islands. He returned to St Malo in 1769, recounting his travels in *Voyage autour du monde*. After further naval service in the American wars, he retired to develop his scientific studies, being appointed field marshal, vice admiral, senator, count and member of the Légion d'Honneur.

De Brosses, Charles. (1709–77). French politician and geographer. His scholarly publication of 1756, *Histoire de la navigation aux terres australes* (*History of Navigation in Southern Lands*), drew on accounts of southern voyages by Binot Palmier de Gonville (1504), Jean-Baptiste **Bouvet de Lozier** and others to revive the concept of **Terra Australis Incognita**, the great southern continent. His views stimulated both French and British interest in possible colonization, helping to justify the exploratory voyages of, among others, **Cook**, **De Bougainville**, and **Bellingshausen**.

Decepción Station. Argentine permanent research station in Fumarole Cove, Deception Island. Established as a year-round station in 1947, it was damaged during the volcanic eruptions of 1967–70 and part-restored for summer-only use. Research at the station centres on vulcanology.

Deception Island. 62°57′S, 60°38′W. Island of volcanic origin in the South Shetland Islands, ringed by cliffs but with an entry on the southeast side (Neptune's Bellows) to an extensive flooded inner crater. Charted in 1820, the inner harbour of Port Foster was used by generations of nineteenth-century sealers, and from the early twentieth

Remnants of the abandoned whaling factory, Whalers Bay, Deception Island. Photo: BS

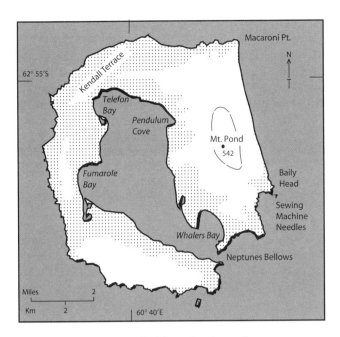

Deception Island, so-called from its deceptive appearance. The island contains an inner harbour much appreciated by generations of sealers and whalers, despite frequent volcanic eruptions that radically alter the inner shoreline. Height in m

century by whalers, who valued the secure anchorage and fresh water. A Chilean-Norwegian whaling station (Hektor Whaling Co.) operated in Port Foster during 1912–31. The island was a port of call for almost every scientific expedition to the South American sector of Antarctica. Port Foster provided a base for the **Wilkins–Hearst Antarctic Expeditions 1928–30, Ellsworth's Antarctic Expeditions 1933–39**, and the **Falkland Islands and Dependencies Aerial Survey Expeditions 1955–57**.

The old whaling station became Base B of **Operation Tabarin**, later the Falkland Islands Dependencies Survey. Fumarole Bay, a shallow bay on the west side of the crater harbour, accommodates the Argentine research station **Decepción** and the Spanish station **Gabriele de Castilla**, both currently summer-only. Pendulum Cove, on the northeast side, contains the burnt-out remains of the Chilean station Presidente Pedro Aguirre Cerda. **Volcanic activity** has been reported throughout the island's history. Major eruptions occurred in 1967, 1969 and 1970, reshaping the shore, creating a new islet in the harbour, and causing fires and mud slides that destroyed or damaged many of the installations. Hot springs at low tide create clouds of steam in Pendulum Cove and Port Foster, and warm the sea water enough to kill plankton and allow tourists to bathe. Five locations around the crater harbour, with a total area of $1.7\,km^2$ ($0.7\,sq$ miles), are collectively designated SSSI No. 21 (Deception Island, parts of). Two small areas of harbour sea floor, totalling $1.9\,km^2$ ($0.7\,sq$ miles) are designated SSSI No. 27 (Port Foster). Both designations facilitate long-term research on biological recolonization following the volcanic eruptions.

Deception Island (Base B).
$62°59'S$, $60°34'W$. British research station at Whalers Bay, Deception Island, South Shetland Islands. The station, the first to be opened by British Naval Operation Tabarin 1943–45 (later Falkland Islands Dependencies Survey and British Antarctic Survey), was established in February 1944 in barracks of the old whaling buildings. It transferred from one (Bleak House) to a second (Biscoe House) in September 1946 after a disastrous fire. The station was evacuated in December 1967 during a volcanic eruption, and again, finally, in 1969 after further eruptions.

De Gerlache de Gomery, Baron Adrian Victor.
(1866–1934). Belgian polar explorer. Born into a wealthy family in Hasselt, Belgium, he joined the Royal Belgian Navy, taking particular interest in scientific research. In 1895 he began planning, organizing and raising funds for a southern polar expedition, eventually leading the **Belgian Antarctic Expedition 1897–99**. Beset by pack ice in the Bellingshausen Sea, this became the first to winter south of the Antarctic Circle. After the expedition de Gerlache led further scientific expeditions to the Persian Gulf, Greenland and northern Siberia. He did not return to Antarctica, but maintained polar interests and helped others, including J.-B. Charcot and Sir Ernest Shackleton, with preparations for polar work.

De Haven, Edwin Jesse.
(1816–65). American naval officer and explorer. Born in Philadelphia, he entered the US Navy aged 13. As Acting Master in 1839 he joined Lt. Charles **Wilkes** in USS *Vincennes*, flagship of the **United States Exploring Expedition 1838–42**, in a voyage that took him to both polar regions, and in which he navigated a considerable length of the coastline of East Antarctica. De Haven Glacier, Porpoise Bay, Banzare Coast, is named for him. After service with the fleet in the Mexican wars, and ashore at the US Naval Observatory, De Haven was commissioned in 1850–51 to lead an expedition of two ships to the Canadian Arctic in search of Sir John Franklin's missing expedition. He found no traces of the expedition, but spent a winter beset by pack ice, drifting for over 1000 miles (1600 km) before breaking free and returning to New York. De Haven left the navy due to ill health in 1862 and died three years later.

Dell, James William.
(d. 1968). British seaman and explorer. Joining the Royal Navy in 1895, he served with the **British National Antarctic Expedition 1901–4** as

a crewman aboard *Discovery*. He used sailmakers' skills in producing canvas ration bags and harnesses for men and dogs, and took part in several sledging journeys. He served with the navy throughout World War I, retiring as a chief petty officer in 1921 in time to join the **Shackleton-Rowett Antarctic Expedition 1921–22** as boatswain and electrician. Later he was employed as resident engineer and electrician on a country estate, becoming a founder member of the Antarctic Club. He died on 21 January 1968, aged 87. (CH)

Detaille Island (Base W). 66°52′S, 66°30′W. British research station in Lallemand Fjord, Loubet Coast, Graham Land. The station was built in February 1956 and operated for three years for local survey, geology and meteorology. It was closed in March 1959. The hut remains in good order. A satellite refuge hut installed at Orford Cliff, on the mainland nearby, was removed in 1997.

Detroit Plateau. 64°10′S, 60°00′W. Extensive ice plateau of central Graham Land, 1500–1800 m. high, observed in a flight of December 1928 by Sir Hubert Wilkins and named for the Detroit Aviation Society.

Dickason, Harry. (d. 1943). British polar seaman. A Royal Navy petty officer, he joined the British Antarctic (*Terra Nova*) Expedition, serving with the Northern Party that wintered first at Cape Adare, then in a snow cave on Inexpressible Island.

Diego Ramirez, Islas. 66°30′S, 68°45′W. Part of Chile, these are the southernmost islands of South America: a group of two main islands and several outlying rocks, 56 km (35 miles) southwest of Cape Horn, standing on the edge of the continental shelf. Biologically akin to the coastal islands of southwestern Chile, they are ringed by steep cliffs, with rolling tussock-covered moorlands. There is no resident human population: they are the breeding ground of many sea birds, including Magellanic penguins, black-browed albatrosses and smaller petrels.

Dion Islands. 67°52′S, 68°43′W. Group of small, rocky, low-lying islands 14 km (8.7 miles) south of Adelaide Island in Marguerite Bay. Discovered and named during the Second **French Antarctic Expedition (*Pourquoi Pas?*) 1908–10**, they were resurveyed in 1948 and found to accommodate the only-known colony of emperor penguins on the west coast of Antarctic Peninsula. An area approximately 6 km^2 (2.3 sq miles) including all the islands and channels, has been designated SPA No. 8 (Dion Islands) to protect the colony.

Discovery Bay. 62°29′S, 59°43′W. Bay on the northern coast of Greenwich Island, South Shetland Islands, Guesalaga Peninsula, within the bay, is the site of the Chilean permanent research station Arturo Prat. The bay includes two small areas of benthic habitat, of total area 0.8 km^2 (0.3 sq miles), that have been designated SSSI No. 26. (Chile Bay (Discovery Bay)) to protect long-term research.

Discovery Investigations (1925–51). A long-term series of oceanographic cruises, coupled with inshore hydrographic surveys and laboratory-based studies, providing information on which could be based the rational management of the whaling and sealing industries. Following the development of Antarctic whaling in South Georgia from 1904, and its rapid expansion through the maritime Antarctic region, the Government of the Falkland Islands sought to control the industry through licensing. It recognized, however, that too little was known of the biology of whales or seals for controls to be effective. An Interdepartmental Committee, established in 1918, was required to consider what scientific investigations and other steps were needed to preserve the whaling industry and develop other industries in the Falkland Islands and Dependencies. Reporting to Parliament in April 1920, the committee recommended that a thorough hydrographic and oceanographic survey of the area be undertaken, coupled with both ship-borne and land-based investigations of whale biology. The recommendations were accepted. RRS *Discovery*, Capt. R. F. Scott's first expedition ship, was commissioned and refitted for hydrographic cruising. A scientific group, the Discovery Committee, was set up to develop the research programme, which became Discovery Investigations under the leadership of Dr Stanley W. Kemp.

In January 1925 Discovery House, a laboratory with living quarters, was set up on King Edward Point, South Georgia. This accommodated scientists and technicians who worked on whale carcasses that were being processed at nearby Grytviken whaling station. In September 1925 RRS *Discovery* (Capt. J. R. Stenhouse) sailed from Dartmouth for South Georgia, arriving on 20 February 1926 and immediately started a programme of plankton and water sampling around the island. The first cruise ended in May, with *Discovery*'s departure for the Falkland Islands and Simonstown, South Africa. There she was joined by RRS *William Scoresby* (Capt. G. M. Mercer), a smaller and faster steam trawler specially commissioned for whale marking. *Discovery* returned to South Georgia in early November 1926, where she and *William Scoresby* made extensive surveys of the whaling grounds. In February 1927 the work was extended to the South Orkney and South Shetland Islands and Antarctic Peninsula. The end of this cruise marked the end of the first two years' research, the decommissioning of *Discovery* and closure of Discovery House.

Results from the investigations fully justified a continuing programme, though it was realized that a ship far hardier than *Discovery* would be needed. The Committee commissioned a successor, *Discovery II*, a purpose-built oil-fired research vessel of 1036 tons, steel-hulled but reinforced for work in ice, with winches and laboratory space for oceanographic research. During 1928–30 a naval hydrographic unit led by Cdr John M. Chaplin RN, with the motor launch *Alert*, charted the harbours and coasts of South Georgia. RRS *Discovery II* began work in late 1929, and for the next ten years operated in parallel with *William Scoresby*, in separate but linked cruises that covered much of the southern oceans. *Discovery II*'s first commission, December 1929 to June 1931, commanded by Capt. William M. Carey with Dr Kemp as chief scientist, explored mainly in the Dependencies but extended also to the southern Atlantic Ocean and Bellingshausen Sea. The second commission, October 1931 to May 1933 (Capt. W. M. Carey, Dr D. Dilwyn John) included work in the Dependencies but also involved more extensive cruises in the Southern Ocean and a complete circumnavigation. The third commission, October 1933 to June 1935 (Capt. Andrew L. Nelson, Dr Neil A. Mackintosh) studied seasonal variations along the meridian 80°W and extensive cruises in the Atlantic and Pacific sectors of the Southern Ocean, reaching 71°25′S in pack ice off the Ruppert Coast. During the fourth commission, October 1935 to May 1937 (Capt. Leonard C. Hill, Dr George E. R. Deacon), *Discovery II* was diverted from a circumnavigation in December 1935 to visit the Bay of Whales, searching for missing aviators Lincoln **Ellsworth** and Herbert Hollick-Kenyon. She explored in the Ross Sea, then completed oceanographic surveys in the Indian and Atlantic Ocean sectors and around the Dependencies. The fifth commission, October 1937 to May 1939 (Capt. L. C. Hill, Dr N. A. Mackintosh and Dr Henry F. P. Herdman), included research along the edge of the pack ice in the Indian and Pacific Ocean sectors, research in the northern Ross and Weddell Seas, and visits to South Georgia and Bouvetøya. Intensive surveys in waters south of South Africa took the ship to within sight of Dronning Maud Land. *Discovery II* was diverted to other activities during World War II, but in 1950–51 made a sixth and final cruise (Capt. J. F. Blackburn, Dr H. F. P. Herdman) under management by the Institute of Oceanography, effectively rounding off the work of Discovery Investigations.

Meanwhile *William Scoresby* undertook similar commissions involving trawling, dredging and whale marking. The first commission of 1926–27 (Capt. G. M. Mercer) involved whale marking off South Georgia and trawling off the Falkland Islands. The second commission of 1927–30 (Capt. H. de G. Lamotte) devoted the first season to oceanography and hydrography in the Dependencies, and the second (Capt. R. L. V. Shannon) to supporting Sir Hubert **Wilkins**'s continuing explorations of West Antarctica by air. The third commission of 1930–32 (Capt. J. J. C. Irving, Capt. T. A. Joliffe) involved whale marking around South Georgia and the Weddell Sea, oceanographic research off Peru and trawling on Burdwood Bank. The fourth commission of 1934–35 (Capt. Claude R. U. Boothby) was devoted largely to whale marking off South Africa. The fifth commission of 1935–36 and the sixth of 1936–37, both under the command of Capt. Boothby, began as whale marking voyages in the southern Indian Ocean sector but also headed south toward the continent: the fifth charted the Lars Christensen and Mawson coasts of East Antarctica from Mackenzie Bay to Edward VIII Gulf, and penetrated to Enderby Land. The seventh commission of 1937–38 (Capt. Ronald C. Freaker) was a whale marking voyage in the Scotia and Bellingshausen Seas. In 1939 *William Scoresby* was diverted to war work, later returning to Antarctica as part of **Operation Tabarin**.

Itineraries and scientific results of Discovery Investigations cruises appear in the extensive series of Discovery Reports, which were published irregularly throughout the period of investigations. In 1951 Discovery Investigations was incorporated into the Institute of Oceanography.

Further reading: Hardy (1967); Ommanney (1938, 1971).

Discovery Sound. 64°31′S, 63°01′W. Narrow channel off Briggs Peninsula, Anvers Island, Palmer Archipelago, discovered by and named for RRS *Discovery* in 1927.

Diving in birds and mammals. Though the diving abilities of penguins, seals and whales have long been recognized, until recently little was known of depths and durations of dives, or of the physiological mechanisms involved. Recent field research in the Antarctic region has shown that such small penguins as gentoos, chinstraps and macaronis weighing 4–6 kg can reach depths of 100 m, staying submerged for 2–5 min. Larger king and emperor penguins weighing 12–35 kg often dive to 50–100 m, occasionally to 300–500 m, and can submerge respectively for 8–15 min. For every 10 metres of descent, hydrostatic pressure on a diving animal's body increases by one atmosphere. If air remains in the lungs, pressure forces its gases, including nitrogen, the main constituent, into solution in the blood, and nitrogen poisoning may result. In a rapid ascent the gases bubble out into the tissues, causing in humans the painful condition 'the bends'. Seals avoid these problems mainly by exhaling before diving, reducing the air in the lungs and allowing the air passages to collapse. Whales retain air in their lungs during diving, but the quantity of nitrogen absorbed by an animal as large as a whale is unlikely to cause harmful effects. It is not clear how the lungs and air sacs of emperor and king penguins cope with deep dives.

Antarctic fur seals seldom dive deeper than 100 m or stay down longer than 3 min. In contrast Weddell seals often dive to 400 m and for 20–25 min: they can reach 600 m and last up to 70 min. submerged. Southern elephant seals are the most accomplished divers of all pinnipeds. During trips to sea lasting several months they are submerged for 90 per cent of the time, periods at the surface rarely exceeding ten minutes. Dives are characteristically deep and long, to recorded maxima of over 2000 m and 120 minutes. Sperm whales have been tracked on echo sounders to 2250 m (though with dive recorders only to 1200 m), and dive durations of at least one hour have been recorded. Baleen whales, by contrast, are surface-living creatures, seldom diving deeper than 170 m or for longer than 17 minutes. In deep-diving species, bouts of intensive diving or long deep dives are succeeded by lengthy rest periods, or tend to occur at reduced frequency with longer surface intervals.

Diving ability depends mainly on the amount of oxygen capable of being stored in tissues, and by responses to the effect of pressure at depth. Penguins, seals and whales have larger body oxygen stores than non-diving birds and mammals of similar size. Weddell seal blood has higher haemoglobin concentrations than human blood and carries 1.6 times more oxygen. Weight for weight, the blood volume is double that in man, and the myoglobin concentration in muscle about ten times greater. During dives, blood flow to many internal organs and muscles, though not to the brain, is greatly reduced, with the result that, in Adélie penguins a body oxygen store lasting 2–3 minutes at resting rates ashore would last 5–6 minutes under water. Marine mammals can extract oxygen from blood at lower gas tensions than terrestrial mammals; i.e. more of the stored oxygen can be used.

Further reading: Schmidt-Nielsen (1997); Kooyman (1989)
(JPC)

Diving petrels. Oceanic flying birds of the order Procellariiformes, family Pelecanoididae, of which two species are prevalent in southern oceanic waters. The smaller petrels, with body length about 20 cm (8 in) and wingspan 32 cm (12.5 in), skim over the sea surface with rapid wing beats, diving into and emerging through wave crests in search of the small crustaceans on which they prey. South Georgian diving petrels *Pelecanoides georgicus* are distinguished from very similar common diving petrels *Pelecanoides urinatrix* by white wing bars and slightly heavier bills. Both species breed on islands in the sub-Antarctic and cool temperate zones: *P. urinatrix* has a wider range, a subspecies *exsul* extending to the warm temperate zone and beyond.

Dobrowolski, Antoni Bolesaw. (1872–1954). Polish polar meteorologist, hydrologist and glaciologist. Educated in physical sciences at the universities of Zürich and Liège, he joined the **Belgian Antarctic Expedition 1897–99** as assistant meteorologist, making particular studies of cloud formations and atmospheric ice. These became the focal points of his later studies, on which he published extensively during a distinguished academic career. Dobrowolski Island, close to Anvers Island, Palmer Archipelago, is named for him. He died on 27 April 1954.
(CH/BS)

Dobson unit. Unit of measurement of the concentration of ozone in the atmospheric column, expressed in milli-atmo-centimetres (see **Ozone hole, Antarctic**). The name is derived from the Dobson spectrophotometer, the instrument used to monitor density of ozone in the atmosphere. Employing a photomultiplier and optical wedge, the spectrophotometer compares intensities of incident ultraviolet radiation at two pairs of frequencies within the band $0.30\,\mu m - 0.33\,\mu m$, one of which is absorbed strongly by ozone, the other only weakly. An increase in the proportion of the strongly-absorbed frequency denotes depletion of stratospheric ozone. The amount of ozone present is expressed in Dobson units: 100 DU is equivalent to a layer of ozone 1 mm thick at sea level.

Dome Fuji Station. 77°30′S, 37°30′E. Japanese station on peak of Valkyrjedomen, an ice dome over 3600 m (11 808 ft) high on the icecap of southeastern Dronning Maud Land. Established in February 1995, it is used intermittently for summer parties.

Dominion Range. 85°20′S, 166°30′E. A range at the northern end of Queen Maud Mountains, flanked by Beardmore and Mill glaciers. It was named by Shackleton for the Dominion of New Zealand, which strongly supported his **British Antarctic (*Nimrod*) Expedition 1907–9**.

Doorly, Gerald Stokely. (1880–1956). British expedition seaman. Born in Trinidad, in 1895 he joined HMS *Worcester* to train for the merchant navy. In 1902 he became third officer in *Morning*, making two voyages south to relieve *Discovery*, the ship of the **British National Antarctic Expedition 1901–4** that was beset by ice in McMurdo Sound. Settling in New Zealand, he continued in the merchant navy, achieving command and serving throughout World War I. In 1925 he moved to Melbourne, Australia, becoming a Port Phillip pilot. He returned to New Zealand in 1951 and died in Dunedin on 3 November 1956.

Dotson Ice Shelf. 74°24′S, 112°22′W. Ice shelf between Bear Peninsula and Martin Peninsula, Walgreen Coast, Marie Byrd Land. The shelf is named for. Lt. W. A. Dotson, USN, an officer who specialized in ice reconnaissance.

Douglas, Eric. (1902–70). Australian polar aviator. Born in Victoria on 6 December 1902, he joined the Australian Flying Corps as an air mechanic, qualifying as a pilot, Royal Australian Air Force, in 1927 and receiving his commission in 1929. With Flt-Lt S. Campbell RAAF, he joined the **British, Australian, and New Zealand Antarctic Research Expedition 1929–31**, flying a Gipsy Moth floatplane over the coast of East Antarctica. In 1935 he returned to Antarctica to search for the missing aviators Lincoln Ellsworth and Herbert Hollick-Kenyon. He retired in 1948 with the rank of group captain, and died on 4 August 1970.

Douglas, George Vibert. (1892–1958). Canadian geologist who served with the **Shackleton-Rowett Expedition 1921–22**. Born in Montreal on 2 July 1892, he qualified as a geologist before serving in World War I. His work with the expedition included a survey of the geology of South Georgia. Later he lectured at Harvard University, and in 1932 became Professor of Geology at Dalhousie University. He died in October 1958.

Doumer Island. 64°51′S, 63°34′W. A small island between Anvers and Wienke Islands, Palmer Archipelago. Charted by J.-B. Charcot in 1904, it was named for Paul Doumer, a prominent French statesman. It accommodates the Chilean refuge **Yelcho**. An area of about 1 km^2 (0.4 sq miles) of coastal and sub-tidal sea floor on the southern end of the island has been designated SSSI No. 28 (Doumer Island, South Point), to facilitate long-term marine ecological studies.

Dovers, George Harris Sargeant. (d. 1971). Australian polar surveyor. A surveyor in government service, he joined the **Australasian Antarctic Expedition 1911–14**, and was assigned to the western base. He took an active part in the sledging programme, with two companions forming the Western Coastal Party that explored the coast westward to Gaussberg and discovered Haswell Island. After the expedition Dovers returned to surveying. He died on 7 July 1971, aged 84.
(CH)

Drake, Francis. (*c*. 1540–1596). English navigator and explorer. Born of a poor agricultural family in Devonshire, he went to sea aged 13. After apprenticeship in the North Sea, he joined the fleet of John Hawkins, quickly rising to command. In 1572, licensed by Queen Elizabeth as a privateer, he led a successful expedition to Panama. From December 1577 to September 1580, in *Pelican* (later renamed *Golden Hind*), he led a fleet of five ships on an expedition to the Pacific Ocean. Traversing the Strait of Magellan in August 1578, he was carried southward by contrary winds to discover open water south of Tierra del Fuego (now called **Drake Passage**). Drake sacked Spanish ports in South America, explored north to the latitude of Vancouver and returned to England across the Pacific and Indian Oceans – the second world circumnavigation. He made several later voyages, and in 1588 played a prominent role in defeating the Spanish Armada. He died at sea off Portobello, Panama.

Drake Passage. Sea passage 1000 km (625 miles) wide between southern Tierra del Fuego and the South Shetland Islands, dominated by strong westerly winds and the powerful eastward-flowing Antarctic Circumpolar Current. Discovered by Francis **Drake** in 1578, it was first navigated by Willem **Schouten** in 1615. Cape Horn forms a prominent landmark on its northern flank. With the Strait of Magellan, it was an important sea route between the Atlantic and Pacific Oceans throughout the late eighteenth, nineteenth and early twentieth centuries, until the opening of the Panama Canal in 1914.

Dronning Fabiolafjella. 71°30′S, 35°40′E. (Queen Fabiola Mountains). Easternmost range of mountains in Dronning Maud Land, rising to 2400 m (7872 ft) from a high ice plateau. They were discovered in 1960 during a Belgian reconnaissance flight, and named for a newly-married Queen of Belgium.

Dronning Maud Land. 76°13′S, 13°00′E. (Queen Maud Land). Wide sector of East Antarctica south of the Indian Ocean between latitudes 20°W and 45°E, bounded in the west by Coats Land and in the east by Enderby Land. It includes five ice-fronted coasts named for Kronprinsesse Märtha, Prinsesse Astrid, Prinsesse Ragnhild, Prins Harald and Prins Olav, and several major ranges of ice-mantled mountains, surrounded by icy plains that rise gently toward the South Pole. The land subsequently designated Dronning Maud Land was first seen on 15 and 16 January 1930, during flights by Capt. Hj. Riiser-Larsen from *Norvegia*, the research ship of the early **Norwegian (Christensen) Whaling Expeditions**. The name honoured Queen Maud, wife of the reigning Norwegian sovereign. Following further discoveries by Christensen and other Norwegian whalers, the whole sector between existing British claims was brought under Norwegian sovereignty by a royal decree of 14 January 1939. The wording of the decree specified the mainland coast and land lying with the coast: a government map published simultaneously illustrated a triangular sector extending from latitude 50°S to the South Pole. The inland ice and mountains of western Dronning Maud Land were overflown by photo-reconnaissance aircraft of the **German Antarctic (*Schwabenland*) Expedition 1938–39**, and explored from air and ground by the **Norwegian-British-Swedish Antarctic Expedition 1949–52**. Much of the rest has been explored extensively by Belgian, South African and Japanese expeditions from coastal stations.

Druzhnaya stations. A series of Soviet and Russian ice shelf research stations. Druzhnaya I (77°34'S, 40°13'W), was established in 1976 on the Filchner Ice Shelf. The station was lost in 1986–87 when the ice shelf calved. Druzhnaya II (75°36'S, 57°52'W) was established in 1980 on the Ronne Ice Shelf, Druzhnaya III in 1987 near Kap Norvegia, Kronprinsesse Märtha Kyst, and Druzhnaya IV in 1987 on Amery Ice Shelf. All were used mainly for summer glaciological research.

Drygalski, Erich von. (1865–1949). German geophysicist and explorer. Born on 9 February 1865 at Königsberg, he studied the natural sciences at Bonn, Leipzig and Berlin, and at the Geodätisches Institut in Potsdam. Becoming interested in glaciers and their movements, he led an expedition of the Berlin Gesellschaft für Erdekunde to West Greenland in 1891–93. He was appointed Professor of Geography and Geophysics at Berlin, and in 1900 was invited to lead the **German South Polar Expedition 1901–3**. The expedition ship *Gauss* spent an enforced winter in the pack ice 95 km (60 miles) off the Wilhelm II Coast. Expedition members were confined to small-scale local studies of sea ice, marine biology and the nunatak Gaussberg, their nearest point of land. Nevertheless Drygalski produced a general account of the expedition and an impressive series of scientific reports, of which the last appeared in 1931. In 1906 he became Ordinarius für Geographie at Munich, holding the appointment until he retired in 1934. He took part in expeditions to Svalbard and Siberia, studying glaciers and periglacial features. He died in Munich on 10 January 1949.

Drygalski Fjord. 54°49'S, 36°00'W. Narrow spectacular fjord in the southeastern end of South Georgia, that was used by whale catchers as a mooring and refuge in stormy weather. It was charted by the **German Antarctic (*Deutschland*) Expedition 1911–13** and named for Prof. **E. von Drygalski**, the German scientist and explorer who had led the earlier **German South Polar Expedition 1901–3**.

Ducks and geese, southern oceanic. A few species of waterfowl (family Anatidae) have reached the southern oceanic islands and become endemic, feeding in fresh water along the shore and occasionally inland. Coastal waters of the Falkland Islands support two species of maritime steamerducks, the endemic flightless species *Tachyeres brachypterus*, and the volant *T. patachonicus*, which occurs also on South American coasts. South Georgia's pintail *Anas georgica*, an endemic island subspecies, feeds intertidally and nests mainly on offlying islets. A similar species, Eaton's pintail *A. eatoni*, is endemic to Iles Kerguelen and Crozet. The speckled teal *A. flavirostris* of South America and the Falkland Islands has recently become an established breeder on South Georgia. Pacific Black ducks *A. superciliosa* of southern Australia and New Zealand are resident on the Auckland Islands, Campbell Island and Macquarie Island.

Dufek, George J. (1903–77). US naval officer and Antarctic explorer. Born on 10th February 1903 in Rockford, Illinois, he entered the Naval Academy in 1921 and was commissioned in 1925, training both in submarines and aviation. He first visited Antarctica with the **United States Antarctic Service Expedition 1939–41** as navigating officer aboard USS *Bear*. After active service in World War II and the Korean War, he returned to Antarctica in command of the eastern task force of **United States Navy Antarctic Development Project 1946–47 (Operation Highjump)**. In 1954–59 he commanded the US Naval Support Force **(Operation Deep-Freeze)** that established McMurdo, Amundsen-Scott, Little America V, Byrd, Wilkes, Hallett and Ellsworth stations. In October 1956 he became the first American aviator to land at the South Pole. Rear Admiral Dufek retired in 1959 and died on 10th February 1977.

Dufek Coast. 84°30'S, 179°00'W. Part of the coast bordering the Transantarctic Mountains, East Antarctica, facing the Ross Ice Shelf and extending from Liv Glacier in the east to Airdrop Peak, the eastern portal of Beardmore Glacier, in the west. The coast consists of steep mountains rising inland to the polar plateau, interspersed with glaciers

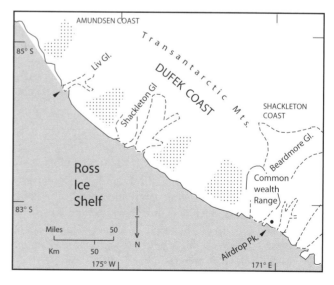

Dufek Coast

that descend and merge into the ice shelf. Dufek Coast was defined and named for Rear Admiral George J. **Dufek** USN.

Dufek Massif. 82°36′S, 52°30′W. Northeastern massif of the Pensacola Mountains, ice-covered and rising to over 1800 m (5904 ft). It was explored by a traverse party from Ellsworth Station in 1957–58, and named for Rear Admiral G. J. **Dufek**. An area of approximately 390 km^2 (28 sq miles), between Brown Nunataks, Cox Nunatak and Forlidas Ridge, is designated Specially Reserved Area No. 1 (North Dufek Massif) for its outstanding geological, geomorphological, aesthetic, scenic, and wilderness values, which management plans will seek to protect while allowing access and multiple use to scientists and other visitors. The reserved area includes SPA No. 23 (Forlidas and Davis Valley ponds).

Dumont d'Urville, Jules Sébastien César. (1790–1842). French naval officer, navigator and explorer; leader of the **French Naval Expedition 1837–40**. Born in Normandy to a patrician but impoverished family, Dumont d'Urville entered the French navy at the age of 17. Gaining an education in classics and natural history as well as maritime subjects, he was singled out for expeditions involving survey and exploration. During his first expedition, to the eastern Mediterranean and Black Seas, he discovered on the Greek Island of Milos the statue of Venus that has since graced the Paris Louvre. In 1826 he was given command of a three-year scientific expedition to explore and chart the coasts of Australia, New Zealand and the western Pacific Ocean. In 1837–40 he led a further scientific expedition to explore the south Pacific islands and the Southern Ocean, with mapping magnetic variation an important objective. With two corvettes, *Astrolabe* and *Zélée*, he sailed south to the South American sector, where in February 1838 he discovered 'Terre Louis-Phillippe' and 'Terre Joinville', now known to be large islands off the tip of the Antarctic Peninsula. After wintering in Indonesia and the western Pacific Ocean, he revictualled in Hobart. In early January 1840 *Astrolabe* and *Zélée* again headed out to sea, this time southeastward toward the assumed position of the Magnetic Pole. Though unable to reach the pole, he discovered and claimed for France a sector of the Antarctic mainland coast, that he named Terre Adélie for his wife. Dumont d'Urville returned to France in November 1840, where he was fêted, promoted to rear admiral, and returned to naval duties. In May 1842 he was killed, together with his wife and son, in one of France's earliest railway accidents. A 23-volume account of his travels, *Voyages vers la Pôle Sud et Oceania*, appeared at intervals between 1841 and 1854.

Dumont d'Urville Sea. 65°S, 140°E. (Mer Dumont d'Urville). Coastal sea off East Antarctica, bordered by latitudes 132°E and 136°E, the latitudinal limits of Terre Adélie, East Antarctica. There is no definitive northern boundary. Fast ice and pack ice extend off the coast in winter: in summer pack ice and bergs move westward, driven by easterly currents and offshore winds. The name commemorates the leader of the **French Naval Expedition 1837–40**, who was the first to explore this stretch of the East Antarctica coast.

Dumont d'Urville Station. 66°40′S, 140°01′E. French permanent year-round station on Ile de Pétrels, Adélie Coast, East Antarctica. Established in 1956, it replaced a smaller station built in 1952, which in turn replaced an earlier station built in 1950 at nearby Port Martin.

Dundee Whaling Expedition 1892–93. In early September 1892 four steam whaling ships from Dundee, *Balaena* (400 tons; Capt. A. Fairweather), *Active* (340 tons; Capt. T. Robertson), *Diana* (340 tons; Capt. R. Davidson) and *Polar Star* (216 tons; Capt. J. Davidson), sailed south to investigate possibilities for whaling in the South American sector of the Southern Ocean. Financed privately,

Jules Sébastien César Dumont d'Urville (1790–1842). Photo: Scott Polar Research Institute

mainly by Scottish ship owner Robert Kinnes, the ships carried scientific equipment on loan from the Royal Scottish Geographical Society and the Meteorological Office, in the charge of two surgeon-naturalists, William Bruce and C. W. Donald. Also aboard *Balaena* was a marine artist, William Burn Murdoch. The fleet left Dundee in early September 1892, reaching the Falkland Islands independently in early December and departing quickly for the south. Through late December and January they explored waters north of the Weddell Sea, searching particularly for right whales, earlier reported by Ross and other explorers, that would yield valuable baleen. Encountering only rorquals which, even if caught, would yield only poor baleen, they contented themselves with an easier harvest of 16 000 seal skins and blubber, secured mainly from the pack ice. Capt. Robertson in *Active* searched Erebus and Terror Gulf and neighbouring waters, discovering and landing on Dundee Island. On 24 January 1893 the ships met the Norwegian whaler *Jason* (Capt. C. A. Larsen: see **Norwegian (Sandefjord) Whaling Expeditions 1892–94**). The surgeon-naturalists were disappointed in finding few opportunities for landings or research, and the sponsors of the expedition, unlike their Norwegian counterparts, saw no reason to repeat the experiment. However, William Bruce gained the knowledge and experience that encouraged him, a few years later, to lead his own **Scottish National Antarctic Expedition 1902–4**. *Further reading*: Murdoch (1894); Bruce (1896); Donald (1896).

Dusky dolphin. *Lagorhynchus obscurus.* Slender, elegant dolphins up to 2 m (6.5 ft) long, dusky dolphins are found only in warm or cooler southern waters, commonly off South Africa, New Zealand, and southern coasts of Australia and South America. They have been recorded off the Falkland Islands, Iles Kerguelen and the cool temperate islands south of New Zealand, and, indeed, are widespread in southern waters north of the Antarctic Convergence. They appear to move south toward the Convergence in summer and north into warmer waters in winter.

Distinguishing features are the white belly, grey and white stripes extending from the forehead almost to the tail flukes, tall recurved dorsal fin (sometimes with pale trailing edge), and long, pointed flippers. They move in groups of up to 20, occasionally more, which are often attracted to ships or small boats. Many observers have recorded spectacular leaps and acrobatics: indeed, this is how schools of dusky dolphins are first located. They feed on fish, locating schools and gathering round them, often in association with flocks of sea birds. Calves are seen with their mothers from mid-winter onward.

Dyer Plateau. 70°30′S, 65°00′W. Ice plateau of central Palmer Land, named for American surveyor J. Glenn Dyer, leader of the United States Antarctic Service Expedition sledging party that discovered it in 1940.

E

Earthquakes and earth movements. The single crustal plate that carries East Antarctica, and the complex of micro-plates that carry West Antarctica, show very little current tectonic activity. Seismological observatories operating in Antarctica since the 1950s have recorded only minor quakes on the continent and neighbouring islands, for example **Deception Island**, South Shetland Islands, and **Ross Island**, Victoria Land. In both, earth movements are usually associated with overt volcanic activity. By contrast the semi-continuous submarine ridges between the plates, on the Southern Ocean floor, are sites of intense, continuing volcanic activity and earthquakes: indeed, their positions can be determined from seismological records of earthquake epicentres. The **Scotia Arc** provides several current examples of intense plate activity. Meteor Deep, for example, on the convex side of the **South Sandwich Islands**, is an area of continuing subduction. Elsewhere, molten rock extruded from the ridges spreads symmetrically on either side.
(RJA)

East Antarctica. The larger of Antarctica's two provinces, lying mainly within the eastern hemisphere. Presenting a huge, complex ice dome, East Antarctica rises from ice shelves and ice-covered coastal plains to a high plateau exceeding 4000 m (13 100 ft). Along its eastern flank run the Transantarctic Mountains, a range some 300 km (187 miles) wide and 4800 km (3000 miles) long. Representing the uplifted edge of an underlying continental bloc, this forms a massive dam that holds back the ice of the high plateau, penetrated by glaciers that flow into the Ross and Ronne-Filchner ice shelves. Toward the Indian and Pacific Ocean coasts are several minor ice domes and lower plateaux, many penetrated by mountain ranges. Other ranges are exposed in other coastal areas of the province. Radar studies have revealed at least one very large mountain range, the **Gamburtsev Subglacial Mountains**, and many lesser ranges underlying Antarctica's high ice dome.

Geological studies from many scattered localities reveal that East Antarctica consists of a basement complex of ancient gneisses, schists and other metamorphic rocks, with massive intrusions. These are overlain by sediments of Lower Cambrian to Permian age, reflecting a complex climatic history from glacial conditions to tropical forests and deserts. Structurally the province shows common origins with South America, South Africa, India and Australia, with all of which it was for long united in the Gondwana supercontinent. Evidence from mid-oceanic ridges surrounding Antarctica indicates that the supercontinent began to fragment some 150 million years ago, allowing East Antarctica to drift gradually to its present polar position, where it acquired its ice cap. East Antarctica includes the sites of the southern **Geographical pole**, **Geomagnetic pole** and **Pole of inaccessibility**. The southern **Magnetic Pole** has migrated across it in recent decades from Victoria Land to Terre Adélie and is currently located offshore.

Edholm, O.G. (1910–85). British polar physiologist. With strong research interests in human performance under climatic stress, Edholm was from 1949 based in Hampstead, London, directing the Division of Human Physiology of the British Medical Research Council. He advised many polar expeditions, and wrote several books of polar interest, notably *Man in a Cold Environment* (with A. Burton, 1955) and *Man – Hot and Cold* (1978).
Further reading: Burton and Edholm (1955); Edholm (1978).

Edward VII Peninsula. 77°40′S, 155°00′W. Extensive peninsula, mountainous and ice-covered, that forms the northwestern corner of Marie Byrd Land. Including parts of the Shirase and Saunders coasts, and abutting onto the Ross Ice Shelf, it was discovered by the British National Antarctic Expedition and named Edward VII Land for the contemporary British monarch.

Edward VIII Plateau. 66°35′S, 56°50′E. Ice-covered peninsula close to Edward VIII Bay, Kemp Coast, East Antarctica. It was identified by Norwegian and British cartographers in the mid-1930s and named for the contemporary British monarch.

Eielson Peninsula. 70°35'S, 61°45'W. A peninsula of the Wilkins Coast, eastern Palmer Land, named for the American aviator Carl B. Eielson who in 1928 overflew the area with Sir Hubert Wilkins.

Eights, James. (1798–1882). American physician, artist and naturalist. Born in Albany, New York, his early life is obscure. He was appointed naturalist on the United States Southern Exploring Expedition 1829–30, a government-supported sealing and scientific expedition that visited Tierra del Fuego, Staten Island and the South Shetland Islands. He appears to have sailed in *Annawan* (Capt. N. B. Palmer). In the period 1833–52 he published five papers recording observations from the expedition, including illustrated first descriptions of the marine isopod *Glyptonotus antarctica*, the malacostracan crustacean *Brongniartia trilobitoides*, and the pycnogonid *Decolopoda australis*, all elegantly illustrated by the author. He also wrote a treatise on icebergs and their origins from glaciers, and a geological and general description of the South Shetland Islands, with notes on the plants, mammals, birds and invertebrates. Eights applied to join the **United States Exploring Expedition 1838–42** as a naturalist and appears to have been short-listed, but did not sail with them. He worked for a time on the Natural History Survey of New York, and wrote reports on mines of North Carolina. Little else is known of his life: he died in obscurity in Ballston, New York.
Further reading: Hedgepeth (1971).

Eights Coast. 73°30'S, 96°00'W. Part of the coast of Ellsworth Land, West Antarctica, facing the Bellingshausen Sea and extending from Cape Waite in the west to Pfrogner Point in the east. The coast is lined with the broad Abbot Ice Shelf, linking it with Thurston Island and other smaller islands, and often inaccessible because of heavy pack ice and bergs. It is named for James Eights, who served as naturalist with the United States Southern Exploring Expedition 1829–30.

Eights Station. 75°14'S, 77°10'W. Former American research station on the icecap of southwestern Palmer Land. Established as 'Sky-hi' in November 1961, it was occupied as a base for seismic land traverses and studies of glaciology and upper atmosphere physics. The station was closed in November 1965.

Eklund, Carl Robert. (1909–62). American polar biologist. On graduating in biology from Carlton College, he joined the Shenandoah National Park, Va. and later the United States Fish and Wildlife Service. In 1939 he joined the **United States Antarctic Service Expedition 1939–41**,

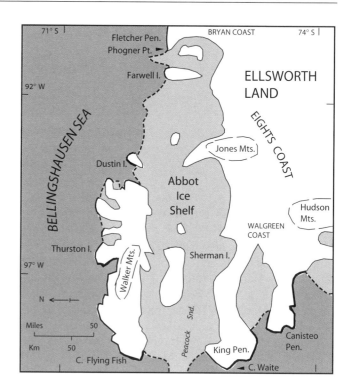

Eights Coast

serving at East base and taking part in a 1900 km (1200-mile) dog-sledging journey to the southern end of George VI Sound. The Eklund Islands, at the southern end of the Sound, were named for him. During the International Geophysical Year 1957–58 he was scientific leader at Wilkes Station. On his return he was appointed Chief of the Polar and Arctic Branch of the Army Research Office. Eklund made major contributions to international co-operation in Antarctic biology. He died on 4 November 1962.
(CH)

Eklund Islands. 73°16'S, 71°50'W. A group of small islands within the ice shelf of the southern arm of George VI Sound, off English Coast, Palmer Land. The largest rises to 410 m (1345 ft). The islands were discovered by a sledging party of the United States Antarctic Service Expedition 1939–41 and named for the expedition biologist, Carl R. Eklund.

Ekström, Bertil. A. W. (1919–51). Swedish expedition mechanic. Born on 20 October 1919 in western Sweden, he trained as a tank mechanic with the Swedish Army. Joining the **Norwegian-British-Swedish Antarctic Expedition 1949–52**, with particular responsibility for servicing the Weasels, he played a vital role in maintaining the sledging programme. He was killed in a tractor accident near Maudheim on 23 February 1951.

Ekströmisen. 71°00′S, 8°00′W. (Ekström Ice Shelf). Ice shelf on Kronprinsesse Märtha Kyst, Dronning Maud Land. The shelf includes an ice port, Atkabukta, the port of access to the German research station Georg von Neumayer. The shelf was named for Swedish mechanic B. Ekström, a member of the **Norwegian-British-Swedish Expedition 1949–52**.

Elephant Island. 61°10′S, 55°14′W. A large, rugged, ice-capped island, 38 km (24 miles) long and 19 km (12 miles) across, forming an eastern outlier of the South Shetland Islands. It was discovered and charted by Edward Bransfield RN in 1820, and named for the many southern elephant seals that graced its beaches.

Elephant seal. See **Southern elephant seal**.

Ellefsen Harbour. 60°44′S, 45°03′W. Harbour between Christoffersen Island and Michelsen Island on the south coast of Powell Island, South Orkney Islands, named in 1821 by early nineteenth-century sealers.

Ellsworth, Lincoln. (1880–1951). US aviator and explorer, whose flights over Antarctica included a first crossing of the continent. Born to a wealthy family in Chicago, Ellsworth studied engineering and survey at Columbia and Yale universities, in preparation for inheriting his father's mining interests. For several years he worked as an engineer in railway construction. He trained as an aviator in World War I, afterwards returning to mining with an expedition to the Peruvian Andes. In 1925 he gave way to a growing interest in exploration, using family money to team with Roald **Amundsen** in an unsuccessful bid to fly two Dornier Wal seaplanes across the north polar basin. In the following year he funded and took part in a flight with Amundsen and Umberto Nobile in *Norge*, an Italian-built dirigible, from Ny Ålesund, Svalbard to Teller, Alaska, successfully passing over the North Pole. In 1931 he took part in flights by *Graf Zeppelin*, a German dirigible, over Franz Josef Land. In the same year he contributed funding to an attempt by the Australian explorer George Hubert **Wilkins** to take a submarine under the Arctic pack ice. This started a lasting association: between 1933 and 1935, in association with Wilkins, he made three attempts to cross Antarctica by air (see **Ellsworth's Antarctic Expeditions 1933–39**). He succeeded in the third flight after four forced landings and a 24 km (15-mile) walk to Little America, abandoned station of the first and second Byrd expeditions. In 1938–39, again with Wilkins, he explored parts of the East Antarctica coast by air and sea. Ellsworth died in New York in 1951.
Further reading: Ellsworth (1938).

Ellsworth Land. Area of West Antarctica bounded to the west by Marie Byrd Land, to the north by Antarctic Peninsula, and to the east by Ronne Ice Shelf. Much of it is a high ice-covered plateau, from which emerge the Ellsworth, Hudson, Jones, Behrendt, Merrick, Sweeney and Scaife mountains. The land was named for US aviator Lincoln Ellsworth, who in 1935 overflew the area in his trans-continental flight. Politically, it forms part of the unclaimed sector of Antarctica.

Ellsworth Mountains. 78°45′S, 85°00′W. Narrow bloc of mountains 380 km (237 miles) long forming the south-western coast of Ellsworth Land. Discovered by US aviator Lincoln Ellsworth on 23 November 1935, they include two major ranges separated by the Minnesota Glacier system. The southern Heritage Range includes several striking peaks above 1800 m (5900 ft). The northern Sentinel Range has several peaks above 4000 m (13 120 ft), rising to the central bloc of Vinson Massif (4901 m, 16 075 ft), Antarctica's highest mountain.

Ellsworth Station. 77°43′S, 41°08′W. United States International Geophysical Year research station, established on the Filchner Ice Shelf in January 1957. From 1959 it was operated jointly with Argentina. The station was closed in December 1962.

Ellsworth's Antarctic Expeditions 1933–39. Between 1934 and 1936 Lincoln **Ellsworth** made three bids to fly across Antarctica. Teaming with Sir Hubert **Wilkins**, whom he had met in 1931, he bought as an expedition ship a Norwegian herring boat *Fanefjord* (400 tons), refitting and renaming it *Wyatt Earp*. For his first venture, he planned to fly with pilot Bernt Balchen (who had recently flown with Richard Byrd's first Antarctic expedition) from Little America, Byrd's old expedition base on the Ross Ice Shelf, to the South Shetland Islands. *Wyatt Earp* left Dunedin, New Zealand on 10 December 1933, successfully negotiating the pack ice to reach Bay of Whales in early January. Shortly after the maiden flight of *Polar Star*, the Northrop Gamma monoplane selected for the long flight, the sea ice on which it was parked overnight broke up, and the aircraft was damaged beyond repair. The expedition returned to New Zealand. In the following year Ellsworth and Wilkins determined that it would be safer to fly in the opposite direction, so *Wyatt Earp* sailed with the rebuilt Northrop Gamma to Deception Island, arriving in October 1934. The aircraft was re-assembled and prepared for a test flight. However, during the engine run-up a connecting rod snapped and caused severe damage, which took over a month to repair. By the time the engine had been stripped and rebuilt, it was late November – too late in the season to fly from Deception Island. *Wyatt Earp* embarked on a voyage, first south along Antarctic Peninsula, then into the Weddell Sea, to find a suitable take-off site. A snow-covered runway was found on the ice shelf of Snow Hill Island. The aircraft was re-assembled and test-flown, and on 3 January 1935 Ellsworth and his pilot Bernt Balchen took off for the long flight. After an hour Balchen refused to fly further, complaining that the flight was suicidal. The expedition was again aborted: *Wyatt Earp* broke with difficulty from the pack ice and returned to Deception Island.

In November 1935 Ellsworth and Wilkins returned south, this time with two Canadian pilots, Herbert Hollick-Kenyon and J. H. Lymburner, both of Canadian Airways. By good fortune they found a landing strip on the shelf ice of Dundee Island. Ellsworth and Hollick-Kenyon made a first attempt at the long flight on 20 November, but returned after an hour with minor engine trouble. On the following day they flew south beyond the mountains that Wilkins had discovered and called 'Hearst Land', discovering in 69°S a further cluster of peaks rising over 3600 m (12 000 ft) that Ellsworth called 'Eternity Mountains'. Here bad weather intervened, and the fliers returned to Dundee Island after a flight of over ten hours. On 22 November Ellsworth and Hollick-Kenyon took off for a third time, flying in good weather beyond the Eternity Range, and beyond a further range of sedimentary mountains to a wide, featureless plain. They had planned a flight of 3700 km to Little America, hoping to complete it in about 14 hours with a single stop for refuelling. After 13 hours in the air, when dead reckoning calculations indicated that they should be approaching the Ross Sea, they landed to take a sun sight. To their surprise they found themselves in 79°S, 104°W, far short of their calculated position, with some 650 miles still to go. Now out of radio contact with the ship, they could only fly on. *Polar Star* made three more landings, each enforced by poor visibility, and each resulting in delay – one of three days and one of seven – due to bad weather. Finally, they ran out of fuel some 24 km (15 miles) south of their destination. It took them a further six days to find the station, now under snow with only roofs and radio masts showing. Arriving on 15 December, they broke in and found plenty of food and fuel to sustain them. Meanwhile Wilkins in *Wyatt Earp* headed first for South America, then to the Bay of Whales. News of the missing fliers brought *Discovery II* also to the Bay: on 16 January 1936 a party from the ship walked inland to find them safe and well. *Wyatt Earp* arrived four days later. Ellsworth left in *Discovery II*, while Hollick-Kenyon and helpers returned to *Polar Star*, refuelled it and flew back to the Bay, to be taken aboard *Wyatt Earp*.

Though two of his three attempts to cross Antarctica were aborted and the third came close to disaster, Ellsworth's expeditions filled an important geographical gap, linking the southern end of the Antarctic Peninsula with the earlier discoveries of Byrd's expeditions. Both the largest mountain range and the surrounding land now bear his name.

In 1938 Ellsworth and Wilkins made a fourth expedition to Antarctica in *Wyatt Earp*, intending to explore inland by air from a point on the coast of Enderby Land to the Bay of Whales. Leaving Cape Town on 29 October, the ship encountered heavy pack ice, finally reaching Ingrid Christensen Coast on 2 January 1939. Ellsworth and Lymburner made several local flights around the Rauer Islands and Vestfold Hills, and a longer flight over an almost featureless ice cap to a point in 72°S, 79°E, where the ice surface was 1830 m (6000 ft) above sea level. The expedition reached Hobart on 4 February 1939.

Further reading: Ellsworth (1938).

Emperor Island. 67°52′S, 68°43′W. A small island of the Dion Islands group, standing at the northern entrance to Marguerite Bay. In 1948 it was found to accommodate a small colony of emperor penguins, the only colony along the western flank of Antarctic Peninsula. The island is included in Specially Protected Area No 8 (Dion Islands).

Emperor penguin. *Aptenodytes forsteri*. Largest of the living penguins, emperors stand up to 95 cm (3 ft) tall and weigh up to 45 kg (100 lb) in premoult fat. They breed exclusively on the coast of Antarctica: some 40 colonies are known, mostly in East Antarctica, ranging in size from hundreds to tens of thousands of pairs. Distinguished by their vivid orange auricular patches, narrow purple or orange bill plates, and yellow breasts, they are seldom found far from continental Antarctica. Emperors are unique in

Emperor penguins on sea ice at Cape Royds, Antarctica. Photo: BS

laying their single eggs in winter: males undertake the full incubation of about 64 days during the coldest months of June, July and early August. Parents then take turns to feed the chicks, which form creches from September, growing to about three-quarters size and moulting into juvenile plumage during December and early January. Emperors feed mainly on fish and squid, which they catch at depths down to 500 m (1640 ft). The world population is estimated at 195 000 pairs.

Enderby Land. Mountainous area of East Antarctica between 45° and 55°E, forming the westernmost sector of Australian Antarctic Territory. Bounded to the west by Dronning Maud Land and to the east by Kemp Land, it includes an extent of coast, the Nye, Scott and Napier mountains, and an extensive interior ice plateau extending to the South Pole. The coast was discovered in 1831 by British sealing captain John Biscoe and named for Enderby Bros. of London, owners of his vessel *Tula*.

Enderby, Messrs. A firm of eighteenth- and nineteenth-century ship owners with long-standing interests in whaling, sealing and exploration of the Southern Ocean. The firm was established about 1750 at Paul's Wharf, Lower Thames St., London, by Samuel and George Enderby, Americans of Quaker stock who had returned from America to Britain. After the founders died in 1829, Samuel's sons Charles, Henry and George took over, moving to Great St Helens, in the City of London. Over many decades Enderby ships made profitable voyages to both northern and southern whaling and sealing grounds. Charles, a founder member of the Royal Geographical Society, in particular encouraged their masters (including John **Biscoe,** Matthew **Brisbane,** Abraham **Bristow** and John **Balleny**) to discover new grounds and provide accurate charting. Toward the mid-century, as the southern sealing grounds became less profitable, the fortunes of the firm declined. Charles Enderby's enterprising but ill-starred **Enderby Settlement** of 1849–52, a land-based whaling station at Port Ross, Auckland Islands, overstrained the firm's resources and brought about its liquidation.

Enderby Settlement. A whaling station and colony established in 1849–52 on Main Island, Auckland Islands, by Messrs **Enderby**, the London firm of oil merchants, whalers and sealers. Inspired by recommendations from Sir James Clark **Ross**, who had spent three weeks on the islands in 1840, Charles Enderby obtained the support of the British government in establishing a settlement 'for the purpose of the whale fishery, as a station at which to discharge the cargoes and refit vessels'. The site chosen was 'Sarah's Bosom', the harbour discovered by Abraham Bristow on the north end of Enderby Island, and renamed Port Ross. Appointed Lieutenant Governor of the small colony, Charles Enderby established the subsidiary Southern Whale Fishing Company to manage the enterprise. Three ships from England, loaded with settlers and stores, arrived at Port Ross in December 1849, to find a group of about 70 Maoris, fugitives from the Chatham Island, already in residence. Land was cleared in Erebus Cove for a village of 18 prefabricated buildings, including housing and a church. Opening officially on 1 January 1850, the village was named for the Earl of Hardwicke, Governor of the company. The colonists cleared more land for farming, dockyard and harbour installations. However, the cold, damp climate and acid soils made agriculture impossible, and the eight whaling ships attached to the station caught very few whales. Dispirited colonists left for Australia and New Zealand, and the expense of maintaining Hardwicke strained the company's resources beyond limits. The Enderby settlement was dismantled and closed on 5 August 1852.
Further reading: Fraser (1986).

England, Rupert G. (d. 1942). British polar seaman. A merchant navy officer and Royal Naval Reservist, he was first officer in the relief ship *Morning* on two voyages (1902–3 and 1903–4) to relieve *Discovery*, expedition ship of the **British National Antarctic Expedition 1901–4**, in McMurdo Sound. In 1907–9 he commanded the ship of the **British Antarctic (*Nimrod*) Expedition 1907–9**, on the voyage that took Ernest Shackleton back to McMurdo Sound. Following disagreements with Shackleton, he was relieved of command on return to Lyttelton: the relief voyage of 1908–9 was under command of Capt. Frederick Pryce Evans.

English Coast. 73°30'S, 73°00'W. Part of the west coast of Palmer Land, West Antarctica, facing Bellingshausen Sea. Bounded by Rydberg Peninsula in the west

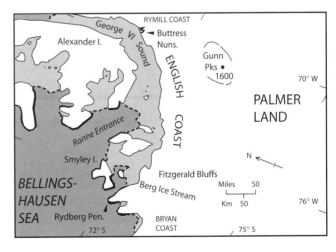

English Coast

and Buttress Nunataks in the east, the coast consists mainly of a featureless ice plain descending to the ice shelves of Strange Sound and the southern entrance to George VI Sound. It was identified by the **United States Antarctic Service Expedition 1939–41**, and named for Capt. Robert A. J. English USN, who commanded the expedition's East Base. A ground traverse party from Eights Station explored the hinterland in 1961–62.

English Strait. 62°27′S, 59°38′W. Passage between Greenwich and Robert Islands, South Shetland Islands. The name, first used in 1821, may record the presence of British sealers in the area.

Enterprise Island. 64°32′S, 62°00′W. A small, rugged island 2.5 km (1.5 miles) long in Wilhelmina Bay, lying at the NE end of Nansen Island in Wilhelmina Bay, off Danco Coast, Graham Land. Foyn Harbour, in the south side, was one of several bays and channels in the area used by whalers for mooring and anchoring factory ships during early decades of the twentieth century.

Equilibrium line. Zone on the surface of a glacier below which there is a net loss of mass (ablation), and above which there is net gain in mass (accumulation). See **Firn** line, **Snow line**.

Eratosthenes of Alexandria. (*c.* 200 BC). Alexandrian geographer whose measurement of Earth led to the concept of **Terra Australis Incognita**. Eratosthenes noted that in the town of Syene (modern Aswan, 80 km (50 miles) north of the Tropic of Cancer), the midday sun in midsummer stood vertically overhead: simultaneously in Alexandria, the sun's rays made an angle of just over 7° from the vertical. Estimating the distance between the two towns at 5000 stadia, he showed by simple proportion that the circumference of the circle on which both were standing must be about 250 000 stadia (approximately 46 000 km, 28 750 miles). This is some 15 per cent higher than we now know it to be. The main source of error was Eratosthenes's estimate of the distance between the towns, which he had no means of measuring accurately, and was about 11 per cent too high. However, it indicated a large Earth, of which the known world immediately surrounding the Mediterranean Sea was but a small fraction, and prompted other geographers to speculate on climatic zones, northern and southern polar regions, and a large southern polar continent.

Erebus, Mt. 77°32′S, 167°10′E. Volcanic mountain rising to 3795 m (12 448 ft), forming the central peak of Ross Island, and overlooking McMurdo Sound. The sole remaining active crater in a complex of Cenozoic volcanoes, it is formed of an olivine-basalt base with Beacon sandstone and dolerite inclusions, capped by kenyites and other younger rocks. When discovered in January 1841 by Capt. J. C. Ross RN (and named for his ship), it was in mild eruption, venting steam and ash. When first climbed by a party from the **British Antarctic (*Nimrod*) Expedition 1907–9**, it was releasing clouds of vapour: indeed, it usually emits enough vapour to indicate the direction of wind at the summit. In 1982 it became more active, and positively boisterous in September to December 1984, when explosions shook the surrounding area: the crater emitted plumes of coloured vapour and incandescent volcanic bombs.

Escudero Station. 62°12′S, 58°58′W. Chilean permanent year-round research station on King George Island, South Shetland Islands, formerly Teniente Rodolfo Marsh Station.

Esperanza Station. 63°24′S, 57°00′W. Argentine permanent year-round research station at Hope Bay, Trinity Peninsula. Built in 1952, close to the site of an earlier British station (Base D), and the still-earlier stone hut of the **Swedish South Polar Expedition 1901–4**, it has operated continuously since then.

Esther Harbour. 61°55′S, 57°59′W. Harbour in Venus Bay, on the north coast of King George Island, South Shetland Islands. William Smith landed in the harbour on 16 October 1819, when he took formal possession of the group for King George III. It is named for an early US sealing vessel.

European Project for Ice Coring in Antarctica (EPICA). International programme to drill and analyse ice cores from Antarctica, with the objective of studying current and past climates. The first major core is to be drilled at **Concordia Station**: a second core will be taken from Dronning Maud Land, at a site to be determined.

Evans, Edgar. (1876–1912). British polar seaman. Born at Rhossili, in Wales, he enlisted in the Royal Navy in 1891. Ten years later, a seaman petty officer, he joined the British National Antarctic Expedition 1901–4, taking part in many of the sledging operations. Impressed by his strength and intelligence, Scott invited him to join the **British Antarctic (*Terra Nova*) Expedition 1910–13**. Again Evans proved a stalwart sledger, and was included in the party of five that reached the South Pole. On the return journey he was the first to show signs of strain. He died on 17 February 1912.

Evans, Edward Ratcliffe Garth Russell. (1881–1957). British naval officer and explorer. Born in London on 25 October 1881, he joined the Royal Navy in 1896. As a junior officer in *Morning* in 1902 he took part in the relief of the **British National Antarctic Expedition 1901–4**. In 1910 he was appointed navigator and second-in-command of the **British Antarctic (*Terra Nova*) Expedition 1910–13**. He took part in all the major sledging journeys, and commanded the expedition after Scott's death. He had a distinguished career in World War I, and was decorated for action in HMS *Broke*. Evans was knighted in 1935, appointed admiral in 1936, and retired from the Navy in 1939. During World War II he was London's regional commissioner for civil defence, and received a barony in 1945. He died in 1957.
Further reading: Evans (1921).

Evans, Hugh Blackwall. (1874–1975). British polar biologist. Born on 19 November at Aylburton, Gloucestershire, at age 16 he left for Saskatchewan, Canada to study and practise agriculture. In 1896 he joined a cousin in a sealing venture to Iles Kerguelen, and on return joined the **British Antarctic (*Southern Cross*) Expedition 1898–1900** as assistant zoologist. One of the first to winter on Antarctica, he collected avidly and took part in the very limited sledging operations. After the expedition he returned to Canada and to farming in the Vermilion River Valley. Evans died in Vermilion on 8 February 1975.

Evans, Cape. 77°38′S, 166°24′E. Rocky, ice-free cape on the west side of Ross Island, named during the **British Antarctic (*Terra Nova*) Expedition 1910–13** for Lt **E. R. G. R. Evans**, RN, the expedition's second-in command. The expedition hut, and a nearby cross from the **Imperial Trans-Antarctic Expedition 1914–17**, are scheduled respectively as HSM 16 and 17, and a surrounding area of approximately 0.5 km² (0.2 sq miles) has been designated SPA No. 25 (Cape Evans) to provide for a management plan covering these sites.

Sorting stores outside the living hut, Cape Evans. Photo: Scott Polar Research Institute

Evans Ice Stream. 76°00′S, 78°00′W. Major ice stream draining southeastward from Ellsworth Land into the southwest corner of Ronne Ice Shelf. The stream was identified during radio echo-sounding overflights in 1975 and named for British physicist Stanley Evans, who developed the echo-sounding technique.

F

Falkland Harbour. 60°44′S, 45°03′W. Harbour on the southwest coast of Powell Island, South Orkney Islands, charted in 1912 and named for a Norwegian whale factory ship that grounded there in that season.

Falkland Islands. 52°00′S, 59°00′W. Archipelago of cool-temperate islands in the southern Atlantic Ocean, 600 km (375 miles) west of central Patagonia. The group includes two large islands, East and West Falkland, and some 350 lesser islands, generally of rolling moorlands. Steep cliffs invest many of the western islands, elsewhere sandy beaches alternate with low headlands. The highest peak is Mt Usborne (705 m, 2312 ft) in East Falkland. The rocks are ancient sediments of palaeozoic and mesozoic age, much folded, metamorphosed, eroded, and smoothed by a long period of former glaciation. The climate is damp and windy, with mean annual temperature 5.8 °C, range of mean monthly temperatures 7.1 °C, annual precipitation 60 cm and strong westerly winds. Frosts are common and snow and sleet fall frequently, though snow seldom settles for long. Thin soils and thicker peat deposits cover much of the ground, supporting heath and scrub close to the coast and fellfield on higher ground. Over 160 native flowering plants are recorded, including tussock and other grasses, rushes and shrubs, with many algae, lichens, mosses and ferns in a variety of habitats. The native flora now competes with over 90 introduced species. Native mammals include herding elephant seals, fur seals, sea lions, and occasional leopard seals. The Falklands support many species of native land bird and waterfowl, and huge populations of breeding seabirds, including five species of penguin.

Discovered in 1592 by the British mariner John Davis, the islands were first settled, though not formally annexed, in 1690. The publication in 1756 of *Histoires des navigations aux terres australes*, by the French explorer Charles de Brosses, stimulated French interest in the southern region. In 1763–64 the French scientist and explorer Louis-Antoine de Bougainville established a small settlement, Port Louis, on East Falkland, introducing cattle and sheep, and encouraging the settlers to exploit the fur seals that bred on the islands. A British naval expedition under Capt. John Byron set up a similar colony at Port Egmont early in 1766. Byron claimed the islands for Britain, and advised the French to leave. However, in April 1767 France formally transferred sovereignty of Bougainville's settlement to Spain, which had also laid claim: Port Louis became Puerto de la Soledad, a Spanish colonial ranching outpost and convict settlement. In June 1770 a Spanish task force from Buenos Aires evicted the British from Port Egmont. Threats of war led to reinstatement of the colonists just over a year later, but the settlement could not pay its way and the British withdrew for economic reasons in 1774. Puerto de Soledad closed for similar reasons in 1811.

The islands continued to be visited by sealers, notably from the United States and Britain, but remained unoccupied and without formal government. In 1820 Daniel Jewitt, a US adventurer in the pay of the Government of the Provincias Unidas de Sud-América (successors to the Spanish colonial government), landed at the site of Puerto de Soledad and asserted his employers' claim to the islands, one that the government did not follow up. British, French and Spanish adventurers hunted seals and raised cattle, sheep and crops, but later abandoned the islands. In 1829 the government of the Provincia of Buenos Aires asserted a claim to the islands and appointed a German adventurer, Louis Vernet, to the governorship. Vernet, who established the settlement of Puerto de la Soledad on the site of the former Port Louis, had already been granted land rights and exclusive fishing rights on the islands. In late July 1831 Vernet confiscated three US sealing ships, which he claimed were operating illegally. In retaliation, on 31 December of the same year an armed party from the US naval sloop *Lexington* dispersed the small garrison and declared the islands free of government. Buenos Aires sent a replacement for the governor, whom the disordered troops of the garrison murdered. On 20 December 1832 Capt. John James Onslow RN, commanding HMS *Clio*, reasserted British sovereignty over the islands, returning some of the garrison to South America and establishing a British settlement in its place. In August 1833 five members of the British settlement were murdered, and an officer from the survey ship HMS *Challenger* was put ashore to restore order. Eventually civil government was restored: the Falkland Islands were again a British colony.

The settlement made use of peat for domestic fuel, growing potatoes, raising cattle, pigs and horses, and cropping sheep wool for export. Burning and grazing made severe inroads into the original vegetation, including the arcades of tall tussock grass that were once a feature of the coasts. Fast-growing conifers were introduced to create windbreaks and provide fencing. A single species of land animal, a native fox (worrah), was destroyed in the interests of sheep rearing. Rats, mice, rabbits, hares, Patagonian foxes and domestic dogs and cats were at various times introduced. During the mid-to-late nineteenth century Port Stanley, with its capacious harbour, became an important naval base and a focal point for shipping *en route* for Cape Horn and the west coast of the Americas. With the opening of the Panama Canal both maritime functions declined. The population, by then numbering about 2000, found itself relatively isolated and impoverished, subsisting mainly on exports of wool.

The Argentine invasion of April 1982 brought war to the Falklands, to be followed by a wave of world attention and relative prosperity. A large military airport was built and permanently garrisoned, and investment in infrastructure supported offshore fishing and tourism. Though the islanders have never developed a fishing fleet of their own, they currently license other nations to fish in their waters, and Stanley provides facilities for both fishing boats and Antarctic-bound tour ships. A Crown Colony of the United Kingdom, the islands are governed by a Governor-in-Council and locally elected Executive Council. Current population (2002) is about 2000, excluding the garrison.
Further reading: Christie (1951); Shackleton, Lord (1982).

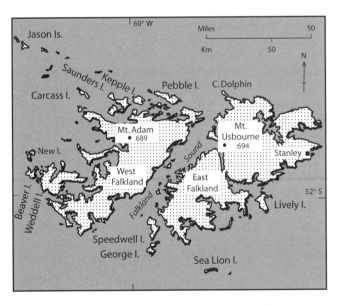

Generally low-lying, with steep western cliffs, the Falkland Islands are intermittently snow-covered in winter, but have no permanent ice. Heights in m

Falkland Islands and Dependencies Aerial Survey Expeditions 1955–57 (FIDASE). Airborne expedition that over two summers surveyed the Antarctic Peninsula and neighbouring islands south to 68°S. In 1955–57 the British Government contracted with Hunting Aerosurveys Ltd for aerial surveys of the **Falkland Islands** and the **Falkland Islands Dependencies**. This required a major two-year expedition, involving a transport ship *Oluf Sven*, two Canso amphibian aircraft, two helicopters, and a field party of 22; led by Peter G. Mott, and based in the old whaling station at Whalers Bay, in the crater of Deception Island. (FIDASE) in two successive summers surveyed a wide area including the South Orkney and South Shetland Islands and the Antarctic Peninsula south to the northern end of Marguerite Bay, the limit of the Cansos' range.

Ground control was provided both by the expedition's own surveyors, who were deployed by helicopter and ship, and by reports from the many survey parties of the **Falkland Islands Dependencies Survey** that had by this time operated in the area. Taking advantage of a very limited number of days of weather clear enough for photography, the aircraft flew a total of 240 hours, logging over 10 000 exposures. FIDASE provided the basis for some of the most detailed and accurate large-scale maps to emerge from Antarctic surveys, and the Survey's photographs continue to provide source material for the smaller scale maps that are now required for more detailed work.
Further reading: Mott (1986).

Falkland Islands Dependencies. British territory in the South American sector of Antarctica. The territory was defined by two British Letters Patent. The first, dated 21 July 1908, declared that 'the group of islands known as South Georgia, the South Orkneys, the South Shetlands, and the South Sandwich Islands, and the territory known as Graham Land, situated in the South Atlantic Ocean to the south of the 50th parallel of South latitude, and lying between the 20th and 80th degrees of West longitude, are part of our Dominions, and it is expedient that provision should be made for their government', and that the said islands and territory should become Dependencies of the Colony of the Falkland Islands. The second Letters Patent, dated 28 March 1917, defined the area more precisely, to include 'all islands and territories whatsoever between the 20th degree of West longitude and the 50th degree of West longitude which are situated south of the 50th parallel of South latitude; and all islands and territories whatsoever between the 50th degree of West longitude and the 80th degree of West longitude which are situated south of the 58th parallel of South latitude'. The Letters Patent thus assumed that the territories were already recognized as British possessions, and were making provision for their government, in particular for the regulation of a fast-growing whaling industry. The Dependencies so defined included all the islands of the Scotia Arc, plus a wedge-shaped sector of Antarctica mainland including the Antarctic Peninsula, and the eastern coast of the Weddell Sea (then recently named Coats Land by William Bruce), south to the South Pole.

In 1962, following Britain's accession to the Antarctic Treaty, a British Order in Council of 26 February split the Dependencies, providing for a new colony within the Treaty area (i.e. south of 60°S), to be called **British Antarctic Territory** and to be administered by a High Commissioner, usually the governor of the Falkland Islands. This included Graham Land, Coats Land, Alexander Island and the South Shetland and South Orkney Islands. The remaining areas, South Georgia, the South Sandwich Islands, Clerke Rocks and Shag Rocks, have, since 1985, been administered separately by the Government of South Georgia and the South Sandwich Islands.

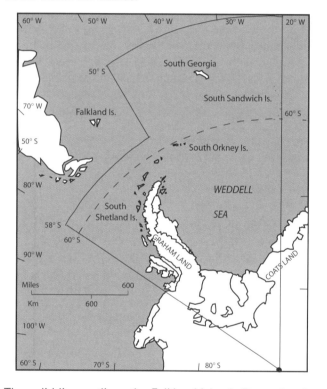

The solid line outlines the Falkland Islands Dependencies as defined in 1917. In 1962, when British Antarctic Territory was created south of 60°S (pecked line), the Dependencies were redefined to include only South Georgia and the South Sandwich Islands

Falkland Islands Dependencies Survey (FIDS) 1945–61. British permanent expedition to occupy, administer and survey the sector of Antarctica claimed by Britain as Falkland Islands Dependencies. The expedition arose from **Operation Tabarin 1943–45,** a continuation of a naval operation under the civilian management of the British Colonial Office. In 1947, when Dr V. E. Fuchs became leader, five bases were in operation (Cape Geddes, **Deception Island**, **Hope Bay**, **Port Lockroy** and **Stonington Island**.) Emphasis was on routine meteorological observations, undertaken at all the bases, coupled with local topographical, geological and biological survey at the smaller bases, more ambitious dog-sledging surveys at Hope Bay and Stonington Island. In 1947 were added **Argentine Islands**, and **Signy Island**, and in 1948 **Admiralty Bay**. On the return of Dr Fuchs from Stonington Island in 1950, to take charge of operations at a newly opened secretariat in London, objectives of the survey were redefined and scientific programmes established. Responsibility for meteorology was transferred to a newly formed Falkland Islands and Dependencies Meteorological Service, and new bases were planned to extend both the scientific scope and the geographical spread of the field work. From 1955 were added **Anvers Island** and **Horseshoe Island**, from 1956 **Danco Island** and **Detaille Island**, from 1957 **Prospect Point**, from 1959 **Halley Bay**, and from 1961 **Fossil Bluff**. The introduction of larger, more efficient resupply ships, and of aircraft capable of supporting and if necessary transporting field parties, introduced a new dimension of logistic efficiency, showing the way to new developments with the inauguration of British Antarctic Survey.

Falla, Robert. (1901–79). New Zealand polar ornithologist. Born in Palmerston North and educated at Auckland University, he became a lecturer in natural history at Auckland Teachers' Training College. In 1929 he became assistant naturalist on the **British, Australian, and New Zealand Antarctic Research Expedition 1929–31**, with special responsibility for bird studies. On return he joined the staff of Auckland War Memorial Museum, later becoming director of Canterbury Museum, Christchurch (1937–47). During World War II he served with the Cape Expedition coast-watching on the Auckland Islands 1943–44. In 1947 he became director of the Dominion Museum (now National Museum), Wellington, promoting several expeditions to New Zealand's outlying islands before retiring in 1966. He served on the Ross Sea Committee from 1955, and was chairman of the New Zealand Nature Conservation Council 1962–74. Falla was knighted in 1973. He died on 23 February 1979.

Fallières Coast. 68°30′S, 67°00′W. Part of the west coast of Graham Land, Antarctic Peninsula, facing Marguerite Bay. Bounded by Bourgeois Fjord in the north and Cape Jeremy in the south, deeply indented with bays and fjords, it consists of high coastal mountains rising to the central plateau, and glaciers descending to narrow piedmont ice shelves. The coast was charted in 1909 by J.-B. Charcot during the French Antarctic (*Pourquoi Pas?*) Expedition 1908–10, and named for C. A. Fallières, who was then president of France.

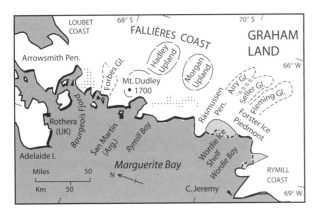

Fallières Coast

Faraday Station (Base F). 65°15′S, 64°16′W. British research station on Marina Point, Galindez Island, Argentine Islands, Biscoe Archipelago. The first Argentine Islands station (Wordie House), built in January 1947 on Winter Island, was occupied by the Falkland Islands Dependencies Survey until May 1954, when operations were transferred to a new building (Coronation House) on Galindez Island. This was maintained as a year-round station until February 1996, when ownership was transferred to the government of Ukraine. The station continues to operate as Akademik Vernadsky. Wordie House, which was re-occupied briefly in the winter of 1960, is now designated an Historic Monument.

Fast Ice. Coastal sea or lake ice that forms in close attachment to the shore, and remains after dispersal along its outer edge. It is sometimes held in place by islands, rocks or grounded icebergs.

Fast ice over 1 m (3.3 ft) thick along the Weddell Sea coast provides a safe substrate for wintering emperor penguins and their chicks. Photo: BS

Faure Islands. 68°06′S, 68°52′W. A group of bare rocky islands and reefs standing in isolation in central Marguerite Bay, off the Fallières Coast of Graham Land. Discovered during the French Antarctic (*Pourquoi Pas?*) Expedition 1908–10, they were named by J.-B. Charcot for the French academic and statesman Maurice Faure.

Ferrar, Hartley Travers. (1879–1932). British polar geologist. Born in Dublin, he read natural sciences at Cambridge. He joined the **British National Antarctic Expedition 1901–4** as geologist and seawater chemist, taking part in many of the sledging journeys, including those to the western mountains: Ferrar Glacier, Victoria Land, is named for him. He surveyed in Egypt, became a schoolmaster in New Zealand, and served with the New Zealand Army in World War I. In 1919 he joined the New Zealand Geological Survey, serving with them until his death.

Filchner, Wilhelm. (1877–1957). German traveller and explorer. Born in Munich, Germany, he trained in a military academy and was commissioned in the German army. Interested in survey and geophysics, he spent as much time as possible travelling in Russia and central Asia. After meticulous study of all the available geographical evidence he planned and led the **German Antarctic (*Deutschland*) Expedition 1911–13**, intending to cross the Antarctic from the Weddell Sea to the Ross Sea via the as-yet-unattained South Pole. Successfully penetrating the Weddell Sea, he landed a station hut in the shelf ice but had to withdraw hurriedly when the ice broke away. Later caught in the pack ice, Filchner and his small scientific team spent a relatively unproductive winter, breaking free in November 1912. Though the expedition failed in its primary aim, Filchner's encouragement ensured that his team produced excellent biological and oceanographic results. In later life he continued to explore in warmer climates of central Asia and Nepal. He died in Switzerland on 7 May 1957.

Filchnerfjella. 72°03′S, 7°40′E. Range of ice-capped mountains forming the south-western corner of Orvinfjella, central Dronning Maud Land, named for German expedition leader Wilhelm Filchner.

Filchner Ice Shelf. 79°00′S, 40°00′W. Extensive shelf bordering the Weddell Sea, between Berkner Island and the Luitpold Coast. From January 1955 it accommodated the Argentine research station General Belgrano, and from January 1957 to January 1962 the United States research station Ellsworth. In January 1956 it became the starting-off point for the **Commonwealth Trans-Antarctic Expedition**. The shelf was named for W. Filchner, leader of the **German Antarctic (*Deutschland*) Expedition 1911–13**.

Fildes Peninsula. 62°12′S, 58°58′W. Peninsula forming the southwestern end of King George Island, South Shetland Islands, extending into Fildes Strait. Peninsula and strait were named for Capt. Robert Fildes, an early nineteenth-century Liverpool sealer, who surveyed and wrote sailing directions for the area. The peninsula accommodates Russian, Chilean, Argentine, Uruguayan and Chinese research stations. Two areas totalling about 1.8 km^2 (0.7 sq miles) were together designated SSSI No. 5 (Fildes Peninsula) to protect fossil ichnolites that appear in outcropping Tertiary strata.

Fildes Strait. 62°14′S, 59°00′W. Passage separating King George Island and Nelson Island, South Shetland Islands. The strait was named in about 1822 for the British sealer Robert Fildes.

Fimbulisen. 70°30′S, 00°10′W. (Fimbul Ice Shelf). Extensive ice shelf bordering Kronprinsesse Märtha Kyst, Dronning Maud Land. It includes Trolltunga, a remarkable ice tongue that periodically breaks off and floats away. From January 1957 to December 1959 it was the site of Norway Station, a Norwegian research station. In January 1960 this was transferred to South Africa and re-named **SANAE**, which has been renewed periodically since that date. The Norwegian name of the shelf translates as 'giant ice'.

Fin whale. *Balaenoptera physalus.* Growing to lengths of about 25 m (82 ft), and weighing up to 40 tonnes, fin whales are second-largest of the rorquals, exceeded in size only by **blue whales**. Named for their prominent recurved dorsal fin, they are dark grey above and pale grey or white under the throat, belly, flippers and tail, usually with a distinctive patch of pale grey or white on the right-hand side of the face. The anterior baleen plates on the right are also white, the remainder dark grey or striped. Fin whales occur in all the world's oceans, most commonly in warm waters, but extending north and south into polar waters in summer. Southern stocks breed in tropical and warm temperate waters, moving south during the spring and summer into the Southern Ocean, including the northern edge of the pack ice. Often they move in small pods of three or four. On rich feeding grounds they may gather in much larger numbers, feeding independently on surface and near-surface shoals of krill or small fish. During the years of Southern Ocean whaling (1905–87) they were by far the commonest whales. During the 1950s and early 1960s, after the devastation of blue whale stocks, they became the most heavily hunted, to be devastated in turn and replaced by **sei** and **minke** whales. Southern stocks appear to be recovering slowly: a world population of 70 000–80 000 is estimated.
Further reading: Watson (1981); Evans (1987).

Firn. Snow in process of consolidation to ice by thawing and re-freezing. Snow becomes firn with the development of coarse crystals, porous texture and a density of 400 kg per m^3 or greater: also called névé. The firn line on a glacier surface is the boundary between firn and ice at the end of a melt season.

First-year ice. Floating ice on a sea or lake surface that formed within the year, i.e. is less than one year old.

Fishes, Southern Ocean. Coastal and shallow-water fishes of the Southern Ocean are unusual in that more than 60 per cent of the total species and over 90 per cent of all individuals belong to the single suborder, Notothenioidei. At greater depths the fishes are mainly representatives of widespread deep-water families, including rat-tails, snailfishes, eel pouts and dragon fish. The notothenioids have undergone a remarkable radiation in shallow Antarctic waters. Lacking a swimbladder, most species live on or close to the bottom. A few species have evolved mechanisms for increasing their buoyancy (for example, by reducing the skeleton and laying down fat), and have been able to exploit the midwaters. However, the Southern Ocean essentially lacks the large schools of plankton-feeding pelagic fish (herring and mackerel, for example) so characteristic of Arctic and temperate northern waters. Most fish, like other vertebrates, use the red pigments haemoglobin and myoglobin in carrying and storing oxygen. All Southern Ocean notothenioids contain less red pigment than temperate species, and one family, the Channichthyidae or ice fish, are remarkable in almost completely lacking either pigment. In consequence, their blood and tissues are colourless, but the higher solubility of oxygen at low temperatures means that sufficient can be carried by the blood plasma alone. The blood of fishes is more dilute than seawater and would normally freeze at the sub-zero temperatures found when ice is present in the Southern Ocean. Indeed, temperate-water fish cooled to these temperatures quickly freeze and die. All cold-water notothenioids produce an antifreeze glycoprotein (AFGP), synthesized in a variety of molecular sizes and found in virtually all the body fluids. It works by preventing ice nuclei from growing large enough to cause tissue damage. Arctic fish have also evolved antifreezes, though these are more usually proteins (rather than glycoproteins) and some species use them only in winter when the sea is coldest.
Further reading: Eastman (1992); Kock (1992).
(ACC)

Fitzsimmons, Roy. (*c.* 1916–45). American polar geophysicist. He served as a geophysicist with the MacGregor Arctic Expedition 1937, and later joined the **United States Antarctic Service Expedition 1939–41**, when he worked at West Base on magnetometry, auroral studies and seismology in the Rockefeller Mts. In World War II he served with the US Army Air Force, and was killed on 5 May 1945 while on active service in Cuba.

Fleming, William Launcelot Scott. (1906–90). British polar geologist and cleric. Born in Edinburgh on 7 August 1906, he read natural sciences at Cambridge University and completed a master's degree in geology at Yale University. While reading for the priesthood at Westcott House, Cambridge, he took part in two undergraduate expeditions to the Arctic. In 1933 he was appointed to a chaplaincy and fellowship at Trinity Hall, Cambridge. In the following year he joined the **British Graham Land Expedition 1934–37** as chaplain and geologist, taking part in exploratory sledging journeys along the southwest coast of Graham Land, and studying particularly the geology and glaciology. During World War II he served as a chaplain in the Royal Navy. Returning to Cambridge, he resumed his fellowship and for two years became part-time director of the Scott Polar Research Institute. He was appointed successively Bishop of Portsmouth (1949), Bishop of Norwich (1959) and Dean of Windsor (1971). He was knighted in 1976, the year of his retirement, and died on 30 July 1990.

Flight in Antarctic seabirds. Seabirds probably spend more time in flight than most birds. Those in Antarctic habitats have to contend with extreme conditions of wind and weather. Adaptations for coping with demands of long-distance flight have been extensively studied in Southern Ocean seabirds. The range of size in the Antarctic petrels, from 25 g storm-petrels to 12 kg albatrosses, is the greatest of any order of birds and offers unique opportunities for understanding how body size relates to aerodynamic properties. Diving-petrels, counterparts of the familiar northern hemisphere auks, provide a functional link between flight in air and the underwater 'flight' of penguins.

Soaring

The characteristic soaring flight by which albatrosses and petrels quarter the oceans involves a long sweep down across the wind into a wave trough, and a steep turn to windward in the direction of the oncoming wave. The updraught of faster-moving air above the wave lifts the bird so that the manoeuvre can be repeated. Procellariforms maintain this gliding flight endlessly in the constant winds of the Southern Ocean. Such large species as albatrosses and giant petrels that rely on gliding keep their wings outstretched by locking tendons. Smaller petrels lack this adaptation, presumably because it is not compatible with flap-gliding (bouts of flapping flight and short glides) which is their characteristic flight. In soaring, albatrosses travel much faster than small petrels, though much slower in proportion to body size. Wilson's storm-petrels, among the lightest of the petrels, can stay airborne at speeds as low as 4 m per second: albatrosses require 12 m per second at least. To remain stationary in light winds storm petrels dip one foot into the sea as an anchor and allow the wind to lift them; in this position they can easily pick tiny food items from the sea surface. Studies tracking albatrosses by using devices to transmit their positions to satellites have revealed that, between successive visits to feed their chicks, they may cover up to 10 000 km, often averaging 1000 km per day at speeds of 80 km per hour. The pattern of the tracks also shows the influence of low pressure weather systems.

Flap-gliding and diving

Storm-petrels, which mainly use flap-gliding, fly at speeds close to their maximum. Soaring, for them a relatively slow means of travel, would not result in overall energy savings where time is a constraint e.g. when needing to return to feed chicks. For the larger birds, that have little extra power available from muscles above that required to fly, flapping is energetically expensive. In calm air they drop to the sea surface and wait for the wind to rise. Seabirds that dive have relatively shorter and stubbier wings, to meet the requirements of a medium 800 times denser than air. Wings of diving-petrels, cormorants and medium-sized shearwaters are compromises between the demands of flight in air and of swimming underwater. Penguins take wing reduction to an extreme. For a macaroni penguin to fly, it would need to cruise at about 40 m per second, requiring a wing-beat frequency of 50 Hz, similar to that of hummingbirds.
Further reading: Pennycuick (1987).
(JPC)

Floe. Fragment of floating ice on the surface of a sea or lake that originated as part of a larger ice sheet.

Flood Range. 76°03′S, 134°30′W. An extensive range of ice-covered mountains behind Hobbs Coast, northern Marie Byrd Land, identified in 1934 during Richard Byrd's second expedition and named for his uncle, US Senator Henry ('Hal') D. Flood.

Flora, Mt. 63°25′S, 57°01′W. Prominent conical peak rising to 520 m (1706 ft) behind Hope Bay, Trinity Peninsula, named for a rich Jurassic fossil flora in a band of sandstones and siltstones underlying a volcanic capping. A location of approximate area 0.6 km² (0.2 sq miles) on its upper slopes has been designated SSSI No. 31 (Mt Flora, Hope Bay) to protect the fossil-bearing outcrops.

Flowering plants, Antarctic. Two species of native flowering plant have been found in Antarctica and the neighbouring Antarctic islands: Antarctic hair grass *Deschampsia antarctica*, and Antarctic pearlwort *Colobanthus quitensis*. Both are restricted to the maritime sector, including Antarctic Peninsula south to about 68°S, and

Unusually dense meadow of Antarctic hair grass, South Shetland Islands. Small patches of Antarctic pearlwort can be seen close to the film carton. Photo: BS

the Scotia Arc. Nowhere plentiful, they are patchily distributed on coastal flats and lowlands, forming small clusters 3–5 cm (1.5–2 in) high, usually growing from moist soils in association with mosses. Both form seeds during favourable summers in the warmer parts of their range.

Forbidden Plateau. 64°47′S, 62°05′W. Ice plateau crossing Graham Land from Charlotte Bay to Flandres Bay, named for difficulties experienced by sledging parties in approaching it.

Ford, Charles Reginald. (1880–1972). British expedition seaman. Born in London, he joined the Royal Navy at 15, and volunteered for the **British National Antarctic Expedition 1901–4** as a steward. He served also as R. F. Scott's secretary and accountant. In 1903 he took part in a depot-laying sledging trip. On return to Britain he became responsible for the expedition's financial affairs. In 1906 he settled in New Zealand, where he qualified as an architect, forming a partnership that designed several public buildings in Auckland. Ford published a short account of his experiences in Antarctica and a book on English ceramics. He died in Auckland on 19 May 1972.
(CH)

Ford Massif. 85°05′S, 91°00′W. Snow-capped massif rising to 2810 m (9216 ft) in the Thiel Mountains, southern Ellsworth Land. The massif was named for United States geologist Arthur B. Ford, who led many survey parties into the Ellsworth Land mountains.

Ford Ranges. 77°00′S, 144°00′W. A complex of mountain ranges behind the Ruppert and Saunders coasts, northwestern Marie Byrd Land, discovered in 1929 during Richard Byrd's first expedition and named for Edsel Ford, a supporter of the expedition.

Forde, Robert. (c. 1877–1959). British expedition seaman. As a Royal Navy petty officer he joined the **British Antarctic (Terra Nova) Expedition 1910–13**, spending two years ashore at Cape Evans. He became an expert sledger, in the second year taking part in the major western journey. He died in Cobh, County Cork, Ireland on 13 March 1959.

Forlidas Pond. 82°27′S, 51°21′W. One of several freshwater ponds in a dry valley area at the eastern end of Dufek Massif. Similar ponds are found in a similar area along the edge of the ice cap in neighbouring Davis Valley. As these are the most southerly-known bodies of seasonally recurring fresh water, an area of approximately 9.8 km^2 (3.8 sq miles) surrounding them is scheduled as SPA No. 23 (Forlidas Ponds and Davis Valley Ponds), protecting them from contamination by human activity.

Forrestal Range. 83°00′S, 49°30′W. Northernmost range of the Pensacola Mountains, southern Coats Land. Mainly snow-covered, it includes several peaks of over 1000 m (3280 ft), with glaciers draining through an eastward-facing escarpment into the massive Support Force Glacier.

Forster, Johann Georg Adam. (1754–94). Prussian naturalist who, with his father J. R. **Forster**, sailed with James **Cook** on his second expedition to the Southern Ocean. Born near Gdansk, Georg Forster moved with his family first to St Petersburg, Russia, then to England. At the age of 18 he accompanied his father as assistant naturalist in HMS *Resolution*, making substantial collections of plants and animals at the many southern localities visited. *Observations Made During a Voyage Round the World*, largely by Reinhold but published in Georg's name to avoid difficulties with the British Admiralty, established his credentials as a field naturalist and taxonomist. There followed academic appointments at Kassel, Vilnius and Mainz. He died in Paris in January 1794.

Forster, Johann Reinhold. (1729–98). Prussian naturalist who, with his son J. G. A. **Forster**, sailed with James **Cook** on his second expedition to the Southern Ocean. Born in Prussia to an expatriate British family, he became a minister of the Reformed Church. After a spell in Russia he moved to northern England, where he taught languages and natural sciences in a Warrington school. In 1772 Forster came to the notice of James Cook, and was taken aboard HMS *Resolution* as naturalist, with his son as assistant. He collected assiduously, illustrating, describing and naming many specimens of birds, mammals and other animals from the southern regions. In addition, he studied ocean temperatures, made deep-sea soundings, and collected anthropological data on Pacific island populations. In 1777 a description of his travels, *Observations Made*

During a Voyage Round the World, was published under his son's name. Returning to Prussia, he became director of the Halle botanical gardens and taught at the local university.

Fosdick Mountains. 76°32′S, 144°45′W. A range of mountains forming a northern flank of the Ford Ranges, Marie Byrd Land, discovered in 1929 during Richard Byrd's first expedition and named for Raymond B. Fosdick, an expedition supporter.

Fossil Bluff Station (Base KG). 71°20′S, 68°17′W. British research station at Fossil Bluff, George V Sound, Alexander Island. The station (Bluebell Cottage) was built in February 1961 by British Antarctic Survey, as an advance field station for operations from other stations in Marguerite Bay, being maintained mostly by air transport. It was occupied year-round in 1961–63 and 1969–75, thereafter only in summer.

Foster Plateau. 64°43′S, 61°25′W. Ice plateau between Drygalski and Hektoria glaciers, Graham Land, named for a British station leader R. A. Foster.

Foster, Port. 62°57′S, 60°39′W. Harbour formed by the breached volcanic crater of Deception Island, South Shetland Islands. Previously called 'Deception Harbour' and 'Yankee Harbour', it was finally named for Capt. H. Foster RN, who visited in HMS *Chanticleer* in 1829 to make scientific observations.

Foundation Ice Stream. 83°15′S, 60°00′W. Major ice stream draining from the southern Pensacola Mountains north-north-eastward into the Ronne Ice Shelf. Identified from aerial photographs, it was named for the US National Science Foundation, which had supported glaciological studies in the area.

Foyn Coast. 66°40′S, 64°20′W. Part of the east coast of Graham Land, Antarctic Peninsula, facing the Weddell Sea. Bounded by Cape Alexander in the north and Cape Northrop in the south, the coast consists of high coastal mountains rising to the central plateau, and glaciers descending to the broadest extent of Larsen Ice Shelf. The northern section was sighted by Capt. C. A. Larsen during the Norwegian (Sandefjord) Whaling Expeditions 1892–94. Larsen named it for Svend Foyn, the Norwegian doyen of whaling and inventor of the explosive grenade harpoon, who had supported the expedition.

Foyn Harbour. 64°33′S, 62°01′W. Harbour between Nansen Island and Enterprise Island, in Wilhelmina Bay, on the Danco Coast of Graham Land. Used extensively by whaling ships from about 1912 onward, the harbour was named for the factory ship *Svend Foyn* that moored there during the early 1920s.

Framnes Mountains. 67°50′S, 62°35′E. Group of coastal mountains, including the Casey, Masson and David ranges, along the Mawson Coast, Mac.Robertson Land, rising to peaks of about 1500 m (4920 ft). Though individual ranges were identified by the British, Australian and New Zealand Antarctic Expedition in 1931, the coast was overflown and photographed by the Norwegian (Christensen) Whaling Expeditions in 1937, and the group named for Framnesfjellet, a locality near their home port of Sandefjord, Norway.

Francis, Samuel John. (d. 1983). British polar surveyor. After training with the Ordnance Survey he joined the Royal Engineers and served through World War II. In 1945–48 he was a surveyor with the Falkland Islands Dependencies Survey, undertaking field surveys and mapping at Base D (Hope Bay). In 1947–48 he and three companions sledged along the east coast of Graham Land, resurveying 400 km (250 miles) previously charted by the **Swedish South Polar Expedition 1901–4**, and adding a further 200 miles (320 km) of new coast along the Larsen Ice Shelf, before crossing the plateau to reach Base E (Stonington Island). From 1848 he was chief surveyor to the Worcestershire County Architect. He died on 11 September 1983.

Franklin Island. 76°05′S, 168°19′E. A steeply flanked island 11 km (7 miles) long, standing 122 km (76 miles) north-northeast of Ross Island in the Ross Sea. Ice-capped, and surrounded by fast ice or pack ice for much of the year, the island offers near-shore accommodation to a colony of emperor penguins. The island was discovered and charted by the **British Naval Expedition 1839–43**. Its name honours Sir John Franklin, a celebrated Arctic explorer, who as Governor of Van Diemen's Land (Tasmania) had provided hospitality for the expedition in Hobart.

Frazier, Russell G. (1893–1968). Born in 1893, he was the chief medical officer of the **United States Antarctic Service Expedition 1939–41**, during which he undertook

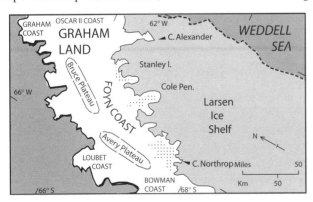

Foyn Coast

a study on acclimatization and the effects of cold on the human body. From 1919 until 1951 he was physician to the Utah Copper Mines, leading expeditions into remote areas of the Rocky Mountains. He died on 11 January 1968.

Frazil ice. Small crystals and flakes of ice formed at or near the surface of an ocean, lake or river, which float to create a distinctive surface layer. Frazil forms in the open ocean when cold air chills the surface, and there is enough turbulence to cause stirring. As it accumulates, it thickens the water to a porridge-like consistency, eventually becoming a semi-solid layer 60 cm (2 ft) or more deep. This may break into platelets (pancake ice), or solidify into continuous sheets or floes. Interstices may contain living organisms of phyto- and zooplankton, which become incorporated into the ice sheet. Individual crystals forming in the sea include only fresh water, leaving a residue of enriched brine which, being denser than the surrounding water, drains gradually away.

Frei Station. See **Presidente Eduardo Frei Station.**

French Antarctic (*Français*) Expedition 1903–5. The first expedition of Dr J.-B. **Charcot** to Antarctic Peninsula. Charcot, a wealthy Parisian medical practitioner, spent much of his leisure time sailing off northwestern Europe. In 1900 he financed the construction of a three-masted schooner, *Français* (245 tons), planning to explore to Greenland and beyond. In 1903, hearing that the Swedish explorer Otto Nordenskjöld was missing in Antarctica, he determined on an expedition that would perhaps rescue Nordenskjöld, but would anyway explore south along the coast of Antarctic Peninsula, from which de Gerlache's Belgian expedition had recently returned. Gaining the support of French scientific societies, the government and the public, Charcot left Le Havre on 27 August 1903, accompanied by a small scientific team. Adrien de Gerlache, who had helped in planning the expedition, sailed with the expedition as far as Brazil. Arriving in Buenos Aires on 16 November Charcot learnt that Nordenskjöld's expedition had been rescued, leaving him free to explore. He met the returning Swedish team, and took aboard five of their Greenland huskies.

Français sailed from Orange Harbour, Tierra del Fuego on 27 January 1904, heading south down Gerlache Strait to Flanders Bay, where the engineers repaired an ailing steam engine. Exploring south, Charcot discovered Port Lockroy (which he named for the French Minister of Marine who had helped the expedition). *Français* reached the edge of the pack ice at Wandel Island (now Booth Island) where Charcot found a small, comfortable harbour (now Port Charcot) to winter the ship. There physicists André Matha and Raymond Rallier du Baty set up astronomical, meteorological and magnetometric equipment, and biologist J. Turquet and geologist Ernest Gourdan collected specimens. During the winter Charcot led surveying expeditions by sledge and boat to nearby Hovgaard Island and Cape Tuxen, on the peninsula coast. In late December *Français* broke free of the ice, and throughout January and early February Charcot forged southward through pack ice, discovering and charting the Loubet Coast, Biscoe Islands and southern Adelaide Island, then heading toward Alexander Island. On 15 February the ship struck a rock, seriously damaging the keel. Charcot made his way slowly to Port Lockroy to make temporary repairs, then returned to Argentina. In Buenos Aires he sold the damaged ship, and he and the expedition staff returned to France in a passenger liner. His first expedition showed Charcot what a small, sturdy ship with a hand-picked crew and dedicated group of scientists could achieve by over-wintering and making full use of two summer seasons. He resolved to return to Antarctica with a stronger and better-founded ship (see **French Antarctic (*Pourquoi Pas?*) Expedition 1908–10**).

Further reading: Charcot (1906).

French Antarctic (*Pourquoi Pas?*) Expedition 1908–10. Dr J.-B. Charcot's second expedition to Antarctic Peninsula. After the limited success of his first **French Antarctic (*Français*) Expedition 1903–5**, and the loss of his first ship, Charcot returned to France and commissioned a new, larger schooner-rigged vessel, *Pourquoi Pas?*, of 800 tons, more comfortable and better equipped than *Français*, with a more powerful and more reliable engine. With the blessing and support of the scientific establishment, government and public, Charcot sailed from Le Havre on 15 August 1908, and from Punta Arenas, Tierra del Fuego on 16 December. At Deception Island, South Shetland Islands, he made friends with the whaling community, whose masters he was pleased to see using the charts published from his first expedition. From there he sailed south on 25 December, heading first for Port Lockroy, then for Port Charcot, his old mooring on Booth Island. *Pourquoi Pas?* was too bulky to winter comfortably in the old harbour, but Charcot and his colleagues rummaged nostalgically among the debris from the earlier expedition. Still seeking a harbour, he made a reconnaissance by motor launch to Petermann Island, where on 1 January 1909 he found a small cove that he named for the day – in the Christian calendar, the Feast of the Circumcision. On 3 January he moved the ship to Port Circumcision, which made an admirable berth. However, there was still time for a summer voyage to the south. *Pourquoi Pas?* headed south and, while making toward Cape Tuxen, struck a submerged rock, sustaining damage to the bow and keel. Despite this Charcot continued south toward Adelaide Island, charting meticulously as he went. Taking advantage of good weather and an absence of pack ice, the ship headed westward around the northern end of Adelaide Island, then south along the island, which turned out to be very much longer than Biscoe, its original discoverer, had indicated. Beyond the southern end Charcot discovered a huge gulf, with a range of mountains extending south toward Alexander Island, which he named Marguerite Bay for his second wife. He considered wintering there, but was running short of coal for the boiler, and felt it safer to return north. On 30 January he headed back toward Petermann Island, and prepared *Pourquoi Pas?* for wintering in Port Circumcision.

The physical scientists set up huts ashore, and spent a profitable season recording magnetometric, meteorological and seismographic observations. On 18 September Gourdan led a party of six for a journey of two weeks onto one of the mainland glaciers. Despite fog and heavy snowfalls they climbed to a height of over 900 m (2950 ft), hoping but failing to find a way onto the high plateau that was visible above. In October the gentoo and Adélie penguins returned, giving biologists Jacques Liouville and Louis Gain opportunities for study.

In November 1909 the ship broke from its harbour and the expedition returned to Deception Island, stocking up with coal from the whalers to provide power for a second long summer voyage. From the whaling manager and his wife, Charcot learnt that his compatriot Blériot had flown the Channel, and that Peary had claimed to have reached the North Pole. A diver from the whaling fleet reported that the damaged keel made the ship unsafe for further exploration, but Charcot decided to continue. Returning south on 7 January 1910, he took advantage of relatively ice-free waters to sweep south past his discoveries of the previous autumn, toward and beyond Alexander Island. On 11 January, an unusually clear day, he saw far to the south the ice-covered rise of a new land, which he called Charcot Land for his father (now Charcot Island). Continuing south and westward into Bellingshausen Sea, three days later he found himself uncomfortably close to Peter I Øy. Remaining close to 70°S, pushing through increasingly dense pack ice, on 22 January *Pourquoi Pas?* achieved 124°W before turning and heading north for Tierra del Fuego.

Pourquoi Pas? returned to France on 4 June 1910 to a well-publicized and warm welcome. The results of Charcot's second expedition far outstripped those of the first, firmly establishing the reputation of a polar explorer of unusual sensitivity and distinction.

Further reading: Charcot (1911).

French Naval Expedition 1837–40.

Two corvettes of the French navy, *Astrolabe* and *Zélée*, commanded respectively by Jules Sébastien César **Dumont d'Urville** and Charles-Hector Jacquinot, left Toulon on 7 September 1837, on a world cruise that would entail two major voyages to the Southern Ocean. The ships undertook charting and survey in the Strait of Magellan, then in January 1838 headed south

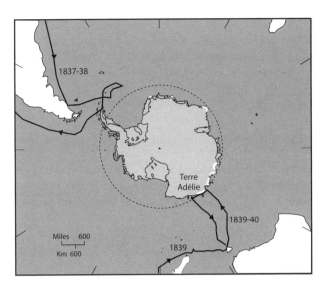

Dumont d'Urville's two voyages of exploration toward Antarctica, the first into the Weddell Sea, the second to and from Hobart, Tasmania

toward the Weddell Sea, where Dumont d'Urville hoped to surpass James **Weddell**'s record southing. Pack ice, fog and foul weather held them in latitudes close to the South Shetland and South Orkney Islands until late February, when they sighted land off the tip of Antarctic Peninsula – a fog-bound coast that Dumont d'Urville named 'Terre Louis Phillipe' for his monarch, separated from the previously discovered 'Palmer Land' (now Trinity Peninsula) by an ice-bound channel, now named Orléans Strait. D'Urville and Joinville Islands are further mementos of their visit. Pack ice restricted the movements of the ships, and the expedition's scientists were unable to land.

Astrolabe and *Zélée* then headed north for an eventful voyage lasting almost two years, mostly in the warmer waters of the southern Pacific Ocean, among islands more familiar and more congenial to Dumont d'Urville. Their second excursion into Antarctic waters began in early January 1840. Now in search of the Southern Magnetic Pole, Dumont d'Urville sailed southeastward from Hobart, encountering an edge of loose pack ice on 16 January. Continuing south through seas relatively clear of ice floes, but dotted with small tabular bergs, on 20 January the two corvettes made a landfall along a stretch of high, icebound coast lined with sheer ice cliffs – a sector of Antarctica's mainland coast.

Clear weather with light breezes allowed the ships to continue eastward along the coast. On the evening of 21 January, parties from both corvettes landed on a small ice-free island, where Dumont d'Urville raised the French flag and claimed the coast for his country, naming it Terre Adélie for his wife. The ice-covered headland first seen along the coast became Cap de la Découverte, the island Ile de la Possession, and a nearby rocky cape Pointe Géologie. The ships continued their eastward exploration to 135°30′E, where a field of dense ice forced them northward. On 29 January, in foul weather and poor visibility, *Astrolabe* and *Zélée* came within sight and almost within hailing distance of USS *Porpoise*, a brig of Wilkes's **United States Exploring Expedition 1838–42**. Minor but unfortunate mishaps forced them apart without speaking. After a further spell in the southern Pacific Ocean, the expedition returned to Toulon in November 1840.

Fridtjof Sound. 63°34'S, 56°43'W. Channel between Andersson Island, Jonassen Island and Tabarin Peninsula, named for one of the relief ships sent in 1903 to look for the missing **Swedish South Polar Expedition 1901–4**.

Friis, Herman R. (1905–89). US polar administrator. Born in 1905 in Chicago, he taught geography at the University of Wisconsin and at Southern Illinois University. He was on several government committees dealing with polar regions, in 1952–59 the Technical Committee on Antarctica, and in 1954–73 the US Advisory Committee on Antarctic Names. Friis Hills, Victoria Land were named for him. He died on 23 September 1989.
(CH)

Frolov, Vyacheslav Vasil'yevich. (*c.* 1902–60). Soviet polar meteorologist and administrator. After a distinguished career in the Arctic, in 1950 he became director of the Arctic Research Institute, Leningrad. Using his considerable influence, he promoted the USSR's active participation in Antarctic research, centred on the re-designated Arctic and Antarctic Research Institute.

Frost smoke. Thin clouds of condensed water vapour accumulating over relatively warmer seas; often associated with newly-forming sea ice.

Fuchs, Vivian Ernest. (1908–99). British polar explorer. Born on 11 February 1908, he read geology at St. John's College, Cambridge. Influenced by his tutor, James **Wordie**, in 1929 he gained his first polar experience on a university expedition to East Greenland. On graduating he joined and led archaeological and geological expeditions to East Africa. During World War II he served as an infantry officer in Africa and Europe. In 1947 Fuchs joined the Falkland Island Dependencies Survey (FIDS) as expedition leader and geologist, serving for two years at Base E (Stonington Island) where he learnt dog-sledging and travel techniques. Appointed director of FIDS, he reorganized operations with increasing emphasis on science, while simultaneously planning a more ambitious exploration, the **Commonwealth Trans-Antarctic Expedition 1955–58** to cross Antarctica from the Weddell Sea to the Ross Sea. Taking leave from FIDS, he planned and undertook the crossing. On return he was knighted and received many honours from scientific societies. With Sir Edmund Hillary he wrote the story of the expedition (1958). Fuchs returned to FIDS in 1958, which under his direction became British Antarctic Survey and expanded massively. Retiring in 1973, he remained active in polar and scientific affairs, serving as president of the British Association for the Advancement of Science 1971, and of the Royal Geographical Society 1982–84. Later books included a history of FIDS and British Antarctic Survey (1982) and an autobiography (1990), both reflecting on the development of Antarctic research during his lifetime. He died in Cambridge on 11 November 1999.
Further reading: Fuchs (1982, 1990); Fuchs and Hillary (1958).

Fuchs Dome. 80°36'S, 27°50'W. Ice-covered dome rising to over 1500 m (4920 ft) in the central Shackleton Range, Coats Land. It was discovered by the Commonwealth Trans-Antarctic Expedition 1955–58 and named for the leader, Sir Vivian Fuchs.

Fuchs Ice Piedmont. 67°10'S, 68°40'W. Piedmont ice sheet extending along the west coast of Adelaide Island, named for the British explorer Sir Vivian Fuchs.

Fulmars, southern oceanic. Petrels of the fulmarine branch of the family Procellariidae, distinguished by the single nasal tube on top of the bill (a family characteristic), chunky build, and soaring and gliding flight. Six species are prominent in the southern oceanic region. Largest by far are Antarctic and Hall's giant petrels, respectively *Macronectes giganteus* and *M. halli*, with bill-to-tail length 90 cm (3 ft), wing span 2.4 m (9–10 ft). The southern (Antarctic) species ranges in colour from almost black through mottled grey to all-white, with yellow and green-tipped bill. The northern species is generally darker, with pink or red-tipped bill. The two are similar in foraging and breeding habits, and overlap considerably in geographical range. They feed by gliding over the sea, settling and dabbling for fish, squid, crustaceans and offal. Unlike albatrosses, they also forage ashore, raiding penguin colonies for chicks, seal colonies for birth membranes, and ravaging carcasses of seals or whales. Hall's giant petrels breed mainly on cool temperate islands, Antarctic giant petrels on sub-Antarctic and Antarctic islands and along the Antarctica coast.

Largest of the smaller fulmarine petrels are Antarctic or southern fulmars *Fulmarus glacioloides*, of body length 45 cm (18 in) and wing span 1.2 m (4 ft). Blue-grey, with grey-brown wing-tip primaries and stout chunky bill, they forage widely in cold waters south of the Antarctic Convergence and along the coast of southern South America, feeding mainly on plankton. Southern fulmars nest in cliff colonies on Antarctica and Antarctic islands. Cape or pintado petrels *Daption capensis* are smaller fulmars, with body length 40 cm (16 in) and wing span 90 cm (36 in). Blue-grey, with distinctive white patches on wings and back, they fly in flocks that often number several hundreds, feeding by settling on the water and dipping for plankton. They nest on cliffs mainly in the Antarctic, sub-Antarctic and cold temperate zones. Antarctic petrels *Thalassoica antarctica*, of similar size, have a predominantly brown head and back, a brown leading edge to the wing, and brown-tipped tail. They forage mainly in colder waters close to the pack ice, and nest on the continent and adjacent islands. Snow petrels *Pagodroma nivea* are the smallest fulmarines, with body length 33 cm (13 in), wing span 75 cm (30 in). Pure white, with black eyes, bill and feet, they forage mostly in waters where ice is present, and nest on continental Antarctica and Antarctic islands. Several of their nesting colonies are on inland cliffs and mountains far from the sea.

Furneaux, Tobias. (1735–81). British naval officer who accompanied James **Cook** on part of his second world voyage into the Southern Ocean. Born in Devon, England, Furneaux joined the Royal Navy, was commissioned, and served as a junior officer in HMS *Dolphin* (Capt. Samuel Wallis), in a voyage of exploration that explored many Pacific Ocean islands and circumnavigated the world (1766–68). In 1772, commanding HMS *Adventure*, he sailed with Cook on his major exploration of the Southern Ocean. After crossing the Antarctic Circle in January 1773, the two ships parted company and Furneaux proceeded independently to Van Dieman's Land (now Tasmania), where he charted much of the coast, then to a rendezvous with HMS *Resolution* in Queen Charlotte Sound, New Zealand. Following a voyage in the South Pacific Ocean and a return to New Zealand, Furneaux again lost contact with Cook, and returned to England alone. Promoted to captain, he saw service in the American wars. He died in his home town of Swilly in 1781.

Fur seals. Of the six closely similar species of southern hemisphere fur seal, two species breed in the Antarctic and southern oceanic regions.

Antarctic or Kerguelen fur seals *Arctocephalus gazella*, which are predominantly russet brown, breed on the South Shetland, South Orkney and South Sandwich Islands, South Georgia, Bouvetøya, Iles Kerguelen and Heard Island. Mature males measure up to 2 m (6.5 ft) and weigh about 120 kg (260 lb): females measure up to 1.6 m (5 ft) and weigh up to 50 kg (110 lb). They have been studied intensively on Bird Island, South Georgia where, like elephant seals, they breed in polygamous groups, favouring rocky rather than sandy beaches. Beachmasters arrive in October to early November, females November to early December. Harem groups rarely exceed ten females. Contests between males involve much physical aggression

Antarctic fur seal resting on tussock grass. Photo: BS

and sometimes severe wounding. Pups, only 60–70 cm (24–27 in.) long at birth, grow slowly and are left in groups after about one week, when the mothers return to feed at sea. Thereafter they are fed on average for two days in every six. Weaned at about four months, they leave the colonies in late summer.

Beachmasters achieve their status at age 7–9 years but rarely hold it for more than three seasons. Females first pup at 3–5 years and live on for another ten years, producing pups in most years. Females and most males feed mainly on krill during the summer: in winter fish and squid are also taken. Each summer feeding trip of three to four days involves several hundred shallow dives. Those at night are rarely deeper than 30 m: in daytime they may reach 40–100 m or below, reflecting the daily vertical migrations of the krill. The South Georgia population, reduced by severe hunting during the nineteenth century to the brink of extinction, survived in very small numbers on Bird Island. From a few hundred individuals in the 1950s, the population has now increased to over three million on South Georgia alone. New populations are slowly developing on other islands, for example the South Orkneys and Iles Kerguelen, where they were formerly plentiful. The world population probably exceeds five million and is still growing.

Sub-Antarctic fur seals *Arctocephalus tropicalis* are predominantly grey-brown with paler underparts, the males with distinctive yellowish chests. They breed on Tristan da Cunha, Gough Island, Marion and Prince Edward Islands and Iles Crozet, Amsterdam and St Paul. There has been no detailed study of any population, though in size, weight, feeding, breeding and general biology, they appear closely to resemble Antarctic fur seals. They are closely similar too in breeding biology, but their offspring grow more slowly and are not usually weaned for 9–11 months.

Further reading: Gentry and Kooyman (1986); Riedman (1990).

(JPC)

G

Gabriele de Castilla. 62°58′S, 60°41′W. Spanish summer-only research station on the western shore of the inner crater of Deception Island, South Shetland Islands.

Gadfly petrels. Small petrels of the family Procellariidae, with broad pointed wings and prion-like flight. Forming large mixed flocks over the southern oceans, they settle to feed voraciously on plankton and other particulate matter. Smallest of the group are blue petrels *Halobaena caerulea*, 30 cm (12 in) long with wing span 68 cm (27 in), which breed on cool temperate islands and forage in cold waters south of the Antarctic Convergence. Five of the remaining six species, all of the genus *Pterodroma*, breed on cool or warm temperate islands. While the white-headed petrel *P. lessonii* and Kerguelen petrel *P. brevirostris* forage over colder waters in summer, the great-winged petrel *P. macroptera*, Atlantic petrel *P. incerta* and soft-plumaged petrel *P. mollis* are more likely to be found in warmer waters north of the Convergence. Mottled petrels *P. inexpectata*, which breed in New Zealand and Australia, forage far south into cold waters in the Pacific and Indian Ocean sectors.

Gaimard, Joseph-Paul. (1796–1858). French naval surgeon and naturalist who, with Jean-René Constant **Quoy**, identified and named many Antarctic and Southern Ocean animals. Born in Provence, Gaimard studied medicine and enrolled in the naval medical service. An early assignment was as assistant surgeon in Louis-Claude de Freycinet's scientific expedition in *Uranie* (1817–20), visiting many islands in the southern oceans, particularly the South Pacific. On board he assisted the already well-established surgeon-naturalists Charles Gaudichaude-Beaupré and Jean-René Constant Quoy in collecting botanical and zoological specimens. *Uranie* was wrecked on the Falkland Islands: the expedition returned to France in a second ship, *Physicienne*. In 1826 Gaimard was appointed surgeon and naturalist in *Astrolabe*, sailing with J. S. C. Dumont d'Urville's expedition to explore and chart the coasts of Australia, New Zealand and the western Pacific Ocean. Again working with Quoy, he collected and named many southern species, providing type-specimens for the French national collection. In 1835–36 he visited Greenland in the French expedition ship *Recherche*. Gaimard became a leading advocate of France's continuing involvement in polar regions, an interest he maintained to the end of his life.

Gamburtsev Subglacial Mountains. 80°30′S, 76°00′E. An extensive range of mountains underlying Dome Argus, the high dome of the central ice cap of East Antarctica. Apart from elevation of the ice cap, there is no indication of its presence at the surface. The range was detected in 1958 by Soviet scientists using seismic methods, and named for Soviet geophysicist Grigoriy A. Gamburtsev (1903–55).

Gaussberg. 66°48′S, 89°11′E. Ice-free coastal mountain of Wilhelm II Land. Emerging from the ice cliffs west of Posadowski Glacier, it rises to a rounded top 370 m (1314 ft) high. It was discovered by the **German South Polar Expedition 1901–3**, and named for their ship *Gauss*.

General Belgrano stations. A succession of Argentine year-round research stations on the southeastern Weddell Sea coast, named for the Argentine patriot General Manuel Belgrano (1770–1820). The first, established on the Filchner Ice Shelf in 77°43′S, 38°04′W, operated from January 1955. Overwhelmed by accumulating snow, it was replaced by General Belgrano II, a new station built from February 1979 in 77°52′S, 34°37′W, on bedrock close to Bertrab Nunatak, Luitpold Coast, Coats Land. The original station drifted out to sea in 1985. General Belgrano II was built on the site of 'Label', an Argentine refuge erected in 1970, and has operated continuously since 1981. General Belgrano III was established in 1981 in 77°54′S, 45°59′W, on the northern end of ice-covered Berkner Island. It operated for three winters before closing in 1984.

General Bernardo O'Higgins Station. 63°12′S, 58°58′W. Chilean permanent year-round research station, named for a nineteenth-century liberator and first president of Chile, and established in February 1948 on Schmidt Peninsula, Cape Legoupil, Antarctic Peninsula. The station was rebuilt after a fire in March 1958, and has since been occupied continuously. An alternative name for the scientific laboratory was 'Luis Risopatron'.

General Ramon Cañas Montalva Station. 63°32′S, 66°48′W. Chilean research station in Duse Bay, Trinity Peninsula, Graham Land. The station, formerly View Point (Base V) of British Antarctic Survey, was transferred to Chilean ownership in July 1996.

General San Martín Station. 68°07′S, 67°06′W. Argentine permanent year-round research station on Barry Island, Debenham Islands, Fallières Coast. Established in March 1951 on the site of the old British Graham Land Expedition base, the station was destroyed by fire in February 1959. Rebuilt in 1976, it has operated continuously since then.

Gentoo penguin. *Pygoscelis papua*. Largest of the three species of pygoscelid (brush-tailed) penguins, gentoos stand 75 cm (30 in) tall, and weigh about 5.5 kg (12 lb): individuals of northern populations tend to be slightly larger than those breeding further south. Distinguished by their orange bill and white flash above the eyes, they live in a circumpolar, wide latitudinal range from the Falkland Islands to about 65°S along the west coast of Antarctic Peninsula, including all the islands of the Scotia Arc, and many of the Southern Ocean and cool-temperate islands. Gentoos feed on shoaling fish and crustaceans, diving to depths of 150 m (500 ft) though usually much less. They breed in colonies of several hundreds to thousands of pairs, building nests of pebbles, moss and tussock grass, and laying two bluish-white eggs, rarely three. Chicks are grey, with distinctive orange bills and feet. The world population is estimated at 300 000 pairs.

Geographic Pole, South. The point, in Latitude 90°S, that marks the southern end of the axis of rotation of Earth. It lies at the southern point of convergence of meridians: from it all distances are north. Its position in relation to the sun shifts slightly as Earth wobbles on its axis. At an elevation of 2835 m (9299 ft) on the South Polar Plateau, it stands on almost the same thickness of ice. First to reach the South Pole were the sledging party of five led by Norwegian explorer Roald Amundsen, on 14 December 1911. Since January 1957 it has been the site of Amundsen-Scott Station, a permanent year-round US research facility.

Geology: geological history and palaeontology. The palaeontological history of Antarctica is reflected in a substantial and growing fossil record from both eastern and western provinces. (For an outline of the geological structure of Antarctica see **Geology: stratigraphy and structure**).

Palaeozoic era

In the Cambrian period (590–505 million years ago), Gondwana including East Antarctica straddled the Equator. The coastline of what is now West Antarctica was inverted from its present position and protruded into the Northern Hemisphere. Vast thicknesses of limestones, sandstones and shales were laid down in shallow tropical seas. East Antarctica's oldest Palaeozoic animal fossils from this period are marine reef-building sponges and other sea-floor invertebrates, contributing to the limestone and long preceding the earliest fossil plants. There is as yet no record of the microscopic marine plants that formed the bases of food chains in those early tropical seas. A common fossil in the limestones is *Archaeocyathus*, a small vase or cup resembling a coral in cross-section, but related more closely to sponges. In some places these were numerous enough to form low mounds or reefs on the sea floor. Living between the reefs on both limey and muddy sea floors were dominant trilobites, relatively rare brachiopods, and small, primitive molluscs.

During the late Cambrian and Devonian periods (480–360 million years ago) Gondwana moved into cooler Southern Hemisphere latitudes. Antarctica's oldest-known fossil land plants are fragmentary remains of a Devonian flora, approximately 360 million years old, mainly from the Transantarctic Mountains. They include ribbed stems of primitive plants akin to modern club-mosses, and spores of an early fern *Archaeopteris*, which probably grew in warm, moist habitats along river banks.

In what is now the Horlick, Pensacola and Ellsworth mountains region, sands and silts were deposited in a large shallow bay. Mid-Palaeozoic strata from this area have yielded brachiopods *Pleurothyrella* and *Australospirifer*, bellerophontid and several kinds of bivalve molluscs, especially burrowing forms, bryozoans and tentaculitids. Trilobites were much less common, but the single species known from the Horlick Mountains, *Burmeisteria antarcticus*, grew to 30 cm long. Similar fossils are widespread across South America and South Africa, and possibly New Zealand. Simultaneously, terrestrial and fresh-water deposits were accumulating in shallow basins between mountains, forming the earliest sediments of the Beacon Supergroup, Transantarctic Mountains. Though fossils are generally rare in these non-marine beds, there are horizons with well-preserved fish fragments, including sharks in southern Victoria Land.

Continued cooling through the late Palaeozoic era produced glacial deposits, the earliest dating from the late Devonian (360 million years ago). In the Permo-Carboniferous period, 280 million years ago, Gondwana was centred about the south polar region much as Antarctica is now. Widespread glaciation produced vast ice sheets over much of what is now East Antarctica, Australia, eastern India, South Africa and southern South America, excluding all forms of life. Though there were warmer interglacial episodes, Antarctica yields only a very poor palaeontological record from late Devonian to Permian times.

Glacial conditions over East Antarctica persisted for at least part of the Permian, 286–248 million years ago, during which Gondwana shifted northward and the climate gradually warmed. By the late Permian a new flora invested the whole Gondwana region, dominated by fern-like plants typified by the genus *Glossopteris*, growing in swampy situations that over many millions of years gave rise to the Permian coal deposits of the Transantarctic Mountains. Continued warming produced drier conditions: by the end of the Palaeozoic much of Gondwana appears to have been a hot desert.

Mesozoic era

Continued warming into the Mesozoic era (about 245 million years ago) culminated in loss of the ice caps from both polar regions, and produced further changes in vegetation throughout Gondwana. In East Antarctica cool-climate glossopterids were replaced by the seed fern *Dicroidium*, of which distinctive forked fronds, spores and pollen are found in Triassic rocks of Victoria Land. This genus appears to have been adapted to dry conditions; a thick, waxy coating covered its leaves, limiting loss of moisture. Stems of horsetails (Equisitales), large fronds of Cycads, and the spores of ferns, lycopods (clubmosses) and bryophytes (mosses) are also preserved in rocks of this period. Along the margins of East Antarctica, thick continental deposits of sandstones and grey-green shales were accumulating in interconnected basins. Now identified as the Victoria Group, they form the topmost unit of the Beacon series. Included among them is the Lower Triassic Fremouw Formation, famous for a fauna of reptiles and amphibians, including the large herbivorous reptile *Lystrosaurus*, which occurs also in South Africa and India.

By Jurassic times, 213–144 million years ago, the islands that form the Antarctic Peninsula were beginning to rise above a warm ocean. Despite their high latitude they were densely vegetated with a flora consisting mainly of ferns (for example, *Cladophlebis*), small scale-leafed conifers such as *Brachyphyllum*, cycads, and *Ginkgo* maidenhair trees. Fossils of *Ginkgo* or related forms occur at several Antarctic Peninsula localities, e.g. at Hope Bay (a particularly well-studied flora), also at Snow Island and Livingston Island in the South Shetland Islands, and on Carapace Nunatak in East Antarctica. Flanking the islands lay deep-sea basins in which volcanic ash and other fine sediments were laid down. The earliest of these was on the SE side of the peninsula, where several species of ammonite and bivalve mollusc are found. Sedimentary Jurassic rocks of Alexander Island, perhaps representing a former seamount, have yielded a small but important fauna including an ammonite resembling *Epophioceras*, the starfish *Protremaster*, a range of gastropod and bivalve molluscs, solitary corals, crinoids, and sea urchins.

Sedimentation continued in this area into the late Jurassic, when a narrower and deeper basin developed on the opposite side of the peninsula. The northern end of the peninsula formed part of a deep basin covering much of the proto-South Atlantic, in which were deposited dark muds rich in radiolaria (minute siliceous animals of the plankton). Late Jurassic marine fossils from the Antarctic Peninsula include ammonites *Virgatosphinctes*, belemnites and bivalve molluscs, for example, *Retroceramus*, as well as brachiopods, serpulid worms, echinoids and occasional fish fragments. Many late Jurassic species identified from Antarctica have been found also in the Malagasy Republic, India, Indonesia, and New Zealand.

The West Antarctica islands remained forest-covered throughout the Cretaceous period (130 million years ago). The Fossil Bluff Formation of Alexander Island includes coals and deltaic sediments, with fossil forests buried in their growth positions. In sediments of similar age on James Ross Island, Seymour Island, and other islands adjacent, trunks of large forest trees, washed into a sedimentary basin on the east side of the land area, are now preserved as petrified logs. These were conifers, mainly ancient relatives of the podocarps and araucarians which are important in Southern Hemisphere forests today. Growth rings in Cretaceous wood are wide and evenly spaced, indicating warm conditions similar to those of modern New Zealand.

Toward the end of the Cretaceous the angiosperms or flowering plants began to diversify, to include broad-leafed trees. Southern beeches of the genus *Nothofagus*, ancestral to the beeches that dominate southern forests today, appeared and spread quickly. The oldest fossil *Nothofagus* traces are wood and pollen from Upper Cretaceous sediments (80 million years ago) from James Ross Island. Leaves are found in younger Tertiary deposits on Seymour Island, along with several other types of angiosperm leaves, including laurel, proteas and myrtles. Podocarp and araucarian conifers were also present. Similar fossil floras occur on King George Island in the South Shetland Islands. Leaf-shape analysis suggests the forests were similar to modern cool- to warm-temperate forests of Australasia. Marine sediments of the northern Antarctic Peninsula area coarsened from muds to sandstones and conglomerates as the islands rose higher above sea level. Early Cretaceous fossils of 119–97 million years ago include distinctive ammonites, belemnites and bivalve molluscs, many with more limited geographical ranges than their late Jurassic counterparts, and possibly restricted to cooler high-latitude seas. Land animals of the late Cretaceous (87–73 million years ago) included dinosaurs, remains of which occur on James Ross Island.

During the late Cretaceous (97–85 million years ago) Gondwana began to split into its component continents, a break-up heralded by widespread volcanic eruptions. The bloc including East Antarctica and Australasia moved southward over the South Pole, and marine faunas along its Pacific margins again began to be isolated. Thick, fine-grained sediments accumulating in a marine basin on the site of James Ross Island have produced some of Antarctica's most fossiliferous strata. The uppermost Marambio Group has yielded a wealth of invertebrates including ammonites, belemnites, bivalves, gastropods, decapod crustaceans, echinoids, asteroids, serpulids, and brachiopods, together with fragmentary remains of large marine reptiles and terrestrial dinosaurs. Though the continuing fragmentation of Gondwana opened up the Atlantic and Indian Oceans, the Antarctic marine fauna remained distinctive, perhaps already isolated by late Cretaceous chilling.

Tertiary era

During the early Tertiary *Nothofagus* and the podocarp conifers were the dominant Antarctic plants. Following the break-up of Gondwana, East Antarctica had started drifting southward into cool-temperate latitudes, encouraging the vigorous growth of trees and other vegetation. Southern beeches of the genus *Nothofagus* and monkey-puzzle trees, *Araucaria* spp., were typical of this period. Exposures on Seymour Island indicate that sedimentation continued in this region into the Cenozoic. Fine-grained sandstones and siltstones of Palaeocene-to-late Eocene age (65–40 million years ago), deposited in a delta or estuary, are richly fossiliferous, with burrowing gastropod molluscs, crabs and brachiopods.

Near the top of this sequence is a remarkable beach deposit crammed with both invertebrate and vertebrate remains. Among the vertebrates are sharks, teleost fish, turtles and other reptiles, birds including giant penguins, and marine mammals including a whale. On land were large flightless birds, which left their footprints in Palaeocene to Eocene volcanic sediments of the South Shetland Islands some 60 million years ago, and a small rodent-like marsupial of the South American family *Polydolopidae*. Thus Antarctica, retaining land links with neighbouring South America and Australia, was the home of some of the earliest mammals.

Fossils are scarce from the uppermost Eocene onwards, possibly due to the destructive effects of later glaciations. Some glacial deposits, which can be traced intermittently from the shores of the Ross Ice Shelf to the South Shetland Islands, contain a shelly fauna including molluscs (especially bivalves), crustaceans, brachiopods and bryozoans. These are almost certainly the product of high sea-level stands during warm inter-glacial spells. Some range back to the Oligocene (38–24 million years ago), while others are only of Pliocene age (5–2 million years ago).

Of the subsequent cooling, the disappearance of the forests and their replacement by tundra and polar desert vegetation, as yet no fossil record has been discovered. Further lateral movement and rotation of the southern continents continued, and has continued to the present day. Now isolated from the other continents by wide expanses of sea floor, Antarctica has returned almost to the south polar position that it occupied as central Gondwana 280 million years ago. In consequence, it has acquired an ice carapace and the accompanying conditions of continental glaciation. *Further reading*: Tingey (1991).
(RJA/JEF/JAC)

Geology: stratigraphy and structure.

Geologically, Antarctica has proved the most difficult of all continents to explore, for thick ice conceals over 98 per cent of its surface. Field observations of early twentieth-century geologists, who man-hauled into an unknown and hostile environment to bring back the first geological information about the continent, are as valid today as when first recorded. However, studies during and since the **International Geophysical Year (1957–58)** have provided a substantial accumulation of geological data, correlations and interpretations. Now almost all of Antarctic bedrock that is visible has been visited and mapped, and geophysical and remote-sensing techniques have helped to resolve the many geological problems disclosed.

The earliest systematic studies were made in the predominantly sedimentary rocks of the Transantarctic Mountains, East Antarctica. Antarctic Peninsula, first studied a few years later, proved to be geologically completely separate and different, and the concept arose of a major rift from the Ross Sea to the Weddell Sea, separating two sub-continents, East and West Antarctica. This concept is still accepted today. East Antarctica clearly originated in close company with southern Africa, India and Australia as part of the Gondwana supercontinent, separating from its former neighbours by progressive spreading of the Southern Ocean floor. West Antarctica in contrast forms part of a circum-Pacific chain of mountains, uplifted by forces of lateral pressure similar to those that lifted the Western Cordillera of North America and the South American Andes.

East Antarctica

East Antarctica forms a continental shield based on a Precambrian metamorphic and crystalline platform.

The basement rocks, of which the oldest known date from 3000 Ma, were folded in several separate events, most recently some 630–1000 Ma (the Precambrian Nimrod orogeny). Overlying are younger Upper Precambrian–Cambrian rocks of the Ross Supergroup, which were folded during the Ross orogeny (475–500 Ma), and in many localities injected with molten rock now identified as Granite Harbour Intrusives. Between Cambrian and Devonian times these rocks were worn down to a surface now represented, throughout the Transantarctic Mountains and elsewhere, by the Kukri Peneplain. On this surface were laid down near-horizontal Devonian–Jurassic sediments, the Beacon Supergroup, named from a prominent mountain in which the succession of sandstones, limestones, shales and coal measures shows clearly. This comprises the bulk of the Transantarctic Mountains. During the Jurassic the Beacon Supergroup was overlain initially by lavas and then intruded by widespread dolerite sills of the Ferrar Supergroup. A much later spell of volcanism is represented by the Cenozoic McMurdo Volcanics. If there was any sedimentation during this period, there is now little evidence of it on land.

In coastal regions the structural grain, defined by fold–axis trends, has been disrupted by large-scale block-faulting. Deep seismic shooting in the mountains of Dronning Maud Land has proved the existence of major upthrust and down-thrust blocks between north–south faults, some virtually undeformed and unmetamorphosed, with throws in the order of 10 km. In Coats Land, at the north-westerly extension of the Transantarctic Mountains, the Shackleton Range and Theron Mountains are horsts defined by major east–west fault systems. In Victoria Land, block-faulting with east–west bounding faults is an important feature, superimposed on the fold systems of the Nimrod and Ross orogenies.

The Precambrian basement rocks, near-horizontal Beacon sediments, and intruded Ferrar dolerite sills of Victoria Land show that major faulting has occurred parallel to the coast. Lesser transverse faulting caused planes of weakness which were frequently occupied by glaciers. This led to the concept of an 'Antarctic Horst' and the 'Inland Plateau', a horst-and-graben system supposedly dominating the structure of East Antarctica. An implied corollary was a supposed Weddell Sea–Ross Sea graben separating East and West Antarctica. However, geophysical studies in Marie Byrd Land have shown beyond doubt that no such graben exists. Instead, a channel extends well below sea level from the head of the Ross Ice Shelf westward to the vicinity of the Amundsen Sea, separating much of Marie Byrd Land as a major island archipelago from continental Antarctica.

In the region of East Antarctica from Dronning Maud Land to the Prince Charles Mountains are exposed some of the oldest rocks in Antarctica. In some high-grade gneisses of the Napier complex in Enderby and Kemp lands, U–Pb data on relict zircon crystals have yielded isotopic ages of 3930 million years, in rocks that were intensely deformed about 3100 million years ago. Several episodes of granite intrusion, migmatization, metamorphism and deformation have been recognized through the Archaean and Proterozoic and into the early Palaeozoic. The youngest rocks of the region are outcrops of sedimentary rocks in Vestfjella, Heimefrontfjella and the Prince Charles Mountains that are equivalent to the Beacon Supergroup in the Transantarctic Mountains.

West Antarctica

In contrast to East Antarctica, West Antarctica bears all the characteristic features of the South American Andes. Nowhere have metamorphic basement rocks akin to those of East Antarctica been found. The rocks forming Antarctic Peninsula were deposited mainly as muds and other fine sediments in a geosyncline (sea-bed trough) from late Palaeozoic to early Mesozoic times, and later metamorphosed by volcanic events and intrusions. The folded Trinity Peninsula Group of greywacke sediments fringe the east coast of northern Antarctic Peninsula, extending southward into central Alexander Island and Marie Byrd Land. At least two major phases of volcanism and several minor ones have been identified. The most important and widespread rocks resulting are the Upper Jurassic andesite–rhyolite volcanic group. At least five main phases of plutonic (deep) intrusion have been dated radiometrically, the youngest and most widespread being the late Cretaceous–early Tertiary Andean Intrusive Suite. Many similar stratigraphical features have been recorded also from Marie Byrd Land, notably the striking Cenozoic volcanic province that includes the Executive Committee Range, showing freshly exposed lava and ash cones, and considerable evidence of subglacial eruption, indicating that volcanism continued after the ice sheet had developed. Similar late volcanism may have occurred on the western extremity of Alexander Island, and possibly also in the South Shetland Islands.

The Ellsworth Mountains, which include Antarctica's highest peaks, stand between the Transantarctic Mountains and Marie Byrd Land as an anomaly in this structural scheme. Stratigraphically they are allied to East Antarctica, but their structural trend, almost east–west, runs at a right angle to the trends of both East and West Antarctica. They form a separate unit which also encompasses the Whitmore and Pensacola mountains.

The arc of the Antarctic Peninsula is related to major Triassic folding which was responsible for the south-west to north-east fold axes in the Trinity Peninsula Group sediments. Similar folds of the same period apparently extended to the Ford Ranges in Marie Byrd Land. In the north-eastern Antarctic Peninsula and Alexander Island, folding possibly during the late Tertiary, with axes trending almost

Table 1 Stratigraphy and structure of East and West Antarctica. Ma = millions of years before present

	East Antarctica	West Antarctica (Peninsula)
Pleistocene	Glaciation	Glaciation
Pliocene		Pecten conglomerate
Upper Miocene	} McMurdo Volcanics	Camptonite dykes (15 Ma) of Alexander Island
Upper–Middle Miocene		James Ross Island Volcanic Group (plus dyke swarms)
Late Miocene		Seymour Island Series [Andean Intrusive Suite (45–70 Ma)]
Upper Cretaceous		Snow Hill Island Series / Cape Longing Series
Middle Cretaceous		[Acid-basic intrusions (90–110 Ma)]
Lower Cretaceous–Upper Jurassic		Fossil Bluff Formation of Alexander Island
Upper Jurassic		Andesite–rhyolite volcanic group [Acid intrusions (130–140 Ma)]
Jurassic	Ferrar Group	Acid intrusions (160–180 Ma)
Triassic		Trinity Peninsula Group
Carboniferous	Victoria Group } Beacon Supergroup / Taylor Group	
Devonian		[Granitic intrusions (370 Ma)]
Carboniferous–Devonian	Admiralty Intrusives (350 Ma) / Kukri Peneplain	
Ordovician–Cambrian	Granite Harbour Intrusives [Ross Orogeny (475–500 Ma)]	
Cambrian	Robertson Bay Group / Byrd Group / Koettlitz Group / Skelton Group — } Ross Supergroup	
Upper Precambrian	Beardmore Group	
[Early Palaeozoic]		Metamorphic complex
	[Nimrod Orogeny (630–1000 Ma)]	
Precambrian	Nimrod Group (1000 Ma) / Wilson Group	

north–south, deformed the Cretaceous sediments, adding complexity to the overall structure. Major block faulting, perhaps also in the late Tertiary, is responsible for much of the present-day topography of Antarctic Peninsula, as it is for that of the Andes. However, the origins of the Antarctic Peninsula plateau and George VI Sound are still uncertain.

Major fracture zones and sea-floor spreading centres can readily be detected by geophysical techniques in the oceans surrounding Antarctica, but they cannot satisfactorily explain the tectonic relationship between East and West Antarctica, the apparently anomalous structural setting of the Ellsworth Mountains, the very existence of the higher Transantarctic Mountains, or the opening of the Weddell Sea. It is generally accepted that the isolated mountains of West Antarctica lie on several micro-plates that separated and moved independently during the break-up of Gondwana. It seems likely that the Ellsworth–Whitmore mountain bloc is a rotated sliver torn from East Antarctica about this time. Evidence for a triple junction in the southern Weddell Sea may help to resolve the complex movement history of this region.

Further reading: Tingey (1991).

(RJA)

Geomagnetic Pole, South. Point on Earth's surface, in 78°30′S, 111°00′E, on the high plateau of East Antarctica, marking the southern end of the axis of Earth's geomagnetic field. From it emerge the lines of force that extend far into space to form the magnetosphere. Convergent lines immediately above the pole create an intense magnetic field within a ring of 23° in the stratosphere: in consequence observers at the geomagnetic pole see a high incidence of displays of the **aurora australis**. The Russian research station **Vostok** is sited close to this pole.

Geophysical techniques. Geophysical surveys in Antarctica involve remote measurement of the physical properties of rocks, mapping geological structures of both surface rocks and those beneath ice or water, and giving information about crustal structure to depths of tens of km. Many techniques can penetrate ice and water, making them particularly valuable in the geological exploration of Antarctica.

Magnetic field measurements

A magnetic sensor towed behind a ship, mounted on a sledge, or attached to an aircraft, at a distance from manmade magnetic disturbances, measures and records the strength of Earth's magnetic field to an accuracy of 1 part in 50 000. Small differences in the recorded field reflect concentrations of magnetic minerals in Earth's crust. A quiet magnetic field suggests sedimentary accumulations with few or evenly distributed magnetic minerals. Large anomalies may be caused by a variety of structures including metamorphic and igneous rocks. Magnetic surveys of ocean floor ridges yield information about age and rate of expansion of the floor (see **magnetic anomalies**), with the youngest stripes closest to the ridges.

Gravity measurements

Gravity surveys make use of sensitive spring balances (gravimeters) accurate to 1 part in one hundred million. Gravimeters can be used in ships or aircraft, but work best in stable conditions on land or satellites. They provide measurements that reflect density differences between different kinds of rock, separating out effects of intervening ice, water and air to reflect only the structure of the rock itself. Light-weight sediments and granites give negative anomalies: denser metamorphic and igneous rock give positive anomalies.

Seismic measurements

Sound shock-waves, generated by a succession of controlled explosions, travel through ice and rock and are reflected or refracted at interfaces. The resulting waves, picked up by receivers, are converted into electrical impulses and stored electronically for later analysis. In the simplest type of seismic reflection survey, depth to bedrock through ice and water is calculated from the travel time of the first reflected wave, using known velocities of sound in those media. A similar technique is used to determine the velocities of sound waves within layers of rock. Sophisticated arrays of receivers are towed behind a ship or by a large over-snow vehicle. Intermittent signals or pulses are transmitted and recorded digitally 500 times per second. Processed by computer, they give a picture approximating to a geological cross-section. Seismic reflection sections show excellent resolution in areas underlain by layered sediments. Refraction surveys use the passage of the shock wave along boundaries. The time taken for energy to reach a distant receiver indicates the velocity of sound along the interface. Small-scale refraction experiments provide information, for example, about compaction in the shallowest layers of snow and ice. Refraction experiments to measure geological structures require large explosive charges and recording arrays tens of km long.

Radiometric dating

Radiometric, isotopic or absolute dating of rocks depends on the capacity of certain unstable elements, contained in crystalline rock minerals, to decay naturally at fixed rates by the emission of radioactive particles. If the rate of decay is known, then the proportion of original or parent element to derived or daughter element, accurately determined by mass spectrometry, is a measure of how long ago the mineral crystals were formed. Some formed during cooling of liquid magma, some by recrystallization during periods of mountain-building, and some by evaporation from solutions: in every case the dating applies to the most recent event in which crystallization occurred. Many elements decay in this way, but for practical purposes the processes most often measured in radiometric dating are the decay of potassium to argon (K–Ar) and rubidium to strontium (Rb–Sr). Potassium is a component of many common minerals, for example, biotite, hornblende and muscovite. Rubidium is rarer, but often found accompanying potassium in feldspars and other crystals. The techniques involved are complex, and results are accurate to between 2 per cent and 5 per cent of the age, e.g. 2–5 million years in a rock 100 million years old, 20–50 million years in one 1000 million years old. By providing time scales of these levels of accuracy, radiometric dating has contributed largely to the unravelling of stratigraphical and tectonic problems of Antarctica. In East Antarctica at least 13 important tecto-magmatic events (mountain-building upheavals and the flow of liquid rocks), each producing rocks of recognizable age, have so far been identified. In the relatively newer rocks of West Antarctica at least five major intrusive phases have been detected.
(SWG)

Georg Forster Station. 70°46'S, 11°50'E. Research station of the former German Democratic Republic in Schirmacheroasen, Prinsesse Astrid Kyst, sharing site and facilities with the Soviet station Novalazarevskaya.

Georg von Neumayer Station. 70°38'S, 8°15'48"W. German permanent year-round research station on Ekströmeisen, Dronning Maud Land. Established in February 1981, it became buried under snow and was replaced in March 1992. The station continues to operate under the name Neumayer Station.

George V Coast. 67°30'S, 145°00'E. Part of the coast of George V Land, East Antarctica, bounded by Point Alden in the east and Cape Hudson in the west. Consisting mainly of ice cliffs with nunataks and emergent islands, the coast is backed by steeply rising ice slopes, dotted with isolated peaks. It includes also the large Ninnis and Mertz glaciers and the Cook Ice Shelf. Both the coast and the interior were explored by the Australasian Antarctic Expedition 1911–14, from their main base at Cape Denison, Commonwealth Bay.

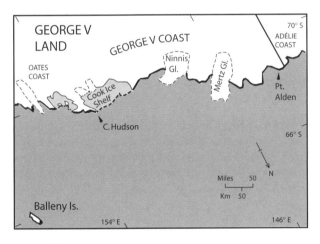

George V Coast

George V Land. Ice-covered area of East Antarctica between Oates Land and Terre Adélie, forming part of Australian Antarctic Territory. It includes George V Coast and part of Oates Coast, steep coastal slopes and an extensive interior ice plateau extending to the South Pole. The coast and interior were explored by the Australasian Antarctic Expedition 1911–14 from their main base at Commonwealth Bay. The land was named in that period for the reigning British monarch.

George VI Ice Shelf. 71°45'S, 68°00'W. Extensive ice shelf filling George VI Sound between Alexander Island and English Coast, Antarctic Peninsula. The sound and shelf were mapped by the **British Graham Land Expedition 1934–37** and named for the contemporary British monarch.

George VI Sound. 71°00'S, 68°00'W. Long ice-filled sound between Alexander Island and English Coast, Antarctic Peninsula, discovered by the **British Graham Land Expedition 1934–37** and named for the contemporary British monarch.

Gerlache Strait. 64°30'S, 62°20'W. Broad strait between the islands of Palmer Archipelago and Antarctic Peninsula. The strait was named for Lt Adrien de Gerlache de Gomery, leader of the **Belgian Antarctic Expedition 1897–99**, who explored and charted it in early 1898.

German Antarctic (*Deutschland*) Expedition 1911–13. A first attempt to cross Antarctica. From 1908 a German army officer, Lt Wilhelm **Filchner**, planned an expedition to the South Pole that would involve crossing from one side of Antarctica to the other. At first he planned for two ships, that would place field parties simultaneously on opposite sides of the continent. Funding, however, limited him to a single venture. Gaining backing from the German government and lottery funds, he bought a sturdy Norwegian ship, renamed it *Deutschland*, and collected together a strong scientific team.

Under the command of Capt Richard Vahsel, the ship left Bremerhaven on 11 May 1911. After a brief stop for bunkering in Buenos Aires, they arrived at Grytviken, South Georgia on 18 October. As yet it was early to challenge the pack ice, so they waited, exploring locally and making a brief excursion to the South Sandwich Islands. On 10 December Filchner headed south toward the Weddell Sea. Encouraged by the earlier success of the Scottish National Antarctic Expedition 1902–4, *Deutschland* moved slowly southward along the Coats Land coast, finding loose pack ice with leads enough to make sound progress. On 30 January Filchner saw land, a high ice cliff backed by steep ice slopes, which he named for Luitpold, the German Prince Regent. Feeling his way westward along the sheer cliffs of a neighbouring ice shelf, he named the shelf for Kaiser Wilhelm II: however, posterity has renamed it the Filchner Ice Shelf. On 9 February they found a landing site with a convenient ramp up onto the shelf, and started to unload ponies, dogs and stores, including prefabricated sections of a wooden hut in which Filchner hoped to winter.

Then disaster struck: the section of ice shelf on which they were building broke away, creating a tabular berg that began to drift north. With great difficulty they retrieved much of the equipment, and began to seek a way out. The pack ice intensified, and in early March ice began to form in the leads. *Deutschland* could no longer manoeuvre, and began to drift with the consolidating pack ice. On board they were comfortable and well-stocked, with a full programme of daily scientific observations to keep them busy. As they drifted northward, Filchner and two of the ship's officers made a winter sledging journey with dog teams to seek 'New South Greenland', a landmass that the American sealer Benjamin Morrell had claimed to have seen in the early nineteenth century. They found no sign of land. In August the ship's master, Capt. Richard Vahsel, sickened and died: his role was taken by the first officer, Alfred Kling. By September wide leads were appearing around the ship, and in late November *Deutschland* was again free. The expedition returned to South Georgia in December, and from there headed back to Germany. Though Filchner failed in his main ambition to cross the continent, his geographical discoveries and the winter studies of his scientific team made it a profitable and worthwhile expedition.

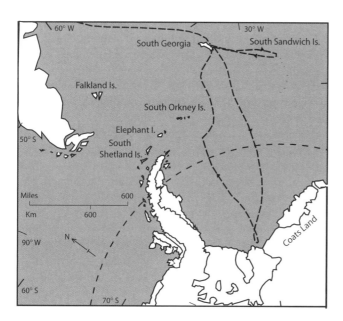

German Antarctic (*Deutschland*) Expedition. From South Georgia *Deutschland* visited the South Sandwich Islands, then headed south. The ship was held for six months in the Weddell Sea pack ice before returning to South Georgia

German Antarctic (*Schwabenland*) Expedition 1938–39.

A bid by the German government to survey and claim land in Antarctica. During the late 1930s Germany returned briefly to Antarctic waters, operating a whaling fleet of up to seven factory ships in the Southern Ocean. In 1938 the German government decided to provide grounds for a formal claim to Antarctic land, within a sector that had briefly attracted the interest of an earlier generation of explorers (see **German Antarctic (*Deutschland*) Expedition 1911–13**). To this end they sponsored an expedition to survey by air some of the mountains, as yet unexplored, that lay behind the East Antarctica coast east of the Weddell Sea. This sector was of particular interest to Norway, having been investigated largely by the Norwegian (Christensen) Whaling Expeditions 1927–37. Led by Capt. Alfred Ritscher, the expedition left Hamburg on 17 December 1938 in *Schwabenland*, an 8500-ton Lufthansa tender and catapult ship. In the Southern Ocean *Schwabenland* headed first for Bouvetøya, then moved south to meet the edge of the pack ice of King Håkon Sea. Between 20 and 23 January *Boreas* and *Passat*, the two Dornier Super Wal seaplanes piloted by Richardheinrich Schirmacher and Rudolf Wahr, flew seven extensive missions, photographing some 250 000 sq km (almost 100 000 sq miles) of ice-covered mountains. They also dropped dozens of aluminium markers, emblazoned with the swastika, staking a claim to the territory that Germany was to call 'Neuschwabenland'. Parties were landed along the mainly ice-mantled coast, but little effort was made to provide ground control for the many hundreds of photographs, substantially reducing their value for accurate mapping. The expedition returned to Hamburg on 10 April 1939. On learning of the German operations, the Norwegian government on 14 January 1939 announced a formal claim to the area now called **Dronning Maud Land**. Many of the expedition records were destroyed in World War II, and names assigned by German cartographers to features discovered by the expedition have not been accepted internationally. Schirmacher Hills, the generally accepted name for a coastal 'oasis' area, commemorates the pilot who discovered it.

German Deep Sea Expedition 1898–99.

A shipborne expedition to the Southern Ocean, sponsored by the German government and led by oceanographer and zoologist Dr Carl Chun. In the chartered steamship *Valdivia* (Capt. Adelbert Krech), the expedition left Hamburg on 1 August 1898 for the Atlantic Ocean, visiting Cape Town in November and heading south to Bouvetøya, the position and magnitude of which were for the first time accurately determined. From there *Valdivia* took a line of soundings southward and eastward along the edge of the pack ice to 58°E. On 16 December the ship achieved a furthest southing of 64°15′S, dredging from a depth of 2540 fathoms (4646 m) erratic boulders of granite, gneiss and sandstone, indicating that Enderby Land, their probable point of origin some 160 km (100 miles) to the south, was of continental rather than volcanic structure. *Valdivia* visited Iles Kerguelen, St Paul and Amsterdam, then continued exploration in warmer latitudes. The expedition returned to Hamburg on 30 April 1899.

German International Polar Year Expedition 1882–83.

The first International Polar Year, a year of coordinated research by scientists of 11 nations at 14 polar research observatories, was inspired by German geophysicist Georg von Neumayer and organized by Lt Karl Weyprecht of the Austro-Hungarian Navy. The German contribution included two research stations, one at Baffin Island in the Arctic, the other at Royal Bay, South Georgia, where a team led by Karl Schrader conducted a comprehensive year-round programme of research. The expedition for South Georgia left Hamburg on 2 June 1882, transhipping on 23 July in Montevideo to the steam corvette *Moltke*, and arriving in South Georgia in mid-August. Bad weather delayed landing and thick snow delayed building, but a station of prefabricated huts, including living quarters, magnetic and astronomical observatories, dark room and laboratories, was set up and operational by 3 September. The team of 11 included physicists, biologists, a meteorologist and a medical officer: they brought with them 3 oxen, 17 sheep, 9 goats and 2 geese. Apart from a rigorous programme of geophysical and astronomical observations, the team (the first scientists to over-winter in the Antarctic region) produced excellent maps of the Royal Bay area and reports on the local weather and natural history. The station was relieved on 3 September 1883 and closed three days later.

Further reading: Headland (1984).

German South Georgia Expedition 1928–29. An expedition led by the German scientist Ludwig Kohl-Larsen to study glaciology and other natural history. Kohl-Larsen, a veteran of the **German Antarctic (*Deutschland*) Expedition 1911–13** and of several subsequent whaling voyages, was accompanied by his wife Margit, a daughter of the whaling entrepreneur Capt. C. A. Larsen, who had visited the island as a child. They were accompanied by Albert Benitz, a cine-photographer who made a photographic record of the expedition and of the island. The expedition was supported by the German Notgemeinschaft der Deutschen Wissenschaft, and the three travelled about the island using ships of the whaling Compañia Argentina de Pesca. During an extended season from mid-September 1928 to mid-May 1929 they explored the island extensively, recording weather, glaciology, geology, limnology, birds, mammals and the whaling industry. Kohl-Larsen published several scientific reports in German journals and a book, *An den Toren der Antarktis* (1930).

German South Polar Expedition 1901–3. An expedition to mainland Antarctica, funded by the Imperial German government and led by Dr Erich von Drygalski, Professor of Geography at Berlin University. One of the national enterprises encouraged by the declaration of the **International Geographical Congress** of 1895, the expedition was provided with a purpose-built wooden barquentine, *Gauss* (1440 tons), commanded by Capt. Hans Ruser, who had previously served in *Valdivia* with the **German Deep Sea Expedition 1898–99**. With a staff of five scientists aboard, *Gauss* left Kiel on 11 August 1901 for Cape Town. In early December they headed southeast, visiting Iles Crozet and Kerguelen, where they spent a month establishing a magnetic observatory. Following a brief visit to Heard Island in late January 1902 the expedition again headed south, seeking a landfall around 90°E along the dimly known coast of East Antarctica. On 12 February they reached latitude 60°S, moving slowly through a zone of icebergs and heavy pack ice. One week later they struck soundings of 130 fathoms (238 m), and on 21 February sighted ice-covered land over 90 km (56 miles) to the south. Almost immediately *Gauss* was immobilized by solidifying pack ice, which held it despite strong winds.

Realizing that they were there for the winter, the scientists set up observatories on the ice and sledged toward the new coast, which they named Wilhelm II Land for their Kaiser. Before the ice cliffs stood a prominent nunatak over 300 m (1000 ft) high and almost ice-free, which they named Gaussberg for their ship. On 29 March 1902, a calm, clear day, von Drygalski launched a tethered hydrogen balloon, ascending in the basket to a height of 457 m (1500 ft) to photograph and make notes on the sea ice. He communicated with the ground through one of Antarctica's earliest telephones.

The team spent a limited but profitable winter studying the geology, flora and fauna of Gaussberg and the biology of the seas around them. In early February 1903 *Gauss* broke free from the surrounding floe: though still beset by ice, and for the first time in almost a year the ship made slow progress along what was now quite clearly an ice-covered coastline. For eight weeks von Drygalski eased slowly westward, hoping to discover more land where his expedition could spend a second winter. On 31 March, with the sea ice threatening to freeze around him, he turned north toward the open sea, reaching Cape Town on 9 June. The expedition returned to Kiel on 24 November 1903.

German Whaling and Sealing Expedition 1873–74. The German steam whaler *Grönland* left Hamburg on 22 July 1873 under the command of Capt. E. Dallmann, an experienced whaling master, for a voyage of commercial exploration in Antarctic waters. Sponsored by the Deutsche Polarschiffahrts-Gesellschaft, a Hamburg-based society for polar maritime exploration, the voyage was funded by the wealthy founder and director of the society, Albert Rosenthal, who had already financed sealing and whaling voyages in Arctic waters. Dallmann reached the South Shetland Islands on 18 November, to find sealing gangs already working on the beaches. His orders allowed him to explore south toward the mainland Antarctic coasts that had earlier been noted by Palmer and Biscoe. On 9 January 1874 he reached 64°45'S. He followed the coast to discover a wide strait which he named for the German Chancellor, Otto von Bismarck. Working northward, he charted a network of lesser channels between the islands of a complex archipelago, now called Palmer Archipelago. There were disappointingly few seals and although whales were plentiful, they were fast-moving rorquals that *Grönland*'s crew were unable to tackle. Still seeking a cargo of oil or skins, Dallmann returned north to the South Orkney Islands for a spell of late-season sealing. He left southern waters in late February, returning to Hamburg on 25 July 1874. The voyage brought no commercial profit, and its sponsors saw no reason to send further ships south. Geographically it provided useful detail of an area that, two decades later, was to be charted more thoroughly by Belgian and French expeditions. The voyage is commemorated in Cape Grönland, at the north end of Anvers Island, and Dallmann Bay, between Anvers and Brabant Islands.

Getz Ice Shelf. 74°15′S, 125°00′W. Extensive ice shelf along the Hobbs and Bakutis coasts of Marie Byrd Land, West Antarctica. The shelf is held in position by Carney Island, Siple Island and other islands. It was discovered during photographic flights by the United States Antarctic Service Expedition 1939–41, and named for an American businessman, G. F. Getz, who supported the expedition.

Giæver, John Schelderup. (1901–70). Norwegian polar explorer. Born in Tromsø, he became a journalist and sealer, gaining experience of northern Norway, Greenland, the White Sea and other Arctic locations. In 1935 he was appointed secretary to the Norges Svalbard-og Ishavs-Undersøkelser (which later became the Norsk Polarinstitutt), with responsibility for organizing annual voyages to northeast Greenland. In World War II his ship was intercepted by British forces and Giæver was interned in Britain. After the German invasion of Norway he was able to join the Royal Norwegian Air Force in Canada, and was eventually involved in the restoration of civil government in central Finmark. In 1948 he became Office Chief of the Polarinstitutt, and was appointed to lead the **Norwegian-British-Swedish Antarctic Expedition 1949–52**. His personal account (1954) testifies to the success of the expedition and of his own role as leader. Ill-health dogged his subsequent retirement: he died in Oslo on 9 November 1970.

Giant petrels. See **Fulmars, southern oceanic**.

Gibbs Island. 61°28′S, 55°34′W. A small, ice-capped island close to Elephant Island in the easternmost group of the South Shetland Islands. The origin of the name is obscure.

Gibson-Hill, Carl Alexander. (1911–63). British ornithologist. Educated at Malvern and Cambridge University, he was especially interested in Malayan ornithology, but also studied seabirds of the North and South Atlantic oceans. He visited South Georgia in 1946 and subsequently published several papers on Antarctic birds. He died on 18 August 1963.
(CH)

Gilbert Strait. 63°38′S, 60°16′W. Strait between Trinity Island and Tower Island, Palmer Archipelago. The strait was named in 1829 by Cdr H. Foster RN, officer commanding the research ship HMS *Chanticleer*, to honour Davies Gilbert, a senior British scientist who planned the scientific programme of the *Chanticleer* expedition.

Girev, Dmitrii Semenovich. (1889–1932). Russian polar explorer. Born on 1 June 1889 in Aleksandrovsk, Sakhalin, the illegitimate son of a convict, he was brought up in Nikolaevsk-on-Amur, where he trained as an electrical engineer. Also an accomplished dog driver and hunter, in 1910 he came to the attention of Cecil Meares, who was in the area buying dogs for Scott's forthcoming **British Antarctic (*Terra Nova*) Expedition 1910–13**. Girev helped Meares to select 33 dogs and, together with groom Anton **Omelchenko**, was himself recruited for Antarctica. Travelling via Vladivostok, Shanghai and Sidney, he joined the ship in Christchurch, New Zealand. Both Russians proved popular and effective members of the expedition. Girev took part in depot-laying expeditions with the dogs, and was one of the group that discovered the bodies of the polar party in November 1912. On return to Siberia he was employed in various capacities in gold mining and dredging. In 1930 he was arrested in an NKVD purge and gaoled in Vladivostok. Released after 18 months, he died of a heart attack on his way home.

Gjelsvikfjella. 72°09′S, 2°36′E. (Gjelsvik Mountains). Range of ice-capped mountains forming part of a chain that extends east-to-west between Fimbulheimen and Hellehallet, parallel to Prinsesse Astrid Kyst, Dronning Maud Land. Several peaks rise above 2600 m (8500 ft). The range was named for Tore Gjelsvik, a former director of the Norsk Polarinstitutt.

Glaciers and ice streams. Streams of ice flowing slowly downhill. Glaciers move between rock walls: ice streams flow as faster zones within an ice sheet. Toward the edges of the Antarctic ice sheet, ice is funnelled through outlets which discharge either directly into the ocean, or alternatively into ice shelves. Outlet glaciers are constrained between rock walls; examples are the massive Beardmore and Byrd glaciers that feed into the Ross Ice Shelf through the Transantarctic Mountains. Ice streams flow between walls of static or slower-moving ice; examples are the Dawson-Lambton and Stancomb-Wills ice streams that flow westward into the Weddell Sea. A few 'rivers of ice' appear first as ice streams and later pass between rock walls, becoming outlet glaciers. One example is the Lambert Glacier of Mac. Robertson Land, probably the world's greatest glacier, that feeds into the Amery Ice Shelf. Some 13 per cent of the Antarctic coastline consists of one or other of these kinds of outflow. Because they are active, their importance in understanding the dynamics of the ice sheets is greater than this proportion suggests. Satellite remote sensing helps us to identify them and estimate their rates of flow.

Outlet glaciers and ice streams result from subglacial valleys which channel the flow. Their starting point on the inland ice is often marked by lines of **crevasses** and long sinuous features resulting from changes in elevation and texture, sometimes termed flowlines, which converge toward the outlet. There may be an abrupt drop in surface height, with arcs of intense crevassing, marking the presence of headwalls under the ice. The elevation of the glacier bed may extend to hundreds of metres below sea level. Ice streams may be separated laterally from adjacent ice by intensely crevassed shear zones, indicating that they receive no significant mass contribution through their side walls. Some outlet glaciers display chaotically broken surfaces throughout, making them impenetrable to studies except by helicopter.

The size of many ice streams and outlet glaciers is in keeping with the huge scale of Antarctica. Some, for example Foundation Ice Stream and Ice Stream D, are well over 200 km long. Typical widths are up to 50 km. Relatively few depths have been measured, but the thickest ice of Byrd Glacier, detected by radio echo sounding, is more than 3000 m. Such outlets are responsible for a significant proportion of the ice sheet's total discharge; 90 per cent of ice accumulation in the coastal zone flows through them. Individual outlets discharge huge volumes of ice. The flux of ice through Byrd Glacier is estimated at 18 km^3 annually, while Ice Stream B may discharge over twice as much. The velocity of ice within outlet glaciers and ice streams is sometimes one to two orders of magnitude higher than in the interior ice sheet. Rates of between 800 and 1500 m per year have been recorded in many outlets. Movement of the seaward front of Stancomb-Wills Ice Stream has been measured at 4000 m yearly. The longitudinal transition from slow to fast flow appears to occur relatively abruptly within a few tens of km of the inland end of the outlet, and builds up to

Glacier front – South Shetland Islands

a maximum close to the grounding line, where the ice starts to float.

Transverse velocity profiles typically show that most of the ice deformation is concentrated at the stream's sides, implying that side-walls impose only very slight frictional restraints on longitudinal motion. Rapid down stream acceleration may in part be due to a change in the mode of ice flow – transition from a flow dominated by internal deformation, where the ice is frozen to the bedrock, possibly to one of basal melting and sliding. The fast flows of some Antarctic outlets are comparable to those attained by valley glaciers during intermittent high-velocity surges. (NmcI)

Glacier Strait. 73°25′S, 169°24′E. Passage between Coulman Island and Cape Jones, on the Borchgrevink Coast of Victoria Land. The strait is named for USS *Glacier*, the icebreaker that in February 1965 first passed through the strait.

Glacier tongue. See **Ice tongue**.

Glaze ice. Transparent coating of clear ice, often formed by rain falling on surfaces at sub-freezing temperatures. Solid and dense, glaze ice may build up on radio antennae and other surfaces, destroying them by its weight.

Glossopteris. A genus of fern-like fossil plants with distinctive tongue-shaped leaves, pronounced mid-ribs and a network of veins, typifying Antarctica's late Permian fossil flora. Antarctic glossopterids were clearly related to similar plants from the other southern continents. Probably evolving from a flora that had survived the Permian ice

age, they grew as forests of shrubs and large trees with woody trunks, the roots having a prominent cross-shaped centre that may have helped with aeration in swampy conditions. They seem to have grown in cool-temperate climates with alpine glaciers still present on the uplands, and were clearly dominant in the forests, as their remains outnumber all other groups in Antarctic fossil beds. Permian coal measures in the Transantarctic Mountains, formed from their accumulated leaves, include fossil stumps in their original growth positions. With them were primitive conifers, ferns and sphenopsids (horse-tails). Pollen and spores from Permian rocks in the Transantarctic and Prince Charles Mountains indicate that the flora was more diverse, but most of the parent species are as yet unknown.
Further reading: Trusswell (1991).
(JEF)

Gold Harbour. 54°37′S, 35°56′W. Bay between Cape Charlotte and Bertrab Glacier, South Georgia, possibly named for the huge king penguin colony that is prominent on the beaches.

Goodale, Edward E. (1903–89). US polar explorer and administrator. Born in Boston in 1903, he joined **Byrd's First Antarctic Expedition 1928–30** while still a student at Harvard. He sledged with L.M. Gould's geology party in the Queen Maud Mountains, providing essential weather reports for Byrd's flight to the South Pole. During World War II Goodale served in the USAAF, establishing search and rescue bases. Later he worked for the US Weather Bureau. In 1955 he revisited Antarctica with the US IGY Antarctic Program. From 1958 to 1968 he worked in Christchurch, New Zealand as representative of the US Antarctic Research Program. Mount Goodale and Goodale Glacier were named for him. He died on 18 January 1989.
(CH)

Goudier Island. 64°50′S, 63°30′W. A small island guarding the harbour of Port Lockroy. Discovered and charted by the **French Antarctic (*Français*) Expedition 1903–5**, it was named for the expedition's chief engineer. The island became a workshop for early twentieth-century whalers and, together with neighbouring Jougla Point, a mooring and repository for coal and other stores. In February 1944 it became the site of Base A of Operation Tabarin, the main hut of which is currently an Historic Site and museum.

Gough Island. 40°19′S, 09°57′W. A warm-temperate island in the southern Atlantic Ocean, Gough Island lies in isolation 350 km (219 miles) south-southeast of Tristan da Cunha. Rectangular and slab-sided, 14 km (8.7 miles) long and 5 km (3.1 miles) wide, it is a volcanic mass some 6 million years old, rising to a plateau at 300 m (984 ft) from which emerge a cluster of higher peaks exceeding 800 m (2624 ft): the highest, Edinburgh Peak, stands central at 910 m (2985 ft). Steep cliffs and narrow beaches provide few points for landing or access to the interior. The climate is mild, overcast, damp and windy, with mean annual temperature 11.5 °C, range of monthly means 5.9 °C, rainfall exceeding 300 cm per year and persistent westerly winds. The island supports lush vegetation including over 60 species of flowering plant and fern. Tall tussock grass grows in exposed coastal areas, scrub and forest in the steep-sided coastal valleys, giving way to wet heath and moorland on the heights. Snow is frequent on high ground in winter. Rockhopper penguins breed in abundance, together with yellow-nosed, sooty and wandering albatrosses, ten species of lesser petrel, skuas, terns and noddies. The island supports also an endemic rail and bunting. Elephant seals and Amsterdam Island fur seals breed on the beaches.

Probably first sighted by Portuguese navigators in 1505, Gough Island was rediscovered, charted and named by Capt. Charles Gough in March 1732. A sea mark for sailing ships *en route* to the Far East, it became well known to navigators, and throughout the nineteenth century was a profitable haven for sealing gangs. House mice, probably introduced by sealers, are the only alien species surviving on the island. Gough Island was investigated by a scientific survey in 1955–56, and from then onward occupied permanently by a South African meteorological station. Since 1938 a British dependency of St Helena, the island has the protected status of a wildlife reserve.
Further reading: Holdgate (1958).

Gough Island Scientific Survey 1955–56. A private British expedition, sponsored by scientific societies and organized and led by surveyor John Heaney, later led jointly by Heaney and Martin Holdgate. A party of eight scientists spent from 13 November 1955 to 13 May 1956 on the island, conducting geographical, biological, geological and topographic surveys from a small station in the Glen. Meteorological observations begun during the expedition have subsequently been maintained by the South African government.
Further reading: Holdgate (1958).

Gough Island Station. 40°21′S, 9°52′W. South African permanent year-round research station on Gough Island, southern Atlantic Ocean, built on land leased from the Government of St Helena. In operation since 1956, the station is concerned mainly with weather reporting.

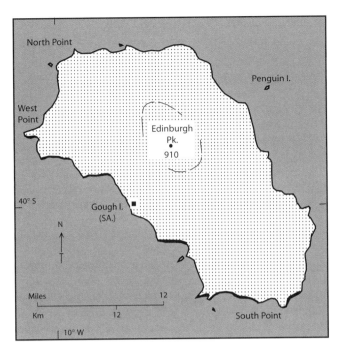

Gough Island. Mountainous and deeply dissected, the island is ringed by steep cliffs, allowing few points of access. There is no permanent ice. Height in m

Gould, Laurence McKinley. (d. 1995). American polar geologist and administrator. Born in Michigan, he studied law and geology at his state university, then joined the US Army Ambulance Service in World War I. On return he resumed geological studies, in 1926 taking part in an expedition to Greenland. In 1928 he joined **Byrd's First Antarctic Expedition**, becoming second-in-command and leading dog-sledging parties to study the geology of the Rockefeller and Queen Maud mountains. In 1932 he established a geology department at Carleton College, Mass., eventually becoming college president. During World War II he was chief of the Arctic section of the US Air Force's Arctic, Desert and Tropic Information Centre. In 1955 he took the lead in planning US Antarctic programmes for the International Geophysical Year. A member of many trustee boards and foundations, Larry Gould was strongly influential in promoting both national and international polar science. In 1962 he was appointed to a chair in geology at the University of Arizona. He died on 20 June 1995.
Further reading: Gould (1931).

Gould, Rupert Thomas. (1890–1948). British seaman, hydrographer and historian. Born in Portsmouth on 16 November 1890, he entered the Royal Navy in 1906. After service with the fleet he was invalided out in 1915. Joining the Hydrographic Department of the Navy, he took particular interest in Antarctic charts, resolving many historical points and issues, and contributing substantially to the first edition of the Admiralty handbook *Antarctic Pilot*. A second interest was marine chronometers, on which he wrote with authority. Further preoccupations included sea serpents, the Loch Ness monster, and many other oddities and enigmas that required intelligent investigation: on all of these he spoke and wrote entertainingly. He died on 5 October 1948.

Gould Coast. 84°30′S, 150°00′W. Part of the west coast of Marie Byrd Land, spanning the junction of East and West Antarctica, and facing onto the southeastern corner of Ross Ice Shelf. Bounded in the south by the western portal of Scott Glacier, the coast includes the foothills of Horlick Mtns and a massive glacier system draining into the ice shelf. The northern boundary lies on a relatively featureless ice plain. The coast is named for geologist Laurence M. Gould, second-in-command of **Byrd's First Antarctic Expedition 1928–30**, and leader of a sledging party that explored in the area.

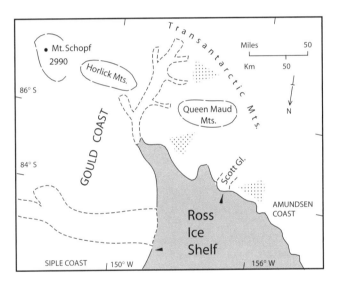

Gould Coast

Gouvernøren Harbour. 62°00′S, 62°60′W. Harbour in the east side of Enterprise Island, Wilhelmina Bay, Danco Coast, Graham Land, which was used extensively by whalers during the 1920s. It contains the half-submerged iron hulk of the whaling transport *Gouvernøren*, that was burnt out while lying at moorings in 1916.

Graham Coast. 65°45′S, 64°09′W. Part of the west coast of Graham Land, Antarctic Peninsula, facing Grandidier Channel, extending from Cape Renard in the north to Cape Bellue in the south. Deeply indented with bays and islands, the coast has steep coastal mountains rising to the central plateau, and glaciers descending to narrow piedmont ice

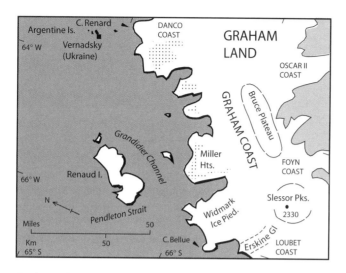

Graham Coast

shelves. It was sighted and delineated in 1832 by the British sealer Capt. John Biscoe, who named it for Sir James Graham, First Lord of the Admiralty at the time.

Graham Land. 66°00′S, 63°30′W. The northern extent of Antarctic Peninsula, south to a boundary line between Cape Jeremy and Cape Agassiz, where it abuts onto **Palmer Land**. Forming a major component of British Antarctic Territory, it is made up of a long, narrow peninsula of alpine mountains and glaciers, flanked by ice shelves and topped by a high ice-covered plateau. The name was applied in 1832 by the British sealing captain John Biscoe, to honour the First Lord of the Admiralty, Sir James Graham.

Gran, Tryggve. (1889–1980). Norwegian polar explorer. Born in Bergen on 20 January 1889, he enrolled as a naval cadet, but on meeting Capt. Robert Scott in Norway, volunteered to join his forthcoming expedition as ski instructor. Serving with the **British Antarctic (*Terra Nova*) Expedition 1910–13**, he took part in Griffith Taylor's exploration of the western mountains, climbed Mt Erebus with Raymond Priestley, and was one of Wright's group that, in November 1912, discovered the bodies of the polar party on the Ross Ice Shelf. On return to Norway he became an aviator, and in World War I flew with the Royal Flying Corps. In 1928 he led the Arctic search for Roald Amundsen, and retired from the Norwegian Air Force with the rank of major. He died in Grimstad, Norway on 8 January 1980, the last survivor of Scott's *Terra Nova* expedition.

Granite Harbour. 76°53′S, 164°44′E. Bay between Cape Archer and Cape Roberts, Victoria Land, named for the glacier-strewn granite boulders that line the shore.

Grearson Oasis. See **Windmill Islands**.

Grease ice. A thick but pliable layer of ice crystals in surface water, giving the appearance of a film of thick grease. This is often the stage after **frazil ice** in the development of new sea ice.

Great Wall Station. 62°13′S, 58°58′W. Chinese permanent year-round station on Fildes Peninsula, King George Island, South Shetland Islands. The station has operated continuously since its establishment in February 1985.

Green, Charles John. (1888–1974). British expedition ship's cook. Born on 24th November 1888, he was apprenticed to bakery, becoming a chef in the merchant navy. In Buenos Aires he joined the **Imperial Trans-Antarctic Expedition 1914–16**, serving as cook in *Endurance*. After the loss of the ship he continued cheerfully to provide hot meals under the rigorous conditions of the pack ice and Elephant Island. Green served in the Royal Navy in World War I and subsequently in the merchant navy. Shackleton recruited him as cook aboard *Quest* in the **Shackleton-Rowett Antarctic Expedition 1921–22**. Retiring from the sea in 1931, he became a baker in Hull, developing skills as a lecturer, especially to school audiences. He died in Beverley on 26th September 1974.

Green Island. 65°19′S, 64°10′W. Northernmost of the Berthelot Islands, Graham Coast of Antarctic Peninsula, remarkable for its wealth of vegetation in a situation so far south. Banks of moss turf overlie moss peat more than 1 m deep, supporting rich stands of Antarctic hair grass. There are also colonies of shags, petrels and skuas. With an area of approximately 0.1 km^2 (0.04 sq miles), the small island is designated SPA No. 9 (Green Island) to protect the flora and fauna from casual interference.

Greene, Stanley Wilson. (1928–89). Irish polar bryologist. Born in Co. Cork, Ireland, he read natural sciences at Trinity College, Dublin, and taught botany successively at the University College of North Wales (1951–55) and the University of Birmingham. His research concentrated on polar mosses: in 1960–61 he spent a season in fieldwork in South Georgia, and in 1964–65 visited Alaska and McMurdo Sound, later developing bi-polar studies for the International Biological Programme on Disko Island, West Greenland, South Georgia and Signy Island. In 1969 he headed the botanical section of the British Antarctic Survey's Division of Life Sciences, moving in 1975 to establish a similar research unit at the Institute of Terrestrial Ecology's Bush Research Station, Penicuik. From 1981 he accepted an honorary readership in botany in the University of Reading, that allowed him to write without academic or administrative responsibilities. He died in the Netherlands in 1989.

Greenhouse effect. Heating of the atmosphere by absorption of infrared energy from Earth's surface. The atmosphere is warmed directly by absorbing some of the high-frequency solar radiation that passes through it *en route* to Earth, and indirectly by absorbing infrared energy re-radiated and conducted from Earth's surface. This surface warming provides the benevolent 'greenhouse' that maintains the temperatures of earth, oceans and atmosphere at life-sustaining levels, and is the most important component in driving global atmospheric circulation. Enhanced concentrations of some gases (e.g. carbon dioxide, methane), produced naturally or industrially, may intensify the greenhouse effect, increasing atmospheric absorption and warming, and raising the mean temperature of Earth's surface. Warming in polar regions may affect distribution and thickness of ice, and possibly accelerate warming elsewhere in the world (see **Climate change**).

Greenwich Island. 62°31'S, 59°47'W. A narrow, ice-capped island 24 km (15 miles) long and up to 10 km (6.2 miles) wide between Livingston Island and Robert Island, in the South Shetland Islands. The origin of the name is obscure. Guesalaga Peninsula, on the northeast side of the island, has since 1947 been the site of the Chilean year-round research station Arturo Prat.

Gressitt, Linsley. (d. 1982). American biogeographical entomologist. On the staff of the Bernice P. Bishop Museum, Honolulu, Gressitt was a dedicated insect fieldworker whose travels took him many times to Antarctica and the southern islands. Concerned not only with the distribution of invertebrate taxa, but with their means of distribution, he pioneered studies of airborne plankton over Antarctica and the southern oceans. His many papers include important contributions to the entomology of Antarctica and the southern islands. He died in an aircraft accident on 26 April 1982.

Grierson, John. (d. 1977). British pioneer polar aviator. Trained as a Royal Air Force pilot, Grierson in 1934 succeeded in a third attempt to cross Greenland in a Fox Moth seaplane. In 1947 he sailed with the whaling factory ship *Balaena*, of United Whalers Ltd., with two Walrus amphibian aircraft. These were used for ice reconnaissance and whale-spotting flights off the coast of Queen Mary Land and Wilkes Land. He died on 21 May 1977, aged 68.
Further reading: Grierson (1949).

Grindley Plateau. 84°09'S, 166°05'E. Ice plateau of Queen Alexandra Range, Victoria Land, named for New Zealand geologist G. Grindley.

Grove Mountains. 72°45'S, 75°E. A scattered group of nunataks emergent from the ice cap on the border of Mac.Robertson and Princess Elizabeth lands, discovered during **Operation Highjump 1946–47** and named for Sqdn. Ldr I. L. Grove, a pilot of the Royal Australian Air Force who later investigated them.

Growler. A small but substantial lump of floating glacier ice, 2 to 5 m (6.5 to 16.5 ft) across, derived by fragmentation from an **iceberg** or **bergy bit**. The term may originate from the colloquial name for four-wheeled cabs, of comparable size, that were common in mid-nineteenth-century London.

Grytviken. 54°17'S, 36°30'W. Whaling station in Cumberland Bay, **South Georgia**, opened in 1904 by an Argentine – Norwegian consortium and operated continuously by the **Compania Argentina de Pesca** until 1965.

Grytviken (Base M). 54°17'S, 36°30'W. British research station on King Edward Point, Cumberland Bay, South Georgia. The station was opened in January 1950 by the Falkland Islands Dependencies Survey for meteorological forecasting and biological research, occupying Discovery House, which had been built in 1924 for Discovery

King Edward Point (foreground) and Grytviken whaling station, Cumberland Bay, South Georgia

Investigations 1924–25. It closed in January 1952, when the meteorological station was transferred to the Falkland Islands Government. The station re-opened in November

1969, British Antarctic Survey taking over all government administrative functions, and occupying the purpose-built Shackleton House and other government buildings. It operated continuously until the Argentine invasion of 1982, when the British staff were removed under duress. British Antarctic Survey returned in 2000 to new, custom-built accommodation, operating the station (now called King Edward Point) as a centre for fisheries research.

Guesalaga Peninsula. 62°29′S, 59°40′W. Peninsula on Greenwich Island, South Shetland Islands, named for a Chilean naval officer, Capt. F. Guesalaga Toro. Since 1947 it has been the site of the permanent Chilean research station Arturo Prat.

Gulls, southern oceanic. Flying birds of the order Charadriiformes, family Laridae, that feed at sea on surface plankton, and on land scavenge or forage in the intertidal zone, in particular for limpets and other molluscs. Most widespread in the southern region is the kelp gull *Larus dominicanus*, a large white gull of body length 54 cm (21 in), wing span 1.3 m (51 in), with yellow bill and black wings and mantle. This species breeds on Antarctic Peninsula and most of the southern oceanic islands, and also in southern Australia, New Zealand, South Africa and South America. Pairs nest in small colonies, spaced to avoid mutual predation but close enough for collective defence: nests are well constructed of seaweed, moss, bones and feathers. Two or three green and brown mottled eggs are laid, usually resulting in one or two chicks. Yearling and second-year immature birds in distinctive mottled plumage are often seen in feeding flocks. Kelp gulls remain year-round on the warmer islands, but leave southern breeding areas for the coldest months of winter. Smaller and more colourful dolphin gulls *L. scoresbii*, red-billed and grey-headed, breed in southern South America and on the Falkland Islands.

Hail. Precipitation of consolidated particles of ice (hailstones), which formed by accretion during their passage through the atmosphere. Usually associated with strong upcurrents of air, hail rarely occurs in the Antarctic or Southern Ocean regions.

Half Moon Island. 62°36′S, 59°55′W. Crescent-shaped island 2 km (1.25 miles) long, standing central to Moon Bay off the eastern end of Livingston Island, South Shetland Islands. Since 1953 it has been the site of the Argentine station Teniente Cámara.

Hallett, Cape. 72°18′S, 179°19′E. Prominent cape on the end of Hallett Peninsula, Borchgrevink Coast, Victoria Land, identified in 1841 and named for Thomas R. Hallett, purser in HMS *Erebus*. **Hallett Station**, a former US and joint US/New Zealand research station, occupies a raised beach on the north side of the cape, close to a large colony of Adélie penguins. A rectangular area of beach of approximately 0.2 km² (0.08 sq miles), including a patch of rich and diverse vegetation and terrestrial fauna, and a rich avifauna, is designated SPA No. 7 (Cape Hallett).

Hallett Station. 72°18′S, 179°16′E. Former US research station at **Cape Hallett**, Borchegrevink Coast, Victoria Land. Built in 1957 for the International Geophysical Year 1957–58, the station in 1959 continued under joint US/New Zealand management until 1964, when fire destroyed the main laboratories. It continued as a summer-only station to 1973. After several years' deterioration, the station was in 1984 reduced in size and refurbished for intermittent use.

Halley Station (Base Z). 74°35′S, 26°30′W. British research station at Halley Bay, Brunt Ice Shelf, Caird Coast. The station began in January 1956 as an International Geophysical Year station operated by the British Royal Society (**Royal Society International Geophysical Year Expedition 1955–57**). In January 1959 it was transferred to the Falkland Islands Dependencies Survey. Because of heavy precipitation, the original hut was replaced in 1961, and again in 1967, 1973 and 1983. The most recent station, Halley V, which has been operational since 1992, stands some 15 km (9.4 miles) from the edge of the floating ice shelf. It consists of four main buildings that can be raised annually above the rising snow level. Research is concentrated on atmospheric sciences, geology and glaciology.

Hanson, Malcolm P. (*c*. 1894–1942). US Navy signals officer and aviation radio pioneer. In 1925–27 Hanson built radio equipment for the Byrd–McMillan Expedition to Greenland, Hubert Wilkins's Arctic expeditions and Byrd's North Polar and trans-Atlantic flights. He served as chief radio engineer on **Byrd's First Antarctic Expedition 1928–30**, studying and measuring the Kenelly-Heaviside Layer, and in January 1929 supervising a historic two-way transmission between Byrd's aircraft above Bay of Whales, Antarctica and Times Square, New York. Hanson was killed in an aircraft crash in Alaska on 10 August 1942.

Hanssen, Helmer Julius. (1870–1956). Norwegian seaman and polar explorer who served on three of Roald **Amundsen's** expeditions. Born in Andøya, northern Norway, he joined the fishing fleet at age 12. By 1898 he had his master's ticket, but served as second mate in *Gjoa*, completing with Amundsen the first transit of the northwest passage. With Amundsen again he sledged to the South Pole in 1911. He was master of *Maud* in Amundsen's 1918–20 voyage through the Northeast Passage, and took part in several other Arctic expeditions. He published three books, including an autobiography. Hanssen died on 2 August 1956.
(CH)

Harbord, Arthur Edward. (1883–1961). British seaman and explorer. Born in Hull of a seagoing family, he entered the merchant navy at 12 and received his master's ticket in sail 10 years later. In 1907 he joined the **British Antarctic (*Nimrod*) Expedition 1907–9** as second officer and navigator, later becoming chief officer. On returning he transferred in 1910 to the Royal Navy, serving in the Hydrographic Department through both world wars. On retiring

Harbord became marine surveyor to the port of Liverpool. He died on 11 October 1961.

Hardy, Alister Clavering. (1896–1985). British polar marine biologist. After serving as an army camouflage officer in World War I, he read zoology at Oxford University. Graduating in 1921, he joined the Fisheries laboratory, Lowestoft. In 1924 he joined the newly-established **Discovery Investigations**. His three years aboard RRS *Discovery*, engaged in research on plankton, fisheries and whales, resulted in many publications and provided a sound foundation both for his own development as a marine biologist, and for the future biological work of Discovery Investigations. Hardy held professorships successively in zoology and oceanography at University College, Hull (1928–43), in natural history at Aberdeen (1942–45), and in zoology and zoological field studies at Oxford (1941–63). Knighted in 1957, he received many scientific honours. An engaging writer and water colour artist, he wrote and illustrated several books on polar and marine science (see below). On retiring in 1963 he founded and directed the Religious Experience Research Unit of Manchester College, Oxford, which continues as the Alister Hardy Research Centre. He died on 22 May 1985.
Further reading: Hardy (1956, 1959, 1967).

Hare, Clarence H. (1880–1967). New Zealand polar seaman. Born in Invercargill on 2 December 1880, he worked as a clerk in New Zealand and Fiji. Recruited to the **British National Antarctic Research Expedition 1901–4**, he served aboard *Discovery* as a steward, but also took part in the sledging programme. On returning to New Zealand he became a professional piano tuner and repairer. He died on 31 May 1967.

Harmony Point. 62°18′S, 59°14′W. Low, ice-free peninsula forming the westernmost point of Nelson Island, South Shetland Islands. Its raised gravel beaches and rocky knolls support an unusually rich and dense vegetation cover, together with an abundance of breeding seabirds including gulls, terns, skuas and giant petrels. An area of approximately 4 km² (1.6 sq miles), including The Toe, a smaller point nearby, has been designated SSSI No. 14 (Harmony Point) to protect and facilitate long-term studies.

Haslop, Gordon Murray. (1922–61). New Zealand expedition pilot. Born in Canada, he moved with his parents to New Zealand, where he trained as a school teacher. Haslop served briefly in the New Zealand Army before transferring to the New Zealand Royal Air Force, in which he trained as a pilot. In 1951 he transferred to the Royal Air Force, where he was selected to fly as second pilot with the **Commonwealth Trans-Antarctic Expedition 1955–58**. He flew many reconnaissance flights from the main base, and toward the end of the expedition accompanied J. H. Lewis on a non-stop flight across Antarctica to McMurdo Sound. Stationed in Singapore with RAF Transport Command, he died on 27 August 1961, the result of a motor accident.
(CH)

Hassel, Helge Sverre. (1876–1928). Norwegian seaman and explorer. Born in 1876, he became a master mariner, serving with Otto Sverdrup in *Fram*. He joined the **Norwegian Antarctic Expedition 1910–12** and was a member of the party that reached the South Pole. On return to Norway he joined the customs service and was in charge of the Grimstad Customs House when he died in 1928.

Haswell Islands. 66°31′S, 93°00′E. A group of islands standing 2.5 km (1.5 miles) off Mabus Point, on Queen Mary Coast, East Antarctica. Discovered by a sledging party from the Western Base of the **Australasian Antarctic Expedition 1911–14**, they were named for William A. Haswell, Professor of Zoology at the University of Sydney. Surrounded by pack ice and fast ice for much of the year, they are notable for a rich flora and avifauna, including five species of petrel. They are also the site of a large emperor penguin colony, the second one to be discovered. Mabus Point in 1956 became the site of the Russian year-round research station **Mirnyy**. An area of approximately 1 km² (0.4 sq miles), including Haswell Island, largest of the group, and its surrounding littoral and sea ice (when present), is designated SSSI No. 7 (Haswell Island), to protect and facilitate research.

Hatherton, Trevor. (1925–92). New Zealand polar geophysicist. Born in Yorkshire, Hatherton studied geophysics at Imperial College, London. In 1953 he moved to New Zealand, where he served for many years with the Geophysics Division of the New Zealand Department of Scientific and Industrial Research, eventually becoming director. In 1957–58 he led the New Zealand International Geophysical Year team at Scott Base, from which developed his lifelong involvement with Antarctic affairs. In 1970 he became vice-chairman of the Ross Dependency Research Committee, and was chairman 1983–88. Hatherton retired in 1989 and died on 2 May 1992.

Heard Island. 53°06′S, 73°30′E. An isolated sub-Antarctic island, 25 km (15.6 miles) across, lying almost equidistant (4550 km, 2844 miles) from Africa and

Australia in the Southern Ocean. An eastern spit and western peninsular extension combine to give it an overall length of 42 km (26 miles). Dominated by the 2745 m (9004 ft) massif of Big Ben, it shares a submarine platform with the **McDonald Islands** and **Iles Kerguelen**. The island consists of laval outpourings and scoria arising from a basement of limestone. About 90 per cent ice-covered, it has a coastline of alternating glaciers, rocky headlands and beaches. Big Ben is volcanically active, frequently producing lava flows and steam from its ice-covered flanks. Heard Island has a cold maritime climate, with mean annual temperature 1.1 °C, range of mean monthly temperatures 4.6 °C, and strong, persistent westerly winds. The flora of eight species of flowering plant includes tussock grasses, the anti-scorbutic Kerguelen cabbage, and a dwarf shrub that grows in sheltered areas. Fur seals and elephant seals dominate the beaches in spring and summer, and there are large colonies of breeding birds. Sealers probably discovered the island as early as 1833, but it remained uncharted until rediscovered by the US merchant captain John J. Heard in November 1853. Repeatedly stripped of fur seals and elephant seals throughout the nineteenth century, it was occupied by Australian scientists from 1947 to 1954. In 1953 it was designated part of Australian External Territory, and is currently administered as a nature reserve; it is also a World Heritage Site. Expeditions visit the island sporadically to study its many interesting biological and geological features and to climb Big Ben.

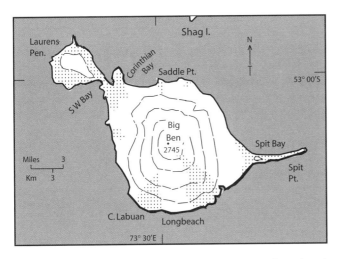

Heard Island. A single high peak, ice-covered and volcanically active, forms the main bulk of the island. Several beaches provide hauling-out grounds for seals and nesting areas for penguins. Height in m

Heard Island Station. 53°06′S, 73°43′E. Former Australian year-round research station in Atlas Cove. Built in February 1947, the station was operated continuously until October 1954, when it was reduced in size to be maintained as a refuge. It has since had frequent but intermittent use.

Hearst Island. 69°25′S, 62°10′W. An ice-covered dome 58 km (36 miles) long, rising to 365 m (1197 ft) in the Larsen Ice Shelf, off Wilkins Coast, Palmer Land. Sighted by Sir Hubert Wilkins on a flight of 20 December 1928, the area was named 'Hearst Land' for newspaper magnate W. Randolph Hearst, who supported the expedition. The name has been transferred to this major feature of the area.

Heimefrontfjella. 74°35′S, 11°00′W. (Home Front Mountains). Chain of three small ice-mantled mountain ranges, Tottanfjella, Sivorgfjella and Milorgfjella, in western Dronning Maud Land. Individual features in the ranges are named for heroes of the World War II Norwegian resistance movement.

Henkes Islands. 67°48′S, 68°56′W. A group of low-lying islands and rocks along the southern coast of Adelaide Island, discovered and named by J.-B. Charcot for a Norwegian director of a Punta Arenas whaling company. The islands are rich in bird life.

Henryk Arctowski Station. 62°09′S, 58°27′W. Polish permanent year-round station on the southwestern shore of Admiralty Bay, King George Island, South Shetland Islands. The station has operated continuously since its establishment in 1977.

Herbert Plateau. 64°32′S, 61°15′W. Ice plateau, part of the dissected central plateau of Graham Land, between Bleriot and Drygalski glaciers. It was named for British explorer Sir Wally Herbert, who sledged and surveyed in the area.

Herdman, Henry Franceys Porter. (1901–67). British polar oceanographer. Born on 11 March 1901 and brought up in Northern Ireland, he graduated in chemistry from Belfast University. In 1924 he joined Discovery Investigations, serving in RRS *Discovery* from 1925–27, later in commissions of RRSs *Discovery II* and *William Scoresby* studying many aspects of oceanography, including whale distribution, pack ice, and the bottom topography of the Southern Ocean. Toward the end of his career he became a recognized authority in the equipping and managing of research ships. He died on 3 September 1967.

Heritage Range. 79°45′S, 83°00′W. Major range of ice-capped peaks forming the southern end of the Ellsworth Mountains, eastern Ellsworth Land. The mountains form a series of northeastward-facing escarpments overlooking the Ronne Ice Shelf. Anderson Massif (2190 m, 7183 ft), a northern outlier, is probably the highest point of the range.

Highjump Archipelago. 66°05′S, 101°00′E. Group of islands within Shackleton Ice Shelf, northwest of Bunger Oasis, at the western end of the Knox Coast, East Antarctica. It was named for Operation Highjump (United States Navy Taskforce 68 1946–47), that provided aerial photographs from which the group was identified.

Hillary Coast. 79°20′S, 161°00′E. Part of the western coast of Victoria Land and its southern extension along the Transantarctic Mountains, facing onto the western Ross Ice Shelf. Bounded by Cape Selborne in the south and Minna Bluff in the north, it consists mainly of steep mountains rising inland to the polar plateau, interspersed with glaciers that descend and merge into the ice shelf. It was identified and named for Sir Edmund Hillary, who in 1953 was one of the first two to climb Mt. Everest, and led the New Zealand party of the Commonwealth Trans-Antarctic Expedition 1955–58.

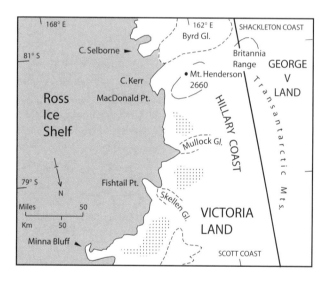

Hillary Coast

Hjort, Johan. (1869–1948). Norwegian fisheries biologist and oceanographer. Born in Christiania (Oslo) on 18 February 1869, he read medicine in Munich, but became interested in marine biology and fisheries, studying in Naples and Jena, and in 1897 taking up the directorship of the Drøbak fisheries biological station. He was a prime mover in securing the Norwegian marine research ship *Michael Sars*, helped to develop the International Council for the Exploration of the Sea, and with Sir John Murray wrote and edited *The Depths of the Ocean* (London 1912), an enduring classic of marine biology. As Professor of Marine Biology at Oslo University he initiated Norwegian studies of whales and the whaling industry, in 1924 becoming chairman of the first Norwegian Whaling Committee, and chairman of the International Whaling Committee 1926–39. He visited Antarctic waters with the whaling fleet in 1929. He died on 7 October 1948.

Hoar frost. A translucent coating of ice crystals formed *in situ* on and near the ground, from condensation and freezing of atmospheric moisture.

Hobbs, William Herbert. (1864–1953). American geologist and polar historian. Born in Worcester, Mass. on 2 July 1864, he trained in industrial design and the natural sciences and for many years taught and undertook research at the University of Wisconsin. In retirement he wrote vehemently and controversially on polar history, championing Peary's priority in reaching the North Pole and – against formidable evidence to the contrary – Palmer's priority in the first sighting of Antarctica. He died in early January 1953.

Hobbs Coast. 74°50′S, 132°00′W. Part of the north coast of Marie Byrd Land, West Antarctica, facing the Amundsen Sea. Bounded by Cape Burks in the west and Dean Island in the east, it consists mainly of nunataks and ice slopes descending to the broad western end of Getz Ice Shelf. The coast was discovered by the United States Antarctic Service 1939–41, and named for the distinguished glaciogist and polar historian Prof. **William Hobbs**, of the University of Michigan.

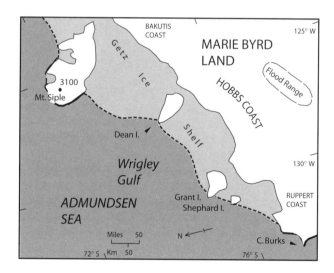

Hobbs Coast

Hoelfjella. 72°00′S, 14°00′E. (Hoel Mountains). Easternmost range of a chain of mountains extending east-to-west between Fimbulheimen and Hellehallet, parallel to Prinsesse Astrid Kyst in central Dronning Maud Land. The range was named for Norwegian geologist and Arctic explorer A. Hoel.

Holland Range. 83°10′S, 166°00′E. A range extending between Robert and Lennox-King glaciers on the Shackleton Coast, East Antarctica, named for New Zealand statesman Sir Sidney Holland.

Hollick-Kenyon Peninsula. 68°35′S, 63°50′W. An ice-covered peninsula on the Bowman Coast, at the junction of Graham Land and Palmer Land. It was named for the Canadian aviator Herbert Hollick-Kenyon, who in 1935 overflew the area with Lincoln Ellsworth.

Hollick-Kenyon Plateau. 78°00′S, 105°00′W. Extensive ice plateau 1200–1800 m. high between the Ellsworth and Crary Mountains, Ellsworth Land, named for Canadian aviator Herbert Hollick-Kenyon, who in 1935 overflew the area with Lincoln Ellsworth.

Home of the blizzard. Name given by Douglas Mawson to Cape Denison, Terre Adélie, where the main base of his **Australasian Antarctic Expedition 1911–14** received some of the strongest and most persistent **katabatic winds** ever recorded. The expedition's recording anemometer logged a mean annual wind speed of 43.4 mph. (19.4 m sec^{-1}) during the two years of the expedition – a world record for a station at sea level. The quietest month was February, with a mean of 26.17 mph (11.7 m sec^{-1}): the stormiest month was July, with a mean of 55.48 mph (24.8 m sec^{-1}). During the stormiest day, 16 August 1912, the mean wind speed was 80.53 mph (36.0 m sec^{-1}). Despite these extraordinary conditions the expedition completed a creditable field programme of sledging, survey and scientific research. *The Home of the Blizzard* became the title of Mawson's two-volume account of the expedition.
Further reading: Mawson (1915).

Hooker, Joseph Dalton. (1817–1911). British naturalist. After training and practising medicine in Glasgow, he was appointed assistant surgeon and botanist to the **British Naval Expedition 1839–43.** Sailing with Capt. James Clark Ross in the flagship HMS *Erebus*, he made a worldwide collection of botanical specimens, including many from southern oceanic islands. The voyage provided material for his classic report *Flora Antarctica* (1844–47), and laid the foundation for a distinguished career in botanical taxonomy and biogeography. Further expeditions followed, to Nepal, Bengal, Morocco and the United States. In 1865 he became Director of Kew Gardens, a position that he held for 20 years.

Hooker's (New Zealand) sea lion. *Phocarctos hookeri*. This species breeds on islands south of New Zealand, notably cool-temperate Campbell Island and Enderby Island in the Auckland Islands group. Non-breeding animals occur on the Snares Islands and New Zealand's South Island, occasionally crossing the Antarctic Convergence to appear on Macquarie Island. Mature males grow up to 3.5 m (11.5 ft) long, probably weighing up to 500 kg (1100 lb): females seldom exceed 2 m (6.5 ft). The fur is dark brown, reddish or tan: large males have a prominent mane. Males appear on the breeding beaches in October: females follow in November and December, giving birth in small nursery groups soon after their arrival. Males dominate the groups, mating with the females as they come on heat soon after the pups are born. The breeding groups disperse after mid-summer; pups grow slowly and are weaned during the following winter. The total population, recovering after severe hunting for skins and oil, is estimated at about 50 000.

Hooper, Frederick J. (*c*.1891–1955). As a steward in the Royal Navy he joined the **British Antarctic (*Terra Nova*) Expedition 1910–13**, serving first on the ship, then with the shore party. He participated in the sledging programmes, including the search for the polar party and the second ascent of Mt. Erebus. He died in Southport, England, on 20 June 1955.
(CH)

Hope Bay (Base D). 63°24′S, 56°59′W. British research station at Hope Bay, Trinity Peninsula, Graham Land. The first station (Eagle House), was built in February 1945 as part of British Naval Operation Tabarin 1943–45. It was destroyed by fire in November 1948, with the loss of two lives. The second station (Trinity House), built in February 1952, operated continuously until February 1964. In December 1997 the building was transferred to Uruguay, to become Station Teniente Ruperto Elichiribehety.

Horlick Ice Stream. 85°17′S, 132°00′W. Major ice stream of the Gould Coast which drains the northern flank of the Wisconsin Range, Horlick Mountains. The stream flows eastward into the lower Reedy Glacier, and ultimately into the southeast corner of Ross Ice Shelf.

Horlick Mountains. 85°23′S, 121°00′W. Group of mountains in southern Ellsworth Land, forming the southeastern end of the visible Transantarctic Mountains, overlooking the Gould Coast of the Ross Ice Shelf. Including the Wisconsin and Ohio ranges, they were investigated by **Byrd's Second Antarctic Expedition 1933–35**, and named for US industrialist William Horlick, who supported the expedition.

Horseshoe Harbour. 67°36'S, 62°52'E. Harbour in Holme Bay, Mac.Robertson Land, that since 1954 has been the site of the Australian Mawson Station.

Horseshoe Island. 67°51'S, 67°12'W. A mountainous island 10.5 km (6.5 miles) long and 5 km (3 miles) across, with peaks rising to 900 m (2952 ft). Standing at the entrance to Square Bay, Fallières Coast of Graham Land, it was named for its distinctive shape. A former British research station, Base Y, now an Historic Monument, stands on the northwest coast.

Horseshoe Island (Base Y). 67°49'S, 67°18'W. British research station in Bourgeois Fjord, Marguerite Bay, Graham Land. The station was opened in March 1955 by the Falkland Islands Dependencies Survey for meteorology and local survey and geology, and closed in 1960 when nearby Stonington Island (Base E) was re-opened. The hut remains in good order, and in 1995 was designated an Historic Monument.

Hoseason Island. 63°44'S, 61°41'W. A mountainous, ice-mantled island of Palmer Archipelago, 9.5 km (6 miles) long, standing alone in Gerlache Strait, west of Trinity Island. It was named for James Hoseason, first mate of the sealing vessel *Sprightly*, that operated in the area in 1824–25.

Hourglass dolphin. *Lagenorhynchus cruciger*. A black dolphin up to about 1.8 m (6 ft) long, with recurved dorsal fin, white chest and belly, and prominent white hourglass-shaped flashes along the flanks. Restricted to cool and cold southern waters, small groups of up to a dozen are often seen in the Southern Ocean, especially around Drake Passage and as far south as Bransfield Strait. Little is known of their biology.

Hovgaard Island. 65°08'S, 64°08'W. An ice-capped, low-lying island 5 km (3.1 miles) long, on the west side of Penola Strait, Graham Coast. It was named by the **Belgian Antarctic Expedition 1997–99** for Capt. P. Hovgaard, an officer and scientist of the Royal Danish Navy who advised the expedition.

Hudson, Hubert T. (1886–1942). British seaman and navigator. Hudson joined the merchant service in 1901, was commissioned in the Royal Naval Reserve in 1913, and in the following year sailed with Shackleton's **Imperial Trans-Antarctic Expedition 1914–17**, as navigating officer in *Endurance*. After the loss of the ship and the drift in the Weddell Sea, he took charge of the lifeboat *Stancomb Wills* for the final journey to Elephant Island. He continued as a master in the merchant service, and as commodore of an Atlantic Ocean convoy was killed in action in June 1942.

Hughes Range. 84°30'S, 175°30'E. A major range of the Queen Maud Mtns, East Antarctica, identified by Richard Byrd in 1929 and named for US Chief Justice Charles E. Hughes.

Human occupation. Antarctica and the southern islands have no indigenous human populations. Though pre-European human settlement extended to the southern extremities of South America and Australasia, as yet there is no evidence of indigenous people living south of Tierra del Fuego or the southern islands of New Zealand. No island south of the Convergence has supported stock or voluntary human occupation for more than a few weeks at a time.

Occupation of the southern islands

The sequence of discovery of islands in the Southern Ocean during the seventeenth to nineteenth centuries is outlined in the **Study guide: Southern Oceans and Islands**. Most were discovered either by early navigators seeking the unknown southern continent, by later explorers or merchantmen pioneering trans-oceanic routes, or by sealers in search of new sealing grounds. Many were occupied temporarily by summer sealing gangs, a few by castaways. Nearly all were rejected as virtually uninhabitable. First of the cold temperate islands to be occupied was **Tristan da Cunha**. Informal occupation of the island by two US mariners in 1813 was followed two years later by the establishment of a small British garrison, placed there to foil possible French attempts to rescue Napoleon from St Helena. The garrison was withdrawn after one year, but permanent settlers remained under British protection. Living mostly by farming, fishing and collecting seabird's eggs, the colony has survived to the present day. The **Auckland Islands**, discovered in 1806, were colonized by a small group of New Zealand natives in 1842, and by a larger settlement of 300 Europeans in 1850. Supported by the London-based sealing and whaling firm of Enderby Brothers, the settlers built the **Enderby Settlement** on Enderby Island, where they intended to make a living from shore-based whaling, ship repairs and farming. The islands' chronically bad weather defeated them, and the survivors returned to civilization in 1852. Iles Kerguelen, and the warmer islands of Amsterdam and St Paul, were from time to time settled by farming or fishing enterprises during the nineteenth and early twentieth centuries.

The first permanent settlement on a Southern Ocean island was Omond House, a meteorological station established on Laurie Island, South Orkney Islands, by the **Scottish National Antarctic Expedition 1902–4**. Offered to Argentina, accepted and renamed Orcadas, it became the Antarctic region's first permanently manned observatory, and has maintained records continuously ever since. The first permanent land-based whaling station was opened in Grytviken, South Georgia, in summer 1904–5.

First landing on Antarctica

The first men to land on the Antarctic continent were probably sealers who, working from the recently discovered South Shetland Islands, may have come ashore on Antarctic Peninsula in 1820–21. Others probably landed in subsequent seasons but, failing to find seals, did not detail their landings. Later in the nineteenth century members of several scientific and sealing expeditions are likely to have landed on offshore islands or ice cliffs. First to make a fully authenticated landing on the continent were members of H. J. Bull's whaling expedition of 1894–95, who in the early morning of 24 January 1895 stepped ashore at Cape Adare, South Victoria Land.

First winterings in Antarctica

While it is possible that survivors of earlier shipwrecks or whaling expeditions wintered unrecorded in the Antarctic region, the first expedition known to have wintered south of the Antarctic Circle was the **Belgian Antarctic Expedition 1897–99,** under Lt Adrien de Gerlache de Gomery, which spent the winter of 1899 in the pack ice of Bellingshausen Sea. First to winter on land were the ten members of Carsten Borchgrevink's **British Antarctic (*Southern Cross*) Expedition 1898–1900**, whose huts, Camp Ridley, still remain at Cape Adare, Victoria Land. The first men recorded as having camped ashore on Antarctica were Savio and Must, the two Norwegian Lapp members of that expedition.

First occupation of Antarctica

For the first four decades of the twentieth century occupation of Antarctica was restricted mainly to the over-wintering expeditions of Scott, Shackleton, Charcot, Mawson, Byrd and other explorers of the 'heroic age'. Several wintered aboard their ships: others established bases ashore. Last in this tradition was the **British Graham Land Expedition 1934–37**, which wintered in huts on the Argentine Islands and Debenham Islands. Throughout the period also the Antarctic population was enhanced by whalers at shore stations on South Georgia, the South Shetlands and South Orkneys, and on factory ships moored each summer at anchorages as far south as Port Lockroy, Cuverville Island and Neko Harbour, on the Danco Coast of Antarctic Peninsula.

Up to the mid-1950s scientists and their supporters accounted for practically all visitors. Since then another category of visitors – tourists – has entered the field, increasing slowly but steadily and recently coming to outnumber all other visitors by an ever-widening margin.

Scientists and support staff

For the first four decades of the twentieth century the resident population of Antarctica and the Antarctic islands totalled four to six individuals – the staff of Orcadas, the meteorological station on the South Orkney Islands that began as a temporary expedition station in 1903 (see above). From 1905 their nearest neighbours were the wintering staff of whaling stations on South Georgia, possibly numbering 100–200. Expeditions from time to time provided an over-wintering population on mainland Antarctica, with peaks of 61 in 1934 (**Byrd's Second Antarctic Expedition 1933–35**) and 60 in 1940 (two bases of the **United States Antarctic Service Expedition 1939–41**). The advent of new research stations during the 1940s brought more wintering staff to Antarctica and the Antarctic islands. In 1945 four stations housed 29 staff. By 1950 11 stations housed 79, and in 1958 (the high point of IGY), 44 stations held a total wintering population of 893. Since then the number of stations operational at any time has varied between 30 and 40, and the number of over-wintering staff has ranged between 650 and 1200.

In addition, each summer has brought a varying influx of scientific and support staff, including summer-only scientists, maintenance staff and ship and aircraft crews. No single authority has responsibility to collect statistics of these influxes, but one reputable source (see *Further reading*) estimates summer support populations to have increased over one hundred-fold from 75 in 1941–42 to 8340 in 1989–90. Wives and families are accommodated at some of the stations (particularly at Argentine and Chilean military stations) and births have been recorded there. However, all who live in Antarctica remain transients, whose first homes are located elsewhere. Figures for resident populations – mostly scientists and support staff remaining overwinter – fluctuate from year to year. Figures for summer-visiting scientists and support staff fluctuate more widely, reflecting the amount of effort and expense that governments from time to time are prepared to devote to Antarctic research.

Further reading: Beltramino (1993).

Hummock. A lump or small ridge in a field of sea ice, usually caused by floes that are raised and displaced by lateral pressure on the ice sheet. (See **pressure ice**.)

Humpback whale. *Megaptera novaeangliae* These are stout, solidly built whales, predominantly black, but often with white patches on the belly, thorax and undersurfaces of the tail flukes and long flippers. The flippers, face and chin carry black knobs, and are often decorated with patches of grey barnacles. The throat pleats, extending from chin to well behind the flippers, are unusually broad: the baleen plates within the mouth are green or black. Measuring up to 19 m and weighing up to 48 tonnes, humpbacks are widespread in both hemispheres. Southern stocks spend their winters in temperate and tropical breeding areas, usually breeding in shallow coastal waters. In summer they move south, often keeping within sight of land, many penetrating to the edge of the pack ice. They feed mainly on plankton and small fish, which they catch by several methods, notably by spiralling up or down within dense shoals, creating nets of small bubbles that concentrate the prey. Humpback whales are notable also for their long, elaborate calls. Slow-moving, predictable, often gathering in groups where food is plentiful, they are among the easiest species to hunt. In the early years of Antarctic whaling around South Georgia, the Scotia Arc and Antarctic Peninsula, they formed the main bulk of the annual catches. Along with all other species, they are currently protected in Antarctic waters.

Further reading: Winn and Winn (1985); Bonner (1989).

Hunter, John George. (1888–1964). Australian Antarctic biologist. After graduating in medicine, biology and geology at Sydney University, he joined the **Australasian Antarctic Expedition 1911–14** as chief biologist, taking part in several sledging journeys from the main base, and serving as marine biologist in *Aurora* during the final survey cruise. He served as an army medical officer in both world wars, practised as a physician, lectured on medical ethics at Sydney University, and became Chairman of the Australian Council of Social Service. He died on 27 December 1964.
(CH)

Hurley, James Francis. (d. 1962). Australian expedition photographer. Leaving school without qualifications he worked in a steel mill and studied at the University of Sydney. A self-taught photographer, he joined the **Australasian Antarctic Expedition 1911–14**, taking part in the sledging programme. Immediately on return he joined the **Imperial Trans-Antarctic Expedition 1914–17**, sailing in *Endurance* and returning with a superb ciné film and still photographs of the expedition, despite having to destroy many plates before entering the lifeboats. He served as an army photographer in World War I, and photographer to several tropical expeditions before returning to the Antarctic with the **British, Australian and New Zealand Antarctic Research Expedition 1929–31**. Hurley was again a war photographer in World War II. He died in Sydney on 17 January 1962, aged 71.

Hussey, Leonard Duncan Albert. (d. 1965). British expedition meteorologist and medical officer. Born and educated in London, he joined the Weddell Sea party of the **Imperial Trans-Antarctic Expedition 1914–17** as meteorologist, in which he was noted for his cheerful disposition and banjo-playing. He served in the army during World War I, then qualified in medicine in time to join the **Shackleton-Rowett Antarctic Expedition 1921–22**. On return to England he practised in London until 1940, when he became a Royal Air Force medical officer. After World War II Hussey continued medical practice in England, and at sea as a ship's surgeon. He died on 26 February 1965, aged 71.
(CH)

Husvik Harbour. 54°10′S, 36°40′W. Harbour on the west side of Stromness Bay, South Georgia, the site of a Norwegian whaling station operated by the Tønsberg Hvalfangeri from 1908 to 1961.

Hut Point. 77°51′S, 166°37′E. Ice-free point on the southern shore of Ross Island, McMurdo Sound, the site of a storage hut erected by the British National Antarctic Expedition 1901–4. The hut is scheduled as HSM No. 18. An area around it of approximately 0.1 km^2 (0.04 sq miles) is designated SPA No. 28 (Hut Point), to allow the application of a management plan to the historic site.

Hutton Mountains. 74°12′S, 62°20′W. A small range flanking Keller Inlet on the Lassiter Coast of southern Palmer Land, named for the Scottish geologist James Hutton (1726–97).

IAATO. See **International Association of Antarctica Tour Operators**.

Ice. Water in solid form. At temperatures below 0 °C water vapour solidifies into crystals of frost or snow: liquid water becomes solid ice, in which a basically crystalline structure is obscured during accretion. Ice crystals are made up of water molecules linked by hydrogen bonds. Each molecule bonds to four others in a tetrahedral pattern, linking to form hexagonal rings: these determine the hexagonal structure of snowflakes and other forms of ice crystal. One gram of ice requires 79.8 calories of heat (latent heat of fusion) to reliquify. One cm^3 of ice weighs 0.919 gm: one cm^3 of water at similar temperature weighs 0.999 gm: thus ice is lighter than water and floats to the surface. Ice crystals forming in ponds or at the sea surface collect as a skin and ultimately a crust, initially in random orientation, later extending in vertical columns (**sea ice formation**).

Pure water freezes at 0 °C. Natural water masses are usually solutions containing ions and molecules, which depress the freezing point. Thus sea water, containing on average 3.5 per cent by weight of solutes, begins to freeze at about −1.8 °C. However, ice crystals that form in aqueous solutions exclude other ions and molecules. So the sea ice that forms annually on the surface of the Southern Ocean is composed only of water, liberating concentrated brine as a by-product.

On freezing to ice, water expands some 9 per cent by volume, creating considerable stresses in damp soils (patterned ground) and in waterlogged crevices between rocks (**weathering**). Pressure reduces the melting point: weight applied to skates, skis and sledge runners provides a self-lubricating film of liquid water. Individual crystals and masses of ice are brittle enough to shatter under impact. Under pressure the layers of ice crystals slide over each other, and the ice mass becomes plastic. Within a glacier, individual crystals under pressure from overlying layers constantly deform and recrystallize, and may even liquify, allowing the ice mass as a whole to flow under its own weight.

Iceberg. A large floating mass of freshwater ice that has broken from a glacier or floating ice shelf: also called berg. Icebergs that calve from glaciers and ice streams tend to be small, up to 50 m (164 ft) long and irregular in shape. These are especially characteristic of Arctic waters, though common off Antarctic coasts too. Floating ice shelves give rise to 'tabular' icebergs, which are often much larger (16 km, 10 miles long is not unusual), flat-topped and square-sided. Large tabular bergs are especially characteristic of Antarctic waters. Single bergs of 5000 km^2 (almost 2000 sq miles) and more have been reported in the Weddell Sea, originating from the Larsen Ice Shelf: even larger ones have from time to time broken from the Ross Ice Shelf.

Weathered and worn iceberg. Photo: BS

Originating from snow that has fallen and accumulated on land, icebergs are composed of ice that is more or less solid, according to its age. Those formed of old ice, for example in the interior of the continent where the ice has been subject to great pressure from overlying layers, contain little air and are of high specific gravity. Those formed in areas of heavy snowfall, where glaciers and ice shelves have a rapid turnover, tend to be soft, contain more air, and in consequence have lower specific gravity. Depending on their density, bergs show between one-seventh and one-fifth of their volume above water. Thus a tabular berg with cliffs rising 40 m (131 ft) above sea level may descend as much as 280 m (918 ft) into the water.

Once calved from their point of origin, bergs may drift for several years. Driven primarily by currents, secondarily by winds, most bergs originating from continental Antarctica drift westward initially under the influence of prevailing easterly currents and winds, then northeastward with the West Wind Drift. Large ones bend and flex in the swell, smaller ones bob in the water. All are subject to constant melting both above and below the water line, which shift their centres of gravity, causing them to rotate and topple, shaping and texturing their surfaces. Eventually all disintegrate into smaller bergs and bergy bits. While most disappear within two or three years, some of the larger tabular bergs, tracked by satellite observations, have retained their identity for five years or more, travelling far into the southern Atlantic and Indian Oceans before finally disintegrating.

Ice blink. Pale glare on the underside of low clouds, reflecting light from an underlying ice sheet. At sea it may indicate the presence of a distant ice field.

Ice cap. Sheet of solid ice, extensive but less than about 50 000 km^2, permanently covering a mountain, mountain range or similar elevated feature. See **Ice sheet**.

Ice core. A cylindrical sample of ice, drilled vertically from an ice sheet or glacier, using a hollow drill that keeps the sample intact. Cores drilled into the upper few tens of metres may show seasonal layering, each layer remaining discrete, often made up of powdery winter snow alternating with a layer hardened by summer melting. Where annual precipitation is high, such layers may be countable to considerable depth. In areas of lower precipitation the layers are thin and tend to merge quickly.

As each layer is overlain and buried by others, pressure on it increases. Deeper cores, recovered by techniques similar to those used in rock drilling, lose all visible signs of layering due to compression and thinning. However, where ice flow does not dramatically deform the layers (for example, towards the centre of ice sheets, and well above the bed where underlying topography is rugged), a chronological sequence may be retained over periods of many hundreds to tens of thousands of years. Apart from ice, each layer contains air – a sample of the lower atmosphere at the time the snow fell, with whatever gaseous components and aerosols it contained – together with continental dust, pollen spores, volcanic ash, sea salts, cosmic particles and other contaminants, both natural and manmade. As the snow is compacted to ice, the air is trapped and held in bubbles, which further pressure compresses to sub-millimetre size. Other impurities remain in tiny but nevertheless detectable amounts, to be revealed only by chemical determination. Physical and chemical analysis of layers within an ice core reveals valuable information on atmospheric and climatic conditions at the time of deposition: see **ice core analysis**.

Ice core analysis. Physical and chemical examination of ice cores, to reveal glaciological, palaeoclimatological and other data contained in their structure. The Antarctic ice sheet keeps deep-frozen a considerable body of information on climate and climate-related parameters. Surface pits reveal the most recent annual layers. Shallow or deep drilling recovers ice cores from greater depths, some descending to the underlying bedrock. The physical structure and chemistry of ice cores are particularly revealing. Their time scales can be established by several techniques, which are generally most precise for recent times but decrease in certainty with depth and age. Isotopic composition of the ice, for example ratios of **oxygen isotopes** ^{18}O to ^{16}O and of deuterium to hydrogen, reveal the temperature of the atmosphere at the time the snow fell. Concentrations of the naturally occurring beryllium isotope ^{10}Be and other chemicals can be used as measures of past rates of accumulation. Inclusions of volcanic dust and other impurities can date particular layers precisely. Air bubbles trapped within the ice allow reconstruction of the past composition of the atmosphere, for example its carbon dioxide and methane content. From the total amount of gas trapped in bubbles the elevation of the ice sheet above sea level at the time of precipitation can be deduced. Much of the information contained in ice cores is relevant not only to Antarctica but also to global environmental changes. Examination of dust and ionic impurities reveals their origins (for example from continents, oceans, volcanoes, cosmic dust or human inputs), as well as wind intensity and transport efficiency from lower latitudes towards Antarctica.

Cores from surface sites and shallow ice drillings have been examined all over Antarctica. Some of the most interesting results have been obtained from deep boreholes penetrating into ice that accumulated during the last ice age. These are located on the high plateau, at Byrd Station in West Antarctica (2164 m or 7098 ft deep, 1968), and at Dome C (905 m or 2968 ft deep, 1978), Dome Fuji (2503 m or 8210 ft deep, 1997) and Vostok (3623 m or 11 883 ft deep, 1998) in central East Antarctica. Results provide important clues toward describing and understanding climatic change. Of particular use in dating recent ice horizons are natural marker layers, i.e. strata marked by particular events. Markers that have been detected in Antarctic ice include debris from such known volcanic eruptions as Krakatoa (1883) and Gonung Agung (1963), and radioactive particles from thermonuclear tests of 1955 and 1965.

Ice cores provide a detailed record of changes in concentration of atmospheric carbon dioxide, showing an increase from about 270 to 280 ppmv (parts per million by volume) in pre-industrial time to over 310 ppmv in the early 1950s. These compare with current values (2001) of over 370 ppmv, determined directly from the atmosphere. Ice cores also show a close association between CO_2 and temperature along glacial and interglacials over the last 420 000 years. Relative dating involves comparison of ice core records with dating from other palaeoclimatic series, for example from ocean floor sediment cores. Evidence for large climatic changes, such as the end of the last ice age some 15 000 years BP, is present in various ice cores and marine records for which radiometric ages are available. The last glacial maximum is characterized by anomalously high concentrations of atmospheric impurities, which are also found in a variety of other kinds of records, and allow possible dating by correlation.

An approximate depth-age relationship for an ice core can be obtained from ice-flow models. Currently this is the most viable approach for deep long-term ice records. Models used are based on velocities obtained from accumulation data, and surface and bedrock topography. They may take into account past accumulation changes, but one of their limitations is that, for simplicity, they assume a constant thickness of the ice sheet which is almost certainly not realistic. Recent results on the Vostok ice core suggest an accuracy of only about 10 per cent over the past 150 000 years. (CL)

Ice edge. Boundary between a floating ice field and open water.

Ice fall. Steep incline in the surface of a glacier or ice stream, usually marked by heavy crevassing and disturbance.

Ice field. An extensive area of ice on sea or land.

Ice fog. Reduced visibility due to ice crystals suspended in still air.

Ice foot. Narrow fringe of coastal ice, forming the landward side of the **tide crack**, that remains for a time after fast ice has dispersed.

Ice front. An ice cliff forming the seaward face of a glacier or ice shelf: see **barrier**.

Ice Island. A large tabular iceberg formed from the breakup of an extensive ice sheet. The term is most often used in the Arctic, where there are few ice shelves large enough to generate large ice masses. More stable than ice floes, Arctic ice islands have been used to support scientific camps over periods of several months. The term is used only rarely in the Antarctic, where large tabular bergs are commonplace, and named as such.

Ice limit. Mean monthly or annual limit of sea ice.

Ice monitoring by remote sensing. The technique of examining Antarctica from spacecraft has revealed much new information about the ice cover of the Antarctic ice sheet (see **satellite imagery**). Before the satellite era, Antarctica's shape and contours were known only from patchy observations. Now glaciologists can outline and contour the ice sheet relatively accurately, including parts that had not previously been surveyed. All-weather day and night observations from satellites are constantly being used to update a surface that is dynamic and changing in response to climatic variations. Studies are fully justified by the importance of the ice sheet in determining present and future world climate. So much detail can be recorded instantaneously in a satellite view that image maps are increasingly used as a substitute for conventional maps.

After the launch of the first satellite of the US Landsat series in 1972, corrections of up to 50 km had to be made to previously sketched parts of the Antarctic coastline and some isolated mountain peaks. The first Landsat image maps lacked contours. This limitation was overcome in 1986 with the launch of Système Probatoire pour l'Observation de la Terre (SPOT), a French civilian satellite series capable of recording stereoscopic pairs of images. This system provided data for the construction of maps with a contour interval of 10 metres. Sun-synchronous satellites, including Landsat and SPOT, follow an inclined orbit that cannot reach further south than latitude 82°S. However, Landsat TM and other passive systems (AVHRR), as well as active radars (ERS SAR and Radarsat SAR) have been used successfully to identify and map a wide range of detailed ice structures. These include current positions of floating ice fronts, grounded ice walls, under-ice coastlines, surface features used to define glacier units, ice streams, direction of ice movement, crevassed areas, ice divides as defined by ridges, superimposed ice, surface equilibrium zones, and ablation areas. The net mass balance of glaciers has been deduced and areas of blue ice, likely to be fruitful both in the search for **meteorites** and in providing landing strips for wheeled aircraft, have been identified.

One reason why periodically surging glaciers, which are common in some other parts of the world, for long remained undetected in Antarctica was the lack of an effective means for monitoring the perimeter of the ice sheet. Satellite imagery now fills the gap. The long-term rate of movement of ice streams and ice fronts has been determined from timed series of images. In 1986 icebergs with an area totalling 4500 sq miles (11 500 km²) were observed newly-calved from Filchner Ice Shelf. A single iceberg almost 1950 sq miles (5000 km²) in area was tracked as it broke and drifted away from Larsen Ice Shelf. Recent imaging systems have achieved a ground resolution of as little as 10 m. An unexpected discovery on infrared images was the extent to which **katabatic winds**, flowing through the Transantarctic Mountains, were able to break down the surface **temperature inversion** over the Ross Ice Shelf.

The problem of clouds between the satellite and the ground was partially overcome by an electrically-scanning microwave radiometer (ESMR) launched in 1972. This instrument detected thermal microwave emissions which can pass through clouds, opening up new possibilities for broad-scale scanning of the ice sheet. Although the pixel (picture element) size of ESMR satellites was about 30 km × 30 km, yielding images of relatively coarse texture, resolution was adequate to reveal unexpectedly stable patterns over the Antarctic ice sheet. These allowed extrapolation of data from the still-sparse network of observations on the ground to the whole area of the continent. Because the mean annual temperature and the accumulation rate of dry polar snow mainly determine the grain size upon which the microwave emission depends, these two parameters can explain the main features of the patterns observed. For snow particle sizes normally encountered, most of the radiation has been shown to emanate from a layer of snow up to 10 m in thickness. Because wet snow emits more than dry snow, it is possible to recognize regions that are affected by summer melting. Extremely low brightness temperatures characterize the dry desert areas of East Antarctica. Again despite their low resolution, satellite images allow glaciologists to detect ice-stream channels extending up to 300 miles (500 km) inland from major outlet glaciers. Some of these features were unknown from any kind of ground or aircraft observations. It was found that the emissivity due to surface temperature, on the one hand, and snow accumulation, on the other, could yield measurements of the rate of snow accumulation to an accuracy of about 20 per cent in most places. Later studies have refined the understanding of these relationships by using dual frequency data from a scanning microwave spectrometer.
(CWS)

Ice piedmont. Shelf of permanent ice along coastal mountains, often covering a narrow strip of coast and terminating in a small ice cliff.

Ice port. Persistent or repeatedly forming bay in an ice front, providing temporary shelter and mooring for ships. Ice ports form in response to topographical features within the ice shelf: thus Bay of Whales, at the eastern end of the Ross Ice Shelf, is caused by the presence of Roosevelt Island some 30 km (19 miles) inland from the ice front.

Ice prisms. Fine needle-shaped crystals of ice precipitating from the atmosphere. They occur in extreme cold, for example on the high plateau, where they are the main form of precipitation.

Ice rind. Thin, brittle crust of hard ice forming on marine or fresh water, usually by freezing of **grease ice**. See **New ice**.

Ice sheet. Extensive area of permanent ice, of continental or sub-continental dimensions, that has formed on land but may spread over a coastline to form a floating **ice shelf**. See **ice cap**.

Ice sheet, Antarctic. The huge mass of **ice** covering over 98 per cent of the surface of continental Antarctica. Roughly circular, centred asymmetrically about the South Pole, the sheet is made up of several interlinked bodies of ice. Most of it lies in East Antarctica, covering an area of some 10.35 million km² (4.04 million sq. miles) and reaching a maximum elevation (**Dome Argus**) of just over 4000 m (13 120 ft). The much smaller West Antarctica ice sheet, with an area of 1.97 million km² (770 000 sq miles), reaches its highest among a small group of nunataks (Mt Woollard) at 2400 m (7872 ft). These sheets are separated by the **Transantarctic Mountains**, which stretch some 2000 km (1250 miles) continuously from Cape Adare in northern Victoria Land to the Wisconsin Mountains, and thereafter as isolated massifs. **Antarctic Peninsula**, a region of complex, mountainous topography with an area of 520 000 km², (203 000 sq miles) supports several highland ice caps, ice shelves, outlet glaciers and ice-covered offshore islands.

Flanking the West Antarctic ice sheet are two large embayments, the Ross Sea and Weddell Sea, which are filled in their inner parts with floating ice shelves. The Ross Ice Shelf covers 540 000 km² (211 000 sq. miles), the confluent Filchner Ice Shelf and Ronne Ice Shelf 530 000 km² (207 000 sq. miles). Elsewhere around Antarctica are many smaller ice shelves, their ice-cliff fronts totalling 14 110 km (8800 miles) long and constituting some 44 per cent of the coastline. Of the remaining coast, 12 156 km (7600 miles, 38 per cent) is characterized by the grounded ice walls of the ice sheet itself, and 3954 km (2470 miles, 13 per cent) by ice streams and outlet glaciers. The remaining 1656 km (1035 miles, 5 per cent) of coastline is ice-free rock.

Ice movement

Ice behaves as a viscous fluid which flows under force of gravity. With a thickness of over several hundred metres, small-scale undulations and irregularities in surface or bed are not critical. The direction of flow of a large ice sheet is down the maximum surface slope, or normal to the regional ice surface contours. From a detailed map of ice surface elevation flowlines may be drawn, manually or by computer, representing the surface movement of packets of ice flowing through the ice sheet. Flowlines radiate from several centres on the ice sheet – either distinctive domes or gently sloping ridges – toward the coast. Depending on the general configuration of the ice sheet and its periphery, ice in some regions diverges and in other regions converges onto narrow zones of discharge. Examples of convergence are the flowlines focusing on the Lambert Glacier and Amery Ice Shelf in East Antarctica, and on the Byrd Glacier in the Transantarctic Mountains. Flowlines extending back from the major zones of ice discharge (outlet glaciers and ice streams) delineate the principal ice drainage basins, equivalent to river catchments. In areas of diverging flow the basins tend to be narrow and indistinct. Where flows converge, the basins are well-defined, almost circular or elliptical.

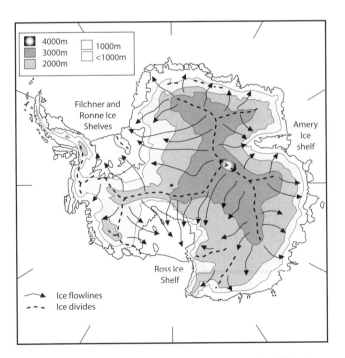

Ice flows downhill to form ice shelves and cliffs along most of the continental coast: underlying topography concentrates a high proportion into the three main bights, forming the massive Ross, Amery and Filchner and Ronne Ice Shelves. Map based on Drewry (1983) and other sources

Ice thickness

Thickness of the ice is measured between sub-glacial bedrock and surface. The most direct method of measuring is by drilling. Shallow depths can be drilled by hand or hand-held mechanical drills. Greater thickness requires drilling rigs similar to those used in drilling through earth and rock. Though accurate, and useful in providing **ice cores** from which data of many kinds can be obtained, drilling is too expensive and slow to be considered a practical means of measuring ice thickness over large continental areas. An alternative method, seismic sounding, was first used in the early 1950s. Sound waves created at the surface by explosive charges travel through the ice and are reflected from the bedrock and picked up by geophones at the surface. When the velocity of sound waves through the ice is also known, the time taken for the echo to return is a measure of depth. Many hundreds of such measurements were taken from sledge-trains during the 1950s and 1960s, providing the first reliable data on ice thickness along transects of the ice cap. The technique could be accurate to about 3 per cent of ice thickness. These techniques have been replaced almost entirely by **radio echo sounding**, in which a source of radio energy is directed towards and propagated through the ice. Energy is reflected from electrical discontinuities in the ice, due for example to changes in density or impurities, and also from the ice–bedrock interface. The travel time, multiplied by the velocity of the radio waves in ice, gives the thickness. Results are accurate to about 1.5 per cent of ice thickness. Undertaken from a moving vehicle or aircraft, the technique provides a much clearer and continuous picture of the internal structure of the ice sheet and of the bedrock surface. It has helped glaciologists to understand the structure and movements of the ice sheet, to detect the presence of lakes under the ice (e.g. **Lake Vostok**), to compute more accurately the total volume of ice, and to model the shape of the underlying continent.

Antarctica's thickest ice is located in major sub-glacial basins of both east and west provinces. The greatest depth to bedrock so far discovered, measured at 4776 m, occurs in the Astrolabe Basin of Terre Adélie. Substantial areas of ice over 4000 m (13 200 ft) thick have been found also in the Aurora sub-glacial basin of East Antarctica, and the Bentley sub-glacial Trench of West Antarctica. Bedrock beneath the Bentley Trench represents the lowest point so far recorded on Antarctica, 2555 m (8430 ft) below sea level. The ice sheet thins towards the coast, but several regions within the continental interior have coverings less than 2000 m (6560 ft) thick, for example, over the **Gamburtsev Subglacial Mountains** and other major sub-glacial massifs of East Antarctica. In West Antarctica detailed radar sounding has revealed a very complex pattern of ice thickness overlying rugged and irregular bedrock. Ice shelves spread and thin rapidly towards the coast, their thickness varying from 500–1000 m (1640–3280 ft) close to the grounding line to

100–200 m (328–656 ft) at the ice front. Exceptions are found in the inner parts of the larger Ross, Filchner, Ronne and Amery ice shelves, where thicknesses reach 1500 m (4920 ft). The total volume of ice in Antarctica is estimated at 30.11 ± 2.5 million km^3 of ice, approximating to 2.8×10^{19} kg in mass, and including 90 per cent of the current volume of ice in the world.

Significance for Earth's climate

Earth's ice sheets play important roles in climatic and oceanic processes and act with varying effects on surface rocks. By far the largest, the Antarctic ice sheet exerts many powerful effects of its own. Climatically it is a heat sink, modulating global climate by controlling rates of transport of energy, mass and momentum between low latitudes and the south polar regions. High and middle latitudes of the southern hemisphere are several degrees C cooler than equivalent latitudes in the north, the ice sheet and the isolation of Antarctica exerting important controlling influences on atmospheric circulation over all the southern continents. For a better understanding of its full effects, physical characteristics and dynamics are being incorporated into general circulation models of the atmosphere. The ice sheet exerts major effects on the world ocean. World sea level rises and falls in response to many factors, ranging from changes in the size and shape of ocean basins to temperature at different levels in the water itself. The major ice sheets represent a volume of water abstracted from the oceans, and capable of returning through warming. Should global warming cause the Antarctic ice sheet to melt, mean sea level would rise, exerting profound effects in coastal areas all over the world. Rapid total melting (a very unlikely possibility) would cause an estimated rise of 80 m (264 ft), which would fall to about 55 m (181 ft) after 1000 years due to sinking of the ocean floors by **isostasy**. Current atmospheric warming seems, at least temporarily, to be exerting an opposite effect, in causing more snow to fall and build up on the Antarctic continent, and a corresponding slight fall in mean sea level. The contrary is true in Greenland.

Floating ice shelves (in Antarctica amounting to about 1.5 million km^2), together with sea ice and icebergs, strongly affect oceanic circulation. They produce very cold, low-salinity but dense water which creeps northward across the bed of the Southern Ocean. So-called 'Antarctic bottom water' is identifiable as a discrete water mass even in the North Atlantic Ocean, its presence affecting both Antarctic and other water masses on a worldwide scale. The ice sheet plays an important role in eroding and shaping the Antarctic landmass. It removes rock material and transports debris over many hundreds or even thousands of km, eventually to be released and deposited either on land beneath the ice or into the sea. Accumulation of marine sediments from glacial debris in particular provides an invaluable record that helps in elucidating past fluctuations of the ice sheet and of climate generally.

The ice sheet is sufficiently large and heavy to generate lateral and vertical stresses in Earth's crust. The upper, low-density and relatively rigid lithosphere or crust sinks into the underlying, more viscous and denser asthenosphere. The amount of sinking is proportional to the ratio of the density of ice (920 kg per m^3) to that of the mantle (of the order of 3300 kg per m^3). Thus an ice thickness of 3000 m causes over 1000 m depression of the underlying surface.
(DJD)

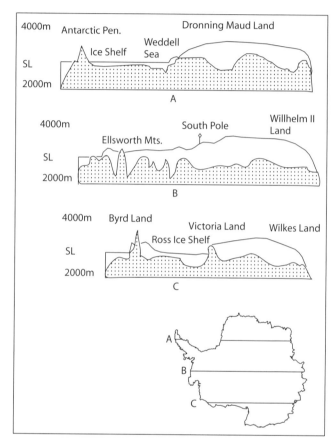

Schematic diagram of ice thickness across three continental transects

Ice sheet, Antarctic: history and development.

The Antarctic continental plate drifted into the south polar position some 100 million years ago (see **plate tectonics**). The resulting cooling alone may have been sufficient to provide conditions for the development of the ice sheet, though two other factors may also have contributed. First, sea passages opened up as the other southern hemisphere continents split away from Antarctica and moved northward. A circumpolar current developed as a result of these plate movements, thermally isolating Antarctica from the rest of the world. Second, the Transantarctic Mountains began to form at the edge of a major rift system around 50 million years ago, raising the land to elevations of 5 km and more. Similarly, other mountain ranges (e.g. the Gamburtsev Subglacial Mountains) may have been uplifted, though these are now buried by the ice sheet. These mountain ranges were the first gathering grounds for local ice fields and valley glaciers that ultimately expanded and coalesced into ice sheets, eventually covering the entire Antarctic continent.

The longevity of glaciation in Antarctica exceeds that of the northern hemisphere by an order of magnitude. The first glaciers reached the sea at least 34 million years ago (at the Eocene–Oligocene transition), as recorded in numerous cores in offshore marine sediments. The record of long-term glaciation on land is scanty in comparison with the marine record, restricted to the flanks of exposed mountain ranges. However, although little reliable evidence exists, it is likely that mountain glaciers existed in the interior of Antarctica well before deposition of the oldest sediments so far recorded offshore.

The most comprehensive record of glaciation has been recovered as a result of drilling on the continental shelf. The earliest offshore drilling was ship-based, and was undertaken in the outer Ross Sea in 1973 by the Deep Sea Drilling Project. This was the first time that scientists found evidence of glaciation that extended back in time far longer than in the northern hemisphere, i.e. to about 25 million years ago, and set the scene for subsequent but elusive searches to define the onset of glaciation. In the late 1970s and 1980s New Zealand scientists pioneered drilling from sea ice in near-shore areas, focusing on McMurdo Sound. In several holes they achieved remarkable recovery of core, culminating in a 702 m (2303 ft) hole in 1986, with the oldest glacial sediments proving to be of late Eocene age (about 36 million years). Next came the visit of the international ship-based Ocean Drilling Program to Prydz Bay in 1987/88, which drilled five holes on a transect across the continental shelf. This research provided a similarly long record of glaciation, at the mouth of one of the main discharge routes from the East Antarctica ice sheet. However, neither these nor the earlier drilling operations actually showed when glaciation started.

Drilling was renewed in 1997, when the international Cape Roberts Project, building on prior New Zealand experience, drilled three holes in the continental shelf of the western Ross Sea (1997–2000). The deepest was 939 m (3080 ft). Cores from these record a progressive cooling of glacial climates. Earliest Oligocene times were near-temperate, with trees (notably southern beech *Nothofagus spp.*) flanking the glaciers discharging into the sea. Miocene climates were similar to those experienced in the Arctic today, chilling to cold polar conditions devoid of vegetation in the Quaternary period (the last two million years). However, the sought-after preglacial/glacial transition remained elusive, because of an unexpected major gap (unconformity) in the stratigraphic record. In 2000 the Ocean Drilling Program returned to Prydz Bay, drilling three sites on the continental shelf, slope and rise. Initial results indicate a transition from fluvial/deltaic deposition to glaciomarine sedimentation around the Eocene–Oligocene boundary, confirming the long-suspected onset of large-scale glaciation at this time. The new Prydz Bay data record also the progressive cooling through the Miocene Epoch that is evident in the Ross Sea cores. Elsewhere, the Ocean Drilling Program has recovered a rather incomplete Pliocene glacial record to the west of Antarctic Peninsula.

Although a clear picture is emerging concerning the offshore glacial record, interpreting the onshore record has proved to be extremely controversial. Throughout the Transantarctic Mountains, glacial sediments are preserved in sections typically up to 100 m (328 ft) thick. Collectively these deposits are known as the Sirius Group, and most have been deposited by terrestrial glaciers, sometimes in association with *Nothofagus* remains, including well-preserved leaf mats (as in the upper Beardmore Glacier region), thus indicating a climate with a mean annual temperature possibly 20 °C higher than that of today. The implications for this association are that there were major fluctuations in the scale and character of the ice sheet at the time of deposition. Based on the presence of microflora (diatoms) in these sediments, and within cores drilled in valley bottoms and a fjord, one school of thought has argued that the Sirius Group is of Pliocene age (1.8 to 5.3 million years), and that this was a time of major ice sheet fluctuations. Another school of thought has refuted the age of these sediments, arguing that a range of volcanic ash dates and cosmogenic dating of land surfaces in the Dry Valleys region of the Transantarctic Mountains indicate that the climate has been cold and stable for at least 15 million years. This debate remains to be resolved.

Similar 'old' glacial deposits occur in several other parts of East Antarctica. Recent studies in the Prince Charles Mountains, bordering the Lambert Glacier that feeds into Prydz Bay, have revealed several sequences of uplifted fjordal sediments, deposited by glaciers under a significantly warmer climate. The oldest sediments, at 1400 m (4592 ft) elevation, are probably of late Oligocene or early Miocene age (*circa* 24 million years), while the youngest are Pliocene and occur close to sea level. The record here tends to support the concept of major fluctuations of the ice sheet to at least Pliocene time, each of which was followed by a phase of uplift.

The question of when the East Antarctic ice sheet switched from being 'dynamic' to the present condition of stability is of more than academic interest. There is a strong correlation between CO_2 composition of the atmosphere and ice cover on Earth. The last time Earth's atmosphere had a composition similar to that projected for the end of this century was during the Pliocene Epoch. Knowing whether the East Antarctica ice sheet was subject to major fluctuations at that time is thus a vital question, given the potential for global sea level rise if the ice sheet melts.

The above lines of evidence point to the following stages of evolution of the East Antarctica ice sheet: (i) growth of a large-scale ice sheet around 34 million years ago, reaching the coast in many areas; (ii) major fluctuations in this sheet until either 15 or 2 million years ago (see above); (iii) progressive cooling of climate from temperate with *Nothofagus*-dominated vegetation to cold-polar by Quaternary times; (iv) a final stage of a stable, cold ice sheet spanning at least the Quaternary Period.

Evidence for the development of the West Antarctica ice sheet is relatively sparse. Small ice masses may have developed over a relatively few elevated areas of the subcontinent at various times from about 30 million years ago. Isotopic records from deep sea cores indicate that growth of a more substantial ice sheet began in mid-to-late Miocene times, 15 to 10 million years ago. There are sites where volcanic rocks erupted subglacially, indicating the presence of ice at different times in the past, notably from the late Miocene onward. Because much of the West Antarctica ice sheet rests on bedrock that is below sea level, it is potentially much less stable than the ice sheet covering East Antarctica, and has probably been subject to major fluctuations through the Quaternary Period to the present day.
(MJH)

Ice sheet movement. Over most of the Antarctic ice sheet there is a net accumulation of snow. Once fallen, snow and ice move downward and outward toward the coast. Several techniques are used to measure these movements. Where there is fixed rock nearby, for example at the edge of a glacier, glaciologists use standard survey techniques (involving theodolites and electronic distance-measuring instruments) to measure the movement of marker stakes. Where there are no rocky fixed points, for example on an extensive ice sheet, satellite-based Global Positioning System (GPS) may be used instead to plot the movement of markers, allowing positions to be calculated quickly and accurately to a few cm over large areas. **Satellite imagery**, both visible (e.g. Landsat or SPOT) and radar (e.g. ERS-1, ERS-2, Radarsat), can be used to measure the travel of individual features over time. Interferometric techniques using synthetic aperture radar (SAR) data give movement patterns over large areas, for example to detect the lifting of ice shelves due to tidal movements, as well as surface velocities.

Where precipitation over the ice sheet is slight, downward movement, i.e. the rate at which objects are buried, is generally very slow – over much of the interior, for example, measuring tens of cm per year. In coastal regions where precipitation may be heavier, downward movement may be measured in metres per year. Only in ablation areas where ice is being removed by melting or sublimation is the vertical component of movement upward, toward the surface. Horizontal velocities are more variable and can be very much higher. At ice divides (the equivalent of watersheds) horizontal movement may be negligible. At the seaward edge of active, floating ice shelves, horizontal velocities can rise to several km per year.

Inland, much of the Antarctic ice sheet is frozen to its bed. Forward movement is by internal deformation, and the highest velocities are at the surface. Toward the coast, ice sheets tend to become channelled into ice streams or outlet glaciers. At some point along the channels the whole body of ice begins to slide over its bed. From there onward velocities rise quickly from tens to hundreds or thousands of metres per year. Processes controlling basal sliding are not well understood, but the base of the ice has to be melting, and the materials and textures of the glacier bed are as important as those of the ice itself. Most ice is lost from the continent by calving of icebergs or melting beneath ice shelves.
(CSMD)

Ice shelf. A floating ice sheet that extends over a coastline. Small areas within the shelf may be grounded on islands or submarine banks. The outer edge, forming an ice cliff (**barrier** or **ice front**) may be grounded or afloat.

Ice stream. See **Glaciers and ice streams**.

Ice tongue. Extension of a glacier or ice stream beyond the coast, in a tongue that retains its shape far out to sea, breaking off and reforming repeatedly. Such tongues occur all around Antarctica. Some of the larger ones extend more than 100 km (63 miles) from their grounding line: an example is the Chelyuskinsky Ice Tongue of the Leopold and Astrid Coast. The Drygalski Ice Tongue in the western Ross Sea extends some 70 km (44 miles) from the coast and 120 km (75 miles) from its grounding line. Smaller ones never exceed a few km before breaking away. Broken fragments of uncertain origin, caught in fast ice, are sometimes distinguished as 'iceberg tongues'. A small, well-studied example is the Erebus Glacier Tongue, in McMurdo Sound, with floating portion up to 12 km (7.5 miles) long and width tapering from 2 km (1.25 miles) at the grounding line to 1 km at the seaward end. Measured velocity and thickness profiles along its length correspond with theoretically expected values, based on treating the tongue as an **ice shelf** with no restricting forces from side walls or from locally grounded areas such as ice rises. Historic records indicate that it calves up to half its length every 30–40 years, though it is not clear why such an unconstrained ice shelf should not calve off as icebergs at the point where it starts to float.
(CSMD)

Imperial Trans-Antarctic Expedition 1914–17. Sir Ernest Shackleton's second Antarctic expedition, in which he planned to cross Antarctica. After Amundsen and Scott had reached the South Pole in 1911–12, Shackleton

Two upturned lifeboats on Pt. Wild, Elephant Island, make a winter home for survivors from the wrecked *Endurance*.
Photo: Scott Polar Research Institute

planned an expedition that revived **Filchner**'s concept of crossing the southern continent at its narrowest point, between the great embayments of the Weddell and Ross Seas. After months of fund-raising and preparation, in August 1914 he launched the Imperial Trans-Antarctic Expedition, effectively the last expedition in the 'heroic' mode of Antarctic exploration. His departure from Plymouth on 8 August 1914, after the start of World War I, was sanctioned by Winston Churchill, then first Lord of the Admiralty. Two ships were involved. S. Y. *Endurance* (Capt. Frank Worsley) would establish a base far south along the Luitpold Coast of Coats Land, on the eastern shore of the Weddell Sea. From there Shackleton would lead a sledging party across the continent. The second ship, *Aurora* (Capt. Æneas Mackintosh), a Newfoundland whaler that had taken Douglas Mawson to Antarctica in 1911, would take a smaller 'Ross Sea' party to McMurdo Sound. There Mackintosh would establish a base in Shackleton's old hut at Cape Royds, with the task of setting up a line of depots across the Ross Ice Shelf to the foot of the Beardmore Glacier – depots on which the polar party would rely to complete their long haul across the continent.

Endurance sailed from South Georgia on 5 December 1914 for the eastern Weddell Sea. Nine days later *Aurora* sailed from Sydney for McMurdo Sound. *Endurance* soon ran into loose pack ice that tightened and solidified as she headed toward the Coats Land coast. By early January 1915 they were beset in ice through which the poorly powered ship could make no headway. By 20 January Shackleton realized that they were there for the winter, and secured the ship accordingly, putting the sledge dogs out onto the ice. *Endurance* continued to drift helplessly in an ever-shifting mass of floes, first southward, then east into the central gyre of the Weddell Sea, and finally north. During October 1915 the ship began to take intense lateral pressure that crushed the wooden hull beyond any hope of repair. Shackleton ordered the 28 men and all movable stores off, together with three lifeboats, creating a makeshift camp of five tents on the ice. *Endurance* finally sank in mid-November. Concerned if possible to reach Paulet Island, where Capt. C. A. Larsen of the **Swedish South Polar Expedition 1901–4** had wintered after the loss of his ship, the party tried marching westward across the rotting ice, but soon gave up. For a further five months the camp continued to drift northward, through the summer and well into autumn. By early April they had reached the northern edge of the pack ice. With open sea in sight, on 6 April they took to the boats, rowing, sailing, and steering northward for Elephant Island, the nearest point of land. On April 14 they stepped ashore at Cape Valentine, a rocky headland on the east coast with narrow cobbled beach. On the following day Wild and four others left in one of the boats to seek a better site for a camp, finding one on another headland just a few miles further along the coast. The weary men made the final journey to what would eventually be called Point Wild. On landing they set up camp, turning two of the lifeboats into living quarters by inverting them over low walls of beach cobbles.

The party was safe, but nobody knew where they were: Shackleton had to seek help. Within three days the carpenters had fitted the third lifeboat, *James Caird*, with a canvas half-deck. On 24 April Shackleton, Frank Worsley, Tom Crean, AB Tim McCarthy, carpenter Harold McNeish and 'Bosun' Vincent left to sail to the nearest settlements, the whaling stations on South Georgia, 1300 km (over 800 miles) away to the east. *James Caird* sailed before the prevailing westerly winds, Worsley navigating with sextant, pocket compass and chronometer. Most of the time the men were wet and near-frozen. Ice built up on the boat and had to be chopped away with axes. After 17 days, on 10 May 1916, they made a landfall in King Håkon Bay, on the deserted southern side of South Georgia. After a brief respite Shackleton decided that, rather than putting to sea again, he, Worsley and Crean would cross the island to the north side, where the whaling stations lay. Starting by moonlight at 2.00 a.m. on 10 May, the three crossed unknown icefields and descended unmapped valleys, to reach Stromness Bay whaling station at tea-time on the following afternoon.

An astonished station manager swung into action, sending a whale-catcher round to King Håkon Bay to pick up the remaining three crew, and a second catcher with Shackleton, Worsley and Crean aboard towards Elephant Island. However, impenetrable pack ice had closed around the island, so the catcher took the three men on to the Falkland Islands. Shackleton made three further bids to reach the island: the third, in the Chilean naval steam tug *Yelcho* (Capt. Luis Pardo), finally broke through the barrier of ice to reach Cape Wild, taking off the 22 scientists and crew on 30 August 1916.

Unknown to Shackleton, the Ross Sea party had also had its share of trouble. In the summer of 1915 fast ice kept *Aurora* from Cape Royds, so they landed instead at Cape Evans, intending to winter the ship immediately offshore. Before unloading was complete, the ship was blown from her moorings and beset by pack ice, making return impossible. So started a ten-month imprisonment. Chief Officer J. R. Stenhouse took command, nursing *Aurora* through the ice until she finally broke free and returned to New Zealand. Meanwhile Cape Evans housed an ill-founded party of ten, lacking most of the stores and equipment they would need to complete their task. With no radios, they had no knowledge that *Endurance* was in difficulties with the polar party still on board. Thus they still faced the awesome responsibility of laying six depots across the Ross Ice Shelf. Mackintosh and his men lost no time. Improvising clothing from old tents, using blubber

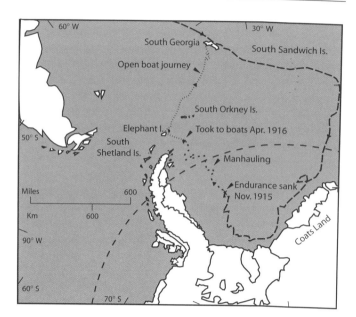

Track of *Endurance* into the Weddell Sea, drift of the stranded party on the pack ice and boat journey to Elephant Island, and the voyage of *James Caird* to South Georgia

stoves to save kerosene for camping, scavenging through rubbish piles, they pulled together makeshift equipment for their sledging programme. During the following months, on meagre rations and in poor, worn-out clothing, dragging patched-up sledges, they doggedly laid their quota of depots across the Ross Ice Shelf as far south as Mount Hope (83°S), providing all that was needed for Shackleton and the trans-polar party. Their efforts cost three lives: Arnold Spencer-Smith died of scurvy on the trail, and Mackintosh and Victor Hayward died while crossing insecure sea ice between Hut Point and Cape Evans. The seven survivors were relieved on 10 January 1917.

Further reading: Joyce (1929); Worsley (1931).

Inaccessible Island. 37°17′S, 12°41′W. A warm temperate island 4 km (2.5 miles) long and 3 km (1.9 miles) across, lies 35 km (22 miles) southwest of Tristan da Cunha, in the southern Atlantic Ocean. A rectangular slab of lava and scoria, tussock-covered and rimmed on all sides by steep cliffs and narrow beaches, it rises to a dissected plateau up to 600 m (1968 ft) high: access is possible only by scrambling up steep water-cut ravines. The climate is similar to that of Tristan da Cunha, though possibly wetter. The dominant vegetation is a tall tussock grass, giving way to shrubs and trees in gullies and on well-drained ridges, and moorland and bog on the uplands. Many millions of seabirds, notably rockhopper penguins and burrowing petrels, nest on the island, which also supports an endemic species of rail. Elephant seals and Amsterdam fur seals breed on the beaches. Sealing gangs and castaways have lived on the island, and Tristan islanders visit to collect seabird eggs, but there is no human settlement. The island is protected as a nature reserve.

Inexpressible Island. 74°54′S, 163°39′E. A low, narrow island 11 km (7 miles) long in Terra Nova Bay, Victoria Land, where six men of the Northern Party, **British Antarctic (*Terra Nova*) Expedition 1910–13** wintered uncomfortably in an ice cave.

Ingrid Christensen Coast. 70°00′S, 75°00′E. Part of the coast of Princess Elizabeth Land, East Antarctica, bounded by Jennings Promontory in the west and the western end of the West Ice Shelf in the east. It consists mainly of ice cliffs with nunataks and emergent islands, but includes a large oasis area, the Vestfold Hills, which is the site of Davis, a permanent Australian research station. The coast was explored from 1935 onward by ships of the Norwegian (Christensen) Whaling Expeditions 1926–37, and named for the wife of Lars Christensen, owner of the whaling fleet.

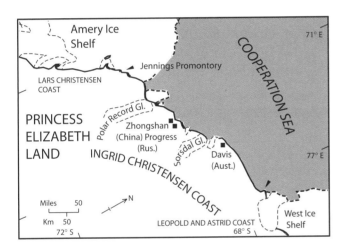

Ingrid Christensen Coast

Institute Ice Stream. 82°00′S, 75°00′W. Ice stream draining northward from Ellsworth Land into the central Ronne Ice Shelf. The stream was identified from radio echo-sounding data and named for the **Scott Polar Research Institute**, Cambridge, UK, from which the research was conducted.

International Association of Antarctica Tour Operators. A member organization of tour operators founded in 1991, with the objective of promoting safe and environmentally responsible private-sector travel to the Antarctic region. With a current (2001) membership of 30 companies in 10 countries, IAATO represents Antarctic tour operators and other travel organisers at meetings of the Antarctic Treaty and other public platforms, seeking to work in accord with the Treaty System and similar international and national agreements. It fosters cooperation among its own members, monitoring its own programmes and activities and coordinating cruise itineraries, and encourages cooperation between the industry and the Antarctic scientific community. Its members encourage research by providing logistical support. IAATO seeks also to ensure that the best qualified staff and field personnel are employed by its members through continued training and education, encouraging and developing international acceptance of evaluation, certification and accreditation programmes for Antarctic personnel.

International Geographical Congress, Sixth. A meeting of professional geographers and other scientists in July 1895, under the presidency of Sir Clements Markham and hosted in London by the Royal Geographical Society, this Congress strongly endorsed further Antarctic exploration, helping to promote during the following decade wave of expeditions under six national flags. Follow research papers by Georg von Neumayer, Carsten Borgrevink, Sir Joseph Hooker, Sir John Murray and others prominent in polar scientific research, the Congress 'recorded its opinion that the exploration of the Antarctic Regions is the greatest piece of geographical exploration still to be undertaken. That in view of additions to knowledge in almost every branch of science which would result from such a scientific exploration the Congress recommends that the scientific societies throughout the world should urge in whatever way seems to them most effective, that this work should be undertaken before the close of the century.' Though the resolution had no direct bearing on government policies, it encouraged individual explorers and entrepreneurs, and geographical and other scientific societies of several nations, to initiate (in several cases to refurbish) plans for expeditions to the Antarctic region, and where possible to seek government support toward their substantial costs. First to take the field were the privately sponsored **Belgian Antarctic Expedition 1897–99**, and the **British Antarctic (*Southern Cross*) Expedition 1898–1900**. Longer in preparation were the **German South Polar Expedition 1901–3**, **Swedish South Polar Expedition 1901–4**, **British National Antarctic Expedition 1901–4**, **Scottish National Antarctic Expedition 1902–4**, and the **French Antarctic (*Français*) Expedition 1903–5**. *Further reading*: Baughman (1994).

International Geophysical Year 1957–58. In 1950 an international group of geophysicists determined that the time had come for a third Polar Year – a year of concerted research on all aspects of polar geophysics, similar to studies that had previously been made in 1882–83 and

1932–33. Plans for a polar year were quickly overtaken by wider-ranging plans for an International Geophysical Year – a year in which geophysical and geographical phenomena would be studied simultaneously all the world over.

An international committee, the Comité Spéciale de l'Année Géophysique Internationale, was established. Some 67 nations were involved worldwide, with special attention paid to polar regions, particularly Antarctica. The period July 1957 to December 1958 was chosen – a period of 18 months spanning two Antarctic summers and a winter, at a time of maximal sun spots and intense solar activity. At a preparatory conference in 1955 it had been agreed that, as the overall aims of IGY were scientific, politics should be set aside and international cooperation encouraged to the full. In the Arctic at the height of the Cold War, this proved almost impossible: most Arctic research was nationally based. In Antarctica questions of sovereignty were temporarily waived, allowing a remarkable degree of international cooperation. Twelve nations operated in or close to the Antarctic region, involving over 5000 scientists and support personnel at 55 stations – by far the largest number ever to operate simultaneously in the area. Claimant nations (Argentina, Australia, Chile, France, New Zealand, Norway, United Kingdom) cooperated fully with each other and with non-claimants (Belgium, Japan, South Africa, Soviet Union, United States of America) on an equal footing. IGY was by no means apolitical: it gave several non-claimants 'rights' to work in Antarctica which might not otherwise have been freely granted. However, it provided grounds for cooperation between such nations as the United Kingdom, Argentina and Chile, whose overlapping claims had for long been a cause of tiresome conflict, and the Soviet Union and USA, which elsewhere in the world found little cause to agree on anything.

While some nations pursued geophysical research at their existing stations, others established new stations expressly for the IGY: some did both. Prominent among the newcomers was the United States, which in two successive seasons established South Pole station (later called **Amundsen-Scott**) at the geographical South Pole, Byrd in the interior of Marie Byrd Land, Wilkes (operating jointly with Australia) on Budd Coast, Ellsworth on Filchner Ice Shelf, Little America V on Ross Ice Shelf, Hallett (operating jointly with New Zealand) in Victoria Land, and Williams Air Operation Facility in McMurdo Sound – later to develop into McMurdo Station. The Soviet Union established Mirnyy on the coast of Queen Mary Land and Oazis in the Bunger Hills, Knox Coast. From Mirnyy tractor trains established inland stations Pionerskaya, Vostok (close to the South Geomagnetic Pole) and Vostok Island, Komsomol'skaya, Sovetskaya, and a transient station at the Pole of Inaccessibility. On a smaller scale Norway established Norway Station on Kronprinsesse Märtha Kyst, Belgium established Roi Baudouin on Prinsesse Ragnhild Kyst, and Japan placed Syowa on East Ongul Island, Kronprins Olav Kyst. New Zealand established Scott Base on Ross Island, and France established Dumont d'Urville and inland Charcot in the narrow strip of Terre Adélie. Australia opened a new coastal station, Davis, in Princess Elizabeth Land. Britain and Argentina used some of their existing stations for IGY work, but each set up a further station in the Weddell Sea, Britain at Halley Bay on the Coats Land coast, Argentina further south on the Filchner Ice Shelf. South Africa investigated Bouvetøya, but finding no safe place to land, settled instead for taking over a station on Gough Island.

On a world scale the IGY achieved many important scientific results. Using recently developed satellites and rockets to study upper-atmospheric phenomena, researchers discovered the Van Allen Belts, and elucidated the forces that drive world climates and weather. New deep-sea techniques were used to study the structure of ocean basins, paving the way for a fuller understanding of plate tectonics. In Antarctica IGY's main achievement was to demonstrate the possibility of international cooperation in scientific research and management. From it developed the Scientific Committee on Antarctic Research (SCAR), which since 1961 has coordinated research throughout the continent. IGY also gave rise to the Antarctic Treaty System, which provided a zone in which peaceful cooperation in scientific research could continue into the post-IGY years.

Further reading: Beck (1986); Fifield (1987).

Inverleith Harbour. 64°32′S, 63°00′W. Sheltered bay between Andrews Pt. and Briggs Peninsula, Anvers Island, Palmer Archipelago, opening onto Discovery Sound. The harbour was used by whalers of the Christian Salvesen fleet whose home base was Leith, Scotland.

Iroquois Plateau. 78°00′S, 105°00′W. Extensive ice plateau of the Pensacola Mountains, named for the Bell UH–1 Iroquois helicopter.

Islands, non-existent. Islands reported to lie in the Southern Ocean and neighbouring seas, which on further investigation proved not to exist. Especially before chronometers came into use for determining longitude, early navigators often reported islands and rocks in positions calculated by dead reckoning, that turned out to be radically wrong. Some were authenticated many years later in their true positions. Bouvetøya, for example, was first reported in 1739, but only in 1898 was its correct position determined. In poor visibility, navigators often mistook icebergs or cloud formations for islands: in these circumstances it was usually inadvisable (with contrary winds, impossible), to approach closer for a more reliable position.

To commercial seamen, the sites of reported islands were generally places to avoid. In the early decades of the twentieth century they attracted the attention of whalers, who at the time were seeking unclaimed lands on which to establish whaling stations that would not be subject to controls or taxation. It was an unrewarding search. Exploring the designated position of each island, the whalers found nothing in sight, and those with efficient sounding gear usually confirmed deep water under the keel, indicating an extreme unlikelihood of land nearby. The islands and rocks here listed are among those that have so far eluded discovery.

Table 1 Non-existent islands

Island(s)	Position	First reported
Aurora Islands	54°S, 45°W	1762
Burdwood's Island	54°S, 59°W	1828
Dougherty's Island	59°S, 120°W	1841
Elizabethides	55°S, 75°W	1578
Emerald Island	57°S, 162°E	1821
Isla Grande	47°S, 49°W	1675
Macy's Island	59°S, 91°W	1824
Middle Island	63°S, 59°W	1821
New South Greenland	63–69°S, 48°W	1821
Nimrod Islands	56°S, 158°W	1828
Pagoda Rock	60°S, 5°E	1845
Royal Company Island	49°S, 142°E	1776
Strathfillan Rock	55°S, 42°W	1927
Swain's Island	59°S, 90°W	1800
The Chimneys	54°S, 3°E	1825
Thompson Island	54°S, 3°E	1825
Trulsklippen	56°S, 24°E	1929
Undine Rock	59°S, 42°W	1916

Note: Data from R. K. Headland, unpublished.

Isostasy. Equilibrium of Earth's crust, manifest in Antarctic regions by the sinking (down-warping) of the continent and adjacent islands under their burden of snow and ice, and recovery when the burden is eased by extensive glaciation. Down-warping extends to a distance typically of the order of 500–600 km from the centre of loading. Degree of down-warping and subsequent uplift can be calculated, taking account of the flexural stiffness of the crust, which has the effect of laterally spreading a load. In Antarctica a complicating factor is uncertainty over the precise thickness of the crust, but values typical of other continental areas (about 100 km) can be adopted. Isostatic movements do not take place instantaneously, as the crust is not truly elastic. Estimates of the time taken to achieve equilibrium, derived from post-glacial uplift rates and a consideration of mantle viscosity, suggest values of the order of 10 000 years or slightly less. In deducing the isostatic influence of the Antarctic ice sheet, ice thickness data indicate that, if full isostatic compensation has been achieved, depression is 950 m in central East Antarctica, and 500 m in West Antarctica. The outer region of the continental shelf is far enough from the ice edge to be unaffected. The Ross Sea and Weddell Sea embayments possess little grounded ice and hence show little warping.
(DJD)

Isotope. Atom of an element that, though chemically identical to other atoms of the same element, has a different atomic weight. Ratios of some isotopes, typically oxygen, may be used to determine palaeoclimates (see **Ice core analysis**).

J

Jack, Andrew Keith. (d. 1966). Australian physicist and polar explorer. Jack joined the **Imperial Trans-Antarctic Expedition 1914–17** as meteorologist and physicist with the Ross Sea Party, taking part in early depot-laying journeys. During World War I he became head of the munitions establishment at Maribyrnong. After the war he remained in munitions, becoming secretary of the Operational Safety Committee of the Australian Department of Munitions Supply. He retired in 1950 and died on 26 September 1966, aged 81.
(CH)

James, Reginald William. (1891–1964). British expedition physicist. Born in London on 9 January 1891, he read physics at London and Cambridge universities. While still a postgraduate student at the Cavendish Laboratory, Cambridge, he joined the Weddell Sea party of the **Imperial Trans-Antarctic Expedition 1914–17**, as a physicist. On return to Britain he served with the Royal Engineers during World War I. After the war he lectured in physics at Manchester University, specializing in X-ray crystallography, and in 1937 became Professor of Physics at Cape Town University. James died in Cape Town on 7 July 1964.

James Ross Island. 64°10′S, 57°45′W. A large island 65 km (40 miles) long, separated by Prince Gustav Channel from the southern coast of Trinity Peninsula, Graham Land. The central peaks rise to 1630 m (5346 ft). Parts of the island were charted in 1842 by Sir James Clark Ross RN, for whom it is named.

Japanese Antarctic Expedition 1910–12. Japan's first venture into Antarctic exploration. Inspired by the successes of European explorers, Lt Nobu (or Choku) **Shirase** of the Japanese Navy in 1910 led a Japanese Antarctic expedition to Victoria Land, hoping to reach the South Pole. His ship *Kainan Maru* (Capt. Naokichi Nomura) left Tokyo on 1 December, arriving in Wellington, New Zealand on 7 February 1911. Though reasonably familiar with cold weather and pack ice, Shirase and his party lacked experience of Antarctic conditions: they had little information about southern conditions, and were hopelessly underfunded. The expedition left New Zealand on 11 February. After pushing through pack ice they reached the Ross Sea, on 6 March sighting the Victoria Land coast. Though they reached as far south as Coulman Island, they were unable to land, and it was too late in the season to attempt further exploration. Shirase returned north, arriving in Sydney, Australia on 1 May. His party wintered in the suburban garden of a kindly Australian host, living at subsistence level in hut and tents. During this time he benefited from discussions with Prof. Edgeworth David, a local veteran of Shackleton's 1907–9 expedition.

On 19 November *Kainan Maru* left again for the south. Now more aware of his own limitations, Shirase aspired only to reach and explore King Edward VII Land. On 16 January 1912 they sighted the Ross Ice Shelf and headed eastward along it to Bay of Whales. There Shirase encountered *Fram*, expedition ship of the Norwegian South Polar Expedition 1910–12, which was waiting to take home the returning polar party. Proceeding further east, Shirase found an iceport or embayment in the shelf ice, Kainan Bay, where he landed a party of his own. A group of five sledged 256 km (160 miles) inland with a small dog team, to a point in 80°05′S, where they raised the Japanese flag. Meanwhile the ship sailed eastward around Cape Colbeck to land a second party on King Edward VII Land, close to the Alexandra Mts in the area now called Sulzberger Bay. Both parties were back on board on 2 February, and the ship headed north, returning to Yokohama on 20 June 1912. Kainan Bay, Okuma Bay (an iceport named for a Japanese statesman who was the expedition's main sponsor) and Shirase Coast are local reminders of the expedition. Though it achieved little geographically, it was the tentative start to Japanese interest in Antarctica, that continues to the present day.
Further reading: Hamre (1933).

Jeannel, René. (d. 1986). French entomologist and biogeographer. During a distinguished academic career he travelled widely, publishing extensively on the morphology, taxonomy, ecology and biogeography of insects. In 1932 he

became Director of the Muséum National d'Histoire Naturel, Paris, and in 1939 organized a collecting expedition in the research ship *Bougainville* to the Indian Ocean. Visiting Marion Island and the French possessions Iles Crozet, Kerguelen, Amsterdam and St Paul, he provided an interesting general account of their history and contemporary ecological status. Jeannel died on 20 February 1965.
Further reading: Jeannel (1941).

Jelbart, John Ellis. (1926–51). Australian expedition physicist. Born in Ballarat on 6 December 1926, he read physics at Queen's College, Melbourne University. In 1947 he joined the Australian National Antarctic Research Expedition and spent over a year at Heard Island, working on cosmic ray physics. Seconded to the **Norwegian-British-Swedish Antarctic Expedition 1949–52** in its second season, he again concentrated on cosmic ray research. He was killed in a tractor accident near Maudheim on 23 February 1951.

Jelbartisen. 70°30′S, 4°30′W. (Jelbart Ice Shelf). Ice shelf on Kronprinsesse Märtha Kyst, Dronning Maud Land. The shelf was mapped by the **Norwegian-British-Swedish Antarctic Expedition 1949–52**, and named for John E. Jelbart, an Australian physicist on the expedition who was killed in a tractor accident in February 1951.

Jenny Island. 67°44′S, 68°24′W. A small, steep island 3.2 km (2 miles) long, rising to 500 m., off the southeast corner of Adelaide Island, Marguerite Bay. It was charted by the **French Antarctic (*Pourquoi Pas?*) Expedition 1908–10**, and named for Jenny Bongrain, wife of the second officer.

Johnson, Charles Ocean. (1867–1949). Swedish-South African fishing, sealing and whaling entrepreneur. Born in Hjälmseryd, Sweden on 21 March 1867, Johnson learnt the fishing business in his home country; then in the early years of the twentieth century he moved to South Africa, taking northern fish-catching technology to southern waters. Co-founder in 1909 of the commercial fishing firm Irvin and Johnson, he formed as a subsidiary the Southern Whaling and Sealing Co., which from 1909 exploited elephant seals for oil on Marion and the Prince Edward Islands, later on Iles Kerguelen. In 1913 he began whaling off South Georgia with *Restitution*, a 3000-ton liner converted to a floating factory. His various sealing and whaling enterprises prospered: in 1930 his Kerguelen Sealing and Whaling Co. bought the 13 000-ton factory ship *Tafelberg*, then the world's largest, which operated in the southern oceans until World War II. Johnson died in Cape Town on 25 June 1949.

Johnston, T. Harvey. (1881–1951). Australian expedition biologist. After a distinguished academic career in zoology and microbiology he became chief biologist on the **British, Australian and New Zealand Antarctic Research Expedition 1929–31**. He sailed on both cruises of RRS *Discovery*, and was later responsible for publication of the biological reports. He died on 30 August 1951.

Johnston, William. (1908–68). British expedition master mariner. Born and educated in Northern Ireland, he qualified as a master in 1933 and became chief officer, later captain, of the Falkland Islands Co. ship *Lafonia*. During World War II he served in the Royal Navy, returning to the Falklands in 1946 to take command of a replacement *Lafonia*. From 1950 to 1965 he commanded in succession three re-supply ships of the Falkland Islands Dependencies Survey, *John Biscoe*, a converted boom-defence vessel, to 1955, *Shackleton* 1955–56 and a new *John Biscoe* 1956–65, making many voyages to the Dependencies and taking an active interest in surveying and improving the navigational charts. Mt Johnston on the Danco Coast and Johnston Passage off the west coast of Adelaide Island, were named for him. He retired in October 1965 and died on 27 February 1968.

Joint Services Expedition to Brabant Island 1983–85. British combined services expedition, led by Cdr J. R. ('Chris') Furse RN, that over-wintered on Brabant Island, Palmer Archipelago, making detailed biological, geological and topographic surveys of the island.

Joinville Island. 63°15′S, 55°45′W. A steep, mountainous island 65 km (40 miles) long off the northeastern tip of Trinity Peninsula, Graham Land. Charted in 1838 by the French Naval Expedition, it was named for a contemporary French prince.

Jones, Sydney Evan. (1887–1948). Australian expedition medical officer. After studying medicine at the University of Sydney he joined the **Australasian Antarctic Expedition 1911–14**, serving as medical officer with the Western Party. He led several major sledging journeys, including the long coastal journey to Haswell Island and Gaussberg. On return to Australia he specialized in psychiatric medicine and occupational therapy. Jones died on 17 February 1948.

Jones Mountains. 73°32′S, 94°00′W. An isolated range overlooking Abbot Ice Shelf on the Eights Coast of Ellsworth Land, identified during flights of the **United**

States Antarctic Service 1939–41 and named for US scientist and administrator Dr Thomas O. Jones.

Joyce, Ernest Edward Mills. (1875–1940). British seaman and explorer. Born in a coastguard cottage in Bognor, Sussex, he became a boy-entrant in the Royal Navy in 1891, serving with the Royal Naval Brigade in the South African War. In Simons Bay, South Africa, he joined the **British National Antarctic Expedition 1901–4** as a seaman on *Discovery*. On return he was rated petty officer, but left the navy by purchase to join the **British Antarctic (*Nimrod*) Expedition 1907–9**. He took part in the sledging programmes of both expeditions. Later he helped to select dogs for Mawson's **Australasian Antarctic Expedition 1911–14**, and ship them to Tasmania. From Sydney, Australia, Joyce joined the **Imperial Trans-Antarctic Expedition 1914–17**, becoming second-in-command of the Ross Sea shore party with the responsibility of laying depots across the Ross Ice Shelf to Mt. Hope. In this he succeeded against extraordinary odds. He was awarded the Albert Medal for his part in rescuing the base leader, Capt. Æneas Mackintosh, who had succumbed to scurvy. He died on 2 May 1940.

Juan Carlos Island. 62°39′S, 60°23′W. Spanish research station on southern Livingston Island, South Shetland Islands. It has operated as a summer station since January 1988.

Jubany Station. See **Teniente Jubany Station.**

June, Harold I. (1895–1962). American pilot and explorer. As a member of **Byrd's First Antarctic Expedition 1928–30** he took part in the first flight over the South Pole. He joined **Byrd's Second Antarctic Expedition 1933–35** as chief pilot and transportation officer making several exploratory flights and participating in two tractor journeys. He died on 22 November 1962.
(CH)

K

Katabatic wind. A strong but shallow downslope wind, usually a stream of air no thicker than the depth of the prevailing **temperature inversion**. Katabatic winds blow with particular strength and persistence in Antarctic coastal regions. For much of the year, especially in winter, the surface of Antarctica loses heat by infrared radiation, cooling the layer of air in contact with it. Surface air, denser than warmer overlying layers, accelerates downslope, slowed by surface friction and diverted to the left by Coriolis force. The result is a katabatic wind directed some 45–60° to the left of the slope. Over gentle slopes of the high plateau katabatic wind speeds rarely exceed 10 m per s: over steep coastal slopes they can build up considerably, in some places reaching hurricane force, particularly where cold air drainage from a wide area converges into a valley. See **Home of the blizzard**.
(JCK)

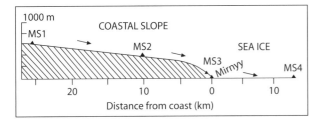

Station	MS1	MS2	MS3	MS4
Distance from coast (km)	25	10	0.7	13.5
Height above sea level (m)	600	400	0	0
Mean wind velocity (m per sec)	9.8	10.8	11.3	4.4

Katabatic wind measured simultaneously at three mobile stations (MS1–3) on the Antarctic icecap above the Russian coastal station Mirnyy, and a station on the sea ice (MS4). The wind accelerates downhill toward the coast, where it reaches its highest velocity, then loses momentum offshore. After Stonehouse 1989: data from Tauber (1960)

Keller Peninsula. 62°05′S, 58°26′W. A hilly peninsula separating Mackellar and Martel inlets, in Admiralty Bay, King George Island, South Shetland Islands. The shore was used as a mooring by whale factory ships during the early twentieth century, and in 1947 became the site of Base G, a research station of the Falkland Islands Dependencies Survey. It is currently the site of Commandante Ferraz, a Brazilian year-round research station.

Kelly Plateau. 81°24′S, 159°30′E. Ice plateau on the eastern flank of the Churchill Mountains, Victoria Land, named for US pilot Cdr G. R. Kelly, USN.

Kemp, Peter. (d. 1834). British sealer and explorer. Details of his early life are unknown. In 1833, a seasoned master with over a dozen southern voyages to his credit, Kemp took command of *Magnet*, a snow-modified brig in the fleet of Daniel Bennett and Son, of Rotherhithe, London, with orders to discover new sealing grounds in the southern Indian Ocean. *Magnet* left the Thames on 15 July, making a non-stop voyage to Iles Kerguelen. After a brief stay Kemp departed south on 26 November, passing close to the position of Heard Island on the following day, and recording a sighting of land, which he did not investigate further. Reaching the edge of the pack ice in early December, he made slow but continuous progress south and west through loosely packed floes for a further three weeks. No new sealing islands were found, but on 27 December Kemp sighted land some 40 miles (64 km) to the south. This proved to be a hitherto unknown sector of the Antarctic coast, east of Enderby Land, the coast discovered two seasons earlier by fellow-sealer **John Biscoe**. Now beset by heavier pack ice, Kemp was unable to approach closer. Needing to justify the voyage commercially, he returned to Iles Kerguelen, where he collected a cargo of elephant seal oil. Heading west toward South Africa, Peter Kemp was lost overboard on 21 April 1834. The voyage was completed under the command of the mate, David Rankin.
Further reading: Jones (1992).

Kemp Coast. 67°15′S, 58°00′E. Part of the coast of Kemp Land, East Antarctica, bounded by Edward VIII Bay in the west, William Scoresby Bay in the east. Consisting mainly of ice cliffs with nunataks and emergent islands, the coast is backed by steeply rising ice slopes, dotted with

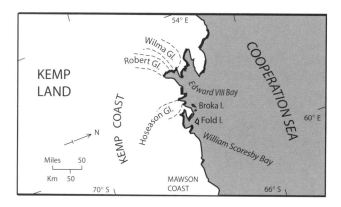

Kemp Coast

isolated peaks. It was first sighted in 1833 by the British sealer Peter Kemp, for whom it is named.

Kemp Land. 67°30′S, 57°30′E. Ice-covered, mountainous area of East Antarctica between meridians 55° and 60°E, bounded to the west by Enderby Land and to the east by Mac.Robertson Land, and forming part of Australian Antarctic Territory. The land includes Kemp Coast, the Framnes and Hansen mountains, and an extensive interior ice plateau extending to the South Pole. It was named for British sealer Capt. **Peter Kemp**, who sighted the coast in 1833. The area has been extensively explored by field parties from the Australian station Mawson.

Keohane, Patrick. (c. 1879–1950). British seaman and explorer. Born in County Cork, Southern Ireland, he enlisted in the Royal Navy, and had reached the rank of seaman petty officer when he volunteered for the **British Antarctic (*Terra Nova*) Expedition 1911–13**. Joining the shore group at Cape Evans, he was involved in many sledging operations, notably the support party that climbed the Beardmore Glacier, and the party that discovered Scott's final camp. On retiring from the navy he returned to Ireland, where he died on 30 August 1950.

Kerguelen, Iles. 49°00′S, 65°00′E. Archipelago in the cold-temperate zone of the southern Indian Ocean, comprising one large, deeply indented island and 200–300 smaller islets. Over 4000 km (2500 miles) east-southeast of South Africa and slightly further west-southwest from Australia, the islands stand close to the Antarctic Convergence, on a submarine platform that they share with Heard Island. All are volcanic, composed of lavas and tuffs representing outpourings over a period of at least 40 million years. Some fumarolic activity remains in the southwestern corner. Grande Terre, the main island, is roughly triangular, measuring approximately 120 km (75 miles) east to west and 110 km (69 miles) north to south, terminating in extensive peninsulas and rising to a cordillera of peaks in the west and south. The highest peak, Mont Grand Ross stands at 1960 m (6431 ft) on the southern seabord. Separate and westward from it rises Calotte Glaciaire Cook (1049 m, 3441 ft), a broad, ice-covered plateau. Deep valleys and fjords indicate that ice formerly extended over most of the island. The lesser islands cluster about Grande Terre mainly to the north and east. The islands are wet, windy and cold. Mean annual temperature is 4.4 °C, with a narrow range of monthly mean temperatures of 5.3 °C. Rainfall exceeds 110 cm per year; the growing season is short, and soils are thin and acid. The islands support 36 species of flowering plant and fern and an unusual wealth of cushion-forming mosses. Meadows of tussock and other grasses predominate at sea level: cliffs hang with mosses and Kerguelen cabbage, and the higher ground supports green moorland of shrubs, herbs and moss.

The islands were discovered in February 1772 by a French naval officer Y.-J. de Kerguelen-Trémarec, who with the support of his king, Louis XV, was searching the Indian Ocean for a southern continent. Fog and strong winds kept his ships from approaching land, but eventually several officers (though not Kerguelen-Trémarec himself) went ashore to claim 'La France Australe' for their nation. On returning to France, Kerguelen-Trémarec delivered a glowing but almost entirely spurious report, claiming to have discovered the central mass of the Antarctic continent, and describing it as a paragon of fertility where in time would be found minerals and other wealth. In 1773 a second government-supported expedition of three ships took out settlers for the new colony, arriving in December and for several weeks seeking sites for landing and establishing farms and settlements. Faced with the reality of a cold, windy climate, dripping peat bogs, and lack of timber or of any other natural resources, they finally returned home, where Kerguelen-Trémarec was court-martialled and disgraced. James Cook, who visited in 1776, at first called them the Islands of Desolation, only later renaming them after their discoverer. Sealers visited frequently during the nineteenth century, hunting for fur seals and elephant seals. Visiting scientists also called, and from 1908 to 1932 the sheltered harbour of Port Jeanne d'Arc provided a base for commercial whaling, fishing and sheep farming.

France claimed the islands formally in 1924, creating a national park and refuge. In 1949 a scientific station, Port-aux-Français, was established on the north coast of Golfe du Morbihan. It has operated continuously since then. Sealers, scientists and settlers introduced black rats, mice, rabbits, cats, dogs, sheep, mouflon and reindeer, stocks of which survive on many of the islands, presenting problems for conservation. Char and salmon introduced more recently in some of the rivers appear to have established themselves, living on freshwater invertebrates. The islands are noted for their huge stocks of seabirds, including king, macaroni and rockhopper penguins, wandering albatrosses, black-browed albatrosses and many species of smaller petrel. Elephant seals are again plentiful: Antarctic fur seals are returning after a long absence to some of the northern islands and beaches.

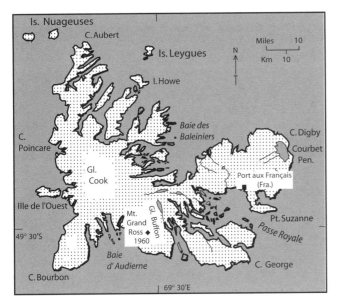

Formerly called 'Desolation Islands', Iles Kerguelen have a damp, windy climate with persistent low cloud cover. The islands are tussock-covered: Glacier Cook is the only remnant of permanent ice. Heights in m

Kernlos effect. A characteristic seasonal pattern of surface cooling at Antarctic plateau stations, in which air temperatures near the ground fall rapidly between February and April, level off during the following winter months, then sometimes fall further in late August or early September immediately before the return of the sun. Literally 'coreless', the effect is due to persistent inversion of temperatures above the snow surface, permitted by lack of wind, that reduce radiational heat losses and prevent temperatures from falling further. In addition, the half-yearly oscillation in atmospheric circulation brings more heat to Antarctica around midwinter, preventing a further fall in temperature on the polar plateau.

Kershaw, John Edward Giles. (1948–90). British Antarctic pilot. Born in Kerala, India on 27 August 1948, he trained as a pilot and in 1974 joined British Antarctic Survey to fly Twin Otters over the Antarctic peninsula region. Though he left in 1979, he flew seasonally with BAS and other Antarctic operations, alternating with more mundane flying as a commercial airlines captain. With his wife Anne he set up Adventure Network International, which specializes in taking parties of scientists, climbers and photographers to out-of-the-way corners of Antarctica. He died in an Antarctic flying mishap on 6 March 1990.

Killer whale. *Orcinus orca*. More formally called orcas, killer whales are large dolphins, measuring up to 9.5 m (31 ft) long and weighing up to 8 tonnes. Mature males are larger than females, with a prominent dorsal fin up to 1.8 m (6 ft) tall. Distinctively piebald, they are mostly black but with prominent white throat and belly, and white patches behind the eyes and dorsal fin. They occur in all the world's linked seas and oceans, usually travelling in what appear to be extended family groups of a dozen or more, including small calves. In the Southern Ocean groups of killer whales are present all the year round, some penetrating far into the zone of pack ice and coastal waters. They hunt in groups, taking fish, seals and penguins, mainly at or near the surface. Often seen 'spy-hopping', i.e. emerging from the water far enough to be able to examine the shore or upper surface of the sea ice, they may act in concert to tip or wash basking seals and penguins from floes and small bergy bits. During the whaling years pods of killer whales frequently attacked carcasses awaiting processing around the stations and factory ships. Though they have been reported to show interest in small boats and in humans standing on pack ice, there are so far no records of attacks on humans in Antarctic waters.
Further reading: Bonner (1989); Evans (1987).

King, Philip Parker. (1792–1856). British naval officer and hydrographer. Born in Norfolk Island, where his father governed the penal settlement, King joined the Royal Navy and served through the Napoleonic wars. In 1817–22 he made four voyages surveying the coasts of Australia, including a detailed survey of Tasmania. From 1826 to 1830, in HMS *Adventure*, he accompanied HMS *Beagle* (Capt. P. Stokes) in similar surveys of the coasts of Patagonia, Tierra del Fuego and the Strait of Magellan. King retired from active service in 1832, settling and pursuing farming interests in Australia.

King George Island. 62°00′S, 58°15′W. Large, ice-covered island, largest of the South Shetland Islands, 70 km (44 miles) long, with a deeply indented southern coast. It was charted in 1820, and named for George III, the contemporary British king. Easily accessible from South America, it is the site of over a dozen research stations.

King Håkon VII Sea. 65°S, 0°. (Kong Håkon VII Hav). Marginal sea off East Antarctica, extending between 20°W and 45°E, the longitudinal limits of Dronning Maud Land, which it borders. There is no definitive northern boundary. It is often ice-filled, even in summer, with bergs and persistent pack ice moving westward and north, under the influence of strong inshore currents and winds, making access to the coast difficult. Explored extensively by whalers of the **Norwegian (Christensen) Whaling Expeditions 1926–37**, the sea was named for the Norwegian ruling monarch. Norwegians regard it as extending to the coast of Dronning Maud Land: Russian cartographers interpose three small, marginal seas along the coast: see **Lazarev Sea**, **Riiser-Larsen Sea** and **Cosmonaut Sea**.

King Leopold and Queen Astrid Coast. 67°20′S, 84°30′E. Part of the coast of Wilkes Land, East Antarctica, bounded in the west by the western end of the West Ice Shelf, and in the east by Cape Penck. Fronted almost

Moulting king penguins: South Georgia. Photo: BS

which they carry on their feet. The chicks, covered in dense brown down, are fed intensively through summer and autumn. Remaining in the colonies during winter, they are fed only intermittently and lose weight. Survivors are fattened during the following spring, entering the sea from November onward. Because the breeding cycle takes over 12 months, parents cannot raise a chick every year, but some may raise two in three years. The world population is estimated at 1 000 000 pairs.

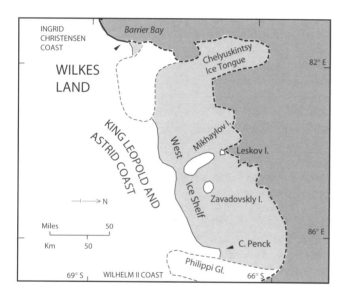

King Leopold and Queen Astrid Coast

entirely by the ice shelf, the coast rises behind the shelf in a featureless ice plain. It was discovered in 1934 by pilots Lt Alf Gunnestad and Capt. Nils Larsen, during flights of the **Norwegian (Christensen) Whaling Expeditions 1926–37**. Christensen named the coast originally for the Norwegian Princess Astrid: it was later re-named for King Leopold III of Belgium and Queen Astrid.

King penguin. *Aptenodytes patagonicus*. Second-largest of the living penguins, kings stand up to 85 cm (33 in) tall and weigh up to 16 kg (35 lb) in full fat. Distinguished by their orange auricular patches and broad purple or orange bill plates, they breed on islands within the sub-Antarctic and cool temperate zones, in colonies that range in size from a few hundred to many thousands of pairs. They feed mainly on fish and squid. Kings lay a single egg,

King Sejong Station. 62°13′S, 58°47′W. Permanent year-round station of the Republic of Korea on Barton Peninsula, King George Island. The station has operated continuously since its establishment in February 1988.

Kirkwood, Harold. (1910–77). British expedition ship's officer. Born in 1910, he joined the Royal Navy, serving as Third Officer in RRS *Discovery II* in 1933–35 and as Second Officer in 1935–38, taking part in many Southern Ocean cruises. He served with distinction in World War II. In 1948–50 he commanded RRS *John Biscoe*, servicing scientific bases of the Falkland Islands Dependencies Survey, and in 1955–58 commanded the same ship in her role as HMNZS *Endeavour*, resupplying the New Zealand Antarctic station. He retired from the Royal Navy in 1960 and died on 25 September 1977.
(CH)

Knox Coast. 66°30′S, 105°00′E. Part of the coast of Wilkes Land, East Antarctica. Bounded by Cape Hordern in the west and the Hatch Islands, Vincennes Bay, in the east, it consists mainly of featureless ice slopes descending to steep ice cliffs: in the west, massive ice streams converge to form the eastern end of Shackleton Ice Shelf. Explored by

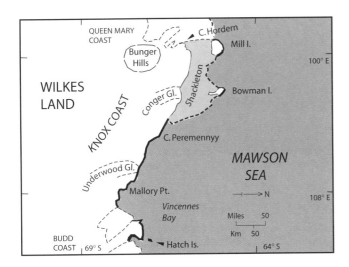

Knox Coast

ships of the **United States Exploring Expedition 1838–42**, the coast was named for Lt Samuel Knox, acting master of the expedition flagship USS *Vincennes*.

Kohl Plateau. 54°14′S, 36°57′W. Ice plateau rising to 760 m. in central South Georgia, identified by and named for German medical officer and botanist L. Kohl-Larsen.

Kohler Range. 75°05′S, 114°15′W. A range flanking Smith Glacier behind Bakutis Coast, Marie Byrd Land, identified during **Byrd's Second Antarctic Expedition 1933–35** and named for US industrialist and expedition supporter Walter J. Kohler.

Komsomol'skaya Station. 74°05′S, 97°29′E. Soviet research station on the plateau of Queen Mary Land, East Antarctica, at an elevation of 3497 m (11 470 ft). Opened for the International Geophysical Year on 6 November 1957, it was occupied sporadically as one of a network of staging posts for tractor trains exploring the high plateau, *en route* from Mirnyy to Vostok Station.

Kraulberga. 73°10′S, 13°45′W. (Kraul Mountains). Range of scattered nunataks and ridges emerging from the ice cap of Kronprinsesse Märtha Kyst, western Dronning Maud Land. It was named for the German whaling scientist Otto Kraul.

Krebs, Manson. (d. 1963). US naval officer who played a major role in developing air support operations during US Operation Deepfreeze between 1960 and 1962, particularly the work of Air Development Squadron VX-6. Mt Krebs, in the Prince Olav Mts, is named for him. He died on 10 April 1963.
(CH)

Krill, Antarctic. A species of surface-living crustacean, *Euphausia superba*, endemic in parts of the Southern Ocean. The term krill is used loosely to include other euphausiid crustaceans, and occasionally even more loosely to include Antarctic zooplankton generally. Antarctic krill dominate the macrozooplankton in many areas of the Southern Ocean, especially in the lower latitude ice-free regions. Because its biomass is so high, and because it forms schools or swarms in summer, krill is both a key organism in the Southern Ocean food web and an important target species for commercial fisheries. *Euphausia superba* is the largest of the euphausiids. Individuals grow to about 40 mm length in two to three years, and up to 60 mm by the end of life, which can extend up to seven years. Krill have well-developed swarming behaviour, comparable with the schooling of small fish, associated with both feeding and reproduction and a defence against predators. Swarms can extend up to 500 m across, and contain krill at densities as high as 3000 per m^3. In summer krill feed primarily on phytoplankton, although they also take such microzooplankton as tintinnids, choanoflagellates and small copepods. Winter populations are found under sea ice, browsing on the algae: some may also feed at depth over the continental shelves by stirring up bottom sediments. Laboratory experiments have shown that adult Antarctic krill can withstand long periods of starvation by using their muscle tissue as a reserve, shrinking in size in the process.
Further reading: Everson (2000); Knox (1994).
(ACC)

Kronprins Olav Kyst. 69°30′S, 36°00′E. Part of the coast of Dronning Maud Land, East Antarctica, bounded in the southwest by the eastern entrance of Lützow–Holmbukta and in the east by Shinnan Glacier. Consisting mainly of ice cliffs and glaciers, with nunataks and emergent islands, the coast was discovered in 1930 by Capt. Hj. Riiser-Larsen, in flights of the Norwegian (Christensen) Whaling Expeditions, and named for the Norwegian crown prince. The interior ice cap was explored

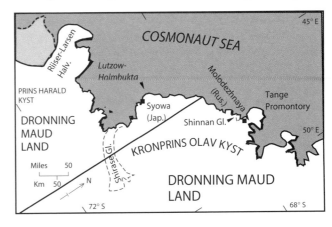

Kronprins Olav Kyst

in the 1960s by traverses from Syowa, a permanent Japanese research station.

Kronprinsesse Märtha Kyst. 72°00'S, 07°30'W. Part of the coast of Dronning Maud Land, East Antarctica, extending between Stancomb–Wills Ice Stream in the west and longitude 5°E. Facing King Håkon VII Sea, the coast consists mainly of ice-covered peninsulas and islands lined by the wide ice shelves Ekströmisen, Jelbartisen and Fimbulisen. It was discovered in flights by Capt. Hj. Riiser-Larsen in 1930, during the **Norwegian (Christensen) Whaling Expeditions**, and named for the Norwegian crown princess. The interior ice cap and mountains were explored extensively by the **Norwegian-British–Swedish Expedition 1949–52**: the coast is also the site of **SANAE**, a permanent South African research station.

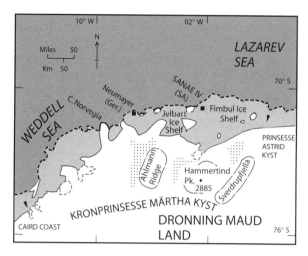

Kronprinsesse Märtha Kyst

L

Laclavère Plateau. 63°27′S, 57°47′W. Ice plateau between Misty Pass and Theodolite Hill, Trinity Peninsula, Graham Land, named for French engineer, cartographer and administrator G. R. Laclavère.

Lagotellerie Island. 67°53′S, 67°24′W. Small island in Marguerite Bay, Fallières Coast, discovered and named by the **French Antarctic (*Pourquoi Pas?*) Expedition 1908–10**. The beaches and lower slopes of the island support an unusually diverse flora, with well-developed moss and lichen communities, and broad stands of both species of flowering plant, that in some seasons produce viable seeds. The invertebrate fauna is rich and diverse, including apterous midges at a southern limit of their range, and a diverse avifauna includes the southernmost breeding colony of blue-eyed shags. An area of approximately 1.2 km^2 (0.5 sq miles) has been designated SPA No. 19 (Lagotellerie Island)

Lakes, Antarctic. Lakes, ponds and streams are widely distributed across the ice-free areas of Antarctica, forming on ice sheets, alongside glaciers, and in basins eroded by ice. Some are permanent, fed by seasonal streams: others are ephemeral melt-water pools. While large lakes in high latitudes thaw only round their edges in summer, smaller lakes and pools may thaw completely, allowing both for thermal and saline stratification and for internal circulation. The biogeographical isolation of Antarctica is reflected in the diversity of the freshwater flora and fauna, which is best developed in maritime Antarctic lakes, and is very limited in most continental water bodies.

The nutrient status of the water-bodies varies widely from ultra-oligotrophic (very pure water) to hypersaline (more salty than the Dead Sea). Some of the saline water bodies are almost sterile, supporting only a few species of bacteria, and constituting some of the most hostile habitats on Earth. A unique feature of lakes in the dry valley areas of the Ross Desert, Victoria Land, is that almost all losses of water, gases and biological material occur through the thick ice cover. Algal felts melt upwards through the ice and are finally distributed by wind to provide the principal carbon source for the micro-invertebrates living in these valleys.

The nutrient status of a lake is related to its origin. Many of the lakes in the Ross Desert and Vestfold Hills contain saline water derived either from the isolation of pockets of sea water during uplift of the land, or by the continual slow evaporation of fresh water containing salts weathered from the rocks of the surrounding catchment. The mixture of salts in each lake is unique and gives valuable clues to its evolution. Some lakes are strongly and permanently stratified with warm, anoxic saline deep water containing only bacteria, overlain by colder layers of brackish and fresh water with different groups of phytoplankton and bacteria in each layer.

By contrast, freshwater lakes are normally much younger and usually with a diverse algal flora, often with aquatic mosses in the maritime Antarctic, but with no flowering plants. The fauna is limited to small invertebrates (protozoa, rotifers, nematodes, tardigrades and crustaceans), with the largest animal a fairy shrimp 2 cm long. The strongly seasonal nature of open water and the effects of winter snow on the penetration of light through the ice turn the lakes into closed systems in winter. Many of the deeper ones in winter develop strongly anoxic layers, inhibiting nutrient cycling. While small pools of melt water may occur anywhere, the Ross Ice Shelf has a very large number of freshwater melt pools with remarkable well-developed cyanobacterial mats and associated bacteria and protozoa. These simple communities, believed by some to be very similar to those in which life first began, may offer a unique insight into the first stages of the evolution of life on Earth. One valuable feature is the ability of a lake to act as an integrator of the changes in the surrounding catchment. Thus, the changes in vegetation and in local climate are reflected in the inflows to the lake and preserved in lake sediments. The present interest in patterns of change shown by previous climates has highlighted the value of these lake sediments for palaeoclimatic reconstructions for the last 10 000 years.
Further reading: Vincent (1988).
(DWHW)

Lakes beneath the ice sheet. Ice under great pressure may melt, reverting to water. Deep down at the ice–rock interface, usually where the ice is at least 2000 m (6560 ft) thick, there are zones at temperatures close to or at the pressure melting point. They occur only where the surface accumulation rate is low, so that atmospheric cold is not carried deep into the ice sheet. The presence of lakes of melt-water under the Antarctic ice sheet was first confirmed by a drilling made through the ice sheet near Byrd Station in 1968. When the drill reached the base of the ice at 2164 m (7140 ft), water flooded into the drill hole to a height of about 60 m (198 ft), showing that the base of the sheet was at melting point. The temperature at the bottom of the borehole was $-1.6\,°C$. Since then, radio echo soundings from aircraft have been used to detect the presence of water under ice sheets. This depends on the fact that water reflects radio waves very strongly. To obtain a reflection, the water layer has to be of a critical depth, termed the 'skin depth', which depends on the conductivity of the water and the radio frequency used. For fresh water at the frequency of ice radar sounders, skin depth is of the order of a few metres, so quite shallow lakes can be detected.

Such lakes occur in areas of low surface slope and low ice velocity, where they typically occupy hollows in the bedrock. Many have been detected, most with dimensions of a few kilometres but some very much larger. Lake **Vostok**, close to the Russian Vostok Station on the high plateau of East Antarctica, has an area of $14\,300\,km^2$, more than three times that of the State of Delaware, USA, and a depth of 125 m (412 ft). The existence of extensive areas of the ice sheet (estimated at 10 per cent of its base) at the pressure melting point, underlain by lakes, has important implications for ice sheet flow and stability, with observable effects on internal ice deformation and surface slopes. Buried as they are under ice which will exert a normal pressure of about 40 MPa., they will certainly contain dissolved gases, atmospheric impurities and dissolved minerals from glacial rock flour. It is not known whether they contain flora or fauna.
(DJD)

Lallemand Fjord. 67°05′S, 66°45′W. Spectacular fjord over 48 km (30 miles) long separating Arrowsmith Peninsula from Loubet Coast, Graham Land, opening into Crystal Sound. It was named by J.-B. Charcot for a French geographer, Charles Lallemand, who was helpful to his second expedition.

Lamb, Ivan Mackenzie. (1910–90). British polar lichenologist. Born in London on 10 September 1910, he read natural science at Edinburgh University. In 1935 he was appointed assistant keeper in the natural history section of the British Museum, where he studied the collections of Antarctic lichens from early twentieth century expeditions. In 1943–45, as botanist to **Operation Tabarin**, Lamb spent two years studying lichens in two of the areas – Port Lockroy and Hope Bay – whence the collections had originated. In later years he studied lichens at Tucuman University, Argentina, the National Museum, Ottawa, and the Farlow Herbarium, Harvard University. He revisited Antarctica with US Operation Deep-Freeze in 1960, and again in 1964–65 to study marine algae *in situ*. Lamb retired in 1972, and died in Braintree, Mass., on 27 January 1990.

Land birds, southern oceanic island. A few species of land birds are endemic on southern oceanic islands: several other species of neighbouring mainlands have become residents or recurrent vagrants. The Falkland Islands, offering a rich variety of habitats some 480 km (300 miles) downwind from South America, support over 65 species of breeding bird, of which two thirds feed on land or fresh water (see **Ducks and geese, southern oceanic**). Populations of more remote island groups reflect a narrower choice of habitats and fewer opportunities for recruitment. Of the Tristan da Cunha group's 32 species of breeding bird, only 10 feed on land: Macquarie Island and South Georgia, each with 25 species of breeding bird, muster respectively only five and four species that feed on land. The endemic South Georgia pipit *Anthus antarcticus* appears to have been derived from a similar species, *A. correndera*, of southern South America and the Falkland Islands. Of body length 16 cm (6.5 in) and wing span 23 cm (9 in), with brown and buff striped plumage, South Georgia pipits forage mainly among dense coastal tussock grass, feeding mainly on insects from the soil and grass, and also on marine copepods and small planktonic animals washed up by the tides. Though still plentiful on the main island, they breed almost entirely on offlying islets, which alone remain free of introduced rats. From late November they weave cup-shaped grass nests among the bases of tussock stems, laying three to five greenish-brown eggs. The Tristan da Cunha group has the resident Tristan thrush *Nesocichla eremita* and two species of bunting *Nesospiza acunhae* and *N. wilkinsi*, now found mainly on the outlying islands. Gough Island has an endemic bunting *Rowettia goughensis*, and an endemic moorhen *Gallinula nesiotis* that it formerly shared with Tristan da Cunha. An endemic rail *Atlantisia rogersi* breeds on Inaccessible Island, and wekas *Gallirallus australis*, endemic to New Zealand, breed successfully on Macquarie Island. The islands south of New Zealand have attracted several species from Australasia, including introduced European species. Thus starlings *Sturnus vulgaris* are resident on Macquarie Island, redpolls *Acanthis flammea* on the Auckland Islands, Campbell Island and Macquarie Island, and red-fronted parakeets *Cyanorhamphus novaezelandiae* thrive alongside penguins and petrels on the Antipodes Islands.

Lands, Antarctic. In Antarctica as elsewhere, explorers have traditionally named their major discoveries for royalty, patrons, other explorers or loved ones. Antarctic gazetteers list 17 lands, 13 in East Antarctica, four in West Antarctica. Of these six are named for members of European royal families, five for explorers, four for patrons, and two for the wives of explorers. Not all names or identities are universally accepted: some lands in the Antarctic Peninsula area have alternative Spanish names proposed by Chile and Argentina. In East Antarctica Alberts (1995) disallows George V, Kemp, Oates, Princess Elizabeth, Queen Mary and Wilhelm II lands, and Terre Adélie, assigning the names of all but Princess Elizabeth to coastal stretches of Wilkes Land. For further details see individual entries.

Lands of East Antarctica

Coats Land
Dronning Maud Land
Enderby Land
George V Land
Kemp Land
Mac.Robertson Land
Oates Land
Princess Elizabeth Land
Queen Mary Land
Terre Adélie
Victoria Land
Wilhelm II Land
Wilkes Land

Lands of West Antarctica

Ellsworth Land
Graham Land
Marie Byrd Land
Palmer Land

Langhovde. 69°14′S, 39°44′E. (Long knoll). Area of bare rocky hills and islands along the east shore of Lützow-Holmbukta, Prins Olav Kyst, Dronning Maud Land. At a point where there is no fringing ice shelf, the coastal ice cap has receded to expose some 170 km² (66 sq. miles) of rocky headlands and islands. The area was discovered and photographed from the air by the **Norwegian (Christensen) Whaling Expeditions** in 1926–37. In January 1957 **Syowa**, a permanent Japanese research station, was established on East Ongul Island, at the northeastern end of the area.

Larkman, Alfred Herbert. (d. 1962). British expedition ship's engineer. He joined the **Imperial Trans-Antarctic Expedition 1914–17** as chief engineer in *Aurora*, the vessel of the Ross Sea Party. After World War I he returned to New Zealand, where he became head of the engineering department of Wanganui Technical College. He died on 15 July 1962.

Lars Christensen Coast. 69°00′S, 69°00′E. Part of the coast of Mac.Robertson Land, East Antarctica. Bounded in the west by Murray Monolith and in the east by the southern end of Amery Ice Shelf, the coast consists mainly of ice cliffs fronting a featureless ice plateau: east of Cape Darnley it turns abruptly south to line the northern end of Amery Ice Shelf. It was discovered during the **Norwegian (Christensen) Whaling Expeditions 1926–37**, and named for the whaling magnate who was the owner of the whaling fleet and instigator of the expeditions.

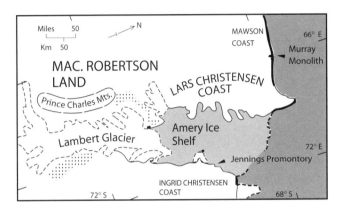

Lars Christensen Coast

Lars Christensentoppen Peak. 68°46′S, 90°31′W. The highest peak of Peter I Øy, rising to 1755 m in the northeastern half of the island. It was named for the instigator of the **Norwegian (Christensen) Whaling Expeditions 1926–37**.

Larsemann Hills. 69°24′S, 76°13′E. A range of low coastal hills emerging from the ice cap, and extending some 15 km (9 miles) along the southern coast of Prydz Bay, Ingrid Christensen Coast. One of the ice-free 'oasis' areas of Antarctica, they include some 35 km² (13.7 sq miles) of ancient metamorphosed rock with over 120 freshwater lakes, ponds and streams. The highest uplands rise to about 300 m (984 ft). Discovered in 1935 by Capt. K. Mikkelsen in *Thorshavn*, a tanker and supply ship of the **Norwegian (Christensen) Whaling Expeditions**, they were first occupied by Australian scientists, who in 1986 opened Law Base, a small summer-only station (later expanded for overwintering). In the same season the Soviet Antarctic Expedition opened Progress Station, for personnel servicing a nearby ice airstrip, and in 1989 the Chinese Antarctic Research Expedition opened Zhong Shan, also for year-round occupation.

Larsen Harbour. 54°50′S, 36°01′W. Narrow harbour off Drygalski Fjord, at the southern end of South Georgia, that was used as a haven and mooring by whale catchers and named for Norwegian whaler Capt. C. A. Larsen.

Larsen Ice Shelf. 67°30′S, 62°30′W. Extensive ice shelf bordering the east coast of Graham Land, Antarctic Peninsula, extending south from Cape Longing to Hearst Island. It was discovered in December 1893 by the Norwegian whaler and explorer Capt. C. A. Larsen, leader of the **Norwegian (Sandefjord) Whaling Expeditions 1892–94**.

Laseron, Charles Francis. (1886–1959). Australian expedition scientist. Born in the United States, he was educated at Sydney Technical College. He joined the **Australasian Antarctic Expedition 1911–14** as assistant biologist, taking part in sledging journeys from the main base along the coast and towards the South Magnetic Pole. On returning to Sydney he served in World War I, then became a collector on the staff of the Technological Museum. Laseron published several works on Australian geology, and a book, *South with Mawson* (London 1947), recounting his Antarctic experiences. He died on 27 June 1959. (CH/BS)

Lashly, William. (1868–1940). British polar seaman. Born in Hampshire, Lashley enlisted in the Royal Navy. Joining the **British National Antarctic Expedition 1901–4** as a stoker, he took part in several sledging operations. He was sought by Capt. Robert Scott to take part in the **British Antarctic (*Terra Nova*) Expedition 1910–13**, in which his experience as a sledger proved invaluable: for saving the life of Lt E. R. G. R. Evans, he and his companion Tom Crean were awarded the Albert Medal. He served with the fleet in World War I, then retired from the navy to join the Customs and Excise Service.

Lassiter Coast. 73°45′S, 62°00′W. Part of the east coast of Palmer Land, Antarctic Peninsula, facing onto the Weddell Sea. Bounded by Cape Mackintosh in the north and Cape Adams in the south, the coast consists mainly of mountain blocks alternating with deeply indented inlets, with glaciers feeding into narrow but extensive ice shelves. It was delineated and photographed during flights by the **Ronne Antarctic Research Expedition 1947–48**, and named for Capt. James W. Lassiter, USAAF, the expedition's chief pilot.

Latady Island. 70°45′S, 74°35′W. A low, ice-covered island 56 km (35 miles) long south of Charcot Island, in the Bellingshausen Sea. It was identified during photo-reconnaissance flights of the **Ronne Antarctic Research Expedition 1947–48**, and named for the expedition photographer William R. Latady.

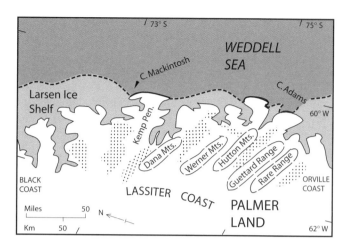

Lassiter Coast

Latitude and longitude distances. As lines of latitude run parallel east to west across the world, the degrees, minutes and seconds separating them are for navigational purposes the same length in equatorial or polar regions. Lines of longitude (meridians), running north and south, converge toward the poles, so degrees, minutes and seconds of longitude shorten. One minute (1′) of latitude spans 1.858 km (1.16 statute miles, or 1.0 nautical miles) throughout the world. One minute of longitude spans: in 55°S, 1.07 km (0.67 statute miles); in 60°S, 0.93 km (0.58 statute miles); in 65°S, 0.79 km (0.49 statute miles); and in 70°S, 0.64 km (0.40 statute miles). To obtain the length of one degree (1°), multiply by 60: to obtain the length of one second (1″), divide by 60.

Laubeuf Fjord. 67°20′S, 67°50′W. Broad waterway between Adelaide Island and Arrowsmith Peninsula, Fallières Coast, named by J.-B. Charcot for French marine engineer M. Laubeuf, who supervised construction of the engine for the expedition ship *Pourquoi Pas?*.

Laurie Island. 60°44′S, 44°37′W. A mountainous, ice-mantled island 20 km (12.5 miles) long, easternmost of the South Orkney Islands. Discovered and charted by sealers George Powell and Nathaniel Palmer in 1821, it was named for a contemporary American sealing captain.

Law Base. 69°23′S, 76°23′E. Australian permanent research station, established in Larsemann Hills for summer research parties. It was named for Dr Philip Law, who was for long the Director of Australian National Antarctic Research Expeditions.

Law Dome. 66°53′S, 113°15′E. A large ice dome rising to 1395 m (4576 ft), 110 km (67 miles) inland from Budd Coast. The peak is the site of an Australian ice drilling and coring programme, serviced from Casey Station.

Lazarev Ice Shelf. 69°37′S, 14°45′E. Ice shelf between Leningradskiy Island and Verblyud Island, Prinsesse Astrid Kyst, Dronning Maud Land. This section of shelf was explored from the Soviet research station Lazarev, which operated on Leningradskiy Island from March 1959 to February 1961: station and shelf were named for the nineteenth-century Russian explorer Capt. Mikhail P. Lazarev.

Lazarev Sea. 68°S, 7°E. Coastal sea off East Antarctica, extending between Trolltunga and Sedovodden, and bordering Prinsesse Astrid Kyst, Dronning Maud Land. At its western end is Leningradbukta, the inlet into the Russian station Novolazarevskaya. There is no definitive northern boundary. Fast ice and pack ice extend off the coast in winter: in summer pack ice and bergs move westward, driven by easterly currents and offshore winds. Designated by Russian cartographers to honour the Russian explorer Capt. Mikhail Lazarev, it occupies a near-shore region of the broader sea named earlier for King Håkon VII of Norway.

Lazarev Station. 70°00′S, 13°00′E. Former Soviet research station on Prinsesse Astrid Kyst. Established in March 1959 on the coast, it was replaced in February 1961 by Novolazarevskaya, 100 km (63 miles) inland to the south.

Lead. Narrow stretch of open water in floating ice. In ice navigation, leads often provide a welcome way forward through a field of fast or pack ice.

Lecointe Island. 64°16′S, 62°03′W. A steep, ice-mantled island 6.5 km (4 miles) long and rising to 700 m (2296 ft), standing east of Brabant Island in Palmer Archipelago, Gerlache Strait. Discovered and charted by the **Belgian Antarctic Expedition 1897–99**, it was named for Georges Lecointe, second-in-command and hydrographic officer of the expedition.

Leith Harbour. 54°08′S, 36°41′W. Harbour on the west side of Stromness Bay, South Georgia, the site of a whaling station that was operated from 1909 to 1965 by Salvesen and Co. of Leith, Scotland.

Lemaire Channel. 65°05′S, 63°59′W. Narrow passage between Booth Island and Graham Coast, Graham Land. The channel was sighted by the **German Whaling and Sealing Expedition 1873–74**, but was first navigated and charted by the **Belgian Antarctic Expedition 1895–98**, and named by them for Capt. Charles Lemaire, a prominent Belgian explorer of Africa.

Léonie Islands. 67°36′S, 68°17′W. A cluster of small islands in Ryder Bay, on the southeastern flank of Adelaide Island, Marguerite Bay. The group was charted by the French Antarctic (*Pourquoi Pas?*) Expedition 1908–10: the largest island was named for Léonie, a lady whose identity was not disclosed.

Leopard seal. *Hydrurga leptonyx*. Leopard seals are plentiful both in pack ice and in open water, ranging north to the sub-Antarctic islands, especially in winter. Slender and lithe, large individuals measure over 3 m (10 ft) and weigh up to 270 kg (600 lb). The fur is silver grey, dark above and paler below, liberally marked with darker spots. Leopard seals are thought to be monogamous, but little is known of their mating, and few young leopard seal pups have been observed. Most pupping is likely to occur in early spring among the pack ice; however, some females have given birth on South Georgia about August. Leopard seals are opportunistic predators taking krill, penguins, seals and fish. They are remarkable for skinning penguins by beating them on the sea surface before eating them. A population of up to 50 000 is estimated.
Further reading: Bonner (1982); Laws (1993).
(JPC)

Leskov Island. 56°40′S, 28°10′W. A small semi-circular volcanic island, 900 m (984 yds) long, rising to 190 m (623 ft), westernmost in the Traversay Islands at the northern end of the South Sandwich Islands chain. About 50 per cent ice-covered, made up of interbedded lavas and scoria, it occasionally shows fumarolic activity. The island was discovered in 1819 by the **Russian Naval Expedition 1819–21**, and named for a junior officer in the flagship *Vostok*.

Lester, Maxime Charles. (1891–1957). British seaman and surveyor. A merchant navy officer in World War I, he became a member of the advance party of the **British Imperial Antarctic Expedition 1920–22** to Graham Land. When the leader (Dr. J. L. Cope) and second-in-command (Sir Hubert Wilkins) withdrew from the expedition, Lester stayed on with geologist Thomas Bagshawe, wintering in a waterboat in Paradise Harbour and completing a year's

meteorological, ornithological and tidal observations. Lester returned to the merchant navy, but in 1926–27 spent more time in Antarctica with *Discovery* Expeditions, in RRS *William Scoresby*. He died on 3 March 1957.
(CH/BS)

Levick, George Murray. (1876–1956). British naval surgeon and explorer. Levick qualified in medicine at St Bartholomew's Hospital, London, and joined the Royal Navy on qualifying in 1902. Specializing in physical training, he served with the fleet, then joined Scott's **British Antarctic (*Terra Nova*) Expedition 1910–13**. Assigned to the Northern Party, he wintered with Victor Campbell first at Cape Adare, then on Inexpressible Island, finally manhauling back to the main base at Cape Evans. While at Cape Adare he studied the behaviour and breeding biology of Adélie penguins, which became the material for a scientific report and popular book on their natural history. Levick remained in the navy throughout World War I, when he became Fleet Surgeon. He rejoined for World War II, serving mainly in Naval Intelligence, with particular interests in commando and Arctic warfare training. In between he worked in the Ministry of Pensions Hospital, concerned particularly with physiotherapy and rehabilitation medicine. In 1932 he founded and developed the Public Schools Exploring Society, which gave schoolboys opportunities to experience camping in the wilds of Scandinavia and Canada. He died on 30 May 1956.
Further reading: Levick (1914).

Lewis Bay. 77°38′S, 166°25′E. Ice-lined bay in the north coast of Ross Island, McMurdo Sound. The ice-covered slopes of Mt Erebus behind the bay were in November 1979 the site of an air disaster, and have been declared a tomb, to protect the unrecovered bodies from disturbance. An area of 16 km² (6.25 sq miles) is designated SPA No. 26 (Lewis Bay), to allow the application of a management plan to Historic Site 73, a memorial cross for the passengers and crew who died in the accident.

Lewthwaite Strait. 60°42′S, 45°07′W. Passage between Coronation Island and Powell Island, South Orkney Islands. The strait was discovered in December 1821 by Capt. George Powell and Capt. Nathaniel Palmer, who discovered and surveyed the islands, and named for a friend of Powell.

Lichens. Plants made up of a symbiotic association of algae (usually green) and fungi (usually Ascomycetes or Basidiomycetes), forming plant bodies (thalli) that are well adapted to living in cold, dry climates. Without roots or vascular system, they attach themselves to rocks by tough hairs, absorbing moisture and minerals through their surfaces. Rocky areas of Antarctica that at first glance appear sterile often support a limited flora of lichens. Though many are pink, orange, yellow, grey or black, all contain photosynthetic pigments within their algal component. Crustose lichens form a thin, leaflike thallus (body) closely apposed to the rock surface, 5–10 cm (2–4 in) across. They are often dense on coastal rocks that are subject to spray. Dendritic lichens, in the form of tiny branching shrubs up to 5 cm (2 in) high, are often plentiful in areas of high humidity, e.g. at cloud level. Both kinds form associations with mosses and other plants. Of some 15 000 species identified worldwide, about 130 have so far been recorded from continental Antarctica, about 150 species from the maritime sector.
Further reading: Longton (1988).

Dendritic and encrusting lichens: South Shetland Islands. Photo: BS

Lidke Ice Stream. 73°30′S, 76°30′W. Ice stream draining northward, flowing into the Sange Ice Shelf, Stange Sound, English Coast. It was identified from aerial photographs, and visited in 1985 by a field party which included the American geologist D. J. Lidke.

Liège Island. 64°02′S, 61°55′W. A mountainous, icemantle island 14.5 km (9 miles) long, off the northeast corner of Brabant Island, Palmer Archipelago. Charted by the Belgian Antarctic Expedition 1897–99, it was named for a province of the expedition's homeland.

Linnaeus Terrace. 77°36′S, 161°07′E. An ice-free terrace at the east end of Asgaard Range, Victoria Land, where Beacon Sandstones are exposed and subject to fierce **weathering**. An area of approximate area 3.2 km² (1.25 sq miles) is designated SSSI No. 19 (Linnaeus Terrace) to facilitate study of weathering phenomena, and of a rich community of endolithic alga and lichens found in the rocks.

Lion Island. 64°41'S, 63°08'W. A small, ice-covered island off the eastern coast of Anvers Island, in Gerlache Strait. Shaped vaguely like a reclining lion, it guards the northern entrance to Neumayer Channel.

Lion Sound. 64°40'S, 63°09'W. Narrow channel between Lion Island and Anvers Island, Palmer Archipelago.

Lions Rump. 62°08'S, 58°07'W. Prominent cape marking the southern entrance to King George Bay, King George Island, South Shetland Islands. The beach and cliff lining the bay support a rich flora of algae, lichens, mosses and flowering plants, twelve species of nesting bird, breeding elephant seals and resting fur seals. An ice-free area of approximately 1.3 km^2 (0.5 sq miles) is designated SSSI No. 34 to facilitate long-term monitoring and studies.

Litchfield Island. 66°16'S, 64°06'W. Rocky Island in Arthur Harbour, southern coast of Anvers Island, Palmer Archipelago, that supports an unusually rich assembly of marine and terrestrial life, including six species of breeding bird. Its total area of approximately 2.7 km^2 (0.9 sq miles) is designated SPA No. 17 (Litchfield Island) to protect 'an outstanding example of the natural ecological system of the Antarctic Peninsula area'.

Little America stations. A series of stations operated by US expeditions in the region of Kainan Bay, on the western end of Ross Ice Shelf. The original Little America, established by **Byrd's First Antarctic Expedition 1928–30**, was close to the site of Roald Amundsen's station Framheim, which by then had disappeared under accumulating snow. Little America II was the base of **Byrd's Second Antarctic Expedition 1933–35**: Little America III was the unofficial name of West Base of the **United States Antarctic Service Expedition 1939–41**. The **US Navy Development Project 1946–47** (Operation Highjump) established Little America IV, and Little America V was the main base and depot of the first Operation Deepfreeze 1955–57.

Livingston Island. 62°36'S, 60°30'W. A large, mountainous and heavily glaciated island 75 km (47 miles) long, between Snow Island and Greenwich Island: second-largest of the South Shetland Islands. Seen and charted in 1819 by William Smith, it was possibly the first of the group to be recorded. A likely source of the name is Capt. Andrew Livingston, a sealer of Glasgow, Scotland.

Lockroy, Port. 64°49'S, 63°30'W. Harbour on the west side of Wiencke Island, Palmer Archipelago, charted by J.-B. Charcot during the **French Antarctic (*Français*) Expedition 1903–5**. It was named for French statesman E.-A.-E. Lockroy, who helped to finance the expedition. Used extensively by whaling ships, in 1944 it became the site of Base A, Operation Tabarin. The refurbished station hut is now an Historic Site and museum.

Longhurst Plateau. 79°23'S, 156°20'E. Ice plateau rising to 2200 m between Darwin and McCleary glaciers, Victoria Land, named for neighbouring Mt Longhurst.

Loranchet, Jean. (d. 1966). French naval hydrographer. In 1908–9 he helped to establish the Franco-Norwegian whaling station at Port Jeanne d'Arc, Iles Kerguelen, later returning with R. Rallier du Baty to make hydrographic surveys and ornithological observations in the archipelago. A career naval officer, he served at sea throughout World War I and in the Ministère de la marine Marchande in peacetime. He died on 14 November 1966, aged 78 years.

Loubet Coast. 67°00'S, 66°60'W. Part of the west coast of Graham Land, Antarctic Peninsula, most of which faces Crystal Sound. Bounded in the northeast by Cape Bellue and in the southwest by the north shore of Bourgeois Fjord, the coast is deeply indented with bays and islands, with steep mountains rising to the central plateau of Graham Land, and glaciers descending to narrow piedmont ice shelves. It was delineated by J.-B. Charcot's **French Antarctic (*Français*) Expedition 1903–5**, and named for Emile Loubet, then President of France.

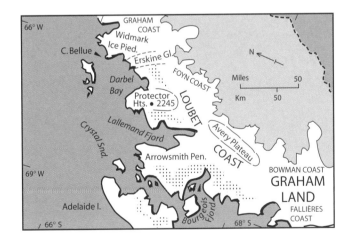

Loubet Coast

Louis Philippe Plateau. 63°37'S, 58°27'W. Ice plateau rising to 1370 m. in central Trinity Peninsula, Graham Land, named to commemorate the name given to the whole peninsula by the **French Naval Expedition 1837–40**.

Luitpold Coast. 77°30′S, 32°00′W. Part of the coast of Coats Land, East Antarctica, facing onto the Weddell Sea and bounded by Hayes Glacier in the north and Filchner Ice Shelf in the south. The coast consists mainly of ice cliffs fronting extensive ice shelves, with featureless ice plains rising inland: at the southern end are the exposed rocks of Moltke Nunataks. It was discovered and charted by the **German Antarctic (*Deutschland*) Expedition 1911–13**, and named for Prince Regent Luitpold of Bavaria.

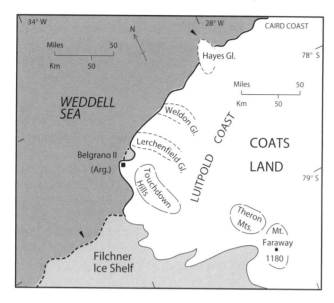

Luitpold Coast

Lynch Island. 60°40′S, 45°38′W. A small island in Marshall Bay, southern Coronation Island, South Orkney Islands, remarkable for its fertile mosses, soils, and an extensive stand of Antarctic hair grass. An area of approximately 0.1 km^2 (0.04 sq miles) is designated SPA No. 14. (Lynch Island) to protect this rare ecological system.

Lystad, Isak. (*c*.1896–1945). American seaman. Born in Kristiansand, Norway, Lystad went to sea as a fisherman from the age of 14. At 22 he emigrated to the USA, fishing off Alaska, trading with Siberia, and commanding *Boxer*, a supply ship of the Department of Indian Affairs. In 1939 he commanded *North Star*, one of the ships of the **United States Antarctic Service Expedition 1939–41**. Lystad was commissioned (Lt Cdr) in the US Naval Reserve. He died in Seattle on 24 May 1945.

M

Mabus Point. 66°33′S, 93°01′E. Rocky point on the coast of Queen Mary Land, facing the Haswell Islands. From February 1956 it has been the site of the permanent Soviet research station Mirnyy.

Macaroni penguin. *Eudyptes chrysolophus.* Crested penguins standing 70 cm (28 in) tall and weighing about 5 kg (11 lb), macaronis are distinguished by brilliant golden yellow crests that extend on either side above the eyes from the midline of the forehead. They breed on Antarctic, sub-Antarctic and cool temperate islands, penetrating furthest south of all the crested penguins, and feeding mainly on euphausiid and other shoaling crustaceans, small fish and squid. Macaronis breed in open colonies on cliffs or coastal flats, laying two eggs usually of dissimilar size, incubating only the second, larger one, and rearing a single chestnut-brown chick. The world population is estimated at 11 800 000 pairs.

Macfie Sound. 67°22′S, 59°43′E. Channel between Islay Island and Bertha Island in the William Scoresby Archipelago, Kemp Coast. The sound was surveyed in 1936 during a cruise by RRS *William Scoresby*, and named for the hydrographic officer, Lt A. F. Macfie, RNR.

Machu Picchu Station. 62°05′S, 58°28′W. Peruvian permanent research station on Crespin Point, Admiralty Bay, King George Island, South Shetland Islands. Established in February 1989, it has since been used intermittently for summer research parties.

Mackellar Islands. 66°58′S, 142°40′E. A group of small, low islands on the approaches to Cape Denison, George V Coast. Discovered by the Australasian Antarctic Expedition 1911–14, they were named for C. D. Mackellar, an expedition patron.

Mackenzie, Kenneth Norman. (1897–1951). British seaman and explorer. Born in Oban, Scotland, he became a merchant navy officer. Joining the **British, Australian and New Zealand Antarctic Research Expedition 1929–31**, he served as chief officer in RRS *Discovery* in 1929–30, and in the following season replaced Capt. J. K. Davis as master. Mackenzie Sea (now Mackenzie Bay) was named for him. He died on 29 September 1951.
(CH)

Mackintosh, Neil Alison. (1900–74). British polar oceanographer. Born in Hampstead, London on 19 August 1900, he read zoology at Imperial College, London. In 1924 he joined the embryo staff of **Discovery Investigations**, establishing the marine laboratory at King Edward Point, South Georgia in 1924–25 and spending a further period there in 1926–27. There followed a whale-marking voyage in RRS *William Scoresby*. Appointed chief scientific officer in 1929, he completed three oceanographic commissions in RRS *Discovery II* (1929–31, 1933–35 and 1937–39). From 1936 he was editor of *Discovery Reports*, an enormous and painstaking task which he fitted in with his own research. In 1949, when Discovery Investigations merged with the new National Institute of Oceanography, Mackintosh was appointed deputy director. From 1946 he was actively involved in the newly-founded International Whaling Commission, and from 1952 served as chairman of its scientific committee. He was also a member of the small sub-committee that advised the Colonial Office during the early days of the Falkland Islands Dependencies Survey. In 1961 he founded the NIO's Whale Research Unit. His research spanned a wide range of biological topics centering on commercial whales, from their life histories to zooplankton and the distribution of pack ice. Mackintosh retired in 1968, and died in London on 9 April 1974.

Macklin, Alexander Hepburne. (d. 1967). British expedition medical officer. After a short spell as a deckhand he read medicine at Manchester University. As a surgeon he joined the **Imperial Trans-Antarctic Expedition 1914–17**, serving with the Weddell Sea party. In World War I he served with distinction in the Italian campaign, and later in Russia with Ernest Shackleton. Repatriated and demobilized, he became surgeon to the **Shackleton-Rowett Expedition 1921–22**, treating Shackleton in his final illness. Thereafter he held hospital appointments in Dundee and Aberdeen. He died on 21 March 1967, aged 76.

Macquarie Island. 54°35′S, 158°55′E. An isolated cold-temperate island standing close to the junction of the Indian, Pacific and Southern oceans, 1760 km (1100 miles) southeast of Tasmania and 850 km (531 miles) southwest of Stewart Island, New Zealand. In the form of a narrow ridge 34 km (21 miles) long, up to 6 km (3.7 miles) wide, it rises to a rolling plateau 430 m (1410 ft) above sea level. Formed mainly of oceanic crustal rocks, it is the small visible section of a much longer submarine ridge extending in a north-to-south arc between New Zealand and the Balleny Islands. An elongate central plateau dotted with small lakes and low peaks is bordered by a coastline of steep cliffs and small, boulder-lined bays. The climate is cool temperate, damp and cold, with annual mean temperature of 4.6 °C and range of monthly means 3.7 °C. Winds blow constantly from the west, and annual rainfall exceeds 90 cm. The flora includes 40 species of shrub, herbs and ferns and over 50 species of moss. A green sward of tussock grass covers the clifftops and spreads into the valleys, growing from mature loams. The moorlands above, poorly drained and acid, support fellfield and bog. There is evidence of recent glacial activity but no permanent ice. King, gentoo, rockhopper and royal penguins, four species of albatross and many smaller petrels breed on the island, together with shags, skuas, terns and gulls. Elephant seals and two species of fur seal breed on the beaches. Black rats, house mice, domestic cats and rabbits, formerly abundant, are currently being brought under control.

The British sealing captain Frederick Hasselburg discovered the island in 1810, naming it after the contemporary governor of New South Wales. From 1825 it was administered as part of Van Dieman's Land, now Tasmania. Sealing gangs stripped it repeatedly of fur seals and elephant seals, later making inroads into the stocks of penguins for oil. A temporary research station was established in Hasselburg Bay on the northern end in 1911–14, and a permanent station has occupied the site since 1947. The island is currently a nature reserve administered by the Wildlife and Heritage Department, Tasmania, with the additional status of an IUCN Biosphere Reserve and a Strict Nature Reserve. Small numbers of tourists visit each year under permit from the Tasmanian government.

Further reading: Selkirk *et al.*. (1990).

Macquarie Island Station. 54°30′S, 158°57′E. Australian year-round research station at Atlas Cove, on the northern end of the island. Standing close to the site of the earlier station established by the **Australasian Antarctic Expedition 1911–14**, it has operated continuously since 1947.

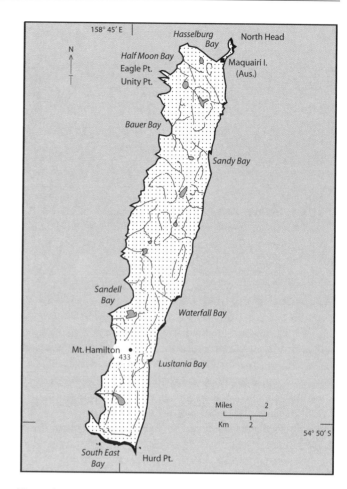

Though snow-covered for several months in winter, Macquarie Island has no permanent ice. Height in m

Mac.Robertson Land. 70°00′S, 65°00′E. Ice-covered, mountainous area of East Antarctica between meridians 60° and 73°E, bounded to the west by Kemp Land and to the east by Princess Elizabeth Land, and forming part of Australian Antarctic Territory. The land includes Mawson and Lars Christensen coasts, the Prince Charles Mountains, the massive Lambert Glacier and Amery Ice Shelf, and a high interior plateau extending to the South Pole. Identified by the **British, Australian and New Zealand Antarctic Research Expedition 1929–31**, it was named for Australian industrialist Sir MacPherson Robertson, whose support made the expedition possible. The area has been extensively studied by field parties from the permanent Australian coastal station Mawson.

Madigan, Cecil Thomas. (*c.*1890–1947). Australian geologist and explorer. A graduate in mining engineering from Adelaide University and a former Rhodes Scholar, Madigan joined Mawson's **Australasian Antarctic Expedition 1911–13** as geologist and meteorologist. Spending two winters at Cape Denison, he took an active role in the sledging programme, leading the Eastern Coastal Party that explored some 480 km (300 miles) of the George V Land coast toward Cape Freshfield. Madigan served in World War I with the Royal Engineers. After the war he was appointed assistant geologist with the Sudan Geological Survey. He returned to Australia to take up a lectureship in geology at the University of Adelaide, a post that gave him opportunities for geological exploration in central Australia. During World War II he served with the army as chief instructor in the School of Military Field Engineering. He died on 14 January 1947.

Magellan, Strait of. Strait linking the southern Atlantic and Pacific oceans through Tierra del Fuego. Discovered and first navigated in 1520 by Ferdinand Magellan, this waterway became an important sea-route for Spanish explorers *en route* to the west coast of South America. Contemporary maps show that its southern shore was still assumed to be part of **Terra Australis Incognita**. In 1577–80 Francis **Drake**, blown by storms south of Tierra del Fuego, found open water in what is now called **Drake Passage**. Having sailed through the Strait of Magellan he raided Spanish colonial settlements along the South American coast. Fearful of further piracy, the Spaniards in 1581–87 tried to fortify the Strait and establish settlements along it, but remoteness, hostile natives and a miserable climate defeated all their efforts. Early English and Dutch expeditions followed, the English attempting to explore south and perhaps repeat Drake's successful privateering, the Dutch seeking a westward passage to their own rapidly developing empire in southeast Asia. A Netherlands expedition of 1598–1600 under Jacques de Mahu made the first accurate survey. The Strait attained considerable importance as a shipping lane during the eighteenth to early twentieth centuries, offering a welcome alternative to the stormy Drake Passage. Though its importance declined after 1914, when the Panama Canal opened, it remains an important southern trade route between the Atlantic and Pacific Oceans.

Magellanic penguin. *Spheniscus magellanicus*. Standing 70 cm (28 in.) tall and weighing up to 4.5 kg (10 lb), Magellanic penguins are distinguished by their black cheeks encircled with white, and two broad black bands across the chest. Widely distributed in southern Argentina and Chile, they breed also on the Falkland Islands. The southernmost representative of the burrowing spheniscid penguins, Magellanics nest in burrows that they dig with bill and feet in sandy soil among coastal tussock grass, laying two eggs and raising one or two chicks. They feed mainly on fish and shoaling crustaceans. The world population is estimated at 1 000 000 pairs.

Magnetic anomalies. Local deviations of Earth's magnetic field, usually caused by local concentrations or irregular distributions of magnetic minerals in bedrock. Anomalies are detected by magnetometer surveys, that determine, first, the local magnetic field due to Earth's dipole, and, second, any deviations from the expected smooth pattern. Anomalies may indicate concentrations of particular minerals, of prime interest to prospectors. Those occurring at constructive boundaries of crustal plates (i.e. where the plates are drifting apart, allowing long bands of molten magma to emerge: see **Plate tectonics** in Antarctica) are of special interest to geophysicists, in providing a record of when the magma solidified, and a means of ascribing a time scale.

The strength and polarity of Earth's magnetic field change, even reverse, from time to time, in ways that are well understood and documented. As magma emerges and cools in an eruption event, for example at a mid-oceanic ridge, magnetic minerals within it take up the prevailing magnetic orientation. Thus each new band of rock acquires a distinctive magnetic signature, and retains it after solidification. Successive eruptions provide a succession of bands, with the youngest central to the ridge, older ones progressively further away. Precise magnetometric measurements across the ridge reveal the succession of signatures, which can be compared with the known record of secular magnetic changes.

Magnetic Pole, South. Known also as the 'dip pole', this is the point in the southern hemisphere where a dip needle (a bar magnet suspended freely on a horizontal axis) points toward Earth's centre. The position of the South Magnetic Pole shifts rapidly from year to year. When first reached in January 1909 by a sledging party from the **British Antarctic (*Nimrod*) Expedition 1907–9**, it lay in 72°25′S, 155°16′E among the mountains of South

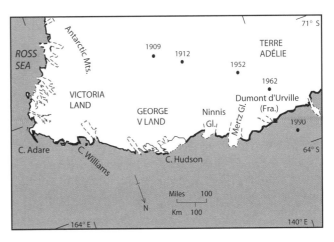

Approximate twentieth-century positions of the South Magnetic Pole. Formerly in King George V Land, it currently lies off the coast of Terre Adélie

Victoria Land. So rapidly was it moving that the sledgers considered waiting for the magnetic pole to come to them. In December 1912 a sledging party from the **Australasian Antarctic Expedition 1911–14** estimated that it had moved to 71°S, 150°E. In 1990 it lay off the coast of Terre Adélie in approximately 65°S, 139°E.

Magnetism, terrestrial. Scientific appraisal of Earth's magnetic field – a study that, particularly in the early and mid-nineteenth century, acted as a considerable stimulus to Antarctic and Southern Ocean exploration. That Earth to some degree mirrors the properties of a bar magnet, possessing North and South Magnetic Poles towards which suspended lodestones or artificial bar magnets are attracted, has long been a subject of scientific study. Even longer has it been a matter of practical application in navigation, in particular the close but varying relationship between true and magnetic north and south, as revealed by the magnetic compass. William Gilbert (1544–1603), among the first to study terrestrial magnetism, recognized that magnetic deviation or variation (the angular difference between true and magnetic north in a horizontal plane) varies with time. Astronomer and practical navigator Edmund Halley (1656–1742) plotted lines of equal variation on a terrestrial globe. Explorer Friedrich Humboldt (1769–1859) studied terrestrial magnetism in South America and Siberia, setting up chains of observatories to measure secular changes in variation. Norwegian geophysicist Christopher Hansteen (1784–1873) summarized contemporary early nineteenth-century knowledge in his book *Magnetism of the Earth*, published in 1819. Carl Friedrich Gauss (1777–1855) provided a mathematical formula by which the three magnetic elements (declination or departure from true north and south, dip or inclination of a horizontally suspended needle to the horizon, and intensity or strength of magnetic force) could be calculated for any point on Earth's surface. The formula appeared to work well for the many points in the well-explored northern hemisphere, but remained relatively untested in the south.

Development of Southern Ocean shipping routes, and increasing use of iron in ship construction, combined to make more detailed studies of terrestrial magnetism imperative, particularly in the empty spaces of the southern oceans. At the 1835 annual meeting in Dublin of the British Association for the Advancement of Science, in which several papers on geomagnetism were presented, a recommendation was made on the need for an Antarctic expedition 'for the purpose of making observations and discoveries in various branches of Science, as Geography, Hydrography, Natural History and especially Magnetism, with a view to determine precisely the place of the South Magnetic Pole or poles, and the direction and inclination of the magnetic force in those regions'. Already the US and French governments were considering major expeditions that would include these objectives. Not until 1838, following a further appeal from the British Association, was the British government moved to provide for an expedition involving the Royal Navy. So developed the three great mid-century scientific expeditions that, though motivated by research into terrestrial magnetism, between them established beyond doubt the existence of Antarctica: the **French Naval Expedition 1837–40** led by Dumont d'Urville, the **United States Exploring Expedition 1838–42** led by Charles Wilkes, and the **British Naval Expedition 1839–42** led by James Clark Ross.

Further reading: Mills (1905).

Maitri Station. 70°46′S, 11°44′E. Indian permanent year-round research station in Schirmacheroasen, Prinsesse Astrid Kyst, Dronning Maud Land. The station was opened on 9 March 1989 to replace the earlier ice-shelf station Dakshin Gangotri.

Marie Byrd Land. 80°00′S, 120°00′W. Mountainous, ice-covered area of West Antarctica, bounded to the east by Ellsworth Land and to the west by the Ross Ice Shelf and Ross Sea. The land includes the Walgreen, Bakutis, Hobbs, Ruppert, Saunders, Shirase, Siple and Gould coasts, the Executive and Ford ranges, and the Rockefeller Plateau. The area was explored particularly by the Byrd expeditions, and named for the wife of Rear Admiral R. Byrd USN.

Marion du Fresne, Marc Macé. (d. 1772). French naval officer and explorer. Sailing south in the Indian Ocean from Ile de France (now Mauritius) for Tahiti, Marion du Fresne reached the Prince Edward Islands on 13 January 1772. Believing them to be undiscovered, he named them 'Terre d'Esperance'. Marion Island, the largest island of the group, is now named for him. Ten days later, further south and east, he discovered Iles Crozet, landing on Ile de la Possession and claiming the group for France. The islands were named for Jules Marie Crozet, his second-in-command. Marion du Fresne was killed during a fracas on Tahiti in June 1772.

Marion Island. See **Prince Edward Islands**.

Marion Island Station. 37°50′S, 77°34′W. South African permanent year-round station on Marion Island, in the Prince Edward Islands group. Established in December 1947, shortly after South Africa took control of the islands, it has operated continuously ever since.

Maritime Antarctic. A climatic and ecological province of the Antarctic region including Antarctic Peninsula (south to about 69°S on the west coast, 64°S on the east coast), and the South Orkney, South Shetland and South Sandwich Islands. Though the land is permanently ice-mantled, and surrounded by sea ice in winter, the province is notably milder and damper than continental Antarctica, with longer summers and correspondingly richer flora and fauna. For temperature data and climatic descriptions of representative stations see **Climatic zones**.

Markham Plateau. 82°56′S, 161°10′E. High ice plateau in northern Queen Elizabeth Range, Victoria Land, named for British scientific administrator Sir Clements Markham.

Marr, James William Slesser. (1902–65). British expedition biologist. Born in Aberdeen on 9 December 1902, Marr was a student at Aberdeen University when in 1920, as a scout, he was selected by Sir Ernest Shackleton to join the **Shackleton-Rowett Expedition 1921–22**. Serving as a general assistant, he became interested in biology and, on return to Scotland, completed degrees in classics (1924) and zoology (1925). In 1925 he took part in a summer expedition to Svalbard, and in 1927 joined Discovery Investigations as a marine biologist, serving in *William Scoresby* (1928–29). In 1929–31 he was seconded to the **British, Australian and New Zealand Antarctic Research Expedition**, returning to complete two further commissions in *Discovery II* (1931–33, 1935–37). At the outbreak of World War II he sailed in a whale factory ship to see if whale meat could be made palatable for human consumption. On return in 1940 he was commissioned in the Royal Naval Volunteer Reserve, and in 1943 was selected to lead **Operation Tabarin**, to establish permanent bases in the Falkland Islands Dependencies. At the end of the war he returned to Discovery Investigations, later to the National Institute of Oceanography, where he completed and published a major long-term study of krill. He died on 29 April 1965.

Marsh Station. See **Teniente Rodolfo Marsh Station**.

Marshall, Edward Hillis. (1885–1975). British expedition ship's surgeon. After distinguished service in the British Army in World War I, he became surgeon and bacteriologist in RRS *Discovery* 1925–27, later serving as an observer in a whaling factory ship (1928–29) and as surgeon in RRS *Discovery II* 1929–31. He died on 28 February 1975. He was an older brother to Eric Stewart Marshall, who served with Shackleton's **British Antarctic (*Nimrod*) Expedition 1907–9**.
(CH/BS)

Marshall, Eric Stewart. (1897–1963). British expedition surgeon. After studying medicine at St Bartholomew's Hospital, London, he joined the **British Antarctic (*Nimrod*) Expedition 1907–9** as medical officer and cartographer with the shore party. Taking part in the sledging operations, he was one of the party that ascended the Beardmore Glacier and man-hauled to within 155 km (97 miles) of the South Pole. In 1909–11 he was medical officer on a British expedition to New Guinea. He served with distinction in World War I, later farming in Kenya. Retiring to the Isle of Wight, he died on 26 February 1963. He was brother to **Edward Hillis Marshall**, who served with Discovery Investigations.
(CH/BS)

Marshall Archipelago. 77°00′S, 148°30′W. Group of large ice-mantled islands within Sulzberger Ice Shelf, on Saunders Coast, Marie Byrd Land, West Antarctica. The group was identified during several of the expeditions led by Rear Admiral Richard Byrd, and named by him for General George C. Marshall, who supported the expeditions.

Marston, George Edward. (1882–1940). British expedition artist. Born in Southsea on 19 March 1882, he attended art schools in London and qualified as a school teacher. As an artist he joined the **British Antarctic (*Nimrod*)Expedition 1907–9**, taking part in several sledging journeys including an ascent of Mt Erebus. He contributed several lithographs to *Aurora Australis*, a limited-edition book produced at Cape Royds, and paintings to *The Heart of the Antarctic*, Sir Ernest Shackleton's two-volume account of the expedition. Highly valued by Shackleton, he was an early recruit to the **Imperial Trans-Antarctic Expedition 1914–17**, again as official artist. He sketched life on the pack ice and on Elephant Island, again contributing to Shackleton's official account of the expedition. On return to Britain Marston taught at Bedales School. In 1925 he joined the Rural Industries Bureau as Handicrafts Adviser. Appointed director in 1934, he died in office in 1940.

Martin, Port. 66°49′S, 141°24′E. Harbour off Cape Margerie, Adélie Coast, Terre Adélie. The site of the first research station of the French Antarctic Expedition 1950–52, it was named for a member of the expedition, A.-P. Martin, who died *en route* to Antarctica.

Martin-de-Viviès Station. 37°50′S, 77°34′E. French research station that has operated year-round on Ile Amsterdam since 1949.

Mason, Douglas P. (1920–86). Born in 1920 he joined the Falkland Islands Dependencies Survey in 1946, serving at Base E, Marguerite Bay, for two years and taking part in

several major sledging operations. Graduating in engineering science at Oxford University he was appointed a lecturer in the Department of Surveying and Photogrammetry, University College, London in 1962 after he had worked with the Sudan Survey Department and Shell International. He retired in 1985 and died on 29 October 1986.
(CH)

Matha Strait. 66°34′S, 67°30′W. Passage between Adelaide Island and the southern end of the Biscoe Islands, named for Lt A. Matha, second-in-command of the **French (*Français*) Antarctic Expedition 1903–5**.

Mather, John Hugh. (1887–1957). British expedition seaman. Born at Stroud Green, he enlisted as a rating in the RNVR. While helping to load the expedition ship in London he applied successfully to join the **British National Antarctic Expedition 1901–4**, serving for three years as a petty officer, helping particularly with clerical work and becoming adept at taxonomy. Later commissioned, he served throughout World War I, retiring with the rank of Commander. In 1929 he inaugurated the Antarctic Club, an annual dining club for former members of British expeditions, and was its enthusiastic Honorary Secretary until his death on 10 April 1957.

Matthews, Leslie Harrison. (1901–86). British polar zoologist. Graduating in natural history at Cambridge, in 1924 he joined the Discovery Investigations as a biologist, serving for two seasons in the marine laboratory on South Georgia. He held an academic post at Bristol University before becoming Scientific Director of the Zoological Society of London 1951–66. His publications include *South Georgia: The Empire's sub-Antarctic Outpost*. He died on 27 November 1986.
(CH)

Mauger, Clarence Charles. (d. 1963). British polar seaman. A merchant navy shipwright, he joined the Ross Sea party of the **Imperial Trans-Antarctic Expedition 1914–17**. Serving aboard *Aurora* during the long drift in the ice pack, he used his skills to provide a jury rudder for the damaged ship. After service in both world wars he worked with the Otago Harbour Board, Dunedin, New Zealand. He died in Dunedin on 13 October 1963.

Mawson, Douglas. (1882–1958). Australian geologist and explorer. Born in Bradford, England on 5 May 1882, he emigrated with his parents to Australia. Graduating in geology at the University of Sydney, he became a lecturer in geology at Adelaide University. From there he joined Shackleton's **British Antarctic (*Nimrod*) Expedition 1907–9**, taking an active part in the sledging programme: with fellow-Australian Edgeworth David he climbed Mt Erebus and reached the South Magnetic Pole. On return to Australia he immediately began planning his own **Australasian Antarctic Expedition 1911–14**, a most successful enterprise that explored virtually unknown areas of the East Antarctica coast from two centres. Knighted, he served in Europe in World War I, returning to take up a chair in geology at Adelaide University. During the late 1920s he lobbied strongly for Australia's further involvement in Antarctic affairs, culminating in the **British, Australian and New Zealand Antarctic Expedition 1929–31**. A shipborne expedition, this enabled his adopted country to lay claim to over half the total area of the continent. He was active also in geological fieldwork in Australia. Mawson died on 5 May 1958.

Douglas Mawson (1882–1958)

Mawson Coast. 67°40′S, 63°30′E. Part of the coast of Mac.Robertson Land, East Antarctica. Bounded by William Scoresby Bay in the west and Murray Monolith in the east, the coast consists mainly of ice shelves with emergent nunataks and offshore islands, with mountains rising close inland. Discovered and explored during the **British, Australian and New Zealand Antarctic Research Expedition 1929–31**, it was named for the expedition leader, Sir Douglas Mawson. Since February 1954 the coast has been the site of the permanent Australian research station Mawson.

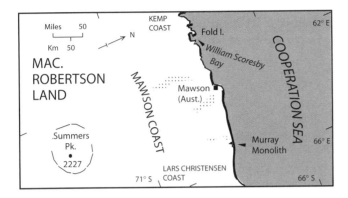

Mawson Coast

Mawson Escarpment. 73°05′S, 68°10′E. A prominent rock escarpment along the eastern flank of Lambert Glacier, Mac. Robertson Land, rising to over 1000 m (3280 ft), and backed by the Law Plateau. It was discovered in a reconnaissance flight of 1956 and named for the doyen of Australian explorers, Sir Douglas Mawson.

Mawson Sea. 65°S, 105°E. Coastal sea off East Antarctica, between Law Dome and the eastern end of Shackleton Ice Shelf, bordering Knox Coast, Wilkes Land. There is no definitive northern boundary. Fast ice and pack ice extend off the coast in winter: in summer, pack ice and bergs move westward, driven by easterly currents and offshore winds. The sea was designated by Russian cartographers to honour the Australian explorer Sir Douglas Mawson.

Mawson Station. 67°36′S, 62°52′E. Australian permanent year-round research station at Horseshoe Harbour, on the coast of Mac.Robertson Land, East Antarctica. Established in February 1954, it holds the longest record for a continuously operated station south of the Antarctic Circle.

McCarthy, Mortimer. (d. 1967). British seaman. Born in Kinsale, Co. Cork, older brother to **Timothy McCarthy**, he served in the Royal Navy during the Boer War, and as an able seaman with the **British Antarctic (*Terra Nova*) Expedition 1910–13**. He settled in Lyttelton, New Zealand, working mainly in shipping, and died in a fire at his home on 11 August 1967.

McCarthy, Timothy. (d. 1917). British polar seaman. Born in Kinsale, Co. Cork, younger brother to **Mortimer McCarthy**, he served in the merchant navy and Royal Naval Reserve. As an able seaman he joined the **Imperial Trans-Antarctic Expedition 1914–17**, and was one of the party that sailed in the lifeboat *James Caird* from Elephant Island to South Georgia. Returning to sea as a gunner in World War I, he was killed in action on 16 March 1917.

McDonald Islands. 53°03′S, 72°36′E. Group of sub-Antarctic islands 2 km (1.25 miles) long and half as wide, with a total area of about 260 ha., lying 40 km (25 miles) west of Heard Island. McDonald Island, largest of the group, consists of two steep-sided platforms rising to about 230 m (754 ft), joined by a narrow isthmus. The rocks are volcanic tuffs intruded by veins of harder phonolite. Two smaller islets flank the main island. Standing on the Kerguelen submarine platform, the group is probably a remnant of a once-larger volcanic mass. The islands were discovered in 1854 by a British sealing captain, William McDonald. Too steep to accommodate many fur seals, they were seldom visited and appear to have remained relatively untouched by man for over a century. Australia assumed sovereignty in 1947. The first recorded landing was a brief visit by helicopter from the French research ship *Galliéni* in January 1971, and in 1980 an Australian party worked for four days on and around the islands. Five species of flowering plant were recorded, including tussock grass, Kerguelen cabbage and small shrubs. Macaroni penguins and several species of petrel were noted, together with breeding Antarctic fur seals and southern elephant seals. There is no permanent ice. The islands are currently scheduled as a strict scientific reserve: landings are permitted only for scientific purposes.

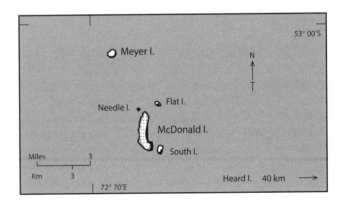

Covered with tussock grass and low shrubs, the McDonald Islands support large populations of sea birds, including burrowing petrels. The highest point rises to about 230 m

McFarlane Strait. 62°32′S, 59°55′W. Passage between Greenwich Island and Livingston Island, South Shetland Islands. The name appears on a chart of 1822 by British sealer Capt. George Powell, honouring Capt. Robert McFarlane, master of the London sealer, *Dragon*.

McIlroy, James Archibald. (1897–1968). British expedition surgeon. Born in Ireland on 3 November 1897, he studied medicine at Birmingham University. He served as surgeon with Shackleton's **Imperial Trans-Antarctic Expedition 1914–17**, sailing with *Endurance* and surviving the landing on Elephant Islands. Badly wounded while serving with the British army in World War I, he recovered sufficiently to join the **Shackleton-Rowett Antarctic Expedition 1921–22**. In cooperation with L. Hussey, McIlroy contributed an appendix on meteorology to Frank Wild's *Shackleton's Last Voyage* (London, 1923). Later he became a ship's surgeon on passenger liners, serving throughout World War II. He died on 30 July 1968.
(CH)

McKenzie, Edward A. (1888–1973). British polar seaman. Born in Norfolk on 20 April 1888, he joined the Royal Navy in 1903, and was rated leading stoker when he volunteered for service with the **British Antarctic (*Terra Nova*) Expedition 1910–13**. On return from the expedition he joined the London Metropolitan Police. After service in both the army and the navy in World War I he re-joined the police until invalided out in 1942. He died on 21 June 1973.

McKinley, Ashley C. (d. 1970). American expedition photographer. As official photographer to Byrd's first Antarctic expedition of 1927–29, he took part in the flight to the South Pole in 1929. During World War II McKinley was concerned with testing and developing cold weather equipment and he also invented a pneumatic aircraft float. He died on 11 February 1970.
(CH)

McMurdo Ice Shelf. 78°00′S, 166°30′E. Westward extension of the Ross Ice Shelf into the area between Ross Island, White Island and Brown Island (some authorities include the area south to Minna Bluff), and draining into McMurdo Sound. This shelf provided access to the Ross Ice Shelf for British explorers Robert F. Scott and Ernest Shackleton in their attempts to reach the South Pole.

McMurdo Oasis. See **Victoria Land Dry Valleys.**

McMurdo Sound. 77°30′S, 165°00′E. Deep bay between Ross Island and Victoria Land, discovered by Capt. J. C. Ross in 1840 and named for Lt A. McMurdo, HMS *Terror*. An important gateway to the Ross Ice Shelf and the route to the South Pole, the Sound was used by British expeditions under Scott, who based his first polar expedition at Hut Point and his second at Cape Evans, and Shackleton, who was based at Cape Royds. It is currently the site of US research station McMurdo, and of New Zealand's station Scott Base.

McMurdo Station. 77°51′S, 166°40′E. American permanent year-round research station at Hut Point, Ross Island, McMurdo Sound. Established in 1955 as a cluster of temporary buildings, it has undergone several stages of development and remodelling to form Antarctica's largest settlement, a township with roads, laboratories, stores, recreational facilities and accommodation for about 1200 seasonally resident and transient personnel. It serves as the main research and logistics centre for US Antarctic operations (other than those at Palmer Station, Antarctic Peninsula), providing port and airstrip facilities for stores and equipment destined for **Amundsen-Scott** and other inland research stations.

Meinardus, William. (1867–1952). German polar geographer and climatologist. An academic who taught at Berlin, Bonn and Göttingen universities, he prepared the meteorological reports of the **German South Polar Expedition 1901–3**, drawing attention for the first time to the discontinuity in surface waters now called the Antarctic Convergence or Polar Front. Later he wrote the section 'Klimakunde der Antarktis', a comprehensive account of Antarctic climates, in W. Köppen and R. Geiger's *Handbuch der Klimatologie*. He died on 28 August 1952.

Melba Peninsula. 66°31′S, 98°18′E. A low, ice-covered peninsula extending onto Shackleton Ice Shelf, Queen Mary Coast. It was discovered by the **Australasian Antarctic Expedition 1911–14** and named for the Australian soprano Dame Nellie Melba.

Melbourne, Mt. 74°21′S, 164°42′E. Quiescent volcano rising to 2730 m. (8954 ft) above the southern end of Borchgrevink Coast, Victoria Land. Fumaroles and warm surface vents produce patches of locally enriched vegetation. A layer of ash in ice nearby suggests that the most recent eruption occurred about 200 years ago. The mountain was discovered in 1841 by Capt. J. C. Ross RN, who named it for the contemporary British prime minister. An area of approximately 8.4 km^2 (3.3 sq miles), including all the ground above the 2200 m contour (*c.* 7200 ft) surrounding the main crater, is designated SSSI No. 24 (Summit of Mt Melbourne) to facilitate and protect long-term research.

Melchior Islands. 64°19′S, 62°57′W. A cluster of low, ice-covered islands in Dallmann Bay, between Anvers Island and Brabant Island, Palmer Archipelago. Noted first by Dallman in 1873–74, later by the **French Antarctic (*Français*) Expedition 1903–5**, they were named for a French vice-admiral. Individual islands bear the names of Greek letters (Eta, Gamma, Lambda, Omega, etc.). Gamma Island is the site of Melchior, an early Argentine research station occupied from 1947 to 1961.

Melchior Station. 64°20′S, 62°59′W. Argentine permanent research station, built on Gamma Island, Melchior Islands, Palmer Archipelago in 1947. Argentina's first research facility in the Antarctic Peninsula area, it was for several years used for wintering parties, later as a summer-only station.

Melsom, Henrik Govenius. (1870–1946). Norwegian sealer and whaling captain. Born in Sandar, Norway, he began his sea career hunting seals and bottle nose whales. From local fisheries he became a noted deep-sea whaler, in his later career commanding factory ships in both polar regions. He took a leading part in developing pelagic whaling, and in the invention and use of the stern slipway to haul whale carcasses aboard for flensing. He died in Sandar on 19 August 1946.

Meteorites. Meteorites are fragments of rock that have fallen to Earth from inter-planetary space and are considered to be primordial material from disintegrated planetary bodies, now represented by the asteroid belt between Mars and Jupiter. To geologists, meteorites give valuable information on the likely composition of Earth soon after it was formed. In the field they are difficult to distinguish from terrestrial rocks, and are easily overlooked. Most are classified as *stones*, composed of the minerals olivine, pyroxene and feldspar, outwardly similar to some types of terrestrial igneous rocks. Many are *chondrites*, distinguished by the presence of small spherical inclusions of silicate material called chondrules. Much rarer meteorites called *irons* are formed of iron alloys with up to 20 per cent nickel. *Stony-irons* are mixtures of iron-nickel alloys and silicate minerals. All meteorites yield radiometric ages of about 4600 Ma.

In 1969 Japanese scientists working on the southern side of Dronning Fabiolafjella (Yamato Mountains, Dronning Maud Land), made the first discovery of a concentration of meteorites on a blue-ice field, far from the nearest rock outcrop. Since then searches of similar sites have proved extremely fruitful. Antarctic meteorites now outnumber all others collected elsewhere on Earth. Most are 1–10 cm in diameter, though larger ones have been found. Antarctic meteorites have provided a wealth of research material. Some were types new to science, but they have proved particularly valuable for other reasons. Excellent preservation in the ice sheet has ensured their almost total freedom from terrestrial contamination. The fusion crust, formed by frictional heating as the meteorite fell through the atmosphere, can be dated accurately so that the Earth residence time can be calculated. Ages up to 700 000 years have been obtained in this way, providing essential information on the age of the Antarctic ice and helping glaciologists to model the behaviour of the ice sheet.
Further reading: Cassidy (1991).
(PDC)

Michigan Plateau. 86°08′S, 133°30′W. Ice plateau rising to 3000 m on the western flank of Reedy Glacier, Wisconsin Mountains, named for the University of Michigan.

Microclimate. Climate in small areas, as opposed to the macroclimate that is measured under standard conditions by meteorologists. Those studying microclimate are usually interested in conditions for life in small environments, e.g. near the ground, in soils or burrows, among rocks, inside moss clumps or dense vegetation. In such microhabitats, especially in polar regions, microclimate differs from macroclimate in ways that favour small plants and animals. Winds that blow strongly 2 m above the surface of ground or snow are usually lighter at surface level, and among rocks and vegetation. Still or slowly-moving air is readily warmed by contact with sun-warmed rock. For small plants and animals, rock crevices provide life-sustaining moisture and shelter from wind-blast. Temperatures and humidity may be several degrees higher among pebbles and rocks, in tufts of moss and immediately under the surface of snow, than in the open.

Microhabitat. Small, highly local environment in which plants and animals find congenial living conditions. Many polar organisms live in microhabitats in which the local **microclimate** surrounding them is less severe than the more extreme macroclimate of the environment at large. Favourable Antarctic microhabitats include burrows occupied by petrels in soils or scree slopes; north-facing crevices among rocks, sun-warmed, sheltered from winds and moisture-retaining; minute spaces behind translucent rock crystals, occupied by endolithic lichens; the interior of moss clumps, home to collembolae and mites, and bases of feathers of many species of bird, infested by parasitic lice and ticks.

Migration, seasonal. Seasonal movement of animals that enables them to survive adverse changes in climate and food supply. Unlike the Arctic, the Antarctic region does not provide expanses of summer tundra vegetation attractive to long-distance migrant waterfowl and waders. However, Antarctic surface waters are rich in summer plankton, attracting some 42 species of seabird to breed, and a few northern-breeding species (for example, Arctic terns, and Manx and other shearwaters) to feed. At the end of summer most species, breeding and non-breeding, leave impoverished southern waters. Adélie and chinstrap penguins migrate away from the continent to the newly formed fast ice or pack ice: though cold, these are much warmer than the land. Flying birds migrate to spend winters over ice-free areas of the Southern Ocean or warmer seas further north. Giant petrels are recorded off South Africa, Australia and South America, Wilson's petrels in the north Atlantic Ocean, southern species of skua off the coasts of Britain and Japan. Emperor penguins, largest and best-insulated of the penguins, are unusual in making a reverse migration, moving south in autumn to breed on the newly-formed fast ice, and north in spring when the ice disperses. Among mammals, Weddell seals and some **minke**, **killer** and **bottlenose whales** remain close to the continent in winter. Most other species migrate northward in autumn and return south in spring.

Mikkelsen Harbour. 63°54′S, 60°47′W. Harbour in the south coast of Trinity Island, Palmer Archipelago, used by whalers for mooring factory ships and named for Norwegian whaler Capt. Klarius Mikkelsen.

Miller, Joseph Holmes. (d. 1986). New Zealand polar surveyor. Born in Waimate, South Canterbury, 'Bob' Miller served with the New Zealand Army in North Africa before training as a surveyor at Victoria University, Wellington. In 1950 he visited the Bounty and Antipodes Islands, in 1955 joined the **Commonwealth Trans-Antarctic Expedition** as surveyor, later a second-in-command under Sir Edmund Hillary. From October 1957 to January 1958 Miller and George Marsh completed a surveying journey of over 4000 km (2500 miles) from Scott Base to the Queen Alexandra Range, Antarctica's longest dog-sledging journey. In 1963–64 he led the northern party of a New Zealand expedition that explored the mountains of Oates Land and Victoria Land from the Pennell Glacier to the Tucker Glacier. Miller chaired the Ross Dependency Research Committee 1981–84, gained international credit in his profession of surveyor, and was knighted in 1979. The Miller Range, inland from Shackleton Coast, is named for him. He died on 7 February 1986.

Miller Range. 83°15′S, 157°00′E. A range inland behind Shackleton Coast, flanked by the polar ice cap. It was identified and surveyed by the **Commonwealth Trans-Antarctic Expedition 1955–58** and named for the New Zealand expedition surveyor **J. Holmes Miller**.

Minerals and mining in Antarctica. Coal seams discovered in the Transantarctic Mountains during Shackleton's **British Antarctic (*Nimrod*) Expedition 1907–9** were the first evidence of the continent's mineral potential. Clearly too far from the sea to be even remotely considered exploitable, their main value was comparability with similar deposits in southern Africa. Their presence confirmed Antarctica's central position among the southern continents, and made it seem likely that more minerals would be found in deposits similar to those of Africa, India, Australia and South America. Traces of economically important minerals have indeed since been found in Antarctica's meagre exposures of bare rock, and also at sea, including offshore oil and gas. However, no occurrence so far discovered has proved of commercial interest, and more research is needed before sensible predictions of resource potential become possible.

Green-stained cliffs and mountains on the Antarctic Peninsula alerted several geologists to the presence of copper mineralization; a very early report even claimed discovery of native copper. However, all copper indications so far investigated are weathering products of other copper minerals, forming impressive but thin veneers across large expanses of exposed rock. Only in one or two places has the source of the copper been traced to abundant primary minerals, and these sources are small. Traces of iron and molybdenum minerals have been found, and more rarely nickel, chromium, cobalt, lead, and zinc. Gold and silver assays on Antarctic Peninsula rocks yielded values respectively of 0.3–2 ppm and 1–10 ppm, only 100–1000 times the average crustal abundance, and not commercially significant. The Antarctic Peninsula clearly has metalliferous resources but their potential for mining is still far from proven.

The Transantarctic Mountains offer a few resource possibilities. Permian (250 Ma) coal is exposed in seams of varying thickness and extent, ranking in quality from bituminous to semi-anthracitic. This coalfield is unusual because much of the coal has not formed *in situ*: the original vegetable matter was transported from a distance by water before being deposited, buried and converted to coal. The seams are therefore discontinuous. The presence of more coal in the Prince Charles Mountains, 2000 km (1250 miles) away, may indicate that coal-bearing strata could be continuous across East Antarctica. However, the thick overlay of ice between them makes this no more than speculation. Sandstones of the coal-bearing sedimentary sequences have been investigated for radioactive thorium and uranium minerals. Airborne gamma-ray surveys have revealed higher-than-average background levels, but the highest levels relate to small radioactivity anomalies associated with dykes and veins. Occurrences of various other minerals have been reported, though none of compelling interest.

In all Antarctica, the layered Dufek Massif of the Pensacola Mountains has provided the most intense speculation, though on the least evidence. The exposed upper part of the intrusion bears strong chemical similarities with the upper part of the Bushveld Igneous Complex of South Africa, the world's most important platinum-producing deposit. It seems possible that the lower, unexposed part of the intrusion may contain an equivalent of the Merensky Reef in the Bushveld Igneous Complex. However, the vast difference in age between the two regions (1900 Ma) negates any realistic comparison. A better comparison with the Insizwa Igneous Complex, of similar age and provenance, would suggest a rather poor nickel-copper prospect. Nevertheless an economic feasibility study of mining this intrusion has been made (see below).

East Antarctica's ancient craton or shield, exposed mainly in marginal coastal mountain ranges, largely comprises metamorphic rocks up to 3800 Ma old. The most important mineral occurrences have been recorded in pegmatites (very coarse crystalline granites) in Dronning Maud Land and in a sedimentary iron formation near the Prince Charles Mountains. Many pegmatites contain large crystals of mica, beryl and quartz, all of which have wide industrial applications. Some gneisses and schists in these areas contain significant concentrations of graphite, and iron in the form of magnetite. However, the sedimentary banded ironstones contain an average of 34 per cent iron, and aeromagnetic surveys have indicated an extension beneath the ice of 600–1800 km^2. This occurrence, although worthy of further investigation, is of lower potential than similar, currently producing deposits in Australia.

The oil and gas potential of the Antarctic is very great. Little indication of suitable environments has been seen on land, but drillings on Antarctica's continental shelves have yielded traces of both gas and oil. Estimated reserves of 45×10^9 barrels of oil and 115×10^{12} cubic feet of gas have been made, based on worldwide averages, but these are meaningless without the support of data on source rocks, heat flow, reservoir rocks, and trap structures. Other natural resources in and around Antarctica with limited practical potential for local use include such building materials as slate, limestone and aggregate. Industrial materials such as fluorite and sulphur are also known to occur. Geothermal energy could be available at a very few sites. One further Antarctic product – icebergs – has been considered in detail: schemes have been suggested for towing them north to provide fresh water for the world's arid regions.

Mineral occurrences have no economic significance unless the minerals can be extracted, processed and marketed profitably. There is little at present to justify the enormous investment necessary for initiation and operation of Antarctic mineral extraction, when many mines and oil wells in more accessible situations worldwide are closed. To prospectors and miners Antarctica presents problems of remoteness, lack of existing infra-structure, hostility of climate, access to operational sites over difficult terrain, and technical difficulties of operating under ice on land, or drilling in very deep, iceberg-strewn waters offshore.

Considerable attention has been paid to where Antarctica's most valuable minerals are likely to be found. Different authors have divided Antarctica into provinces where particular minerals may be present. Others have summarized geophysical data to indicate where geological structures may trap hydrocarbons, catalogued known mineral occurrences in Antarctica, discussed possible impacts of minerals exploration and exploitation, and made an economic feasibility study of mining the Dufek Intrusion, assuming the presence of platinum.

In considering the pros and cons of Antarctic mining, political and legal questions arise concerning operations in international territory. Currently the **Antarctic Treaty System** provides the only generally accepted rules of conduct. International conservationist groups have consistently expressed strong opposition to any form of mineral operations in Antarctica, and a **Convention on the Regulation of Antarctic Mineral Resource Activities** (CRAMRA), agreed under the Antarctic Treaty System in 1989, was not adopted. In its place, the **Protocol on Environmental Protection to the Antarctic Treaty** was agreed in 1991 and entered into force on 14 January 1998. Article 7 of the Protocol prohibits all mineral resource activities, other than scientific research, for a period of 50 years. Thus mining activities are unlikely to occur in the Antarctic Treaty area within the foreseeable future.

(PDC)

Minke whale. *Balaenoptera acutorostratus.* Also called piked whales (from their sharp pointed rostrum or fore-end), these are the smallest of the rorquals, measuring up to 9 m and weighing up to 10 tonnes. Dark grey above, paler below, they are distinguished mainly by their small length and girth. Some have a prominent white mark on the right upper jaw: nearly all have a white band or patch on the upper side of the right flipper. Plentiful in all the world's oceans, they are most often seen in temperate and cool waters, but are also summer visitors to both polar oceans. They travel in groups of three or four: though shy, they can readily be approached by small boats and occasionally show curiosity. On rich feeding grounds there may be much larger assemblies of several dozen, even hundreds, feeding together but independently. In the Southern Ocean they appear mostly to be summer visitors, though some are known to winter among the pack ice close to Antarctica. In northern waters they are hunted

by small catchers, and are small enough to be processed on deck. During the early and middle years of Southern Ocean whaling (1905–65) they were ignored: having an oil value of only one-sixth of a blue whale, they were deemed hardly worth the trouble of catching and processing by the large factory ships. Following the destruction of **blue**, **fin** and **sei whale** stocks, minkes became the mainstay of Southern Ocean hunting. They are still plentiful in southern waters in summer. A world population of about 200 000 is estimated.
Further reading: Evans (1987); Bonner (1989).

Mirnyy Station. 66°33′S, 93°01′E. Russian permanent year-round station close to Haswell Island on the Davis Coast. Established in February 1956 for the International Geophysical Year, it formed a port of entry and point of departure for stations inland, and continues to service the plateau station Vostok.

Mizuho Plateau. 70°41′S, 44°54′E. Ice plateau east of Dronning Fabiolafjella, Dronning Maud Land, identified by the Japanese Antarctic Research Expedition and named with a traditional name for Japan. A Japanese research facility, Mizuho Station 70°41′S, 44°54′E, was established in July 1970 and has been used intermittently since then.

Moe Island, South Orkney Islands. 60°45′S, 45°41′W. Small island off the southwest coast of Signy Island, South Orkney Islands, of area approximately 1.3 km² (0.5 sq miles) which supports a representative sample maritime Antarctic ecosystem, including moss peat, breeding chinstrap penguins, several breeding species of petrel and Weddell, leopard and fur seals. The island is designated SPA No. 13 (Moe Island) to ensure its protection as a control area for future comparisons with Signy Island.

Mollymawks. Seamen's name for the smaller albatrosses of the genus *Thalassarche*, of which four species forage and nest in the southern oceanic zone. All have body length up to 90 cm (35 in), and wing span 2 m (79 in): all are distinguished from wandering and royal albatrosses by size, and by their black back and black-tipped tail. Black-browed albatrosses *T. melanophris* have a pale head, yellow bill and conspicuously dark ridge above the eye. Grey-headed albatrosses *T. chrysostoma* have a markedly darker head and dark grey bill. Both breed on islands in the sub-Antarctic and cool temperate zones, usually in cliff colonies numbering hundreds of nests. Yellow-nosed albatrosses *T. chlororhynchos*, distinguished by a dark bill with yellow central culmen band and tip, breed mainly in the warm temperate zone, singly or in large colonies on windswept headlands. Shy albatrosses *T. cauta* breed on the Auckland, Bounty and Snares Islands.

Molodezhnaya Station. 67°40′S, 45°51′E. Russian permanent research station in Alasheyev Bight, Enderby Land. Opened in February 1962, it was maintained year-round until 1999, when it reverted to summer-only use.

Moltke Harbour. 54°31′S, 36°04′W. Harbour on Royal Bay, South Georgia. The **German International Polar Year Expedition 1882–83** wintered there, naming the harbour for their expedition ship.

Montagu Island. 58°25′S, 26°20′W. A kite-shaped island 25 km (15.5 miles) long standing between Bristol Island and Saunders Island in the central section of the South Sandwich Islands. About 97 per cent ice-covered, it rises to a central dome of Mt Belinda, 1370 m (4494 ft) high. Mt Oceanite forms a lower and smaller cone in the southeast corner of the island. The island was discovered and charted in 1775 by Capt. J. Cook RN, and named for John Montagu, the contemporary Earl of Sandwich and First Lord of the Admiralty. There is no evidence of recent volcanic activity.

Moraine. A deposit of rock debris (till) resulting from the activity of ice masses. Often massive, the debris consists of angular and striated clastic rocks, well compacted with ungraded, ill-sorted finer materials. *Ground moraines* form within the glacier bed, *lateral moraines* along its flanks, *medial moraines* at the junctions of two or more glacier flows, and *end* or *terminal moraines* at the glacier foot.

Morgan, Charles Gill. (1906–80). American polar geologist and geophysicist. Graduating from Southern Methodist University in 1928, he joined **Byrd's Second Antarctic Expedition 1933–35**. With E. Bramhall he sledged by tractor train from Little America to Rockefeller Plateau, Marie Byrd Land, where he undertook seismic and magnetic surveys. After service with the US Army in World War II he became a consultant in geophysics and geology and president of the Atlas Land Company, Dallas. He died on 8 August 1980.
(CH)

Morton Strait. 62°42′S, 61°14′W. Passage between Snow Island, Rugged Island and Livingston Island, South Shetland Islands. The name appears on an 1825 chart by British sealer Capt. James Weddell.

Moss mires. Wet, moss-dominated vegetation. In the maritime Antarctic and on some sub-Antarctic islands, waterlogged habitats around lake, pool and stream margins, seepage slopes below melting icefields and glaciers and in rock depressions, may be dominated by several moss species that form mire communities. Including *Sanionia uncinata*, *Warnstorfia fontinaliopsis* and *W. sarmentosa*, they occur typically near sea level and close to the coast where the underlying glacial till has a high moisture and nutrient regime, and may periodically be inundated by melt water. On deeper mineral soils in the maritime Antarctic, permafrost is usually present 50–100 cm below the surface. Such wet moss-dominated soils are usually mildly acidic, becoming mildly alkaline in areas of sandstone and calcareous rocks. Mires are often a favourable habitat for a diverse and active microflora and microfauna. Decomposition of the moss shoots is relatively rapid (up to 25 per cent dry weight loss per annum). This results in only slight **moss peat** accumulation, comprising a thin mantle of decaying moss shoots seldom deeper than 15–20 cm (6–8 in). The lower, moribund layer of shoots is yellowish and usually smells of hydrogen sulphide, indicating chemical-reducing action. (RILS)

Moss peat. Peat-like soil developed from the accumulation of partially decomposed moss. In the Antarctic region, continuous stands of moss are rare. They are restricted to coastal localities where, during the brief growing season (two months in continental Antarctica, up to four months in the maritime Antarctic), daytime temperatures rise marginally above 0 °C and moisture is briefly available. Only a few species are involved in the development of moss peat. In continental regions *Bryum argenteum*, *B. pseudotriquetrum*, *Ceratodon purpureus* and *Schistidium antarctici* accumulate up to 10 cm (4 ins) depth of undecayed moss, though rarely over areas exceeding 100 m² (120 sq yards). Much deeper and more extensive accumulations, occasionally covering up to 0.5 ha (1.23 acres) or more, occur in the milder and wetter conditions of the maritime Antarctic. Here, annual growth of shoots varies from 1–2 mm (0.1 in) in dry-habitat (xeric) mosses and 5–15 mm (up to 0.6 in) in wet-habitat (hydric) mosses. Because of the low summer temperatures, bacterial decomposition is very slow: moss stands lose only 1–2 per cent of mass per year, thus accumulating organic matter faster than it decays. The result is deposits of acidic moss peat, achieving depths of up to 2 m (6.5 ft), and in one locality an exceptional 3 m (9.8 ft).

The deepest moss peat deposits occur in the South Orkney and South Shetland Islands and northwestern Antarctic Peninsula. Two species of moss, *Chorisodontium aciphyllum* and *Polytrichum strictum*, with an upright growth habit, give rise to *moss turf* – banks of dense moss. These dominant mosses do not extend into the drier and colder southern Peninsula region. Deposits develop, not by the infilling of wet hollows, but by the upward growth of the tall and compact turf-forming moss shoots, which build up raised mounds of fibrous moss peat, capped by living moss and associated epiphytic lichens. In several places

Moss turf bank composed entirely of two mosses (*Chorisodontium aciphyllum* and *Polytrichum strictum*); the peat accumulation is 1.8-2 m thick, and stands above the level of the local terrain. Peat from the base of the bank has been radiocarbon dated as ca. 5500 years old. Signy Island, South Orkney Islands. [Photo. R.I. Lewis Smith]

peat moss mounds are large enough to become landscape features. Marginal erosion caused by seasonal snow drifts and freeze–thaw activity often causes vertical edges to form, but in general the banks are compact, their stability resulting from the dense root-like rhizoids of the mosses. These moss turf banks tend to remain permanently frozen except for the upper 25 cm (10 in), which thaws during the summer. They support very little animal life: the biomass of herbivorous invertebrates in moss peats is an order of magnitude lower than in most other Antarctic plant communities. Dove prions burrow into moss peat beds, and brown skuas often scoop out nests on the surface. *Chorisodontium*-dominated turfs, with their less-matted shoots, are looser and more easily damaged than the more compact banks.

Radiocarbon dating of moss peat from the base of the deeper mounds indicates ages of up to about 5500 years since development began. The milder, wetter climates and longer growing seasons on sub-Antarctic and cool temperate islands between 45° and 55°S permit comparatively diverse floras and soil faunas, in which true peats develop (see **Peats and peat formation; Moss mires**).
(RILS)

Moyes, Morton Henry. (1886–1981). Australian polar explorer. Born on 29 June 1886 at Koolunga, South Australia, he read mining engineering at Adelaide University, studying geology under Douglas Mawson. Prompted by Mawson, he joined the **Australasian Antarctic Expedition 1911–14** as meteorologist in the western party. In World War I he was commissioned in the Royal Australian Navy, but was seconded in 1916 to assist in relieving the Ross Sea party of the Imperial Trans-Antarctic Expedition. He returned again to Antarctica as cartographer during the first year of the **British, Australian and New Zealand**

Antarctic Research Expedition 1929–31. Moyes retired from the navy with the rank of instructor captain in 1946. He died in Sydney on 20 September 1981.

Mühlig-Hofmannfjella. 72°00'S, 5°20'E. (Mühlig-Hofmann Mountains). Range of ice-capped mountains between Gjelsvikfjella and Orvinfjella, part of a chain extending east-to-west between Fimbulheimen and Hellehallet, parallel to Prinsesse Astrid Kyst, central Dronning Maud Land. This range includes the highest mountains of the chain, with several peaks rising above 2800 m (9200 ft). It was photographed from the air by the **German Antarctic (*Schwabenland*) Expedition 1938–39**, and named by Norwegian cartographers for an official of the German Air Ministry.

Mule Peninsula. 68°39'S, 77°58'E. Rocky peninsula in Vestfold Hills, Princess Elizabeth Land. The peninsula includes 'Marine Plain', an important locality for vertebrate fossils, including a unique fossil dolphin, and 'Burton Lake', formerly an arm of the sea. An area of approximately 19.9 km^2 (7.8 sq miles) is designated SSSI No. 25. (Marine Plain) to safeguard research at this site.

Mulock, George Francis Arthur. (1882–1963). British naval officer and polar explorer. Born in 1882, he qualified in marine surveying while serving in HMS *Triton*. In 1902 he joined the **British National Antarctic Expedition 1901–4**, initially as sub-lieutenant in the relief ship *Morning*, but in 1903 taking the place of Ernest Shackleton in the shore party. In November 1903 he participated in a ten-week southern journey to Barne Glacier. Mulock served with distinction in the Royal Navy in World War I. Retiring in 1920, he became Marine Superintendent for the Asiatic Petroleum Co. in Shanghai. He saw further naval service in World War II. Mulock died on 26 December 1963. (CH)

Murphy, Robert Cushman. (1887–1973). American ornithologist. Born in Brooklyn, New York on 29 April 1887, he became an assistant at the American Museum of Natural History and student at Brown University, where he graduated in biology in 1911. Appointed curator of mammals and birds, he embarked in 1912 in the whaling brig *Daisy* for a year-long cruise to the Southern Ocean and South Georgia, collecting specimens and information. From this arose his long-term interest in Southern Ocean seabirds, and an informal but extremely informative book on the natural history of South Georgia (1947). Though Murphy himself returned only briefly to southern waters (including a 1960 guest visit to McMurdo Sound), he sponsored a succession of expeditions on behalf of the museum that resulted in an unparalleled collection of seabird skins and data. Results were summed up in his two-volume monograph 'Oceanic birds of South America' (1936): a liberal interpretation of 'South America' extends it to Antarctica and many of the southern islands. He remained with the museum throughout his life, serving as Curator Emeritus from 1955. He died on 19 March 1973.
Further reading: Murphy (1936), (1947).

Murray Harbour. 64°21'S, 61°35'W. Harbour on the north coast of Murray Island, Danco Coast, Graham Land, used by whalers for shelter during the 1920s. The name commemorates Sir John Murray, the scientist who organized the collections and reports of the *Challenger* Expedition.

Mushroom Island. 68°53'S, 67°53'W. A small ice-covered island lying 16 km (10 miles) west-southwest of Cape Berteaux, off the Fallières Coast of Graham Land. Charted by the **British Graham Land Expedition 1934–37**, and named for its mushroom shape, it marks one of the southernmost points at which Antarctic hairgrass *Deschampsia antarctica* is found.

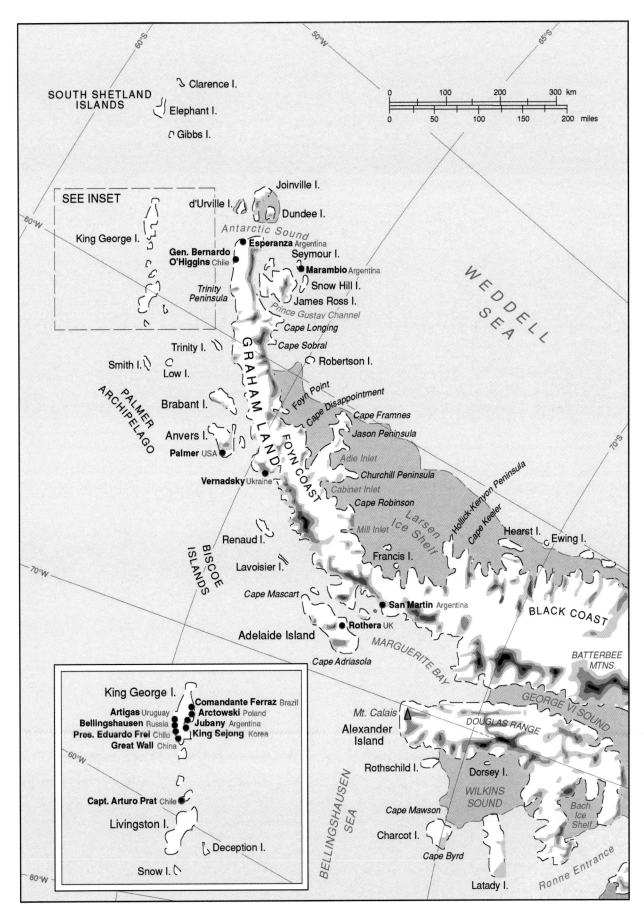

Plate 1 The Antarctic Peninsula

N

Nansen, Fridtjof. (1861–1930). Norwegian polar explorer and statesman. Trained in zoology, Nansen made his first Arctic voyage to Spitsbergen in 1882, and subsequently explored Greenland and the Arctic basin. Though never in the Antarctic region, he strongly influenced southern polar exploration by his methodical approach to techniques of expedition camping, travelling and ship-handling. His carefully designed sledges, cookers and tents were used by many subsequent expeditions in both polar regions: his ship *Fram*, designed for use in the Arctic, proved equally successful as an expedition ship and platform for oceanic research in southern waters. Nansen's professional approach to exploration strongly influenced Roald Amundsen, who in turn provided a role model for Byrd and other, later Antarctic explorers.

Nansen Island. 64°35′S, 62°06′W. A large island in Wilhelmina Bay, Danco Coast of Graham Land. Charted by the Belgian Antarctic Expedition 1897–99, it was named for the Norwegian Arctic explorer Fridtjof Nansen. With neighbouring Enterprise Island it provided shelter and safe moorings for early twentieth-century whalers.

Napier Mountains. 66°30′S, 53°40′E. A scattered group of peaks and nunataks rising to 2300 m (7544 ft) in northern Enderby Land. Discovered in January 1930 during flights by the British, Australian and New Zealand Antarctic Research Expedition 1929–31, they were named for J. M. Napier, a South Australian Supreme Court judge.

Nares, George Strong. (1831–1915). British naval explorer. Born in Aberdeen, he entered the Royal Navy in 1846. In 1852 he was second-in-command of HMS *Resolute*, a ship engaged in the search for the Franklin Arctic expedition, on a voyage that took him far into the pack ice of the Canadian Arctic. In 1872, now a captain, he commanded HMS *Challenger* during the first years of a four-year oceanographic voyage (see **Challenger Expedition 1872–76**), in the course of which he crossed the Antarctic Circle south of the Indian Ocean. Nares was recalled to command an Arctic expedition of two ships, HMSs *Discovery* and *Alert*, which attempted unsuccessfully to reach the North Pole. He was knighted and promoted to rear admiral in 1892.

Nash Range. 81°55′S, 162°00′E. A range at the southern end of Churchill Mountains, flanking the Shackleton Coast, between Dickey and Nimrod glaciers, named for New Zealand statesman Walter Nash (1882–1968).

National expeditions. Until the late 1930s most Antarctic expeditions were non-governmental, i.e. they were organized by individuals and scientific bodies with varying levels of government support. Planned as short-term campaigns, most extended over two summers and the intervening winter: a few remained in the field for a second winter and third summer. A notable exception is the Argentine meteorological station, **Orcadas**, on the South Orkney Islands. Taken over by the Argentine government as an operational station from the **Scottish National Antarctic Expedition 1902–4**, it has since operated continuously with annual changes of personnel, providing the longest unbroken meteorological record for the whole Antarctic region.

In 1939 the United States set up a government-run operation, the **United States Antarctic Service Expedition 1939–41**, to establish the first-ever long-term scientific programme on the Antarctic continent. In 1940 East Base on Stonington Island, Fallières Coast, and West Base at Little America, on the Ross Ice Shelf, became the first two stations to be set up with this intent. However, the threat of World War II overtook the programme: after a single winter both stations closed in 1941, and the USAS was disbanded.

In 1943–44 the UK government embarked on a similar enterprise, **Operation Tabarin 1943–45**, using ships of the Royal Navy to establish long-term stations in the South American sector of Antarctica. The intention was mainly to confirm Britain's political claim to the area in the face of Argentine and Chilean counter-claims and US interest. With Orcadas, these bases became the Antarctic region's first permanent settlements. Soon after the end of World War II, responsibility for Operation Tabarin was transferred to the

British Colonial Office, and transformed into the **Falkland Islands Dependencies Survey**, which continued the policy of establishing permanent stations.

Argentina extended its national Antarctic operation to the Peninsula region, setting up **Melchior**, its second permanent station, on Gamma Island, Melchior Islands, in January 1947, and six further stations by 1951. Chile too set up its national operation, establishing Arturo Prat Station on Guesalaga Peninsula, Greenwich Island, South Shetland Islands, in February 1947, **General Bernardo O'Higgins** at Cape Legoupil, Trinity Peninsula in 1948, and **Presidente González Videla** in 1951.

France established a national presence in Terre Adélie with Port Martin in 1950, replacing it (after a disastrous fire) with **Dumont d'Urville** in 1952. Australia built **Mawson** in 1954, as the first of three stations in its extensive Antarctic territories. Thereafter national expeditions and long-term stations became the norm. This policy has been encouraged by the terms of the **Antarctic Treaty**, under which contracting parties establish and maintain their credentials for membership by 'conducting substantial research activity there, such as the establishment of a scientific station'. Establishing a station has become an accepted, though not essential, qualification for an aspiring state to join the Treaty.

National expeditions in the Peninsula region vied with each other to establish 'refuges' and summer-only stations as well as wintering stations, and to set up navigational beacons and other indications of management. All these became manifestations of increased presence and effective occupation, helping to substantiate political claims. Over the years many of the early stations, refuges and beacons fell into disrepair or became derelict, reproaching the chauvinism that created them, and putting expeditions to the considerable trouble and expense of refurbishing or removing them. See **Study guide: National Interests in Antarctica**.

Neko Harbour. 64°50′S, 62°33′W. Bay in the ice cliffs of the Andvord Bay, Danco Coast, Graham Land, named for the whale factory ship *Neko* of the Salvesen fleet, which moored there for many seasons in the 1920s.

Nelson, Andrew Laidlaw. (1904–58). Polar navigator and hydrographer. Born in Scotland, he joined the merchant navy and, between 1929 and 1935, served as navigator, first officer and finally captain of *Discovery II*. He was responsible for surveys in the South Sandwich, South Orkney and South Shetland Islands, and many hydrographic surveys of the Southern Ocean. After service in the Royal Navy in World War II he retired with the rank of captain, and joined the oceanography staff of Lamont Geological Observatory, Columbia University. He died on 26 August 1958.

Nelson Island. 62°18′S, 59°03′W. An ice-capped island 19 km (12 miles) long, southwest of King George Island in the South Shetland Islands. It was charted by Edward Bransfield in 1820. The name may commemorate a sealing ship that operated locally about that time.

Neptune Range. 83°30′S, 56°00′W. Central range of the Pensacola Mountains, southern Coats Land. It includes several peaks of over 1500 m (4920 ft) and the massive west-facing Washington Escarpment, through which glaciers drain into the Foundation Ice Stream.

Neumayer Channel. 64°47′S, 63°27′W. Narrow passage between Wiencke Island and Anvers Island, Palmer Archipelago. The channel was noted by the **German Whaling and Sealing Expedition 1873–74**, but first navigated and charted by the **Belgian Antarctic Expedition 1895–98**. It was named for the German geographer Georg von Neumayer.

Névé. See **firn**.

New College Valley. 77°14′S, 166°23′E. Ice-free valley rising from Caughley Beach, close to Cape Bird, Ross Island, notable for its luxuriant algae, mosses and lichens and associated soil microflora and microfauna. An area of approximately 0.1 km^2 (0.04 sq miles) is designated SPA No. 20 (New College Valley), to maintain it inviolate as a control area for neighbouring SSSI No. 10, which surrounds it on three sides.

New Harbour. 77°36′S, 163°51′E. Bay between Cape Bernacchi and Butter Pt., Victoria Land, discovered and first used during Robert F. Scott's **British National Antarctic Expedition 1901–4**. The harbour provides a way in to Taylor Valley, one of the dry (ice-free) valleys of Victoria Land.

New ice. Recently formed floating ice. Earlier stages identified in ice formation include **frazil ice**, **grease ice**, slush, **shuga**, **ice rind**, **nilas** and **pancake ice**.

New Zealand Cape Expeditions 1941–45. A wartime operation to report on possible enemy activity in New Zealand's southern islands. Concern that German raiders were using harbours on some of the islands led the New Zealand Government in 1941 to deploy parties of three to five observers in small, hidden huts, two in the Auckland Islands at Port Ross and Carnley Harbour, and one at Perseverance Harbour, Campbell Island. Equipped

with radio, the observers watched for shipping through the daylight hours, also making regular meteorological observations. Though traces of earlier activities were found, no enemy shipping was reported. The Auckland Islands' stations were closed in 1944 and 1945. The meteorological work of the Campbell Island station was continued after the war; it became a meteorological station that operated continuously until its closure in 1995. The expeditions involved several naturalists, who later contributed reports to the series *Scientific Results of Cape Expedition Bulletins 1941–45*, published by the New Zealand Department of Scientific and Industrial Research.
Further reading: Eden (1955); Fraser (1986).

Nickerson Ice Shelf. 75°45′S, 145°00′W. Extensive ice shelf off Ruppert Coast, Marie Byrd Land, fed from the Siemiatkowski Glacier and locked in place by Newman Island. The shelf is named for Cdr H. J. Nickerson USN, a staff officer during Operation Deep-Freeze 1966.

Nightingale Island. 37°25′S, 12°29′W. An island 2.5 km (1.6 miles) long and 1.5 km (0.9 miles) wide, rising to 400 m (1312 ft) some 38 km (24 miles) south-southwest of Tristan da Cunha, in the South Atlantic Ocean. Nearby are the much smaller Stoltenhoff Island and Middle Island, together with many stacks and isolated rocks. Oldest and most eroded of the Tristan group, the islands are composed of lava and trachyte, rugged and rimmed by cliffs and narrow beaches. Climate and vegetation are similar to those of nearby Inaccessible Island. Seabirds including rockhopper penguins and two species of albatross nest in abundance: the main island has a breeding population of great shearwaters estimated at several millions. Sub-Antarctic fur seals breed on some of the beaches. There is no permanent human habitation. The islands are protected as nature reserves, though Tristan islanders are allowed several visits per year to collect eggs, young birds and guano.

Nilas. Thin pliable floating ice, consolidating from **grease ice** in the formation of **new ice**. Up to about 10 cm thick, it is solid enough to raft under lateral pressure, though unlikely to support the weight of a man on skis.

Nilsen Plateau. 86°20′S, 158°00′W. Ice plateau rising to 3940 m (12 900 ft) between Amundsen and Scott glaciers, Queen Maud Mountains, named for Capt. T. Nilsen, Norwegian master of the expedition ship *Fram*.

Nordenskjöld Coast. 64°30′S, 60°30′W. Part of the east coast of Graham Land, Antarctic Peninsula, facing the Weddell Sea, from which it is separated by the northern section of Larsen Ice Shelf. Bounded by Cape Longing in the north and Cape Fairweather in the south, the coast consists mainly of steep mountains rising to the central Graham Land plateau, with glaciers descending and merging into the ice shelf. It was sighted during the Dundee and Norwegian whaling expeditions of 1893, and charted by Otto Nordenskjöld, leader of the **Swedish South Polar Expedition 1902–4**, for whom it is named.

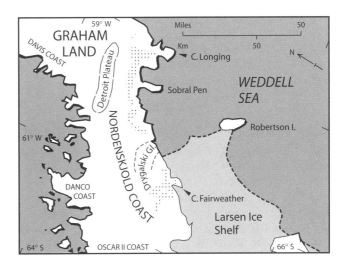

Nordenskjöld Coast

Normanna Strait. 60°40′S, 45°38′W. Passage between Signy Island and Coronation Island, South Orkney Islands, named for the factory ship *Normanna*, of the Normanna Whaling Co., based in Sandefjord, Norway, that operated in the area for three seasons in 1912–15.

Norsk Polarinstitutt. Norwegian scientific institution based in Tromsø, forming a government-funded centre for polar research. The Institute was founded in March 1948 under the Industridepartementet (Ministry of Commerce), as a successor to the Norges Svalbard-og Ishavsundersøkelser, with a remit to include Antarctic regions in its sphere of influence. The first director was H. U. Sverdrup. The Institute publishes *Norsk Polarinstitutt Skrifter* and *Norsk Polarinstitutt Meddelelser*.

Norway Station. 70°30′S, 2°32′W. Norwegian International Geophysical Year research station on Fimbulisen, Kronprinsesse Märtha Kyst, operated from January 1957 to December 1959.

Norwegian Antarctic (*Brategg*) Expedition 1947–48. Marine biological and hydrological expedition to the South Pacific Ocean. Funded by the Federation of Norwegian Whaling Companies, the expedition was commanded by Capt. Nils Larsen. The ship, M/S *Brategg*, was a cargo vessel of 500 tonnes equipped with an additional sounding winch for hydrological work, carrying a crew of 17 and four scientists led by zoologist H. Holgersen. *Brategg* left Sandefjord, Norway, on 22 October 1947 and passed through the Strait of Magellan to the Pacific Ocean. Hydrological observations were conducted at three transects along longitudes 90°W, 120°W and 150°W and a series of stations along the edge of the pack ice. On 10 February 1948 the ship visited Peter I Øy, taking observations and soundings close by. *Reports on the Scientific Results of the Brategg Expedition 1947–48* cover an expedition narrative and observations on fishes, ornithology and hydrology.

Norwegian-British-Swedish Antarctic Expedition 1949–52. An international expedition that undertook a two-year research and survey programme in Dronning Maud Land. Conceived by the Swedish glaciologist H. W. Ahlmann during World War II, the expedition was planned during the immediate post-war years, mainly at the **Norsk Polarinstitutt**, Oslo, but sponsored jointly by major scientific bodies in each of the three countries. The area chosen for study, western Dronning Maud Land, had not previously been explored from the ground, though parts had been photographed from the air by the pre-war **Norwegian (Christensen) Whaling Expeditions 1926–37** and the **German Antarctic (*Schwabenland*) Expedition 1938–39**. The main objectives of the expedition were topographic and seismic survey of the glacial cover, geology and meteorology. The leader was Capt. **John Giæver**, a Norwegian air force reserve officer with Arctic experience.

The expedition ship *Norsel* (Capt. Guttorm Jakobsen), a converted sealer, left Oslo on 17 November and London on 23 November 1949, calling at Cape Town *en route* south. Much of the equipment, including 60 sledge dogs and three weasels (tracked snow vehicles) travelled south in the whaling factory ship *Thorshøvdi*, and was transferred to *Norsel* at the edge of the Scotia Sea pack ice. With the help of two Royal Air Force reconnaissance Auster floatplanes the expedition landed at a bay, Norselbukta, in the ice cliff coast close to Kapp Norvegia, Kronprinsesse Märtha Kyst, and established a station, Maudheim, 3 km (1.9 miles) inland. *Norsel* departed on 20 February 1950, leaving a party of 15 to over-winter. Maudheim was occupied for two winters and three summers. On 26 July 1950 the return of the sun made it possible to start field activities. From 8 September two dog teams reconnoitred a route inland for the weasels, which through October and November established depots, and an advanced base 140 miles east-southeast of Maudheim. From this in both seasons sledging parties fanned out into the neighbouring mountains, undertaking glaciological, geological and seismic surveys. *Norsel* returned in January 1951 to resupply the station, which by this time had almost disappeared under accumulated snow. A Norwegian air unit with two aircraft made a number of local flights, severely hampered by bad weather. On 24 February 1951 three members of the party were killed when one of the weasels drove over the edge of the ice cliff in fog. During the 1951–52 field season the expedition accomplished detailed surveys of the inland ice and mountains as far east as the Sverdrupfjella and southeast onto the ice cap 600 km (375 miles) from Maudheim. An important aspect of their work was the first chain of seismic soundings showing the complex and mountainous topography under the ice. *Norsel* returned on 22 December with a Swedish air unit. Both ship and aircraft took advantage of good weather to survey the coast, and the aircraft took survey photographs far inland. By 6 January 1952 all the field parties had returned to Maudheim, and the expedition left on 14 January. *Further reading*: Giæver (1954).

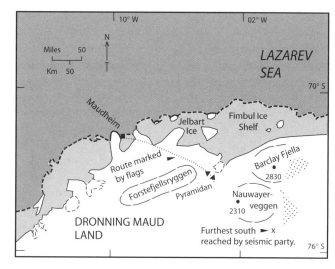

Area of western Dronning Maud Land explored by the Norwegian-British-Swedish Antarctic Expedition

Norwegian (Christensen) Whaling Expeditions 1927–37. Annual expeditions in Antarctic waters organized by Consul **Lars Christensen**, making use of the ships of his whaling fleet. The constant search for new hunting grounds, free of actual or potential political and fiscal controls, led Christensen, a Norwegian whaling magnate, in 1927 to begin a series of research expeditions in the Southern Ocean, using ships of his whaling fleet. Prompted by British and French territorial claims in Antarctica, Christensen was particularly concerned to establish and promote Norwegian interests, and give Norwegian scientists opportunities to work in Antarctica. In 1927 he sent *Odd 1*, a whale catcher commanded by Capt. Eyvind Toft, to explore for whaling opportunities in the Bellingshausen Sea area. Toft visited Peter I Øy, but was unable to land,

and also investigated Palmer Archipelago, the South Shetland Islands and South Georgia. From the following season onward Christensen's ships worked in Antarctic waters, both whaling and undertaking a systematic programme of research.

For the 1927–28 season Christensen bought and refitted a wooden sealer, renamed *Norvegia*, to serve as a research vessel. During its first commission, commanded by Capt. Harald Horntvedt, *Norvegia* (1927–28) visited Bouvetøya, taking 800 fur seals, surveying the island and setting up a small hut with emergency supplies. The intention was for *Norvegia* to explore further south toward Antarctica, but the ship struck a rock and spent the rest of the summer under repair in South Georgia. Meanwhile Christensen's other ships assisted geologist Otto Holtedahl and biologist Ola Olstad to work in the South Shetland Islands and at Port Lockroy.

In 1928–29 *Norvegia*, commanded by Capt. Nils Larsen, revisited Bouvetøya with the intention of landing a meteorological station. However, the hut previously landed had disappeared, and no safe site for a new station could be found. Larsen searched unsuccessfully for 'Thompson Island' (see **Islands, non-existent**), then moved into the Bellingshausen Sea. He landed a party on Peter I Øy, making a quick survey and establishing a refuge hut. Later he searched for 'Dougherty's' or 'Swain's' Island, but found only deep water in the reported position.

In 1929–30 *Norvegia* (Capt. N. Larsen) undertook a more ambitious programme of exploration under the direction of an army aviator, Capt. **Hjalmar Riiser-Larsen**. The ship called first at Bouvetøya, where they erected another small hut, this time on nearby Larsøya. Then they met the Christensen whaling factory ship *Thorshammer*, which transferred two aircraft and other equipment. Returning to Bouvetøya, Riiser-Larsen spent three days taking aerial photographs and surveying the island. Moving on to Antarctica, they reached Enderby Land in early December 1929, making several exploratory flights. Riiser-Larsen landed on a small islet southwest of Cape Ann and claimed the area for Norway. On 14 January 1930, again close to Cape Ann, *Norvegia* met RRS *Discovery*, the ship of the **British, Australian and New Zealand Antarctic Research Expedition 1929–31**, engaged on a similar mission of exploration. After convivial exchanges of views, the expedition leaders agreed to continue their researches in opposite directions, the Norwegians exploring along the unknown coast to the west. Riiser-Larsen and his co-pilot F. Lützow-Holm discovered the stretch of ice-mantled coast that is now called the Kronprins Olav Kyst. Bad weather forced them north, but they returned south in about 8°W of the Greenwich meridian and continued westward, on 18 February making further landfalls along what is now the Kronprinsesse Märtha Kyst. Thus the 1929–30 expedition defined respectively the eastern and western ends of a new coastline – a sector of the Antarctic continent that Christensen called Dronning Maud Land for the Norwegian queen.

In October 1930 *Norvegia* (Capt. N. Larsen) left Cape Town on the fourth Christensen expedition, an eastward circumnavigation of Antarctica under the leadership of Major Gunnar Isaacson. They visited Bouvetøya, searched for 'Dougherty's Island' and the equally elusive 'Truls Island' and 'Nimrod Islands', and headed toward Dronning Maud Land, hoping to find more of the coast. Later in the season Riiser-Larsen replaced Isaacson as leader, and discovered from the air a new stretch of coast that is now Prinsesse Ragnhild Kyst. By mid-season much of the pack ice had drifted away from the land, and other ships of the Christensen fleet headed south to join in the search, making landfalls at different points along the newly-discovered coast and in Enderby Land. Capt. Klarius Mikkelsen in the catcher *Torlyn* landed at the southern end of Mackenzie Bay, naming his discovery 'Lars Christensen Land'. Part of Mac.Robertson Land in the Australian sector, this is now Lars Christensen Coast.

The season 1930–31 produced a glut of whale oil, much of which remained unsold. In 1931–32 most of the whaling fleet stayed in port. Activities resumed in 1932. Christensen's fifth expedition visited Tristan da Cunha and Gough Island and in early March 1933 Riiser-Larsen made an unsuccessful bid to land on and sledge along part of the new coast. In 1933 the expedition ship *Norvegia* was crushed by sea ice and sunk during an Arctic voyage, and Christensen took immediate steps to replace her with a new steel ship, *H. J. Bull*.

Operations were fully restored in the two seasons 1933–34. There was as yet no dedicated expedition ship, but the tanker *Thorshavn*, brought south to relieve the factory ships of their accumulated oil, circumnavigated the continent with Lars Christensen aboard. A floatplane piloted by Lt Gunnestad flew within sight of a further stretch of ice cliff in 85°E that is now the King Leopold and Queen Astrid Coast of Princess Elizabeth Land. Continuing eastward, he discovered an extensive ice shelf in 131°W, part of a hitherto undiscovered coast of Marie Byrd Land, West Antarctica.

The seventh expedition of 1934–35, again led by Capt. Mikkelsen in *Thorshavn*, similarly discovered what is now the Ingrid Christensen Coast (named for Christensen's wife), further east in Princess Elizabeth Land. On 20 February 1935 he landed in an ice-free 'oasis' area which he called Vestfold Hills, for the district of southern Norway that includes Sandefjord, home port of the Christensen expeditions. With Mikkelsen was his wife Karoline, who thus became the first woman to land on the Antarctic continent. The eighth expedition of 1935–36 featured the new ship, *H. J. Bull*, a strongly built whalecatcher of 500 tons, that made a whaling reconnaissance in the Amundsen Sea.

On the ninth and final expedition of 1936–37 Christensen sailed from Cape Town in *Thorshavn* on 28 December 1936, making for West Ice Shelf off the King Leopold and Queen Astrid Coast. From there between 25 and 28 January 1937, in flights by Viggo Widerøe and Nils Romnaes, the Ingrid Christensen Coast and the Lars Christensen Coast were photographed westward to 66°E. On 30 January Christensen made his first landing on Antarctica at Scullin Monolith. *Thorshavn* continued westward. On 31 January photographs were taken along the Mawson and Kemp coasts. By 4 February the ship had passed longitude 45°E, soon to be defined as the eastern boundary of a Norwegian sector of Antarctica, and lay off Kronprins Olav Kyst. Flights westward discovered a new length of coast running northeast–southwest, which was photographed and named for Prins Harald Kyst, the infant son of the Crown Prince and Princess. On the following two days more photographic flights were made over Prinsesse Ragnhild Kyst, and over an extensive range of inland mountains, later named Sør Rondane as a reminder of southern Norway's Rondane mountains.

The nine Christensen expeditions, undertaken as adjuncts to industrial operations, secured for Norway an unequivocal interest in the two remote islands of Bouvetøya and Peter I Øy, and the huge sector of Antarctic coast between Coats Land and Enderby Land. As a result of their visits and investigations, the Norwegian government felt justified in annexing Bouvetøya on 23 January 1928, Peter I Øy on 1 May 1931, and defining a Norwegian sector of Antarctica, Dronning Maud Land, between 20°W and 45°E, on 14 January 1939.

Further reading: Christensen (1935).

Norwegian (Sandefjord) Whaling Expeditions 1892–94. In 1892–93 the Norwegian whaling ship *Jason*, owned by Christen Christensen of Sandefjord, southern Norway and commanded by experienced northern whaler Capt. C. A. Larsen, made a solitary voyage to Antarctic waters in search of opportunities for whaling. Larsen explored the South Orkney Islands, then headed for Seymour Island and the western Weddell Sea. Landing on Seymour Island, he collected specimens of sedimentary rocks that included fossils – the first recovered from the East Antarctica mainland. He caught no whales, but secured enough seal oil to justify the cruise. Larsen impressed Christensen with the commercial possibilities for sealing and whaling, and returned in the following season, accompanied by two whaling colleagues, Capt. Julius Evensen in *Hertha* and Capt. Morten Pedersen in *Castor*. Leaving Sandefjord in August 1893, the three ships spent the southern summer exploring widely on both flanks of Antarctic Peninsula. Larsen penetrated to 66°S in loose pack ice of the western Weddell Sea, discovering land on the eastern flank of Graham Land that he named for the Norwegian king. Now called Oscar II Coast, this is lined by an extensive ice shelf named for Larsen himself, and includes the prominent Mount Jason, named for his ship. Continuing south to 68°10'S, he discovered what is now the Foyn Coast, named for the Norwegian whaler Svend Foyn. On returning northward he named Christensen Island (now Christensen Nunatak) for his owner. Meanwhile Evensen in *Hertha* explored the western flank of Antarctic Peninsula to 69°10'S.

Hertha, *Castor* and *Jason* met in Cumberland Bay, South Georgia, in early April 1894, preparing for the homeward voyage. Further south they had secured cargoes of sealskins and oil, mainly from the pack ice and beaches, and seen plenty of rorquals. Larsen and his colleagues quickly saw that South Georgia, by contrast, was well stocked with the slow-moving right and humpback whales that would immediately repay commercial exploitation. On returning to Sandefjord in July, Larsen lost no time in investigating possibilities for establishing a whaling station on South Georgia, the step that founded the southern whaling industry (see **Whaling in the Southern Ocean**).

Norwegian South Polar Expedition 1910–12. The expedition in which Roald **Amundsen** and his four colleagues became the first men to reach the South Pole. For several years Amundsen, already an accomplished Arctic explorer, planned to make a bid for the North Pole in 1910–11. Almost single-handedly he raised the money, assembled a group of reliable, experienced Arctic travellers, first-rate equipment and dog teams, and arranged with Nansen to use the explorer's specially built polar ship *Fram*. Three months before his scheduled date of departure (January 1910), news broke of two rival claims to have achieved the North Pole. Robert Peary's claim to have reached it on 6 April 1909 was closely followed by a claim from Frederick Cook to have got there on 21 April 1908, almost a year earlier. As one or other was likely to be right, Amundsen saw no reason to follow. In a swift and secret change of plans he determined to go to the South Pole instead. Delaying sailing until early August, he headed for Madeira as though still bound for the Arctic. There he notified his shipmates of the change of plans, wrote letters home to his sponsors, and, in a telegram sent by his brother, notified Capt. Scott (who was also bound for the South Pole) of his intention. From Madeira he sailed directly to Bay of Whales, on the Ross Ice Shelf, arriving in January 1911. He immediately set up his base hut Framheim and a group of store tents 3 km (2 miles) inland from the ice edge.

Using superb dog teams, Amundsen and his companions, all highly skilled in ice and snow travel, began immediately to set up a string of three large depots southward across the ice shelf. As Scandinavians, all were familiar with long

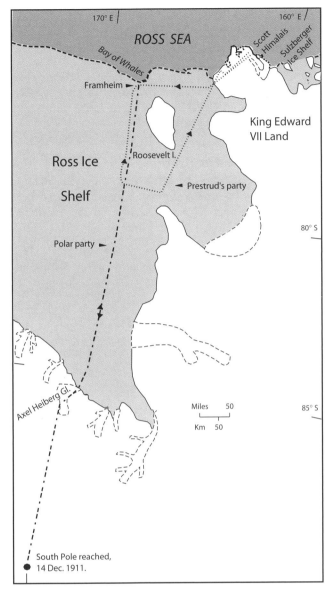

Norwegian South Polar Expedition 1910–12

winter nights: in preparing equipment and loads for the polar journey, the Antarctic winter passed quickly for them. On 8 September Amundsen made an early start for the Pole, but was forced back by dangerously low temperatures and strong winds. After a quick reorganization and a wait for better weather, they made a second start on 20 October. Led by Amundsen, the polar party included Olav Bjaaland, Helge Hassel, Helmer Hanssen and Oscar Wisting, plus four sledges, each hauled by a team of 13 Greenland huskies in excellent condition. Crossing the shelf ice from one depot cairn to the next, they maintained a daily distance of 32 km (20 miles), usually within about five hours, building intermediate cairns to guide their return journey. *En route* they killed ten of the dogs, feeding the meat to their companions. On 11 November the peaks of a magnificent range of mountains appeared ahead: Amundsen named the range for Queen Maud of Norway. On 17 November the party began to ascend the Axel Heiberg Glacier, a steep and by no means easy route through the mountains, named for a patron of the expedition. In five days they drove the dogs, with heavily laden sledges, a distance of 71 km (44 miles) over heavily crevassed ice, climbing some 10 000 ft (3000 m), through rough weather with strong, biting winds. At the top of the glacier they killed 24 more of the dogs, leaving 18 for the next stage – a rapid and relatively uneventful run across the polar plateau. On 14 December 1911 they reached the Pole.

Amundsen ensured that all five planted the Norwegian flag together. They spent three days at the Pole, determining the accuracy of their position. Knowing that they were first, they left a tent and messages for Scott's party. From there it was a relatively easy journey back along their track, across the plateau, down the glacier and across the ice shelf to Framheim, which they reached on 25 January 1912.

During the polar journey a second sledging party consisting of Hjalmar Johansen, Jörgen Stubberud and leader K. Prestrud, with two teams of seven dogs, left Framheim on 8 November 1911 to explore south, east and northward into King Edward VII Land. Only the cook, Lindström, remained at the base. The sledgers found themselves in an area of heavy snowfall: during an enforced lie-up of three days their tents were buried under more than 2 m (6.5 ft) of wind-packed snow. Only a few rock outcrops appeared toward the northeastern extremity of the journey: in 77°S, 154°W they visited for the first time a group of nunataks that Robert Scott had noted early in 1902 from the decks of RRS *Discovery*. Prestrud collected rocks and lichen specimens, naming the group Scott Nunataks in honour of their discoverer. The party returned to Framheim on 16 December. Meanwhile at sea the expedition ship *Fram*, commanded by Lt Thorvald Nilsen, had completed an extensive oceanographic cruise in the southern Atlantic Ocean, and was heading back across the Indian Ocean to Framheim. The expedition left Bay of Whales on 30 January, reaching Hobart on 7 February 1912.

Amundsen reached his main objective, the South Pole, with an efficiency born of his hard-earned Arctic experience. Though his small team at Framheim lacked scientific acumen, the two main sledging journeys broke new ground and provided useful geological information, and the winter cruise of *Fram* made full use of her capacity for oceanographic research. The surviving dogs returned almost immediately to Antarctica with Mawson's **Australasian Antarctic Expedition 1911–14**.

Further reading: Amundsen (1912).

Norwegian (Tønsberg) Whaling Expedition 1893–95. On September 20 1893 the whaling ship *Antarctic* (Capt. Leonard Kristensen), left Tønsberg for a voyage to the Ross Sea. Financed by the Norwegian whaling entrepreneur Capt. Sven Foyn, and led by H. J. Bull, a Norwegian businessman resident in Australia, the purpose of the expedition was to investigate possibilities for whaling in the sector of the Southern Ocean south of New Zealand and Australia. After a successful hunt for elephant seals on Iles Kerguelen, and a less successful winter hunt for whales off Campbell Island, *Antarctic* headed south from southern New Zealand in late November 1894. Caught for several weeks in the pack ice during December and early January, the ship broke free on 14 January 1895 and continued southward through open water into the Ross Sea, seeing a few whales but lacking the speed to chase them. On 22 January in 74°S *Antarctic* turned northward, and on 24 January Bull and a small party including Kristensen and seaman naturalist Carsten. E. **Borchgrevink**, landed at Cape Adare. At the time this was thought to be man's first landing on the Antarctic continent. After more desultory sealing and whaling, *Antarctic* returned to Australia on 12 March and to Tønsberg five months later. Commercially a failure, the voyage was of interest mainly in showing that the polynya discovered over 50 years earlier by James Clark Ross still afforded entry to high latitude in the Ross Sea, and in inspiring Borchgrevink, who later led the **British Antarctic (*Southern Cross*) Expedition 1898–1900** to winter in the same area. *Further reading*: Bull (1896).

Novolazarevskaya Station. 70°46′S, 11°50′E. Russian research station operating year-round in Schirmacheroasen, Prinsesse Astrid Kyst. Opened in early 1961, it replaced Lazarev, an earlier station on the coastal ice shelf 100 km (63 miles) to the north.

Nunatak. A rocky crag or mountain entirely surrounded by an ice sheet.

O

Oases. In the Antarctic context an oasis is a sizeable area (e.g. at least 10 km^2, 3.9 sq. miles) of ground that is free of permanent ice, though surrounded or limited by permanent ice sheets, or by the sea. Oases occur away from the mean tracks of depressions, which bring heavy precipitation where they cross the coast. In consequence there is little snowfall: local melting and ablation exceed precipitation, and no ice forms. Unlike tropical oases, they are arid and inhospitable to terrestrial flora and fauna, though they offer more possibilities for colonization than the contiguous ice sheets. Local conditions determine which of those possibilities are realized, providing for a wide range of ecological development. In topography they vary widely, from formerly glaciated valley systems to smooth coastal expanses with moraines. Some are crossed by seasonal glacial torrents, ending in outwash fans: some have extensive lake systems. Most show evidence of recent isostatic uplift, resulting in saline lakes, and raised beaches and benches. Where there is colonization, the newest areas are closest to the receding ice edge. Some currently show evidence of recent expansion, due to accelerated glacial retreat. Some of the largest show evidence of many changes over histories of several millennia.

Scientifically they are of great interest, first because they are areas of exposed rocks in a continent of which over 98 per cent is ice-covered, secondly because they reveal glacial and post-glacial processes, including weathering, soil formation and plant and animal colonization under extreme conditions of cold and aridity. Seven continental localities in particular have been studied: the inland **Victoria Land Dry Valleys**, and six coastal oases, **Bunger Hills**, **Langhovde**, **Schirmacheroasen**, **Thala Hills**, **Vestfold Hills** and **Windmill Islands**. For details, see the table below and individual entries. However, many other continental sites qualify for inclusion under the term. Some ecologists recognize as oases almost any ice-free land areas, including many in the maritime province, i.e. along the west coast of Antarctic Peninsula and on the southern islands of Scotia Arc. Individually these may be smaller than the classical continental oases: cumulatively they are much more extensive. Oases in the maritime area often have substrates of thin organic soils and patches of continuous vegetation, indicating possibly longer periods of exposure, certainly more favourable conditions for soil formation and plant recruitment and growth. Including them as oases emphasizes a factor common to all ice-free Antarctic areas: all are likely to have emerged relatively recently (i.e. within the past few thousand years or less) from under land ice, and are in active process of colonization by plants, animals and man.

Oasis areas that are open to the sea are likely to be colonized by penguins, other seabirds and seals, which appreciate the low precipitation and bare ground for resting and breeding. Perhaps representing the only ice-free ground for many kilometres on either side, they are colonized also by man: each of the seven sites listed has supported one or more research stations.
Further reading: Pickard (1986); Beyer and Bölter (2001).

Oates, Lawrence Edward Grace. (1880–1912). British soldier and polar explorer. Born in Putney, London on 17 March 1880, he was commissioned in 1900 in the Iniskilling Dragoons, a cavalry regiment. During the Boer War he saw active service in South Africa, and received a leg wound. Further service followed in Ireland, Egypt and India. After ten years of regimental life Capt. Oates volunteered to join the **British Antarctic (*Terra Nova*) Expedition 1910–13**. In Christchurch, New Zealand he took charge of the 19 Manchurian ponies, noting despondently their poor condition. With the Russian stable assistant Anton Omelchenko he cared for them throughout the expedition, until the final demise of the last ten during the march to the South Pole. Included in the five-man polar party, Oates suffered gravely from frostbite during the return from the Pole. On 16 March 1912, concerned that he was reducing his companions' chances of survival, he ended his life by walking out into the blizzard.

Oates Coast. 69°30′S, 159°00′E. Part of the coast of George V Land and Oates Land, East Antarctica, and bounded by Cape Hudson (Mawson Peninsula) in the west and Cape Williams in the east. Like the neighbouring Pennell Coast, it is deeply indented, backed by spectacular ranges of mountains, carved by sweeping glaciers and lined

Table 1 Seven Antarctic oasis areas

	Area (km²)	Highest point (m)	Stations	Location, topography
Victoria Land Dry Valleys	>1000	>2000	Vanda	West of McMurdo Sound. Deep glacier-cut valleys between tabular mountains.
Bunger Hills	1000	172	Oasis, Edgeworth David	Lakes, islands and low hills inland from Shackleton Ice Shelf.
Vestfold Hills	411	157	Davis	Ingrid Christensen Coast. Lakes, ponds, low hills.
Windmill Islands	400	75	Wilkes, Casey	Budd Coast. Low scattered islands and lakes.
Schirmacheroasen	50	200	Novolazarevskaya Georg Forster	Prinsesse Astrid Kyst. Lakes and low hills.
Ongul Islands	100	500	Syowa	Prins Olav Kyst. Rocky peninsulas and islands.
Thala Hills	10	>100	Molodezhnaya	Prins Olav Kyst. Low hills and lakes.

Source: Data from Pickard (1986), Bakayev (1966) and other sources. For details see individual entries.

by piedmont ice shelves. The eastern end of the coast was delineated in 1911 by Lt Harry Pennell RN, of the British Antarctic (*Terra Nova*) Expedition 1910–13, who named it for Capt. Lawrence Oates, one of the polar party of the expedition who died on the return journey from the South Pole.

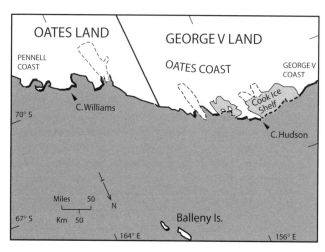

Oates Coast

Oates Land. 72°S, 159°30′E. Ice-covered and mountainous area of East Antarctica between meridians 153°E and 160°E, bounded to the east by Victoria Land and to the west by George V Land, and forming part of Australian Antarctic Territory. The land includes part of Oates Coast, Usarp Mountains, and an extensive ice plateau. The coast was surveyed by the **British Antarctic (*Terra Nova*) Expedition 1910–13**, and named, together with its hinterland, for expedition member Capt. L. E. G. Oates, who died with Robert Falcon Scott on the return from the South Pole.

Oazis Station. 66°16′S, 100°44′E. Soviet research station in Bunger Oasis, Knox Coast. Established in October 1956 as an International Geophysical Year station, it operated to November 1958. In the second season a Polish group joined the Russians: in January 1959 the station was transferred to Poland and renamed Antoni Dobrowolski. Since then it has been used intermittently for summer parties.

Ocean Harbour. 54°20′S, 36°16′W. Sheltered fjord harbour on the northern coast of South Georgia, named for the Ocean Whaling Co. which occupied a shore station there 1909–20.

Odbert Island. 66°22′S, 110°33′E. A small, rocky island of the Windmill Islands, Vincennes Bay, Budd Coast, noted for its rich avifauna, including nesting Antarctic petrels and Antarctic fulmars. Together with neighbouring Ardery Island, which is similarly endowed, it has been designated Specially Protected Area No. 3 (Ardery Island and Odbert Island). The total area is approximately 1.9 km² (0.75 sq miles).

Oddera, Alberto J. (d. 1965). Argentine naval officer, leader of the Argentine First Antarctic Naval Expedition, which in 1942 surveyed parts of the South American sector of Antarctica by sea and air. On behalf of the Argentine Government he took formal possession of the sector between 25° and 68°34′W, south of latitude 60°S. In 1947–49 he served as Naval attaché in London, and in 1956 became secretary of the Instituto Antártico Argentino. He died on 24 September 1965.
(CH)

O'Higgins Station. See **General Bernardo O'Higgins Station**.

Ohio Range. 84°45′S, 114°00′W. Western range of the Horlick Mountains, south Ellsworth Land. The highest peak, Mt. Schopf, rises to 2990 m (9807 ft). The range was surveyed in 1960–62 and 1961–62 by geologists of the Ohio State University Institute of Polar Studies, for which it is named.

Ohridiskii Station. See **St Kliment Ohridiski Station**.

Olympus Range. 77°29′S, 161°30′E. A range of steep-sided mountains behind Scott Coast, Victoria Land, almost entirely ice-free and overlooking the oasis area of dry valleys. The range was identified by New Zealand surveyors in 1958–59.

Omelchenko, Anton. (1883–1932). Ukrainian polar explorer. Born in Bat'ki, near Poltava, Ukraine into a peasant farming family, he became a groom on a local estate, eventually excelling as a jockey. In 1909, employed as a jockey in Vladivostok, he was recruited by Cecil Meares to help in selecting Manchurian ponies for the **British Antarctic (*Terra Nova*) Expedition 1910–13**. With fellow-countryman Dmitrii **Girev**, who was escorting 33 sledge dogs, he helped in transporting the ponies to Christchurch, New Zealand and sailed with them to Antarctica. As assistant to 'Titus' Oates, who was in charge of the ponies, he proved a popular and hard-working member of the expedition, taking part in several sledging journeys in which the ponies were involved. Back in Russia he fought with the army in World War I, later with the Red Army, then returned to Bat'ki where he helped to establish a collective farm. He was killed by lightning in 1932.

Ommanney, Francis Downes. (1903–80). British polar marine biologist and author. Born in 1903, he graduated in zoology at the Royal College of Science, London. Appointed Assistant Lecturer at Queen Mary College, London in 1926, three years later he joined the staff of Discovery Investigations, serving on several cruises of RRS *Discovery II*. After service with the Royal Navy in World War II, he joined the Mauritius-Seychelles Fisheries Service and later became reader in marine biology at the University of Hong Kong. He wrote several books, including *South Latitude* (1938) covering some of his work with Discovery Investigations. He died on 30 June 1980.
(CH)

Ongul Island. 69°01′S, 39°32′E. A low, partly ice-capped island in Lützow-Holm Bay, discovered from aerial photographs by the Norwegian (Christensen) Whaling Expeditions, but named by Japanese scientists in 1957, to whom the shape suggested a fishhook. The eastern end of the hook, on which they established their research station Syowa, proved to be a separate island, now called East Ongul Island. The area including islands and neighbouring coasts is relatively ice-free, and regarded as one of Antarctica's **oases**.

Operation Deep-Freeze. See **United States Navy Antarctic Expeditions 1955–98**.

Operation Highjump. See **United States Navy Antarctic Development Project 1946–47**.

Operation Tabarin 1943–45. A World War II operation of the Royal Navy to establish permanent research stations in the **Falkland Islands Dependencies** sector of Antarctica. Up to the start of World War II, British presence in the Dependencies had been due mainly to **Discovery Investigations**, involving hydrographic or marine biological surveys, and the **British Graham Land Expedition 1934–37**, a two-year land-based operation. In 1943, following the wartime withdrawal of RRSs *Discovery II* and *William Scoresby*, and in response to reports of Argentine activities in the area, the British government decided to re-establish a presence. Following the precedent of the **United States Antarctic Service Expedition 1939–41**, it planned to install permanent land stations, that would demonstrate the continuity of occupation deemed necessary for supporting territorial claims. Combining naval with Colonial Office resources, Operation Tabarin (named somewhat obscurely for a Parisian night-club) involved HMS (formerly RRS) *William Scoresby* (Capt. V. A. J. B. Marchesi) and SS *Fitzroy* (Capt. K. A. J. Pitt) a cargo ship of 600 tons normally deployed between the Falkland Islands and South America. Under the command of Lt. Cdr. J. W. S. **Marr** RNVR, the expedition left Stanley on 31 January 1944. Arriving at Whalers Bay, Deception Island (South Shetland Islands) on 3 February, they established Base B in the barracks of the abandoned whaling station, with a party of four including a geologist and meteorologist. The ships proceeded toward Hope Bay, where Marr hoped to set up a larger main base, but were stopped by ice in Antarctic Sound. Exploring southward along the west coast of Antarctic Peninsula, they finally landed the base hut and stores at Port Lockroy, which became Base A. The ships departed for Stanley on 17 February, leaving a party of ten. The two stations quickly became operational, transmitting regular weather reports to the Falkland Islands, and undertaking local surveys which continued throughout the winter.

In late January 1945 *William Scoresby* and *Fitzroy*, accompanied by the wooden Newfoundland sealer *Eagle* (Capt. C. S. Sheppard), relieved Deception Island, and *Fitzroy* went on to relieve Port Lockroy. *Fitzroy* then headed eastward to Coronation Island, South Orkney Islands, to land materials for a third station (Base C) at Moreton Point, Sandefjord Bay. The hut was built in the middle of a large penguin colony, but never occupied except by enterprising penguins. *Eagle* meanwhile proceeded to Hope Bay, where Base D, a larger station, was established on a site close to the shore. From mid-February this was occupied by a party of 13, with enough dogs and equipment to support a considerable sledging programme. Thus three stations operated during the winter of 1945. In September 1945, when World War II ended, administrative responsibility for Operation Tabarin was transferred from the Admiralty to the Colonial Office, and the operation continued as the **Falkland Islands Dependencies Survey**.

Operation Windmill. See **United States Navy Second Antarctic Development Project 1947–48**.

Orca. See **Killer whale**.

Orcadas Station. 60°44′S, 44°44′W. Argentine year-round research station on Laurie Island, South Orkney Islands. Established in 1903 as a meteorological observatory by the **Scottish National Antarctic Expedition 1902–4**, the station was passed to the Argentine government in February 1904, and has been maintained continuously ever since. It provides the longest unbroken meteorological record of any Antarctic station.

Orde-Lees, Thomas Hans. (*c.* 1879–1958). British motor expert with the **Imperial Trans-Antarctic Expedition 1914–17**. Born in Germany of Anglo-Irish stock, he was educated in England and commissioned in the Royal Marines. A skier and physical fitness expert, he was seconded by the navy to Shackleton's expedition, and assigned to the Weddell Sea party, taking charge of the tractors (including some of his own design) and stores. After the expedition he joined the Royal Flying Corps, developing parachutes and ballooning. Retiring as a Lt Colonel, he lived in Japan, where he lectured at Kobe University, and New Zealand. He died in Wellington on 2 December 1958.

Orléans Strait. 63°50′S, 60°20′W. Passage between the northeastern end of Palmer Archipelago and Antarctic Peninsula; charted by the **French Antarctic Expedition 1837–40**, and named for Louis Philippe, King of France and former Duc d'Orléans.

Orne Harbour. 64°37′S, 62°32′W. Glacier-lined bay in the Danco Coast, Graham Land, used as a shelter and anchorage by Norwegian whalers, and probably named for ships of the Norwegian whaling company Ørnen.

Orville Coast. 75°45′S, 65°30′W. Part of the east coast of Ellsworth Land, West Antarctica. Facing the Ronne Ice Shelf, it is bounded by Cape Adams in the northeast and Cape Zumberge in the southwest. Backed by high mountains rising to the interior ice plateau, it presents huge glaciers that sweep down to merge with the ice shelf. The coast was delineated by the **Ronne Antarctic Research Expedition 1947–48**, and named for Capt. Howard T. Orville USN, of the Naval Aerological Service, who assisted in planning the expedition's meteorological programmes.

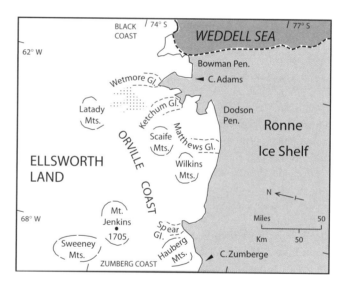

Orville Coast

Orvinfjella. 72°00′S, 9°00′E. (Orvin Mountains). Range of mountains extending for about 100 km (62 miles) between Mühlig-Hofmannfjella and Wohlthat Massivet, Prinsesse Astrid Kyst, central Dronning Maud Land. The range, which includes several peaks over 2500 m (8200 ft), was named for A. K. Orvin, a former director of the Norsk Polarinstitutt.

Oscar II Coast. 65°45′S, 62°30′W. Part of East Graham Land coast of Antarctic Peninsula, facing the Weddell Sea across a wide extent of Larsen Ice Shelf. Bounded by Cape Fairweather in the north and Cape Alexander in the south, it consists mainly of steep mountains rising to the central Graham Land plateau, with glaciers descending and merging into the ice shelf. The coast was sighted by Capt. C. A. Larsen in the **Norwegian (Sandefjord) Whaling Expedition 1893–94**, and named by him for the ruling Norwegian king.

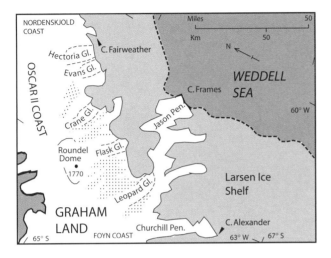

Oscar II Coast

Outlet glacier. A glacier that drains an inland ice field, usually discharging through a gap in a mountain range.

Oxygen isotope ratios. Oxygen occurs naturally in two stable forms, the normal isotope ^{16}O, and a slightly heavier isotope ^{18}O, each atom of which contains two additional neutrons. The two isotopes occur in varying proportions in the atmosphere, in water vapour and snow, and in all compounds of oxygen. Natural water contains about 99.8 per cent of ^{16}O and 0.2 per cent of ^{18}O, and this proportion is taken as standard. Because ^{18}O evaporates less readily than ^{16}O and is generally less chemically mobile, less of it in proportion to ^{16}O appears in the atmosphere, water vapour, snow and compounds of oxygen precipitated from solution in water. Differences in proportion of the two isotopes are called $\delta^{18}O$ (delta 18 O) and expressed in parts per thousand. Differences vary with the temperature at which the change from water occurred. For example, vapour evaporating from water (and the ice crystals that later form from that vapour), or salts precipitating from sea water to form animal shells, will have less ^{18}O in proportion to ^{16}O at lower environmental temperatures than at high, and correspondingly lower $\delta^{18}O$ ratios.

Thus the ratio of oxygen isotopes in a sample of snow or sea shell, carefully interpreted, is a measure of mean atmospheric or sea temperature at the time the sample was formed. Similarly, ice recovered from deep inside an ice sheet by coring will, on melting, release oxygen and carbon dioxide, representing samples of the environmental atmosphere at the time of the original snowfall. From both gases further isotope ratios and environmental indicators may be obtained.

Ozawa, Keijiro. (1922–71). Japanese polar oceanographer. Born on 28 January 1922, he studied fisheries at the Imperial Fisheries Institute. In 1955–57 he was First Officer in *Umitaka-Maru*, oceanographic research ship of Tokyo University of Fisheries, during its first Antarctic cruise. In subsequent southern oceans voyages of 1961–62, 1964–65 and 1966–67 he commanded the ship and supervised many aspects of the research. In 1968 he was appointed professor at the University of Fisheries. He died on 14 October 1971.

Ozone hole, Antarctic. Area of the stratosphere over the Antarctic region from which ozone is seasonally depleted, resulting in enhanced reception of solar radiation on the ground beneath. Ozone, a tri-atomic form of oxygen (O_3), is found almost exclusively in a shallow layer of atmosphere at heights from 12 to 25 km above Earth's surface, depending on latitude and season. It is formed when ultraviolet radiation (wavelengths less than 242 nm) splits O_2, releasing oxygen atoms that quickly attach themselves to other O_2 molecules. Constant production of O_3 is balanced by natural destruction, re-forming O_2.

Continuous observations of the amount of ozone within the column of atmosphere above Halley Station, Antarctica, revealed a sudden and dramatic disappearance of ozone each spring, from background levels of over 300 DU (see **Dobson units**) in the late 1950s to only 180 DU in 1984, when the discovery was first reported. Continuing observations showed a continuing fall, to values of around 40 per cent of normal throughout the 1990s. Further monitoring, from the ground, from aircraft and from stratospheric balloons, showed that, during the sunless southern winter, a vortex of cold, descending air forms in the stratosphere over Antarctica, sustained by strong westerly winds that separate the air over the continent from that at lower latitudes. Temperatures within the vortex fall below −80 °C, allowing icy clouds to form. Chlorine and other halogens, released from compounds including chlorofluorocarbons (CFCs), is present in the form of 'reservoir' molecules which, though not themselves reactive with ozone, may be partially activated through reactions on the cloud surfaces. Complete activation occurs when the sun returns in spring, and the chlorine radicals so produced react with and destroy ozone. Bromine released from other industrial halocarbons is similarly involved, to a lesser extent. Thus the vortex above Antarctica provides a localized area of stratosphere where ozone is destroyed suddenly and dramatically – the so-called 'ozone hole'. These conditions last until late spring or early summer, when the vortex breaks down and horizontal stratospheric circulation is resumed, bringing in ozone-rich air from lower latitudes.

The unusual conditions over Antarctica dramatize a more general situation over the rest of the world. Stratospheric ozone is being depleted by man-made chemicals, with loss of a shield against forms of high-energy radiation that are known in certain circumstances to cause cancers, and otherwise interfere with life processes. Discovery of the ozone hole led in 1987 to the adoption of the Montreal Protocol on Substances that Deplete the Ozone Layer, which has been strengthened in subsequent amendments.
(AJ)

P

Pack ice. Floating sea ice in the form of more or less consolidated floes. *Open pack ice* is made up mainly of separate floes with extensive leads between, giving a total ice cover of four to six tenths. *Close pack ice* is composed of floes mostly in contact, totalling seven-tenths to nine-tenths cover. In *consolidated pack ice*, the floes are frozen together, providing complete or near-complete cover.

Palmer, Nathaniel Brown. (1799–1877). American sealing master and navigator, possibly the first to see and explore Antarctic Peninsula. Born on 8 August 1799 in Stonington, Conn., the son of a shipbuilder, Palmer started his seagoing life as a ship's boy on New England coasters. At 18 he was given his first command, an inshore schooner. At 19 he sailed as second mate aboard the sealing brig *Hersilia* (Capt. J. A. Sheffield) to seek new sealing islands in waters south of Cape Horn. While replenishing stores in the Falkland Islands in 1819, Palmer may have persuaded his captain to follow a British sealing brig, *Espiritu Santo*, to the newly-discovered South Shetland Islands, where they secured a substantial cargo of skins. In the following season Palmer commanded a sloop, *Hero*, assisting a Stonington-based fleet of five brigs and two schooners in a return to the islands. Under the overall command of Capt. Benjamin Pendleton, *Hero*'s role was to carry skins and stores, and explore for new islands while the rest of the fleet hunted. In the course of a busy season, Palmer and Pendleton sighted land to the south of Deception Island, which Palmer visited, exploring both south and west but finding no seals in the area now called Palmer Archipelago. Returning to the South Shetland Islands he met the ships of the Imperial Russian Naval Expedition 1819–21 off Deception Island, and discussed his findings with the leader, Fabian von Bellingshausen. In a further search eastward, in company with the British sealer George Powell, on 7 December 1821 he participated in the discovery and exploration of the South Orkney Islands.

Palmer turned his attention to more profitable trading in northern waters, returning south for a final sealing voyage in the brig *Annawan* in 1829–30, in company with his brother Capt. A. C. Palmer in the brig *Penguin*. Thereafter he became a prosperous ship owner, contributing substantially to the design and development of clipper ships trading to Europe, the Antipodes and China. He retired from the sea in 1850, maintaining his interests in sailing, ship design and long-distance maritime routes. He died in San Francisco on a return journey from China in 1877.

Palmer Archipelago. 64°15′S, 62°50′W. Group of large ice-covered islands bordering Gerlache Strait, off the northeast coast of Graham Land, West Antarctica. The group, which includes Anvers, Brabant, Liège, Hoseason, Trinity and Tower Islands, was named for US sealer and explorer Capt. Nathaniel B. Palmer.

Palmer Coast. See **Davis Coast**.

Palmer Land. 71°30′S, 65°00′W. Southern part of Antarctic Peninsula, bounded in the north by Graham Land (along a line joining Cape Jeremy and Cape Agassiz), and in the south by a line joining Cape Adams and a point in 72°W on the English Coast. The western half is a broad, ice-covered plain sloping gently toward a shore formed by George VI Sound: the eastern half includes the mountainous, glaciated and deeply indented Wilkins, Black and Lassiter coasts. Forming part of British Antarctic Territory, the land was named for US sealing captain Nathaniel B. Palmer, who explored the northern Antarctic Peninsula area in November 1820.

Palmer Station. 64°46′S, 64°03′W. American year-round research station on the southern end of Anvers Island, Palmer Archipelago. Built close to the site of a former British research station, it was opened in 1965 and has since then operated continuously as the main US centre for Antarctic biological research.

Pancake ice. Small floes of new ice up to 3 m across, formed by fragmentation of grease ice, shuga or slush due to slight movement in the water. Movement rubs the floes together, rounding them and providing their characteristic raised edges.

Paradise Harbour. 64°51′S, 62°54′W. Extensive bay in the Danco Coast, Graham Land, sheltered by Lemaire and Bryde Islands. It was used and named by whalers in the 1920s, later becoming the site of Argentine station Almirante Brown and Chilean station González Videla.

Parker, Alton N. (*c*.1895–1942). US pilot and explorer. After enlisting in the US Navy in 1917, he flew as a pilot with the US Army Air Force, and in 1926 joined Byrd's expedition to the North Pole. He remained with the same leader for **Byrd's First Antarctic Expedition 1928–30**, flying as his pilot and sharing his discovery of the Edsel Ford Mountains, for which Parker was awarded the Distinguished Flying Cross. In 1930 he became a commercial pilot. He died in Miami, Florida on 30 November 1942.

Patterned ground. Polygonal patterns, circles or stripes formed in the active layer of **permafrost**. Water in the active layer of damp soil increases some 9 per cent in volume during freezing. Ice crystals, lenses and needles form then disrupt the soil structure: repeated freezing and thawing causes soil particles to be sorted into different sizes. Coupled with lateral movements due to expansion and contraction of soil blocks, this results in the formation of stripes on sloping ground and polygons on the flat. Soils with low water content, for example in the dry valleys, have no active periglacial processes, but show fossil polygons in which the ice has been replaced by sand.
(DWHW)

Patuxent Ice Stream. 85°15′S, 67°45′W. Major ice stream draining northward from the Patuxent Range of Pensacola Mountains, joining the Foundation Ice Stream and ultimately the Ronne Ice Shelf. It was identified by ground parties in 1961–62 and subsequently delineated from aerial photographs.

Patuxent Range. 85°15′S, 67°45′W. Southernmost range of the Pensacola Mountains, southern Coats Land. It includes several peaks of over 1000 m (3280 ft), with glaciers draining south and west into the Academy Glacier and Foundation Ice Stream.

Paulet Island. 63°35′S, 55°47′W. A roughly circular island about 1.6 km (1 mile) across, 4.8 km (3 miles) south of Dundee Island off the eastern tip of Trinity Peninsula. Made up of scoria cones on a platform of lavas, probably formed within the last 1000 years, it was discovered by Capt. J. C. Ross RN in 1842 and named for a fellow naval captain. The stone hut in which Capt. C. A. Larsen and the crew of *Antarctic* wintered in 1903 stands on a raised beach on the south side. The island is home also to one of Antarctica's largest colonies of Adélie penguins.

Peacock Sound. 72°45′S, 99°00′W. Broad ice-filled sound between Thurston Island and Eights Coast, Ellsworth Land, named for USS *Peacock*, one of the ships of the **United States Exploring Expedition 1838–42**.

Peale's (blackchin) dolphin. *Lagenorhynchus australis*. A species known mainly from cool South American waters, likely to be seen around the Falkland Islands. Peale's dolphins measure up to 2 m (6.5 ft) long. Distinguishing features are a tall recurved dorsal fin, prominent white chin, belly and flanks, with broad grey flashes extending along the sides of the thorax. They travel in small groups of up to four or five, and are seen mostly in shallow waters. Little is known of their biology.

Peats and peat formation. Peats are organic soils formed largely of accumulated, part-decayed plant material. Within the southern polar region, inland continental Antarctica is currently too dry and cold for closed stands of peat-forming mosses to survive. In coastal continental Antarctica, habitats receiving a regular supply of melt water occasionally support stands of a few species of moss extending to tens of square metres and capable of accumulating a thin mantle of undecayed moss shoots – proto-peat or **moss peat**. In the maritime Antarctic as far south as about 68°S, several turf- and carpet-forming mosses and, to a very limited extent the Antarctic hair grass (*Deschampsia antarctica*), accumulate similar types of peat moss. On well-drained hillsides the tall turf-forming mosses *Chorisodontium aciphyllum* and *Polytrichum strictum* accumulate peat to a thickness of 2 m (6.5 ft) or more. On wetter ground, mat- or carpet-forming mosses (*Sanionia uncinata*, *Warnstorfia* spp.) develop a thin mantle of peat seldom more than 15–20 cm (6–8 in) thick.

True peat is found only on islands north of the northern limit of pack ice, where milder, wetter climates and longer growing seasons encourage floral and microfaunal diversity. Bogs and mires dominated by grasses, rushes, mosses and liverworts occur on the poorly drained plains, valley floors and depressions. A relatively high rate of primary production, with slow decomposition and no permafrost, produces accumulations of true peat to depths of 5–6 m (16–20 ft) in suitable areas on most of the islands; Iles Kerguelen features deposits up to 10 m (33 ft) deep. Blanket bog formations normally develop a typical peat under conditions of waterlogging and de-oxygenation, inducing the production of humic acids which modify the mineral soils below. Peat at the base of several deep bogs has been radiocarbon dated at between 9000 and 10 500 years, suggesting that plant colonization and peat formation began shortly after the end of the last major glaciation. Palynological (pollen and spore) analyses of peat cores on several islands have indicated minor changes in the abundance of the principal plant species in response to fluctuations in climate, but there is no evidence of change in floristic composition during the past 10 000 years.

Fossil deposits testify to peat formation in Antarctica, in both the remote and the nearer past. Petrified peat deposits containing well-preserved vascular plant remains of Permian age (about 260 million years ago, just before the break-up of the Gondwana supercontinent) appear in association with coal deposits in the Transantarctic Mountains. In the same area accumulations of partially fossilized mosses, dated possibly as recent as two million years ago, indicate a peat-forming flora associated with swamps on former glacial outwash plains.
(RILS)

Pedro Vicente Maldonado. 62°27′S, 59°43′W. Ecuadorian research station, built in February 1990 at Spark Point (incorrectly called 'Fort William'), on the southern coast of Greenwich Island, South Shetland Islands. The station has been occupied by summer parties.

Pendleton Strait. 66°00′S, 66°30′W. Passage between Rabot and Lavoisier Islands, Biscoe Islands; named for US sealer Capt. Benjamin Pendleton, whose sealing fleet explored the area early in 1821.

Penfold, David. (1913–91). British polar hydrographer. Born in 1913, he joined the Royal Navy in 1938 and served throughout World War II. As a hydrographic officer in successive naval cruises between 1948 and 1952 he made surveys of Deception Island, Port Lockroy, and Wiencke Island. Retiring in 1967, he served for nine years in the Hydrographic Office, Ministry of Defence. Penfold Point, Deception Island is named for him. Penfold died on 20 April 1991.
(CH)

Penguin Island. 62°06′S, 57°54′W. An oval island about 1.6 km (1 mile) across, off Turret Point, King George Island, in the South Shetland Islands. A red scoria cone, Deacon Peak (160 m, 525 ft), resting on a platform of lavas, and other features suggest volcanic activity within the last 300 years. Charted by Edward Bransfield in 1820, the island was named (as were many others) for its penguins, mostly nesting chinstraps and Adélies. Apparently the earliest to receive the name, it has been granted priority.

Penguins, southern oceanic. Flightless marine birds (order Sphenisciformes, family Spheniscidae), widely distributed in the southern oceans. Darkly feathered on head and back, with white undersides, penguins carry most of their distinguishing features on the head and neck. Wings are adapted as flippers for underwater propulsion, and used also in fighting. Feet trail in the water, combining with the tail to form a rudder. On land penguins walk upright over rocky or snow-covered surfaces: in soft snow they toboggan on their stomachs, pushing forward with feet and flippers. On hard snow or ice the strong claws become crampon spikes. The smaller species hop with feet together, and climb using bill, claws and flippers. Body size is related to underwater ability: the larger the penguin, the deeper and more prolonged its diving capability. Penguins reach their highest numbers and densest concentrations in cold temperate, subpolar and polar areas, where nine of the 18 living species (representing five of the six genera) occur as breeding birds. Most Antarctic, sub-Antarctic and temperate islands have several resident species (see **Birds, southern oceanic**). Most species produce clutches of two eggs in open nests of stones, moss, grass etc. Magellanic penguins raise two eggs in burrows; emperors and kings raise only single eggs, carrying them on their feet. Nine of the 18 living species breed within the Antarctic and southern oceanic regions: see entries for individual species.
Further reading: Williams, 1995.

Young gentoo penguins. Photo: BS

Pennell Coast. 71°00′S, 167°00′E. Part of the north coast of Victoria Land, East Antarctica, bounded by Cape Williams in the west and Cape Adare in the east. It is a deeply indented coast, backed by spectacular ranges of mountains, carved by sweeping glaciers and lined by piedmont ice shelves. The narrow coast consists mainly of ice-covered peninsulas and islands lined by the wide ice shelves. It was charted in 1911 by Lt Harry Pennell RN, on

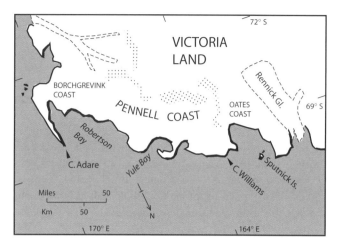

Pennell Coast

one of the voyages of the **British Antarctic (*Terra Nova*) Expedition 1911–14**, and subsequently named for him.

Penola Strait. 65°10'S, 64°07'W. Passage between the Argentine Islands, Petermann Island and Hovgaard Island and the Graham Coast, named by the British Graham Land Expedition 1934–37 for the expedition ship *Penola*. This was also the name of the family sheep farm in Australia of the leader, John Rymill.

Pensacola Mountains. 83°45'S, 55°00'W. Narrow, elongate range of mountains trending north-northeast to south-southwest at the southern end of Filchner Ice Shelf, southern Coats Land. Over 400 km (250 miles) long, they include, from north to south, the Forrestal, Neptune and Patuxent ranges, and the isolated Dufek Massif. The mountains were named for the United States Naval Air Station, Pensacola, Florida, to honour its role in training naval aviators.

Periantarctic islands. A term used variously to describe (a) all the islands of the Southern Ocean surrounding Antarctica and related islands north of the Antarctic Convergence, or (b) islands that stand between the mean northern limit of pack ice and the Antarctic Convergence. To avoid possible confusion, the term is not used in this Encyclopedia.

Permafrost. Permanently frozen ground. Polar and alpine soils freeze in winter, and summers are too cold and short to allow them to thaw out completely. Only an upper, so-called 'active layer', thaws, leaving the lower layers permanently frozen. Permafrost ranges in depth from a few centimetres to many metres. In the Antarctic region within the northern limit of pack ice it probably occurs in all exposed soils. On Southern Ocean islands it occurs mainly close to glaciers. See **patterned ground**.
(DWHW)

Peter I Øy. 68°51'S, 90°35'W. An Antarctic island 19 km (11.9 miles) long and 11 km (6.9 miles) across, isolated in semi-permanent pack ice in the Bellingshausen Sea. With a central volcanic peak named for the whaling magnate consul Lars Christensen, rising to about 1750 m (5740 ft), the island is formed of successive flows of basaltic lava. Almost entirely ice-covered, it is ringed by steep cliffs with few accessible landing points. Very little vegetation, or wildlife other than breeding seabirds, has been reported. The Russian explorer Capt. Fabian von Bellingshausen discovered the island in 1821, naming it for the founder of his navy. To approach with certainty in most summers requires an icebreaker, however, ice-strengthened ships have reached it in exceptional summers when the pack ice has cleared. Not surprisingly, the island is seldom visited and little studied. Scientists have lived ashore for a few days at a time, but none has wintered there. Currently the most frequent visitors are tourist cruise ships attracted by the island's remoteness. The island was claimed by Norway on 1 May 1931 and is administered as a scientific reserve. Landing is allowed only under permit.

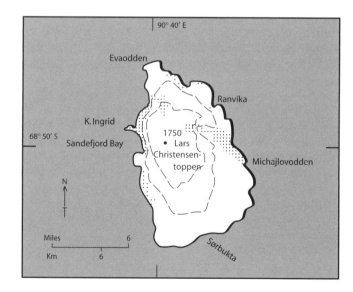

Peter I Øy. Isolated in Bellingshausen Sea, the island is mostly ice-covered and usually surrounded by pack ice. Height in m

Petermann Island. 65°10'S, 64°10'W. A small, ice-capped island about 1.5 km (1 mile) long off the northern Graham Coast, facing Penola Strait: part of the Wilhelm Archipelago. It was discovered by the German Whaling and Sealing Expedition 1873–74 and named for August Petermann, a distinguished German geographer. Port Circumcision, a small bay on the southeastern side, provided a safe over-winter mooring for the ship *Pourquoi Pas?* during J.-B. Charcot's second expedition. A cairn on top of the island is scheduled as an Historic Monument.

Peters, Nikolaus. German biologist and historian who, from the Reichstelle für Walforschung in Hamburg, became his country's leading authority on the whaling industry. His study of German whaling, *Der neue deutsche Walfang* (1938), includes information on pre-war German southern hemisphere whaling operations.

Petersen, Carl, O. (d. 1941). American polar photographer and radio operator. He took part in both of Byrd's Antarctic expeditions, in 1928–30 as a radio operator, and in 1933–35 as radio operator and photographer. He was involved in several sledging operations and climbed in the Edsel Ford Mts. He died on active service with the US Navy on 10 November 1941.

Petrel Station. 63°28′S, 56°17′W. Argentine permanent station at Petrel Cove, southern Dundee Island. The station began in 1952 as a refuge, which was expanded and reopened in February 1967 as a naval air station, and occupied continuously until 1977. Since then it has been used intermittently for summer parties.

Petrels, southern oceanic. Flying birds of the order Procellariiformes. All petrels are seabirds, with nasal salt glands and nostrils more or less encased in tubes. Most breed and live in the southern hemisphere, many in the cool-temperate and cold environments south of the southern oceans. All produce a single white egg, large in proportion to body weight, with long incubation periods and correspondingly slow chick growth. Petrels secrete crop oil, which both adults and chicks may eject defensively. Largest of the order are wandering and royal albatrosses, with wing spans of up to 3.7 m (12 ft), body length 1.2 m (4 ft) and weighing up to 8.2 kg (18 lb). The smallest are storm-petrels with wing span 40 cm (16 in), body length 32 cm (12.5 in), weighing 50–60 gm (2 oz). The order includes four families: Diomedeidae (**albatrosses** and **mollymawks**), Procellariidae (**fulmars, prions, gadfly petrels** and **shearwaters**), Hydrobatidae (**storm-petrels**) and Pelecanoididae (**diving-petrels**), all of which are well represented in Antarctic waters. *Further reading*: Warham (1990), (1996).

Phleger Dome. 85°52′S, 138°24′W. Dome-like mountain rising prominently to 3315 m. (10 873 ft) above sea level from Stanford Plateau, on Watson Escarpment, at the southern end of the Queen Maud Mtns. The dome is named for H. Phleger, a US delegate to early discussions on the Antarctic Treaty.

Pickersgill Islands. 54°37′S, 36°45′W. A small, isolated group of islands off the southern coast of South Georgia, discovered by Capt. Fabian von Bellingshausen in 1819. Bellingshausen thought that the largest of the group was an island named by Capt. James Cook for Lt. Pickersgill, one of his officers, and wrongly ascribed the name to this group. The island that Cook originally named is now **Annenkov Island**.

Piedmont glacier. The lower end of a valley glacier, expanded to form a plain of shallow glacier ice.

Pieter J. Lenie Station. 62°11′S, 58°27′W. US research station ('Copacabana') on the western shore of Admiralty Bay, established in 1985 and used mainly for summer studies of seabirds and elephant seals in SSSI No. 8.

Piked whale. See **minke whale**.

Pionerskaya Station. 69°44′S, 95°30′E. Soviet research station on the plateau of Queen Mary Land, East Antarctica, at an elevation of 2741 m (8104 ft). Opened on 27 May 1956, one of a network of staging posts for tractor trains *en route* from the coastal station Mirnyy, it housed a small party of technicians who became the first to winter on the polar plateau. The station was closed on 15 January 1959.

Pirie, James Hunter Harvey. (1877–1965). Scottish expedition medical officer and scientist. Born near Aberdeen and educated at Edinburgh University, he joined the **Scottish National Antarctic Expedition 1902–4** as medical officer, bacteriologist and geologist. He wintered in 1902, taking part in the sledging programme, and remained in the station on Lauric Island in late 1903, joining the subsequent voyage to the Weddell Sea. He published scientific reports on bacteriology, deep sea deposits and the geology of Gough Island and the geology and glaciology of the South Orkney Islands, contributing also to joint zoological papers, maps and reports. On return he took up private practice in Edinburgh, then served as a government pathologist in East Africa. During World War I he served in the army, afterwards moving to South Africa where he spent the rest of his life in medical research. He died in Johannesburg on 27 September 1965.

Plate tectonics. Plate tectonics has been a dominant process in the geological evolution of Antarctica. The *lithosphere* or outermost layer of Earth, approximately 100 km thick, is divided into plates, of which the outer parts form the crust. Oceanic crust is thin (10 km) and dense, and has formed within the past 180 million years. Continental crust is thicker (30–50 km), lighter, more complex in structure, and may be up to 4500 million years old. The plates float on a viscous underlying *aesthenosphere* or mantle, moving over it at rates of a few centimetres per year.

Boundaries between plates are usually sites of geological activity, topographic complexity and earthquake generation. *Constructive* boundaries, usually mid-ocean ridges, are sites where two plates are moving apart and new lithosphere is being formed by the intrusion or eruption of basaltic igneous rocks. New oceanic lithosphere is magnetized as it cools, and the resulting magnetic anomalies may be used to date the ocean floor (see **magnetic anomalies**). *Destructive* boundaries, characterized by very deep ocean trenches, are sites where one plate is subducted beneath another. Dense, cold lithosphere, forced downward at an angle of approximately 45°, can be traced by earthquake records to depths of up to 700 km, where it fragments and is destroyed. Melting of the subducted slab and loss of water from it may result in dramatic volcanism on the over-riding plate, and large gravity anomalies. At *transcurrent* boundaries crustal plates move laterally in relation to each other, causing earthquakes but rarely volcanic activity. Transcurrent sites may be marked by ocean-floor ridges or shallow troughs alongside continental flanks.

Over 200 million years ago East Antarctica formerly formed part of Gondwana, a much larger continent including South America, Africa, India and Australia. Forces within the mantle caused Gondwana to split into separate continents, each on its own plate. The plates drifted apart, constructive boundaries forming ocean floor between them. East Antarctica, which appears to rest on a single plate, drifted toward West Antarctica, a more complex formation standing on a group of micro-plates, which was simultaneously separating from South America. The Antarctic Peninsula, like the Andes, was a site of active subduction in the Tertiary but is now relatively quiescent. Antarctica as a whole is unique among continents in being almost entirely surrounded by constructive plate boundaries, and ringed by an active mid-ocean ridge. Magnetic anomalies indicate that the entire sea bed of the Southern Ocean formed within the past 180 Ma.

Both parts of Antarctica show evidence of earlier splitting and drifting, East Antarctica some 1000 million years ago, West Antarctica during the early stages of the fragmentation of Gondwana. Similarities between ancient rocks at the Pacific margin of West Antarctica and those at current destructive plate boundaries indicate that subduction may have occurred for over 200 million years. The Transantarctic Mountains include volcanoes that have been active in recent times but the lack of associated deep earthquakes within the continent suggests that they are not associated with current plate boundaries, though they may mark sites of earlier boundary activity.
Further reading: Tingey (1991).
(SWG)

Plateau Station. 79°28′S, 40°35′E. Former US research station, opened in December 1965 on the high plateau of East Antarctica at an elevation of 3625 m (11 890 ft). Sited on a gentle slope below the highest crest of the plateau, surrounded entirely by ice, it was for three years the coldest inhabited place on Earth, and an important laboratory for investigating solar radiation and climate. Each year the sun disappeared about 24 April and reappeared about 21 August. The highest temperature (−18.5 °C) was recorded in January, the lowest (−84.2 °C) in August, three days after the first re-appearance of the sun. In the intensely dry atmosphere, moisture produced by human respiration formed a permanent plume of ice fog downwind. The station was closed in 1968. Had it remained open longer, it would almost certainly have proved colder than Vostok Station, which is 137 m (450 ft) closer to sea level.

Pléneau Island. 65°06′S, 64°04′W. A small island less than one mile long in Wilhelm Archipelago. Discovered by the German Whaling and Sealing Expedition 1873–74, it was charted erroneously as a peninsula of neighbouring Hovgaard Island by the **French Antarctic (*Français*) Expedition 1903–5**, and named for their photographer, Paul Pléneau. It is one of Graham Land's southernmost localities for moulting elephant seals.

Plumley, Frank. (1875–1971). British polar seaman. Born in Clevedon, Somerset on 5 May 1875, he served an apprenticeship as a blacksmith and wheelwright before enlisting in the Royal Navy. He joined the **British National Antarctic Expedition 1901–4** as a stoker, proving useful in building and repairing the expedition's sledges and hardware, and taking part in several sledging journeys. He remained in the navy until retirement.
(CH)

Pobeda Ice Island. 64°40′S, 98°50′E. An ice island (large tabular iceberg) grounded 200 km off Shackleton Ice Shelf, Queen Mary Coast, East Antarctica. Almost 70 km (44 miles) long, 36 km (22 miles) wide and up to 300 m (984 ft) thick, it was discovered by Soviet scientists in 1960. From 9 May to 12 August 1960 it became the site of a small research station, where scientists investigated the ice structure.

Pointe Géologie, Archipel de. 66°39′S, 139°55′E. Group of small rocky islands, islets and rocks off the ice-cliff coast of Terre Adélie, East Antarctica. The group includes Iles Doumoulin, on one of which Capt. J. Dumont d'Urville landed in January 1840, to claim the area for France. The French research station Dumont d'Urville is close by. The group is rich in botanical and bird life, which has been intensively studied by French biologists. During the late 1980s several of the islands were incorporated into an aircraft runway, which was never completed. An area of approximately 2 km^2 (0.8 sq miles), including Iles Jean Rostand, Alexis Carrel, Lamarck and Claude Bernard, the Nunatak du Bon Docteur, and a breeding colony of emperor penguins, is designated SPA 24 (Archipel de Pointe Géologie).

Pole of Inaccessibility. 85°50′S, 65°47′E. A point on the ice cap of Antarctica that is farthest from any point on the coast. It is located in an area toward the southern end of Mac.Robertson Land, at an elevation of about 3700 m (12 136 ft) above sea level. A temporary Soviet International Geophysical Year research station was established nearby between 14 and 26 December 1956.

Polynya. A persistent area of open water within a field of floating ice. Geophysicists distinguish *sensible heat polynyas*, caused by constant upwelling of relatively warm water over reefs or submarine banks, from *latent heat polynyas* caused by persistent winds or currents that sweep the water clear of ice crystals as they form and re-form, however, many polynyas seem to include elements of both. Most Antarctic polynyas form close to the coast, maintained by strong systemic or katabatic winds. By providing open water throughout winter, and especially in early spring, they are often of great significance for marine life, allowing whales, seals and seabirds to remain in high-latitude polar waters. They are also an invaluable aid to navigation. The

Coastal polynya off the ice cliffs of Coats Land, eastern Weddell Sea. 'Sea smoke' arises from the open water. Photo: BS

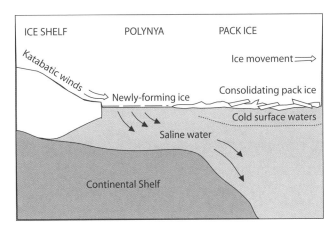

Offshore polynyas, for example that in the eastern Weddell Sea, are formed by strong offshore winds, usually katabatic, that persistently sweep newly-formed ice away from the land. Salt rejected as the new ice forms enhances the salinity and density of underlying sea water, which cascades down the continental slope

occurrence of open water in spring and summer off Victoria Land, in the otherwise congested Ross Sea, enabled Sir James Clark Ross and later explorers to penetrate south to McMurdo Sound and navigate along a polynya close to the 'barrier' or cliff of the Ross Ice Shelf. Similarly a recurrent polynya off the Coats Land and the Luitpold Coast enabled explorers to penetrate almost as far south in the Weddell Sea.

Pomona Plateau. 60°35′S, 45°55′W. Ice plateau about 300 m. high on Coronation Island, South Orkney Islands, named by Capt. J. Weddell in 1822 for one of the Scottish Orkney Islands.

Port aux Français. 49°21′S, 70°15′E. French permanent year-round research station on Iles Kerguelen. The station has operated continuously since 1951.

Port Lockroy (Base A). 64°49′S, 63°30′W. British research station on Goudier Island, Wiencke Island, Palmer Archipelago. The station, Bransfield House, was built in February 1944, the second to be opened by British Naval Operation Tabarin 1943–45 (later Falkland Islands Dependencies Survey and British Antarctic Survey). It was occupied until April 1947: reopened in January 1948, it operated intermittently as a year-round station until January 1962, mainly for ionospheric research, then as an intermittent summer-only station. Between January and March 1996 the hut was restored and the site cleaned up. Currently listed as an Historic Site, it is manned each

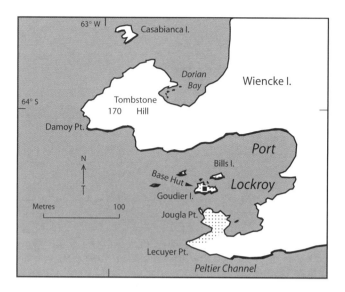

Port Lockroy, southwest Wiencke Island, a harbour much used by early twentieth century whalers. Base A, Britain's first permanent station in Antarctica, is on Goudier Island

summer by the UK Antarctic Heritage Trust, attracting cruise-ship tourists and other summer visitors.

Portal Point Refuge. 64°40′S, 61°00′W. British refuge hut established on Reclus Peninsula, Danco Coast, in December 1956 for use by survey parties from Base O, Danco Island. It remained disused after April 1958. In April 1997 it was removed to the Falkland Islands Museum, Stanley, where it is currently on exhibition.

Porthos Range. 70°25′S, 65°50′E. The central range of three that together make up the northern ranges of Prince Charles Mountains, Mac.Robertson Land. All three were named for Alexander Dumas's three musketeers (see **Athos Range**, **Aramis Range**).

Possession Islands. 71°56′S, 171°10′E. A group of small islands off Cape McCormick, Borchgrevink Coast of Victoria Land. The islands were discovered and named by Capt. J. C. Ross RN, who landed on 12 January 1841 and claimed the area for Britain.

Potter Peninsula. 62°15′S, 58°37′W. Peninsula forming the east flank of Maxwell Bay, King George Island, South Shetland Islands. An area of approximately 1.9 km^2 (0.74 sq miles), between Mirounga Point and the east side of Stranger Point, occupying the coastal zone up to 500 m from the shore, is designated SSSI No. 13 (Potter Peninsula) to preserve its vegetation and birds and seal populations for long-term scientific studies.

Pourquoi Pas Island. 67°41′S, 67°28′W. A steep mountainous island 27 km (17 miles) long off the Fallières Coast of Graham Land. Discovered and charted by the French Antarctic (*Pourquoi Pas?*) Expedition 1908–10, it was named for the expedition ship.

Powder snow. Thin, dry snow formed from separate ice crystals, often rounded and redistributed by strong winds.

Powell, George. (*c.*1796–1824). British sealing captain who, with American sealer N. Palmer, discovered the South Orkney Islands. Born in England, possibly in London, Powell entered the maritime record in August 1818, as master of *Dove*, a small smack of 58 tons, owned by the Wapping-based firm of Daniel **Bennett** and Son. In *Dove* (1818–19) and *Eliza*, a larger sloop of 132 tons (1819–21), Powell made successful voyages to the southern sealing grounds, including the newly-discovered South Shetland Islands. In July 1821, again in *Eliza*, he left England for a third southern voyage. Now an experienced sealer and accomplished navigator, by late November of that year he was again hunting for fur seals on the South Shetland Islands, competing intensively with both British and American sealing vessels. After working with little success around Elephant Island, easternmost of the South Shetland Islands, he joined forces with an American sealer, Nathaniel B. **Palmer**, to explore unknown seas further eastward. On 7 December 1821, in a sea strewn thickly with icebergs and brash ice, he sighted small islands close by, and a larger, ice-covered landmass further east – a hitherto unknown archipelago. Powell and Palmer spent six days exploring and charting the islands. On returning to Britain Powell published a chart of the South Shetland Islands, including details of the new archipelago. At first known as 'Powell's Islands', these were later named the South Orkney Islands.

In January 1823 Powell became master of *Rambler*, a larger whaler of the Enderby fleet, in which he sailed on a whaling voyage to Sydney, Australia. In April 1824, still whaling, he was killed in a fracas with natives of the Friendly Islands.
Further reading: Jones (1992).

Powell Island. 60°41′S, 45°03′W. A narrow, slab-sided island 11 km (7 miles) long between Coronation Island and Laurie Island, in the South Orkney Islands. It was discovered and charted by sealing captains Nathaniel Palmer and George Powell in December 1821. An area of approximately 18 km^2 (7 sq miles), including part of Powell Island, neighbouring Fredriksen, Michelsen, Christoffersen and Grey Islands, and adjacent unnamed islands, is designated SPA No. 15 (Powell Island south and adjacent islands). The purpose is to protect from human interference local vegetation, birds and mammals, which are currently at risk from expanding breeding colonies of fur seals.

Prebble, Michael. (d. 1998). New Zealand polar geographer and administrator. 'Mike' first visited Antarctica with the 1960–61 expedition to restore the historic huts at Cape Royds and Cape Evans. In 1961–62 he was a field assistant at Scott Base: in 1964–65 he was appointed deputy leader, and in 1965–66 base leader. After completing a master's thesis in geography at Victoria University, Wellington, he took a year's study-leave at the Scott Polar Research Institute, Cambridge. He again led Scott Base in 1979–80. From 1989, as an officer of the Ministry for the Environment, Prebble was involved in planning New Zealand's Antarctic conservation policies and legislation, and attending meetings of the Antarctic Treaty: he also revisited Antarctica as an observer and government representative on cruise ships. He died on 18 April 1998.

Presidente Eduardo Frei Station. 62°11′S, 58°57′W. Chilean meteorological station on Fildes Peninsula, King George Island, South Shetland Islands. Established in February 1969 by the Chilean Air Force, since 1980 it has been engulfed by the more extensive **Teniente Rodolfo Marsh Station** that has grown around it. The name remains attached to the meteorological facility.

Presidente González Videla Station. 64°49′S, 62°51′W. Permanent Chilean research station on Waterboat Point, Paradise Harbour. Built in 1951 as a Chilean air force meteorological station, it operated continuously until 1964, intermittently thereafter as a summer-only station.

Pressure ice. A floating ice sheet in which lateral pressures have caused **rafting, hummocks** and **pressure ridges**.

Pressure ridge. An extended ridge in a field of floating ice, caused by lateral pressure at a right-angle to the direction of the ridge. Ridges extend both upwards and downwards: a visible ridge may be matched by an underwater ridge (keel) two to three times deeper.

Priestley, Raymond Edward. (1886–1972). British polar scientist and explorer. After reading botany and geology at Bristol University College, he joined the **British Antarctic (*Nimrod*) Expedition 1907–9** as a geologist. He took part in the sledging programme, working in the field with Prof. Edgeworth David, under whose guidance at Sydney University he later wrote up his expedition reports. Almost immediately he joined the **British Antarctic (*Terra Nova*) Expedition 1910–13**, as geologist to the northern party. Priestley spent a winter at Cape Adare, then a second winter in a snow cave on Inexpressible Island, ultimately reporting on both geology and glaciology. He served with distinction as a signals officer in World War II, then returned to Cambridge to write a history of the Signal Service and complete his Antarctic reports. From Cambridge he was appointed vice-chancellor of Melbourne University 1935–38, and vice-chancellor of Birmingham University 1938–52. Knighted in 1949, he became acting-director of the Falkland Islands Dependencies Survey in 1955–59, and president of the Royal Geographical Society in 1961–68. He died on 24 June 1972.

Primavera Station. 64°09′S, 60°57′W. Argentine permanent year-round research station, established on Cierva Point, Danco Coast, Antarctic Peninsula in March 1977. It became a summer-only station after 1982.

Primero de Mayo Station. See **Decepción Station**.

Prince Albert Mountains. 76°00′S, 161°30′E. A major block of mountains overlooking Scott Coast, extending between the Priestley and Ferrar glaciers. Discovered during the British Antarctic Expedition 1839–43, it was named for Prince Albert, consort of Queen Victoria.

Prince Andrew Plateau. 83°38′S, 162°00′E. Ice plateau of Queen Elizabeth Range, Victoria Land, named in 1961–62 for the infant British prince.

Prince Charles Mountains. 72°00′S, 67°00′E. A major arcuate formation of mountain ranges and scattered peaks, about 400 km (250 miles) long, aligned along the western flank of Lambert Glacier, Mac.Robertson Land, East Antarctica. Discovered during aerial reconnaissance flights of United States Operation Highjump 1946–47, they have been surveyed and explored geologically by field parties from the Australian National Antarctic Research Expedition. They are named for the heir-apparent to the British throne.

Prince Charles Strait. 61°05′S, 54°35′W. Passage between Cornwallis Island and Elephant Island, in the South Shetland Islands. Though known to sealers from 1821, and traversed by Lt C. Ringgold in the brig USS *Porpoise* during the United States Exploring Expedition in 1839, the strait remained unnamed until 1948, when it was named for the recently born British Prince of Wales.

Prince Edward Islands. 46°54′S, 37°45′E. A group of cool temperate islands in the southern Indian Ocean, 1800 km (1125 miles) southeast of South Africa, including Marion Island and the smaller Prince Edward Island. Both are volcanic islands, set 22 km (14 miles) apart on a submarine platform that marks the western edge of the Prince Edward–Crozet submarine ridge. Marion Island is roughly circular and about 20 km (12.5 miles) in diameter: Prince Edward Island is a rectangle 9 km (5.6 miles) long and 5 km (3.1 miles) across. Each takes the form of a low, irregular dome, composed of lavas and tuffs that have emerged from several volcanic cones over a period of half a million years. Marion Island rises to 1230 m (4034 ft). Its high plateau has fragments of a formerly extensive ice sheet, and there is evidence of glaciation that must formerly have covered most but not all of the island. Prince Edward Island rises only to 672 m (2204 ft) and may not have borne an ice-cap. Steep cliffs and rocky beaches line the coasts: inland are rolling plains crossed by steep lava ridges, covered with moorland and scrub vegetation. The islands are damp and humid, with mean annual temperature 5.1 °C, range of monthly mean temperatures 3.7 °C, almost constantly overcast skies, persistent westerly winds, and 250 cm of rain per year.

Though the islands were probably recorded by the Dutch navigator Barend Barendszoon Lon in 1663, their discovery is usually credited to the French explorer Marc Macé **Marion du Fresne**, who in January 1772 named them Terre d'Espérance. The British explorer Capt James Cook gave them their present names in 1775. Both islands were frequented by sealers from the early nineteenth century onward. In 1948 the group was annexed by South Africa, which in the following year established a scientific station at Transvaal Cove, Marion Island. South African scientists have worked there ever since, visiting Prince Edward Island occasionally. Marion Island has 24 native species of flowering plant and over a dozen introduced species: Prince Edward Island shares many of the native species, and both are well endowed with algae, lichens and mosses. Both islands support breeding fur seals, elephant seals, and seabirds including penguins, albatrosses and lesser petrels. Both are managed as nature reserves: tourists are occasionally allowed ashore on Marion Island, but not on Prince Edward Island.

Further reading: van Zinderen Bakker *et al.* (1971).

Prince Olav Harbour. 54°04′S, 37°09′W. Fjord harbour of South Georgia, the site of a whaling station managed by the Southern Whaling and Sealing Company 1911–32, now derelict. The harbour was named by Norwegian whalers for a member of their royal family.

Prince Olav Mountains. 84°57′S, 173°00′W. A major range of the Queen Maud Mountains, fronting Dufek Coast between Shackleton and Liv glaciers. Discovered in 1911 by Roald Amundsen, they were named for the contemporary Norwegian crown prince.

Princess Elizabeth Land. 68°30′S, 80°00′E. Ice-covered area of East Antarctica between meridians 73° and 86°E, bounded to the west by Mac.Robertson Land and to the east by Wilhelm II Land. Forming part of Australian Antarctic Territory, the land includes Ingrid Christensen Coast, King Leopold and Queen Astrid Coast, and an inland ice plateau extending to the South Pole. The coast was identified during the **Norwegian (Christensen) Whaling Expeditions** in 1934, and further explored during the second voyage of the **British, Australian and New Zealand Antarctic Research Expedition 1929–31**, who named the land for the young British princess who later became Queen Elizabeth II.

Prins Harald Kyst. 69°30′S, 36°00′E. Part of the coast of Dronning Maud Land, East Antarctica, bounded by Riiser–Larsenhalvøyer in the northwest and the southern end of Lützow–Holmbukta in the southeast. Almost entirely ice-covered, it consists mainly of sheer ice cliffs backed by gently-rising ice plains. The southern section, in Lützow-Holmbukta, has several emergent rocky capes and islands. The coast was discovered by the **Norwegian (Christensen) Whaling Expeditions** in early February

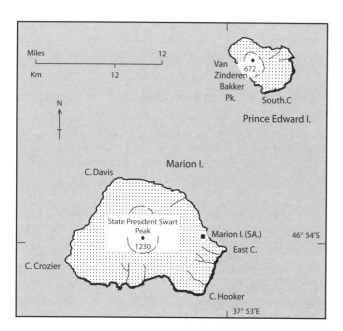

Prince Edward Islands. Though Marion Island is much larger, with a permanent research station, the group takes its name from its smaller island. There is no permanent ice. Heights in m

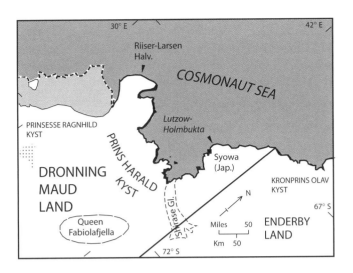

Prins Harald Kyst

1937, during a flight by Viggo Widerøe and Nils Romnaes, and named for an infant prince of Norway.

Prinsesse Astrid Kyst. 70°45′S, 12°30′E. Part of the coast of Dronning Maud Land, East Antarctica, bounded by the meridians 5°E and 20°E. It consists mainly of ice-covered peninsulas and islands lined by wide ice shelves, but includes the ice-free oasis area Schirmacher Hills. The coast was first sighted in March 1931 by Capt. H. Halvorsen, in the whaling factory ship *Sevilla*. It was revisited in January 1934 by Lars Christensen in the tanker *Thorshavn* and overflown and photographed by Lt Gunnestad. The coast was named for a princess of Norway. In 1939 it was again overflown and photographed by pilots of the **German Antarctic (*Schwabenland*) Expedition**, who discovered the ice-free area at the junction of land and ice shelf. The coast is the site of the former Soviet research station Lazarevskaya, and of the permanent Russian station Novolazarevskaya.

Prinsesse Ragnhild Kyst. 70°30′S, 22°00′E. Part of the coast of Dronning Maud Land, East Antarctica, extending from Riiser-Larsenhalvøyer in the east to a western limit in longitude 20°E. The coast consists mainly of ice-covered peninsulas and islands, fronted by wide ice shelves. It was overflown and photographed on 16 February 1931 by Capt. Hj. Riiser-Larsen, pilot of the **Norwegian (Christensen) Whaling Expeditions**, and named for a Norwegian princess. From December 1957 to January 1961 it was the site of Roi Baudouin, a Belgian research station from which the inland Sør Rondane Mountains were explored.

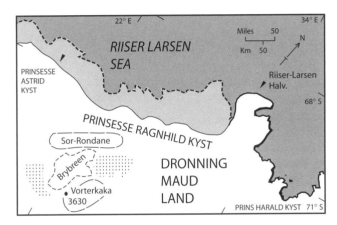

Prinsesse Ragnhild Kyst

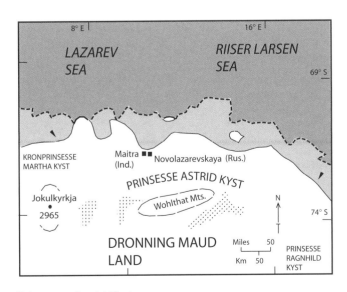

Prinsesse Astrid Kyst

Prions, southern oceanic. Blue-grey petrels of the family Procellariidae, genus *Pachyptila*, that feed in large flocks in surface waters of the southern oceans. Also called whalebirds, they often flock in areas where whales are feeding, pecking vigorously to pick up disturbed krill and other particulate foods. In length they range from 23–31 cm (9–12 ins), in wing span 48–60 cm (19–24 in). Underparts are usually pale grey or white, upper wings and mantle blue with dark grey or brown diagonal bars. The dark grey-brown bills vary in shape, perhaps indicating slightly different food preferences, and providing the most positive interspecific differences. Largest are the broad-billed prions *P. vittata*, with bills containing lamelli that strain plankton from the sea water. These and the slightly smaller Salvin's prions *P. salvini* breed and forage mainly in the warm temperate zone. Antarctic prions *P. desolata* breed on cold temperate, sub-Antarctic and Antarctic islands and forage mainly south of the Antarctic Convergence. Slender-billed prions *P. belcheri* occupy an intermediate zone. Fulmar prions *P. crassirostris* breed and feed around the temperate islands of the Indian Ocean and southern New Zealand:

fairy prions *P. turtur* are most often seen in the temperate New Zealand sector, and also around the Falkland Islands and Prince Edward Islands.

Proclamation Island. 65°51′S, 53°41′E. A small island 2.5 miles (4 km) off Cape Batterbee, Enderby Land, discovered by the **British, Australian and New Zealand Antarctic Research Expedition 1929–31**. Expedition leader Sir Douglas Mawson landed on 13 January 1930 and claimed the area for Britain.

Progress Station. 69°23′S, 76°23′E. Russian year-round station in the Larsemann Hills, Ingrid Christensen Coast, built in January 1989 to provide accommodation for a nearby icecap airstrip. It has since been occupied intermittently.

Prospect Point (Base J). 66°00′S, 65°21′W. British research station at Prospect Point, Ferin Head, Graham Land. The station was built in February 1957 and occupied for geological and survey work until June 1959, since when it has remained closed.

Protocol on Environmental Protection to the Antarctic Treaty. Instrument of the Antarctic Treaty System dedicated to the comprehensive protection of the Antarctic environment and its dependent and associated ecosystems. Measures for protecting the environment were among the first manifestations of activity from early Antarctic Treaty consultative meetings (see **Agreed Measures for the Conservation of Antarctic Fauna and Flora**). However, by the late 1980s concern was being expressed by visitors to the region, including scientists, tourists and environmentalists, that many sites of human operation had already suffered severe environmental degradation. Though most of this was clearly due to scientific and support activities following the International Geophysical Year (1957–58), there were implied threats of further damage from the development of tourism and the possibility of mineral extraction. The presentation and rejection of CRAMRA (see **Convention on the Regulation of Antarctic Mineral Resource Activities**), focusing attention as never before on environmental issues, made it apparent that the Treaty's environmental measures required repackaging into a coherent system of mandatory obligations. This became the business of the Eleventh Antarctic Treaty Special Consultative Meeting, which opened in Viña del Mar, Chile in late 1990 and held its final session in Madrid on 3–4 October 1991. The Protocol on Environmental Protection (known also as the Madrid Protocol), that was adopted in the final meeting included many principles adapted from the defunct Convention.

Summary of the Protocol

The Protocol consists of a preamble, 26 articles, a schedule and four (later five) annexes: for the full text see **Appendix D**. The preamble recalls 'the designation of Antarctica as a Special Conservation Area and other measures adopted under the Antarctic Treaty System to protect the Antarctic environment and dependent and associated ecosystems'. Article 1 lists seven definitions: in Article 2 the Treaty parties commit themselves to the comprehensive protection of the Antarctic environment and its dependent and associated ecosystems, and 'designate Antarctica as a natural reserve, devoted to peace and science'.

Article 3 (1) specifies that the protection of the Antarctic environment 'and the intrinsic value of Antarctica, including its wilderness and aesthetic values and its value as an area for the conduct of scientific research, in particular research essential to understanding the global environment, shall be fundamental considerations in the planning and conduct of all activities in the Antarctic Treaty area'. Article 3(2) and ensuing articles require all activities in the Treaty area to be planned and conducted in ways that limit adverse impacts on the environment, avoiding adverse effects on climate or weather patterns, air or water quality, significant changes in atmospheric, terrestrial, glacial or marine environments, detrimental changes in distribution, abundance or productivity of fauna or flora. Activities must not increase the jeopardy of endangered or threatened species, or hazard areas of biological, scientific, historic, aesthetic or wilderness significance. They must be planned and conducted on the basis of information sufficient to allow prior assessments of, and informed judgements about, their possible environmental impacts. They must provide for regular and effective monitoring to ensure that no adverse changes are occurring, accord priority to scientific research, preserve the value of Antarctica as an area for the conduct of research, and be modified, suspended or cancelled if they prove detrimental. Article 6 requires parties to cooperate in planning and conducting Antarctic activities, and share information arising from them.

Article 7 prohibits any activity relating to mineral resources, other than scientific research. Articles 8 and 9 cover requirements for environmental impact assessments, referring to four (later five) annexes. The remaining articles cover mainly procedural matters, including (Articles 11 and 12) the establishment of a Committee for Environmental Protection and its functions. The purpose of the committee is to advise and make recommendations on its own effectiveness, and provide for a system of inspection. Parties are required to provide prompt and effective responses to emergencies arising from their activities, including those that endanger the environment. They agree to make rules and set procedures covering liability for damage arising from their activities, and to report annually on steps they have taken to implement the Protocol. The

Protocol may be modified or amended at any time, and reviewed on request by one of the parties after 50 years. The five annexes cover procedures for environmental impact assessment, conservation of Antarctic fauna and flora, waste disposal and management, prevention of marine pollution, and area protection and management.

Environmental protection

Where many of the earlier environmental measures arising within the Treaty system could be interpreted as safeguarding scientific research first and the Antarctic environment only secondarily, the Protocol gives higher priority to environmental protection, and has at least the potential for being applied with equal rigour and impartiality to all who visit Antarctica, for whatever purpose. Critics have been quick to identify shortcomings, broadly, that the Protocol reinforces the Treaty's scientific elitism, and concerns itself with minutiae but fails to set standards or provide for monitoring, the practicalities of inspection or enforcement of standards. In the longer term, scientists and administrators are concerned at the effort and expense that will be required from national expeditions to meet all the requirements of the Protocol.

Further reading: Chaturvedi (1996).

Prudhomme, André. (1920–59). French polar meteorologist. Prudhomme served on two French expeditions in Terre Adélie. He disappeared, presumed drowned, during a blizzard on 7 January 1959 while making meteorological observations at Dumont d'Urville station.
(CH)

Publications Ice Shelf. 69°38′S, 75°20′E. Rectangular ice shelf in Prydz Bay, Ingrid Christensen Coast, fed by three glaciers that were named for polar publications: Polar Record, Polar Times and Polar Forschung.

Q

Quar, Leslie Arthur. (1923–51). British expedition radio technician. Born in London on 27 March 1923, he trained as a radio technician with the Royal Air Force, serving throughout World War II. In 1949 he joined the RAF Antarctic Flight, which in summer 1949–50 was attached to the **Norwegian-British-Swedish Antarctic Expedition 1949–52**. When the flight returned home, he remained with the expedition as radio technician and handyman, playing an important role in maintaining electrical and mechanical equipment. He was killed in a tractor accident near Maudheim on 23 February 1951.

Quarisen. 71°20′S, 11°00′W. (Quar Ice Shelf). Small ice shelf east of Kapp Norvegia, Kronprinsesse Märtha Kyst, Dronning Maud Land: site of Maudheim, the station of the **Norwegian-British-Swedish Antarctic Expedition 1949–52**. The ice shelf was surveyed by the expedition and named for Leslie Quar, a British radio and electrical engineer of the party who was drowned in a tractor accident.

Quartermain, Leslie Bowden. (1895–1973). New Zealand polar historian. Born in Hororata on 10 June 1895, he became a school teacher of English in Christchurch and Wellington. Contacts with Antarctic expedition ships passing through New Zealand ports inspired him and several colleagues to found the New Zealand Antarctic Society, which for many years helped to develop and promulgate Antarctic interest in New Zealand. In 1950 he founded and became first editor of the Society's bulletin *Antarctic*. In 1957, and again in 1960–61 and 1968 he visited Antarctica, on his second foray as leader of a task-force that dug out and stabilized the historic huts at Cape Royds and Cape Evans. On retiring from teaching he became Information Officer for the Antarctic Division of the New Zealand Department of Scientific and Industrial Research, a post that gave him opportunities for historical research and writing. He died on 28 April 1973.

Queen Alexandra Range. 84°00′S, 168°00′E. A major range extending between Beardmore and Lennox-King glaciers, merging on its southwestern flank into the polar plateau. Discovered in 1908 by Ernest Shackleton, it was named for the consort of King Edward VII.

Queen Elizabeth Range. 83°20′S, 161°30′E. A major range of the Transantarctic Mountains bounded by Marsh, Nimrod and Law glaciers, and merging on its southwestern flank into the polar plateau. Identified by a New Zealand sledging party of the **Commonwealth Trans-Antarctic Expedition** in 1958, it was named for the contemporary British monarch.

Queen Mary Coast. 66°30′S, 95°00′E. Part of the coast of Queen Mary Land, East Antarctica, bounded by Cape Filchner in the west and Cape Hordern in the east. In an area of heavy precipitation, it consists mainly of ice cliffs, large glaciers and ice shelves backed by steeply rising ice slopes, with few nunataks or emergent islands. It includes the permanent year-round Russian research station Mirnyy, and the western half of Shackleton Ice Shelf. Both the coast and the interior were explored by the **Australasian Antarctic Expedition 1911–14**, from their Western Base on the ice shelf.

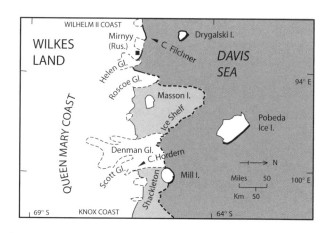

Queen Mary Coast

Queen Mary Land. Ice-covered area of East Antarctica between meridians 91° and 102°E, bounded to the east by Wilkes Land and to the west by Wilhelm II Land. Forming part of Australian Antarctic Territory, it includes Queen Mary Coast, much of Shackleton Ice Shelf, the Bunger Hills oasis area, and an inland ice plateau extending to the South Pole. The coast was identified and explored by the **Australasian Antarctic Expedition 1911–14**, and named for the contemporary British queen. From the permanent coastal station Mirnyy the inland ice was traversed during the 1950s and early 1960s by Soviet tractor trains and aircraft, which established several temporary stations and also Vostok, a permanent inland station close to the geomagnetic pole. A Soviet coastal station Oazis operated in the Bunger Hills oasis from October 1956 to November 1958.

Queen Maud Land. See **Dronning Maud Land**.

Queen Maud Mountains. 86°00′S, 160°00′W. A complex group of mountains forming the southernmost section of the Transantarctic Mountains, extending from Reedy Glacier to Beardmore Glacier, and including many peaks of over 2500 m (8200 ft). Forming a massive barrier some 400 km (250 miles) long, they stand between the polar plateau and the Amundsen and Dufek coasts. The Norwegian explorer Roald Amundsen in 1911 passed through them via the Axel Heiberg Glacier, *en route* to the South Pole, naming them for the contemporary queen of Norway.

Quoy, Jean-René Constant. (1790–1869). French naturalist who, with J.-P. Gaimard, collected and described many animal species of the Southern Ocean. Born near Rochefort, Bay of Biscay, Quoy trained as a surgeon and naturalist, and in 1808 joined the French naval service. After service in the West Indies and Indian Ocean, in 1817–20 he joined L.-C. de Freycinet's expedition in *Uranie*, on a scientific voyage to explore Australia, South Pacific islands and the southern oceans. In 1826 he was appointed surgeon and naturalist in *Astrolabe*, sailing with J. S. C. Dumont d'Urville's expedition to explore and chart the coasts of Australia, New Zealand and the western Pacific Ocean. In both expeditions he worked closely with Gaimard, collecting and naming many southern species, and providing type-specimens for the French national collection. In later life he became Professor of Medicine at the Rochefort naval school, where he wrote extensively of his voyages and collections.

R

Radio echo sounding. Airborne technique that has provided the most rapid, continuous and accurate method of measuring thickness of ice sheets and glaciers, notably the Antarctic ice sheet. Using such long-range aircraft as Lockheed C-130 Hercules, and smaller Twin Otters, Dornier 228s, Pilatus Porters and various helicopters, over half of the Antarctic ice sheet (about 2.7 million sq miles, 7 million km^2) has been surveyed by radio echo sounding, with an average line spacing of 30–60 miles (50–100 km). Ice is a highly resistive protonic semi-conductor – a medium which scatters, absorbs and refracts radio signals. Velocity of propagation, about 168 m per micro-second, depends upon ice density but is largely independent of ice crystal fabric, pressure and temperature. Impurities within the ice influence the number and location of point defects in the ice crystal lattice, which in turn help to determine the electrical properties of the ice, and hence its ability to transmit radio energy. One practical result is that echoes are returned, not only from the bedrock, but also from 'interfaces' between the layers of ice, providing much information on the internal structure of the ice sheet.

Glaciers contain many inhomogeneities, for example, ice lenses, water pockets, cracks and rock debris, from which extensive polar ice sheets are relatively free. Their presence causes attenuation of signals and gives rise to scatter or clutter echoes which obscure the bottom return. The colder the ice, the less radio energy is lost by absorption. These factors combine to make the technique work better with the polar ice sheets of Antarctica and Greenland than with glaciers. Frequency and design parameters can be chosen to achieve maximum penetration with the highest possible resolution. Over Antarctica, for example, glaciologists require depth sounding of up to 5 km with an accuracy of the order of 10 m. To investigate internal glacier structures they need high definition of certain reflecting horizons on a scale of the order of 10–20 cm. With frequencies at or beyond 1 GHz (30 mm wavelength) very high absorption of the radio signal is the limiting factor. At low frequencies around 1 MHz resolution is poor, measurable in tens of metres, and the physical size of the antennas becomes a dominant consideration. An optimum range has been found between 30 and 400 MHz (10 to 0.3 m wavelength).
(DJD)

Radiometric dating. A technique that uses the known decay rate – the half-life – of a particular radioactive isotope of an element to determine the time elapsed since its formation. The half-life of an isotope is the time taken for half the mass of the isotope to decay to a daughter isotope with the loss of some radiation, such as the decay of uranium to lead. Within a nuclear reactor, the radiation lost can be harnessed to provide nuclear power. Several well-known radioactive decay processes can be used to determine the age of a mineral or of a rock. These include potassium to argon (K–Ar), rubidium to strontium (Rb–Sr), uranium to lead (U–Pb), and samarium to neodymium (Sm–Nd). Such techniques are now routinely used on a variety of different rock types to determine their ages. Usually they determine the time when the rock crystallized from a melt, as in the case of lavas and plutonic rocks. They can be used also to determine the time when a metamorphic rock last underwent a major recrystallization, without necessarily having passed through a molten stage. The initial ratios of ^{87}Sr and ^{86}Sr isotopes can also help to indicate the source of an original magma.

Radiometric dating can be particularly useful in Antarctica, where rock exposures are rarely continuous for any distance. Do the lavas exposed on two nunataks a few kilometres apart belong to the same volcanic episode or even the same volcano? Radiometric dating provides a positive answer. Similarly, many of the granitic plutons in the Antarctic Peninsula are similar in composition, having been similarly formed by the subduction of the oceanic floor of the Pacific Ocean. However, subduction has proceeded over some 200 million years. By revealing the ages of the plutons, radiometric dating provides them with identities and a chronology of their intrusive history.

Originally the experimental method needed large quantities of a rock or a mineral to ensure a homogeneous sample for analysis in a mass spectrometer. However, modern instrumentation, for example the ion probe, can be used to find the age of part of a single crystal. This has had far-reaching implications, particularly for the pre-intrusive history and origin of some plutonic rocks, and for the sequential history of some metamorphic rocks. The mineral zircon can frequently survive high-grade metamorphism and crustal melting, and is particularly useful

where it is now possible to date the growth stages of an individual crystal through several thermal events. Whole-rock K–Ar dating of gneisses around much of the East Antarctica coast showed a major metamorphic event around 450–550 million years ago. This proved to be only the most recent imprint on rocks that were actually much older: their true age was revealed by mineral U–Pb dating of zircon crystals. This technique showed them to be among Antarctica's oldest rocks, formed around 3800 million years ago, with further periods of metamorphism 3100 and 2500 million years ago.
(PDC)

Rafting. A feature of young floating ice sheets, in which lateral pressures force adjacent parts of the sheet to override each other, doubling the effective thickness.

Rankin, Arthur Niall. (d.1965.) Scottish traveller and polar photographer. Experience as a natural history photographer with the Oxford University Arctic Expedition in 1924 led Rankin, over 20 years later, to take his own small expedition to South Georgia for a similar objective. A capable yachtsman, he took down *Albatross*, a converted lifeboat, and with two Shetland Island companions spent the summer of 1946–47 making a unique record of South Georgia's wildlife. He died on 7 April 1965, aged 77.
Further reading: Rankin (1951).

Rasmussen, Johan. (1878–1966). Norwegian whaling entrepreneur. Born in Stavanger on 11 January 1878, he qualified as a lawyer and became first a solicitor, later a partner in the extensive whaling interests of Peder Bogen, based in Sandefjord. In 1914 he took charge of the company, which by the end of World War I had contracted to two operations, A/S Vestfold, operating the factory ship *Vestfold* and Stromness whaling station, South Georgia, and A/S Sydhavet, operating *Svend Foyn* around the South Shetland Islands. In 1923 he added a new company, A/S Rosshavet, which pioneered whaling operations in the Ross Sea with the factory ships *Sir James Clark Ross* and *C. A. Larsen*. Rasmussen was chairman of Den Norsk Hvalfangerforening 1919–28 and of the international Association of Whaling Companies 1936–38. Much involved in the international politics of whaling, he had many other shipping, financial and business interests in Norway.

Rawson Plateau. 85°52′S, 164°45′W. Extensive ice plateau rising to 3400 m (11 152 ft) in Queen Maud Mountains, Victoria Land. It was explored by a sledging party of **Byrd's first Antarctic Expedition 1928–30**, and named for K. L. Rawson, who supported the expedition and participated in its 1932–34 sequel.

Raymond, John East. (d.1977). British expedition carpenter. After service in World War II with the Royal Engineers, John Raymond was in 1950 employed by the Public Works Department of the Falkland Islands. Seconded to the Falkland Islands Dependencies Survey, he supervised the construction of expedition huts at Hope Bay, the Argentine Islands, Anvers Island, Signy Island, and Admiralty Bay, wintering twice with the Survey. He died on 15 April 1977, aged 62.

Reece, Alan. (1921–60). British expedition geologist. Born in London, he served as a meteorological officer in World War II. In 1944 he joined Operation Tabarin, soon to become the Falkland Islands Dependencies Survey, serving at Deception Island, Hope Bay and Admiralty Bay. On return to Britain he graduated in geology, then spent two further years in Antarctica with the **Norwegian-British-Swedish Antarctic Expedition 1949–52**. Later he worked as a geologist in Uganda, Greenland and Canada. Reece was killed in an aircraft crash near Resolute, northern Canada, on 28 May 1960.

Richards, Richard Walter. (1893–1985). Australian polar physicist and explorer. Born in Bendigo, Australia on 14 November 1893, he was educated at Melbourne University. In 1914 he joined the Ross Sea party of the **Imperial Trans-Antarctic Expedition 1914–17**, sailing in *Aurora* to McMurdo Sound. He quickly found himself involved in sledging stores across the Ross Ice Shelf, laying depots for the trans-polar party that never came. As one of the few who remained capable of hauling during the privations of the main southern journey, he was instrumental in saving the lives of several of his companions, an achievement for which he was awarded the Albert Medal. Only later did he collapse, taking five months to recover. On return to Australia he taught in technical schools, retiring in 1958. He died on 8 May 1985.

Right whale, southern. *Eubalaena australis*. Southern representatives of a cosmopolitan genus, southern right whales are found mainly in warm subtropical and cool temperate waters. Stout, stocky whales, seldom exceeding 16 m (52.5 ft) in length, in full fat they may weigh up to 100 tonnes. Uniformly black, except for white or grey abdominal flashes, they often carry large white patches of barnacles, or a 'bonnet' of pale, horny skin on nose and chin. The huge head with recurved lower jaw contains baleen plates up to 2.5 m (8.2 ft) long. They swim slowly and deliberately, usually singly or in small groups: only rarely are more than three or four seen together. During annual migrations from warm to cooler, richer waters, they frequently come close to land, females entering harbours

and shallows to produce their single calves. Not surprisingly, right whales were among the first to be hunted commercially by man, mainly from land stations in southern South Africa, Australia, New Zealand and South America. Close to extinction, their stocks were afforded international protection in 1935. Recovery has been very slow: except in known haunts off eastern South America and southern Australia, southern right whales are seldom seen. Only rarely are they recorded from waters south of the Antarctic Convergence: in recent years they have been sighted off South Georgia, the South Orkney and South Shetland Islands, and as far south as Gerlache Strait, Antarctic Peninsula.
Further reading. Brownell *et al.* (1986).

Riiser-Larsen, Hjalmar. (d. 1965). Norwegian aviator and explorer. Joining the Royal Norwegian Navy in 1912, he became one of its earliest pilots. In 1921 he became chief secretary to the Aviation Council of the Ministry of Defence. In 1925 he joined Roald Amundsen and Lincoln Ellsworth in their attempt to fly two seaplanes to the North Pole, and in the following year piloted the airship *Norge* on its historic trans-Arctic flight. In 1929–31 he spent two summers in the Antarctic with the Norwegian (Christensen) Whaling Expeditions, photographing Bouvetøya, Enderby Land and parts of the future Dronning Maud Land from the air. In 1932–33 he returned with a sledging party that tried unsuccessfully to survey from the sea ice. During World War II he served as naval attaché in Washington, in 1944–46 commanding the Royal Norwegian Air Force in Britain. After the war he became involved in civil aviation planning, from 1948 managing Norwegian Air Lines. He died on 3 June 1965, aged 73.

Riiser-Larsen Halvøya. 68°55′S, 34°00′E. (Riiser-Larsen Peninsula). Extensive peninsula forming the west side of Lützow-Holmbukta, Prinsesse Ragnhild Kyst, Dronning Maud Land. It was discovered during the **Norwegian (Christensen) Whaling Expeditions** and named for the expedition pilot, Capt. Hjalmar Riiser-Larsen.

Riiser-Larsen Sea. 68°S, 25°E. Coastal sea off East Antarctica, between Sedovodden and Riiser-Larsen Halvøya, and bordering Prinsesse Ragnhild Kyst. There is no definitive northern boundary. Fast ice and pack ice extend off the coast in winter: in summer pack ice and bergs move westward, driven by easterly currents and offshore winds. Designated by Russian cartographers to honour the Norwegian aviator and explorer, it occupies a near-shore strip of the broader and earlier named King Håkon VII Sea (Kong Håkon VII Hav).

Riiser-Larsenisen. 72°40′S, 16°00′W. (Riiser-Larsen Ice Shelf). Elongate ice shelf off the western end of Kronprinsesse Märtha Kyst, Dronning Maud Land, named for aviator Capt. H. Riiser-Larsen, who overflew it in the 1930 **Norwegian (Christensen) Whaling Expedition**.

Rime. Accretionary deposit of fine crystalline grains of ice, due to rapid freezing of water droplets from the atmosphere.

Ripamonti (Luis Ripamonti). 62°12′S, 58°53′W. Chilean refuge or small research station on Ardley Island, South Shetland Islands, used for summer biological studies.

Risopatron. 62°22′S, 59°40′W. Chilean refuge or small research station in Coppermine Cove, Robert Island, South Shetland Islands, established in 1952 and used mainly for summer biological studies.

Ritscher, Alfred. (1879–1963). German polar explorer. Born in Bad Lauterberg on 23 May 1879, he became a mercantile marine officer, transferring to the Imperial German Navy in 1911. In 1912 he was master of *Hertzog Ernst*, expedition ship of an ill-fated attempt to explore northern Spitsbergen, of which several members including the leader, Lt. H. Schröder-Stranz, were lost through inexperience. Later he led the **German Antarctic (*Schwabenland*) Expedition 1938–39**, commanding a seaplane tender with two seaplanes that overflew and photographed extensive areas of East Antarctica, in the sector east of Coats Land in which Norway already had a firm research commitment. Named 'Neu-Schwabenland' and claimed by Germany, the area had already been claimed by Norway as part of Dronning Maud Land. Though the flying and photography were competent, ground control was lacking. Many of the photographs and records were lost during World War II. Ritscher salvaged the remains, some of which proved useful in planning the **Norwegian-British-Swedish Antarctic Expedition 1949–52**. Ritscher died in Hamburg on 30 March 1963.

Robert Island. 62°24′S, 59°30′W. An ice-covered island 17.5 km (11 miles) long between Nelson Island and Greenwich Island in the South Shetland Islands. Charted in 1819, it was possibly named for the sealing brig *Robert*. Coppermine Cove, in the northwest corner of the island, is the site of a Chilean refuge.

Roberts, Brian Birley. (1912–78). British polar geographer, explorer and administrator. Interested in polar exploration when still at school, Roberts read geography at

Cambridge University, where he organized undergraduate expeditions to Iceland and Greenland. He joined the **British Graham Land Expedition 1934–37** as an ornithologist, spending the austral winter of 1935 at the Argentine Islands and a further winter on South Georgia, where he studied gentoo penguins. In 1944 he was appointed to the Foreign Office Research Department, and in 1945 became Secretary of the UK Antarctic Place-names Committee. He was also a committee-member of the British Glaciological Society, editing *Journal of Glaciology*. In 1946 he joined the Scott Polar Research Institute, Cambridge, as a part-time Research Fellow. Through his work at the Foreign Office he was instrumental in gaining government support for the **Norwegian-British-Swedish Antarctic Expedition 1949–52** and continuing support for the Falkland Islands Dependencies Survey. He was involved in drafting the Antarctic Treaty of 1959, and in 1961 became official UK observer on the United States Naval 'Operation Deep-Freeze'. At the Scott Polar Research Institute he undertook general supervision of library and information activities, developed the Universal Decimal Classification for use in polar libraries, and made the library collections accessible worldwide by publication of its catalogue. He died on 9 October 1978.
(CH)

Roberts, Cape. 77°02′S, 163°12′E. Prominent cape forming the southern entrance to Granite Harbour, Scott Coast. Between 1997 and 2000 sea ice off the cape has been the site of a US/New Zealand joint seabed drilling programme, raising cores of sediment to provide information on the glacial and climatic history of the region.

Robertson, MacPherson. (d. 1945). Australian industrialist and philanthropist. Generous with his fortune (based on manufacture of confectionery), and interested in Antarctica, he provided funding that made possible Sir Douglas Mawson's **British, Australian and New Zealand Antarctic Research Expedition 1929–31**. Mac.Robertson Land, discovered during the expedition, was named for him. He died in Melbourne on 20 August 1945.

Rockefeller Mountains. 78°00′S, 155°00′W. A scattered group of low ridges and nunataks forming the interior of the western end of Edward VII Peninsula. Discovered by the first Byrd expedition in 1929, it was named for John D. Rockefeller, Jr., an expedition supporter.

Rockefeller Plateau. 80°00′S, 135°00′W. Extensive ice plateau 1000–1500 m. high in central Marie Byrd Land, named for John D. Rockefeller, Jr., who supported the Second Byrd Antarctic Expedition 1933–35.

Rockhopper penguin. *Eudyptes chrysocome*. Smallest of the crested penguins, standing 50 cm (20 in) tall and weighing up to 2.8 kg (6 lb), rockhoppers are distinguished by their black forehead and yellow and black crests over the eyes. They live mainly on the temperate islands: those of the warm temperate islands (*E. c. mosleyi*) have longer and fuller crests than the southern nominate subspecies. Rockhoppers have a distinctive hopping gait that allows them to climb steep cliffs and scree slopes. They feed mainly on shoaling crustaceans, small fish and squid. Like other crested penguins, rockhoppers breed in open colonies, often on tussock-covered cliffs that they share with albatrosses and other petrels. They lay two eggs of different sizes, usually incubating only the second, larger one, and rearing a single brown chick. The world population is estimated at 3 700 000 pairs.

Rockhopper penguin, representative of the long-crested northern subspecies *Eudyptes chrysocome moseleyi*. Photo: BS

Rogers, Allan Frederick. (1918–90). British polar medical officer. A graduate in medicine from Bristol University, where he was also a lecturer in physiology, Rogers joined the **Commonwealth Trans-Antarctic Expedition 1955–58** to undertake a programme of physiological research. At Shackleton Base he measured consumption of food and drink, energy expenditure, sleep rhythms and acclimatisation to polar conditions, and was one of the party that crossed the continent to the Ross Sea. On return to Britain he resumed his lectureship, studying low blood pressure syndrome in neonates and infants. Mt Rogers in the Shackleton Range is named for him. Rogers died in June 1990.
(CH)

Roi Baudouin Station. 70°26′S, 24°19′E. Former Belgian and Belgian/Dutch research station built in January 1958 for the International Geophysical Year 1957–58, and abandoned in 1961. Revisited in 1963, the buildings had disappeared under 5 m of snow. A new station, built near the original site by a Belgian-Dutch team, was occupied for three years 1963–66.

Rongé Island. 64°43′S, 62°41′W. A mountainous island forming the west side of Errera Channel, Danco Coast of Graham Land. Discovered by the **Belgian Antarctic Expedition 1897–99**, it was named for Mme. de Rongé, a supporter of the expedition.

Ronne, Finn. (1899–1980). American polar explorer. Born in Horten, Norway, son of the polar explorer Martin Ronne, he trained as a mechanical and marine engineer and naval architect. In 1923 he emigrated to the United States, becoming a mechanical engineer with Westinghouse Corporation. He joined **Byrd's Second Antarctic Expedition 1933–35**, taking part in the sledging programme as a skier and dog driver, and the **United States Antarctic Service Expedition 1939–41**, in which he sledged from East Base to the southern end of King George VI Sound. After service in World War II with the US Navy, he returned to East Base with a private expedition, the **Ronne Antarctic Research Expedition 1947–48**. During the International Geophysical Year 1957–58 he commanded the US base Ellsworth Station. Ronne retired from the navy in 1962 with the rank of captain. He died in Bethesda, Md. on 12 January 1980.

Ronne Antarctic Research Expedition 1947–48. Private American expedition based on Stonington Island, Fallières Coast, that undertook aerial and ground survey on both sides of Antarctic Peninsula. Cdr Finn **Ronne**, USNR, who had previously served with the second Byrd expedition and at the East Base of the **United States Antarctic Service Expedition 1939–41**, reoccupied the old expedition hut on Stonington Island on 12 March 1947. His party of 22 including three pilots, a group of scientists, and two women, his wife Edith (Jackie), and Jennie Darlington, the wife of chief pilot Harry Darlington III, who became the first women to winter in Antarctica. The expedition ship *Port of Beaumont*, commanded by Cdr Isaac Schlossbach USN (Ret.), wintered in the bay immediately behind the island. The expedition cooperated with a British party of 11, based a short distance across the island, at Base E of the **Falkland Islands Dependencies Survey**. US geologists Robert Nichols and Robert Dodson made a 100-day sledging journey south along the coast of Marguerite Bay to King George VI Sound. Joint Anglo-American survey parties led initially by British surveyor Douglas Mason, later by the British base leader Kenelm Butler, sledged across the peninsula and south along the east coast to Mt Tricorn. These provided a degree of ground control for extensive photographic flights by US pilots Chuck Adams and Jim Lassiter across the shelf ice south of the Weddell Sea (part of which is now the Ronne Ice Shelf). Photographic flights were made also over Alexander Island and Charcot Island. The expedition ship was released from the bay ice by two visiting icebreakers of Operation Windmill on 19 February 1948, and started for home on the following day.
Further reading: Ronne (1949); Darlington (1949).

Ronne Ice Shelf. 78°30′S, 61°00′W. Extensive ice shelf (Antarctica's second largest) bordering the Weddell Sea, between the east coast of Palmer Land, Antarctic Peninsula and Berkener Island, extending south to Ellsworth Land. The shelf was named for Cdr Finn Ronne USNR, leader of the **Ronne Antarctic Research Expedition 1946–48**, which overflew and photographed it.

Rookery Islands. 67°37′S, 62°33′E. Group of islands in Holme Bay, Mac.Robertson Land, remarkable for its rich avifauna of penguins and petrels. An area of approximately 30 km^2 (11.7 sq miles) is designated SPA No. 2 (Rookery Islands).

Rooney, Felix. (d. 1965). A member of the ship's crew during the **British Antarctic (*Nimrod*) Expedition 1907–9**, Rooney remained in New Zealand on the return of the expedition, serving as a seaman in trans-Tasman and coastal ships. He died in Christchurch, New Zealand, on 4 November 1965.
(CH)

Roosevelt Island. 79°25′S, 162°00′W. A totally ice-covered island, about 130 km (80 miles) long and 65 km (40 miles) wide, forming a prominent bump of up to 550 m (1800 ft) towards the eastern end of Ross Ice Shelf. The island lies immediately south of Bay of Whales, and probably accounts for its constant re-formation. Discovered by the Second Byrd Antarctic Expedition 1933–35, it was named for Franklin D. Roosevelt, the contemporary United States president.

Rosamel Island. 63°34′S, 56°17′W. A circular island about 1.5 km (1 mile) across ringed by steep basalt cliffs, rising to 435 m. (1427 ft). It stands prominently in Antarctic Sound, between Andersson Island and Dundee Island, off Tabarin Peninsula, Graham Land. Discovered by the **French Naval Expedition 1837–40**, it was named for

Admiral Claude de Rosamel, the French Minister of Marine who sanctioned the expedition.

Rosita Harbour. 54°01′S, 37°27′W. Sheltered harbour at the west end of Bay of Isles, South Georgia, used by whalers and named for the whale catcher *Rosita*.

Ross, James Clark. (1800–62). British naval officer and polar explorer. Born in London of a Scottish family, James Clark joined the Royal Navy in 1812, under the patronage of his uncle Capt. John Ross. In 1818 he sailed on his first Arctic expedition, as a midshipman in HMS *Isabella* under his uncle's command, to explore Baffin Bay for an entrance to the legendary but as yet undiscovered Northwest Passage. Between 1819 and 1827 he took part in four Arctic expeditions under Sir William Parry, attempting to explore northward from eastern Canada and Svalbard. In 1829–33 he was second-in-command to John Ross on a private expedition, again seeking the Northwest Passage. He led several overland sledging parties, on May 31 1831 reaching the North Magnetic Pole. Promoted to captain, between 1835 and 1838 Ross pursued magnetometry studies in and around Britain.

In 1839 he was appointed to command an expedition of two stoutly built monitors, HMSs *Erebus* and *Terror*, to investigate magnetic variation in the southern hemisphere, if possible to penetrate the pack ice to locate and reach the South Magnetic Pole. The **British Naval Expedition 1839–43,** with Ross in HMS *Erebus* and Capt. F. M. R. Crozier commanding HMS *Terror*, left England in September. Establishing magnetic observatories in St. Helena, Cape Town, and Iles Kerguelen, and taking running observations *en route*, the two ships reached Hobart, Tasmania early in 1840. Later in the same year Ross and Crozier headed south into the Southern Ocean, crossing the Antarctic Circle on 1 January 1841 and continuing south through heavy pack ice into the huge bight now called the Ross Sea. Ten days later the expedition sighted the mountains of a hitherto unknown coast, which he called Victoria Land and claimed for Britain. Continuing south, he discovered Ross Island, with twin peaks that he named Erebus and Terror, and the huge ice shelf that also bears his name. Less to his own satisfaction, he located the South Magnetic Pole behind the mountains of Victoria Land, out of reach either by boat or by sledging. After wintering in Australia, he returned to the Ross Sea in December 1841, then proceeded to the Falkland Islands and a further foray into the South American sector of Antarctica. On return to England he was knighted. In 1848–49 he led a further expedition to the Arctic in search of Sir John Franklin's lost expedition. He retired from the Navy in 1856 with the rank of rear admiral, and died at his home in Aylesbury in 1862.

Capt. James Clark Ross RN. Photo: Scott Polar Research Institute

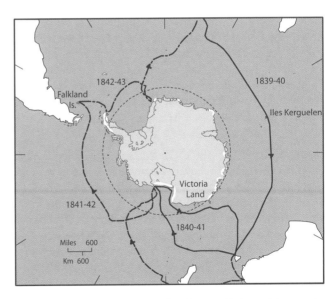

The voyages of Ross and Crozier to Hobart, Tasmania; from Hobart to Victoria Land; from Hobart via Sydney and the Ross Sea to the Falkland Islands; and from the Falkland Islands to the eastern Weddell Sea

Ross Archipelago. 77°30′S, 167°00′E. Extended group of volcanic islands in Victoria Land, in the southwestern corner of the Ross Sea. The islands include Beaufort Island, Ross Island, the Dellbridge Islands, Black Island and White Island. Brown Peninsula marks the southern end of the geological formation.

Ross Dependency. Triangular sector of Antarctica bounded by 160°E (bordering Australian Antarctic Territory) and 150°W, and extending from 60°S to the South Pole. The territory was defined by a British Order in Council of July 1923, provision being made for its administration by the Government of New Zealand. The territory includes Victoria Land, most of the Transantarctic Mountains, the Ross Ice Shelf, and the northwestern corner of Marie Byrd Land.

Ross Island. 77°30′S, 168°00′E. A large T-shaped volcanic island 80 km (50 miles) long forming the eastern coast of McMurdo Sound, off the Scott Coast of East Antarctica. Partly ice-capped, it emerges from the western end of Ross Ice Shelf. It incorporates a complex of three volcanoes, Mt Erebus (3795 m, 12 448 ft), Mt Terror (3230 m, 10 594 ft) and Mt Bird (1765 m, 5789 ft). Only Mt Erebus is currently active. The island was discovered in 1841 by the **British Naval Expedition 1839–43** and named for its commander, Capt. J. C. Ross RN. Hut Point, Cape Evans and Cape Royds, on its eastern flank, are historic sites associated with British expeditions of the period 1901–13: current year-round research stations are McMurdo (United States) and Scott Base (New Zealand), respectively at Hut Point and Pram Point.

Ross Ice Shelf. 81°30′S, 175°00′W. Extensive ice shelf (Antarctica's largest) between Marie Byrd Land and Victoria Land, bordering the Ross Sea. The frontal ice cliff (later termed the Barrier) was discovered and explored by the **British Naval Expedition 1839–43**, and named for its leader, Capt. J. Clark Ross RN.

Ross Sea. 77°S, 180°. Coastal sea occupying a deep bight between East and West Antarctica, bounded by Cape Colbeck and the Shirase Coast in the east, the Ross Ice Shelf in the south, and the coast of Victoria Land in the west. There is no definitive northern boundary. The sea was named for Capt. James Clark Ross RN, who discovered and first penetrated it in 1841. Much of the eastern half is ice-filled throughout the year. A polynya that forms every year along the Victoria Land Coast and Ross Ice Shelf (or Barrier) allows summer access to McMurdo Sound and the 800 km (500 mile) long cliffs (barrier) of the shelf. The Ross Sea therefore provides Antarctica's most reliable marine route to high latitudes.

Ross seal. *Ommatophoca rossi*. Ross seals live singly or in small groups, favouring the densest areas of pack ice. Up to 2.3 m (7.5 ft) long and weighing up to 200 kg (440 lb), they are the least-known of the Antarctic seals. The fur is dark grey on top, silver grey underneath. Huge eyes and needle-like teeth suggest that they feed deep, probably on squid. They pup in November and December. A population of up to 100 000 is estimated.
Further reading: Bonner (1982); Laws (1993).
(JPC)

Rothera Point. 67°34′S, 68°06′W. Headland of Square Peninsula, at the southeastern corner of Adelaide Island. An area of approximately 0.1 km^2 (0.04 sq miles) is designated SSSI No. 9 (Rothera Point) to facilitate studies of the effects of man (including the proximity of Rothera Station airstrip) on an Antarctic fellfield ecosystem.

Rothera Station. 67°34′S, 68°07′W. British permanent year-round station in Ryder Bay, Adelaide Island. Built in 1976 on a rocky promontory at the southern end of Wormald Ice Piedmont, and since extended, the station includes a dock, scientific laboratories and an aviation facility. The crushed rock runaway, 900 m (2952 ft) long, accommodates both intercontinental and local flights, making Rothera the logistics centre for British Antarctic Survey operations in the Antarctic Peninsula area.

Rothschild Island. 69°25′S, 72°30′W. An ice-mantled island 27 km (17 miles) long off the northwest corner of Alexander Island, in the entrance to Wilkins Sound. It was discovered and charted by the **French Antarctic (*Pourquoi Pas?*) Expedition 1908–10**, and named for Baron Edouard-Alphonse de Rothschild, a prominent French banker.

Royal penguin. *Eudyptes schlegeli*. Crested penguins, markedly similar to **macaroni penguins** but slightly larger, with golden yellow crests and white chin. Some ornithologists regard them as a subspecies (*E. chrysolophus schlegeli*). They breed in extensive colonies, mainly on Macquarie Island, where macaronis are rarely seen: the breeding cycle is similar to that of macaronis. The world population is estimated at 850 000 pairs.

Royal Society International Geophysical Year Antarctic Expedition 1955–57. British scientific expedition that in January 1956 established a research station on the Brunt Ice Shelf, Caird Coast, specifically for two years' geophysical studies. Led in the first year by Surg. Lieut-Cdr David. G. Dalgliesh RN, in the second by Lt Col Robert A. Smart RAMC, the expedition was fortunate in finding a site that justified continuing research beyond the two years originally intended. The station at 'Halley Bay' (an indentation in the ice shelf named for Edmund Halley, an eighteenth-century British Astronomer Royal), continued operations for a further two years, and in 1959 was taken over by the Falkland Islands Dependencies Survey (see **Halley Station**).

Royal Society Range. 78°10′S, 162°40′E. A massive range of mountains forming a barrier between the polar plateau and the Hillary and Shackleton coasts. Several peaks rise over 3000 m (9840 ft): Mt Lister, the highest, rises to 4025 m (13 202 ft). The range was identified and explored via its major glacier by sledging parties of the British National Antarctic Expedition 1901–4, and named for the prominent British scientific society that had supported the expedition.

Royds, Charles William Rawson. (1876–1930). British naval explorer. Born on 1 February 1876, he entered the Royal Navy from school. After varied service with the fleet he joined the **British National Antarctic Expedition 1901–4** as first lieutenant in *Discovery*. He was also meteorologist (for which he received special training in the Ben Nevis Observatory), and took part in the sledging programme, leading a sledging journey of exploration across the Ross Ice Shelf. He served throughout World War I, became Director of Physical Training and Sports, and commodore of Devonport Royal Naval Barracks, retiring in 1926 with the rank of rear admiral. In retirement Royds was appointed Deputy Commissioner of the Metropolitan Police, promoted to vice admiral and knighted in 1929. He died on 31 Dec 1930.

Royds, Cape. 77°33′S, 166°08′E. Low-lying rocky cape north of Barne Glacier, Ross Island, named for Lt **C. W. R. Royds**, meteorologist of the **British National Antarctic Expedition 1901–4**. Chosen by Ernest Shackleton as the shore base of the **British Antarctic (*Nimrod*) Expedition 1907–9**, the cape accommodates the expedition hut and also Antarctica's southernmost colony of Adélie penguins. An area of approximately 4.6 km^2 (1.8 sq miles), including an area of sea ice extending 500 m (1640 ft) offshore, was designated SSSI No. 1 (Cape Royds) to protect these features. The hut is scheduled as HSM 15, and the area immediately surrounding it is designated SPA No. 27, to allow the application of a protective management plan.

Rudmose-Brown, Robert Neal. (1879–1957). British expedition botanist and geographer. Born in London, he read natural sciences at Aberdeen University and lectured in botany at University College, Dundee. Making the acquaintance of Dr W. S. Bruce, he joined him in the **Scottish National Antarctic Expedition 1902–4**, wintering in *Scotia* at the South Orkney Islands, where he made extensive biological collections. Later, as a lecturer in geography at Sheffield University, he spent several seasons as a field botanist in Svalbard. Rudmose-Brown

Prefabricated hut at Cape Royds, providing living quarters for Shackleton's **British Antarctic (*Nimrod*) Expedition 1907–09**. Photo: Scott Polar Research Institute

was appointed Professor of Geography at Sheffield in 1931, retiring in 1948. He died on 27 January 1957.

Ruppert Coast. 75°S, 142°W. Part of the south coast of Marie Byrd Land. Bounded by Cape Burks in the east and Brennan Point in the west, the coast consists of ice-covered capes and ridges, with steep glaciers falling from the interior plateau to merge into ice shelves or protrude as glacier tongues. The coast was identified by pilots of the Second Byrd Antarctic Expedition 1933–35, and named for Col Jacob Ruppert, a supporter of the expedition.

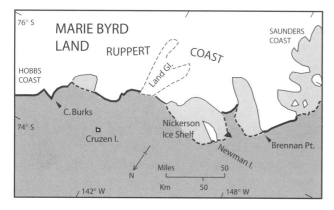

Ruppert Coast

Russkaya Station. 74°42′S, 136°51′W. Soviet research station on Hobbs Coast, Marie Byrd Land. After early efforts to establish this station from 1972–73, which failed due to persistently difficult ice conditions, it was eventually established in 1979. It operated year-round until 1989.

Russian Naval Expedition 1819–21. Russia's first major contribution to Antarctic exploration was a lengthy expedition of two ships, the corvette *Vostok* (900 tons) and transport *Mirnyy* (530 tons), scheduled to run simultaneously with an equally extensive north polar expedition. Its commander, Fabian Gottlieb von **Bellingshausen**, an Estonian-born German serving in the Imperial Russian Navy, was an able and determined navigator. His second-in-command, Mikhail Petrovich **Lazarev**, was also widely travelled and experienced. Bellingshausen planned his expedition thoughtfully to complement the high-latitude voyages of James **Cook**. Where Cook had been forced northward by winds or ice, Bellingshausen sought to continue south, so completing the survey of the Southern Ocean and increasing the likelihood of discovering **Terra Australis Incognita**. Sailing from Kronstadt in July 1819, the two ships called at Portsmouth to pick up navigational equipment, and Rio de Janeiro. Their first Antarctic landfall was South Georgia, where they added substantially to Cook's survey, and to the South Sandwich chain where they charted and added Russian names to several of the northern islands. Moving southward and east along the ice edge during January and February 1820 they twice crossed the Antarctic Circle. On 16 February they penetrated the pack ice to 69°06′S, and later in the month penetrated again further east, reaching 66°53′S. At both points the Russian ships stood among tabular icebergs close to the ice cliffs of the southern continent, but neither then nor later did Bellingshausen claim to have sighted Antarctica.

The two ships made their way independently to Sydney and wintered in the Pacific Ocean. Leaving Sydney again in November 1820 they visited Macquarie Island, where Bellingshausen met and spoke to sealers engaged in elephant seal hunting. From there they sailed south to the ice edge and headed eastward across the Pacific Ocean sector below 60°S, continuing their high-latitude circumnavigation of the southern continent. On 10 January 1821, in 92°38′W, having crossed the Antarctic Circle for the sixth time, they pressed through pack ice to reach their southernmost point, 69°53′S. On the same day Bellingshausen sighted the dark rocks of an island or headland over 30 miles away. On further investigation this proved to be a lonely, ice-covered island with steep rocky cliffs, about 9.5 miles long, rising to over 4000 ft. They named it for Peter I, founder of the Russian navy (now Peter I Øy). Continuing eastward, one week later they saw at a distance the dark cliffs and peaks of a more extensive land, which they named Alexander I Land (now Alexander Island) after their current Tsar. On 24 January Bellingshausen sighted Smith Island, westernmost of the South Shetland Islands. On the following day, close

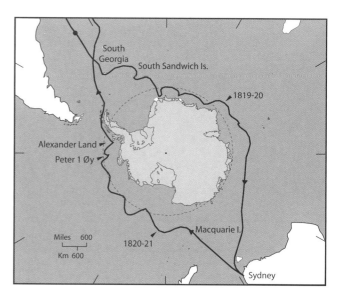

Bellingshausen's two voyages in the Southern Ocean. Between these voyages his two ships explored in the southern Pacific Ocean

to Deception Island, he met British and American sealers, among them Nathanial **Palmer**, with whom he discussed sealing. Continuing eastward, during the following three days he completed a running survey of the southern coasts of all the islands, ascribing Russian names to many geographical features that, unknown to him, had already been named in the surveys of Edward **Bransfield** and James **Weddell**. The expedition returned to Kronstadt in early July 1821.

Like Cook, Bellingshausen and Lazarev had sailed close to Antarctica, probably in sight of it, without the satisfaction of positive discovery. However, they encompassed more than two-thirds of the world in latitudes south of 60°S, employing meticulous navigational skills and scientific acumen. In so remarkable a voyage, the final dispatch of Terra Australis Incognita was a relatively minor achievement.
Further reading: Debenham (1945); Rubin (1982).

Rutford Ice Stream. 79°00′S, 81°00′W. Major ice stream draining the eastern flank of Ellsworth Mountains south-southeastward into the Ronne Ice Shelf. It was identified from aerial photographs and named for the distinguished American geologist and administrator Robert H. Rutford.

Rymill, John Riddoch. (1905–68). Australian polar surveyor and explorer. Born on 13 March 1905, he was educated at Melbourne University. His first visit to a polar region was with an expedition of the Cambridge University Department of Archaeology and Anthropology to Arctic Canada. There followed an invitation to join the British Arctic Air Route Expedition 1930–31 to Greenland, in which he learnt dog-sledging, navigation and kayaking, and a short follow-up expedition in 1933. With companions from these expeditions forming a trained and experienced nucleus, Rymill then organized and led the **British Graham Land Expedition 1934–37**, a small, economical but highly successful scientific expedition that explored the west coast of Graham Land and Palmer Land by aircraft and dog sledge, south to 72°S. Rymill received the Founders Medal of the Royal Geographical Society. After service with the Royal Australian Navy in World War II he returned to manage his family sheep farm in South Australia.

Rymill Coast. 67°30′S, 145°00′E. Part of the west coast of Palmer Land, West Antarctica, bounded by Cape Jeremy in the north and Cape Buttress Nunataks in the south. Facing onto George VI Sound, it consists mainly of ice cliffs, glaciers and ice shelves, backed by ice slopes rising to Dyer Plateau. The sound and hinterland were explored by sledging parties of the **British Graham Land Expedition 1934–37** (led by **John Rymill**, for whom the coast is named) and later by the **United States Antarctic Service Expedition 1939–41**.

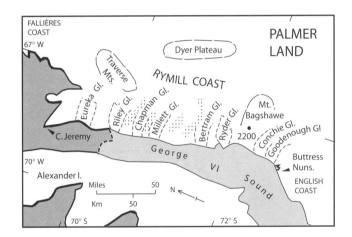

Rymill Coast

S

Sabrina Coast. 67°S, 119°E. Part of the coast of Wilkes Land bounded by Cape Poinsett in the west and Cape Southard in the east. Ice-covered mountains were seen first in or close to this position by John Balleny in March 1839, later by Charles Wilkes in 1840. The coast was finally identified from the air by Mawson in 1931. Almost completely ice-covered, terminating in ice cliffs, and rising to ice-covered slopes inland, it was named for the cutter *Sabrina* which, commanded by Capt. H. Freeman, accompanied Balleny on his voyage.

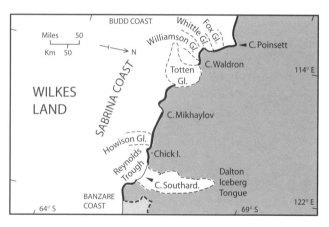

Sabrina Coast

Sabrina Island. 66°57′S, 163°17′E. A small, mainly ice-covered island of the Balleny Islands, standing 2 km (1.25 miles) south of Buckle Island. Discovered in 1839 by sealing captains J. Balleny and H. Freeman, it was named for Freeman's cutter *Sabrina*. With an area of approximately 0.4 km^2 (0.16 sq miles), the island was designated SPA No. 4 (Sabrina Island), as representative of a group that forms the most northerly Antarctic land in the Ross Sea region, with a flora and fauna that reflect circumpolar distributions in this latitude.

Saffery, John Hugh. (d. 1985) British polar aviator. After distinguished service in World War II Sqdn Ldr Saffery retired from the Royal Air Force to become chief pilot, later flying manager, of Hunting Aero Surveys. In this capacity he served as flying manager and deputy leader of the **Falkland Islands Dependencies Aerial Survey Expedition 1955–57**, which in two successive summers photographed some 90 000 km^2 (35 000 sq miles) of Antarctic peninsula and neighbouring islands. For this achievement Saffery and his aircrews were awarded the Johnston Memorial Trophy of the Guild of Aviators.

Salvesen, Harold Keith. (d. 1970). Scottish whaler and industrialist. Born into the prominent maritime industrial family **Salvesens of Leith**, he served with the Indian Army in World War I, then read economics at New College, Oxford. Joining the family firm in 1928, he concentrated on the whaling operations, gaining hands-on experience by frequently visiting Antarctica with the whaling fleet. He maximized efficiency and profits by pioneering more economical use of carcasses, development of by-products, and maintaining good relations with staff and crews. Highly respected within the industry, Salvesen became its spokesman in matters relating to national and international management, frequently despairing of what he regarded as effete or ill-advised policies based on inadequate scientific research. Predicting accurately the demise of Antarctic whaling in the early 1960s, he played an important role in re-directing the company's interests both before and afterwards. He died on 1 February 1970, aged 72.

Salvesen Range. 54°40′S, 36°07′W. A major snow-covered range forming much of the southeastern end of South Georgia, named for Capt. Harold Salvesen, a director of the Scottish whaling company Christian Salvesen and Co.

Salvesens of Leith. A family of Scottish-Norwegian shipowners, prominent in both Arctic and Antarctic whaling. In 1846 Johan Theodore Salvesen, a Norwegian businessman, established a shipbroking and forwarding agency in Leith, the port of Edinburgh. In 1855, with George V. Turnbull and others, he founded the firm of Turnbull and

Salvesen, developing trade based on importing Norwegian timber and exporting Scottish coal. Salve Christian Fredrik (born 1827), a younger brother of Johan Theodore, joined the firm, but left it in 1872 to found his own company, Christian Salvesen and Co., trading in coal, timber, fish oils, and products of Norwegian whaling, notably oils and meat meal. During the 1880s Christian took into partnership three of his sons; Johan Thomas (born in 1854), Frederick Gulov (1855) and Theodore Emile (1863). The company owned tramp steamers and cargo liners that traded as far afield as Malta and Alexandria.

In 1891 Christian Salvesen and Co. took shares in a small Arctic whaling schooner. From that beginning it moved gradually into whaling, first in Iceland, later in the Faroe and Shetland Islands, and in 1907 starting operations in the southern hemisphere, opening their first station on New Island, in the west Falkland Islands. In 1909 they secured a lease to operate on South Georgia, erecting a station in Stromness Bay which they called Leith Harbour after their home port. To obtain a second lease the Salvesen brothers set up a second company, South Georgia Co. Ltd., which operated also from Leith Harbour. In 1911 they began operations on the South Shetland Islands with two floating factory ships, *Neko* and *Horatio*, both converted from cargo ships. Stations and factory ships each required two or three whale catchers to keep them busy, and cargo ships, later tankers, to provide fuel and ship out the oil. By 1911 Christian Salvesen and Co. had become the world's largest whaling operators, feeding a fluctuating but expanding market with oil and meal products.

During the 1920s a third generation, notably Theodore's sons Noel and Harold Salvesen, developed the company's whaling interests. More of the whaling operations were conducted from purpose-built factories with stern slipways and greatly improved processing plants, working in the open sea, especially along the edge of the pack ice (see **Whaling in the Southern Ocean**). Salvesens prospered through the years of expansion, when over 40 such units were operating in Antarctic waters, and held on through the years of depression when whale products glutted the market and production almost ceased. The company prospered again during the difficult years after World War II, when the International Whaling Commission introduced a quota system to partition the rapidly declining Antarctic stocks of whales. From the late 1950s Christian Salvesen and Co. wound down its whaling operations, closing Leith Harbour, last of the land stations, in 1961, and selling off its remaining factory ships and catchers in 1962 and 1963. From 1964 the company developed its shipping interests, and diversified into fishing and fish processing, frozen foods, property development, brickmaking and other enterprises. South Georgia's Salvesen Range is named for Harold, who chaired the firm during its final years in southern whaling.

Further reading: Vamplew (1975).

Sandefjord Bay (Base C, later Base P). 60°37′S, 46°02′W. British research station at Moreton Point, Coronation Island, South Orkney Island. Built in February 1945, the third station of **British Naval Operation Tabarin 1943–45**, the hut was ill-founded in the middle of an extensive penguin colony, with no access to fresh water. It was never used, and quickly became derelict. Confusingly, the name Base C was later applied to a second station, built in 1946 at Cape Geddes, Laurie Island, South Orkney Islands. Still later, the Sandefjord Bay hut, by then completely derelict, was re-designated Base P.

SANAE Stations. A succession of South African research stations operating year-round on Kronprinsesse Märtha Kyst, Dronning Maud Land. The current station, built in 70°41′S, 2°50′W. on a nunatak 140 km (94 miles) from the coast, is the latest in a series that have occupied successive sites on Fimbulisen from January 1960 onward.

San Martin Station. See **General San Martin Station**.

San Telmo Island. 62°28′S, 60°49′W. A small ice-mantled island on the western side of Shirreff Cove, on the northwest coast of Livingston Island, South Shetland Islands. The island was named for a Spanish sailing ship which was lost while trying to round Cape Horn in 1819. Wreckage attributed to the ship was found by sealers two years later on nearby Half Moon Beach.

Sarie Marais Station. 72°03′S, 2°49′W. South African field station established in 1982–83 on Grunehogna, Ahlmannryggen, Dronning Maud Land, for use mainly by biological and geological field parties.

Sastrugi. Ridges formed by wind deposition and erosion and on a snow surface. Sastrugi 30 to 60 cm (1–2 ft) high are not unusual: strong, persistent winds may produce sastrugi over 1 m (3.3 ft) high, which can interfere seriously with sledging operations.

Satellite imagery. A technique used widely in monitoring the parameters and behaviour of the Antarctic ice sheet (see **Ice sheet**). Satellites carry sensors that receive and record electromagnetic radiation from Earth's surface, returning the information in a form that can be analysed and used for mapping and other purposes. Different portions of the electromagnetic spectrum – visible, near infrared, thermal infrared, microwave and other wavelengths yield different kinds of information, to be used in different

Sastrugi: a snow surface packed and eroded by strong winds. Photo: BS

ways. Some sensors, for example radar, are termed 'active' because they both transmit and receive pulses of radiation. Others, for example radiometers, are 'passive' because they simply receive reflected or naturally emitted radiation from surface features.

The television cameras and scanning radiometers of weather satellites, designed to cover whole weather systems rather than detail, give relatively low ground resolution. However, they record an almost instantaneous view over very large areas and monitor different properties of snow and ice over a range of wavelengths, an improvement over the visible wavelength sensors of early weather satellites. Synthetic Aperture Radars mounted on satellites, especially ERS-2 and the Canadian Radarsat, have enabled the construction of mosaics which display large-scale spatial variations in radar brightness. These relate to zones of melting and re-freezing, ice divides and surface textures associated with flow of the ice sheet. Radar altimeters are used to construct detailed contour maps of the Antarctic ice sheet, and to obtain an instantaneous view of the position of the ice margin. A repeatability of better than ±1 m has been claimed for elevation measurements at a given point. This is an improvement over any other method of height determination, except for those involving instruments on the ground. Precision is lost as the slope of the surface or the amplitude of surface undulations increases. Given the extremely high cost of polar field work, remote sensing from spacecraft provides a valuable alternative for providing basic data about the ice sheet.

Further reading: Massom (1991).
(CWS)

Saunders, Harold E. (1890–1961). American polar cartographer. A career officer in the United States Navy, Saunders specialized in naval architecture and ship design, but was also a distinguished cartographer with special interests in developing maps of Antarctica from aerial photographs. He was chiefly responsible for the cartography arising from Richard Byrd's expeditions of 1928–30 and 1933–35. In 1943 he was invited to join the newly-formed Special Committee on Antarctic Names of the US Board on Geographical Names, a committee that he subsequently chaired for 14 years, developing principles for international agreement on geographical nomenclature in Antarctica. Saunders Mountains and the Saunders Coast, Marie Byrd Land, are named for him. He died in Tacoma Park, MD. on 11 November 1961.

Saunders Coast. 77°45′S, 150°00′W. Part of the coast of Marie Byrd Land and Edward VII Land, West Antarctica. Bounded by Brennan Point (Block Bay) in the east and Cape Colbeck in the west, the coast presents a complex of massive mountain ranges, peninsulas and islands, set among the permanent ice of Sulzberger Ice Shelf. The western section was sighted by the **British National Antarctic Expedition 1901–4**, and members of the **Japanese Antarctic Expedition 1910–12** made a brief landing and sledging journey inland. The coast was overflown and photographed by aircraft of **Byrd's First Antarctic Expedition 1928–30**, and named for Capt. Harold Saunders USN, cartographer to the expedition.

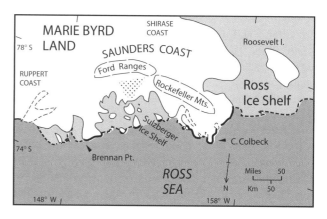

Saunders Coast

Saunders Island. 57°47′S, 26°27′W. A crescent-shaped island 8 km (5 miles) across, standing between Candlemas Island and Montagu Island in the central South Sandwich Islands. About 85 per cent ice-covered, it rises to a central peak, Mt Michael, 990 m (3247 ft) high. The horns of the crescent are low plains of lava and scoria. Steam is often reported to issue from the summit craters. The island was discovered in 1775 by Capt. J. Cook, RN and named for Sir Charles Saunders, contemporary First Lord of the Admiralty.

SCALOP. See **Standing Committee on Antarctic Logistics and Operations**.

SCAR. See **Scientific Committee on Antarctic Research**.

Schirmacheroasen. 70°45′S, 11°40′E. (Schirmacher Oasis). A group of low, ice-free hills, alternating with morainic shingle flats and shallow lakes. Lying between the edge of the continental ice plain and the coastal ice shelf, this is an **oasis** or anomalously ice-free area over 20 km (12.5 miles) long, with an area of about 50 km^2 (20 sq. miles). Its presence was noted during a photo-reconnaissance flight of the **German Antarctic (*Schwabenland*) Expedition 1938–39**, by pilot R. Schirmacher, for whom it was subsequently named. In 1960 it was visited by Soviet researchers from nearby Lazarev Station, and in February 1961 the permanent Soviet station Novalazarevskaya was established at the eastern end of the oasis. Later was added the German (Democratic Republic) research station Georg Forster. The oasis has been the subject of intensive multi-disciplinary studies by German and Soviet scientists (see *Further reading*).
Further reading: Borman and Fritzsch (1996).

Schouten, Willem Korneliszoon. (*c.* 1567–1625). Dutch navigator and explorer. Born in Hoorne, Schouten sailed with the Dutch East India Company. In 1615, commanding an expedition sponsored by merchant

Table 1 Membership of SCAR*, with dates of joining

State	Associate membership	Full membership
Argentina		3 February 1958
Australia		3 February 1958
Belgium		3 February 1958
Brazil		1 October 1984
Bulgaria	5 March 1995	
Canada	5 September 1994	27 July 1998
Chile		3 February 1958
China, People's Republic		23 June 1986
Colombia*[1]	(23 July 1990)	
Ecuador	12 September 1988	15 June 1992
Estonia	15 June 1992	
Finland	1 July 1988	23 July 1990
France		3 March 1958
Germany[2]		22 May 1978
India		1 October 1984
Italy	19 May 1987	12 September 1988
Japan		3 February 1958
Korea (South)	8 December 1987	23 July 1990
Netherlands	20 May 1987	23 July 1990
New Zealand		3 February 1958
Norway		3 February 1958
Pakistan*	15 June 1992	
Peru	14 April 1987	
Poland		22 May 1978
Russia[3]		3 February 1958
South Africa		3 February 1958
Spain	15 January 1987	23 July 1990
Sweden	24 March 1987	24 April 1984
Switzerland	16 June 1987	
Ukraine	5 September 1994	
United Kingdom		3 February 1958
United States of America		3 February 1958
Uruguay	29 July 1987	12 September 1988

Notes: *Austria, Colombia, Cuba, Czech Republic, Denmark, Greece, Guatemala, Hungary, Korea (North), Papua New Guinea, Romania, Slovakia, Turkey and Venezuela are signatories to the Antarctic Treaty, but not members of SCAR. Pakistan is an associate member of SCAR, but not a signatory to the Treaty.
[1] Colombia withdrew from SCAR 3 July 1995.
[2] German Democratic Republic joined 9 September 1981: Germany was unified 3 October 1990.
[3] Formerly the Soviet Union: represented by Russia from December 1991.
Source: R. K. Headland, unpublished.

adventurer Jakob le Maire, he sailed in *Eendracht*, accompanied by his brother Jan in *Hoorne*, to establish a southwesterly passage to the East Indies. *Hoorne* was destroyed by fire in December 1615: in the following month *Eendracht* became the first ship to navigate Drake Passage. Schouten identified Le Maire Strait, on the western flank of Tierra del Fuego, and named Cape Horn (Cabo de Hornos), the southern tip of South America, for his brother's lost ship and his home town.

Scientific Committee on Antarctic Research (SCAR). A non-governmental committee established by the International Council of Scientific Unions (ICSU), for the purpose of initiating, promoting and coordinating scientific activity in Antarctica. Through the ICSU network, the committee also coordinates Antarctic research with research in other parts of the world. Members of SCAR are the national scientific academies of countries with active and continuing programmes of Antarctic research, i.e. with consultative party status under the Antarctic Treaty. Such countries normally appoint national Antarctic committees to provide their research programmes and advise their delegates in scientific matters. Associate membership of SCAR is granted to national organizations which seek participation for scientific reasons but do not qualify for full membership, and Union membership is available for other ICSU organizations that seek to monitor, or to participate in, Antarctic research. Delegates are mostly scientists who are active in Antarctic research. SCAR's ability to coordinate research and offer advice is based mainly on its working groups of scientists covering the major disciplines, and groups of specialists in specific research fields, which provide the necessary up-to-date scientific and technological knowledge. SCAR has a further important role in providing scientific advice on request to Antarctic Treaty Consultative Meetings. It publishes *SCAR Bulletin*, a quarterly report available separately or as part of each issue of *Polar Record*. The table shows current associate and full membership in January 2002.

Scotia Arc. Oceanic ridge looping eastward between Tierra del Fuego and Antarctic Peninsula, including submarine **Burdwood Bank** and a succession of islands including **Shag Rocks**, **South Georgia** and the **South Sandwich**, **South Orkney** and **South Shetland Islands**. Geologically and topographically similar to the Antarctic Peninsula, the islands form a series of inter-continental stepping stones, with submarine contours revealing their true relationship to each other. Almost all the islands are heavily glaciated down to sea level. Those of the southern and eastern arc, surrounded by sea ice in winter, are climatically similar to western Antarctic Peninsula. South Georgia is by contrast almost entirely free of sea ice, with a warmer climate throughout the year. The ridge forms on the periphery of an active plate where subduction and geological uplift are occurring. Several of the islands are volcanically active. Deception Island, in the South Shetlands, erupted most recently during the 1970s, and eight of the 11 South Sandwich Islands show steaming vents and fumaroles. The arc was named for the SY *Scotia*, expedition ship of the Scottish National Antarctic Expedition 1902–4.

Scotia Sea. 58°S, 35°W. Sea bounded by the submarine Scotia Ridge, South Georgia, the South Sandwich Islands, the submarine Bruce Ridge, and the South Orkney and South Shetland Islands. In the west it merges with Drake Passage, in the south with the Weddell Sea. The mean northern limit of pack ice runs diagonally across from southwest to northeast. Though the sea is normally free of fast ice, except around the southern islands, the southern half is often blocked in winter, spring and early summer by pack ice and fleets of ice bergs, driven northeastward from the cold, clockwise gyre of the Weddell Sea. The sea was first investigated by William Bruce of the **Scottish National Antarctic Expedition 1902–4**, and named for his expedition ship SY *Scotia*.

Scott, Robert Falcon. (1868–1912). British naval polar explorer. Born near Devonport, the son of a brewer, Scott became an officer cadet in the Royal Navy at the age of 13. He specialized as a torpedo officer, following a steady but uneventful early career. With no previous experience in polar work and no predilection for it, he was in 1900 selected to lead the **British National Antarctic Expedition 1901–4**, Britain's response to the contemporary demand for Antarctic scientific research. Recruiting a company of Royal Navy and merchant navy seamen aboard the purpose-built expedition ship *Discovery*, and with a strong scientific team, Scott led a conspicuously successful expedition to McMurdo Sound, with achievements in both exploration and scientific research.

Continuing his naval career, he was promoted to captain, decorated and returned to more routine fleet commands. Seeing a more interesting professional future in further Antarctic work, he planned an expedition to the South Pole. Ernest Shackleton, who had served and sledged with him in the *Discovery* expedition, in 1909 led the **British Antarctic (*Nimrod*) Expedition 1907–9**, locating the Pole on the featureless high ice plateau, and coming within 97 miles of it. This did not deter Scott: in 1910 he returned to McMurdo Sound with the **British Antarctic (*Terra Nova*) Expedition**, again fielding a strong scientific team, both naval and civilian, that included several companions from his previous expedition. The expedition wintered at Cape Evans. Spurred by a rival bid

Robert Falcon Scott (1868–1912). Photo: Scott Polar Research Institute

from Roald Amundsen (**Norwegian South Polar Expedition 1910–12**), who was equipped with excellent dog teams, Scott started out for the Pole in late October 1911. He followed his own earlier route across the Ross Ice Shelf, then took the route up the Beardmore Glacier that Shackleton had pioneered. Experimental tractors, dog teams of doubtful reliability, and ponies hauled depots across the ice shelf, successively giving way to the man-hauling on which Scott primarily relied for the final march to the Pole. The five-man polar party successfully traversed the plateau, reaching the Pole on 17 January 1912, to find that Amundsen's group had arrived and departed almost five weeks earlier. Unseasonably bad weather delayed the returning party and ultimately defeated them. Scott died with the last of his companions late in March. Controversy over Scott's personality, leadership and tragic death has tended to mask the creditable exploratory and scientific achievements of his two major Antarctic expeditions.

Further reading: Huntford (1979); Solomon (2001).

Scott Base. 77°51'S, 166°46'E. New Zealand permanent year-round research station on Pram Point, Ross Island, McMurdo Sound. Built in 1956–57 to accommodate the New Zealand contingent of the Commonwealth Trans-Antarctic Expedition, and a five-man International Geophysical Year team, it has operated continuously since then as the main centre for New Zealand's Antarctic research.

Scott Coast. 76°30'S, 162°30'E. Part of the coast of Victoria Land, East Antarctica, facing the Ross Sea. Bounded in the north by Cape Washington and in the south by Minna Bluff, this is a deeply indented coast of mountains that rise in ranks to the polar plateau, glaciers and piedmont ice shelves. Discovered by Capt. James Clark Ross in January 1841, the coast was explored by sledging parties of three early twentieth-century British polar expeditions and named for Capt. R. F. Scott RN, who led two of them.

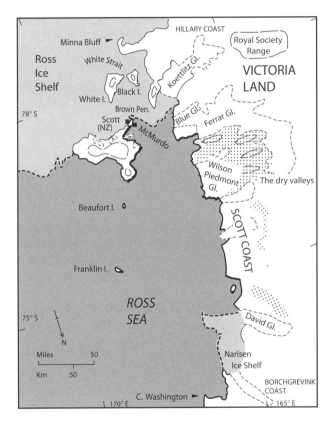

Scott Coast

Scott Island. 67°30'S, 180°00'W. An Antarctic island standing in isolation 700 km (440 miles) east of the Balleny Islands, in the Ross Sea. The island is a slab-sided cone of phonolitic rock 54 m (177 ft) tall and 3 km (1.9 miles) across, rising from a much broader submarine plateau. It appears to be the remnant of a fairly recent (Cenozoic) volcano. Discovered by Capt. W. Colbeck in December

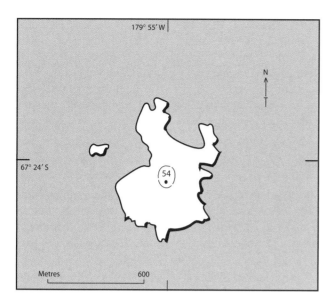

An isolated volcanic column, steep sided, heavily eroded and without beaches, Scott Island provides limited accommodation only for cliff-nesting seabirds. Height in m

1902 and named for Robert Falcon Scott, it is usually surrounded by pack ice, difficult to reach and almost impossible to land on except by helicopter. Sparse vegetation is reported from the top: breeding gulls, skuas and petrels invest the cliffs. The island is claimed by New Zealand as part of the Ross Dependency.

Scott Mountains. 67°30′S, 50°30′E. A scattering of mountain peaks and nunataks emerging from the continental ice cap south of Amundsen Bay, northern Enderby Land. Discovered during flights of the **British, Australian and New Zealand Antarctic Research Expedition 1929–31**, they were named for the British explorer Robert Falcon Scott.

Scott Polar Research Institute. Scientific institute in Cambridge, UK, a sub-department of the Department of Geography, University of Cambridge. The institute was founded in November 1920 on the initiative of Frank Debenham and Raymond Priestley, both of whom had served with Capt. R. F. Scott and sought to preserve his memory in a centre for fostering and encouraging polar research. Financial support was provided by a grant of £6000 from the Polar Research Fund of the Captain Scott Memorial Mansion House Fund, which had been raised by the Lord Mayor of London from public subscription after the death of Scott in 1912. Though based originally in the attic of the Sedgewick Museum of Geology, the Institute had no formal link with the university until May 1925, when a committee of management was appointed and Debenham became the first director. The university provided Lensfield House to accommodate a growing library and archive, until the Institute acquired its own building on Lensfield Road in 1934. The directorship was part-time until 1957, when Dr G. Robin became the first full-time director. The Institute houses an important polar library and archive, and since 1931 has published the journal *Polar Record*.

Scottish National Antarctic Expedition 1902–4. An entirely Scottish expedition that wintered in the South Orkney Islands, explored the Scotia Sea and started the longest-running Antarctic meteorological record. Responding to the 1895 appeal of the Sixth **International Geographical Congress**, Scottish surgeon and oceanographer William **Bruce** planned an expedition to the northern Antarctic Peninsula and Weddell Sea, an area that he had previously visited with the **Dundee Whaling Expedition 1892–93**. Eclipsed by Scott's **British National Antarctic Expedition 1901–4**, the smaller and more modest Scottish National Antarctic Expedition was with great difficulty and economy funded almost entirely from Scottish sources. Bruce sailed from Troon, western Scotland on 2 November 1902 in the Norwegian-built wooden whaling ship *Scotia* (Capt. Thomas Robertson), with a small scientific party that included botanist R. N. Rudmose-Brown, meteorologist Robert C. Mossman, and bacteriologist and medical officer Harvie Pirie, who was also a competent geologist. *Scotia* reached the Falkland Islands on 6 January 1903, and from there made for the South Orkney and South Sandwich Islands. Bruce and Robertson attempted to enter the Weddell Sea, but had left it too late in the season. Extricating themselves from heavy pack ice, the expedition returned to the South Orkney Islands, where on Laurie Island they found a small sheltered bay (quickly named Scotia Bay) and allowed the fast ice to form around the ship. Ashore the party built a stone hut, Omond House, which became the meteorological observatory, operating from April 1903. Using *Scotia* as a base, the scientists made sledging journeys with their small dog team, completing local surveys and collecting biological and geological specimens.

The ice in Scotia Bay broke up late in November. *Scotia* returned north to Buenos Aires for a refit. Perceiving the value of the meteorological station, Bruce offered it first to the British government, which declined to support it, then to the Argentine government, which was happy to accept it. *Scotia* returned south with three Argentine observers, who took over the station under Mossman's guidance. In late February the expedition packed up and departed for a second attempt to explore the Weddell Sea. This time they were more successful. Pushing through loose pack ice, on 3 March they sighted an ice-cliff that fronted an ice-covered coast – the eastern shore of the Weddell Sea. Bruce named his discovery Coats Land, for James and Andrew Coats, two Scottish brothers who had strongly supported his expedition. Again *Scotia* was almost beset

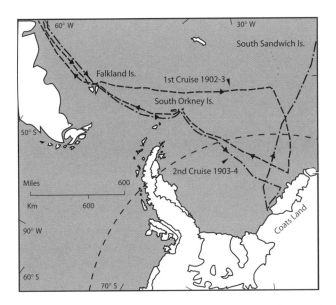

The two summer cruises of *Scotia* to and from the South Orkney Islands

by ice: again Robertson managed to escape, and the ship made her way to open water, and ultimately to the open sea. The expedition returned to Britain in July 1904.
Further reading: Bruce (no date).

Scrimshaw, Antarctic and southern. The name now given to the wide variety of functional and decorative objects, in the Western tradition, made as a pastime by people involved with whaling. Many small things were made, the commonest materials being baleen (whalebone), bone from whale mandibles and sperm whale teeth. The characteristic decoration consists of pictures and designs resembling engravings cut into polished surfaces with a point or blade and accentuated by colour. Most old scrimshaw dates from 1830–70. Engraved sperm whale teeth with Yankee associations are the most common, but inscriptions are rare and most scrimshanders anonymous.

Southern Ocean whaling developed from the late eighteenth-century search for fur and elephant seals. Sealers used the latter's teeth with albatross wing bones to make tobacco pipes. Scrimshaw from nineteenth-century exploration is rare, though a baleen cosh and three whale teeth with images of Patagonia survive from the 1826–36 surveying voyages of HMSs *Beagle* and *Adventure*. By the 1830s mixed whaling and sealing voyages were common, and whale jawbone was used for seal clubs and knife handles. They are usually impossible to localize, but a rare seal club of whale tooth and rope, with inscriptions including 'Croisetts ... 1843' is at Mystic Seaport Museum. A few decorated elephant seal teeth of similar age have been identified.

Southern Ocean scrimshaw is mainly mid-twentieth century, done by men from shore-based stations on South Georgia, and on factory ships and floating factories. Sperm whales were a minor catch, but their teeth remained the commonest material, generally sculpted or relief-carved into penguins, sometimes other birds and whales. In Norway, after World War II, such carved teeth were sold for the Red Cross charity. Modern materials such as vulcanite and coloured plastics were also used. With rorquals the main catch, and the whole carcass processed, new products were available: their humanoid 'earbones' (tympanic bullae) were painted as comic faces. Norwegians made whale eyes into small lamps: when dried and hardened, the translucent cornea became the shade and the tough sclerotic capsule held the bulb holder. Capsules were also turned or carved into ash trays and small bowls: if de-greased they darkened less and resembled onyx. Broken saw blades made excellent knives and model harpoon guns were filed from scrap metal. Expedition members and scientists on Antarctic shore bases also developed forms of modern scrimshaw: one specialized in making rings from the teeth of leopard seals.
Further reading: West and Credland (1995).
(JW)

Sea ice: Southern Ocean. Sea ice is formed when surface waters of sea or ocean freeze: for details of freezing processes see **Study Guide: Southern Oceans and Islands**. Sea ice is present year-round in the Southern Ocean. During the summer (February) minimum it occupies some 4 million km^2 (1.56 million sq. miles). Through autumn and winter it expands five-fold, toward a winter maximum area of up to 20 million km^2 (7.8 million sq. miles) in August and September. Its annual spread and decline provide one of Earth's most marked seasonal surface changes. The winter extent of sea ice surrounding the continent more than doubles the effective size of Antarctica as a reflector of solar radiation. Sea ice inhibits both the absorption of solar radiation and exchanges of energy and mass between ocean and atmosphere, with important implications for the planet's climate. It also influences the biological productivity of the surrounding waters (**Southern Ocean: biology**) and the ecology of the Antarctic islands.

As sea water freezes, salt is rejected from the ice crystalline structure and the salinity and density of the water below increase. Conversely, during sea ice melt in spring and summer, fresher water is released to the surface ocean. In the Southern Ocean these seasonal injections of salt and freshwater lead to extensive modifications of the water mass, strongly influencing the circulation of surface and deep waters around Antarctica.

Antarctic sea ice takes either of two forms. *Fast ice* (literally 'landfast' ice) forms continuous immobile sheets attached to the coast, that may extend many kilometres from the shore. *Pack ice* consists of floes packed tightly

Helicopter landed on a floe in recently-broken fast ice of the Weddell Sea. Photo: BS

or loosely together. Pack ice may result from the break-up of fast ice or of large offshore ice sheets: the break-up is usually caused by invading swell. The floes are mobile, drifting freely with winds and currents. Individual floes vary widely in thickness and size, depending on their origins and history. Their concentration too is highly variable, depending on the degree to which they are packed together by winds and currents. During the winter maximum, the outer edge of the pack ice may extend as far as 2200 km (1375 miles) from the Antarctic coastline.

The spring break-up and decay of Antarctic sea ice, most rapid in November and December, is aided by the Antarctic Circumpolar Current, which disperses the ice northward to melt in relatively warmer waters. By February, when the extent is minimal, there remains only an asymmetric pattern of fast and pack ice around the continent. Except for small bays and channels, most of East Antarctica's coastline is ice-free in summer, as is the northern tip of Antarctic Peninsula. In three large areas – the eastern Ross Sea, the Bellingshausen and Amundsen seas (60–140°W) and the western Weddell Sea – remnants of pack ice survive successive summers to become multi-year or perennial ice. These regions are relatively little affected by coastal currents. In the Weddell Sea, a strong clockwise circulation provides a semi-enclosed gyre, in which accumulated ice moves south, west and eventually north along the eastern coast of Antarctic Peninsula. Caught within this gyre in 1915, Shackleton's ship *Endurance* was trapped by pack ice under considerable lateral pressure (see **Imperial Trans-Antarctic Expedition 1914–17**). After the ship was crushed and sunk, members of the expedition continued to drift westward and north with the pack ice for five months, finally reaching Elephant Island.

When the pack ice reaches its maximum winter extent, about 18–21% of its total area remains ice-free. **Leads** and **polynyas** (persistent or recurring openings in sea ice) within the pack comprise areas where considerable heat exchanges occur between ocean and atmosphere. A very large polynya, detected by satellite observations, occurred in the Weddell Sea during three consecutive winters in 1974–76. Measuring 1000 × 350 km (625 × 219 miles), the polynya is thought to have been maintained by upwelling of warm subsurface waters over the nearby seabed feature Maud Rise. However, it has failed to reappear since the mid-1970s. More typical are smaller coastal polynyas maintained by local winds or currents that remove newly forming ice. A recurring polynya in Terra Nova Bay in the western Ross Sea, where strong offshore winds from the continent keep a small area of 50 × 50 km (31 × 31 miles) open throughout the winter, is estimated to produce 10% of the sea ice formed in the Ross Sea. Coastal polynyas regularly occur along the fronts of the Ross, Filchner, Ronne and Amery ice shelves, where **katabatic winds** blow directly off the continent over the front–ocean boundary.

Much of Antarctic sea ice remains less than 0.5 m (20 in) thick: about 25% of the pack ice is no thicker than 0.3 m (13 in) and only rarely does fast ice in coastal areas reach 2.0 m (6.5 ft) thick. By contrast, in the Arctic much of the ice is several years old, and may be up to 4 m (13 ft) thick in level areas, even thicker where it is ridged. These differences are attributable to geography. Antarctic pack ice at its northern limit forms a wide, unconstrained belt, continually opened and closed by winds and ocean currents during its constant shift eastward. In consequence, very little survives for more than one season. The enclosed Arctic basin retains significantly more multi-year ice in its centre, and is peripherally neither so mobile nor variable in extent. The melting process also is different in the two regions. Arctic fast ice and floes initially melt at the surface to form ponds or pools; Antarctic ice, in contact with a more open and warming ocean, melts more rapidly from the bottom and sides.

Passive microwave satellite observations of Antarctic and Arctic sea ice began in 1973. Since then several research teams have studied the interannual variability of sea ice extent in both regions, seeking to interpret trends in the context of global climate change. While significant changes have been observed in the extent and thickness of Arctic sea ice, satellite data have so far revealed no comparable trends in Antarctic sea ice.

Further reading: Gloersen et al. (1992); Nicol and Allison (1997); Wadhams (2000).
(LWB)

Seal Islands. 60°59'S, 55°23'W. Small islands and skerries in Sealers Passage, off Yelcho Point, the northwest corner of Elephant Island, South Shetland Islands. Seal Island the main island, together with all land and rocks within 5.5 km of its highest point, are scheduled under the **Convention on the Conservation of Antarctic Marine Living Resources** as Ecosystem Monitoring Site No. 1.

Sealing in the Southern Ocean. The vast stocks of **fur seals** and **elephant seals** that bred on the Southern Ocean islands were the first major Antarctic resource to suffer human exploitation. Capt. J. Cook's visit to South Georgia of 1775 reported fur seals in abundance on its beaches, stimulating British, American and other sealers to visit the island. Earliest records of sealing date from 1786, though earlier voyages may have gone unrecorded. The harvest lasted only a few decades. Sealer James Weddell calculated that by 1822 at least 1.2 million fur seals had been taken from South Georgia, and few remained. The search for fur seals spread rapidly to other Southern Ocean island groups, each of which was stripped of its stocks soon after discovery. By 1850 there were few southern fur seals left, but elephant seals had begun to provide an alternative resource. Though their skins were of limited worth, oil from their blubber was equivalent in value to whale oil. Ashore or afloat, the life and activities of sealing parties were brutish and dangerous. Though some sealers deplored the unrestricted slaughter on the beaches, recommending more controlled and sustainable systems of harvesting, none of the southern islands had local administrations capable of enforcing control measures. A brief revival of sealing in the 1850s was followed by a lull, and a subsequent brief revival in the 1870s with elephant seals the main prey and the few remaining fur seals as a bonus. American sealers from New England were predominant, often combining elephant oiling with whaling in lower latitudes.

Antarctic fur seal stocks took over a century to recover. Elephant seal stocks, never as drastically reduced, continued to support an uncertain harvest into the early twentieth century. In 1910 the British administration on South Georgia issued licences for elephant sealing to whaling companies then operating, which processed the oil in their factories before the start of the whaling season. The catch was limited to mature males. Breeding females and young were absolutely protected, and for many years an annual quota of 600 bulls was sustained with no noticeable decline in stocks. In 1948 world demand for oil prompted an increase in hunting, followed by a stock decline. From a biological study of the seals was devised a scientifically-based harvesting regime, which reversed the decline without decreasing oil production. Elephant sealing on South Georgia continued until 1964, when whaling ceased.

Fur seal stocks remained low until the mid-twentieth century, when they began to show a remarkable recovery throughout their former range. Both elephant seals and fur seals are once again plentiful on many Antarctic and Southern Ocean islands: South Georgia's breeding stocks of fur seals may now exceed their original size.

Tooth of elephant seal, 125 mm long, inscribed "Desolation 1883/Sea Elephant" from a New London sealing expedition. Private collection. Photo: JW

Sealing zones and reserves. The annex of the **Convention for the Conservation of Antarctic Seals** provides that each of six zones within the Antarctic region shall be closed in numerical sequence to all sealing operations involving the scheduled species, during the sealing season (1 September to the last day of February inclusive). The zones lie between: (1) 60°W and 120°W; (2) 0° and 60°W (including that part of the Weddell Sea lying west of 60°W); (3) 0° and 70°E; (4) 70° and 130°E; (5) 130°E and 170°W; and (6) 120°W and 170°W. In addition, killing or capturing seals is forbidden in the following reserves, which are seal breeding areas or sites of long-term research: (1) The area around the South Orkney Islands between 60°20′S and 60°56′S, 44°05′W and 46°25′W; (2) The area of the southwestern Ross Sea south of 76°S and west of 170°E; (3) The area of Edisto Inlet south and west of a line drawn between Cape Hallett (72°19′S, 170°18′E) and Helm Point (72°11′S, 170°00′E).

Seals, southern oceanic. Seals (Order Pinnipedia) live mainly in the cold oceans. Living species are divided into three families, the Phocidae ('earless' or true seals, 18 species), Otariidae ('eared' seals including fur seals

and sea lions, 12 species), and Odobenidae (walruses, one species). 'Earless' and 'eared' are misleading; all seals have ears and good hearing, but only the 'eared' seals have an external flap or pinna. Only phocids and otariids are represented in southern waters: there are no southern walruses. Phocid seals have fine hair rather than fur (see below). Zoologists divide them into two subfamilies, the Phocinae, which include all the northern polar seals, and the Monachinae, which include the warm-water genus *Monachus* (monk seals of the Mediterranean, Caribbean and Hawaiian Islands) and all five species of Southern Ocean seal. Thus southern phocids appear to have evolved from a northern warm-water stock and diverged in the southern hemisphere, quite independently of the Arctic seals and probably much more recently. Otariid seals are similarly divided into two subfamilies, the Arctocephalinae (fur seals) and Otariinae (sea lions), which evolved in the north Pacific Ocean and spread southward to become circumpolar in the southern hemisphere. Though dense otariid fur with its even denser undercoat might be expected to be beneficial in very cold water, only one of the seven species of fur seal breeds on Antarctic islands subject to sea ice, and all fur seals leave southern oceanic islands in winter. Sea lions, with coarse fur lacking an undercoat, are restricted to temperate and warmer waters: only one species occasionally ventures to cross the Antarctic Convergence. Phocid seals found in the southern oceans include Weddell, crabeater, Ross, leopard and southern elephant seals. Southern oceanic otariids include the Antarctic (Kerguelen) and sub-Antarctic **fur seals**, and South American, Australian and New Zealand (Hooker's) sea lions. Elephant seals, fur seals and sea lions were in the past severely hunted on many southern ocean islands (see **Sealing in the Southern Ocean**). All species of seal within the Antarctic Treaty area (i.e. south of the Antarctic Convergence) are currently included in the provisions of the **Convention for the Conservation of Antarctic Seals**, under which Ross seals and fur seals are afforded complete protection.
Further reading: Bonner (1982); King (1983).

Seas, Antarctic. Within the Southern Ocean surrounding the Antarctic continent, 13 local seas have been designated as separate entities, all but one (Scotia Sea) along the continental coast. The two largest, the Weddell Sea and Ross Sea, occupy unequivocally the large bights between East and West Antarctica. Lesser seas, usually identified for the convenience of marine biologists and hydrographers, are less clearly defined. Though all designations are to some degree useful, not all are generally accepted. Some areas designated as seas by earlier explorers (for example, Mackenzie Sea) have been reduced to bays. Three designated in the 1960s by Russian cartographers (Lazarev, Riiser-Larsen and Cosmonaut) overlap with the earlier-named, more extensive and more generally-accepted King Håkon VII Sea. Three from the same origins (Cooperative, Riiser-Larsen and Mawson) appear no longer to be supported even by their originators, in that they do not appear in the 1998 *Composite Gazetteer of Antarctica* (see **Study Guide: Information Sources**). For further details see individual entries.

Seasonality. Climatic variation between seasons. Antarctica's seasonal shifts of climate are strongly continental, with winters generally much colder than summers, and short, transient springs and autumns. The formation of winter fast ice in a broad belt surrounding the continent extends these harsh continental influences far beyond the coast. Though the Southern Ocean is generally much warmer than the continent, fast ice and fringing pack ice from early winter to late spring effectively keep the sea from warming the land. Their effects are felt for three to four months in the Antarctic Peninsula and Scotia Arc areas, and eight to ten months or longer in the far south. On shores of islands toward the northern edges of the pack ice zone, winters remain harsh but summers tend to be milder, especially in years when the ice disperses early. The relatively ice-free western flank of Antarctic Peninsula has milder summers than the icebound east coast, and still milder conditions prevail on the Scotia Arc islands, which are usually ice-free for several months each year. The South Orkney Islands, though slightly north of the South Shetlands, tend to be cooler year-round because of chilling influences of air and ocean from the icebound Weddell Sea, with harsher winters. Beyond the northern limit of pack ice both winters and summers are milder. The periantarctic and cool temperate islands have equable climates strongly dominated by oceanic influences, with relatively weak seasonal changes. See **seasons**.

Seasons. Climatic divisions of the year, varying most intensively in polar regions. Seasons occurs because Earth's axis is tilted some 23°28′ from vertical in relation to the ecliptic – the plane of rotation of Earth about the sun. In consequence all latitudes receive varying amounts of illumination throughout the year. Differences in day length between summer and winter are negligible in equatorial regions, stronger in temperate regions, and strongest of all in polar regions. Areas poleward of the Arctic and Antarctic circles receive 24-hour illumination around midsummer, none at all around midwinter (see **day length**). At the North and South Geographic Poles, days and seasons converge: days lasting six months alternate with six months of night. Levels of incident radiation are particularly high during the long polar summer days. Of its total annual incidence of solar radiation, **Vostok** (78°S) receives over 82 per cent during the four summer months of November to

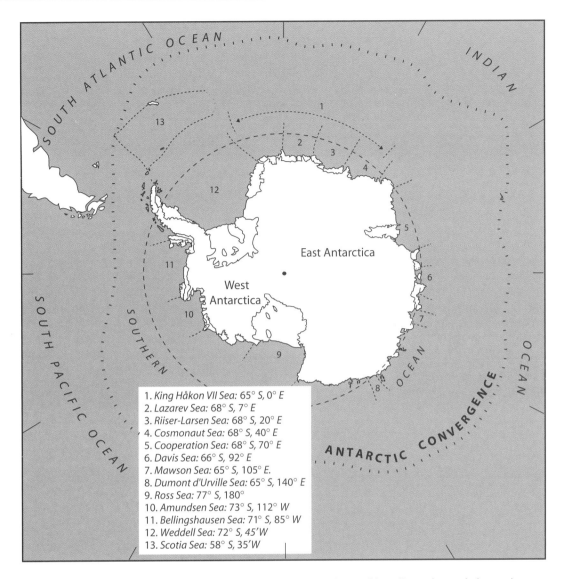

The 13 Antarctic seas. All but one (Scotia Sea) are inshore to the continent. Not all continental shores have seas ascribed to them. King Håkon VII Sea overlaps with three others

February, and none during the four winter months of May to August.

Second-year ice. Floating ice that has failed to melt during its first year, and has persisted through a second winter. While most Antarctic **sea ice** disperses and melts after a single year, patches tend to persist in particular areas, notably the bights of the Weddell and Ross Seas, and in the Amundsen and Bellingshausen Seas off the coast of West Antarctica. Here may be found third-year and older ice, characterized by greater thickness and rugged profile.

Sei whale. *Balaenoptera borealis*. Third largest of the rorquals, sei whales are long and slender, growing to lengths of 20 m and weights of 15 tonnes. Predominantly dark grey, with patches of paler grey under the throat and belly, they are distinguished from other species by the single longitudinal ridge on the nose. Like other rorquals they feed mainly on krill and small fish. Southern stocks live mainly in temperate and cool waters, penetrating south in summer. During the early decades of Antarctic whaling they were largely ignored, because they yielded far less oil than other species. (in terms of blue whale units, a sei rated only one-sixth of a blue whale). As the more profitable species declined, catches of seis increased, especially after World War II. They peaked briefly in the mid-1960s after the decline of fin whale stocks, but were never plentiful enough to form more than a small proportion of the total annual catch in Antarctic waters.

Further reading: Bonner (1989); Evans (1987).

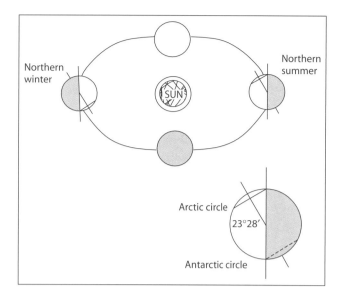

Earth's axis of rotation is tilted about 23°28′ from the vertical or 66°32′ from the ecliptic or plane of rotation about the sun. In the southern summer and northern winter (left) the whole area poleward of the Antarctic Circle is exposed to the sun throughout each 24 hour period: in the northern summer and southern winter (right) the area receives no direct illumination

Sentinel Range. 78°10′S, 85°30′W. A major mountain range forming the northern bloc of the Ellsworth Mountains, eastern Ellsworth Land. Over 160 km (100 miles) long and up to 48 km (30 miles) wide, it forms a series of escarpments overlooking the Ronne Ice Shelf. Its highest peaks rise over 4000 m (13 120 ft): Vinson Massif (4901 m., 16 075 ft), an outlier towards the southern end, is Antarctica's highest peak.

Seymour Island. 64°17′S, 56°45′W. A steep, ice-mantled island 16 km (10 miles) long forming the southern shore of Erebus and Terror Gulf, discovered by the **British Naval Expedition 1839–43** and named for a contemporary British admiral. In 1892–93 Capt. C. A. Larsen discovered rich seams of fossil plants and animals, which have since been extensively studied. The island is the site of a major landing strip for Argentina's Marambio Station.

Shackleton, Ernest Henry. (1874–1922). British mariner and explorer who pioneered the route to the South Pole. Born in Kilkea, rural Ireland, Shackleton was one of six children of Anglo-Irish parents. Aged six, he moved with his family to Dublin, where his father Henry studied medicine. On qualifying four years later, Henry took up a practice in south London, sending his son to the local day-school, Dulwich College. Gifted with charm and intelligence, Shackleton worked hard in subjects that interested him, notably literature and poetry. At 16 he began training for the merchant marine, sailing in square-riggers and tramp steamers. In 1896 he took his First Mate's ticket, and in 1900 was appointed third officer in RRS *Discovery*, the newly-built ship that Lt Robert Scott was preparing for the British National Antarctic Expedition 1901–4. Shackleton helped to provision and load the ship, and prepare her for the long voyage.

Ernest Shackleton (1874–1922). Photo: Scott Polar Research Institute

Sailing in August 1901, *Discovery* crossed the world to New Zealand, then lumbered south to Antarctica, arriving in McMurdo Sound in early February 1902. Shackleton was a popular crew member and got on well with Scott. In autumn and early winter he learned to ski, drive dog teams and camp in the snow, finding time also to edit and write poems for the expedition newspaper, *South Polar Times*. Scott's long spring sledging journey with Edward Wilson and Shackleton into the interior of the continent was a gruelling march over the featureless Ross Ice Shelf, with ill-trained and underfed dogs, inadequate rations and poorly designed equipment. On their return from 82°15′S the party suffered from scurvy, and Shackleton showed symptoms of heart strain which left him unable to do more than walk beside the sledge. On return to the ship he remained an invalid for over a month. Against Shackleton's own wishes, Scott sent him home on the relief ship.

Shackleton married and settled in Edinburgh, becoming secretary of the Royal Scottish Geographical Society. Ever restless, he stood unsuccessfully for Parliament and became involved in money-making enterprises that failed to enrich him. In 1906 he announced his intention of mounting a private expedition to the South Pole. Though he himself inspired confidence, he had no official support. However, in August 1907 on the eve of his departure, King Edward VII inspected *Nimrod*, his tiny expedition ship, signifying approval of the operation. Shackleton finally left New Zealand with the **British Antarctic (*Nimrod*) Expedition 1907-9** in early January 1908.

From their base at Cape Royds, starting in late October, the expedition's main polar journey took Shackleton, Marshall, Wild and Adams along the previously explored route to 82°S in less than a month. By Christmas 1908 they had found a route up the formidable Beardmore Glacier. On inadequate rations, and already half-starving, they man-hauled across mile after mile of featureless polar plateau. On 9 January 1909 in 88°23'S the party abandoned the Pole and turned back. A hard struggle brought them to the coast on 4 March, just as their relief ship was about to sail for home.

Shackleton returned to Britain a hero, receiving a knighthood but unfortunately little income on which to sustain his new social position. After both Amundsen and Scott reached the South Pole, Shackleton determined on another expedition – one based on Filchner's concept of crossing Antarctica from one side to the other between the Ross and Weddell Seas. One ship, *Endurance*, would penetrate the Weddell Sea, landing a party that would approach the Pole from that side. Meanwhile a second ship, *Aurora*, would land parties in McMurdo Sound to lay depots that would help the polar party complete its crossing. The **Imperial Trans-Antarctic Expedition 1914-17** sailed in August 1914, returning after a succession of disasters and triumphs that added substantially to Shackleton's reputation, and rounded off an era in Antarctic exploration.

Shackleton returned to a war-torn Europe. He lectured in the USA on behalf of Britain, and served briefly with the army in Russia. After the war he picked up the threads of his former life, again joining a succession of business enterprises that began with hope and ended in disappointment. In 1920 Shackleton planned an expedition to Arctic Canada, which came to nothing. One year later, desperate to explore again, he planned an expedition to the Southern Ocean, funded largely by John Rowett, a wealthy friend. On 21 September 1921 the **Shackleton-Rowett Antarctic Expedition 1921-22** sailed from London in *Quest*, a small whale-catcher. Ill health dogged Shackleton across the South Atlantic Ocean. Arriving in South Georgia in early January, he suffered a series of heart attacks and died at Grytviken. He was buried in the whalers' graveyard overlooking the bay.

Further reading: Fisher and Fisher (1957); Huntford (1985).

Shackleton Coast. 82°00'S, 162°00'E. Part of the coast of the Ross Dependency, East Antarctica. Bounded by Airdrop Peak in the south and Cape Selborne in the north, and facing the Ross Ice Shelf, it comprises steep mountains rising inland to the polar plateau, interspersed with glaciers that descend and merge into the ice shelf. The coast was first visited by a sledging party seeking a route to the South Pole, led by the British explorer Ernest Shackleton, for whom it is named.

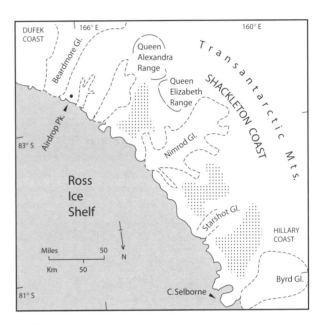

Shackleton Coast

Shackleton Ice Shelf. 66°00'S, 100°00'E. Extensive ice shelf on the Knox and Queen Mary coasts of East Antarctica. About 400 km (250 miles) long and up to 180 km (112 miles) wide at its western end, it is fed by a series of glaciers and ice streams, of which the Roscoe, Denman and Scott are the largest. Discovered by Lt. Charles Wilkes in 1840, it was explored by the Western Party of the **Australasian Antarctic Expedition 1911-14**, and named for the British explorer Sir Ernest Shackleton.

Shackleton Range. 80°30'S, 25°00'W. Isolated range of mountains of central Coats Land, overlooking the Filchner Ice Shelf. The range is bounded to the north by Slessor Glacier, which separates it from the Theron Mountains, and to the south by Recovery Glacier. Central to the group, Fuchs Dome rises to 1525 m (5002 ft):

some surrounding peaks rise to over 1800 m (5904 ft). The range was explored by the **Commonwealth Trans-Antarctic Expedition 1955–58** while searching for a route south. It was named for the British explorer Sir Ernest Shackleton.

Shackleton–Rowett Antarctic Expedition 1921–22. Sir Ernest Shackleton's final expedition to the Southern Ocean. Following unsuccessful attempts to raise funding for more ambitious projects, in 1921 Shackleton planned a modest expedition to search for some of the doubtful and unauthenticated islands of the Southern Ocean, to visit and explore some of the lesser known groups, and explore previously unvisited stretches of the continental coast. With financial backing from John Q. Rowett, a wealthy philanthropist friend, the expedition left London on 17 September 1921 in *Quest*, an elderly, 125-ton wooden sealer. Ship's captain and second-in-command was former colleague Cdr. Frank Wild RNVR: other Antarctic veterans aboard included the doctors, Alexander H. Macklin and James McIlroy, meteorologist Leonard Hussey and able seaman Thomas McLeod, all of whom had served in the **Imperial Trans-Antarctic Expedition 1914–17**. The Australian adventurer Hubert Wilkins joined the expedition as photographer, naturalist and pilot of a small floatplane that the expedition planned to pick up in Cape Town. James Marr, who later became a prominent marine biologist and expedition leader, was a Boy Scout on attachment to the party. Engine troubles delayed the old ship, and Cape Town was struck from the itinerary. After a stormy crossing from South America, *Quest* reached South Georgia on 3 January 1922, some five weeks behind schedule. In the early morning of 5 January Shackleton suffered a heart attack and died.

While a freighter carried his body to Montevideo for return to Britain, *Quest* continued the voyage under Frank Wild. Leaving South Georgia on 18 January, they visited Zavodovski Island in the South Sandwich group, then made two attempts to approach the continent in about 16°E, meeting heavy pack ice in 68° to 69°S. Returning westward across the Weddell Sea, they disposed of 'New South Greenland', a large landmass supposedly seen by early nineteenth-century sealer Benjamin Morrell. However, in 45°W *Quest* became dangerously entrapped in pack ice, emerging with difficulty in late March. Continuing eastward they made a running survey of Elephant Island, South Shetland Islands, returning to South Georgia on 6 April.

At the request of his widow, Shackleton's body had been returned to Grytviken and was already buried in the whalers' graveyard. His shipmates built a cairn and erected a cross on King Edward Point. The expedition's final calls were made at Gough Island, Tristan da Cunha and Cape Town.

Further reading: Marr (1923); Wild (1923).

Shackleton Station. 77°59′S, 37°09′W. Temporary British research station established on Filchner Ice Shelf in January 1956 by the **Commonwealth Trans-Antarctic Expedition 1955–58**. The crossing party left on 24 November 1957, and the station was closed on 27 December 1957.

Shag Rocks. 53°33′S, 42°02′W. An isolated cluster of six tall, triangular rocks and several smaller reefs 260 km (162 miles) west-northwest of South Georgia, forming part of the northern arm of Scotia Arc. The highest rises to 70 m (230 ft). Shags roost on the peaks: the steep sides support lichens, but there are few other signs of plant life. The sea close by is a rich feeding ground for seabirds and Antarctic fur seals.

Shcherbakov, Dimitriy Ivanovich. (d. 1966). Soviet geologist and academician. As Academician-Secretary of the Division of Geological and Geographical Sciences of the USSR Academy of Sciences 1953–1963, and chairman of the successive committees responsible for Soviet scientific work in the Antarctic, he played an important role in determining the course of research from 1955 until his death on 25 May 1966. Shcherbakov Range, in Orvinfjella, Dronning Maud Land, is named for him.
(CH)

Shearwaters. Medium-sized petrels of the family Procellariidae with long, narrow wings and a soaring and gliding mode of flight. They forage widely over the warmer oceans, spreading southward in summer but usually avoiding ice. Often solitary, seldom in groups of more than a dozen, they settle to feed on cephalopods, fish and crustaceans. White-chinned petrels ('shoemakers') *Procellaria aequinoctialis* – dark brown to black, with body length 50 cm (20 in), wing span up to 1.4 m (55 in) – breed in burrows on cool temperate and sub-Antarctic islands. Gray petrels *P. cinerea* and sooty shearwaters ('mutton-birds') *Puffinus griseus* favour warmer islands for breeding: the latter are particularly prominent in the New Zealand sector. Greater shearwaters *P. gravis* forage close to their breeding grounds on the Falkland Islands, Tristan da Cunha and Gough Island. Little shearwaters *P. assimilis* too breed on Tristan da Cunha and Gough Island, but seldom appear in southern waters. A small colony of flesh-footed shearwaters *P. carneipes* maintains a tenuous hold on Ile St. Paul, at the southern edge of the species' mainly tropical range.

Sheathbills. Flying birds of the order Charadriiformes, family Chionidae, that forage among penguin and cormorant colonies on many southern islands. Plump white birds with pink wattled faces and unwebbed feet, they feed also along the shore on tidal debris. They nest in sheltered corners under rocks, laying clutches of two to four white speckled eggs. American or snowy sheathbills *Chionis alba* nest on islands of the Scotia Arc and Antarctic Peninsula south to the Danco Coast. Most are summer-only residents, migrating in late autumn to South America: a few may remain year-round if food is plentiful. Black-faced sheathbills *C. minor* nest on cool temperate islands of the Indian Ocean, where they are year-round residents.

Shimizu Ice Stream. 85°11′S, 124°00′W. Narrow ice stream draining from the polar plateau between the Ohio and Wisconsin Ranges of the Horlick Mountains. The stream flows northwestward into Horlick Ice Stream and ultimately into the Ross Ice Shelf. It was identified from aerial photographs and named for the glaciologist H. Shimizu, who worked with American field parties in the area.

Shirase, Choku. (*c.* 1860–1946). Japanese Antarctic explorer. As a lieutenant in the Japanese navy, Shirase led the **Japanese Antarctic Expedition 1910–12**, which visited the Ross Sea and in January 1912 made short excursions by dog sledge across the Ross Ice Shelf and into Edward VII Land. In 1933, on the founding of the Nippon Polar Research Institute, Shirase became first president. He died at Nagoya on 10 September 1946.

Shirase Coast. 78°30′S, 156°00′W. Part of the coast of Marie Byrd Land. Bounded by Cape Colbeck in the north, it meets Siple Coast on a featureless ice plain in 80°00′S. The southern section faces the Ross Sea, presenting ice cliffs of piedmont shelves that rise steeply inland to the Rockefeller Mts.. North of Prestrud inlet heavily crevassed ice rises merge into the Ross Ice Shelf. The coast is named for Lt Choku Shirase, leader of the Japanese Antarctic Expedition 1910–12.

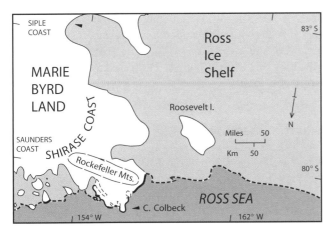

Shirase Coast

Shirreff, Cape. 62°27′S, 60°41′W. Peninsula close to the western end of northern Livingston Island, South Shetland Islands. Antarctic fur seal, elephant seal and penguin colonies make this an important area for biological research. An area of approximately 6.3 km^2 (2.5 sq miles) was in 1967 scheduled SPA No. 11, later rescheduled SSSI No. 32, to protect long-term research on these species. The penguin and seal colonies, and krill fisheries within the foraging range of these species, make this a critical site for inclusion in the ecosystem monitoring network. An area of approximately 3.5 km^2 (1.4 sq miles) between the cape and nearby Telmo Island is accordingly scheduled Ecosystem Monitoring Site No. 2.

Shore Lead. A persistent stretch of open water between a field of floating ice and a shore, usually caused and maintained by offshore winds.

Shuga. Lumps of sea ice a few cm in diameter, formed by agitation within grease ice or slush.

Signy Island. 60°43′S, 45°38′W. A rectangular island 6.5 km (4 miles) long immediately south of Coronation Island, South Orkney Islands. It was charted in 1825 by James Weddell, but remained unnamed until 1912–13, when Norwegian whaler Capt. P. Sorlle resurveyed the island and named it for his wife. A whaling station operated briefly in Factory Cove from 1920–21: part of the site is currently occupied by Signy, a British Antarctic Survey research station.

Signy Station (Base H). 60°43′S, 45°36′W. British research station at Factory Cove, Borge Bay, Signy Island, South Orkney Islands. The first station (Clifford House) was built by the Falkland Islands Dependencies Survey at Berntsen Point in March 1947. In 1955 the station was re-sited closer to the sea in Tønsberg House, Factory Cove, and the old buildings were destroyed. The station was extended from time to time, becoming an important centre for marine and terrestrial biological research. It operated year-round until 1996, thereafter only in summer.

Simpson, George Clarke. (1878–1965). British expedition atmospheric physicist. Born in Derby, he graduated in science from Manchester University and studied atmospheric physics at Göttingen. He returned to Manchester as a lecturer in physics, then in 1906 joined the India Meteorological Department. From India he joined the **British Antarctic (*Terra Nova*) Expedition 1910–13** as meteorologist, contributing outstanding reports on the weather and physics of the atmosphere. He returned to India, but in 1920 moved back to Britain as Director of the Meteorological Office. He retired in 1938, but returned during World War II to become superintendent of Kew Observatory. He died on 1 January 1965.

Siple, Paul Allman. (1908–68). American Antarctic explorer and administrator. Born in Montpelier, Ohio, he completed a first-year's study of biology and geology at Allegheny College, Meadville, PA. At the end of his first year he was selected by Richard Byrd to represent the Boy Scout movement on **Byrd's First Antarctic Expedition 1928–30**. Willing and intelligent service as a general assistant recommended him to the leader who, after he had completed his degree, recruited Siple to **Byrd's Second Antarctic Expedition 1933–35**, this time as a personal assistant and field scientist. Siple led an important dog-sledging traverse of Marie Byrd Land, and published papers on botany, ornithology and bacteriology. Completing a doctorate at Clark University, Worcester, Mass., he was appointed by Byrd to organize the logistics of the United States Antarctic Service Expedition 1939–41, and to be leader of the West Base, on the site of his two earlier expeditions. On return to the US, Siple wrote up the reports of his highly successful year's leadership. In World War II he became an army specialist on clothing for use in polar and tropical climates, and in 1946, now a Lieutenant Colonel, joined the Army General Staff as a research geographer and scientific advisor. In that capacity he took part in the US naval Operation Highjump in 1946–47, and Operation Deep-Freeze 1955–56, in preparation for the International Geophysical Year. During the IGY 1957–58 he commanded the South Pole station, again with conspicuous success. An influential figure in polar diplomacy during the development of the Antarctic Treaty, in 1963 he became the US Scientific Attaché to Australia and New Zealand. A crippling stroke in 1966 restricted his activities: he died on 25 November 1968.
Further reading: Siple (1931; 1936, 1959).

Siple Coast. 82°00′S, 155°00′W. Part of the west coast of Marie Byrd Land, West Antarctica, abutting into the Ross Dependency along meridian 150°W. Intermediate between Gould Coast (from a boundary in 83°S) and Shirase Coast (from 80°S), the coast marks an almost indefinable rise from the Ross Ice Shelf to the ice plain of Marie Byrd Land. It was named for American scientist Paul Siple, a member of several Byrd expeditions and first leader of Amundsen-Scott Station.

Siple Island. 73°39′S, 125°00′W. A large ice-covered island 112 km (70 miles) long embedded in the Getz Ice Shelf, off Bakutis Coast, Marie Byrd Land. Discovered

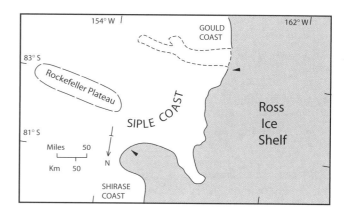

Siple Coast

from aerial photographs, it was named for Dr Paul A. Siple, biologist and leader in many United States expeditions.

Siple Station. 75°33′S, 83°33′W. Three United States research stations occupied successively on the icecap of Ellsworth Land. Though difficult to reach, the site was chosen as optimal for the study of Very Low Frequency waves in the upper atmosphere. Established by air from McMurdo Station in November 1969, Siple was enlarged in December 1972 to Siple I, which was occupied continuously except in winter 1976. In the three summers from 1976 to 1979 a replacement station, Siple II, was built in 75°56′S, 84°15′W, again by air shuttle from McMurdo. Siple I was closed, and Siple II was occupied continuously from January 1979 to 1981, then intermittently until its closure in 1989.

Sites of Special Scientific Interest (SSSIs). Within the Antarctic Treaty system, SSSIs are designated 'to protect sites where scientific investigations are being carried out or are planned, and there is a demonstrable risk of interference which would jeopardise those investigations, or to protect sites of exceptional scientific interest'. SSSIs are protected for a specified time period, which can be reviewed and extended at Antarctic Treaty Consultative Meetings. See **Study Guide: Protected Areas under the Antarctic Treaty**.

Skelton, Reginald. (1872–1956). British naval engineer officer and explorer. Born in Lincolnshire, Skelton was a senior engineer in the Royal Navy when he joined the **British National Antarctic Expedition 1901–4** as the ship's chief engineer. He supervised the fitting-out of *Discovery*: on the expedition he took part in several sledge journeys and became an accomplished photographer. In World War I Skelton took charge of dockyard facilities

at Archangel, Russia. He retired from the Navy as Engineer Vice-Admiral in 1932, becoming a director of John I. Thorneycroft and Co., shipbuilders. He died on 5 September 1956.
(CH)

Skottsberg, Carl Johan Fredrik. (1880–1963). Swedish expedition botanist. As a recent graduate in biology he joined the **Swedish South Polar Expedition 1901–4**. Aboard the expedition ship *Antarctic* he missed no opportunities of studying local floras in Tierra del Fuego, the Falkland Islands, the South Shetlands, South Georgia and Antarctic Peninsula. Extensive collections, including the marine algae that were his particular interest, were lost when the ship was crushed in the ice of the Weddell Sea. However, his studies on this expedition formed the basis of a long-term interest in southern biogeography. Academic posts in Uppsala and Göteborg provided bases from which he travelled extensively, in Tierra del Fuego and southern Chile (1907–9), Juan Fernandez and Easter Island (1916–17) and later throughout the world, producing a succession of publications that remain a foundation for all southern temperate and Antarctic botanical studies. He died on 14 June 1963.

Skuas, southern oceanic. Flying birds of the order Charadriiformes, family Stercorariidae, that forage on fish and plankton at sea: on land they scavenge, prey on small petrels, and raid penguin colonies for eggs and chicks. Skuas of the southern oceanic area are confusingly similar gull-like birds, with plump body 50 cm (20 in) long, and wing span up to 130 cm (51 in). Plumage is predominantly grey to golden-brown, with prominent white wing bars, dark brown bill and black webbed feet with distinctive curved talons. Some authorities include them with the northern great skua in the single species *Catharacta skua*. Others distinguish northern from southern skuas, subdividing the southern forms into separate species with overlapping ranges and considerable hybridization. All are noted for their strident calls and vigorous diving attacks on intruders close to their nest sites. Skuas are summer migrants, appearing to breed in October or November, and disappearing at the end of the breeding season, to winter in warmer waters as remote as the northern temperate zone. Pairs scoop nests in moss or gravel and lay two green-brown speckled eggs, usually raising only one chick. Southernmost in geographical range is the south polar or McCormick's skua *C. maccormicki*, a grayish, short-legged form that breeds mainly on continental and peninsular Antarctica. The larger brown skua *C. antarctica* has a much wider breeding range, extending from Antarctic Peninsula (where it hybridizes with the south polar skua) to most of the southern oceanic

Incubating brown skua. Photo: BS

islands. A closely-related variant breeds on the South American mainland and the Falkland Islands.

Slava Ice Shelf. 68°49′S, 154°44′E. Ice shelf on the Oates Coast, in a deep bay between Mawson Peninsula and Cape Andreyev. The shelf was photographed from the air in 1947 during Operation Highjump, and visited in 1958 by the Soviet whale ship *Slava*.

Sledge dogs. Large dogs of strong build and dense fur that are used to haul sledges in polar regions. Also called 'huskies', they are trained to accept harness and run in teams. Breeders recognize strains including long-legged malmutes of Alaska, stocky Asian Samoyeds and Siberians, and heavier, more powerful Eskimo huskies from Greenland and Labrador. However, working sledge dogs show wide varieties of size, body proportions and colour. Antarctica has no native dogs. Its first sledge dogs were introduced from Siberia and Greenland by the **British Antarctic (*Southern Cross*) Expedition 1898–1900**, but limited access from Cape Adare inhibited sledging in any direction. The **Swedish South Polar Expedition 1901–4** made better use of its dogs, running small teams of five or six with light sledges, mainly on short survey journeys. Their longest journey covered 380 miles in 38 days. Dogs on later British Antarctic expeditions fared badly. The **British National Antarctic Expedition 1901–4** landed 23 huskies which proved sadly ineffective. Of 19 that left Hut Point on a journey across the Ross Ice Shelf in November 1902, five died of malnutrition during the first six weeks, and the rest died or were killed before the end of the journey. Shackleton's **British Antarctic (*Nimrod*) Expedition 1907–9** took only nine dogs in a last-minute decision, relying instead on Manchurian ponies and manpower. The **British Antarctic (*Terra Nova*) Expedition 1910–13** used dogs only marginally: Scott, like Shackleton, relied on ponies and man-hauling on his polar journey, to his ultimate defeat and death. Amundsen's **Norwegian South Polar Expedition 1910–12** by contrast used sledge dogs with conspicuous, if ruthless, success. Leaving Maudheim with 55 dogs, he drove to the South Pole and back, a distance of 1860 miles in 89 days, killing 41 dogs along the way to feed the 14 that brought him home. Mawson's **Australasian Antarctic Expedition 1911–14** used dogs effectively to explore one of Antarctica's windiest and most uncomfortable corners. **Byrd's First Antarctic Expedition 1928–30** and **Byrd's Second Antarctic Expedition 1933–35** brought aircraft and aerial survey to Antarctica, but included teams of huskies with experienced drivers to provide ground control and collect scientific specimens. For the first expedition Byrd favoured Alaskan dogs, for the second graduating to heavier stocks from Labrador and Greenland.

On a more modest scale, the **British Graham Land Expedition 1934–37** used mainly Labrador huskies in their surveys of western Antarctic Peninsula. Techniques of travel and dog-driving developed on that expedition were used by sledgers during British Naval **Operation Tabarin 1943–45** and its successor the **Falkland Islands Dependencies Survey**. From thence, in the years following World War II, they spread to a succession of British, Australian, New Zealand and other long-term national expeditions, which introduced, exchanged and bred from Greenland and Labrador stocks. Over the years there developed an 'Antarctic' strain of husky, typically short-limbed, muscular, weighing 32–40 kg (70–90 lb), and selected over many generations for intelligence, amiability and willingness to work. These dogs ran in teams of seven to nine, trained to fan or centre trace, following leaders that answered to voice commands. Fed mainly on seal meat at the base, requiring 1 lb of pemmican per day on the trail, each dog could haul more than its own weight day after day for journeys of several weeks. Using a Nansen sledge on firm snow surfaces, a nine-dog team could haul up to 545 kg (1200 lb), averaging 16 km (10 miles) per day.

This high plateau in the history of Antarctic sledge dogs did not last long. From the 1970s onward tractors, skidoos and other motorized vehicles began to replace dogs at Antarctic stations. Dog-driving became recreational, dog-handling merely therapeutic, rather than essential crafts. Finally, in response to pressure from environmental groups, states signatory to the Antarctic Treaty removed all sledge dogs from the Treaty area by 1 April 1994.

Further reading: Walton and Atkinson (1996).

Sleet. Simultaneous precipitation of snow flakes and rain.

Smith Island. 63°00′S, 62°30′W. A steep, narrow island 29 km (18 miles) long, standing alone at the southwestern end of the South Shetland Islands. Its high, ice-covered ridge includes Mt. Foster (2100 m, 6889 ft) and Mt. Pisgah (1860 m, 6100 ft). Discovered in 1819 by William Smith, it was named for him in later surveys.

Snow. Snow consists of ice crystals which form when water vapour condenses on suitable nuclei. They grow in the upper atmosphere at temperatures between −10° and −20 °C, and range in size from one hundredth of a mm to a few mm. Snow crystals have complicated shapes, with underlying hexagonal or six-fold symmetry. Precise shape is determined by external conditions while the crystal is growing, principally the temperature of the air and the degree to which it is supersaturated with water vapour. Most common are columnar (pencil-shaped) crystals, plate-like hexagons, and the well-known fern-like structures. In still

Sledge dog of mixed Greenland and Newfoundland stock, typical of many used widely by Antarctic expeditions. Photo: Scott Polar Research Institute

air the crystals fall at between 0.3 and 1 m s⁻¹, depending on size. In rising air they descend more slowly or are forced upward. Single snow crystals suspended in the atmosphere, because of their regular shape, give rise to many meteorological-optical phenomena such as halos. At temperatures above $-10\,°C$ individual crystals adhere to one another. At temperatures only a little below freezing they produce the characteristic large clusters called snow flakes.
(WGR)

Snow algae. Several species of single-celled plant that live on or near the surface of snow banks, at times imparting bright red, yellow or green tints to the snow. Species identified in Antarctic snows include *Stichococcus bacillaris*, *Chlamydomonas nivalis*, *Raphidonema nivale*, *Hormidium subtile* and *Chlorosphaera antarctica*. Concentrations vivid enough to attract attention appear most often in the maritime sector, on islands of the Scotia Arc and along the west coast of Antarctic Peninsula. They have also been found along the Prins Harald Kyst, and could be more widespread around the continent along coasts where air temperatures are close to or above freezing point during the warmest month. Thin layers of algae persist from year to year close to sea level, in areas where sea spray may provide nutrients in summer. Masked by winter snows, they appear during the early summer melt, intensify throughout the season, then disappear when snow settles again in autumn.
Further reading: Vincent (1988).

Snow bridge. Arch of wind-packed snow forming over, and sometimes hiding, the mouth of a crevasse.

Snow drift. A mass of wind-blown snow, usually accumulated in the lee of an obstruction.

Snow field. An extensive area of snow that persists for most of the year and is renewed annually.

Snow Hill Island. 64°28′S, 57°12′W. A narrow island 32 km (20 miles) long standing southeast of James Ross Island, and separated from it by Admiralty Sound. Almost entirely snow-covered, it was discovered in 1843 by the British Naval Expedition 1839–43. A raised beach at its eastern end accommodates the main base hut of the **Swedish South Polar Expedition 1901–4**, now an Historic Monument.

Snow Island. 62°47′S, 61°23′W. An ice-mantled island 16 km (10 miles) long standing southwest of Livingston Island, South Shetland Islands.

Snow line. Altitudinal or latitudinal limit beyond which snow persists year-round. On glaciers, the snow line is the boundary between an upper zone of no melting, and a lower zone where some surface melting occurs, allowing water to percolate into the snow.

Snow petrels. See **Fulmars, southern oceanic**

Sobral, José Maria. (1880–1961). Argentine Antarctic explorer. Entering the Argentine Navy in 1898, he was commissioned and in 1901 was seconded to join the **Swedish South Polar Expedition 1901–4**. He served at Snow Hill Island, assisting the scientific staff and taking part in the long sledging journey across the Larsen Ice Shelf. On return to Argentina he left the navy and studied geology at the universities of Buenos Aires and Uppsala, later becoming director of the department of mines and a petrologist in the department of oil resources. He served also in the Argentine consular service in Norway.
(CH/BS)

Sobral Station. 81°05′S, 40°30′W. Argentine temporary research station established in April 1965 on the eastern edge of Filchner Ice Shelf. The station was closed in 1969.

Soils. Mixtures of rock particles, living organisms, gases and water-based solutions, fundamental to the development of ecosystems on land. The simplest ahumic soils, i.e. containing little or no organic material, are composed of rock fragments, sand and rock flour, the products of physical and chemical **weathering**, redistributed and sorted by water or wind. Chemically inert and hostile to life, these are characteristic of the cold desert areas of Antarctica. Simple humic soils, in which rock particles blend with organic materials both living and dead, attract moisture and provide a substrate for complex chemical interactions. Weak humic acids washed from dying and dead plant material help further to break down rock particles, releasing ions that provide nutrients, and making the soils more hospitable to plants. Humification can occur only in the presence of liquid water and living material, and is reduced when, in the absence of decomposing bacteria, dead vegetation fails to break down. Ahumic and simple humic soils characterize some of the more hospitable ice-free Antarctic coastal areas, the most complex and advanced humic soils occurring in the warmer, damper regions of the maritime Antarctic and sub-Antarctic islands.

More mature protoranker soils, peats and loamy brown soils, comparable to those of temperate regions, occur only on islands beyond the northern limit of pack ice,

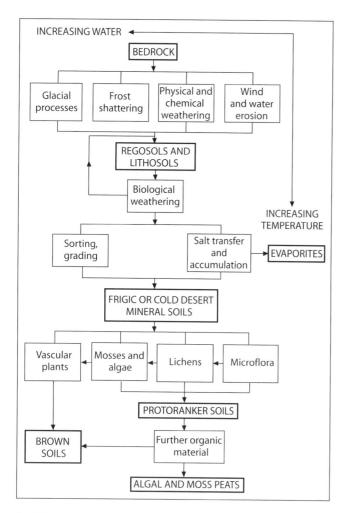

Soil formation in polar environments. In the cold, dry conditions of the continent ahumic frigic or cold-desert mineral soils predominate. Warmer and damper conditions on the southern islands give rise to protorankers, brown soils and moss peats. After Stonehouse (1989)

where moisture and summer warmth encourage a wider variety of soil bacteria, microfauna and more advanced plants and contribute more to soil formation. Only ferns and flowering plants, with their mats of penetrating roots, ensure development to brown soils and beyond. These are unknown on continental Antarctica and rare even in the Maritime Antarctic, where soils in consequence fail to develop beyond protoranker or moss-peat stages. Their full effects are seen in the rich brown soils of warmer Southern Ocean islands beyond the limits of pack ice.
Further reading: Campbell and Claridge (1987).
(DWHW)

Solander, Daniel Carl. (1733–82). Swedish botanist who sailed on James **Cook**'s first expedition to the Southern Ocean. Born in Pitea, northern Sweden, Solander became a student at Uppsala University, studying natural history under Carl Linneus, the distinguished botanist and taxonomist. After expeditions to the Arctic and Canary Islands, in 1763 he joined the staff of the British Museum, and in 1768–71 accompanied Joseph **Banks** as assistant naturalist on Cook's first voyage of exploration. Returning to the British Museum, he was appointed curator of the natural history collection, spending virtually the rest of his life in sorting, classifying and preserving the thousands of specimens brought back from his earlier travels.

Solar radiation. Radiation emitted in a range of frequencies by the sun, part of which is received by Earth and atmosphere. Of the total solar energy radiated to space, the portion impinging at the outer surface of Earth's atmosphere is received at a rate of about 1 kW per m^2. To reach and warm Earth's surface, rays must first pass through the atmosphere, where part of their energy is reflected back into space, part absorbed by cloud and gases. At Earth's surface more of their energy is reflected (see **albedo**), leaving only a fraction to be absorbed. Polar regions are colder than tropical regions because they receive their radiation obliquely, spread it over a wider area, lose more of it to the atmosphere, and reflect more of it away. A shaft of $1 m^2$ shining on an equatorial region passes through a thin layer of atmosphere to warm $1 m^2$ of ground beneath. A similar shaft striking a polar region passes obliquely through a greater thickness of atmosphere, to spread obliquely over a wider area of Earth's surface, where a higher proportion is reflected from the prevailing ice and snow.

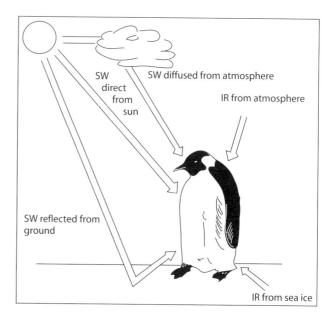

The penguin receives direct, diffuse and reflected short-wave (SW) radiation from the sun, and long-wave infra-red (IR) radiation from the atmosphere and substrate. After Stonehouse (1989)

Table 2 Monthly and yearly radiation budgets for Antarctic polar stations Mirnyy, Oazis and Vostok. All values are kcal cm^{-2}; to obtain in MJ^{-2} multiply by 41.8. Data after Rusin (1964)

Month	Jul	Aug	Sep	Oct	Nov	Dec	Jan	Feb	Mar	Apr	May	Jun	Year
Mirnyy													
DirSWR	0.0	0.8	2.2	5.7	10.3	12.8	10.6	7.5	4.0	0.9	0.1	0.0	54.9
DifSWR	0.1	0.7	2.9	5.3	7.6	9.1	9.1	6.4	3.6	1.7	0.2	0.0	46.7
TSWR	0.1	1.5	5.1	11.0	17.9	21.9	19.7	13.9	7.6	2.6	0.3	0.0	101.6
RSWR(−)	0.1	1.4	4.6	9.3	15.0	17.7	15.8	11.7	6.6	2.2	0.3	0.0	84.7
ESWR	0.0	0.1	0.5	1.7	2.9	4.2	3.9	2.2	1.0	0.4	0.0	0.0	16.9
GLWR(−)	10.7	10.5	11.5	11.5	12.6	13.8	13.8	12.7	12.0	11.5	11.1	10.8	142.5
ALWR	8.8	8.6	9.5	9.5	11.2	12.3	11.9	11.7	9.6	9.6	9.4	8.5	120.6
ELWR(−)	1.9	1.9	2.0	2.0	1.4	1.5	1.9	1.0	2.4	1.9	1.7	2.3	21.9
NRB	−1.9	−1.8	−1.5	−0.3	1.5	2.7	2.0	1.2	−1.4	−1.5	−1.7	−2.3	5.0
Oazis													
DirSWR	0.0	0.6	2.7	4.8	7.3	9.2	8.7	6.2	3.0	1.1	0.4	0.0	44.0
DifSWR	0.0	0.7	2.8	4.7	8.0	9.5	8.4	5.8	3.6	1.5	0.2	0.0	45.2
TSWR	0.0	1.3	5.5	9.5	15.3	18.7	17.1	12.0	6.6	2.6	0.6	0.0	89.2
RSWR(−)	0.0	0.5	2.3	2.5	3.1	3.5	2.7	2.6	2.1	0.9	0.3	0.0	20.5
ESWR	0.0	0.8	3.2	7.0	12.2	15.2	14.4	9.4	4.5	1.7	0.3	0.0	68.7
GLWR(−)	9.8	9.9	10.3	11.4	13.3	14.7	14.9	13.7	12.3	11.6	10.7	9.0	141.6
ALWR	8.0	8.0	8.1	11.3	8.6	11.4	11.0	11.1	10.5	8.5	8.6	7.0	112.1
ELWR(−)	1.8	1.9	2.2	0.1	4.7	3.3	3.9	2.6	1.8	3.2	2.1	2.0	29.5
NRB	−1.8	−1.1	1.0	6.9	7.5	11.9	10.5	6.8	2.7	−1.4	−1.8	2.0	39.2
Vostok													
DirSWR	0.0	0.0	1.8	8.7	20.0	25.6	22.3	14.8	4.2	0.4	0.0	0.0	97.8
DifSWR	0.0	0.0	0.8	3.3	4.1	4.7	5.8	2.0	1.6	0.0	0.0	0.0	22.3
TSWR	0.0	0.0	2.6	12.0	24.1	30.3	28.1	16.8	5.8	0.4	0.0	0.0	120.1
RSWR(−)	0.0	0.0	2.5	10.3	20.0	24.6	23.1	13.8	5.1	0.3	0.0	0.0	99.7
ESWR	0.0	0.0	0.1	1.7	4.1	5.7	5.0	3.0	0.7	0.1	0.0	0.0	20.4
GLWR(−)	4.5	4.0	4.4	5.2	7.0	8.5	8.8	6.8	5.3	4.6	4.7	4.3	68.1
ALWR	3.5	2.8	3.4	3.2	3.3	3.6	4.4	3.7	3.7	3.4	4.1	3.5	42.5
ELWR(−)	1.0	1.2	1.0	2.0	3.7	4.9	4.4	3.1	1.7	1.2	0.6	0.8	25.6
NRB	−1.0	−1.2	−0.9	−0.3	0.4	0.8	0.6	−0.1	−1.0	−1.1	−0.6	−0.8	5.2

Though daily totals of incoming radiation during Antarctica's long summer days are among the world's highest, there is no daily income during the sunless winters (see table). Throughout the year Antarctica as a whole receives only about 40 per cent as much radiant energy as the tropics. High albedo reflects away on average over 80 per cent of what it receives: in addition it radiates some of its own meagre store of heat outward to space, which is even colder. Overall Antarctica loses more energy each year than it gains from solar radiation. It would become progressively colder but for heat imported annually in poleward flows of relatively warm air and atmospheric moisture from temperate regions, plus a little from ocean currents.

Even weak Antarctic sunshine can be felt as warmth, and is strong enough to warm rocky environments, melt snow and stimulate plants into photosynthesis. Of the relatively small amount of energy that reaches and is absorbed at the surface, much becomes involved in melting ice and snow and evaporating the resulting water. There is consequently very little left for warming. However, a person or animal standing in the snow on a clear sunny day feels warmth from the several kinds of radiation received directly or indirectly.

Over the world as a whole some 30–50 per cent of the incident radiation is reflected back by atmosphere and earth, leaving 50–70 per cent to warm the earth/atmosphere system. In polar regions a very much higher proportion is lost by reflection. Though the Arctic region ends each year with a small net positive balance, the Antarctic region has a net deficit over the year as a whole. Antarctica's energy balance and surface temperatures are maintained by an annual inflow of heat, mostly atmospheric, from lower latitudes.

The table provides complete radiation budgets for three Russian Antarctic stations: Vostok, a high plateau station, Mirnyy, a coastal station in an area where the ground is snow-covered for much of the year, and Oazis, a station in an almost snow-free coastal oasis area.

Further reading: Rusin (1964); Stonehouse (1989).

Somov, Mikhail Mikhaylovich.

(d. 1973). Soviet polar oceanographer and administrator. After training in oceanography, he joined the Arctic Research Institute of the

Northern Sea Route Administration and was for many years involved in Arctic research and shipping. In 1951 he was appointed deputy director of the Arctic Research Institute, Leningrad, and four years later was selected to lead the first Soviet expedition to the Antarctic, establishing Mirnyy, Oazis and Pionerskaya stations. He was the first Soviet delegate to the Scientific Committee on Antarctic Research, and in 1963 led the ninth Soviet Antarctic Expedition. Somov died in Leningrad on 30 December 1973, aged 65.

Sooty albatrosses. Small, dark-plumaged albatrosses of the genus *Phoebetria*, widespread in the southern oceanic region. Similar in size to the **mollymawks**, with body length up to 90 cm (35 in), wing span 2 m (79 in), they are distinguished by smoky-brown plumage and dark bills. Sooty albatrosses *P. fusca* breed only on the temperate islands and forage mainly in temperate and warmer waters of the Indian Ocean. Light-mantled sooty albatrosses *P. palpebrata*, distinguished by their pale fawn back, breed mainly on colder temperate and sub-Antarctic islands, and range freely throughout the west wind zone. Both breed individually on cliffs or in small colonies of five or six pairs. Both feed mainly on fish, crustaceans and squid, caught at or near the surface.

Sør Rondane. 72°00′S, 25°00′E. Major range of ice-clad mountains 200 km inland from Prinsesse Ragnhild Kyst, Dronning Maud Land. Discovered in reconnaissance flights by the Norwegian (Christensen) Whaling Expeditions on 6 February 1937, the group was named for Rondane, a mountain massif of southern Norway. The highest peaks, toward the south, rise above 3400 m (11 152 ft). The mountains were re-photographed during **Operation Highjump 1946–47**, and surveyed from the ground by Belgian survey parties operating from Roi Baudouin Station in 1958–60.

South American sea lion. *Otaria byronia*. This species is found along southern coasts of South America from about 12°S in Brazil to 4°S in Peru, including southern Argentina and Chile. Most breeding colonies lie south of 30°S, usually on inaccessible stretches of coast. Hundreds of thousands were reported to breed on the Falkland Islands up to the mid-twentieth century: the current population is estimated at 30 000. Mature males grow to a length of 2.8 m (9 ft) and weigh up to 300 kg (660 lb): females are smaller and more slender. The fur is dark brown, shading to tan underneath: males have a prominent yellowish mane. Mature males take up territories on beaches and rock flats close to the high-water mark in late November, rounding up pregnant females as they appear, to form harems or breeding groups of up to a dozen. Most pups are born in December and early January. They grow slowly, remaining on the breeding grounds until June or July. The species was formerly hunted for its tough hides. The world population is estimated at 275 000.

South Georgia. 54°17′S, 36°30′W. A sub-Antarctic island forming part of the northern arm of the **Scotia Arc**. Almost 200 km (125 miles) long and 35 km (22 miles) across, deeply indented by fjords and bays, South Georgia forms a banana-shaped ridge of alpine mountains and glaciers in the northeastern corner of the Scotia Sea. Of its total area of 3750 km^2 (1465 sq miles), about 60 per cent is ice-covered. The highest peak, Mt Paget, rises to 2934 m (9623 ft), and several other peaks of the central Allardyce Range rise above 2000 m (1250 ft). The rocks are mostly metamorphosed volcanic shales and tuffs. The island has a wet, windy and cold climate: mean annual temperature about 1.9 °C, range of mean monthly temperatures 6.8 °C, annual precipitation 140 cm of rain-equivalent. It supports 20 species of flowering plant including tussock and other grasses, rushes and shrubs, giving rise to stands of dense tussock along the coast and fellfield on higher ground. Lichens and mosses are also abundant. At least a dozen species of alien plant have been introduced. Norway rats and mice introduced by sealers now flourish among the tussock. Whalers who lived on the island from 1904 onward introduced cattle, dogs, cats, sheep, rabbits, horses and reindeer, but fortunately only the reindeer survive: several herds browse freely on lowland vegetation and currently appear to be spreading.

Spectacular colonies of king, gentoo and macaroni penguins, together with albatrosses, mollymawks, and huge flocks of smaller petrels, breed among the tussock grass and scree slopes. Southern elephant seals and Antarctic fur seals breed on many beaches, both having recovered from nineteenth-century devastation. Fur seals have made a remarkable comeback and are now estimated to number over 3 million.

South Georgia: Tussock grass in the coastal zone. Photo: BS

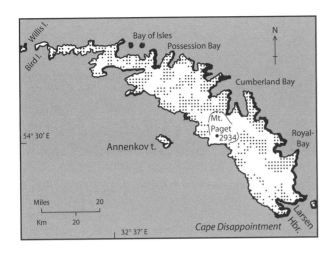

South Georgia. Though standing in a latitude equivalent to that of northern Britain, South Georgia is heavily ice-covered, with glaciers down to sea level. Its sheltered northern fjords provided harbours for several whaling stations. Height in m

The island was probably discovered during the seventeenth century, but was charted, named and claimed for Britain by Capt. James Cook in 1775. From about 1790 it was exploited by sealers, whose hunt for fur seals and elephant seals continued sporadically throughout the nineteenth and twentieth centuries. In 1904 the first whaling station was opened at **Grytviken**, and a British government station was built on nearby King Edward Point, augmented in 1925 by a whale research station operated by the *Discovery* Committee. Whaling stations were established also in Cumberland Bay and other harbours: several closed before World War II, but those at Grytviken, Husvik Harbour and Leith Harbour persisted until the 1960s, when whaling from the island ceased. British Antarctic Survey operated a station at King Edward Point from 1969 to 1982. It continues to operate a year-round station for bird and seal studies on Bird Island, and has recently (2001) opened a fisheries research laboratory at King Edward Point. Meteorological records have been maintained almost continuously since 1905.

Claimed by both Argentina and Britain, South Georgia was invaded by Argentine forces in April 1982 and fought over briefly. It is administered by the Government of South Georgia and the South Shetland Islands on behalf of the United Kingdom. Wildlife and magnificent scenery combine to make South Georgia a popular tourist venue.
Further reading: Headland (1984); Leader-Williams (1988).

South Georgia Survey Expeditions 1951–57. A succession of privately sponsored expeditions led by Duncan Carse (a veteran of Discovery Investigations and the **British Graham Land Expedition 1934–37**), which, in the four summers 1951–52, 1953–54, 1955–56 and 1956–57, undertook a comprehensive topographic and geological survey of South Georgia. Carse relied on volunteer surveyors, geologists and assistants, some of whom had gained previous experience with the Falkland Islands Dependencies Survey. Helped logistically by catchers and buoy boats of the whaling fleet, the expeditions involved man-hauling, back-packing and camping for travel about the island. The topographic map 610 South Georgia, sheet ix.1958 embodies many details arising from these expeditions.

South Orkney Islands. 60°40′S, 45°15′W. A compact archipelago of mountainous Antarctic islands in the southern Scotia Arc. Their land area, totalling about 600 km^2 (234 sq miles) is 90 per cent ice-covered. The largest island, Coronation Island, rises to 1265 m (4149 ft): smaller islands include Powell Island, Laurie Island and Signy Island. The islands are made up of ancient metamorphosed gneisses, schists and limestones, overlain by greywackes, shales and conglomerates. Alternating piedmonts, glaciers, rocky headlands and narrow beaches form the coast. Fast ice forms between the islands in winter, and pack ice often invests them to November or later. The climate is windy, overcast and wet, with mean annual temperatures −4 °C, annual range 10 °C. Thick snow invests the beaches for six to eight months each year, and rain is frequent in summer. Vegetation is sparse, with algae, lichens and mosses predominating; the two Antarctic flowering plants, hair-grass and pearlwort, grow abundantly in sheltered places. Breeding seabirds, notably chinstrap and macaroni penguins, petrels, skuas, Dominican gulls and sheathbills, are abundant. Antarctic fur seals and elephant seals breed on most beaches in early summer.

The group was first sighted by sealers George Powell and Nathaniel Palmer in December 1821, and named for the latitudinally similar Orkney Islands of North Britain. They were first charted and claimed for Britain by sealer Capt. James Weddell in 1822. Marginally colder than the

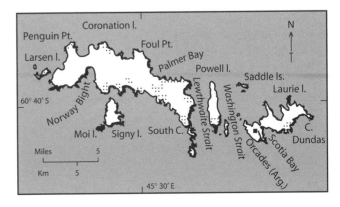

The South Orkney Islands are mountainous, steep and heavily glaciated, with abundant winter snow

South Shetland Islands, more often invested with sea ice and correspondingly more difficult to access, they were nevertheless quickly stripped of their fur seals during the following decade, and again intermittently throughout the nineteenth century. In 1903 the Scottish National Antarctic Expedition wintered their ship SY *Scotia* in Scotia Bay, Laurie Island, establishing a meteorological station ashore. The station, taken over in the following year by the Argentine Meteorological Service and renamed **Orcadas**, has maintained observations ever since, providing the longest uninterrupted record of Antarctic climate. From about 1910 the waters surrounding the islands were hunted by whalers. A small land station established in 1920 at Factory Cove, Signy Island, operated for a few seasons but proved unsuccessful. In 1947 the Falkland Islands Dependencies Survey established Base H close to the remains of the station. Enlarged and extended, it became **Signy Station**, for long British Antarctic Survey's main centre for terrestrial biological research. Currently claimed by Argentina and Britain, the islands are administered as part of British Antarctic Territory.

South Polar Times. Monthly journal produced by and for members of the **British National Antarctic Expedition 1901–04**. Text consisted of articles and poems submitted by expedition members, edited by Ernest Shackleton and typed by Charles Ford; illustrations included pencil sketches and watercolours by Edward Wilson. A single copy of each issue was produced and passed around. On the return of the expedition, a limited edition was printed commercially, bound between hard covers in three volumes. A second limited edition has recently (2000) been published to celebrate the centenary of the expedition.

South Sandwich Islands. 58°00′S, 27°10′W. Chain of 11 Antarctic islands 350 km (219 miles) long, extending north to south and forming the eastern limit to the Scotia Sea. The total land area of about 300 km² is 85 per cent ice-covered. The chain is formed of two northern groups, the Traversay Islands and Candlemas Islands, and a southern group, the Southern Thule Islands, with Saunders, Montagu and Bristol Islands standing alone in mid-chain. All but one, Leskov Island, stand along the submarine ridge forming the eastern loop of the Scotia Arc. A few kilometres to their east lies the 8000 m (26 240 ft) deep South Sandwich Trench. Leskov Island stands 56 km (35 miles) west of the main arc. The chain is a classic example of an island arc in an early stage of evolution. All the islands are complex volcanic cones of basalt and scoria, arising from submarine eruptions that began some 3 million years ago, producing fragmental basal rocks on which were deposited lavas and ashes, and subsequently fluid basaltic lavas. Each island currently consists of one or more cones of basalt and andesite, eroded by wave action to form high cliffs on all sides. Only Visokoi, Montagu, Bristol and Cook Islands have peaks rising above 1000 m (3289 ft): the rest have broad upland plateaux and lower peaks. All present formidable headlands and wave-cut cliffs, with few beaches or landing points.

Seven islands – Bellingshausen, Bristol, Saunders, Candlemas, Visokoi, Leskov and Zavodovski – show fumaroles, warm springs or other evidence of contemporary or recent activity. In 1956 a major eruption formed a large crater on the eastern side of Bristol Island, ejecting black cinders. In March 1962 a submarine eruption of pumice 56 km (35 miles) northwest of Zavodovski Island formed a sea mount, Protector Shoal, rising to within 27 m (89 ft) of the surface. Floating pumice carried by the West Wind Drift at a speed of about 20 km (12.5 miles) per day was subsequently distributed widely on beaches of many southern islands.

Pack ice invests the chain for several winter months, persisting longer around the southern islands. Mean annual temperatures are close to freezing point. Windswept and bare, the islands support little mature soil or vegetation. Algae, lichens and mosses predominate, and the two Antarctic flowering plants, hairgrass and pearlwort, grow

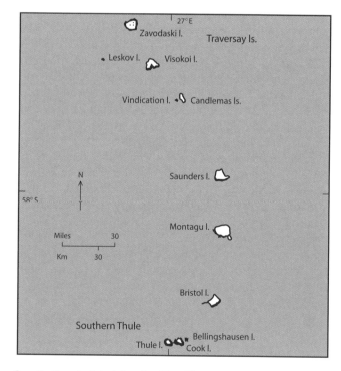

South Sandwich Islands. The Traversay Islands are the three northernmost islands, discovered and named by von Bellingshausen. The Candlemas Islands include only Candlemas and Vindication Islands and neighbouring rocks. Southern Thule includes Bellingshausen, Thule and Cook Islands. Most of the islands stand tall enough to be visible from their immediate neighbours on clear days

in sheltered places. Breeding elephant seals are plentiful: Antarctic fur seals breed on several of the islands, probably in increasing numbers. Breeding seabirds, notably penguins, petrels, skuas, Dominican gulls and sheathbills, are abundant.

The eight southern islands were sighted by Capt. J. Cook RN in 1775, and named for the Earl of Sandwich. The three Traversay Islands were named by Capt. F. von Bellingshausen in 1819. Sealers visited the islands during the nineteenth century and whalers and scientists are more recent visitors. There are no harbours and currently no settlements. An Argentine scientific station built on Thule in 1976 was destroyed by Britain in 1982 following the Argentine occupation of South Georgia. The islands are claimed by both Britain and Argentina, and currently administered by Britain under the Government of South Georgia and the South Sandwich Islands.

Further reading: LeMasurier and Thomson (1990).

South Shetland Islands. 62°00′S, 59°00′W. A chain of 11 large and many smaller Antarctic islands over 500 km (312 miles) long, between Drake Passage and Bransfield Strait. The main western group comprises, west to east, Smith, Snow, Livingston, Deception, Greenwich, Robert, Nelson and King George Islands: the smaller eastern group includes Gibbs, Elephant and Clarence Islands. The islands are over 90 per cent covered with permanent ice. Ice piedmonts and glaciers form much of their coastline, alternating with headlands and narrow beaches. Forming part of the Scotia Arc ridge, they are composed of volcanic lavas, agglomerates and tuffs, some including fragments of fossil wood, dating from the Jurassic onward, contorted, metamorphosed, faulted and punctuated by volcanic plugs. Deception Island was reported active at intervals throughout the nineteenth and early twentieth centuries, and erupted spectacularly on three occasions during the period 1967–70. Bridgeman Island is reported to have erupted in historic times, and Penguin Island, a small island with red scoria cone off the southern flank of King George Island, is clearly of recent origin. Fast ice forms around the islands in winter, and pack ice may make access difficult in November and December. The climate is windy, overcast and wet, with mean annual temperatures around freezing point. Thick snow invests the beaches for six to eight months each year. Rain is frequent in summer. Vegetation comprises mainly algae, lichens and mosses. The two Antarctic flowering plants, hairgrass and pearlwort, grow well in sheltered places. Breeding elephant seals are plentiful and Antarctic fur seals are increasing in numbers: breeding seabirds, notably penguins, petrels, skuas, Dominican gulls and sheathbills, are abundant.

Livingston Island, with its highest peak rising almost to 2000 m (6560 ft), was sighted by William Smith in February

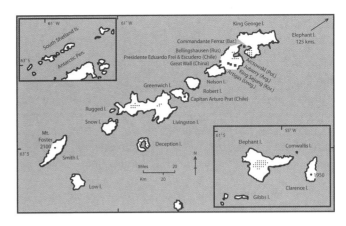

Forming a long north-to-south chain, the South Shetland Islands are steep and heavily glaciated. The highest peaks stand at either end of the chain: between them, Elephant Island rises to 975 m, Livingston Island to 750 m and King George Island to 700 m

Extensive moss beds at Hannah Point, Livingston Island, South Shetland Islands. Photo: BS

1819. The first recorded landings were made at Turret Point and Penguin Island later in the same year, when the group was charted and claimed for Britain. Almost immediately it became a hunting ground for sealers from Britain and the United States, who within a few seasons stripped it of most of its fur seals. Later the islands became a centre for elephant sealing, and in the early decades of the twentieth century several of the sheltered bays accommodated whale factory ships. Accessibility throughout summer has made the islands centres for scientific work and tourism. King George Island especially supports the research stations of a dozen different nations, among them the Chilean station Teniente Rodolfo Marsh which is equipped with an airstrip capable of accommodating heavy transport aircraft. Claimed by Argentina, Britain and Chile, the islands are administered by Britain as part of British Antarctic Territory.

Southern elephant seal. *Mirounga leonina.* The world's largest seals, these breed on Antarctic Peninsula, South Georgia, Iles Kerguelen, Heard and Macquarie Islands, and on other islands in the cool temperate zone. Immature animals and moulting bulls are found as far south as the continental shore in summer. Large males measure up to 6.5 m (21 ft) nose-to-tail and weigh up to 4 tonnes: females rarely exceed 3.5 m (11.5 ft) in length and 1 tonne. After wintering at sea, mature beachmaster bulls, usually 8–9 years old, arrive in early spring, hauling up on traditional sandy breeding beaches just before the main influx of pregnant cows. During the pupping season and for some three to four weeks later when the cows come on heat, the beachmasters seek to guard their harems, usually groups of 20–30 cows but occasionally as many as 100, from the attentions of younger bulls. In consequence over 95 per cent of matings are performed by about 3 per cent of the bulls. Physiological stress for beachmasters is considerable: it is not surprising that few retain this status for more than two years. Females give birth first at age 3–5 years and on average live another 8–10 years. Little is known of elephant seals at sea, though their main foods are fish and squid, probably caught at great depth. The world population is estimated at about 30 000.
Further reading: Le Boeuf and Laws (1994).
(JPC)

Southern islands. Collective term for over 20 remote island groups and archipelagos of the Southern Ocean and the southern Atlantic, Indian and Pacific oceans, occupying a latitudinal span of over 25° and ranging in climate from polar to warm-temperate. The southernmost **Southern Ocean islands**, standing south of the Antarctic Convergence and thus in the Southern Ocean, are cold and ice-capped with desert or semi-desert vegetation. Groups north of the convergence, sometimes called 'sub-Antarctic' or 'subtropical' but here called **Southern temperate islands**, span climates from cool to mild, and support a range of vegetation from subpolar desert to forest. Though geographically marginal to the limits of this encyclopedia, Southern temperate islands are included because of strong historical links and other affinities with Southern Ocean islands.

All the southern islands are small compared with the surrounding oceans, which control their daily, seasonal and annual mean temperatures. Most are of volcanic origin, some of very recent formation. All tend to be cool or cold in winter, wet in summer, and windy throughout the year, their weather dominated by successive eastward-moving depressions. Only a few of the larger islands or groups, e.g. South Georgia, the Falkland Islands and Iles Kerguelen, are large enough to generate significant local climates. All are the haunts of seals and of petrels, penguins and other seabirds that breed ashore and feed in the rich waters close at hand.

These islands were discovered during voyages of exploration between the fifteenth and nineteenth centuries, some by naval or scientific explorers, many by sealers in search of new hunting grounds. None is known to have supported aboriginal populations. Indigenous South American Indians and Australasians may from time to time have ventured or been blown southward into higher latitudes. Maori legends speak of at least one such journey to an ice-covered land south of New Zealand, and stones pierced by holes discovered on the sea bed off the South Shetland Islands have been interpreted by one authority as primitive anchor stones. Aboriginals capable of colonizing Tierra del Fuego and Stewart Island could almost certainly have survived on some of the milder islands, e.g. South Georgia or Macquarie Island with their fur seals and year-round seabird populations. The first men known to have lived for any length of time on southern islands were shipwrecked mariners and sealing gangs of European origin, for which there are well-authenticated records dating from the late eighteenth or early nineteenth centuries. Many mariners spent months on some of the islands: a few that were lost or forgotten spent several years, living mainly on seals and birds. Some left on southern temperate islands contrived to grow crops from potatoes and other vegetables that were left with them for food.

Settlement of the southern temperate islands began on the **Falkland Islands** in 1690 (though early occupations were transitory), and on **Tristan da Cunha** in 1816. Both groups were (and continue to be) settled by farming and fishing communities. The **Auckland Islands** were colonized by native New Zealanders in 1842 and by Europeans in 1850, though neither settlement survived more than a few years. **Campbell Island** was farmed in successive summers from 1895. **South Georgia** gained its first whaling station in 1904, and **Iles Kerguelen** supported a whaling station and stock farm from 1908. Of the colder Antarctic islands, first to be occupied was Laurie Island, **South Orkney Islands**, in 1903 by a Scottish, later Argentine, meteorological station. The **South Shetland Islands** gained their first long-term settlement, a British magistrate's hut, on Deception Island in 1909, and their first whaling station in 1912. Several of the islands that stood in the shipping lanes of the southern oceans were stocked with goats, pigs, sheep, rabbits and other domestic animals, as food for castaways. Almost all, through shipwreck and the landing of stores, accidentally received populations of rats and mice. Though most have been ravaged by their introduced mammals and other man-induced changes, all show remnants of unique terrestrial flora and fauna, and remain ecologically interesting by virtue of their isolation. Some have been well studied, others neglected. A very few, mostly peripheral islets, have escaped human attention altogether and remain almost pristine. Several are currently sites of permanent scientific stations.

Sovereignty of some of the islands is equivocal. The South Shetland, South Orkney and South Sandwich Islands, South Georgia and the Falkland Islands are subject to rival claims by Argentina and Britain: Chile also claims the South Shetland Islands. Lying poleward of latitude 60°S, the South Shetlands and South Orkneys fall within the Antarctic Treaty area, under which rival claims are at present shelved. The Falkland Islands and South Georgia, claimed and for long administered by Britain, were invaded in 1982 by Argentine forces and subjected to brief warfare. Within a few weeks a British task force restored the status quo, for good measure evacuating and destroying an Argentine research station on Thule, South Sandwich Islands. All other southern islands and groups, whether occupied or unoccupied, fall unequivocally under the sovereignty of one of six nations – Australia, Britain, France, New Zealand, Norway or South Africa – which thereby assume responsibility for their welfare. The World Conservation Strategy of the International Union for the Conservation of Nature and Natural Resources identifies many of these islands as priority regions for protective measures. Despite their remoteness, none is completely isolated from man, for the southern oceans have recently become busy with fishing boats, tour ships, private yachts and research vessels. For a table showing all the islands in their ecological zones see **Study Guide: Southern Oceans and Islands**. For more detailed accounts see entries for individual islands and groups.

Southern lights. See **Aurora australis**.

Southern Ocean. The ocean surrounding Antarctica. Identified and named by Capt. James Cook in the 1760s, its discovery pre-dated that of Antarctica, and British oceanographers tend to follow his lead in calling it the Southern Ocean. Some authorities call it the Antarctic Ocean (counterpart to the Arctic Ocean of the far north) or the Southern Oceans: others deny its existence, referring only to the southern extremities of the Atlantic, Indian and Pacific oceans. Bounded in the south by the shores of Antarctica, the ocean extends northward to the Antarctic **Convergence**, where its surface waters are over-ridden by warmer northern waters. The total area so defined is about 28 million km² (10.9 million sq miles), roughly twice that of continental Antarctica. In comparison the Pacific Ocean is roughly six times as extensive, the Atlantic Ocean three times, and the Arctic Ocean half. At the ocean's narrowest point, Tierra del Fuego lies 1200 km (750 miles) from the tip of Antarctic Peninsula. New Zealand and Tasmania lie 2600 km (1625 miles) from East Antarctica, South Africa 4000 km (2500 miles). To reach the continent ships must cross water more than 2000 m (6560 ft) deep. The mean depth of the ocean is close to 4000 m (13 120 ft): the greatest depths (South Sandwich Trench or Meteor Deep) exceeds 8000 m (26 240 ft).

The ocean floor

Antarctica is surrounded by a continental shelf, generally 50–100 km (31–62 miles) wide around East Antarctica, up to three times as wide off the Ross Ice Shelf and West Antarctica. The shelf lies mainly between 350 and 500 m (1150–1640 ft) deep, i.e. two to three times the depth of continental shelves elsewhere, a difference imposed by the weight of the ice cap. A steep continental slope, cut by canyons and littered with sediment from glacial activity, descends to a ring of ocean basins up to 5000 m (16 400 ft) deep, floored by oceanic crust on which have accumulated thick layers of sediment. Off the Amundsen and Bellingshausen Seas lies the Southeast Pacific Basin: off East Antarctica lie the South Indian and Atlantic-Indian basins, separated by a prominent rise, the Kerguelen Plateau. A similar plateau surrounds the Balleny Islands, extending toward Macquarie Island and New Zealand, and the broken ridge of the Scotia Arc loops between Antarctic Peninsula and South America. Beyond the basins, mainly between latitudes 50° and 65°S, the ocean floor rises to a complex but almost complete ring of mid-oceanic ridges, the site of active sea-floor spreading.

Winds and currents

The Southern Ocean is distinguished from neighbouring oceans by its strong, wind-driven circumpolar currents and distinctive water masses, including cold surface waters (see **Currents, convergences and divergences**). Immediately around the continent, prevailing east and southeasterly winds drive surface waters westward with a northerly component. Further north, beyond the boundary of the Antarctic divergence, surface waters are driven strongly eastward by the prevailing west wind drift. Below the surface, currents of differing strength and direction prevail according to depth.

Like all other oceans, the Southern Ocean is made up of water masses containing inorganic salts, in solutions of remarkably constant chemical composition. Ionic constituents in order of prevalence are chloride, sodium, sulphate and magnesium: next commonest are calcium, potassium, bromide and inorganic carbon. In addition, there are varying concentrations of organic substances, notably compounds of phosphorus and nitrogen, derived from living materials. Salinity, usually determined from electrical conductivity, is expressed as grams per kg of solution, equivalent to parts per thousand (ppt), or one-tenth of percentage by weight. An overall mean is 35 ppt. The discrete water masses making up the ocean are distinguished from each other by very small differences in salinity and temperature, resulting in differences in density. As each mass is

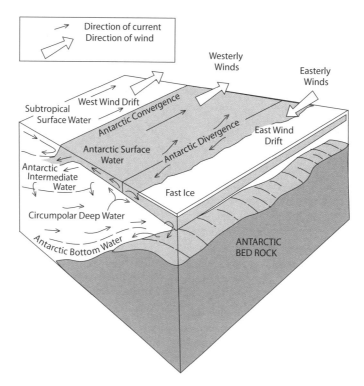

Major winds and currents in surface and deep waters of the Southern Ocean

large in relation to its surface area, adjacent water masses tend to retain their identities on contact, rather than mixing freely. Cold sea water is denser (heavier per unit of volume) than warm water of the same salinity. Similarly, highly saline water is denser than less saline water at the same temperature. When sea water cools, for example in autumn, or in passage along a poleward-flowing current, it becomes denser. If it is diluted, for example by an inflow of fresh water from rivers or melting around an iceberg, it becomes less dense – though often in calm conditions the fresh water rides on the surface rather than immediately mixing. When sea water freezes, fresh water is withdrawn from it in the form of ice crystals, leaving a solution of higher salinity and greater density.

Cold surface waters spreading northward from the continent form a discrete surface layer called Antarctic Surface Water that, at the Antarctic Convergence or Polar Front, sinks below the warmer, less dense mass of Subtropical Surface Water that flows southward from the subtropics. Similarly a mass of cold, very saline water of Antarctic origin, called Antarctic Bottom Water, sinks to the ocean floor close to the continent and flows slowly northward at great depth, retaining its identity well beyond the equator. When forced to the surface by upwelling, for example off the coast of Peru, it emerges as an anomalously cold current in a tropical area. These two major outflows of water away from Antarctica are countered by inflows of Antarctic Intermediate Water and Circumpolar Deep Water at intermediate depths. The Antarctic Convergence form the northern limit of the Southern Ocean, and is a generally accepted northern ecological boundary to the **Antarctic region**.

Sea ice

The southern two-thirds of the ocean's surface freeze in winter to depths of 1–2 m (3–6 ft) or more, forming floating **sea ice**, some attached to the land (fast ice), the rest drifting with wind and currents (pack ice). The ice forms gradually from about March onward, spreading northward to cover up to 20 million km^2 (7.8 million sq miles) of ocean by late winter. The northern boundary is the mean northern limit of pack ice. Like the open sea in summer, the ice in winter is driven by the prevailing winds, parting to form leads and polynyas, rafting and deforming into pressure ridges. Snow builds up on it, pressing down and thickening the original floes. In early spring the process slowly reverses: the ice sheets break and disperse, producing loose pack ice which continues to shift in patterns determined by wind and currents. Most of this gradually melts and disperses, reducing to a summer area of about 4 million km^2 (1.6 million sq miles). The remnant that persists through summer is incorporated in the following season's ice sheet, becoming second-, third- or multi-year ice. For details of sea ice formation and distribution see **Sea ice: Southern Ocean**; for its effect on biota see **Southern Ocean: biology**.

The Southern Ocean contains also drifting ice bergs and 'ice islands', which break from continental glaciers and ice cliffs. These drift first westward about the continent, then eastward and north, driven partly by winds but mainly by currents that bear on their relatively enormous keels. As they move northward they gradually break up and melt, contributing the products of their disintegration – bergy bits, growlers and brash ice – to surface waters. The smaller fragments disappear quickly, the larger bergs and ice islands last longer and travel further under the influence of wind and current.

Study techniques

Almost every expedition that penetrated the Southern Ocean from the eighteenth to the early twentieth centuries used contemporary methods to record sea temperature, currents, winds, and distribution of sea ice and biota. From 1925 to 1939 the annual expeditions of **Discovery Investigations** used the most up-to-date shipborne techniques, contributing substantially to knowledge of water mass movements, chemistry and biology. During the second half of the twentieth century, the number of dedicated oceanographic expeditions decreased, but rate of data acquisition was enhanced, for example by deployment of continuously-recording conductivity-temperature-depth (CTD) probes, current meters, and rosettes of water bottles that took samples of water from pre-determined depths, for analysis by rapid electronic techniques. Shipborne observations were augmented by data collected from automatic stations, free-floating or placed on sea ice: data were logged and recovered during subsequent visits, or transmitted at intervals by radio. Since the 1960s, satellites carrying radiometers and other remote-sensing equipment contributed substantially to year-round knowledge of the Southern Ocean. TIROS, ARGOS, LANDSAT, SPOT, Nimbus, SEASAT and other systems have recorded sea ice and ice berg distribution, wind speed and direction, surface temperatures, chlorophyll density, wave height and many other variables, on a scale and with continuity that was never possible from ships.

Further reading: Deacon (1984); Wadhams (2000).

Southern Ocean: biology.

After annual variations in solar radiation, the seasonal factor that most strongly influences the biology of the marine environment is variation in the distribution of sea ice, especially pack ice.

Sea ice and annual productivity

At its maximum extent, usually September or October, sea ice may cover up to 20 million km^2. At the minimum, usually around February, only 3–4 million km^2 remain covered, and large areas of inshore seas, including the Weddell and Ross Seas, are open. This variation induces strong seasonal changes in the biology of the water column.

The presence of sea ice reduces gaseous exchange and heat flow at the sea–air interface, and increases stability of the water column. Thin sea ice is translucent, and an important habitat both for phytoplankton and for an associated animal community. When the floes thicken or are covered with snow, light penetration is reduced. When the ice begins to melt in spring, starting in low latitudes and extending to high latitudes later in the season, phytoplankton held within it is released, and there is often an intense bloom along the ice edge. There is also a clear seasonal increase in open water and inshore (neritic) primary production. The bloom tends to be more intense and more sharply defined inshore, more diffuse and less concentrated in offshore waters.

Seasonal production is reflected throughout Southern Ocean ecosystems. Most herbivorous zooplankton and benthic suspension feeders feed only in summer when production is high, so most of their growth and reproduction is confined to this period. For zooplankton which have high energy demands because of swimming activity, winter energy comes mainly from fat reserves, which also fuel the production of eggs so that larvae can be released during the short spring bloom. Benthic (sea bed) organisms that do not swim have lower metabolic rates, which can be met by relatively small reserves of energy. Their larvae are usually released during or before the spring bloom, sometimes after an extended period of brooding by the adult during the preceding winter. Some demersal fish are seasonal breeders, despite an apparently constant supply of food throughout the year.

Biological communities

Living organisms were first found in, around and under sea ice in the mid-nineteenth century by Joseph Hooker, but only recently have different communities of organisms been associated with different sea ice zones. The communities gather in different ways, depending on how the ice forms. **Frazil ice**, for example, concentrates particulate matter (including phytoplankton) from the surrounding seawater during its formation. Diatoms are dominant in sea-ice communities, which include such other primary producers as flagellates. Also present are bacteria, colourless flagellates, cilates and nematodes. These microbial organisms cycle a significant fraction of the total fixed carbon within the communities. Physical conditions in the brine channels which harbour the communities are difficult to measure precisely, but seem to vary widely. Standing crops of chlorophyll can be very high, sometimes exceeding 1 g carbon per m^3. Production, also very difficult to measure, may contribute at least 4 gm carbon per m^3 to pelagic and benthic systems. The sea ice microbial community attracts macrozooplankton, including amphipods and other crustaceans which feed on it, and are in turn preyed upon by fish, especially *Pagothenia borchgrevinki* which is rarely found elsewhere. The microbial community may also provide an important food source for Antarctic krill in winter.

Food chains and webs

Oceanic food chains, like those in all other ecosystems, begin with primary producers, the organisms that fix inorganic carbon from the atmosphere to produce organic matter. In oceans they range from microscopic bacteria and phytoplankton, drifting in surface waters, to the large, attached algae familiar along the shore as kelps and seaweeds. The latter are important only in coastal areas: in the open ocean primary production is due partly to bacteria and very much more to phytoplankton, mostly single-celled or colonial algae. Primary producers are browsed by consumers of various kinds, mostly in the zooplankton, including arrow-worms, polychaetes, crustacea, and the larvae (juveniles) of many organisms from hydroids to fish. Zooplankton is in turn fed on by **fishes**, **birds**, **seals** and **whales**, so the energy fixed by the primary producers passes along the food chain. A proportion of primary producers sink below the level of the plankton, to be taken by a host of predators on the sea bed.

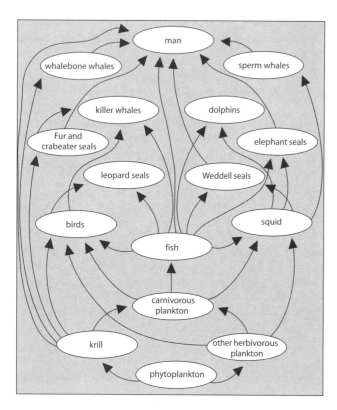

Food web of the Southern Ocean. Phytoplankton and zooplankton, notably krill, form the bases of all the major food chains. Man currently takes very few seals or whales, more krill and fish

Phytoplankton

Phytoplankton cells are classified according to size, specifically the size of mesh of the net, or the pores of filters, that will catch them. Those retained by a 20-micron filter are microplankton, or sometimes net phytoplankton, since these are taken in traditional fine phytoplankton nets. Smaller ones retained by a 2-micron filter are nanoplankton, and organisms smaller than about 0.2 micron are picoplankton. Early studies of Southern Ocean phytoplankton concentrated on the larger forms such as diatoms and dinoflagellates. Only recently has the importance of the smaller forms, that slip through standard nets and filters, been realized. In some areas at some times nanoplankton and picoplankton are responsible for much of the biomass and most of the production. Southern Ocean phytoplankton is made up of over 100 species of diatom and more than 60 species of dinoflagellate, with diatoms generally the more abundant. Important diatom genera include *Chaetoceros*, *Corethron*, *Nitzschia*, *Thalassiosira* and *Thalassiothrix*. Typical Southern Ocean dinoflagellates include *Ceratium*, *Dinophysis* and *Protoperidinium*.

Silicoflagellates are important in the Antarctic, for example the haptophyte *Phaeocystis*, which exists both as small flagellates and as extensive mucus-bound colonies in very high population densities.

Primary production is strongly seasonal. Very low densities of plankton in winter, particularly under sea-ice, make it impossible for many zooplankton species (see below) to feed at all for several months each year. This winter standing crop begins to proliferate in spring, and by summer gives rise to a large biomass, often dense enough to colour the water green or red in patches. Because phytoplankton includes such a variety of organisms, the most practical way of estimating its biomass is to take a sample of known volume, extract the photosynthetic pigment chlorophyll and measure its concentration. From winter concentrations too low to measure at all by this method, peak biomass during late spring or summer in rich inshore areas may exceed 10 mg per m^3 of chlorophyll, equivalent to roughly 800 mg per m^3 of carbon, or 1.6 g per m^3 of organic matter. In offshore waters representative concentrations are 0.2 to 0.6 mg per m^3, rising to maxima usually closer to 1 mg per m^3 or more of chlorophyll.

A substantial biomass of phytoplankton develops within sea-ice. Thick ice, which does not transmit sunlight, has very little phytoplankton beneath it, but concentrations immediately under thin ice can be high, greatly exceeding those in the water column below. A complex association of diatoms and other microscopic plants develops, together with a fauna of bacteria, rotifers, nematodes, crustaceans and other organisms. The overall contribution to Southern Ocean primary production by these sea-ice communities is unknown, but present estimates suggest that it may be substantial. In most seas the spring and summer increase in primary production results in a rapid depletion of such important growth elements as nitrogen, phosphorus and silicon, often to levels where cell growth is inhibited. In the open Southern Ocean spring nutrient levels are usually high, and although some nutrient depletion occurs, it rarely reaches levels that would be limiting. Yet phytoplankton production peaks and falls, presumably in response to some non-nutrient limiting factor: it is not at present clear what this might be. The most intense blooms occur either with periods of calm weather at sheltered inshore sites, in areas of upwelling (where nutrient-rich water is constantly moving up to the surface), and in the spring marginal ice zone where sea-ice is melting rapidly.

Zooplankton and other consumers

The standing crop of phytoplankton at any time is a balance between the rate of primary production and rates of loss by grazing and sedimentation. Grazers include both microzooplankton (minute animals only a few times larger than the cells on which they are feeding) and macrozooplankton, animals usually visible to the naked eye, of which the impact is very much greater. Southern Ocean macroplankton includes a wide range of species including medusae (jellyfish), ctenophores (comb-jellies), siphonophores, chaetognaths (arrow-worms), polychaet worms and pteropods, together with a wide spectrum of crustacean and fish. Dominant in most summers are salps and two groups of crustacean, copepods and shrimp-like euphausiids. The relative importance of these three groups varies with area and season, but in many areas (for example Bransfield Strait) the single species of euphausiid, Antarctic krill *Euphausia superba*, appears to be the dominant grazer for much of the summer. Elsewhere this role is taken by copepods or, closer to the continent, by another species of euphausiid, *E. crystalorophias* (sometimes called ice-krill). On occasion salps can be very important as grazers, but our understanding of their role in the ecosystem is limited by problems with obtaining accurate quantitative estimates of biomass. In the ice-free zone of the West Wind Drift or Antarctic Circumpolar Current, zooplankton is dominated by herbivorous copepods, salps and small euphausiids. In the zone of seasonal ice-pack, covered by ice in winter but largely ice-free in summer, and dominated by the East Wind Drift or Antarctic Coastal Current, *Euphausia superba* is dominant, but the complex food web also includes salps, copepods, fish larvae and chaetognaths. Among the permanent pack-ice close to the Antarctic continent, including the two large embayments of the Weddell Sea and Ross Sea, zooplankton biomass appears to be low, and the smaller *Euphausia crystalorophias* largely replaces *Euphausia superba*.

All levels of the Southern Ocean food web are influenced by the seasonal production. Calanoid copepods, typifying herbivores that are tied more or less directly to the annual pattern of primary production, feed and grow only during the summer when phytoplankton is available. In winter they cease feeding and sink to lower depths, living off a store of fat synthesised the previous summer. This store is used also for the production of eggs in late winter, timed so that newly-emerged larvae can take advantage of the spring bloom of phytoplankton. In some widely-ranging species the period of maximum growth is later in high-latitude stocks, tracking precisely the later period of primary production in these latitudes. Carnivorous and omnivorous organisms are less strictly coupled to seasonality. The largest predators in the food webs, among them many birds, seals and whales, respond to seasonality of food availability and harsh winter conditions by migrating to lower latitudes. Microplankton, Southern Ocean phytoplankton and zooplankton smaller than about 200 microns, has so far been little studied, but is likely to account for a high proportion of total productivity and energy transfer. The microbial community includes representatives of the following groups: bacteria, Bacillariophyta (diatoms); Haptophyta (including coccolithophorids; Dinoflagella; Chrysophyta (including silicoflagellates); Chlorophyta; Euglenophyta; Zoomastigina (including choanoflagellates and bicoecids); Ciliophora (ciliate protozoans); Foraminifera and Actinopoda (including radiolarians). Many groups of organisms (for example dinoflagellates) can be both autotrophic primary producers and heterotrophic consumers: in these small organisms the traditional distinction between plants and animals breaks down. Within the microzooplankton a considerable fraction of the primary production is grazed immediately by protozoan consumers. This so-called microbial loop is an important pathway for energy flow, separate from the better-known route to large zooplankton grazers such as salps, copepods and euphausiids. The loop is likely to be particularly important among sea ice communities of microplankton.

Further reading: Knox (1994).
(ACC)

Southern Ocean islands. Groups of **Southern islands** that stand south of the Antarctic **Convergence** in the Southern Ocean. This category excludes those that stand on the continental shelf, and are geographically close enough to come directly under continental influences (cf. **Antarctic fringe islands**). The broad zone of the Southern Ocean is divided into two latitudinal sub-zones by the mean northern limit of **pack ice**. Islands within the southern sub-zone are invested with fast ice and pack ice for several winter months each year, and are grouped as **Antarctic islands**. They include the South Shetland Islands, South Orkney Islands, South Sandwich Islands, Peter I Øy, Balleny Islands and Scott Island: for details see individual entries. Those of the northern sub-zone, between the mean northern limit of pack ice and the convergence, are called **sub-Antarctic** islands. They include South Georgia, Bouvetøya, Heard Island and the McDonald Islands. Fast ice and pack ice are normally absent from their coasts, though thin, transitory fast ice may appear in sheltered harbours during cold winters. For a table see **Study Guide: Southern Oceans and Islands**.

Southern Ocean Sanctuary. Area of the Southern Ocean and adjacent oceans in which whaling is prohibited under the rules of the International Whaling Commission. The Commission's first sanctuary for Antarctic whales was established in 1948. Baleen whales were protected from killing by pelagic whale catchers (i.e. those from factory ships) in a broad zone of the South Pacific and Southern oceans extending south of 40°S, and between longitudes 70°W and 160°W. A broader proposal to set up a sanctuary covering all oceanic waters south of 40°S was considered by the Commission in 1992, endorsed in 1993, and finally accepted with modifications by majority vote in May 1994. The resulting Southern Ocean Sanctuary, within which all commercial whaling is prohibited, extends from the coast of Antarctica northward to an irregular northern boundary (see map), along 60°S (that of the **Convention for the Conservation of Antarctic Marine Living Resources**) in the southeastern Pacific and southwestern Atlantic oceans, and to 55°S in the Indian Ocean between 20°E and 130°E, where for much of its length it runs contiguously with the pre-existing Indian Ocean Sanctuary (established in 1979). In the central Atlantic and western Pacific oceans, and south of Australia, the sanctuary extends further north to 40°S to accommodate feeding grounds of sei and fin whales. An open area immediately east and south of South America allows for local hunting. Japan objected to the establishment of the sanctuary, and continues to hunt limited numbers of minke whales within its boundaries. The prohibition is subject to review after ten years and at succeeding ten-year intervals.

Southern temperate islands. Groups of **Southern islands** that lie north of the Antarctic **Convergence** in the southern Atlantic, Indian and Pacific oceans. Those lying between the convergence and the 10°C isotherm for the warmest month are called **Cool temperate islands**. They include the Falkland Islands, Marion Island, the Prince Edward Islands, Iles Crozet, Iles Kerguelen, Macquarie Island, Campbell Island and the Antipodes Islands. Those lying north of the 10°C isotherm are called **Warm temperate islands**. They include the Tristan da Cunha group, Gough Island, Ile St. Paul, Ile Amsterdam, and the Auckland Islands and Bounty Islands.

On the cool temperate islands mean monthly temperatures near sea level always exceed freezing point in winter and remain below 10°C in summer. Precipitation exceeds 100 cm annually, falling most often as sleet or rain: snow that falls in winter seldom settles for long. Only the main island of Iles Kerguelen carries a remnant glacial cap, formerly more extensive. Predominant vegetation at sea level is tussock grassland, with high meadows of shorter grasses, low scrub and fellfield. On the warm temperate islands there are no permanent snowfields and few signs even of former glaciation. Mean monthly temperatures in winter rarely fall below 5°C. Snow is rare: rainfall usually exceeds 100 cm annually. Characteristic vegetation close to sea level is low, wind-sculptured forest, giving way to grassland in higher or more exposed parts, though Iles St Paul and Amsterdam have lost their forests to burning and grazing, and the Bounty Islands are mostly bare and sea-washed.

Southern Thule. 59°30'S, 27°00'W. A cluster of three islands, Thule Island, Cook Island and Bellingshausen Island, standing south of Forsters Passage and forming the southernmost group of the **South Sandwich Islands**. The group was discovered and charted by Capt. J. Cook, who considered that it might be part of a large land mass. For details, see entries for individual islands.

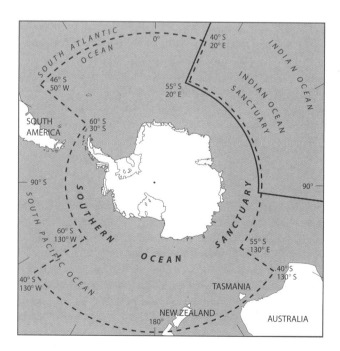

Boundaries of the Southern Ocean Sanctuary, within which commercial whaling is prohibited. The Indian Ocean Sanctuary lies contiguous along part of its northern boundary

Southwest Anvers Island and vicinity. The only area of Antarctica so far (2001) designated a Multiple-use Planning Area. Situated between 64°41'30"S and 65°S and between 63°40′W and 64°35′W, it includes some 1535 km² of ice shelf, sea and islands: Palmer Station, SPA 17 (Litchfield Island) and SSSI 20 (Biscoe Point) all lie within it. It was designated because the area, already well studied biologically, is likely to grow in importance for long-term studies of natural variability, effects of humans on Antarctic communities, and possible effects of global change.

Sovereignty in the Antarctic region. Though explorers and navigators of the seventeenth to nineteenth centuries from time to time claimed newly-discovered lands in the Antarctic region for their sovereign or state, by the late nineteenth century claims based solely on discovery carried little credibility. What mattered were occupation and administration: while the lands remained unoccupied, ownership was hardly an issue. The onset of sealing in the late nineteenth century (**Sealing in the Southern Ocean**), and of whaling in the early twentieth century (**Whaling in the Southern Ocean**) led several nations to conclude that the Antarctic region had taxable resources and was thus no longer politically negligible. Political consciousness led to the filing of national territorial claims, at first small-scale and restricted in scope, later extended to include parts of the continent itself.

Though sovereignty in the region at the start of the twentieth century was by no means clearly established, Norwegian whalers, the prime movers in the newly-developing whaling industry, sought on behalf of their own newly-established state to maintain good international relations with Britain and other maritime nations. They were thus prepared to recognize British claims to sovereignty initially on South Georgia, later on other islands and coasts within the Maritime Antarctic sector, and seek permits for their operations from the nominal British authority, the Government of the Falkland Islands and Dependencies. Their requests gave Britain opportunity in 1906 to provide whaling ordinances, issue permits and collect taxes, and in 1908 to issue Letters Patent (amended and reissued in 1917: see **Falkland Islands Dependencies**) reasserting its claim to the whole sector, including the Scotia Arc and a wedge-shaped portion of the continent between 20°W and 80°W, extending from the coast to the South Pole. To provide an administrative authority, British magistrates were installed in 1909 on South Georgia and in 1910 at Deception Island, South Shetland Islands, where moored whaling ships were operating. In 1912–13 the Deception Island whaling station was built, licensed on a 21-year lease by the Government of the Dependencies. Simultaneously the whalers were working southward along Antarctic Peninsula, finding sheltered moorings where fresh water was plentiful for their factory ships, and sending off fast steam-driven catchers to hunt among the many icebound islands and bays.

The development in 1912–13 of pelagic whaling – catching and processing whales along the edge of the pack ice far from land – encouraged the industry to spread to other areas of the Southern Ocean. In 1923, responding to a request from Capt. C. A. Larsen to begin whaling in the Ross Sea sector of Antarctica, the British Government promulgated an Order in Council defining a new Ross Dependency, and placing it under the administration of the Dominion of New Zealand. So began New Zealand's political involvement in Antarctica. Anticipating a British bid to annex more of Antarctica, France in 1924 renewed a long-standing claim to Terre Adélie, a narrow wedge of East Antarctica extending from the coast to the South Pole. American influence and interest grew with the expeditions of Richard Byrd inland from the Ross Ice Shelf. Byrd staked claims to some of the areas that his expeditions discovered, but failed to persuade his government to ratify them. The **British, Australian and New Zealand Antarctic Research Expedition 1929–31** explored a lengthy stretch of the East Antarctica coast, presaging a 1934 British claim to two broad wedges of the continent flanking Terre Adélie, to be administered by the government of Australia as Australian Antarctic Territory. From 1926 onward Lars Christensen, a Norwegian whaling and shipping magnate, used his whaling fleet to explore widely along the continental coasts south of the Southern Ocean's richest whaling grounds. The **Norwegian (Christensen) Whaling Expeditions 1926–37** substantiated Norway's claims to **Bouvetøya** (1928), **Peter I Øy** (1931) and **Dronning Maud Land** (1939), the latter claim narrowly heading-off a rival bid by Germany, based on the activities of the **German Antarctic (*Schwabenland*) Expedition 1938–39**, and a possible Japanese bid based on the activities of their whaling fleets in Antarctic waters.

Argentine claims to a sector of Antarctica immediately south of continental South America were implied in many statements of the nineteenth and early twentieth centuries, centred latterly on the permanent meteorological station Orcadas on Laurie Island, South Orkney Islands. In July 1939 a presidential decree set up an enquiry into Argentina's role in Antarctica, and a further decree of April 1940 established a National Antarctic Committee, to be 'responsible for the consideration and handling of all matters connected with the defence and development of Argentine interests in the Antarctic or in the Antarctic continent'. Semi-official sources asserted claims to a sector, as yet ill-defined, south of mainland Argentina, and an Argentine naval transport visited the Peninsula area in 1942 and 1943, posting notices of claims in prominent situations. A British protest at these incursions, into what Britain regarded as the Falkland Islands Dependencies, invoked a ministerial Argentine reply of February 1943, reaffirming a claim to all Antarctic lands and dependencies south of 60°S and between meridians 25°W and 68°W.

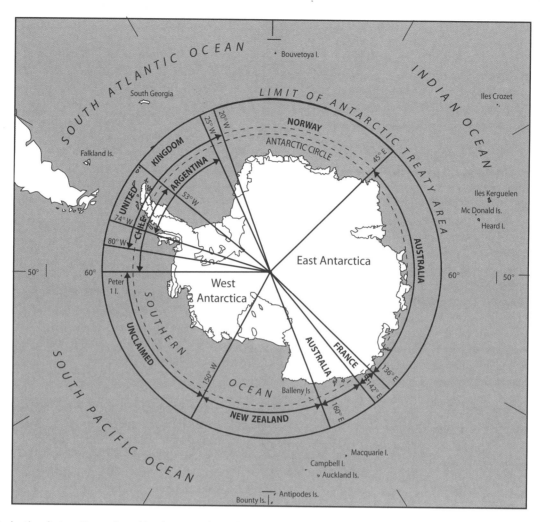

Sovereignty in the Antarctic region. Northern and southern limits of the Norwegian sector are not specified

Chilean claims to a related sector of Antarctica were stated most clearly in Presidential Decree 1747 of 6 November 1940, which asserted that all lands, islands, islets, reefs, rocks and glaciers, already known or to be discovered, and their respective territorial waters, in the sector between longitudes 53° and 90°W, constitute the Chilean Antarctic or Chilean Antarctic Territory. Subsequently Chilean Supreme Decree No. 844 of 19 May 1945 divided the coastline of the Republic into three naval zones, of which the southernmost was defined as from parallel 47°00′S to the most southerly point of the Republic. Chilean naval operations in the area began in 1947.

Currently, therefore, seven states claim sovereignty over parts of the Antarctic continent, adjacent seas and neighbouring islands. Claims to four of the areas are not disputed between the claimants. Claims to the remaining three overlap considerably, and are or have in the past been the cause of political friction between Argentina, Chile and the United Kingdom, the states concerned. A large area of West Antarctica, significantly the most difficult area to approach by sea, remains unclaimed. The degree to which claims are recognized by other states varies. The United States and Russia neither make claims nor recognize the claims of others, but both reserve the right to claim at some time in the future. Other states that are not themselves claimants either tacitly or overtly recognize the validity of at least some existing claims.

The Antarctic Treaty adopts a position widely interpreted as 'freezing' all sovereignty claims. Article IV asserts first: 'that nothing in the Treaty shall be interpreted as (a) a renunciation by a Contracting Party of previously asserted rights of or claims to territorial sovereignty in Antarctica, or (b) a renunciation or diminution by any Contracting Party of any basis of claim to territorial sovereignty in Antarctica which it may have whether as a result of its activities or those of its nationals in Antarctica, or otherwise, or (c) prejudicing the position of any Contracting Party as regards its recognition or non-recognition of any other State's rights of or claim or basis of claim to territorial sovereignty in Antarctica.' Second, 'No acts or activities

taking place while the present Treaty is in force shall constitute a basis for asserting, supporting or denying a claim to territorial sovereignty in Antarctica or create any rights of sovereignty in Antarctica. No new claim, or enlargement of an existing claim, to territorial sovereignty in Antarctica shall be asserted while the present Treaty is in force.' It was instructive to note that, during the conflict between Argentina and the United Kingdom over sovereignty of the Falkland Islands and South Georgia, hostilities extended to those areas and to the South Sandwich Islands, but not to areas covered by the Treaty.
Further reading: Dodds (1997).

Sovetskaya Station. 78°23′S, 87°32′E. Soviet research station at 3650 m (11 972 ft) on the plateau of Princess Elizabeth Land. Opened on 16 February 1958, it was one of a network of staging posts for tractor trains crossing the plateau, *en route* between Mirnyy and the Pole of Inaccessibility. The station closed on 3 January 1959.

Soyuz Station. 70°35′S, 68°47′E. Russian permanent research station established inland from Amery Ice Shelf, by Beaver Lake, in the Aramis Range of Prince Charles Mountains, Mac.Robertson Land. It operated as a summer station for traverses and glaciological work between September 1982 and February 1989.

Spaatz Island. 73°12′S, 75°00′W. An ice-domed island 80 km (50 miles) long and 40 km (25 miles) wide off the coast of Ellsworth Land, embedded in ice shelves between George VI Sound and Stange Sound. Surveyed from the ground by a sledging party from the United States Antarctic Service Expedition 1939–41, and from the air by the Ronne Antarctic Research Expedition 1946–48, it was named for Gen. Carl Spaatz, USAF, who supported the Ronne expedition.

Specially Protected Areas (SPAs). Within the Antarctic Treaty system SPAs are designated 'to preserve both unique and representative examples of the natural ecological systems of areas in Antarctica which are of outstanding scientific interest'. Entry is prohibited without a permit: permits can be issued by appropriate national authorities for their own nationals, and only for compelling scientific purposes that cannot be served elsewhere. Vehicles of all kinds are forbidden.

Sperm whale. *Physeter catadon.* Largest of the toothed whales, sperm whales occur in all the world's oceans. There are separate northern and southern hemisphere stocks. Bulls in the southern hemisphere grow to 18 m and weigh up to 55 tonnes. In winter all are found in tropical, subtropical and temperate waters. In summer cows and calves remain in warm or cool waters, but many bulls, probably those that are not involved in breeding, swim south to feed in Antarctic waters. They are usually seen alone or in small groups, penetrating to the edge of the pack ice, where they feed mainly on fish and squid, often at great depths. Many thousands were taken in Antarctic waters by whalers, usually in a short season preceding or following the main whaling season. Their waxy oil, including valuable spermaceti from the head, was processed and marketed separately from that of other whales. Sperm whales are included in the general ban on whaling in Antarctic waters. The world population is estimated at well over 300 000.
Further reading: Watson (1981); Bonner (1989).

Spring warming. Period during which the return of the sun brings the first post-winter warming to the Antarctic. In early spring sufficient light may penetrate shallow snow and ice layers to permit photosynthesis. Partial melting under the snow produces miniature 'greenhouses', in which algae, lichens and mosses have adequate water and light and are protected from wind and temperature fluctuations, reaching temperatures several °C higher than that of the air outside. Melting and refreezing of the snowpack increase its transparency by the formation of long ice crystals which act as light pipes. Snow melting begins mainly on north-facing slopes, as they receive most sunshine. Melt water trickling into crevices moistens soils and their microflora and fauna. With further warming, all biological activity increases. Lichens and mosses become rehydrated and, in the lengthening days, start to photosynthesize. Freezing and thawing alternate rapidly: surface temperatures of dark rock may reach over 40 °C in direct sunlight, but fall within minutes to ambient air temperatures well below freezing point when the sun is obscured by cloud.
(DWHW)

Standing Committee on Antarctic Logistics and Operations (SCALOP). A subcommittee of the **Council of Managers of National Antarctic Programmes** (COMNAP) that investigates and initiates research on operational problems, arranges subgroups of experts to discuss and foster advances in technology, and exchanges information on logistics and operations. SCALOP and its parent body COMNAP arose in 1988 to provide management bodies in which the requirements of scientists could be reconciled with the practicalities of running national programmes.

Stanford Plateau. 85°57′S, 140°00′W. Ice plateau over 3000 m high in the Queen Maud Mountains, Ross Dependency, named for Stanford University.

Stange Ice Shelf. 73°15′S, 76°30′W. Extensive ice shelf occupying Stange Sound, between Smyley Island, Spaatz Island and the English Coast, and facing Bellingshausen Sea. It was first identified from aerial photographs by the Ronne Antarctic Research Expedition 1946–48, and named for Henry Stange, a supporter of the expedition.

Stefansson Sound. 69°28′S, 62°25′W. Ice-filled passage between the Wilkins Coast and Hearst Island, in the southwestern Antarctic Peninsula, named for the US Arctic explorer Vilhjalmur Stefansson. Sir Hubert Wilkins gave the name Stefansson Strait to an ice-filled channel in about 70°S that he identified from the air during his pioneering flights, thinking that it crossed the Peninsula, naming it for his former leader on an Arctic expedition. The channel proved non-existent, so the name has been transferred to a lesser feature nearby.

Stenhouse, Joseph Russel. (1887–1941). British polar seaman and explorer. Born on 15 November 1887 in Dumbarton, Scotland, he joined the merchant navy in 1903, ultimately qualifying as a master in sail. In 1914 he joined the **Imperial Trans-Antarctic Expedition 1914–17**, initially as chief officer of *Aurora*, the ship carrying the expedition's Ross Sea party, later as master replacing Æneas Mackintosh, who joined the shore party at Cape Evans. Stenhouse commanded *Aurora* through the long period of drifting in the ice, returning eventually to New Zealand in 1916. Already commissioned in the Royal Naval Reserve, he served during World War I first in 'Q' (mystery) ships, later with Ernest Shackleton in northern Russia, and was twice decorated. After the war he was engaged by the Discovery Committee, fitting out RRS *Discovery* and taking command during the 1925–27 commission. In 1931–32 he worked with an international travel service that tried unsuccessfully to promote shipborne travel and tourism, in particular to historic sites in Antarctica. Recalled to the fleet in World War II, he saw action in northern waters and the Gulf of Aden. He was lost at sea, missing presumed drowned, on 12 September 1941.

Stephenson, Alfred. (1908–99). British polar surveyor and explorer. Born in Norwich on 25 November 1908, Stephenson read geography at Cambridge University. On graduating in 1930 he became chief surveyor to the British Arctic Air Route Expedition, taking part in boat and sledging journeys to survey and chart parts of the coast of east Greenland. In 1932–33 he explored and surveyed in the Great Slave region of central Canada with the British PolarYear Expedition to Canada. In 1934 he joined several former Greenland colleagues to form the **British Graham Land Expedition 1934–37**, in which his skills as both dog-driver and surveyor were fully employed. On return he was appointed lecturer in surveying at Imperial College, University of London. During World War II he served with distinction in the Royal Air Force, developing photo-interpretation techniques. After the war he returned to teaching at Imperial College, retiring in 1972. Active for many years in the Royal Geographical Society, 'Steve' served on the expeditions committee, supervised survey equipment and helped to found the Young Explorers' Trust. From 1956 to 1996 he was honorary secretary of the Antarctic Club. On his retirement the club established an annual prize for expedition work within the trust, in his honour naming it the Stephenson Award. He died on 3 July 1999.

Stevens, Alexander. (1886–1965). British polar geologist. Born in Scotland, he graduated in arts (1907) and geology (1913) at Glasgow University. He joined the **Imperial Trans-Antarctic Expedition 1914–17**: assigned to the Ross Sea party, he became chief of the scientific staff, and took part in the depot-laying operations. On returning to Britain he saw wartime service in the Royal Engineers, then returned to teach geology at Glasgow University where he was appointed Professor in 1947. He retired in 1953 and died on 20 December 1965.
(CH)

Stewart VII, Duncan. (1905–69). Antarctic petrologist. A graduate in geology of the University of Michigan (1933) and Brown University (1930), he joined the staff of Carlton College in 1933, where he remained until his death. Developing an interest in Antarctic petrology, he spent the austral summer of 1960–61 in fieldwork in the Ross Sea area, and also studied rocks from earlier expeditions. Stewart Hills, north of Ford Massif in the Thiel Range, were named for him. He died on 5 November 1969.
(CH)

Stillwell, Frank Leslie. (1888–1963). Australian expedition geologist. Born in 1888, he graduated in engineering from Melbourne University in 1911. Joining the **Australasian Antarctic Expedition 1911–14**, he served at the main base, taking part in the sledging programme and leading two geological sledging surveys east of the base. On his return he wrote up his petrological studies and, after war service and a spell with the Commonwealth Advisory Council of Science and Industry (1916–19) later he became a mineralogist and petrologist with the Council of Scientific and Industrial Research (1929–53). He died on 8 February 1963.
(CH)

St Kliment Ohridiski Station. 62°38′S, 60°22′W. Bulgarian permanent summer-only research station on 'Bulgarian Beach', Livingston Island, South Shetland Islands. Begun as a refuge in April 1988, it was subsequently expanded and inaugurated as a station in December 1993. The station is named for a tenth-century Bulgarian bishop and scholar.

Stonington Island. 68°11′S, 67°00′W. A low, rocky island off the Fallières Coast, southern Graham Land. Formerly joined to the mainland by a tongue of Northeast Glacier, it was selected by the **United States Antarctic Service Expedition 1939–41** as the site of their East Base, from which they could gain direct access to the Graham Land plateau. They named it for the whaling port of Stonington, Conn. In 1945–46 it became the site of Base E, Falkland Islands Dependencies Survey, and in 1947–48 East Base was reoccupied by the **Ronne Antarctic Research Expedition**. The ice tongue has since retreated, cutting off access to the glacier.

Stonington Island Station (Base E). 68°11′S, 67°00′W. British research station on Stonington Island, Marguerite Bay, Graham Land. The first station (Trepassy House), was built by the Falkland Islands Dependencies Survey in February 1946 and operated until February 1950. It was reoccupied from March 1958 to March 1959 and again from August 1960. A new, two-storey hut was built in March 1961 and occupied until February 1975. Trepassy House was destroyed in 1974. The two-storey hut remains, currently in good order, and is designated an Historic Monument.

Storm-petrels. Oceanic flying birds of the order Procellariiformes, family Hydrobatidae, of which five species breed and are widespread in the southern oceanic region. They form loose flocks of a few dozen birds, often mixed with prions and other species, that forage over the ocean surface, fluttering and dipping for food particles, seldom settling except in very calm weather. They breed on islands and mainland throughout the region, usually in colonies, nesting in cavities in soil, under rocks or in screes. Often they emerge from and return to their nests only at night, to avoid predation by gulls and skuas. Commonest and most widespread are Wilson's storm-petrels *Oceanites oceanicus*, 18 cm (7 in) long, with wing span 40 cm (16 in), distinguished by their white rump, dark underbelly and pale grey or fawn wing coverts. They occur over cold northern hemisphere waters during the non-breeding season. Black-bellied storm-petrels *Fregetta tropica*, similar but slightly larger, with wing span 48 cm (19 in), are white underneath with a broad black line along the chest and belly, breeding mainly on Antarctic islands. White-bellied storm-petrels *Fregetta grallaria*, similar again but with white midriff, breed on some of the warm temperate islands. Grey-backed storm-petrels *Garrodia nereis*, smaller and greyer, nest mainly on cold temperate islands. White-faced storm-petrels *Pelagodroma marina*, with conspicuous white face and throat, nest mainly in the warm temperate zone.

St Paul, Ile. 38°43′S, 77°33′E. A warm temperate island of the southern Indian Ocean, 110 km (69 miles) from **Ile Amsterdam**, its larger neighbour. The semicircular remnant of a volcanic cone 5 km (3.2 miles) across, it is an island that has lost its northeastern half, probably by downfaulting. A central peak 270 m (886 ft) high overlooks the caldera, flooded to form a sheltered yacht harbour less than 3 km (1.9 miles) across. Steep cliffs on all sides of the island are made up of interbedded tuffs and lavas. Fresh-looking scoria cones and active hot springs in the harbour echo a record of eruptions and natural fires in historic times. The island has a mild and windy climate similar to that of Ile Amsterdam. Introduced goats and sheep, together with natural and man-made fires, have ensured that little of the original vegetation remains. The exposed western slopes carry a sparse covering of tussock and other grasses, with patches of scrub and sedge: the more sheltered crater is lined with tall tussocks. Rabbits and house mice are still active. Together with outlying stacks, the island supports rockhopper penguins, yellow-nosed and sooty albatrosses, five species of smaller petrel, and small stocks of terns and skuas. Amsterdam Island fur seals breed in the crater. Long used as a summer base for fishing boats, the harbour

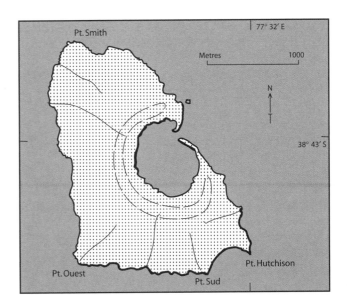

Ile St Paul is a tiny, steeply-flanked crater island of volcanic origin, covered with tussock grass and shrubs. The crater provides a sheltered harbour for small boats

contains the ruins of a canning factory that was active in the 1920s. Claimed by France in 1843, the island has the protected status of a French national park.

Strand crack. A hinge line marking the junction between a fixed inland ice sheet and a floating ice shelf.

Stratton, David George. (d. 1972). British polar surveyor. After service in the Royal Navy, he read geography at the University of Cambridge. In 1951 he spent some months studying glaciology in Swedish Lapland, then joined the Falkland Islands Dependencies Survey as assistant surveyor and dog driver. On return to Britain in 1954 he was appointed stores officer of the **Commonwealth Trans-Antarctic Expedition 1955–58**. Later second in command, he was one of the party that crossed the continent from Shackleton Base on the Weddell Sea to Scott Base in McMurdo Sound. In 1959 Stratton joined British Petroleum Ltd., serving in Europe and North Africa. In 1970 he contracted poliomyelitis. He died on 22 May 1972, aged 45.

Stromness Harbour. 54°09′S, 36°41′W. Fjord harbour off Stromness Bay, South Georgia, named by whalers for the capital of the Orkney Islands. From 1908 it was the site of a whaling station, in 1916 the destination of Sir Ernest Shackleton following his boat journey from Elephant Island to South Georgia (see **Imperial Trans-Antarctic Expedition 1914–17**). The station was later used as an engineering depot and store, closing in the early 1960s.

Sturge Island. 67°28′S, 164°38′E. A narrow ice-mantled island about 32 km (20 miles) long, the southernmost and largest island of the Balleny Islands. Discovered and surveyed in 1839 by sealer Capt. J. Balleny, it was named for one of the expedition supporters.

Sub-Antarctic. A term used variously to describe islands and other localities and phenomena in the area surrounding Antarctica. It has been employed indiscriminately over many years: to avoid possible confusion, is not used in this encyclopedia. The term **sub-Antarctic islands** is used to describe islands within the northern zone of the Southern Ocean: see also **Southern islands.**

Sub-Antarctic islands. Islands of the Southern Ocean that stand between the mean northern limit of **pack ice** and the Antarctic **Convergence**. They include South Georgia, Bouvetøya, Heard Island and McDonald Islands: for details see individual entries. Though milder at sea level than **Antarctic islands**, all but the McDonald Islands are cold enough to support permanent upland ice caps, that descend to sea level in glaciers. Coastal plains are snow-covered in winter, but their mature soils support rich stands of tussock grasses. Coastal uplands have a thin moorland vegetation of herbs and small shrubs. Monthly mean air temperatures close to sea level range from +6 °C in summer to 0 °C or just below in winter.

Sulzberger Ice Shelf. 77°00′S, 148°00′W. Ice shelf on the Saunders Coast of Marie Byrd Land and Edward VII Land. Occupying Sulzberger Bay, the shelf is fed by glaciers from the Ford Ranges, and held in position by many large islands. It was identified by Byrd's first Antarctic Expedition 1928–30 and named for Arthur H. Sulzberger, publisher of the *New York Times* and a supporter of the expedition.

Svarthamaren. 71°44′S, 5°12′E. An ice-free mountain area some 200 km (125 miles) inland from the coast in the Mühlig-Hoffmannfjella, Dronning Maud Land. Northeast-facing cliffs and screes on its flank are remarkable for accommodating a large colony of Antarctic petrels, snow petrels and southern skuas. An area of approximately 6.4 km^2 (2.5 sq miles) of the breeding grounds is designated SSSI No. 23 (Svarthamaren) to facilitate research. The Norwegian research station **Tor** has been established close by for this purpose.

Svea Station. 74°35′S, 11°13′W. Swedish research station in the Heimefrontfjella, Dronning Maud Land, opened in 1987–88 for summer glaciological and geological studies.

Sverdrup, Harald Ulrik. (1888–1957). Norwegian meteorologist, oceanographer and polar administrator. Born on 15 November 1888, Sverdrup studied meteorology with V. Bjerknes. He became chief scientist with the Norwegian North Polar Expedition 1918–25, held academic appointments in Norway and USA, was chief scientist on Sir Hubert Wilkins's *Nautilus* expedition (1931), and joint leader of the Norwegian–Swedish Expedition to Northwest Spitsbergen (1934). He directed the Scripps Institution of Oceanography (1936–48), and was Director of the Norwegian Norsk Polarinstitutt from 1948. Sverdrup chaired and inspired the international committee of the Norwegian-British-Swedish Antarctic Expedition 1949–52, visiting the expedition station Maudheim in 1951. He died on 21 August 1957.

Sverdrupfjella. 72°20′S, 1°00′E. (Sverdrup Mountains). Westernmost of a chain of mountain ranges extending east-to-west between Fimbulheimen and Hellehallet, parallel to

Prinsesse Astrid Kyst, Dronning Maud Land. Several peaks rise above 2600 m (8500 ft). The mountains were named for Norwegian glaciologist H. U. Sverdrup.

Swedish South Polar Expedition 1901–4.

Geological and biological survey expedition to northern Graham Land. Responding to the 1895 declaration of the London **International Geographical Congress**, geologist Otto Nordenskjöld of Sweden spent over four years in lobbying and fund-raising for a Swedish expedition to the Antarctic. Private benefactors eventually provided most of the money. He bought *Antarctic*, the whaling ship that had previously carried Bull's **Norwegian (Tønsberg) Whaling Expedition 1893–95** to Cape Adare, securing Capt. C. A. Larsen as master, and a team of eight scientists. Already familiar with the geology of southern South America, and aware that Larsen had found interesting fossils on Seymour Island, Nordenskjöld settled on Antarctic Peninsula as his study area. *Antarctic* left Göteborg on 16 October 1901, reaching the South Shetland Islands on 10 January 1902. From there the expedition explored the islands and channels off the tip of Antarctic Peninsula, then continued south into the Weddell Sea. Trying to reach the coast of King Oscar II Land, which Larsen had discovered on his earlier voyage, *Antarctic* met impenetrable pack ice. On 12 February, unable to proceed further south, the expedition landed on Snow Hill Island, close to Seymour Island, setting up a small hut where Nordenskjöld and five companions could spend the winter in comfort. The ship returned north, completing a programme of oceanographic research and survey *en route* for the Falkland Islands.

Despite bad weather the wintering party explored widely. In October 1902 Nordenskjöld with companions José Sobral and Ole Jonassen took a small team of dogs on a sledging journey southwestward to 66°S along the Larsen Ice Shelf, collecting rock samples from a prominent nunatak

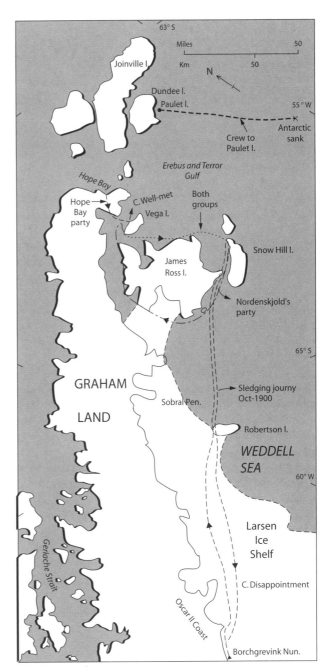

Routes of the Swedish South Polar Expedition

Stone hut of the Swedish South Polar Expedition, Hope Bay. Photo: BS

that they named for Carsten Borchgrevink. In December they explored Seymour Island, finding fossil plants that told of the region's temperate, even subtropical history, and penguin bones, including some from species larger than any alive today. After wintering in the Falkland Islands *Antarctic* moved to South Georgia, picking up a new member of the expedition, scientist Gunnar Andersson, and in early November headed south to relieve the Snow Hill base. Heavy pack ice barred the way. Larsen spent time charting in the Orléans Strait area, then moved south into Antarctic Sound, again hoping to reach Snow Hill Island. Again the ship met solid pack ice. On 29 December Gunnar Andersson, Samuel Duse and Toralf Grunden were put ashore at Hope Bay, with a tent, sledge and rations, to walk the 200 miles over the ice to Snow Hill Island. This turned out to be far more difficult than anticipated, so they returned to Hope Bay and camped, expecting the ship to pick them up within a few weeks.

Meanwhile *Antarctic* sailed northward around Joinville Island, trying yet again to enter the pack ice and push southward. This time the ship was gripped by the ice. After a month of struggling off the southeastern coast of Dundee Island, relentless pressure severely damaged the stern and allowed the sea to pour in. The crew unloaded spare timbers, boats, sails and stores, and on 14 February 1903 abandoned *Antarctic*, which disappeared through the ice. The 22 men under Larson's leadership made their way across 40 km (25 miles) of rough sea ice and wide leads of open water to Paulet Island, the nearest land. There they built a stone hut, roofing it with spars and sail canvas, and settled in for a crowded and uncomfortable winter.

On Snow Hill Island Nordenskjöld and his party, knowing nothing of the fate of the rest of the expedition, resigned themselves to a second winter. At Hope Bay Andersson's party of three waited in vain for the ship to return for them, then made the best of their situation, building a stone wall round their tent and roofing it with sledges to protect it from the violent winter winds. All three parties stockpiled penguins and seals to see them through the lean months. On Paulet Island one of the seamen, Ole Wennersgaard, died on 7 June, the only death recorded on the expedition.

On 29 September 1903 Andersson, Duse and Grunden began walking southwest from Hope Bay toward Snow Hill Island. By 9 October, approaching Vega Island, they were surprised to meet Nordenskjöld and two companions, who were sledging with a dog team from their winter base towards Hope Bay. The two groups together returned to Snow Hill Island. On 31 October, as soon as the ice had opened in Erebus and Terror Gulf, Larsen and five of the *Antarctic* crew set out in a rowing boat from Paulet Island towards Hope Bay, threading a hazardous passage between shifting ice floes. Arriving at Hope Bay on 4 November, they found a message to say that Andersson's party had left for the south. Larsen rowed on for a further three days toward Snow Hill Island, leaving the boat at the edge of the fast ice and walking the last 24 km (15 miles) to the hut. They arrived on 7 November, almost simultaneously with a party from *Uruguay* (Capt. J. Irizar), a relief ship sent out by the Argentine navy. *Uruguay* cleared the Snow Hill Island base, then went on to rescue the Paulet Island castaways, and returned the whole expedition safely to Buenos Aires. Despite their difficulties the Swedish scientists produced exceptional results, publishing a series of reports that set a pattern for later expeditions to follow.

Further reading: Nordenskjöld *et al.* (1905).

Syowa Station. 69°00'S, 39°35'E. Permanent Japanese research station operating year-round on East Ongul Island, Prins Olav Kyst, Dronning Maud Land. It opened in January 1957 for the International Geophysical Year and was abandoned in February 1958. Re-opened in 1959, it has since operated continuously.

T

Tabarin Peninsula. 63°32′S, 57°00′W. Mountainous peninsula forming the southeastern tip of Trinity Peninsula, Graham Land. It was named for Operation Tabarin 1943–45, members of which explored the area from their research station, Base D, at Hope Bay.

Tasman, Abel Janszoon. (c. 1603–59). Dutch navigator and explorer. Born near Groningen in the northeastern Netherlands, he sailed for many years on trading and exploratory voyages with the Dutch East India Company. In 1642 Anton van Dieman, governor-general of the Dutch East Indies, commissioned him to explore southward from Batavia (now Jakarta). Sailing in *Heemskerck*, accompanied by Gerrit Janszoon in *Zeehaan* and the Dutch cartographer Frans Jakobszoon Visscher, Tasman crossed the southern Indian Ocean, on 24 November discovering and naming 'Van Dieman's Land' (now Tasmania). Continuing eastward across the Tasman Sea, on 13 December 1642 he discovered and named 'Staten Land' (now New Zealand). Heading northward back to Batavia he discovered the Friendly and Fiji Islands. In a second voyage of 1644 Tasman navigated along the north and west coasts of Australia. His voyages effectively isolated New Guinea from Australia, and Australia and New Zealand from the hypothetical **Terra australis incognita**.

Taylor, Thomas Griffith. (1880–1963). British expedition geologist and geographer. Born in Essex on 1 December 1880, he moved in 1893 with his family to Australia, where he graduated in geology and physics at the University of Sydney. As a postgraduate research scholar in Cambridge, he was appointed senior geologist to the **British Antarctic (*Terra Nova*) Expedition 1910–13**. He explored extensively among the mountains and glaciers west of McMurdo Sound. In later life he held chairs in geography in the universities of Sydney (1917–18), Chicago (1929–35) and Toronto (1936–51). He died in Sydney on 5 November 1963.

Taylor Rookery, Mac.Robertson Land. 67°26′S, 60°50′E. Moraine beach on the east side of Taylor Glacier, Mawson Coast, the site of a land-based colony of emperor penguins. An area of approximately 0.4 km² (0.15 sq miles) is designated SPA No. 1 (Taylor Rookery), to protect the colony.

Taylor Valley. 77°37′S, 163°00′E. Extensive ice-free valley in Prince Albert Mountains, Victoria Land. Upper Taylor Glacier, which carved the valley, has retreated to leave a broad U-shaped valley with several lakes and streams. The valley is open at the seaward end. The upper end was discovered by a sledging party led by R. F. Scott in December 1903. The valley was explored and surveyed during Scott's British Antarctic (*Terra Nova*) Expedition 1910–13, by geologist Griffith Taylor, for whom both valley and glacier are named. The valley forms part of the **Victoria Land Dry Valleys**.

Temperature inversion. A condition of the atmosphere close to the ground in which air temperature increases anomalously with height. Temperature inversion occurs in calm conditions when radiation cooling chills the ground, and heat is conducted from the layer of air immediately above. It is a common feature of the Antarctic plateau, where air 20–30 m above the ground may be 10–12 °C warmer than at the snow surface. See **kernlos effect**.

Teniente Cámara Station. 62°36′S, 59°54′W. Argentinian permanent research station on Half Moon Island, South Shetland Islands. Opened in April 1953 as a naval station with hangar accommodation for a seaplane, it was named for a naval officer who was killed in a helicopter accident. The station operated year-round until 1960, subsequently for summer-only scientific parties.

Teniente Carvajal Station. 62°45′S, 68°54′W. Chilean research station on southern Adelaide Island. Built by British Antarctic Survey in 1961 and operated as Adelaide (Base T), it was in 1983 transferred to the Chilean Air Force as a summer station for meteorological and flying operations.

Teniente Jubany Station. 62°14′S, 58°40′W. Argentine permanent station, built as a refuge in 1953 on Potter Peninsula, King George Island, and named for a naval officer. It was extended in 1982 and has since operated year-round.

Teniente Matienzo Station. 64°58′S, 60°07′W. Argentine permanent research station on Larsen Nunatak (one of the Seal Nunataks), Larsen Ice Shelf, on the Nordenskjöld Coast of Antarctic Peninsula. Built in 1961, it replaced 'San Antonio', a refuge established two years earlier.

Teniente Rodolfo Marsh (Escudero) Station. 62°11′S, 58°57′W. Chilean permanent year-round station on Fildes Peninsula, King George Island, South Shetland Islands, developed since 1980 to service an air strip of the Chilean Air Force. From 1999 it has become known as Escudero Station.

Terns, southern oceanic. Flying birds of the order Charadriiformes, family Sternidae, terns are slender, slightly-built birds about 36 cm (14 in) long, with wing span 80 cm (32 in), mainly white with coral-red bill, black cap and grey-mantled wings. The wings are sharply pointed, the tail is deeply forked. Terns feed over the sea, seldom settling on the water, most often hovering and dipping for small fish and planktonic animals at the surface. They breed in small gatherings of a dozen or more pairs, nesting on gravel or sand within sight of each other, and rising in noisy concert to mob intruders. Their two or three well-camouflaged brown eggs, flecked with grey and black, hatch into equally well-camouflaged chicks. Antarctic terns *Sterna vittata* breed throughout the Scotia Arc, at points along the mainland coast of Antarctica, and on most of the southern islands. Kerguelen terns *S. virgata*, with darker underwings and tail, breed on the Prince Edward Islands, Iles Crozet and Iles Kerguelen. South American terns *S. hirundinaceae* breed on southern coasts of South America and on the Falkland Islands, where they are called 'swallow-tailed gulls'. Summer visitors from the northern hemisphere include common terns *S. hirundo*, which appear off southern coasts of South America and the Falkland Islands, and Arctic terns *S. paradisea* that are often seen off Antarctic Peninsula and islands of the Scotia Arc.

Terra australis incognita. An imaginary southern continent shown on many medieval world maps, joining southern America, Africa and Asia, and occupying the position of Antarctica and the Southern Ocean. Its origins can be traced to pre-Christian geographers. Eratosthenes of Alexandria (*c.* 200 BC) measured Earth's circumference by a technique based on directions of shadows in Alexandria and Syene (Aswãn), calculating a surface area that showed the known world, bounded by North Africa, the Atlantic seabord, central Europe, and a hazy orient far to the east, to be only a large southern continent that would match and counterpoise Eurasia in the north. Pomponius Mela of Rome (*c.* AD 43) peopled the continent with inhabitants whom he called 'Antichthones'. The earliest map showing a southern continent (*c.* AD 150, ascribed to Ptolemy of Alexandria), linked the southern extremities of Africa and Asia. This image of Terra australis incognita, with variations, persisted for over 1000 years, to be dispelled in stages by voyages of Renaissance navigators and Arab traders.

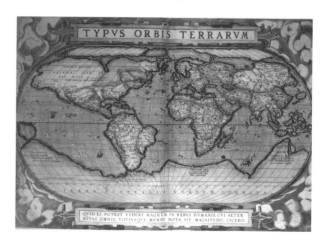

Late 16th century world map by A Ortelius, showing the extent of the unknown southern continent

Terra Nova Bay (TNB) Station. 74°41′S, 164°07′E. Italian permanent research station established in 1985–86 in Terra Nova Bay, Scott Coast of Victoria Land. The station is used for summer studies.

Antarctic tern: Snares Islands, southern New Zealand. Photo: BS

Terre Adélie. Triangular sector of East Antarctica between 136°E and 142°E, bounded on both sides by Australian Antarctic Territory and extending from 60°S to the South Pole. The territory was defined by a French Government Decree of April 1938. It includes a length of coast with limited exposed rocks, steep inland slopes and an extensive plateau. The coast was discovered in 1840 by French Capt. J. Dumont d'Urville and named for his wife. French research stations have operated along the coast since 1950, and temporary stations have from time to time been established inland from the current main base, Dumont d'Urville, at Pointe Géologie.

Terres australes et antarctiques françaises (TAAF). French overseas territories in the southern Indian Ocean and Antarctic. Constituted in August 1955, TAAF includes Terres australes françaises, comprising the three districts of Iles Crozets, Iles Kerguelen and Iles Amsterdam and St Paul, and the Antarctic mainland territory Terre Adélie. Since France's accession to the Antarctic Treaty (1959) Terre Adélie has been administered within the Antarctic Treaty system. The remote islands of Terres australes françaises are administered from the French populated centre, the island territory of Réunion, from which their scientific research stations are visited several times yearly by the research and resupply ship Marion Dufresne II. In 1978 a 200-mile (320 km) exclusive economic zone was declared around the islands, providing a measure of control over local fisheries which are now exploited under licence. Catches include crayfish, squid and several species of fin-fish: krill is regarded as a potential resource. Further revenue is generated by the issue of postage stamps, and from tourist passages on the resupply ship. For details of the island groups and Terre Adélie see individual entries.

Thala Hills. 67°39′S, 45°58′E. Area of low rounded coastal hills between Freeth Bay and Spooner Bay, eastern Enderby Land. In Alashayev Bight, at a site where there is little or no fringing shelf ice, the coastal ice cap has receded to expose about 10 km^2 (4 sq. miles) of rocky headlands and islands, regarded as a small **oasis** area. The area was visited by Australian scientists in 1961, and named for their ship, *Thala Dan*. In January 1962 it became the site of the permanent Russian station Molodezhnaya.

Thermoregulation. Maintenance of constant body temperature. Whatever the environmental conditions, birds and mammals strive to maintain a high and constant central body temperature. Their internal heat is generated by chemical reactions in muscle, liver and other tissues, fuelled by food intake. To maintain constant body temperature, usually within a degree of 37 °C, this must be balanced precisely against heat losses to the environment. Antarctic birds and mammals in summer are likely to experience air temperatures between −10 °C and +10 °C, causing heat losses through their surface to the environment. Those that overwinter may encounter temperatures down to −50 °C or lower. Even light winds accelerate heat losses, and strong winds cause rapid chilling. Entering water at temperatures close to freezing point can be far more demanding, for cold water conducts heat away from the body 25 times more rapidly than air – even faster when an animal is swimming.

Most baleen whales are only, or mainly, summer visitors to the Antarctic. In contrast most seals and some minke, bottlenose and killer whales stay in the Southern Ocean all year round.

Insulation, in the form of feathers, fur or subdermal fat, reduces heat losses considerably. Feathers and fur insulate by entrapping a layer of still air next to the skin. Air has a low specific heat and conductivity, and therefore can easily be warmed to near-body temperature, reducing heat losses. All seabirds have thick plumage, and that of Antarctic seabirds is especially dense, providing a layer of still, insulating air between skin and outer surface. Penguin feathers include a dense under-down: those of emperor penguins are the longest. Seals have insulating fur, though among Antarctic marine mammals, only that of the fur seal provides efficient insulation in air or water. Antarctic fur seals have one of the densest fur coats of all mammals, with an estimated 40 000 hairs per cm^2.

Emperor penguins save energy by being large, and therefore having a low ratio of surface to volume, with thick plumage, relatively short flippers and bills and a highly developed surface vascular heat-exchange system. During their winter incubation and while brooding small chicks in early spring, they huddle tightly together. By forming large dense groups of up to 5000 birds, with 10 in every m^2, they reduce heat losses through their surface by at least 25–50 per cent. This adaptation is feasible only because emperor penguins carry their egg on their feet, protecting it with a fold of abdominal skin, and have none of the aggressive behaviour so characteristic of other penguins during incubation and brooding.

Fur and feathers have serious limitations in water. Though fur is twice as effective an insulator in air as blubber, compression during diving reduces its effective thickness by 50 per cent for each 10 m depth, significantly reducing heat retention. Animals that rely mainly on fur and feathers for insulation tend therefore to be relatively shallower divers. Antarctic fur seals and most penguins rarely dive deeper than 100 m. Emperor and king penguins, however, have been known to dive to 300 m and 500 m respectively. While penguins and fur seals spend relatively long periods ashore, animals that spend nearly all their life in water rely more heavily on blubber for insulation. Blubber is an active tissue containing a significant and variable blood supply. Circulation towards the skin surface can be restricted in cold conditions, or enhanced in warm ones. At times of high muscular activity, when much body heat is produced, an enhanced blood supply flushes the skin and extremities (flippers, feet, etc.), allowing heat to be shed from the body surface. In surface waters blubber is only 25 per cent less efficient as an insulator than fur, and below the shallowest depths it is superior. Phocid seals and whales, therefore, rely on thick blubber layers. In contrast, fur seals are able to shed heat effectively only via the flippers.

If blubber functions solely as insulation we would predict that the smaller species should have proportionately the thickest blubber, but this is not the case. Species like the minke whale, some individuals of which remain in the Antarctic most of the year, have relatively thinner blubber than the large baleen whales which use blubber as a food reserve when they migrate in winter to warmer areas which are less rich in food. Thus blubber has a dual function and the possession of thick blubber has important consequences for reproductive biology and ecology. To function properly, blubber has to be of a minimum thickness, and small animals would be especially vulnerable to inadequacies in this insulation because they have proportionately large surface areas. How then do offspring of the whales and phocid seals cope with birth? Although adult seals are essentially hairless (certainly so far as insulation is concerned), their pups are born with a thick fur coat, which is the ideal insulator for their land- or ice-based existence at this stage. The pups grow fast, acquiring thick blubber and losing their fur when they take to the water. Young whales are born at a relatively large size: most species give birth in warm waters closer to their own body temperature.
Further reading: Schmidt-Nielsen (1997).
(JPC)

Theron Mountains. 79°05′S, 28°15′W. Small mountain range of central Coats Land, separated from the larger Shackleton Range by the Slessor Glacier. The range was discovered and examined from the air by the Commonwealth Trans-Antarctic Expedition 1955–58, and named for the expedition ship.

Thomas, Charles W. (1903–73) US Coastguard captain. Born on 3 September 1903 in Pasadena, California, he entered the US Coast Guard Academy in 1922 and saw a wide range of Arctic service before and during World War II. In 1946–47 he commanded the USCG icebreaker *Northwind* during Operation Highjump, and subsequently played a role in planning Operation Deep-Freeze I and II. He retired from the service in 1957 with the rank of rear admiral. He died following a motor accident in Ushuaia, Argentina, on 3 March 1973.

Thompson, Andrew A. (d. 1970) US polar geophysicist. After graduating in physics and geology at Yale and Columbia Universities, he led the geophysical research team of the **Ronne Antarctic Research Expedition 1947–48**. His particular interests lay in detecting and interpreting microseisms. Later he worked as a research geophysicist in the US Army Ballistic Research Laboratories. He died after a motor accident on 17 September 1970, aged 46.
(CH)

Thomson, Charles Wyville. (1830–82). British oceanographer. As Professor of Natural History at Belfast University, Thomson secured the use of ships of the Royal Navy to trawl and dredge for life in deep waters around Britain. When the navy was seeking a naturalist to head the scientific team of the *Challenger* **Expedition 1872–76**, Thomson was the obvious choice. In the course of the expedition he supervised soundings, trawling and dredging in the Atlantic, Indian and Pacific oceans, and landings to collect specimens on the Prince Edward Islands, Iles Crozet and Kerguelen, and Heard Island. On returning from the expedition he was knighted and appointed to the professorship of natural history at Edinburgh University. He wrote a two-volume account of the voyage (1877) and undertook the immense task of editing the scientific reports, a work ultimately spanning 50 volumes that others continued long after his death.
Further reading: Linklater (1972).

Thule Island. 59°27′S, 27°19′W. A triangular island, about 95 per cent ice-covered: one of three small islands forming the Southern Thule group, at the southern end of the South Sandwich Islands chain. The central peak, Mt Larsen, rises to 725 m (2378 ft). Composed mainly of lava flows and scoria, the island is mildly active: steam occasionally emerges from the central crater. The group was discovered and charted by Capt. J. Cook in 1775; individual islands were identified and named in 1820 by the Russian Naval Expedition 1819–21. Ferguson Bay, in the southeastern corner, in 1976–77 became the site of Corbeta Uruguay, a large Argentine research station. The station was destroyed by a British force in 1982, after its use in the Argentine invasion of South Georgia.

Thule Islands. 60°42′S, 45°37′W. A group of small islands in Borge Bay, Signy Island, South Orkney Islands, named for a floating factory ship of the Oslo-based Thule Whaling Co., which moored nearby in January 1913.

Thurston Island. 72°06′S, 99°00′W. An ice-covered, heavily-glaciated island 216 km (135 miles) long and 88 km (55 miles) across off the Eights Coast of Ellsworth Land, separated from the mainland by part of the Abbot Ice Shelf. Discovered during a reconnaissance flight by the United States Antarctic Service 1939–41, it was named for W. H. Thurston, a benefactor of the expedition.

Tide crack. A persistent line of fracture between an immovable ice foot, and fast ice that is rising and falling with the tide. Tide cracks form along coasts, and surround stranded icebergs. They are of biological importance in providing breathing holes and access to the sea for seals and other animals.

Tierra de O'Higgins. Chilean name for **Antarctic Peninsula**.

Tierra San Martín. Argentine name for **Antarctic Peninsula**.

Till. See **moraine**.

Tomb. A category of Antarctic protected area created by delegates of Antarctic Treaty Consultative Meeting XI (1981), for a site on Mt Erebus, Ross Island, in 77°25′30″S, 167°27′30″E. On 28 November 1979 an Air New Zealand DC-10 aircraft crashed into an ice field on the northern slopes of the mountain, killing all 257 passengers and crew aboard. Under Recommendation XI-3, the site of the accident is declared a tomb, to ensure that the area is left in peace.

Tonkin, John Eliot. (1920–95). British polar explorer. Born in Singapore on 21 July 1920, he was commissioned in the army early in World War II. He joined the Long Range Desert Group and later the SAS, fighting with distinction in Africa, Italy and Germany. In 1945 he joined **Operation Tabarin**, later the Falkland Islands Dependencies Survey, serving for two years at Base E as surveyor and dog driver. He led several sledging journeys across the high plateau of Graham Land, in his second year becoming deputy leader. Tonkin spent much of the rest of his life in industrial management in Australia. He died in Mornington, near Melbourne, on 10 June 1995.

Toothfish. Southern oceans fish of the genus *Dissostichus*. Patagonian toothfish *D. eleginoides* are found mainly in the latitudes of the southern temperate islands (35°–65°S). Antarctic toothfish *D. mawsoni* live closer to Antarctica, mainly south of 65°S. Little is known of their biology. Immature forms live in shallow waters around the islands and continent. Larger mature fish, which in both species grow longer than 1.5 m (5 ft), live mostly in deeper water. Patagonian toothfish especially are commercially important: stocks are currently being fished by long-lining around many of the southern islands, and marketed under several names including Antarctic, Chilean or Australian sea bass. In the wide and poorly patrolled expanses of the southern oceans illegal fishing is rife. Scientists of the **Convention on the Conservation of Antarctic Marine Living Resources** (CCAMLR) estimate that, in the period 1996–99 over 80 000 tonnes were taken illegally from Exclusive Economic Zones around the islands, almost doubling annual legal catches during the same period. A further concern is destruction of seabirds, notably albatrosses and mollymawks, that are caught on the long-line baited hooks. Antarctic toothfish are subject to experimental fisheries, mainly in the Ross Sea under New Zealand control, and thus are far less subject to illegal fishing.

Tor Station. 71°53′S, 5°09′E. Norwegian research station at Svarthamaren, in the Mühlig-Hofmannfjella, Dronning Maud Land, established in March 1993 and used as a summer-only base for ornithological studies in the local SSSI.

Tordesillas, Treaty of. A sixteenth-century treaty allocating the distribution of newly discovered lands between Spain and Portugal. The late fifteenth century saw intense rivalries between Spain and Portugal over discoveries of land in the Atlantic Ocean. Ferdinand and Isabella of Spain in 1493 petitioned Pope Alexander VI to determine a meridional boundary, extending from North Pole to South, separating future discoveries. All lands found west of the line would be Spanish, all to the east Portuguese. Papal bulls were issued selecting a meridian 100 leagues (approximately 550 km) west of the Portuguese-held Cape Verde Islands. Finding this over-restrictive, John II of Portugal proposed an alternative boundary 270 leagues further west, close to the modern meridian 49°W of Greenwich. Diplomats meeting in the northern Spanish town of Tordesillas in June 1494 agreed to this, and the boundary was sanctioned by Pope Julius II in 1506. The new boundary enabled Portugal to claim Brazil on its discovery in 1500 and subsequent exploration, and thus to gain an important foothold in South America.

Torgersen Island. 64°46′S, 64°05′W. A low rocky island off the entrance to Arthur Harbour on the southern coast of Anvers Island, Palmer Archipelago. It was named for Norwegian chief officer T. Torgersen, of the transport *Norsel*, who in 1954–55 made soundings in the harbour.

Tourism, Antarctic. A growing industry that, since the mid-1950s, has brought many thousands of visitors to Antarctica and the Southern Ocean islands. Unsuccessful early attempts to attract tourists to Antarctica included those of Thomas Cook (1910) and J.R. **Stenhouse** (1929). From 1924 a mail steamer of the Falkland Islands Dependencies Government took limited numbers of passengers annually from the Falkland Islands to the whaling stations of South Georgia and the South Orkney and South Shetland Islands, and from time to time non-working passengers had sailed south on whaling and expedition ships. The first dedicated tourists to visit the

Antarctic region were 66 sight-seeing passengers in a Douglas DC 6B aircraft of Linea Aérea Nacional, who on 22 December 1956 enjoyed a four-hour scenic flight over parts of the Antarctic Peninsula and neighbouring islands. One year later, in January 1958, two cruises of the Argentine-operated passenger liner *Les Eclaireurs* brought the first shipborne passengers to the region from Buenos Aires, visiting approximately the same area by sea. Both airborne and shipborne tourism developed slowly, but from 1990–91 summer tourists came to outnumber the scientists and support staff who had previously dominated the population. Currently (2002) between 10 000 and 14 000 tourists land on the continent each year, while several thousand more visit in overflights and non-landing scenic cruises.

Shipborne tourism

Far more tourists travel to the region by ship than by air. Most seaborne visitors are passengers on cruise ships: a very few come by private or chartered yacht (see below). For over four decades Antarctic shipborne tour operators have followed an almost standard pattern of 'expedition' or 'adventure' cruising, inspired by Lars-Eric Lindblad who pioneered regular scheduled cruises to Antarctica in 1966. Currently (2002) about a dozen cruise ships are involved per year, each making several round trips from gateway ports or points of departure in South America, the Falkland Islands, Australia or New Zealand. The tourism season spans the summer months from late October to early March. Cruise ships are usually at least ice-strengthened, with enough engine power and water-line protection to push their way slowly though thin fast ice or loose pack ice. A few are icebreakers (see below). Most carry between 50 and 150 passengers. A recent development has been the advent of larger liners carrying 700 passengers or more.

Most cruises start from one of three gateway ports: Ushuaia (Argentina), Punta Arenas (Chile) or Stanley (Falkland Islands), and spend eight to twelve days in the South Shetland Islands and Antarctic Peninsula. Some slightly longer cruises may include the South Orkney Islands and South Georgia. Alternative cruises start from southern posts of Australia (mainly Hobart) or New Zealand, visiting Australian Antarctic Territory and Ross Dependency, usually calling at Macquarie Island or some of the New Zealand cold temperate islands. These cruises involve greater distances, longer times at sea and correspondingly higher costs.

In the Lindblad pattern of adventure cruising, passengers receive lectures and on-board briefings from experienced and qualified lecturers, and are made aware of their environmental obligations under the Antarctic Treaty. On reaching the Antarctic region they make two or three landings daily from inflatable boats, at sites selected by the cruise operators for scenic attractions or wildlife, accompanied by guides who point out features of interest. Evening talks recapitulate the events of the day. Larger cruise ships carrying 250 to 500 passengers follow a similar pattern with considerable adaptation. Their passengers make fewer landings or none at all, and lectures and briefings take their turn with shipboard entertainment. Ships carrying more than 500 passengers usually visit Antarctic waters for only two or three days, within a much longer round-the-world or round-South-America cruise. Landings are seldom considered, due to lack of inflatable boats and suitable clothing for the passengers.

Though the Lindblad pattern continues to dominate Antarctic cruising, especially for the middle-aged and elderly passengers who form the majority of clients, there have been recent trends toward special-interest tours that allow younger and more active travellers to spend more time ashore (see below). Activity cruises provide a wide range of pursuits including climbing, back-packing, trekking, marathon racing, kayaking and participation in clean-up operations or scientific research. Russian icebreakers adapted for tourism are able to visit landing sites within the pack ice that are inaccessible to the smaller ice-strengthened ships. Some carry helicopters, extending their effective range still further. Passengers can be landed on otherwise inaccessible islands, sea ice, shelf ice or on top of icebergs, and visit research stations and emperor penguin colonies some distance from the open sea.

Shipborne tourists include the small number of visitors who travel in southern waters by private yacht, either as owners or as passengers in chartered vessels. Most visit the southern cold temperate and sub-Antarctic islands: a few each year cross the Drake Passage to spend summer in the Scotia Arc or Peninsula regions. One or two may spend winter in the fast ice of a sheltered bay.

Airborne tourism

Two forms of airborne tourism answer more limited demands. The first is overflights, exemplified by the 1956 first-ever passenger flight from Punta Arenas, Tierra del Fuego, to Antarctic Peninsula. Between February 1976 and November 1979 Australian Boeing 747s and New Zealand DC10s carried passengers on longer scenic flights over Victoria Land. These ceased in November 1979 when a tourist aircraft from New Zealand crashed into the side of Mount Erebus (Ross Dependency), killing 257 passengers and crew. The tragedy drew attention to the absence of an effective infrastructure of safety measures for such operations. However, flights from Australia over East Antarctica began again in January 1995 and are currently proving popular. Commercial overflights are made also from southern Chile and South Africa.

The second form of airborne tourism involves landings and activities on the ground or ice. From 1982 an airstrip operated by the Chilean Air Force at Teniente Rodolfo Marsh Station, King George Island, South Shetland Islands, received C-130s and other aircraft carrying small numbers of tourists, some of whom stayed in a hostel nearby. Since its inauguration the airstrip has been used extensively by commercial tour operators, notably Adventure Network International. These fly small numbers of passengers in ski-equipped aircraft to the Ellsworth Mountains and other locations inland, mainly for climbing and trekking. An Argentine airstrip at Marambio, Seymour Island, was later made available for similar purposes. In 1986 the first experimental landings were made by wheeled aircraft on bare ice, opening up new possibilities for tourist flights direct from South America or South Africa to many points in the Antarctic interior. Using these facilities, ANI and a small number of other operators provide tours and help private expeditions to work in high latitudes. Most who take part are climbers, photographers or trekkers, seeking adventure in the interior of Antarctica. ANI has also carried scientific teams and undertaken emergency and rescue flights that for one reason or another could not be undertaken by national expeditions. Despite constant hazards of poor weather and difficulties inherent in low temperature operations, airborne tourism with landings maintains a high record of achievement and safety. Numbers of tourists making use of it remain low, mainly because of the relatively high costs in comparison with shipborne tourism.

Tourist numbers

Numbers of tourists visiting and landing in the Antarctic region each year have increased from a few hundred in the early 1960s to over 14 000 in the early 2000s. Shipborne tourism continues to be dominant, with most of the passengers travelling conventionally by the smaller cruise ships. Numbers carried by yachts fluctuate from year to year but overall remain small. Numbers carried on non-landing cruises fluctuate because of the size of ships involved. Airborne tourism involving landings and camping has proved consistently more expensive and less popular, accounting for fewer than 5 per cent of tourists annually.

Developments in tourism

Antarctic tourism, particularly shipborne tourism, currently shows every indication of gaining in popularity, and of developing and diversifying. One possible development already being tested is to fly clients from Tierra del Fuego to pick up their cruise ship on the South Shetland Islands, cutting out the unpopular and often uncomfortable Drake Passage, and allowing a higher proportion of cruise time to be spent in Antarctic waters. As younger travellers 'discover' Antarctica, demand for activities ashore is increasing, and tour operators are responding with an ever-increasing diversity of activities. Numbers of big-ship cruises are increasing, attracting a new and different clientele. Gateway ports (see above), particularly Ushuaia (Argentina), Punta Arenas (Chile) and Hobart (Australia) see advantage in investing to improve back-up facilities for this growing industry.

Regulation of tourism

There is no overall authority capable of regulating Antarctic tourism, in the sense of limiting its numbers or controlling its activities. Cruise operators and their clients whose governments are signatory to the Antarctic Treaty are responsible for their activities under the Treaty's **Protocol on Environmental Protection**. This requires a degree of accountability, though far less than is required in many scheduled wilderness areas (for example in national parks) that are subject to tourist incursions. Most operators are members of the **International Association of Antarctica Tour Operators**, which since 1991 has provided environmentally sound guidelines for its members and their clients. Currently numbers of tourists and of visits are small, compared with those visiting other popular tourist venues. Research has so far revealed little evidence of lasting environmental impacts – compared, for example, with those of scientists and support staff operating in the Antarctic region. Whether or not these constraining influences will keep pace with the growing industry remains to be seen.

Further reading: Bauer (2001); Codling (1995); Enzenbacher (1995); Stonehouse (2000).

Tourmaline Plateau. 74°10′S, 163°27′E. Ice plateau central to the Deep Freeze Range, Victoria Land, named for tourmaline–granite found in neighbouring mountains.

Tower Island. 63°33′S, 59°51′W. A tall, narrow island 8 km (5 miles) long, rising to 305 m (1000 ft), the north-easternmost island of Palmer Archipelago. It was charted in Edward Bransfield's survey of 1820.

Tractors, early. The first motor vehicle used in Antarctica was a 15 horse-power Arrol-Johnston motor car taken by Ernest Shackleton on the **British Antarctic (*Nimrod*) Expedition 1907–9**. It performed well: once warmed, the engine gave little trouble, and the leather-lined clutch was successful so long as it remained dry. However, even slight moisture caused it to freeze solid. The wheels gripped well on hard ice, especially when fitted with chains, but in soft snow they sank and spun uselessly. To counter this, mechanic Bernard Day stripped all surplus weight off the chassis, and experimented with a variety of special treads made of timber and rubber. None proved as effective as simple chains. On snow-free surfaces the car reached speeds of 14 mph, dragging loads of up to 1000 lbs. It was used for depot-laying runs, but Shackleton decided not to risk it on long-distance journeys.

Scott on the **British Antarctic (*Terra Nova*) Expedition 1910–13** used two custom-built motor-driven tractors, with caterpillar tracks that gripped both snow and ice better than wheels. Again nursed by Bernard Day, they travelled slowly, easily towing loads of over a tonne. However, they

Motorised sledge used on Scott's *Terra Nova* expedition. Photo: Scott Polar Research Institute

gave constant mechanical trouble, and Scott could not rely on them for anything more than hauling stores about the camp and depot-laying. Mawson's **Australasian Antarctic Expedition 1911–14** took a Vickers REP monoplane which, damaged before it left Australia, was used only as an 'air tractor' – useful on occasions for dragging heavily-loaded sledges, but temperamental, difficult to control and, with whirring exposed propeller, singularly dangerous to operate.

Design improvements during and after World War I produced a generation of more rugged and reliable motor vehicles, several of which were tried out in Antarctica during the 1920s and 1930s. Byrd experimented successfully with modified agricultural tractors, but failed completely with a 'snow cruiser', a huge motor caravan that foundered in the snow and never travelled further than Little America. Edmund Hillary's New Zealand contingent of the **Commonwealth Trans-Antarctic Expedition 1955–58** drove successfully from McMurdo Sound to the South Pole on modified Ferguson agricultural tractors. However, World War II and the ensuing Cold War saw the development of a very wide range of dedicated snow vehicles, many of which are currently in use on Antarctic stations and traverses.

Tramway Ridge. 77°32'S, 167°08'E. A prominent rocky ridge on Mt Erebus, Ross Island, including areas of intense fumarolic activity. A square with sides approximately 100 m (328 ft) long, occupying the lower end of the ridge at an altitude of 3350 m (11 000 ft), is designated SSSI No. 11 (Tramway Ridge), to facilitate research on the warm ground and associated vegetation.

Transantarctic Mountains. 85°00'S, 175°00'W. Continental Antarctica's most extensive visible cordillera, much of it forming the inland flank of East Antarctica. Extending southward in a continuous curve from the northern coast of Victoria Land, along most of its length the cordillera forms a continuous bulwark against the ice of the polar plateau, which cuts deep glacier channels that divide it into a succession of individual ranges. The descending ice forms piedmont glaciers along the Ross Sea coast, the western and southern flanks of the Ross Ice Shelf, and the eastern flank of the Filchner Ice Shelf. Beyond the Horlick Mountains the cordillera disappears under the ice cap, to emerge intermittently as Thiel and Pensacola mountains, Shackleton Range and Theron Mountains.

Traversay Islands. 56°36'S, 27°43'W. Northernmost group of the South Sandwich Islands, including Zavodovski Island, Leskov Island and Visokoi Island. The group was discovered in 1819 by the Russian Naval Expedition 1819–21 led by Capt. F. von Bellingshausen, who named them for the Marquis de Traversay, a French naval officer who was Minister of Naval Affairs at St Petersburg, 1811-31, and a supporter of the expedition.

Trinity Peninsula. 63°37'S, 58°20'W. A peninsula forming the northeastern end of Antarctic Peninsula, beyond a line connecting Cape Kjellman and Cape Longing. It was charted by Edward Bransfield RN in 1820, and named for the 'Trinity Board', a British nautical guild.

Tristan da Cunha group. 37°05′S, 12°15′W. Tristan da Cunha and neighbouring **Nightingale Island** and **Inaccessible Island** form a cluster 4000 km (2500 miles) east of South America and 2700 km (1687 miles) west of South Africa, in the South Atlantic Ocean. Standing at the junction of two submarine ridges, they represent a pinpoint of volcanic activity that began at least 18 million years ago: Nightingale Island has the oldest rocks, Tristan da Cunha some of the youngest. Tristan da Cunha is roughly circular, 12 km (7.5 miles) in diameter, with cliffs on all sides up to 600 m (1968 ft) high. Formed up to 1 million years ago, it is composed mainly of interbedded lava and scoria, rising in steps to a central cone ('The Peak') 2060 m (6757 ft) high. In 1961 its northern flank erupted to provide a small platform of fresh lava close to Edinburgh, the settlement. The climate is mild, damp and windy, with mean annual temperature 14.5 °C, range of monthly means 6.5 °C, rainfall 170 cm per year and constant westerly winds. The Peak, usually shrouded in cloud, is snow-covered in winter. Over 70 native species of flowering plant and fern have been recorded. A wind-sculptured forest of trees and ferns is dominant close to sea level, with tussock meadows, moorland and fellfield above. Farming and animal husbandry since the early nineteenth century have brought in many alien species. Seabirds are abundant, including breeding stocks of rockhopper penguins, yellow-nosed and sooty albatrosses, several species of smaller petrel, and terns, noddies and skuas. Land birds include an endemic thrush and two endemic species of bunting.

Tristan da Cunha was discovered in 1506 by the Portuguese navigator whose name it bears. A seamark for sailing ships bound for the Cape of Good Hope and Indian Ocean, it was often sighted and became the cause of many wrecks. Settled initially by an American seaman in 1811, it was annexed in 1816 by a British garrison, sent there to forestall possible efforts by the French to rescue Napoleon Bonaparte from St Helena. Though the military soon withdrew, the settlement continued under British sovereignty,

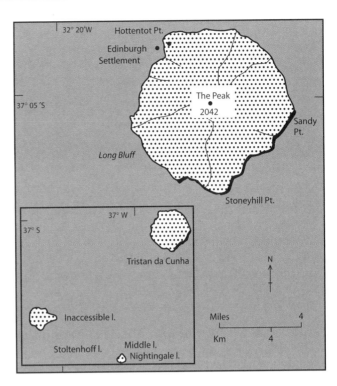

Tristan da Cunha. Mainly ringed by steep cliffs, the islands offer few landing points. Edinburgh, the single settlement, has the only small harbour for landing catches from fishing boats. Heights in m

maintaining a fluctuating population of 50–100 that supported itself by fishing, farming and barter with passing ships. As steam replaced sail, fewer ships visited the island, but the community survived and grew to about 300, supported by more efficient farming and commercial fishing. During World War II it acquired a garrison and meteorological station. The eruption of 1961 close to Edinburgh made it necessary to evacuate the whole population of 290 to the United Kingdom: 250 returned in 1963 to resume their old way of life. The Tristan da Cunha Islands, together with their nearest neighbour Gough Island, are administered by Britain as a Dependency of St Helena.
Further reading: Crawford (1982).

Edinburgh; the settlement on Tristan da Cunha. Photo: BS

Troll Station. 72°01′S, 20°32′E. Norwegian research station in the Gjelsvikfjella, inland from Prinsesse Ragnhild Kyst, at an elevation of 1270 m (4165 ft). Established in 1989–90 by the Norsk Polarinstitutt, it has summer accommodation for 30 scientists.

Tula Mountains. 66°54′S, 51°06′E. A scattering of mountain peaks and nunataks merging into the continental ice cap east of Amundsen Bay, Enderby Land. Discovered during flights of the British, Australian and New Zealand Antarctic Research Expedition 1929–31, they were named for the brig of **Capt John Biscoe**, who first explored the coast in 1831.

U

Undine Harbour. 54°02′S, 37°58′W. Harbour on the south coast of South Georgia, possibly charted by British sealer Capt. J. Weddell in 1823, later used by whalers and named for the sealing ship *Undine*. Undine South Harbour (54°31′S, 36°33′W) is a less sheltered harbour on the south coast of South Georgia.

United States Antarctic Service Expedition 1939–41. A US government-sponsored expedition that for the first time operated 'permanent' stations in Antarctica. Prompted by the contemporary spate of territorial claims to Antarctica, the US government in 1939 established the US Antarctic Service, an inter-departmental agency to continue the geographical exploration of the continent, especially to maintain American interests in the areas where **Byrd**'s two expeditions had operated. The Service was invited to consider a policy of setting up long-term stations, that would be relieved each year with changes of personnel – essentially a first occupation or colonization of Antarctica. With Rear-Admiral Richard Byrd in overall command, the USAS Expedition established two stations. The navy transports USNS *North Star* and USS *Bear* in January 1940 set up West Base in Bay of Whales on the Ross Ice Shelf, close to the site of Byrd's old bases Little America I and II. In early March 1940 the same ships set up East Base on Stonington Island, a small island near Neny Fjord in Marguerite Bay, Antarctic Peninsula. Byrd himself accompanied the ships to West Base, but did not winter. West Base was commanded by Paul A. Siple, East Base by Richard B. Black, and both included members of the earlier Byrd expeditions. From both bases aircraft and dog-sledging parties combined in the kind of exploration, involving aerial survey with competent ground control, that Byrd had pioneered and brought to a high standard of efficiency. From West Base five sledging parties explored Marie Byrd Land, starting where Byrd's earlier expeditions had left off. Long-distance flights explored south to the Queen Maud Range and eastward along the Hobbs and Ruppert coasts to 123°W. From East Base survey aircraft flew south to the Bryan Coast of Ellsworth Land and the southern shore of the Weddell Sea, while sledging parties explored far south down King George VI Sound and eastward across the Peninsula to the Bowman Coast, far extending previous explorations in the area. The US Antarctic Service Expedition might well have continued for several years, but with World War II already raging, the US Congress withdrew funding. The expedition terminated early in 1941, and both bases were evacuated. With the Japanese attack on Pearl Harbor in November of that year, the US entered the war, and the Antarctic Service was disbanded.

Further reading: Polar Record (1941).

United States Exploring Expedition 1838–42. Established by an Act of Congress in 1836, this major US naval expedition was authorized to explore the southern oceans in high latitudes, searching for previously reported and unknown lands, with a further excursion into Japanese and Indo-Chinese waters, with the overall purpose of promoting the safety of US whaling, sealing and commercial shipping operations. Six ships were involved, under the command of Lt **Charles Wilkes**, USN, one of the navy's most experienced hydrographers, with teams of civilian scientists and artists aboard. Through no fault of his own, Wilkes's expedition was under-funded, poorly victualled and ill-equipped, with an assortment of inadequate and ill-prepared ships. These included the flagship USS *Vincennes*, a sloop-of-war of 780 tons, under Wilkes's command, USS *Peacock*, a smaller sloop of 650 tons (Lt W. L. Hudson USN), USS *Porpoise*, a gun-brig of 230 tons (Lt C. Ringgold USN), and two tenders, USS *Sea Gull* of 110 tons (Midshipman J. W. E. Reid USN), and USS *Flying Fish* of 96 tons (Mr S. R. Knox). A store ship, USS *Relief*, accompanied the expedition.

Wilkes and his fleet left Norfolk, Va., on 18 August 1838, reassembling in Orange Harbour, Tierra del Fuego on February 18 1839. From there Wilkes in USS *Porpoise*, accompanied by USS *Sea Gull*, visited the South Shetland Islands and Antarctic Peninsula, encountering foul weather that seriously inhibited survey work. The two ships returned separately to Orange Harbour in late March 1839. Meanwhile USSs *Peacock* and *Flying Fish* were exploring southward, under instruction to revisit James Cook's furthest-south position (71°10′S) and the Bellingshausen

Sea. They too were hampered by bad weather: only *Flying Fish* achieved 70°S. During April the six ships of the fleet left southern waters for Valparaiso. Only five reached their destination: USS *Sea Gull* was lost with all hands off the Chilean coast.

After a long and successful voyage in warmer latitudes of the Pacific Ocean, on 26 December 1839 Wilkes left Sydney in USS *Vincennes*, accompanied by *Peacock, Porpoise* and *Flying Fish*. On 11 January 1940 *Vincennes* and *Porpoise* reached the edge of the pack ice in 64°11′S, to be joined five days later by *Peacock. Flying Fish* reached the ice edge on 21 January after the others had moved on: by this time barely seaworthy, she returned to New Zealand. The three larger ships headed westward through a sea strewn liberally with floes and bergs, close to the ice-cliff coast of what is today called Wilkes Land. From time to time they came within soundings, and all the crews agreed that often behind the cliffs there appeared evidence of high, ice-covered land. Wilkes and his colleagues were clearly sailing close to a continent, and had no hesitation in claiming several sightings. On 24 January *Peacock* was damaged by a collision with the ice: Lt Hudson, in command, deemed the ship unfit for further service and headed back to Sydney. On 30 January *Porpoise* (Lt C. Ringgold) encountered a strange ship, the French corvette *Astrolabe* of Dumont d'Urville's **French Naval Expedition 1837–40**, but by a misunderstanding failed to exchange signals. On 14 February, having lost contact with Wilkes, Ringgold headed north for New Zealand. Wilkes in *Vincennes* continued westward. On the evening of 21 February he encountered an ice cliff that lay across his course, preventing further westward movement without a diversion north. He named it 'Termination Land', because it ended his voyage in high latitudes: today it forms part of the Shackleton Ice Shelf.

Wilkes returned to Sydney well pleased with his accomplishments, notifying his government and announcing to the local press that the expedition had without doubt located a land of continental size. As a mark of courtesy he sent a provisional copy of his chartwork to Capt. James Ross, who was later to return the courtesy by denouncing them as incompetent. The Exploring Expedition ended with a second long Pacific Ocean voyage. Wilkes and the remaining ships of his fleet returned to New York on 2 June 1842, to face recriminations, charges of incompetence, a court of enquiry and a court-martial, though with final exoneration on all but one group of the charges against him. Wilkes's reports of the expedition were published in 1845. Time has done much to mend his reputation: his expedition was indeed the first to delineate a substantial length of Antarctic coastline, and to establish that behind lay a landmass of continental size.

Further reading: Wilkes (1845).

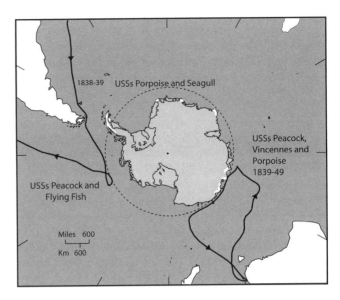

The major voyages of Wilkes's fleet from Orange Harbour, Tierra del Fuego, and from Sydney, Australia. The intervening winter was spent in the southern Pacific Ocean

United States Navy Antarctic Development Project 1946–47. A large-scale United States operation, codenamed Operation Highjump, to deploy naval forces in a cold climate, and explore Antarctica by sea and air. Under the general direction of Rear Admiral Richard Byrd, Operation Highjump was the largest single Antarctic expedition, involving a fleet of 13 ships under the command of Rear Admiral Richard Cruzen USN. Objectives were to test equipment and provide operational experience for about 4000 naval personnel under polar conditions, and to photograph as much as possible of Antarctica from the

Little America IV. Fuel drums, tractors, temporary living quarters and expedition equipment lined up on the Ross Ice Shelf. Official US Navy photo

air. Divided into three separate task forces, the operation explored miles of previously inaccessible coastline and photographed huge areas of the continental coast and interior.

The central task force included the icebreaker USS *Northwind*, headquarters ship *Mount Olympus*, freighters *Yancey* and *Merrick* and the submarine *Sennet*. Cruzen established a temporary land station and airstrip, Little America IV, close to Byrd's old base on the Ross Ice Shelf. This provided facilities for six Douglas R4Ds that flew in from the aircraft carrier USS *Philippine Sea*, on station close to Scott Island. The 'eastern' task force under Capt. George Dufek USN, including *Pine Island*, a seaplane tender with three Martin Mariner (PBM) flying boats, the destroyer *Brownson* and tanker *Canisteo*, cruised eastward around West Antarctica from 120°W to the Greenwich meridian. The 'western' task force under Capt. Charles A. Bond USN, including the seaplane tender *Currituck* with three flying boats, the destroyer *Henderson* and tanker *Cacapon*, cruised westward around East Antarctica from 175°E to 30°E.

The central force experienced heavy pack ice in the Ross Sea, which excluded the submarine almost completely from operations and delayed the start of the flying programme. However, from 29 January 1947 the aircraft flew 29 photo-reconnaissance flights in a wide semi-circular swathe from King Edward VII Land to Victoria Land, including a flight over and beyond the South Pole. This group included also heavy amphibian tractors and other land-based equipment which was under test for cold weather operations. The ships of the central force withdrew early: Little America IV was cleared by the icebreaker USS *Burton Island* on 23 February, leaving the six RD4s on the site. The eastern group, dogged by bad weather, made 12 flights over Marie Byrd Land, West Antarctica, and the area around Alexander Island and Charcot Island. One flying boat crashed among the mountains, with the loss of three lives. The western group made 25 flights along the coast of Eastern Antarctica. In March and April 1947 the three task forces returned independently to the United States. The main objective of conducting naval operations in a polar environment was fully met. The secondary objectives of securing good aerial photographic coverage of Antarctica was achieved insofar as the weather allowed. Little provision had been made for ground control, a defect that was remedied in the following year's operation, the **United States Navy Second Antarctic Development Project 1947–48.**
Further reading: Rose (1980).

United States Navy Antarctic Expeditions (Operation Deep-Freeze) 1955–98. Following a preliminary survey in 1954–55, the US Navy, in cooperation with scientific teams of the National Science Foundation, in 1955–56 established stations on Antarctica in preparation for the International Geophysical Year 1957–58. 'Operation Deep-Freeze', code-name of the naval operation, was later given to the naval and Coast Guard component of

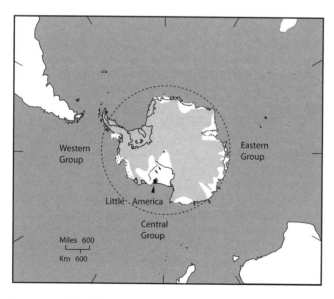

Operation Highjump. Continental areas shown in white were visited and photographed by the Eastern, Western and Central groups

Icebreaker USS *Glacier* cutting consolidated fast ice to reach McMurdo Station in the early years of Operation Deep-Freeze. Official United States Navy photo

all successive annual expeditions. It ceased to apply after 1998, when the Naval Antarctic Support Unit was finally disbanded.

United States Navy Second Antarctic Development Project 1947–48. A United States naval operation, codenamed Operation Windmill, that enhanced the work of the **United States Navy Antarctic Development Project 1946–47** (Operation Highjump), its much larger predecessor. A task force consisting of two icebreakers, USSs *Edisto* and *Burton Island*, under the command of Cdr G. L. Ketchum USN, left Samoa on 5 December 1947 for Antarctic waters. The objective was to penetrate the pack ice at several points around the continent and put ashore small survey parties, to provide astronomical fixes that would tie down the extensive aerial photography of the previous season. The code name 'Windmill' reflected the operational significance of the helicopters used in deploying the parties ashore. The icebreakers entered the pack ice in 92°E, moving toward the coast between Gaussberg and the Shackleton Ice Shelf. They then headed eastward around East Antarctica, entering the Ross Sea on 28 January 1948, and visiting McMurdo Sound and the Bay of Whales. Continuing eastward, they tried but failed to reach the Marie Byrd Land coast, but visited Peter I Øy. In Marguerite Bay by mid-February, they assisted *Port of Beaumont*, the ship of the **Ronne Antarctic Research Expedition 1947–48**, and RRS *John Biscoe*, the relief ship of the Falkland Islands Dependencies Survey, respectively to break out from and break into their bases on Stonington Island. On 23 February they began the journey back to the United States.

Usarp Mountains. 71°10'S, 160°00'E. An extensive mountain range, almost 200 km (125 miles) long, aligned north–south inland from the coast of Oates Land, East Antarctica. Discovered during reconnaissance flights of United States Navy Operation Highjump 1946–47, they were named from the initials of the United States Antarctic Research Program.

V

Valkyrjedomen. 77°30′S, 37°30′E. Large, prominent ice dome rising to 3700 m (12 136 ft) in the southern ice sheet of Dronning Maud Land. The dome is named for the Valkyries of Norse mythology.

Van Oordt, Gregorius Johannes. (1892–1963). Dutch polar ornithologist. Born in Arnem, he became an academic zoologist with special interests in vertebrate endocrinology. An expedition to Svalbard in 1921 resulted in publications on Arctic breeding birds. In 1951–52 he visited the Southern Ocean in the whaling tanker *Barendrecht*, contributing substantially to knowledge of the pelagic distribution of seabirds. He died on 22 April 1963.

Vanda, Lake. 77°32′S, 161°33′E. Freshwater lake in Wright Valley, Victoria Land, identified and named by an exploratory party from Victoria University of Wellington. The lake (named in memory of a sledge dog) is fed mainly from Onyx River, a seasonal meltwater stream from the Lower Wright Glacier. The lake and surrounding area have been intensively studied, mainly by New Zealand and US scientists, operating from a small permanent station from 1968 to 1995.

Vanda Station. 77°32′S, 161°38′E. Former New Zealand research station close to Lake Vanda, in the Victoria Land dry valley system. Established in 1968, it was installed and serviced mostly by air from Scott Base and McMurdo Station. Scientists wintered there in 1969, 1970 and 1974, but it was used mainly by summer parties studying the lakes, streams and surrounding oasis area. The station was decommissioned in 1991 and removed in 1995.

Vega Island. 63°50′S, 57°25′W. A rectangular island 27 km (17 miles) long between Trinity Peninsula and James Ross Island, in the eastern entrance to Prince Gustav Channel. Charted by the Swedish South Polar Expedition 1901–4, it was named by O. Nordenskjöld, the expedition leader, for the ship in which his uncle, Baron A. E. Nordenskjöld, explored the Northeast Passage in 1878–79.

Vegetation, subfossil. Ancient non-silicified plant remains, belonging mainly to the Holocene (i.e. the last 10 000–15 000 years). Besides some poorly preserved mosses in sedimentary rock in the Transantarctic Mountains, possibly as recent as Late Pleistocene (2 million years), the oldest remains of subfossil plant material so far discovered in the Antarctic region are from former lake sediments. A thick deposit of perfectly preserved aquatic moss, *Drepanocladus longifolius*, in the sediment of a former lake bed on James Ross Island, north-eastern Antarctic Peninsula, has a radiocarbon age of 9700 years old. A similar former lake bed deposit in the Vestfold Hills, Princess Elizabeth Land, contains the moss *Bryum pseudotriquetrum* and an alga with radiocarbon ages of 7400 to 8400 years. Small fragments of moss, identical to present-day species, have been found in older lake sediments up to about 15 000 years old. Shoots from deep terrestrial **moss peat** banks on Signy Island, South Orkney Islands, and Elephant Island, South Shetland Islands, have been dated at 5500 years old. On a raised beach on King George Island, South Shetland Islands, a narrow layer of moss peat 50 cm below the surface, also about 5000 years old, contained identifiable remains of moss and both species of Antarctic flowering plant. These mid-Holocene datings indicate that climate and ice cover in the northern maritime Antarctic provided conditions similar to today's, and that areas where they occur have been largely ice-free since then.

In the maritime Antarctic region, current warming that started in around 1950 has caused ice fields to thin and margins to retreat, re-exposing former stands of vegetation and moss peat that were ice-covered for periods of 200 to 2000 years. Mapping and radiocarbon dating of these exposures on Signy Island, South Orkney Island, allow reconstruction of the island's glacial events during the last 1000 years. The living surfaces were covered, though not destroyed, by at least four glacial advances in the ninth, twelfth, sixteenth and eighteenth centuries.
(RILS)

Venable Ice Shelf. 73°03′S, 87°20′W. Ice shelf between Fletcher Peninsula and Allison Peninsula, Bryan Coast, Ellsworth Land. It was identified from aerial photographs and named for Cdr J. D. Venable, USN, a staff officer of United States Naval Support Force 1967–68.

Vernadsky Station. See **Akademik Vernadsky Station**.

Vestfjella. 73°20′S, 14°10′W. (West Mountains). Coastal range of mountains and nunataks overlooking the Riiser-Larsenisen, Kronprinsesse Märtha Kyst, Dronning Maud Land.

Vestfold Hills. 68°33′S, 78°15′E. Area of low rounded coastal hills, lakes and islands north of Sørsdal Glacier, Ingrid Christensen Coast, East Antarctica. In an area where there is no fringing shelf ice, the coastal ice cap has receded to expose 410 km^2 (160 sq. miles) of ice-free land and lakes, making this one of Antarctica's larger **oases**. The area was identified by the Norwegian (Christensen) Whaling Expeditions in 1935: Capt. K. Mikkelsen of the tanker *Thorshavn* landed and named it Vestfold for the county of his home port, Sandefjord. It was further explored by Australian parties in 1954 and 1955. In January 1957 it became the site of Davis, a permanent Australian research station, from which much research on the geology, palaeontology, history and ecology of the region has been undertaken.
Further reading: Pickard (1986).

Vicecomodoro Marambio Station. 64°15′S, 56°45′W. Argentine permanent year-round research station and aviation facility on Seymour Island. Opened in 1969, its runway is used both for intercontinental flights and for local flights servicing other Argentine stations.

Victoria Land. Mountainous and heavily glaciated area of East Antarctica, with a long western coast flanking the Ross Sea and the Ross Ice Shelf. The land extends inland to a high, ice-covered plateau, bordering Oates Land 160°E (part of Australian Antarctic Territory) in longitude 160°E. Extending south to 78°S, it includes the Hillary, Scott, Borchgrevink and Pennell coasts, Ross Island, and the Prince Albert Mountains, northernmost range of the Transantarctic Mountains. Victoria Land is administered as part of the Ross Dependency of New Zealand. Discovered in January 1841 by Capt. J. C. Ross, RN, it was named for the reigning British queen. As a possible route to the South Pole, it was explored extensively by early 20th century British expeditions: it is currently the site for permanent New Zealand, United States and Italian stations.

Victoria Land Dry Valleys. 77°30′S, 162°E. Extensive system of ice-free glacial valleys on the western side of McMurdo Sound, known also as McMurdo Oasis. Extending from Cotton and Debenham glaciers in the north to Ferrar Glacier in the south, they include Taylor, Wright and Victoria valleys, and several smaller tributary valleys, in a roughly rectangular bloc totalling 2500 km^2 (977 sq miles). Taylor Valley was discovered during R. F. Scott's British National Antarctic Expedition 1901–4 and surveyed by Griffith Taylor during the British Antarctic (*Terra Nova*) Expedition 1910–13. Ice-free central sections of Wright and Victoria valleys were detected from aerial photographs in the late 1950s. All have since been explored thoroughly, mainly by New Zealand and US scientists. Broad, steep-sided valleys alternate with high towering ridges, exposing a basement complex of granites and metamorphic rocks, with overlying Beacon-suite shales, sandstones and dolerite sills. Precipitation is low: chemical and mechanical weathering produce sands which blow into dunes and barchans. All three main valleys have remnant 'upper' glaciers, which have retreated to leave broad U-shaped valleys. Wright and Victoria valleys have 'lower' glaciers merging with the Wilson Piedmont Glacier: Taylor Valley opens to the sea. The valleys contain many interesting features, including lakes, summer streams, remnant lateral glaciers, and a meagre but interesting microflora and microfauna. Research was undertaken from **Vanda**, a long-term research station.

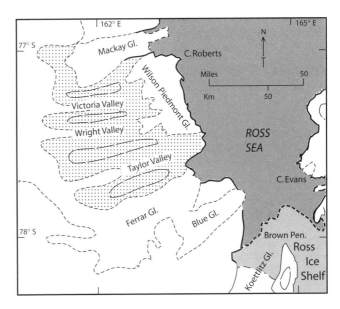

Victoria Land Dry Valleys

Victoria Valley and glacier were named for Victoria University of Wellington, New Zealand, scientists from which have made intensive studies of this and other valleys.

Victory Mountains. 72°40′S, 168°00′E. A rectangular bloc of mountains in northern Victoria Land, extending between Borchgrevink Coast and the polar plateau, bounded north and south by the Tucker and Mariner glaciers. The range was named by New Zealand cartographers: individual peaks within it celebrate naval and military victories.

View Point (Base V). 63°32'S, 66°48'W. British research station in Duse Bay, Trinity Peninsula, Graham Land. The station was occupied intermittently from June 1953 to November 1963 as a satellite station and refuge for survey parties from Hope Bay (Base D). In July 1996 ownership was transferred to the government of Chile, which operates it as General Ramon Cañas Montalva Station.

Vindication Island. 57°04'S, 26°46'W. One of two islands in the **Candlemas Islands** group of the **South Sandwich Islands**. Discovered in February 1775 and charted by Capt. James Cook, this small volcanic island was re-charted and named by Discovery Investigations in 1930.

Vinson Massif. 78°35'S, 85°25'W. Mountain massif forming a southern outlier to the Sentinel Range, Ellsworth Mountains, eastern Ellsworth Land. The summit, rising to 4 901 m. (16 075 ft), is Antarctica's highest peak. First seen on reconnaissance flights from Byrd Station in January 1958, the massif was named for Rep. Carl G. Vinson of Georgia, whose interest and political support encouraged United States' official involvement in Antarctic research.

Vishniac, Wolf V. (d. 1973) American polar microbiologist. Born in Berlin of Latvian parents, he became a US citizen in 1946, and trained as a biologist at Brooklyn College and Washington and Stanford Universities. While on the staff of the Department of Biology, Rochester University, he took particular interest in possibilities of extra-terrestrial life, and in 1971–72 worked in microbiology in the dry valleys west of McMurdo Sound. Returning in 1973 to continue research, he was killed in a fall in the Asgaard Range.

Visokoi Island. 56°42'S, 27°12'W. An oval island 7.2 km (4.5 miles) long and 4.8 km (3 miles) wide, in the South Sandwich Islands. Southernmost of the Traversay Islands, it is about 90 per cent ice-covered: where visible, the rocks are mainly lava flows and scoria. The island was discovered and charted in 1819 by the Russian Naval Expedition 1819–21, and named for its high central mountain (Russian: visokui = high), an ice-mantled dome of 1005 m (3296 ft). The peak was later named Mt. Hodson. The island was reported to be volcanically active in 1830 and again in 1930.

Volcanic activity. Only two sites on mainland Antarctica currently show signs of volcanic activity. **Mt. Erebus**, Ross Island is an active volcano, and **Mt. Melbourne**, Victoria Land, has active fumaroles. Several 'hotspots' (sites of surface warming due to recent volcanic activity) have been identified by infrared scanning, satellite images and photographs in the Flood and Ames ranges of Marie Byrd Land, and the Hudson Mts of Ellsworth Land. The Executive Committee Range, Marie Byrd Land, comprises young volcanoes that were last active about 5 million years ago. The structure of their cones and craters and the material erupted indicate that they formed originally under ice sheets, which melted locally, producing large lakes. Within these lakes, strata of volcanic ash were deposited and flowing lavas were quenched. Remnants of similar subglacial volcanic eruptions have been discovered in south-west Alexander Island.

Elsewhere in the region, **Deception Island** and **Bridgeman Island** in the South Shetland Islands, and several of the **South Sandwich Islands** show varying levels of activity. Lavas and ashes of the South Shetland and South Sandwich Islands contain a high proportion of sodium minerals, which makes them flow readily, slumping and spreading over wide areas. Those of Mount Erebus contain a higher proportion of potassium minerals and are stiffer, building taller cones. The 1969 Deception Island eruption was partly subglacial, a wide chasm opening to release volcanic heat that melted overlying ice.

Many of the Southern Ocean islands are of volcanic origin. Some in the Indian Ocean and the Indian sector of the Southern Ocean show remnants of recent volcanic activity, though only one is currently active. **Heard Island**, ice-capped and frequently enveloped in cloud, has an active central cone, Big Ben. Eruptions were recorded in 1881 and 1910, steam, smoke, ash and glowing vents were often reported during six years of occupation 1948–54, and passing ships recorded spectacular lava flows in 1985 and 1996. **Marion Island**, in the Prince Edward group, erupted in 1980, producing a fissure over 5 miles (9 km) long on its western flank, with a substantial outpouring of lava. Péninsule Rallier du Baty on south-western **Iles Kerguelen** shows fumarolic activity, and steaming springs in the caldera of **Ile St Paul** have for long been hot enough to cook locally caught lobsters. **Bouvetøya**, a tiny, ice-capped island in the Atlantic sector, has a well-defined caldera, clearly of volcanic origin, from which steam is from time to time reported. Between 1955 and 1958 the island developed a new beach, Westwindstranda, on its eastern flank, which may have resulted from volcanic action, but more possibly from a landslip following tectonic activity. (RJA/BS)

Vostok, Lake. A large subglacial lake, detected by geophysical methods some 4 km (2.5 miles) below the surface of the ice sheet close to Vostok Station, on the high plateau of East Antarctica. About 230 km (144 miles) long, it has an estimated area of 14 300 km^2, and a depth that

may exceed 500 m (1640 ft). To avoid polluting its waters (which may contain an interesting biota) drilling for ice cores by an international consortium has stopped short of entering the lake.
Further reading: SCAR Bulletin, (2000).

Vostok Station. 78°28'S, 106°48'E. Russian research station, opened in 1958 on the high plateau of Wilkes Land, East Antarctica, at an elevation of 3488 m (11 441 ft) above sea level. It replaced an earlier station, Vostok Island, set up in the previous year, and is continuously occupied. Sited on an ice plain, it is maintained by summer tractor-trains and aircraft from Mirnyy, the nearest Russian coastal station some 1380 km (863 miles) away. Vostok is currently the world's coldest inhabited place. Temperatures and other weather characteristics are similar to those for the former **Plateau Station**, though perhaps slightly milder because of the lower altitude. Lowest temperatures are recorded each year in July and August. A low temperature of $-88.3\,°C$ recorded on 24 August 1960 remained a world record until 21 July 1983, when the current record of $-89.6°C$ was achieved. Maximum temperatures around $-15\,°C$ occur in January. The station was sited originally for its proximity to the Southern Geomagnetic Pole. A centre for geophysical research, it is also the site of a joint US/Russian ice coring programme.

Voyeykov Ice Shelf. 66°20'S, 124°38'E. Large ice shelf of Banzare Coast, East Antarctica. Emerging from Maury Bay, and fed by three active glaciers, it extends between Paulding Bay and Cape Goodenough. The shelf was mapped by the Soviet Antarctic Expedition 1958, and named for the Russian climatologist Aleksandr I. Voyeykov.

W

Waite, Amory H. (Bud). (1902–85). American polar radar engineer. Born on 14 February 1902 in Newton, Mass., he trained in radio with the US Navy, and completed a degree in electrical engineering at Lowell Technological Institute. A radio engineer and operator on **Byrd's Second Antarctic Expedition 1933–35**, he experimented with the transmission of radio waves through ice. He joined the Institute of Exploratory Research of the US Army Signal Corps (later Electronics Command). Returning to Antarctica as with Operation Deep-Freeze in 1957–58, he used radio waves to sound ice 600 m (1968 ft) thick in an ice shelf close to Wilkes Station, and in 1961 instigated the first airborne survey of the Antarctic ice cap. Waite retired from army service in 1965, and died on 15 January 1985.

Walgreen Coast. 75°30′S, 107°00′W. Part of the coast of Marie Byrd Land, West Antarctica. Facing the Amundsen Sea, it is bounded by Cape Herlacher in the west and Cape Waite in the east. The two ends are mountainous, flanking a mid-section that rises gently to the inland plateau. An area of heavy snowfall, the coast is lined with extensive ice shelves and glacier tongues, and difficult to approach by sea. Discovered during flights by the United States Antarctic Service 1939–41, it was named for Charles Walgreen, president of a Chicago-based drug company who supported both this and the earlier Second Byrd Antarctic Expedition.

Walker, Paul. (1934–59). American polar glaciologist. Graduating in geology from Occidental College, Los Angeles, he gained field experience in Greenland in 1956, then served at Ellsworth Station with the US International Geophysical Year Expedition 1957–58. In 1959 he returned to the Arctic with the Ellsmere Island Shelf Expedition, but became ill and had to withdraw. He died in Pasadena on 9 November 1959.

Warburton, Keith. (1928–59). British mountaineer and expedition medical officer. After training in medicine he served for two seasons (1953–54, 1955–56) with the South Georgia Survey, and later climbed in the Himalayas. He was killed while climbing in the Karakorum mountains in July 1959.

Warm temperate islands. Southern oceanic islands that stand north of the 10 °C summer isotherm, to an arbitrary boundary of 38°S. They include Tristan da Cunha, Inaccessible Island, Nightingale Island and Gough Island in the southern Atlantic Ocean, Ile Amsterdam and Ile St Paul in the southern Indian Ocean, and the Auckland Islands and Bounty Islands south of New Zealand. For a table see **Study Guide: Southern Oceans and Islands**. For more detailed accounts see entries for individual islands and groups.

Wasa Station. 73°3′S, 13°25′W. Swedish summer-only research station in Kraulberga, Dronning Maud Land.

Washington Strait. 60°43′S, 44°56′W. Passage separating Fredriksen Island and Powell Island from Laurie Island, South Orkney Islands. The strait was named in 1821, possibly for George Washington, first President of the United States.

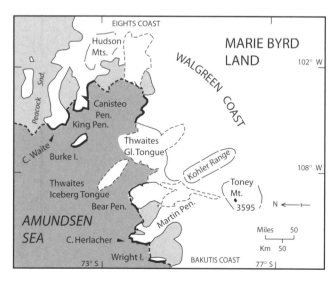

Walgreen Coast

Watson, Andrew Dougal. (1885–1963). Australian polar geologist. Born on 27 June 1885, he graduated in science from Sydney University, and served as geologist with the western party of the **Australasian Antarctic Expedition 1911–14**. He took part in several sledging journeys including the long eastern journey. On return to Australia he became a schoolmaster, retiring in 1949. He died on 18 June 1963.
(CH)

Wauwermans Islands. 64°55′S, 63°53′W. A group of low-lying, snow-capped islands at the northern end of Wilhelm Archipelago, off the Danco Coast of Graham Land. First charted by the **German Whaling and Sealing Expedition 1873–74**, they were re-surveyed by the **Belgian Antarctic Expedition 1897–99** and named for Lt Gen. H. Wauwermans, an expedition benefactor.

Way Archipelago. 66°53′S, 143°40′E. Cluster of small islands forming an extended arc between Watt Bay and Commonwealth Bay, on the George V Coast, East Antarctica. The group, which includes Stillwell Island, was identified by the **Australasian Antarctic Expedition 1911–14**, and named for Sir Samuel Way, who was Chancellor of the University of Adelaide at the time.

Weather, Antarctic. Continental Antarctica has many climates, all of them cold (see **climatic zones**), but with a variety of weather patterns, determined mainly by variations in winds, cloud and precipitation.

Winds

The pattern of average wind flow over and around Antarctica reflects the different mechanisms that force the winds in different regions. Between about 50°S–60°S lies the belt of strong westerlies, driven by the large-scale equator-to-pole temperature gradient. This is also a region of great storminess and winds at a given location will vary greatly from day to day with the passage of individual depressions. Over the continent, winds are forced by the **katabatic** drainage of cold air from the high interior towards the coast. Coriolis force directs winds down and to the left of the large scale topographic gradient, providing katabatic winds that exhibit great constancy in speed and direction. At the coast, the katabatic winds merge with a belt of coastal easterlies, forced by the high pressure that prevails over the continent.

Cloud

Estimating cloud cover and cloudbase height presents a challenge to observers at Antarctic stations. The featureless

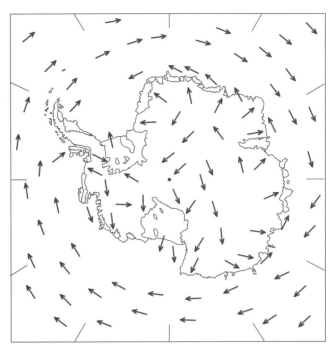

Prevailing wind directions in the Antarctic region. Winds on the plateau are generally light, blowing slightly to the left of the downhill gradient. Near the coast, steeper topography intensifies them and channels them into strong, persistent katabatic winds, with an easterly bias. In lower latitudes westerlies prevail

landscapes of the ice sheets and ice shelves provide few visual clues for estimating cloudbase height and, in winter, permanent darkness makes even the estimation of cloud cover problematic. Instruments such as lidars that provide an automatic record of cloudbase height are being introduced at a few stations and progress is being made with compiling reliable cloud cover climatologies from satellite imagery. However, current knowledge of the climatology of Antarctic clouds is largely based on visual observations.

Clouds at Antarctic coastal stations, such as Halley, are mostly associated with synoptic and mesoscale cyclones. Monthly average cloud cover at such stations (see table) ranges from around 6/8 in the summer months to 4/8 in midwinter. However, on any particular day, cloud cover is most likely to be either zero (0/8) or totally overcast (8/8) and the average figures thus reflect the relative frequency of clear and overcast conditions. The average height of the lowest cloudbase varies from around 500 m in summer to about 1 km in winter. Clouds at the coast are usually of mixed phase, i.e. they consist of a mixture of water droplets and ice crystals. Few cyclones penetrate into the interior of the continent. The atmosphere here is characterized by subsidence and is thus generally less cloudy than at the coast (see South Pole observations in the table). Cloudbase heights are similar to those seen at the coast but the

Table 1 Monthly mean cloud cover (eighths) at Halley and Amundsen-Scott

Month	Halley	Amundsen-Scott
Jan	5.9	5.2
Feb	5.7	5.5
Mar	5.3	5.4
Apr	5.0	3.5
May	4.4	3.9
Jun	4.1	3.5
Jul	4.2	4.4
Aug	4.6	4.3
Sep	4.9	4.1
Oct	5.4	5.6
Nov	5.7	4.7
Dec	5.8	4.3
Mean	5.1	4.5

clouds are optically thinner, consisting almost entirely of ice crystals.

Precipitation

In Antarctica most precipitation falls as snow. Strong winds make measurement by standard gauges unreliable. More consistent are estimates derived from snow stakes, snow pits or ice cores, which measure snow accumulation, i.e. snowfall less losses due to sublimation, blowing snow transport and melt. The loss terms are generally small: over much of Antarctica, accumulation is taken as a reliable guide to precipitation. Around the coasts, cyclones bring relatively high snowfall. The highest values are found on the western side of the Antarctic Peninsula and the coastal regions of Ellsworth Land, where moist airmasses from the Pacific sector of the Southern Ocean rise on encountering the steep topography. Few cyclones penetrate far into the interior of the continent, so snowfall is much lower on the high plateau of East Antarctica. With less than 50 mm water equivalent of snowfall per year, much of this region is technically a desert. The bulk of precipitation over the high plateau results from a near-continuous fall of ice crystals from an apparently clear sky. Known as 'diamond dust', this is formed by the gradual subsidence of (relatively) warm and moist air into the cold surface layers of the atmosphere, and gives rise to spectacular optical effects, such as solar haloes. Total snow accumulation over the grounded Antarctic ice sheets is estimated to be about 1800×10^{12} kg per year. While most falls as snow, rain may occur at any time of year in warmer regions such as the west coast of the Antarctic Peninsula and the Antarctic islands.
(JCK)

Weather buoys. See **Weather stations**.

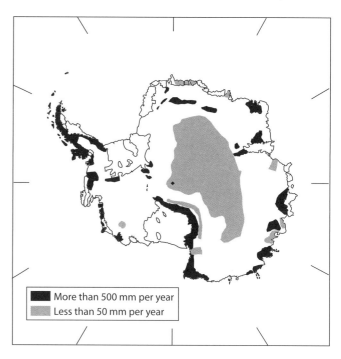

Precipitation over Antarctica. Desert conditions prevail over much of the high plateau. Trains of depressions bring heavier precipitation to many coastal areas, and lighter, less frequent falls to lower plateau areas. Exceptions are the coastal oases, where precipitation is locally very low

Weather data. Meteorological data from Antarctic stations are used primarily for studies of Antarctic climate and the mechanisms that maintain and change it, and also for inclusion in global numerical weather forecasting models, run at several world centres (e.g. the UK Meteorological Office, Bracknell), and in the Antarctic region. To be useful in forecasting, observations must be available with minimal delay. Most stations now use satellite links to transmit their data within a few minutes of observation, and to receive forecast charts and other guidance from meteorological centres outside the Antarctic.
(JCK)

Weather stations. Scientific stations at which weather data are systematically recorded. Most Antarctic scientific stations record weather data at fixed hours, synchronously with observations made worldwide. Temperature, barometric pressure and wind speed are measured using remote-recording instruments, reducing the necessity for three-hourly visits to the thermometer screen in all weathers. However, observers need to check their instruments frequently for snow and ice accretion and other problems. Some meteorological elements, e.g. the form, extent and height of clouds, and form and intensity of precipitation, require direct observation by skilled human observers. Several stations also make once-or-twice daily observations of conditions at heights of up to 20 km, using radiosondes – lightweight balloon-borne packages that transmit data on temperature, pressure and humidity. Upper-level winds can be determined by tracking the radiosonde by radar or directional receivers.

The network of manned meteorological stations in Antarctica is sparse, and further data are collected, mainly from remote locations, using automatic weather stations that operate untended for a year or more. Some 40 stations run by a University of Wisconsin programme measure wind speed and direction, pressure and temperature every 10 minutes, transmitting observations to polar-orbiting satellites. Power is provided by batteries, recharged by solar panels during the Antarctic summer. Data become available to meteorological centres outside the Antarctic within an hour or two of collection. Meteorological and oceanographic data are gathered from instrumented weather buoys, which may be dropped from aircraft or deployed from ships. Equipped also to transmit their position, they provide a continuous record of their drift, provide useful information on pack-ice movement or ocean surface currents.

Further reading: King and Turner (1997).
(JCK)

Weather systems. Most weather phenomena occur within the troposphere and lower stratosphere, the layers of **atmosphere** that are densest and in closest contact with land and ocean. In the troposphere, circulation is both horizontal and vertical, creating the familiar systems of zonal winds that promote exchanges of heat and moisture between earth, atmosphere and ocean, and giving rise to the cyclones and anticyclones that account for much of the world's weather. In the stratosphere the rise in temperature with altitude discourages vertical circulation. Horizontal circulation includes the powerful and predictable jet streams useful in long-distance aviation. Incoming short-wave solar radiation passes through the atmosphere to impinge on Earth's surface, where some is reflected (see **solar radiation, albedo**) and the rest absorbed. Warmed by absorbed radiation, the surface emits energy at longer-wave radiation, some of which is absorbed by water vapour, carbon dioxide, methane and other tropospheric gases, stabilizing surface temperatures and redistributing them between latitudes. In the stratosphere ozone absorbs and is warmed by ultraviolet radiation, accounting for the increase in temperature with height. In doing so ozone provides the shield that protects Earth's surface from the damaging effects of high-energy radiation (see **Ozone hole**). Dominant in determining climatic patterns of the southern hemisphere, Antarctica has become a key area for studying both local and world-scale climates, including climatic changes. It is also a valued platform for ionospheric and magnetospheric research.

Weathering. Natural mechanical, chemical and biological processes that reduce rock to rubble, and ultimately to its component minerals. Four kinds of weathering are distinguished, producing particles of different size and qualities that become basic components in the formation of **soils**.

Physical weathering arises from freezing and thawing of water in surface cracks, thermal expansion and contraction of rock surfaces in sunlight, and abrasion by strong winds loaded with rock dust or snow – all prominent processes in Antarctica. Scree slopes lining the flanks of Antarctic mountains testify to the power of freeze–thaw and expansion–contraction cycles. Sculptured boulders, highly-polished surfaces of rock pavements, and accumulations of wind-blown debris and dunes, show the effectiveness of wind erosion.

Glacial weathering, the pulverizing of rock by moving ice, occurs mainly in glacier beds. Glaciers carry rock debris that scrapes the floor of the glacier channels, producing a fine, powdery glacial flour which washes out in meltwater streams, forming muddy sediment in rivers and lakes. Exposed and dried, the sediment may be blown by wind to settle as ahumic soil, often in the interstices of other rock debris, which holds moisture and becomes a substrate for plants.

Chemical weathering involves erosion and destruction of rocks by chemicals in aqueous solution. Originally thought to be unimportant in polar climates where water is seldom liquid, it is now apparent that water impregnated with salts remains liquid at temperatures well below 0 °C, and chemical weathering is found to be widespread in Antarctica. In highly saline areas, for example dry valleys, can be seen a wide range of chemical weathering phenomena, resulting in the development of xeric (very dry) soils similar to those of hot deserts.

Biological weathering, involving erosion caused by microbes and plants, can break down massive rocks as well as mineral particles. They are eroded also by chemical secretions of microscopic algae and fungi, which attack crystal structures below the surface and cause peeling. Lichens penetrate tiny cracks in the surface of rocks to which they are attached, parts of their thallus (plant body) squeezing between particles and, by shrinking and swelling, forcing them apart. Particles released in this way may be incorporated within the thallus, and further altered by weak organic acids released by the plant. Soft rocks with transparent crystals, for example the Beacon Sandstones, are particularly liable to attack by these 'endolithic cryptogams' living within their surface.
(DWHW)

Webb, Eric N. (1889–1984). New Zealand polar scientist. Soon after graduating in civil engineering from Canterbury University, he joined the **Australasian Antarctic Expedition 1911–14**, based at Cape Denison. With a four-month training at Melbourne Observatory, he specialized in magnetometry, and was one of the party of three that in 1912 attempted to sledge to the South Magnetic Pole. After the expedition he served with distinction in the Australian Army. For much of his later life he travelled widely as a hydro-engineer, servicing operations throughout the world. He retired in Britain in 1959. The last survivor of the AAE, he died on 23 January 1984.

Weddell, James. (1787–1834). British master mariner and polar explorer. Born in Ostend of Scottish parents, orphaned while still a child, Weddell was apprenticed to ships in the North Sea coastal trade. In 1816, while still a seaman, he ill-advisedly knocked down his captain in a dispute, and was transferred to the Royal Navy. There he fared better: his navigational skills secured rapid advancement to midshipman and Master. In 1819, paid off from the navy, he took command of a Leith sealing brig, *Jane* (160 tons), in which he made two successful and profitable voyages to the newly-discovered South Shetland and South Orkney Islands. In 1822 he began a similar and more adventurous voyage. Leaving London on 17 September, accompanied by Capt. Matthew **Brisbane** in the cutter *Beaufoy* (65 tons), he made a routine passage down the South Atlantic Ocean, arriving off the South Orkney Islands on 12 January 1823. There Weddell undertook a running survey and collected specimens of seals for the Edinburgh Museum.

Seeking new islands, the two ships turned south on 22 January, making slow progress in wet, foggy weather, through a sea laced with pack ice and bergs. On 4 February, in better conditions and with favourable winds, Weddell determined to press as far south as possible. By 16 February, having crossed the Antarctic Circle and reached 70°S, *Jane* and *Beaufoy* found themselves in open water, heading southeast under mild west winds, surrounded by flocks of seabirds 'of the blue petrel kind', with many humpback and fin whales in sight. On 20 February 1823 the wind shifted to south, making further progress difficult. Weddell and Brisbane had reached the remarkable latitude of 74°15′S, longitude 34°15′45″W. Having found no new land, and mindful of the need to secure a profitable cargo, Weddell felt it was time to return north. In what must have been an unusually benign season, he had penetrated a huge bight, now called the Weddell Sea, achieving a record furthest-south that took him 214 nautical miles beyond Cook's earlier record. Most remarkably, he had achieved it in an area that is normally full of slowly-circulating pack ice – an area in which several later expedition ships were threatened, beset, damaged or destroyed.

James Weddell (1787–1834). Photo: Scott Polar Research Institute

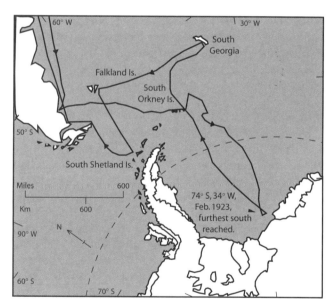

Voyages of James Weddell, 1822–4

After a stop in South Georgia the expedition wintered in the Falkland Islands. In the following summer they returned to the South Shetland Islands and eventually to Tierra del Fuego in constant hunt for a cargo of sealskins and oil. They returned to Britain in May 1824. Weddell published an account of his voyages (see below), including his many carefully recorded observations of weather, tides

and natural history. Little is known of his subsequent life, though he continued sealing, and died in London on 9 October 1834, aged 47.
Further reading: Weddell (1825).

Weddell Sea. 72°S, 45°W. Coastal sea occupying a deep bight bounded by Antarctic Peninsula in the west, Ronne and Filchner ice shelves in the south, and Coats Land and Dronning Maud Land in the east, where it merges with King Håkon VII Sea. Named for the British sealer James Weddell, who discovered and first penetrated it in 1823, the sea is largely ice-filled, and has a clockwise circulation of multi-year ice that makes access to the western side very difficult. A recurring polynya along the Coats Land coast allows relatively reliable summer access to the northeast corner of Filchner Ice Shelf, though several expedition ships have been caught in the ice gyre and involuntarily carried northward.

Weddell seal. *Leptonychotes weddelli*. Weddell seals breed mainly in the fast-ice zone close to the shore of Antarctica and the southern islands: one small group breeds unusually far north in South Georgia. Large males measure up to 3 m (10 ft) and weigh up to 400 kg (880 lb): females are of similar size or slightly larger. The fur is dark grey, mottled with paler spots and patches. Weddells winter under the ice, using tide cracks and leads for breathing and to haul out for basking on calm sunny days. In new ice they punch holes with their head, and keep breathing holes from freezing over by grinding the edges with their teeth. They hunt in darkness under the ice, using musical sonar trills to locate their prey and each other. Feeding mainly on fish, octopus and squid, they are able to dive deep but also hunt just beneath the ice, blowing into it to flush fish. Males compete for underwater territories which include a breathing hole or ice crack, giving access to all females that use the same opening. They fight for dominance in September and October. Pregnant females gather in groups of a dozen or more: in favoured areas several hundred may be seen around a system of cracks. Their pups, about 1 m long at birth, grow quickly on rich milk, and are ready to enter the sea when three to four weeks old. Mating occurs under water about the same time. Females produce their first pups at age 2–6 years and are estimated to live to 10–15 years, a slightly shorter span than other species of the ice. Some are thought to die early due to tooth wear caused by keeping their breathing holes open. The world population is estimated at up to 400 000.
Further reading: Laws (1993); Kooyman (1981).
(JPC)

Welcome Islands. 53°58′S, 37°29′W. A group of steep rocky islands standing 6.5 km (4 miles) west-northwest of Cape Buller, on the north coast of South Georgia. The largest island rises to 90 m (295 ft) and has a natural arch. All are tussock-covered: several of the islands have colonies of macaroni and gentoo penguins, and other breeding seabirds. The islands were charted by Capt. J. Cook RN in 1775: the origin of their name is obscure.

West Antarctica. The smaller of Antarctica's two provinces lies almost entirely in the western hemisphere. Topographically it consists of an undulating ice sheet, much lower than that of East Antarctica, covering a complex of mountainous islands that are geologically similar to Antarctic Peninsula, its visible extension. Deep subglacial trenches and basins separate the islands. Emergent mountains, including those of the Peninsula, indicate that geologically the province is of similar structure to the Andes and the neighbouring islands of the **Scotia Arc.** Standing on a Precambrian basement complex, they consist mainly of Carboniferous, Mesozoic and Tertiary volcanic sediments, including greywackes, siltstones, mudstones and shales, folded and metamorphosed by subsequent tectonics. These are overlain with plant-bearing sediments of mid-Jurassic age, and later Tertiary volcanic sediments, indicating a geological history quite distinct from that of East Antarctica, and generally more disturbed by earth movements. Some sediments were laid down in deep water, others in shallows, suggesting a succession of basins that filled, subsided, refilled, and were repeatedly lifted with accompanying volcanic action. A patchy fossil record indicates warm-to-tropical conditions over the more recent periods of deposition. Unlike East Antarctica, West Antarctica remains tectonically unstable, subject to frequent earth tremors and quakes. Its **Sentinel Range**, immediately west of the Ronne Ice Shelf, includes **Vinson Massif**, Antarctica's highest mountain.

West Ice Shelf. 66°40′S, 85°00′E. Large ice shelf of the King Leopold and Queen Astrid Coast, extending about 300 km (190 miles) between Barrier Bay and

Weddell seal. Photo: BS

Posadowsky Bay. At its western end it includes the massive Chelyuskintsy Ice Tongue. The shelf was identified and named by the German Antarctic Expedition 1901–3.

Wexler, Harry. (1911–62). American expedition meteorologist. Graduating in mathematics from Harvard University in 1932, he joined the United States Weather Bureau in 1934. After service with the Army Air Force in World War II he became head of research at the weather bureau, and in 1957 was appointed chief scientist of the US Antarctic IGY Program, establishing Weather Central at Little America. Active in promoting international scientific exchanges, he participated in planning 'World Weather Watch', coordinating results from US and Soviet weather satellites. Wexler became Vice-President of the American Meteorological Society, and a fellow of the American Academy of Arts and Sciences. He died in Boston on 11 August 1962. (CH/BS)

Whales, Southern Ocean. Whales (Order Cetacea) live in all the world's oceans. Living species are divided into two suborders, Odontoceti (toothed whales, about 64 species including dolphins, porpoises, orcas and sperm whales), and Mysticeti (baleen or whalebone whales, ten species). Both suborders are well represented in the Southern Ocean. Several species breed in tropical or temperate waters and migrate to the southern oceans to fatten on the summer abundance of zooplankton, fish and squid. Some have both Arctic and Antarctic stocks which, though very similar, appear never to mix or interbreed. Individuals within southern stocks tend to be larger and heavier than their northern counterparts, suggesting that southern oceans provide better summer feeding.

Odontocetes are active hunters that feed on a wide variety of prey, including fish and squid (some caught in deep waters), seabirds and seals. Those found in the southern oceans include sperm, southern bottlenose, southern beaked and orca or killer whales. Smaller dolphins are rare in icy waters, but hourglass, Peale's, dusky and Commerson's dolphins are often seen close to or south of the Antarctic Convergence. Mysticetes browse mainly at or close to the ocean surface, feeding mainly on shoals of fish or plankton. Southern Ocean species include southern right, humpback, blue, fin, sei and minke whales. For details of species see individual entries.

During the period when Antarctic whales were intensively hunted (1904–65: see **Whaling in the Southern Ocean**) many thousands of whales that were killed and processed were measured and fully documented. Others were tagged with metal darts that, on recovery, gave information about their seasonal movements: this was mainly the work of Discovery Investigations. From these records, coupled with contemporary and later field observations, it has been possible to discover much about the breeding and life cycles of Antarctic whales, especially of those that were killed commercially. For details, see accounts under individual species.

Mysticetes: baleen whales

The larger rorquals (blue, fin and sei whales) spend only their summers in the Southern Ocean, wintering in temperate or tropical areas. Some minke whales, though probably not all, winter among the pack ice. In high summer blue and minke whales, the largest and smallest species, are found all the way south to the ice edge. Fin whales seldom occur so far south, and sei whales restrict themselves mainly to the northern fringes of the Southern Ocean. The life cycles of baleen whales are closely linked to their seasonal migrations. Mating takes place in temperate and tropical waters. About one year later a single calf is born. Within a few weeks it is ready to migrate with its mother to the Southern Ocean, where it is weaned at the end of its first summer. In most species there is thus at least two years between the start of successive pregnancies. Minkes, smallest of the mysticetes, have a gestation period of only 10 months, so conceptions may occur less than two years apart. Humpback whales suckle their calves for twelve months, weaning them on return to temperate waters. Mysticete whale calves are large at birth: new-born blue whales measure up to 8 m, or 25 per cent of adult length. They grow very rapidly. Blue, fin and sei whale calves grow to about 75 per cent of adult length and 40 per cent of adult weight in their first year, entirely on the abundant milk supplied by the mother, and sustained by her ability to feed on the rich resources of the Southern Ocean.

Blue, fin, minke and humpback whales feed by taking in huge gulps of water and zooplankton, enlarging the mouth by dropping the tongue and expanding the concertina-grooves in the throat. Raising the tongue and contracting the throat expresses the water through the baleen plates, leaving a thick layer of food on the fibrous matting in the roof of the mouth. Krill is the main food of most species, but both smaller crustaceans (copepods, amphipods) and large fish are also taken. Blue whales, with the coarsest baleen, take the fewest small prey. Fin whales with medium-textured baleen take a higher proportion of amphipods and copepods. Southern right whales and sei whales have fine baleen and take large quantities of copepods, as well as krill of all sizes. They sometimes 'skim' the water, swimming with mouth half-open and head raised: when they have gathered sufficient food on the baleen plates, the mouth is closed and the food swallowed. Blue whales need to catch food equivalent to three to five times their body weight each year. As they feed very little or not at all during their months in warm water, they require 2–3 tonnes per day in the Antarctic. Other mysticetes have similar requirements.

Though baleen whales often occur in small groups, male and female or mother-calf pairs are often seen. Whales communicate with each other by vocalizations (songs), which carry great distances through the oceans. The complexity of the songs may be related to more complex social behaviour, especially in humpbacks. They become sexually mature on reaching full body size. During the intensive Antarctic whaling of the 1930s most species were found to have attained maturity at a mean age of 9–11 years. Currently fin, sei and minke whales mature sooner, at 6–7 years, suggesting that growth rates are higher since the number of whales has been reduced.

Odontocetes: toothed whales

Two species of odontocete, sperm whales and killer whales, are common in Antarctic waters. Sperm whales include mainly adult males; females and calves are rarely found in cold waters. They feed almost exclusively on large quantities of squid; nearly 8000 beaks of individual squid have been found in a single sperm whale's stomach. How they hunt and catch these slippery creatures in the darkness of deep waters is unknown, but the slender lower jaws, armed with 20–25 peg-like teeth on each side, are clearly efficient for holding and cutting. Typical prey squid are 2–3 m long and weigh 5–10 kg; the largest probably range up to 39 m long (including tentacles) and weigh 200 kg.

Killer whales are widespread in the Southern Ocean, though curiously rare in some areas, e.g. around South Georgia. Mature bulls are almost twice the weight of adult cows, with a distinctive tall dorsal fin. Highly social, killers live in groups (pods) of 5–40 individuals, usually made up of one bull with a number of cows and calves. Female calves remain in the pod of their birth. Killers hunt cooperatively, pinning fish or penguins against a shore, displacing seals from ice floes by creating waves, or hitting floes to upset them. Fish, squid, penguins and seals are probably of equal importance in their diet. Females reach maturity at 8–10 years, males perhaps not until 16 years old. Females are believed to average only one birth every 5–8 years.

Little is known of southern beaked and bottlenose whales, two species of similar appearance with one or two teeth at the tip of their lower jaw. Usually seen over deep water, they probably hunt for squid in the darkness far below.

Further reading: Bonner (1989); Evans (1987); Gaskin (1982); Simmonds and Hutchinson (1996); Watson (1981). (JPC)

Whaling in the Southern Ocean.

Though the ocean-wide whaling industry had exploited warmer waters of the southern hemisphere during the nineteenth century, the cold waters of the Southern Ocean remained relatively untouched. Explorers and sealers from time to time reported abundant sightings of whales, but these were mostly rorquals – fast-moving species that could not be harpooned by hand from open boats. The **German Whaling and Sealing Expedition 1873–74**, the earliest expedition to investigate commercial whaling in the South American sector, saw few whales and could provide only a negative report for its backers. However, continuing competition for declining stocks of northern whales led to a revival of interest in southern stocks during the 1890s. The four ships of the **Dundee Whaling Expedition 1892–93**, freshly back from northern grounds, explored the Falkland Islands, Weddell Sea and Antarctic Peninsula. In the same and the following season the **Norwegian (Sandefjord) Whaling Expeditions 1892–94**, commanded by experienced whaler Capt. C. A. Larsen, explored around Seymour Island, the western Weddell Sea, the South Orkney Islands and South Georgia. The **Norwegian (Tønsberg) Whaling Expedition 1893–95** under H. J. Bull, explored far south in the Ross Sea sector. These expeditions recouped some of their expenses by sealing, but none was commercially successful.

Whale catcher of the 1950s Salvesen fleet operating from South Georgia. Note the bow-mounted harpoon gun, catwalk from bridge to bow, and crow's nest for reconnaissance. Photo: Scott Polar Research Institute

The start of Antarctic whaling

However, Larsen and his Norwegian colleagues were particularly impressed with possibilities for hunting fin and humpback whales around South Georgia. Returning south with the **Swedish South Polar Expedition 1901–4**, Larsen took the opportunity to convince businessmen in Buenos Aires that whaling from South Georgia would be profitable, and raised capital to form the **Compañia Argentina de Pesca**. Sailing in November 1904 with two supply ships and a whale-catcher to Grytviken, a harbour well known to sealers, he set up the first Antarctic land-based whaling station. So successful was his first season's hunting that other Norwegian whalers quickly became involved, setting up further operations on South Georgia and extending southward to the southern Scotia Sea and Antarctic Peninsula. In 1905–6 the Norwegian whaler Chr. Christensen sent his factory ship *Admiralen* south first to the Falkland Islands, then to Admiralty Bay, King George Island in the South Shetland Islands. In the same season Adolf Amandus Adresen, a Norwegian whaler based in Punta Arenas, southern Chile, formed the Sociedad Ballenera de Magellanes, which sent a factory ship *Gobernador Bories* to work from Deception Island, South Shetland Islands. From then onward increasing numbers of factory ships with their accompanying catchers worked from other harbours around South Georgia and the South Shetland Islands, notably Deception Island. Later operations moved south to harbours and sheltered bays along Antarctic Peninsula. The South Sandwich Islands were investigated in 1909 but found to lack suitable harbours. Floating factories operated at harbours in the South Orkney Islands from 1911, and a small station was built at Signy Island in 1920.

Alert to the growth of a new and potentially profitable industry in dependencies to which it laid claim (see **Sovereignty in the Antarctic region**), the British government issued leases for stations in South Georgia from 1907, and licences to impose controls on land-based and inshore operations elsewhere in the region. Magistrates were installed on South Georgia in 1909 and on Deception Island in 1910. The objectives were to tax and regulate the industry, generally to maintain steady growth and maximize productivity, for example, by encouraging full use of carcasses.

Pelagic whaling

British-imposed regulations and taxes at land stations and moorings soon proved irksome to whaling operators. In 1912–13 three Norwegian whaling factory ships, *Tioga*, *Falkland* and *Thule*, prevented by pack ice from reaching their licensed harbours in the South Orkney Islands, began operations in the open sea, using the ragged edge of the ice field to provide the calm waters they needed. So began pelagic or open-sea whaling. Tax-free, untrammelled by British-imposed regulations, and capable of exploiting huge stocks of whales that were feeding far from land, this form of hunting quickly overtook land-based whaling in Antarctic waters. In the southern summer of 1923–24 the Norwegian whale factory ship *Sir James Clark Ross*, accompanied by catchers and commanded by Larsen, explored the Ross Sea for likely anchorages, making use of the ice-free route south that James Clark Ross himself had discovered in 1841. In the following year the fleet returned for a summer's pelagic hunting. Larsen died on board, but his initiative remained: the Ross Sea became a favoured and profitable hunting ground for many successive seasons.

Like many other factory ships up to this time, *Sir James Clark Ross* was a former liner, converted for whaling by the addition of pressure cookers and oil storage tanks. Its flensers cut up the whales in the water alongside, and the blubber was hauled on board for processing. In 1925 came *Lancing*, the first of a new kind of factory ship, custom-built with a slipway and tackle that allowed the whole whale to be hauled on board. Accompanied by four catchers, *Lancing* spent a season cruising the whaling grounds from South Georgia to Antarctic Peninsula. Despite teething troubles, her success made it clear that the future of whaling lay with ships of this kind.

Regulation and research

In 1917, the penultimate year of World War I, the British government set up an interdepartmental committee to consider all aspects of research and development in the Antarctic whaling industry. In 1920 the committee published a report that stressed the need for research to protect the industry from over-rapid development and the stocks of whales from over-exploitation. Whaling had spread throughout the Falkland Islands Dependencies, where the industry began, and discovered the advantages of tax-free pelagic hunting that could take it to any profitable sector of the pack ice edge. In 1925 the government established **Discovery Investigations**, to gain as much information as possible on Southern Ocean whale stocks, as a basis for regulating the industry.

Reviewing the system

A massive demand for whale oil during the 1920s stimulated rapid growth of the industry worldwide, but particularly in the Southern Ocean. Over-production caused the market to collapse, and in 1931 an international Convention for the Regulation of Whaling was established among the whaling nations, primarily to protect the industry from its own excesses. Initially protection was given to right whales, calves, cows with calves and immature whales: from 1937 grey whales were protected, and hunting seasons defined for commercial species. In the Southern Ocean, humpback whales, slowest and easiest to catch, and always the main prey of the shore stations, had already declined. Blue whales, the largest species and the one most heavily hunted by the deep-sea fleets, were becoming hard to find. Fin whales, the next-largest species, were coming under increasing pressure. While any of the larger species remained, the much-smaller sei whales were barely worth catching. No hunters at that time bothered with minke whales, the smallest rorquals.

Southern Ocean whaling continued throughout World War II from a single station (Grytviken) on South Georgia, but pelagic whaling declined rapidly as the vulnerable ships of the whaling fleet were sunk, captured or converted to other uses. Eleven factory ships operated in Antarctic waters in 1940–41, none in the following two seasons, and only one in 1943–44 and 1944–45. Demand for edible oils increased rapidly after the end of the war, and 1945–46 saw a return to normal business, with three shore stations operating on South Georgia and nine factory ships working in the open ocean.

International Convention for the Regulation of Whaling

Toward the end of the war the representatives of the whaling nations reconvened to establish a new convention – the International Convention for the Regulation of Whaling (1946), with an executive International Whaling Commission (IWC) to provide further controls over the industry. The Commission's first measure was an attempt to limit production under a quota system based on 'Blue Whale Units', a measure already familiar to the industry from its earlier attempts at self-regulation. One unit was equal to one blue whale, two fin whales, two and a half humpbacks or six sei whales – all deemed to be roughly equivalent in terms of oil production. The first annual quota, set at 16 000 BWU, probably bore some relation to the maximum sustainable yield for the combined stocks. However, the system took no account of the different degrees of protection that each species required: while blue and fin whales could still be found, they remained the easiest to catch and process, and thus remained prime targets.

From 1946–47 the three shore stations on South Georgia, and increasing numbers of factory ships operating pelagically, produced annual catches of whales that rose quickly to over 30 000. This output was maintained throughout the 1950s, employing three shore stations and up to 21 factory ships. In 1960–61 over 41 000 whales were taken. Despite the introduction of more rational catch limits and quotas, the IWC found it increasingly difficult to keep a highly capitalized and competitive industry under control, and powerless to halt the overall decline of whale numbers in the Southern Ocean, the industry's main killing ground.

Whaling in decline

Salvation for the remaining stocks of whales came from another quarter. Tropical plantations of vegetable oil crops, established after World War II, increasingly flooded the market during the 1950s, reducing the demand for whale oil. The more enterprising whalers (for example, **Salvesens of Leith**) withdrew from a clearly declining industry, leaving the remnants essentially to Japan and the USSR. By the mid-1960s all three of South Georgia's shore stations had closed. In 1966–67 only nine pelagic factory ships were employed, taking just over 20 000 whales: in 1970–71 only six ships remained in operation, taking just over 12 000 whales. Whaling in the Southern Ocean had become a relatively small operation, taking almost entirely minke whales.

In 1972 the United Nations Conference on the Human Environment, held in Stockholm, expressed special concern at the worldwide decline in stocks of whales, and recommended a 10-year moratorium on whaling that might allow over-taxed stocks to recover. The IWC responded in stages, increasing restrictions during the 1970s, and in 1979 establishing a whale sanctuary in the Indian Ocean. In 1982 it recommended catch limits of zero (i.e. the complete moratorium on whaling) for the 1985–86 season. Japan and the USSR objected, continuing to hunt in the Southern Ocean, taking a few thousand minke whales for meat. Both announced that they would not resume commercial whaling in the 1987–88 season. Japan later resumed catching small numbers of Antarctic whales on somewhat disputed scientific grounds, continuing desultorily to the present day (2002) under relentless opprobrium from conservationists.

In 1994 the IWC announced the formation of the Southern Ocean Sanctuary, a whale sanctuary contiguous with that of the Indian Ocean, from which all commercial whaling activities are excluded. Like the moratorium, this is subject to review under the Commission's current management procedures, that will require substantial justification before legitimate commercial whaling is allowed to re-start.

Derelict whale catchers at the abandoned whaling station, Grytviken South Georgia. Photo: BS

Are whale stocks recovering?

We have no counts or reasonable estimates of whale stocks at the start of whaling, so the true extent to which they were reduced can never be known. We have good information on the breeding biology of the large whales: they breed young and effectively, and their potential for recovery is high, but we can only guess as to their actual recovery and whether former stock levels will be attained. The environment – the

Southern Ocean itself – has changed, and so have many of the component species of its ecosystems. Recent dramatic increases in numbers of fur seals and pygoscelid penguins may result from the reduction of whale populations. They may or may not represent competition for food, which may or may not affect the rate of recovery of whales. Humans continue to draw heavily on the Southern Ocean, fishing mainly for white fish, squid and krill. Anecdotal reports – possibly the best evidence currently available – suggest that during the past 20 years humpback whales have begun to appear more and more frequently at their old hunting ground along Antarctic Peninsula, and that to a lesser extent blue and fin whales may be returning, though none is yet back even to post-World-War-II levels of population.
Further reading: Tønnesen (1970); Tønnesen and Johnsen (1982).

White Island. 78°08′S, 167°24′E. A steep, ice-covered island of Ross Archipelago, 24 km (15 miles) long, set in the Ross Ice Shelf west of Ross Island. Surveyed by the **British National Antarctic Expedition 1901–4**, the island was named for its snow cover. Though some 20 km (12.5 miles) from the sea, the surrounding shelf ice is afloat. A patch of sea ice along its northern shore, with a tide crack and open leads, harbours a small population of Weddell seals, which appear to live permanently in isolation from those in McMurdo Sound. An area of 170 km^2 (66.5 sq miles) extending southward along the northwest coast from Cape Spencer-Smith, and from high water mark to 5 km (3.1 miles) across the Ross Ice Shelf, is designated SSSI No. 18 (Northwest White Island).

Whiteout. Atmospheric condition experienced in snowfields under overcast skies, in which the horizon and shadows disappear, making it difficult for a human observer to orient or perceive distances.

Whitmore Mountains. 82°35′S, 104°30′W. A small group of peaks and nunataks on the southern flank of Hollick-Kenyon Plateau, visited by a US traverse party in 1959 and named for cartographer George D. Whitmore of the US Geological Survey.

Wiencke Island. 64°50′S, 63°25′W. A mountainous, irregular-shaped island 25 km (15.6 miles) long off the southeastern corner of Anvers Island. Discovered and charted by the Belgian Antarctic Expedition 1897–99, it was named for Auguste-Karl Wiencke, a seaman who was washed overboard in a storm of January 1898. Port Lockroy lies on the western side of the island.

Wild, Harry Ernest. (d. 1918). British seaman and explorer. Born in Yorkshire, the brother of **Frank Wild**, he enlisted in the Royal Navy. After service in the Mediterranean and advancement to petty officer, he served with the Ross Sea party of the Imperial Trans-Antarctic Expedition 1914–17, participating in the gruelling sledging programme to lay a chain of depots across the Ross Ice Shelf toward the Beardmore Glacier. He was awarded the Albert Medal for his part in saving the base leader, Æneas Mackintosh, who had succumbed to scurvy. Returning to the navy, he died on active service in March 1918.

Wild, John Robert Francis (Frank). (1873–1939). British seaman and explorer. Born at Skelton, Yorkshire, he joined the merchant navy in 1889, and in 1900 transferred to the Royal Navy. Volunteering for the **British National Antarctic Expedition 1901–4**, he served as an AB in *Discovery*, taking part in the sledging programme. He joined Shackleton's **British Antarctic (*Nimrod*) Expedition 1907–9**, and was one of the party chosen by Shackleton to man-haul up the Beardmore Glacier toward the South Pole. In Mawson's **Australasian Antarctic Expedition 1911–14** he was in charge of the Western Base, wintering and sledging in the difficult, snowbound conditions of Shackleton Ice Shelf. On Shackleton's **Imperial Trans-Antarctic Expedition 1914–17** he was second-in-command, taking charge of the Elephant Island camp while Shackleton sailed for South Georgia. During World War I he was commissioned in the Royal Naval Volunteer Reserve for service in Russia. After the war he farmed in southern Africa, breaking off to serve as second-in-command of the **Shackleton-Rowett Antarctic Expedition 1921–22**, taking over leadership on Shackleton's death. Wild received many geographical awards and was made a Freeman of the City of London in 1923. He returned to southern Africa, engaging in various farming enterprises. He died in Johannesburg on 19 August 1939.

Wilhelm Archipelago. 65°08′S, 64°20′W. Extensive cluster of small, mainly ice-capped islands bordering the Graham Coast, Antarctic Peninsula. The group extends from Bismarck Strait in the north to Southwind Passage in the south, and includes the Wauwermans, Dannebrog and Myriad groups, Booth, Pléneau, Hovgaard and Petermann Islands, the Argentine, Jalour and Betbeder groups and many smaller islets, west to Lumus Rock. Several islands of the group were first sighted by the British sealer John Biscoe. The first charts were produced by Capt. E. Dallman of the German Whaling and Sealing Expedition 1873–74, who named the group for Kaiser Wilhelm I.

Wilhelm II Coast. 67°00′S, 90°00′E. Part of the coast of Wilkes Land, East Antarctica. Bounded by Cape Penck in the west and Cape Filchner in the east, the coast consists

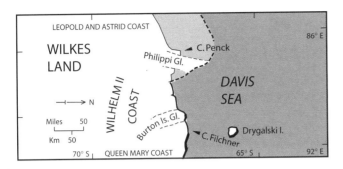

Wilhelm II Coast

mainly of ice cliffs, including the eastern end of the West Ice Shelf. The coast was explored by the German Antarctic Expedition 1901–3, that wintered in the sea ice of Posadowski Bay and worked intensively on the Gaussberg, the only available rock outcrop. The coast was named for the contemporary German Kaiser.

Wilhelm II Land. Narrow ice-covered sector of East Antarctica between longitudes 86° and 91°E, bounded by Princess Elizabeth Land to the west and Queen Mary Land to the east. The area includes the eastern end of King Leopold and Queen Astrid Coast, part of West Ice Shelf, Gaussberg, and a gently sloping ice plateau extending to the South Pole. Discovered by the German Antarctic Expedition 1901–3, the land was named for the German reigning Kaiser. It forms part of Australian Antarctic Territory.

Wilkes, Charles. (1798–1877). American naval officer and explorer, leader of the **United States Exploring Expedition 1838–42.** Born in New York and educated at home, Wilkes joined the US Navy in 1818. Showing outstanding competence as a navigator and hydrographer, he was in 1826 and in 1833 made Director of the US Navy's Depot of Charts and Instruments. Five years later he was appointed to command a fleet of six ships, commissioned on a world voyage of hydrography and exploration that would include forays into the Southern Ocean. Through no fault of his own the expedition was ill-conceived, ill-equipped, and certainly ill-prepared for Antarctic exploration. Leaving Norfolk, Va. in August 1838, Wilkes proceeded to Orange Bay, Tierra del Fuego. During early March 1839, in the sloop USS *Peacock*, he visited the South Shetland Islands and tip of Antarctic Peninsula. In April the squadron headed north to Valparaiso and westward on a prolonged Pacific Ocean cruise, reaching Sydney, Australia in late November. On 26 December the four remaining ships headed south towards the unknown continent. Aboard the flagship USS *Vincennes*, with USSs *Peacock* and *Porpoise*

Charles Wilkes (1798–1877)

in attendance, Wilkes reached the edge of the pack ice in early January 1840, and headed westward through a sea strewn liberally with floes and bergs. Close to the ice-cliff coast of what is today called Wilkes Land, the ships from time to time came within soundings, and there was evidence of higher, ice-covered land behind the cliffs. Wilkes and his colleagues were clearly close to a continent, and had no hesitation in claiming several sightings of ice-covered land. After a most difficult voyage in appalling conditions, covering some 1200 miles of ice-bound coast, Wilkes turned north on 21 February, and on 11 March re-entered Sydney Harbour. The rest of the expedition was conducted in warmer waters of the central and northern Pacific Ocean. Only two of the six original ships, USSs *Vincennes* and *Porpoise*, survived to reach New York in August 1842.

Wilkes thus became the first explorer to delineate a substantial length of Antarctic coastline, enough to establish that immediately behind lay a landmass of continental size. On returning to the US he faced recriminations, charges of incompetence and a court martial for his handling of the expedition. Acquitted on most charges, he was promoted to commodore. During the following years he supervised publication of the scientific results of the expedition – a 20-volume report that, characteristically, was published in

only very limited edition. After service with the Union fleet in the Civil War, he retired from the navy as a rear admiral in 1866.
Further reading: Waldman and Wexler (1992).

Wilkes Coast. See **Clarie Coast**.

Wilkes Land. Broad ice-covered sector of East Antarctica between longitudes 102° and 136°E, bounded by Queen Mary Land to the west and Terre Adélie to the east. The sector includes the Knox, Budd, Sabrina, Banzare and Wilkes coasts, and a gently sloping inland ice sheet extending to the South Pole. Discovered and largely delineated by the United States Exploring Expedition 1838–42, the area was named for the expedition commander, Lt Charles Wilkes USN. The inland ice was explored by tractor traverses from Wilkes Station, a combined United States and Australian research facility that operated in the Windmill Islands oasis area from January 1957 to February 1969. Wilkes was replaced by Casey Station, a permanent Australian research station built nearby.

Wilkes Station. 66°15′S, 110°32′E. Former research station on the Windmill Islands, Clarke Peninsula, Budd Coast, established in 1957 by the United States as a base for overland traverses. In February 1959 its management passed to Australia. Wilkes was abandoned in 1969 when a new Australian station, **Casey**, was built on nearby Bailey Peninsula. The first Casey was occupied between 1969 and 1989, then replaced by a new station built some 500 m away. The sites of Wilkes and 'Old Casey', and of the shore-based rubbish tips to which both gave rise, are currently subjects of intensive research into management and remediation of contaminated ground, of interest and relevance in relation to similar sites elsewhere in Antarctica.

Wilkins, George Hubert. (1888–1958). Australian explorer, aviator and adventurer into both polar regions. Born on a sheep station at Mt Bryan East, 100 miles north of Adelaide, South Australia, Wilkins was the youngest of a family of 13, of whom only seven survived infancy. At 16 he learnt general and electrical engineering in a trade school. At 20 he was running a small business maintaining cinematographs, but stowed away on a cargo ship. Landing in Algeria, he made an adventurous way to England, where he learnt to fly and take aerial photographs. During the Balkans war of 1912–13 he became a photographer and war correspondent, was several times imprisoned and narrowly escaped death by bombardment and firing squad. In 1913 he joined Vilhjalmur Stefansson's expedition to the Canadian Arctic, spending three years in the field as photographer and correspondent for the London *Times*. Returning to Britain in 1916, his ship was torpedoed and sunk. Wilkins was rescued by the Royal Navy, reached England, and joined the Royal Australian Flying Corps as air and field photographer. He saw service in France, was wounded and awarded the Military Cross and Bar. In 1919 he attempted to fly back to Australia as navigator in an air race, but the aircraft crashed in Crete and he returned to England.

Demobilized in 1920, Wilkins joined the British Imperial Antarctic Expedition 1920–21, which initially promised opportunities for flight and aerial photography in Antarctic Peninsula. When the expedition was reduced to a reconnaissance by whale boat, with no aircraft and few possibilities for exploration, Wilkins (together with J. L. Cope, the leader) withdrew. Homeward bound in Montevideo he encountered Sir Ernest Shackleton, who invited him to join the Shackleton–Rowett Antarctic Expedition as naturalist and photographer. Soon after the expedition ship *Quest* arrived in Grytviken, Shackleton died. The expedition continued for a few months under Cdr. Frank Wild, briefly exploring the South Sandwich Islands and the Weddell Sea before returning to Britain. After a spell as a cinephotographer with the Society of Friends, recording relief work in war-devastated Russia and Poland, in 1923–25 Wilkins led a museum-sponsored natural history expedition to northern Australia. In 1926 he joined forces with the North American News Alliance to make a number of exploratory flights in the Arctic, securing for the purpose several aircraft and an entourage of pilots and mechanics. After two preliminary seasons based in Barrow, Alaska, involving flights over unexplored sea ice and mountain ranges, on 16 April 1928 he flew with pilot Carl Ben Eielson across the Arctic basin from Barrow to Green Harbour, Svalbard in a single-engined Lockheed Vega aircraft, a flight of over 4000 km (2500 miles) in just over 20 hours. For this and earlier Arctic achievements Wilkins was knighted and awarded the Patron's Medal of the Royal Geographical Society.

On return to the US Wilkins immediately developed plans for pioneering flights in Antarctica, using the same single-engined aircraft and a small field team. Supported as before by geographical societies and the press, he organized the **Wilkins-Hearst Antarctic Expeditions 1928–30**, which aimed initially to fly from the South Shetland Islands to the Ross Sea, a distance comparable to that achieved on his trans-Arctic flight. Starting from Deception Island, on 16 November 1928 he and pilot Carl B. Eielson made the first-ever flight by powered aircraft in the Antarctic region (just two months ahead of Byrd's first flight over the Ross Ice Shelf). Poor ice conditions and bad weather made the long flight impossible, but on 20 December they flew over 960 km (600 miles) to the base of Antarctic Peninsula. Leaving the aircraft to winter on Deception Island, Wilkins returned to the US, first to become involved in flights of the airship Graf Zeppelin over Europe, second, to marry his fiancée, actress and fellow-Australian Susanne Bennett. After a brief honeymoon he returned to Antarctica for a second season. Again snow, ice and weather conspired to make the long flight to the Ross Sea impossible, but Wilkins and his two pilots, S. A. Cheeseman and P. D. Cramer, achieved several shorter flights along the Peninsula, to Charcot Island and over the Bellingshausen Sea.

In 1931 Wilkins formed a consortium with Lincoln **Ellsworth**, a wealthy US mining engineer, and others, and the support of several scientific bodies, in a plan to operate an ex-US Navy submarine to reach the North Pole under the pack ice of the Arctic Ocean. Though well researched and prepared, the expedition was beset by ill will, even sabotage among the crew, and *Nautilus* made only minor forays under the ice north of Svalbard. In 1932 Ellsworth and Wilkins again formed a partnership for a further attempt to fly across West Antarctica (see **Ellsworth's Antarctic Expeditions 1933–39**). In a series of three expeditions, one from the Ross Sea and two from West Antarctica, Wilkins proved an efficient consultant and manager. In August 1937 he helped in an aerial search of the Arctic basin north of Alaska and Canada, seeking a Russian T4 aircraft reported missing on a trans-Arctic flight from Moscow to Fairbanks, Alaska. Within a week of receiving the report, Wilkins had bought a Catalina flying boat, enrolled pilots H. Hollick-Kenyon and S. A. Cheeseman, and was flying long search flights over the Arctic Ocean from Coppermine, northern Canada. The flights continued for several months, involving about 45 000 miles of searching, but no signs were found of the missing aircraft.

In 1938 Wilkins joined Ellsworth in a fourth Antarctic expedition, this time to Ingrid Christensen Coast of Princess Elizabeth Land. As operations manager Wilkins remained with the ship, taking several opportunities to raise the Australian flag on coastal islands, reinforcing his country's claim to the territory.

During World War II Wilkins remained a civilian but, employed in various capacities as a consultant or observer, characteristically managed to see more action than most servicemen. In later life he continued to travel, visiting the South Pole in 1957 as a guest of the US government, and developing latent interests in religion and the occult. He died of a heart attack on 1 December 1958, aged 70.
Further reading: Grierson (1960).

Wilkins Coast. 69°40′S, 63°00′W. Part of the east coast of Palmer Land, Antarctic Peninsula, facing onto the Weddell Sea. Bounded by Cape Agassiz in the north and Cape Boggs in the south, the coast consists of mountain rising to the high interior plateau, drained by glaciers that descend to the Larsen Ice Shelf, where they are in part blocked by the offshore bulk of Hearst Island. The coast was identified and photographed by Sir Hubert Wilkins on his flight of 20 December 1928.

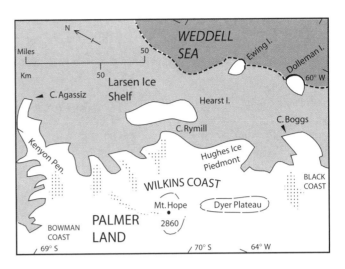

Wilkins Coast

Wilkins-Hearst Antarctic Expeditions 1928–30. Two expeditions in which Sir Hubert **Wilkins** achieved the first exploratory powered flights over Antarctica. On returning to the US after his successful trans-Arctic flight of 1928, Wilkins negotiated with the American Geographical Society, Detroit Aviation Society, the Hearst press and other supporters to sponsor flights of exploration over Antarctica. Equipped with two single-engined Lockheed Vega aircraft, the five-man Wilkins-Hearst Antarctic Expedition left Montevideo in October 1928 in the whaling factory ship *Hektoria*, for Whalers Bay, Deception Island, a sheltered harbour in the South Shetland Islands. Wilkins hoped to fly from this sector of Antarctica to the Ross Sea, involving long-distances that would lie within the range of his Vegas, but only if they could take off fully loaded, preferably on skis or floats to ensure safe landings *en route*. On arrival in early November they found both snow and sea ice too thin for skis or floats. The party levelled a runway in the loose volcanic scoria behind the beach, and on 16 November 1928 pilot Carl Ben Eielson and Wilkins made a wheeled take off and first brief test flight. On 26 November both aircraft were used in flights to seek better landing facilities. Further digging, scraping and raking produced a sinuous runway 730 m (800 yards) long – enough for take-off with light or moderate loads, but inadequate for transcontinental flights. On 20 December Wilkins and Eielson left at 08.20 for their first southern flight along Antarctic Peninsula. Heading first for Trinity Island and Salvesen Bay, they crossed Graham Land plateau at 8000 ft. Again turning south along the broad ice shelf footing the Weddell Sea coast, Wilkins identified headlands and landmarks recorded by Nordenskjöld in the **Swedish South Polar Expedition 1901–4**, and photographed what he interpreted as broad, deep ice-filled channels that carved the peninsula into island blocks. They turned at a point estimated in 71°21′S, 64°15′W, beyond which lay a mountainous land that Wilkins named

Hearst Land for his major sponsor. The largest 'channel' he named Stefansson Strait for the Arctic explorer with whose expedition he had served in 1913–16: lesser ones he named 'Casey' and 'Crane' channels for expedition sponsors. In a second long flight on 10 January 1929 Wilkins sought a suitable landing site further south, but low cloud hampered visibility. No further flights were possible that year: Wilkins wintered the aircraft at Deception Island and returned to the USA.

In the following season the British government, keen to share in Wilkins' discoveries, gave him the use of RRS *William Scoresby*, the small survey ship of Discovery Investigations. Returning to Deception Island in November, Wilkins again found ice and snow conditions that prevented use of skis or floats. Still concerned to make the long cross-continental flight, he moved south to Port Lockroy, Wiencke Island, and on to the southern tip of Adelaide Island, seeking the weather and ice conditions he needed. The expedition returned north to Port Lockroy, and on 19 December 1929 were at last able to fly off from nearby Neumayer Channel. Wilkins and Canadian pilot Silas A. Cheeseman headed southwestward to Leroux Bay, then across the Peninsula. A troublesome engine limited the flight, but Wilkins was able to re-examine 'Crane Channel', one of the deep, ice-filled rifts photographed in the previous year, before returning to base. Later in December, still in search of open water, *William Scoresby* again moved south, this time toward the farthest-south sighting of land made by J.-B. Charcot's **French Antarctic (*Pourquoi Pas?*) Expedition 1908–10**. On 28 December Wilkins and Cheeseman flew toward 'Charcot Land' but were defeated by bad weather. On the following day in better visibility they flew over and around a large ice-covered island, some distance to the west of Alexander Island, with more ice-covered mountains extending away to the south. The island is now Charcot Island, linked to Alexander Island by the extensive Wilkins' Ice Shelf. The ship returned to Deception Island for refuelling, then on 25 January 1930 made a last foray south to 70°10′S, 100°45′W in Bellingshausen Sea. A final flight by Wilkins and pilot Parker D. Cramer on 1 February explored south to latitude 73°S, but saw only cloud and sea ice.

Wilkins' flights drew attention to the great potential offered by aircraft and aerial photography in Antarctic exploration, but also demonstrated their limitations. Visual observations of ice topography from heights above about 600 m (2000 ft), even by an experienced observer and in good visibility, are likely to be misleading, and dead-reckoning surveys without ground control make for inaccurate maps. The 'channels' thought by Wilkins to cross the Peninsula were later found to be deeply incised glaciers flowing down one side or the other from a continuous high escarpment. Later cartographers have with some difficulty re-interpreted his descriptions of geographical features and honoured the names that Wilkins' pioneering flights added to the map.

Further reading: Wilkins (1929; 1930).

Wilkins Ice Shelf. 70°15′S, 73°00′W. Large rectangular ice shelf occupying Wilkins Sound, and linking Charcot Island and Latady Island to the west coast of Alexander Island. It was overflown by the Australian aviator Sir Hubert Wilkins on 29 December 1929, and later named for him.

Wilkins Sound. 70°15′S, 73°00′W. Ice-filled channel between Alexander Island and Charcot Island, named for the Australian explorer and aviator Sir Hubert Wilkins, who discovered it in a pioneering flight of 1929.

Wilkinson, John Valentine. (d. 1986). British naval hydrographer. A career officer in the Royal Navy, Wilkinson served with distinction in World War II. In 1955–57 he commanded HMS *Protector*, guard ship in the Falkland Islands Dependencies. While supporting scientific and logistic operations of the Falkland Islands Dependencies Survey, he made two visits to the South Sandwich Islands, deploying the ship's helicopters in aerial photography and landings on three of the islands. He retired from the navy with the rank of captain, and died on 13 September 1986, aged 72.

William Scoresby Archipelago. 67°20′S, 59°45′E. Group of mainly ice-free islands off William Scoresby Bay, Mawson Coast, Kemp Land, East Antarctica, discovered in 1936 during a cruise of the Discovery Investigations whale-marking ship, RRS *William Scoresby*.

Willis Islands. 54°00′S, 38°11′W. The westernmost group of islands of South Georgia, standing 3 km (2 miles) west of Bird Island. They were charted in 1775 by Capt. J. Cook RN in HMS *Resolution*, and named for Midshipman Thomas Willis RN, who first sighted them.

Wilson, Edward Adrian. (1872–1912). British scientist and explorer who died with Robert Falcon Scott after reaching the South Pole. Born in Cheltenham, Gloucestershire on 23 July 1872, the fifth child of a prosperous family doctor, Wilson studied medicine and natural science at Gonville and Caius College, Cambridge and St George's Hospital, London. In 1898 he contracted pulmonary tuberculosis; treatment requiring recuperation in Britain and Switzerland gave him leisure to develop his interests in wildlife and talents as an artist. Qualifying in

Edward Wilson (1872–1912) with his wife Oriana

medicine in 1900, he practised at Cheltenham Hospital, where in 1901 he was appointed Junior House Surgeon. Later in the same year he was appointed Junior Surgeon and Zoologist to the British National Antarctic Expedition. Shortly before joining the expedition he married Oriana Souper. Spending two winters and three summers in McMurdo Sound, he played an active role in sledging and exploration, invaluable for his skills in accurately illustrating both topography and wildlife. A staunch Christian, he was valued too for his gifts as a physician, mediator and counsellor, and became a close friend and confidant of Scott.

On returning to England in 1904 Wilson wrote up his zoological notes for publication, accepted commissions for illustrating bird and mammal books, and was appointed to investigate disease in grouse on British moorlands. These tasks occupied him fully until 1909, when Scott invited him to join his new venture, the British Antarctic (*Terra Nova*) Expedition, as chief of scientific staff. Now an experienced hand, he became a much-loved 'uncle' to the expedition. He led a winter sledging journey to Cape Crozier to collect emperor penguin embryos, and Scott selected him for the long sledging journey to the South Pole. Wilson shared the triumph of reaching the Pole, the frustration of discovering there the relics of Amundsen's earlier visit, and the long, dispiriting march back toward McMurdo Sound. Together with Scott and 'Birdie' Bowers, he died in a final camp on the Ross Ice Shelf in late March 1912.

Wind chill. Feeling of intensified cold experienced in moving air. Cold winds feel colder than still air at the same temperature, and can more readily cause frostbite. Attempts to quantify wind chill address both the feeling of increased cold and the increased danger of frostbite to exposed flesh. Many factors contribute to the discomfort experienced in cold, windy conditions, including insolation (sunshine), humidity, the quality and amount of clothing worn, and such personal factors as physical fitness, body temperature, metabolic rate and psychological condition. However, the single most important factor is the rate at which heat is lost from an area of exposed skin, and this is governed principally by air temperature and wind speed. The same factor is almost solely responsible for determining if and when frostbite occurs. For dry skin, frostbite occurs when the rate of heat loss exceeds about 1700 W per m^2.

Wind chill is commonly expressed as a subjective feeling of cold, typically ranging from 'cool' through 'cold' and 'very cold' to 'freezing cold'. However, different subjects vary in how they describe their feelings of cold. Alternatively it can be expressed as the air temperature which, in the absence of wind, would produce the same sensation of cold. Here we meet the problem that few people in temperate regions have enough experience of very low temperatures (below $-10\,°C$, for example) to make accurate comparisons.

Most wind chill tables are based on work carried out over 50 years ago, extrapolated using some rather unrealistic assumptions. Use of such tables can lead to exaggerated and essentially meaningless figures for equivalent temperatures. Nevertheless, tables or diagrams showing the wind chill as a function of air temperature and wind speed serve a useful purpose in indicating the approximate severity of

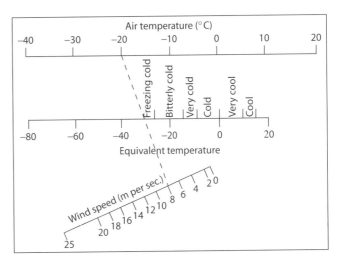

Wind chill. To use this diagram, place a straight edge from the point indicating air temperature to the point indicating wind speed. The intersection of this line with the wind chill axis shows the sensation of cold experienced by average people, and the 'equivalent' air temperature. The dashed line gives an example, showing that an air temperature of $-20\,°C$ and a wind speed of 8 m per second (29 km per hour) would be regarded as 'freezing cold', with an 'equivalent' temperature of $-31\,°C$

the conditions, and the type of clothing that will provide adequate protection. Such a diagram, based on a recent theoretical analysis, appears in the figure above. A similar diagram is used in estimates of wind chill provided by the UK Meteorological Office. The scale is subjective, and should be interpreted by the user with reference to experience. However, the wind chill class 'freezing cold' is not subjective: it corresponds to conditions under which there is a danger of frostbite to exposed dry flesh.
Further reading: Rees (1993).
(WGR)

Windmill Islands. 66°20′S, 110°28′E. A group of ice-free islands some 27 km (17 miles) long and 10 km (6 miles) wide in Vincennes Bay, Budd Coast of Wilkes Land, East Antarctica. The islands were discovered during photographic reconnaissance flights of Operation Highjump 1946–47, and visited to establish ground control during Operation Windmill 1947–48. In January 1957 they became the site of Wilkes, a joint American and Australian IGY research station. Currently they accommodate its all-Australian successor, Casey Station. Regarded as one of Antarctica's **oases** (the area was for a time known as Grearson Oasis), they have been the subject of intensive biological research studies for many years.

Wisconsin Plateau. 85°48′S, 125°24′W. Extensive ice plateau 2800 m. high in the Horlick Mountains, southern Ellsworth Land, named for the University of Wisconsin.

Wisconsin Range. 85°45′S, 125°00′W. The western range of the Horlick Mountains, south Ellsworth Land, bounded in the south by the polar plateau. The highest peak, Faure Peak, rises to 3841 m (12 598 ft). The range was named for the University of Wisconsin, from which many field researchers have worked in Antarctica.

Wisting, Oscar. (1871–1936). Norwegian naval officer. Born in 1871, he joined Roald Amundsen's **Norwegian South Polar Expedition 1910–12**, and was one of the party that successfully reached the South Pole. After the expedition Wisting remained closely associated with Amundsen, accompanying him as master and joint-leader of the *Maud* expedition 1918–25. He retired from the Norwegian navy with the rank of captain, and died in 1936.
(CH)

Wohlthat Massivet. 71°35′S, 12°20′E. (Wohlthat Massif). Cluster of mountain ranges between Orvinfjella and Hoelfjella, toward the eastern end of a chain extending east-to-west between Fimbulheimen and Hellehallet, central Dronning Maud Land. Several peaks rise above 2200 m (7200 ft). The range was photographed from the air by the German Antarctic (*Schwabenland*) Expedition 1938–39, and named by Norwegian cartographers for a H. C. H. Wohlthat, an official who facilitated the expedition.

Worcester Range. 78°50′S, 161°00′E. A small but spectacular range forming the southwestern wall of Skelton glacier on Hillary Coast, overlooking the Ross Ice Shelf. The range was identified by Shackleton's British Antarctic (*Nimrod*) Expedition 1907–9, and probably named for the merchant navy training ship *Worcester*, in which several polar officers received their first nautical schooling.

Wordie, James Mann. (1889–1962). British polar geologist and administrator. Born in Scotland, he read geology at Glasgow and Cambridge Universities. In 1913 he visited the Yukon and Alaska, and in the following year joined the **Imperial Trans-Antarctic Expedition 1914–17** as geologist and chief of scientific staff, serving with the Weddell Sea party. He served with distinction as an artillery officer in World War I. Returning to a tutorship at St. John's College, Cambridge, he ran a series of summer expeditions to the Arctic, including Spitsbergen, Jan Mayen, east and west Greenland and Baffin Island, in which many students received an introduction to polar fieldwork. Successively senior tutor and master of the college, he became also an influential member of many British polar committees, and advisor to the government on polar matters. He died in Cambridge on 16 January 1962.

Wordie Ice Shelf. 69°15′S, 67°45′W. Ice shelf of the southern Fallières Coast, Marguerite Bay, occupying an inlet between Cape Berteaux and Mt. Edgell. It was identified by the British Graham Land Expedition 1934–37 and named for the British geologist and polar explorer J. M. Wordie. Within recent years the shelf has broken back towards the coast and almost disappeared.

World War II in the southern islands. Though Antarctica itself remained untouched by the war, the Southern Ocean and several of the peripheral islands were visited by ships engaged in hostilities. In September 1939 a German steamer, SS *Erlangen*, having left New Zealand with inadequate bunkering and seeking to avoid capture, slipped south to the Auckland Islands and refuelled with timber from Carnley Harbour. HM Submarine *Olympus* in 1939 inspected Iles Crozet and Prince Edward Islands for signs of enemy use, and in late 1940 HMS *Neptune* made a similar search of Marion and Prince Edward Islands and Iles Kerguelen. Naval ships searched the Auckland Islands and Campbell Island for signs of enemy activity. In 1941 New Zealand established the coast-watching stations of the **New Zealand Cape Expeditions** on Auckland and Campbell Islands.

In January 1941 *Pinguin*, a German armed raider, encountered a Norwegian whaling fleet off Dronning Maud Land, capturing two factory ships, a supply vessel and 11 catchers, which were escorted to France. Between January and March 1941 HMS *Queen of Bermuda*, an armed merchant cruiser of the Royal Navy, patrolled with the British southern Atlantic whaling fleet off the Falkland Islands, South Georgia and the South Orkney and South Shetland Islands. In early March the ship entered Port Foster, Deception Island, and destroyed oil tank lines and other installations at the disused whaling station. From January to March 1941 German raiders used Gazelle Basin, Kerguelen, for rendezvous with their supply ship. In November of that year HMAS *Australia* searched and mined the approaches to the main harbours of Iles Kerguelen, frustrating a German plan to establish a secret meteorological station on the islands. Also in 1941, the German raider *Komet*, which had sailed the Northeast Passage with Russian help, visited the Ross Sea sector of Antarctica in an unsuccessful search for whaling ships.

Worsley, Frank Arthur. (1872–1943). New Zealand seaman and navigator. Born in Akaroa, at age 15 he was apprenticed to the merchant navy. He served in sail and steam, was commissioned in the Royal Naval Reserve, and in 1914 joined Shackleton's Imperial Trans-Antarctic Expedition 1914–17 as master of *Endurance*. After the loss of the ship and the long drift on the pack ice of the Weddell Sea, Worsley took charge of the lifeboat *Dudley Docker*, guiding the party to Elephant Island. Ten days later he skippered the lifeboat *James Caird* on the epic 16-day voyage to South Georgia, a remarkable feat of navigation. On arrival, he crossed the island with Shackleton and Crean to seek help from the whalers. Returning to sea after the expedition, he commanded Q ships and was twice decorated for anti-submarine actions and later while serving on the Russian front. After the war he joined the Shackleton-Rowett Antarctic Expedition 1921–22 as sailing master and hydrographer, continuing the expedition briefly after Shackleton's death. In 1925 he became joint leader of an Arctic expedition to Franz Josef Land, and in 1935 joined a treasure-hunting expedition to the Cocos Islands. He retired from the sea in 1939, but continued as an RNR officer and instructor at the Royal Naval College, Greenwich, until his death in February 1943.
Further reading: Worsley, (1931).

Wright, Charles Seymour. (1887–1975). Canadian polar physicist. Born in Toronto, he read physics at the University of Toronto and undertook research on cosmic rays at the Cavendish Laboratory, Cambridge. He joined the **British Antarctic (*Terra Nova*) Expedition 1910–13**, studying the physics of ice and snow, magnetism, gravity and the aurora. Wright took an active part also in the sledging programme, reaching the top of Beardmore Glacier with the final support party, and in November 1912 leading the search that discovered the bodies of Scott and his companions. In World War I he served with distinction as a Royal Engineers officer in wireless intelligence. In 1919 he joined the Admiralty Research Department: from 1934 to 1936 he directed naval research into radar and protection against magnetic mines. Wright was knighted in 1946. From 1947 he retired several times, alternating with research in Canada, the US, Britain and Antarctica. In 1960 he finally retired to Salt Spring Island, off Vancouver. He died in Victoria, BC, on 1 November 1975.

Wright Valley. 77°31′S, 161°50′E. Extensive ice-free valley in Prince Albert Mountains, Victoria Land. Upper Wright Glacier, which carved it, has retreated to leave a broad U-shaped valley with several lakes and streams. Lower Wright Glacier, merging with Wilson Piedmont Glacier, blocks the seaward end. Both valley and glacier were named for Canadian physicist Charles S. Wright. The valley is one of the **Victoria Land Dry Valleys.**

Y

Yalour Sound. 63°34′S, 56°39′W. Sound between Jonassen and Anderson Islands, Trinity Peninsula, named for Lt. Jorge Yalour, who in 1903 served in the Argentine navy ship *Uruguay* that rescued the Swedish South Polar Expedition.

Yankee Harbour. 62°32′S, 59°47′W. Harbour in the southwest coast of Greenwich Island, South Shetland Islands, used by sealers in the early nineteenth century, probably named as a popular venue for the US sealing fleet.

Yelcho. 64°52′S, 63°35′W. Chilean summer-only station on South Bay, Doumer Island, Danco Coast, built in 1962 and named for a Chilean hydrographic survey ship that worked in the area.

Young Island. 66°25′S, 162°24′E. An ice-covered island 30 km (19 miles) long and 6.5 km (4 miles) wide. Rising steeply to 1340 m (4395 ft), it is the northern island of the Balleny Islands, off the north coast of Victoria Land, East Antarctica. The group was discovered and charted in 1839 by sealing Capt. John Balleny, and named for one of the sponsors of his expedition.

Yukidori Valley. 69°14′S, 39°46′E. An ice-free valley in Langhovde, on the east side of Lützow-Holmbukta, Prins Olav Kyst, between a tongue of the Dronning Maud Land ice cap and the sea. Approximately 3.6 km^2 (1.4 sq miles) of the valley has been designated SSSI No. 22. (Yukidori Valley) to facilitate studies of its meagre but representative fellfield ecosystem and populations of breeding birds.

Z

Zavodovski Island. 56°20′S, 27°35′W. A roughly rectangular island 4.5 km (2.8 miles) across, this is the northernmost island of the South Sandwich Islands chain. Almost entirely ice-free, it rises to a near-central peak, Mt Curry, 551 m (1807 ft) high, topped by an active crater. It was discovered by the Russian Naval Expedition 1819–21 and named for the second-in-command of *Vostok*, the expedition flagship. One of the livelier islands of the chain, it is noted for fumaroles, vapours and frequent mild eruptions.

Zhongshan Station. 69°22′S, 76°22′E. Chinese permanent research station established in 1989 and operating year-round in Larsemann Hills, Prydz Bay, Ingrid Christensen Coast.

Zumberge Coast. 77°45′S, 72°30′W. Part of the east coast of Ellsworth Land, West Antarctica. Bounded by Cape Zumberge in the north and Hercules Inlet in the south, it consists mainly of huge ice-covered peninsulas separated by ice streams that feed from the land into the southwestern corner of Ronne Ice Shelf. Inland from

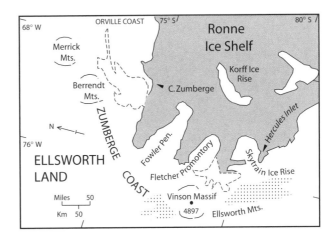

Zumberge Coast

the southern end rise the Ellsworth Mountains. The coast was delineated from aerial photographs, satellite imagery and radio-echosounding data, and named for the distinguished American Antarctic geologist and glaciologist James H. Zumberge.

Appendix A

Agreed Measures for the Conservation of Antarctic Fauna and Flora

The Governments participating in the Third Consultative Meeting under Article IX of the Antarctic Treaty,
Desiring to implement the principles and purposes of the Antarctic Treaty;
Recognizing the scientific importance of the study of Antarctic fauna and flora, their adaptation to their rigorous environment, and their inter-relationship with that environment;
Considering the unique nature of these fauna and flora, their circum-polar range, and particularly their defencelessness and susceptibility to extermination;
Desiring by further international collaboration within the framework of the Antarctic Treaty to promote and achieve the objectives of protection, scientific study, and rational use of these fauna and flora; and
Having particular regard to the conservation principles developed by the Scientific Committee on Antarctic Research (SCAR) of the International Council of Scientific Unions;

Hereby consider the Treaty Area as a Special Conservation Area and have agreed on the following measures:

Article I: [Area of application]

1. These Agreed Measures shall apply to the same area to which the Antarctic Treaty is applicable (hereinafter referred to as the Treaty Area) namely the area south of 60° South Latitude, including all ice shelves.
2. However, nothing in these Agreed Measures shall prejudice or in any way affect the rights, or the exercise of the rights, of any State under international law with regard to the high seas within the Treaty Area, or restrict the implementation of the provisions of the Antarctic Treaty with respect to inspection.
3. The Annexes to these Agreed Measures shall form an integral part thereof, and all references to the Agreed Measures shall be considered to include the Annexes.

Article II: [Definitions]

For the purposes of these Agreed Measures:

a. 'Native mammal' means any member, at any stage of its life cycle, of any species belonging to the Class Mammalia indigenous to the Antarctic or occurring there through natural agencies of dispersal, excepting whales.
b. 'Native bird' means any member, at any stage of its life cycle (including eggs), of any species of the Class Aves indigenous to the Antarctic or occurring there through natural agencies of dispersal.
c. 'Native plant' means any kind of vegetation at any stage of its life cycle (including seeds), indigenous to the Antarctic or occurring there through natural agencies of dispersal.
d. 'Appropriate authority' means any person authorized by a Participating Government to issue permits under these Agreed Measures. The functions of an authorised person will be carried out within the framework of the Antarctic Treaty. They will be carried out exclusively in accordance with scientific principles and will have as their sole purpose the effective protection of Antarctic fauna and flora in accordance with these Agreed Measures.
e. 'Permit' means a formal permission in writing issued by an appropriate authority as defined at paragraph (d) above.
f. 'Participating Government' means any Government for which these Agreed Measures have become effective in accordance with Article XIII of these Agreed Measures.

Article III: [Implementation]

Each Participating Government shall take appropriate action to carry out these Agreed Measures.

Article IV: [Publicity]

The Participating Governments shall prepare and circulate to members of expeditions and stations information to ensure understanding and observance of the provisions of these Agreed Measures, setting forth in particular prohibited activities, and providing lists of specially protected species and specially protected areas.

Article V: [Cases of extreme emergency]

The provisions of these Agreed Measures shall not apply in cases of extreme emergency involving possible loss of human life or involving the safety of ships or aircraft.

Article VI [Protection of native fauna]

1. Each Participating Government shall prohibit within the Treaty Area the killing, wounding, capturing or

molesting of any native mammal or native bird, or any attempt at any such act, except in accordance with a permit.
2. Such permits shall be drawn in terms as specific as possible and issued only for the following purposes:
   a. to provide indispensable food for men or dogs in the Treaty Area in limited quantities, and in conformity with the purposes and principles of these Agreed Measures;
   b. to provide specimens for scientific study or scientific information;
   c. to provide specimens for museums, zoological gardens, or other educational or cultural institutions or uses.
3. Permits for Specially Protected Areas shall be issued only in accordance with the provisions of Article VIII.
4. Participating Governments shall limit the issue of such permits so as to ensure as far as possible that:
   a. no more native mammals or birds are killed or taken in any year than can normally be replaced by natural reproduction in the following breeding season;
   b. the variety of species and the balance of the natural ecological systems existing within the Treaty Area are maintained.
5. The species of native mammals and birds listed in Annex A of these Measures shall be designated 'Specially Protected Species', and shall be accorded special protection by Participating Governments.
6. A Participating Government shall not authorize an appropriate authority to issue a permit with respect to a Specially Protected Species except in accordance with paragraph 7 of this Article.
7. A permit may be issued under this Article with respect to a Specially Protected Species, provided that:
   a. it is issued for a compelling scientific purpose, and
   b. the actions permitted thereunder will not jeopardize the existing natural ecological system or the survival of that species.

Article VII: [Harmful interference]

1. Each Participating Government shall take appropriate measures to minimize harmful interference within the Treaty Area with the normal living conditions of any native mammal or bird, or any attempt at such harmful interference, except as permitted under Article VI.
2. The following acts and activities shall be considered harmful interference:
   a. allowing dogs to run free,
   b. flying helicopters or other aircraft in a manner which would unnecessarily disturb bird and seal concentrations, or landing close to such concentrations (eg within 200 m),
   c. driving vehicles unnecessarily close to concentrations of birds and seals (eg within 200 m),
   d. use of explosives close to concentrations of birds and seals,
   e. discharge of firearms close to bird and seal concentrations (eg within 300 m),
   f. any disturbance of bird and seal colonies during the breeding period by persistent attention from persons on foot.

However, the above activities, with the exception of those mentioned in (a) and (e) may be permitted to the minimum extent necessary for the establishment, supply and operation of stations.

3. Each Participating Government shall take all reasonable steps towards the alleviation of pollution of the waters adjacent to the coast and ice shelves.

Article VIII: [Specially Protected Areas]

1. The areas of outstanding scientific interest listed in Annex B shall be designated 'Specially Protected Areas' and shall be accorded special protection by the Participating Governments in order to preserve their unique natural ecological system.
2. In addition to the prohibitions and measures of protection dealt with in other Articles of these Agreed Measures, the Participating Governments shall in Specially Protected Areas further prohibit:
   a. the collection of any native plant, except in accordance with a permit;
   b. the driving of any vehicle.
   c. entry by their nationals, except in accordance with a permit issued under Article VI or under paragraph 2(a) of the present Article or in accordance with a permit issued for some other compelling scientific purpose;
3. A permit issued under Article IV shall not have effect within a Specially Protected Area except in accordance with paragraph 4 of the present Article.
4. A permit shall have effect within a Specially Protected Area provided that:
   a. it was issued for a compelling scientific purpose which cannot be served elsewhere; and
   b. the actions permitted thereunder will not jeopardize the natural ecological system existing in that Area.

Article IX [Introduction of non-indigenous species, parasites and diseases]

1. Each Participating Government shall prohibit the bringing into the Treaty Area of any species of animal or plant not indigenous to that Area, except in accordance with a permit.
2. Permits under paragraph 1 of this Article shall be drawn in terms as specific as possible and shall be issued to allow the importation only of the animals and plants listed in Annex C. When any such animal or

plant might cause harmful interference with the natural system if left unsupervised within the Treaty Area, such permits shall require that it be kept under controlled conditions and, after it has served its purpose, it shall be removed from the Treaty Area or destroyed.
3. Nothing in paragraphs 1 and 2 of this Article shall apply to the importation of food into the Treaty Area so long as animals and plants used for this purpose are kept under controlled conditions.
4. Each Participating Government undertakes to ensure that all reasonable precautions shall be taken to prevent the accidental introduction of parasites and diseases into the Treaty Area. In particular, the precautions listed in Annex D shall be taken.

Article X: [Activities contrary to the principles and purposes of these Measures]

Each Participating Government undertakes to exert appropriate efforts, consistent with the Charter of the United Nations, to the end that no one engages in any activity in the Treaty Area contrary to the principles or purposes of these Agreed Measures.

Article XI: [Ships' crews]

Each Participating Government whose expeditions use ships sailing under flags of nationalities other than its own shall, as far as feasible, arrange with the owners of such ships that the crews of these ships observe these Agreed Measures.

Article XII: [Exchange of information]

1. The Participating Governments may make such arrangements as may be necessary for the discussion of such matters as:
   a. the collection and exchange of records (including records of permits) and statistics concerning the numbers of each species of native mammal and bird killed or captured annually in the Treaty Area;
   b. the obtaining and exchange of information as to the status of native mammals and birds in the Treaty Area, and the extent to which any species needs protection;
   c. the number of native mammals or birds which should be permitted to be harvested for food, scientific study, or other uses in the various regions;
   d. the establishment of a common form in which this information shall be submitted by Participating Governments in accordance with paragraph 2 of this Article.
2. Each Participating Government shall inform the other Governments in writing before the end of November each year of the steps taken and information collected in the preceding period of 1st July to 30th June relating to the implementation of these Agreed Measures. Governments exchanging information under paragraph 5 of Article VII of the Antarctic Treaty may at the same time transmit the information relating to the implementation of these Agreed Measures.

Article XIII: [Formal provisions]

1. After the receipt by the Government designated in Recommendation I-XIV(5) of notification of approval by all Governments whose representatives are entitled to participate in meetings provided for under Article IX of the Antarctic Treaty, these Agreed Measures shall become effective for those Governments.
2. Thereafter any other Contracting Party to the Antarctic Treaty may, in consonance with the purposes of Recommendation III-VII, accept these Agreed Measures by notifying the designated Government of its intention to apply the Agreed Measures and to be bound by them. The Agreed Measures shall become effective with regard to such Governments on the date of receipt of such notification.
3. The designated Government shall inform the Governments referred to in paragraph 1 of this Article of each notification of approval, the effective date of these Agreed Measures and of each notification of acceptance. The designated Government shall also inform any Government which has accepted these Agreed Measures of each subsequent notification of acceptance.

Article XIV: [Amendment]

1. These Agreed Measures may be amended at any time by unanimous agreement of the Governments whose Representatives are entitled to participate in meetings under Article IX of the Antarctic Treaty.
2. The Annexes, in particular, may be amended as necessary through diplomatic channels.
3. An amendment proposed through diplomatic channels shall be submitted in writing to the designated Government which shall communicate it to the Governments referred to in paragraph 1 of the present Article for approval; at the same time, it shall be communicated to the other Participating Governments.
4. Any amendment shall become effective on the date on which notifications of approval have been received by the designated Government and from all of the Governments referred to in paragraph 1 of this Article.
5. The designated Government shall notify those same Governments of the date of receipt of each approval communicated to it and the date on which the amendment will become effective for them.
6. Such amendment shall become effective on that same date for all other Participating Governments, except those which before the expiry of two months after that date notify the designated Government that they do not accept it.

ANNEXES TO THESE AGREED MEASURES

ANNEX A: Specially Protected Species
All species of the genus *Arctocephalus*, Fur Seals.
Ommatophoca rossii, Ross Seal.

ANNEX B: Specially Protected Areas
[Lists to be promulgated and updated]

ANNEX C: Importation of animals and plants
The following animals and plants may be imported into the Treaty Area in accordance with permits issued under Article IX(2) of these Agreed Measures:

a. sledge dogs
b. domestic animals and plants
c. laboratory animals and plants including viruses, bacteria, yeasts and fungi.

ANNEX D: Precautions to prevent accidental introduction of parasites and diseases into the Treaty Area

The following precautions shall be taken:

1. *Dogs*. All dogs imported into the Treaty Area shall be inoculated against the following diseases:
   a. distemper
   b. contagious canine hepatitis
   c. rabies
   d. leptospirosis (*L. canicola* and *L. icterohaemorragicae*)

   Each dog shall be inoculated at least two months before the time of its arrival in the Treaty Area

2. *Poultry*. Notwithstanding the provisions of Article IX(3) of these Agreed Measures, no living poultry shall be brought into the Treaty Area after 1st July 1966.

Appendix B

Convention for the Conservation of Antarctic Seals

The Contracting Parties,

Recalling the Agreed Measures for the Conservation of Antarctic Fauna and Flora, adopted under the Antarctic Treaty signed at Washington on 1 December 1959;

Recognizing the general concern about the vulnerability of Antarctic seals to commercial exploitation and the consequent need for effective conservation measures;

Recognizing that the stocks of Antarctic seals are an important living resource in the marine environment which requires an international agreement for its effective conservation;

Recognizing that this resource should not be depleted by over-exploitation, and hence that any harvesting should be regulated so as not to exceed the levels of the optimum sustainable yield;

Recognizing that in order to improve scientific knowledge and so place exploitation on a rational basis, every effort should be made both to encourage biological and other research on Antarctic seal populations and to gain information from such research and from the statistics of future sealing operations, so that further suitable regulations may be formulated;

Noting that the Scientific Committee on Antarctic Research of the International Council of Scientific Unions (SCAR) is willing to carry out the tasks requested of it in this Convention;

Desiring to promote and achieve the objectives of protection, scientific study and rational use of Antarctic seals, and to maintain a satisfactory balance within the ecological system,

Have agreed as follows:

Article 1: Scope

1. This Convention applies to the seas south of 60° South Latitude, in respect of which the Contracting Parties affirm the provisions of Article IV of the Antarctic Treaty.
2. This Convention may be applicable to any or all of the following species:
   Southern elephant seal *Mirounga leonina*,
   Leopard seal *Hydrurga leptonyx*,
   Weddell seal *Leptonychotes weddelli*,
   Crabeater seal *Lobodon carcinophagus*,
   Ross seal *Ommatophoca rossi*,
   Southern fur seals *Arctocephalus* spp.
3. The Annex to this Convention forms an integral part thereof.

Article 2: Implementation

1. The Contracting Parties agree that the species of seals enumerated in Article 1 shall not be killed or captured within the Convention area by their nationals or vessels under their respective flags except in accordance with the provisions of this Convention.
2. Each Contracting Party shall adopt for its nationals and for vessels under its flag such laws, regulations and other measures, including a permit system as appropriate, as may be necessary to implement this Convention.

Article 3: Annexed Measures

1. This Convention includes an Annex specifying measures which the Contracting Parties hereby adopt. Contracting Parties may from time to time in the future adopt other measures with respect to the conservation, scientific study and rational and humane use of seal resources, prescribing inter alia:
   a. permissible catch;
   b. protected and unprotected species;
   c. open and closed seasons;
   d. open and closed areas, including the designation of reserves;
   e. the designation of special areas where there shall be no disturbance of seals;
   f. limits relating to sex, size, or age for each species;
   g. restrictions relating to time of day and duration, limitations of effort and methods of sealing;
   h. types and specifications of gear and apparatus and appliances which may be used;
   i. catch returns and other statistical and biological records;
   j. procedures for facilitating the review and assessment of scientific information;
   k. other regulatory measures including an effective system of inspection.
2. The measures adopted under paragraph (1) of this Article shall be based upon the best scientific and technical evidence available.
3. The Annex may from time to time be amended in accordance with the procedures provided for in Article 9.

Article 4: Special Permits

1. Notwithstanding the provisions of this Convention, any Contracting Party may issue permits to kill or capture seals in limited quantities and in conformity with the objectives and principles of this Convention for the following purposes:
   a. to provide indispensable food for men or dogs;
   b. to provide for scientific research; or
   c. to provide specimens for museums, educational or cultural institutions.
2. Each Contracting Party shall, as soon as possible, inform the other Contracting Parties and SCAR of the purpose and content of all permits issued under paragraph (1) of this Article and subsequently of the numbers of seals killed or captured under these permits.

Article 5: Exchange of Information and Scientific Advice

1. Each Contracting Party shall provide to the other Contracting Parties and to SCAR the information specified in the Annex within the period indicated therein.
2. Each Contracting Party shall also provide to the other Contracting Parties and to SCAR before 31 October each year information on any steps it has taken in accordance with Article 2 of this Convention during the preceding period 1 July to 30 June.
3. Contracting Parties which have no information to report under the two preceding paragraphs shall indicate this formally before 31 October each year.
4. SCAR is invited:
   a. to assess information received pursuant to this Article; encourage exchange of scientific data and information among the Contracting Parties; recommend programmes for scientific research; recommend statistical and biological data to be collected by sealing expeditions within the Convention area; and suggest amendments to the Annex; and
   b. to report on the basis of the statistical, biological and other evidence available when the harvest of any species of seal in the Convention area is having a significantly harmful effect on the total stocks of such species or on the ecological system in any particular locality.
5. SCAR is invited to notify the Depositary which shall report to the Contracting Parties when SCAR estimates in any sealing season that the permissible catch limits for any species are likely to be exceeded and, in that case, to provide an estimate of the date upon which the permissible catch limits will be reached. Each Contracting Party shall then take appropriate measures to prevent its nationals and vessels under its flag from killing or capturing seals of that species after the estimated date until the Contracting Parties decide otherwise.
6. SCAR may if necessary seek the technical assistance of the Food and Agriculture Organization of the United Nations in making its assessments.
7. Notwithstanding the provisions of paragraph (1) of Article 1 the Contracting Parties shall, in accordance with their internal law, report to each other and to SCAR, for consideration, statistics relating to the Antarctic seals listed in paragraph (2) of Article 1 which have been killed or captured by their nationals and vessels under their respective flags in the area of floating sea ice north of 60° South Latitude.

Article 6: Consultations between Contracting Parties

1. At any time after commercial sealing has begun a Contracting Party may propose through the Depositary that a meeting of Contracting Parties be convened with a view to:
   a. establishing by a two-thirds majority of the Contracting Parties, including the concurring votes of all States signatory to this Convention present at the meeting, an effective system of control, including inspection, over the implementation of the provisions of this Convention;
   b. establishing a commission to perform such functions under this Convention as the Contracting Parties may deem necessary; or
   c. considering other proposals, including:
      i. the provision of independent scientific advice;
      ii. the establishment, by a two-thirds majority, of a scientific advisory committee which may be assigned some or all of the functions requested of SCAR under this Convention, if commercial sealing reaches significant proportions;
      iii. the carrying out of scientific programmes with the participation of the Contracting Parties; and
      iv. the provision of further regulatory measures, including moratoria.
2. If one-third of the Contracting Parties indicate agreement the Depositary shall convene such a meeting, as soon as possible.
3. A meeting shall be held at the request of any Contracting Party, if SCAR reports that the harvest of any species of Antarctic seal in the area to which this Convention applies is having a significantly harmful effect on the total stocks or the ecological system in any particular locality.

Article 7: Review of Operations

The Contracting Parties shall meet within five years after the entry into force of this Convention and at least every five years thereafter to review the operation of the Convention.

Article 8: Amendments to the Convention

1. This Convention may be amended at any time. The text of any amendment proposed by a Contracting Party shall be submitted to the Depositary, which shall transmit it to all the Contracting Parties.
2. If one-third of the Contracting Parties request a meeting to discuss the proposed amendment the Depositary shall call such a meeting.
3. An amendment shall enter into force when the Depositary has received instruments of ratification or acceptance thereof from all the Contracting Parties.

Article 9: Amendments to the Annex

1. Any Contracting Party may propose amendments to the Annex to this Convention. The text of any such proposed amendment shall be submitted to the Depositary which shall transmit it to all Contracting Parties.
2. Each such proposed amendment shall become effective for all Contracting Parties six months after the date appearing on the notification from the Depositary to the Contracting Parties, if within 120 days of the notification date, no objection has been received and two-thirds of the Contracting Parties have notified the Depositary in writing of their approval.
3. If an objection is received from any Contracting Party within 120 days of the notification date, the matter shall be considered by the Contracting Parties at their next meeting. If unanimity on the matter is not reached at the meeting, the Contracting Parties shall notify the Depositary within 120 days from the date of closure of the meeting of their approval or rejection of the original amendment or of any new amendment proposed by the meeting. If, by the end of this period, two-thirds of the Contracting Parties have approved such amendment, it shall become effective six months from the date of the closure of the meeting for those Contracting Parties which have by then notified their approval.
4. Any Contracting Party which has objected to a proposed amendment may at any time withdraw that objection, and the proposed amendment shall become effective with respect to such a Party immediately if the amendment is already in effect, or at such time as it becomes effective under the terms of this Article.
5. The Depositary shall notify each Contracting Party immediately upon receipt of each approval or objection, of each withdrawal of objection, and of the entry into force of any amendment.
6. Any State which becomes a Party to this Convention after an amendment to the Annex has entered into force shall be bound by the Annex as so amended. Any State which becomes a Party to this Convention during the period when a proposed amendment is pending may approve or object to such an amendment within the time limits applicable to other Contracting Parties.

Article 10: Signature

This Convention shall be open for signature at London from 1 June to 31 December 1972 by States participating in the Conference on the Conservation of Antarctic Seals held at London from 3 to 11 February 1972.

Article 11: Ratification

This Convention is subject to ratification or acceptance. Instruments of ratification or acceptance shall be deposited with the Government of the United Kingdom of Great Britain and Northern Ireland, hereby designated as the Depositary.

Article 12: Accession

This Convention shall be open for accession by any State which may be invited to accede to this Convention with the consent of the Contracting Parties.

Article 13: Entry into Force

1. This Convention shall enter into force on the thirtieth day following the date of deposit of the seventh instrument of ratification or acceptance.
2. Thereafter this Convention shall enter into force for each ratifying, accepting or acceding State on the thirtieth day after deposit by such a State of its instrument of ratification, acceptance or accession.

Article 14: Withdrawal

Any Contracting Party may withdraw from this Convention on 30 June of any year by giving notice on or before 1 January of the same year to the Depositary, which upon receipt of such a notice shall at once communicate it to the other Contracting Parties. Any other Contracting Party may, in like manner, within one month of the receipt of a copy of such a notice from the Depositary, give notice of withdrawal, so that the Convention shall cease to be in force on 30 June of the same year with respect to the Contracting Party giving such notice.

Article 15: Notifications by the Depositary

The Depositary shall notify all signatory and acceding States of the following:

a. signatures of this Convention, the deposit of instruments of ratification, acceptance or accession and notices of withdrawal:
b. the date of entry into force of this Convention and of any amendments to it or its Annex.

Article 16: Certified Copies and Registration

1. This Convention, done in the English, French, Russian and Spanish languages, each version being equally authentic, shall be deposited in the archives of the Government of the United Kingdom of Great Britain and

Northern Ireland, which shall transmit duly certified copies thereof to all signatory and acceding States.
2. This Convention shall be registered by the Depositary pursuant to Article 102 of the Charter of the United Nations. *In witness whereof*, the undersigned, duly authorized, have signed this Convention.
Done at London, this 1st day of June 1972.

Annex

1. Permissible Catch

The Contracting Parties shall in any one year, which shall run from 1 July to 30 June inclusive, restrict the total number of seals of each species killed or captured to the numbers specified below. These numbers are subject to review in the light of scientific assessments:

a. in the case of Crabeater seals *Lobodon carcinophagus*, 175,000;
b. in the case of Leopard seals *Hydrurga leptonyx*, 12,000;
c. in the case of Weddell seals *Leptonychotes weddelli*, 5,000.

2. Protected Species

a. It is forbidden to kill or capture Ross seals *Ommatophoca rossi*, Southern elephant seals *Mirounga leonina*, or fur seals of the genus *Arctocephalus*.
b. In order to protect the adult breeding stock during the period when it is most concentrated and vulnerable, it is forbidden to kill or capture any Weddell seal *Leptonychotes weddelli* one year old or older between 1 September and 31 January inclusive.

3. Closed Season and Sealing Season

The period between 1 March and 31 August inclusive is a Closed Season, during which the killing or capturing of seals is forbidden. The period 1 September to the last day in February constitutes a Sealing Season.

4. Sealing Zones

Each of the sealing zones listed in this paragraph shall be closed in numerical sequence to all sealing operations for the seal species listed in paragraph 1 of this Annex for the period 1 September to the last day of February inclusive. Such closures shall begin with the same zone as is closed under paragraph 2 of Annex B to Annex 1 of the Report of the Fifth Antarctic Treaty Consultative Meeting at the moment the Convention enters into force. Upon the expiration of each closed period, the affected zone shall reopen:

Zone 1 – between 60° and 120° West Longitude
Zone 2 – between 0° and 60° West Longitude, together with that part of the Weddell Sea lying westward of 60° West Longitude
Zone 3 – between 0° and 70° East Longitude
Zone 4 – between 70° and 130° East Longitude
Zone 5 – between 130° East Longitude and 170° West Longitude
Zone 6 – between 120° and 170° West Longitude.

5. Seal Reserves

It is forbidden to kill or capture seals in the following reserves, which are seal breeding areas or the sites of long-term scientific research:

a. The area around the South Orkney Islands between 60°20′ and 60°56′ South Latitude and 44°05′ and 46°25′ West Longitude.
b. The area of the southwestern Ross Sea south of 76° South Latitude and west of 170° East Longitude.
c. The area of Edisto Inlet south and west of a line drawn between Cape Hallett at 72°19′ South Latitude, 170°18′ East Longitude, and Helm Point, at 72°11′ South Latitude, 170°00′ East Longitude.

6. Exchange of Information

a. Contracting Parties shall provide before 31 October each year to other Contracting Parties and to SCAR a summary of statistical information on all seals killed or captured by their nationals and vessels under their respective flags in the Convention area, in respect of the preceding period 1 July to 30 June. This information shall include by zones and months:
   i. The gross and net tonnage, brake horse-power, number of crew, and number of days' operation of vessels under the flag of the Contracting Party;
   ii. The number of adult individuals and pups of each species taken.

When specially requested, this information shall be provided in respect of each ship, together with its daily position at noon each operating day and the catch on that day.

b. When an industry has started, reports of the number of seals of each species killed or captured in each zone shall be made to SCAR in the form and at the intervals (not shorter than one week) requested by that body.
c. Contracting Parties shall provide to SCAR biological information, in particular:
   i. Sex
   ii. Reproductive condition
   iii. Age

SCAR may request additional information or material with the approval of the Contracting Parties.

d. Contracting Parties shall provide to other Contracting Parties and to SCAR at least 30 days in advance of departure from their home ports, information on proposed sealing expeditions.

7. Sealing Methods

a. SCAR is invited to report on methods of sealing and to make recommendations with a view to ensuring that the killing or capturing of seals is quick, painless and efficient. Contracting Parties, as appropriate, shall adopt rules for their nationals and vessels under their respective flags engaged in the killing and capturing of seals, giving due consideration to the views of SCAR.

b. In the light of the available scientific and technical data, Contracting Parties agree to take appropriate steps to ensure that their nationals and vessels under their respective flags refrain from killing or capturing seals in the water, except in limited quantities to provide for scientific research in conformity with the objectives and principles of this Convention. Such research shall include studies as to the effectiveness of methods of sealing from the viewpoint of the management and humane and rational utilization of the Antarctic seal resources for conservation purposes. The undertaking and the results of any such scientific programme shall be communicated to SCAR and the Depositary which shall transmit them to the Contracting Parties.

Appendix C

Convention on the Conservation of Antarctic Marine Living Resources

The Contracting Parties,

Recognizing the importance of safeguarding the environment and protecting the integrity of the ecosystem of the seas surrounding Antarctica;

Noting the concentration of marine living resources found in Antarctic waters and the increased interest in the possibilities offered by the utilization of these resources as a source of protein;

Conscious of the urgency of ensuring the conservation of Antarctic marine living resources;

Considering that it is essential to increase knowledge of the Antarctic marine ecosystem and its components so as to be able to base decisions on harvesting on sound scientific information;

Believing that the conservation of Antarctic marine living resources calls for international co-operation with due regard for the provisions of the Antarctic Treaty and with the active involvement of all States engaged in research or harvesting activities in Antarctic waters;

Recognizing the prime responsibilities of the Antarctic Treaty Consultative Parties for the protection and preservation of the Antarctic environment and, in particular, their responsibilities under Article IX, paragraph 1(f) of the Antarctic Treaty in respect of the preservation and conservation of living resources in Antarctica;

Recalling the action already taken by the Antarctic Treaty Consultative Parties including in particular the Agreed Measures for the Conservation of Antarctic Fauna and Flora, as well as the provisions of the Convention for the Conservation of Antarctic Seals;

Bearing in mind the concern regarding the conservation of Antarctic marine living resources expressed by the Consultative Parties at the Ninth Consultative Meeting of the Antarctic Treaty and the importance of the provisions of Recommendation IX-2 which led to the establishment of the present Convention;

Believing that it is in the interest of all mankind to preserve the waters surrounding the Antarctic continent for peaceful purposes only and to prevent their becoming the scene or object of international discord;

Recognizing in the light of the foregoing, that it is desirable to establish suitable machinery for recommending, promoting, deciding upon and co-ordinating the measures and scientific studies needed to ensure the conservation of Antarctic marine living organisms;

Have agreed as follows:

Article I

1. This Convention applies to the Antarctic marine living resources of the area south of 60° South latitude and to the Antarctic marine living resources of the area between that latitude and the Antarctic Convergence which form part of the Antarctic marine ecosystem.
2. Antarctic marine living resources means the populations of fin fish, molluscs, crustaceans and all other species of living organisms, including birds, found south of the Antarctic Convergence.
3. The Antarctic marine ecosystem means the complex of relationships of Antarctic marine living resources with each other and with their physical environment.
4. The Antarctic Convergence shall be deemed to be a line joining the following points along parallels of latitude and meridians of longitude:
50°S, 0°; 50°S, 30°E; 45°S, 30°E; 45°S, 80°E; 55°S, 80°E; 55°S, 150°E; 60°S, 150°E; 60°S, 50°W; 50°S, 50°W; 50°S, 0°.

Article II

1. The objective of this Convention is the conservation of Antarctic marine living resources.
2. For the purpose of this Convention, the term 'conservation' includes rational use.
3. Any harvesting and associated activities in the area to which this Convention applies shall be conducted in accordance with the provisions of this Convention and with the following principles of conservation:
   a. prevention of decrease in the size of any harvested population to levels below those which ensure its stable recruitment. For this purpose its size should not be allowed to fall below a level close to that which ensures the greatest net annual increment;
   b. maintenance of the ecological relationships between harvested, dependent and related populations of Antarctic marine living resources and the restoration of depleted populations to the levels defined in sub-paragraph (a) above; and

c. prevention of changes or minimization of the risk of changes in the marine ecosystem which are not potentially reversible over two or three decades, taking into account the state of available knowledge of the direct and indirect impact of harvesting, the effect of the introduction of alien species, the effects of associated activities on the marine ecosystem and of the effects of environmental changes, with the aim of making possible the sustained conservation of Antarctic marine living resources.

Article III

The Contracting Parties, whether or not they are Parties to the Antarctic Treaty, agree that they will not engage in any activities in the Antarctic Treaty area contrary to the principles and purposes of that Treaty and that, in their relations with each other, they are bound by the obligations contained in Articles I and V of the Antarctic Treaty.

Article IV

1. With respect to the Antarctic Treaty area, all Contracting Parties, whether or not they are Parties to the Antarctic Treaty, are bound by Articles IV and VI of the Antarctic Treaty in their relations with each other.
2. Nothing in this Convention and no acts or activities taking place while the present Convention is in force shall:
   a. constitute a basis for asserting, supporting or denying a claim to territorial sovereignty in the Antarctic Treaty area or create any rights of sovereignty in the Antarctic Treaty area;
   b. be interpreted as a renunciation or diminution by any Contracting Party of, or as prejudicing, any right or claim or basis of claim to exercise coastal state jurisdiction under international law within the area to which this Convention applies;
   c. be interpreted as prejudicing the position of any Contracting Party as regards its recognition or non-recognition of any such right, claim or basis of claim;
   d. affect the provision of Article IV, paragraph 2, of the Antarctic Treaty that no new claim, or enlargement of an existing claim, to territorial sovereignty in Antarctica shall be asserted while the Antarctic Treaty is in force.

Article V

1. The Contracting Parties which are not Parties to the Antarctic Treaty acknowledge the special obligations and responsibilities of the Antarctic Treaty Consultative Parties for the protection and preservation of the environment of the Antarctic Treaty area.
2. The Contracting Parties which are not Parties to the Antarctic Treaty agree that, in their activities in the Antarctic Treaty area, they will observe as and when appropriate the Agreed Measures for the Conservation of Antarctic Fauna and Flora and such other measures as have been recommended by the Antarctic Treaty Consultative Parties in fulfilment of their responsibility for the protection of the Antarctic environment from all forms of harmful human interference.
3. For the purposes of this Convention, 'Antarctic Treaty Consultative Parties' means the Contracting Parties to the Antarctic Treaty whose Representatives participate in meetings under Article IX of the Antarctic Treaty.

Article VI

Nothing in this Convention shall derogate from the rights and obligations of Contracting Parties under the International Convention for the Regulation of Whaling and the Convention for the Conservation of Antarctic Seals.

Article VII

1. The Contracting Parties hereby establish and agree to maintain the Commission for the Conservation of Antarctic Marine Living Resources (hereinafter referred to as 'the Commission').
2. Membership in the Commission shall be as follows:
   a. each Contracting Party which participated in the meeting at which this Convention was adopted shall be a Member of the Commission;
   b. each State Party which has acceded to this Convention pursuant to Article XXIX shall be entitled to be a Member of the Commission during such time as that acceding party is engaged in research or harvesting activities in relation to the marine living resources to which this Convention applies;
   c. each regional economic integration organization which has acceded to this Convention pursuant to Article XXIX shall be entitled to be a Member of the Commission during such time as its States members are so entitled;
   d. A Contracting Party seeking to participate in the work of the Commission pursuant to sub-paragraphs (b) and (c) above shall notify the Depositary of the basis upon which it seeks to become a Member of the Commission and of its willingness to accept conservation measures in force. The Depositary shall communicate to each member of the Commission such notification and accompanying information. Within two months of receipt of such communication from the Depositary, any Member of the Commission may request that a special meeting of the Commission be held to consider the matter. Upon receipt of such

request, the Depositary shall call such a meeting. If there is no request for a meeting, the Contracting Party submitting the notification shall be deemed to have satisfied the requirements for Commission Membership.
3. Each Member of the Commission shall be represented by one representative who may be accompanied by alternate representatives and advisers.

Article VIII

The Commission shall have legal personality and shall enjoy in the territory of each of the States Parties such legal capacity as may be necessary to perform its function and achieve the purposes of this Convention. The privileges and immunities to be enjoyed by the Commission and its staff in the territory of a State Party shall be determined by agreement between the Commission and the State Party concerned.

Article IX

1. The function of the Commission shall be to give effect to the objective and principles set out in Article II of this Convention. To this end, it shall:
   a. facilitate research into and comprehensive studies of Antarctic marine living resources and of the Antarctic marine ecosystem;
   b. compile data on the status of and changes in population of Antarctic marine living resources and on factors affecting the distribution, abundance and productivity of harvested species and dependent or related species or populations;
   c. ensure the acquisition of catch and effort statistics on harvested populations;
   d. analyse, disseminate and publish the information referred to in sub-paragraphs (b) and (c) above and the reports of the Scientific Committee;
   e. identify conservation needs and analyse the effectiveness of conservation measures;
   f. formulate, adopt and revise conservation measures on the basis of the best scientific evidence available, subject to the provisions of paragraph 5 of this Article;
   g. implement the system of observation and inspection established under Article XXIV of this Convention;
   h. carry out such other activities as are necessary to fulfil the objective of this Convention.
2. The conservation measures referred to in paragraph 1(f) above include the following:
   a. the designation of the quantity of any species which may be harvested in the area to which this Convention applies;
   b. the designation of regions and sub-regions based on the distribution of populations of Antarctic marine living resources;
   c. the designation of the quantity which may be harvested from the populations of regions and sub-regions;
   d. the designation of protected species;
   e. the designation of the size, age and, as appropriate, sex of species which may be harvested;
   f. the designation of open and closed seasons for harvesting;
   g. the designation of the opening and closing of areas, regions or sub-regions for purposes of scientific study or conservation, including special areas for protection and scientific study;
   h. regulation of the effort employed and methods of harvesting, including fishing gear, with a view, inter alia, to avoiding undue concentration of harvesting in any region or sub-region;
   i. the taking of such other conservation measures as the Commission considers necessary for the fulfillment of the objective of this Convention, including measures concerning the effects of harvesting and associated activities on components of the marine ecosystem other than the harvested populations.
3. The Commission shall publish and maintain a record of all conservation measures in force.
4. In exercising its functions under paragraph 1 above, the Commission shall take full account of the recommendations and advice of the Scientific Committee.
5. The Commission shall take full account of any relevant measures or regulations established or recommended by the Consultative Meetings pursuant to Article IX of the Antarctic Treaty or by existing fisheries commissions responsible for species which may enter the area to which this Convention applies, in order that there shall be no inconsistency between the rights and obligations of a Contracting Party under such regulations or measures and conservation measures which may be adopted by the Commission.
6. Conservation measures adopted by the Commission in accordance with this Convention shall be implemented by Members of the Commission in the following manner;
   a. the Commission shall notify conservation measures to all Members of the Commission;
   b. conservation measures shall become binding upon all Members of the Commission 180 days after such notification, except as provided in sub-paragraphs (c) and (d) below;
   c. if a Member of the Commission, within ninety days following the notification specified in sub-paragraph (a), notifies the Commission that it is unable to accept the conservation measure, in whole or in part, the measure shall not, to the extent stated, be binding upon that Member of the Commission;

d. in the event that any Member of the Commission invokes the procedure set forth in sub-paragraph (c) above, the Commission shall meet at the request of any Member of the Commission to review the conservation measure. At the time of such meeting and within thirty days following the meeting, any Member of the Commission shall have the right to declare that it is no longer able to accept the conservation measure, in which case the Member shall no longer be bound by such measure.

Article X

1. The Commission shall draw the attention of any State which is not a Party to this Convention to any activity undertaken by its nationals or vessels which, in the opinion of the Commission, affects the implementation of the objective of this Convention.
2. The Commission shall draw the attention of all Contracting Parties to any activity which, in the opinion of the Commission, affects the implementation by a Contracting Party of the objective of this Convention or the compliance by that Contracting Party with its obligations under this Convention.

Article XI

The Commission shall seek to co-operate with Contracting Parties which may exercise jurisdiction in marine areas adjacent to the area to which this Convention applies in respect of the conservation of any stock or stocks of associated species which occur both within those areas and the area to which this Convention applies, with a view to harmonising the conservation measures adopted in respect of such stocks.

Article XII

1. Decisions of the Commission on matters of substance shall be taken by consensus. The question of whether a matter is one of substance shall be treated as a matter of substance.
2. Decisions on matters other than those referred to in paragraph 1 above shall be taken by a simple majority of the Members of the Commission present and voting.
3. In Commission consideration of any item requiring a decision, it shall be made clear whether a regional economic integration organization will participate in the taking of the decision and, if so, whether any of its member States will also participate. The number of Contracting Parties so participating shall not exceed the number of member States of the regional economic integration organization which are Members of the Commission.
4. In the taking of decisions pursuant to this Article, a regional economic integration organization shall have only one vote.

Article XIII

1. The Headquarters of the Commission shall be established at Hobart, Tasmania, Australia.
2. The Commission shall hold a regular annual meeting. Other meetings shall also be held at the request of one-third of its members and as otherwise provided in this Convention. The first meeting of the Commission shall be held within three months of the entry into force of this Convention, provided that among the Contracting Parties there are at least two States conducting harvesting activities within the area to which this Convention applies. The first meeting shall, in any event, be held within one year of the entry into force of this Convention. The Depositary shall consult with the signatory States regarding the first Commission meeting, taking into account that a broad representation of such States is necessary for the effective operation of the Commission.
3. The Depositary shall convene the first meeting of the Commission at the headquarters of the Commission. Thereafter, meetings of the Commission shall be held at its headquarters, unless it decides otherwise.
4. The Commission shall elect from among its members a Chairman and Vice-Chairman, each of whom shall serve for a term of two years and shall be eligible for re-election for one additional term. The first Chairman shall, however, be elected for an initial term of three years. The Chairman and Vice-Chairman shall not be representatives of the same Contracting Party.
5. The Commission shall adopt and amend as necessary the rules of procedure for the conduct of its meetings, except with respect to the matters dealt with in Article XII of this Convention.
6. The Commission may establish such subsidiary bodies as are necessary for the performance of its functions.

Article XIV

1. The Contracting Parties hereby establish the Scientific Committee for the Conservation of Antarctic Marine Living Resources (hereinafter referred to as 'the Scientific Committee') which shall be a consultative body to the Commission. The Scientific Committee shall normally meet at the headquarters of the Commission unless the Scientific Committee decides otherwise.
2. Each Member of the Commission shall be a member of the Scientific Committee and shall appoint a representative with suitable scientific qualifications who may be accompanied by other experts and advisers.
3. The Scientific Committee may seek the advice of other scientists and experts as may be required on an ad hoc basis.

Article XV

1. The Scientific Committee shall provide a forum for consultation and co-operation concerning the collection, study and exchange of information with respect

to the marine living resources to which this Convention applies. It shall encourage and promote co-operation in the field of scientific research in order to extend knowledge of the marine living resources of the Antarctic marine ecosystem.

2. The Scientific Committee shall conduct such activities as the Commission may direct in pursuance of the objective of this Convention and shall:
   a. establish criteria and methods to be used for determinations concerning the conservation measures referred to in Article IX of this Convention;
   b. regularly assess the status and trends of the populations of Antarctic marine living resources;
   c. analyse data concerning the direct and indirect effects of harvesting on the populations of Antarctic marine living resources;
   d. assess the effects of proposed changes in the methods or levels of harvesting and proposed conservation measures;
   e. transmit assessments, analyses, reports and recommendations to the Commission as requested or on its own initiative regarding measures and research to implement the objective of this Convention;
   f. formulate proposals for the conduct of international and national programs of research into Antarctic marine living resources.
3. In carrying out its functions, the Scientific Committee shall have regard to the work of other relevant technical and scientific organizations and to the scientific activities conducted within the framework of the Antarctic Treaty.

Article XVI

1. The first meeting of the Scientific Committee shall be held within three months of the first meeting of the Commission. The Scientific Committee shall meet thereafter as often as may be necessary to fulfil its functions.
2. The Scientific Committee shall adopt and amend as necessary its rules of procedure. The rules and any amendments thereto shall be approved by the Commission. The rules shall include procedures for the presentation of minority reports.
3. The Scientific Committee may establish, with the approval of the Commission, such subsidiary bodies as are necessary for the performance of its functions.

Article XVII

1. The Commission shall appoint an Executive Secretary to serve the Commission and Scientific Committee according to such procedures and on such terms and conditions as the Commission may determine. His term of office shall be for four years and he shall be eligible for re-appointment.
2. The Commission shall authorize such staff establishment for the Secretariat as may be necessary and the Executive Secretary shall appoint, direct and supervise such staff according to such rules and procedures and on such terms and conditions as the Commission may determine.
3. The Executive Secretary and Secretariat shall perform the functions entrusted to them by the Commission.

Article XVIII

The official languages of the Commission and of the Scientific Committee shall be English, French, Russian and Spanish.

Article XIX

1. At each annual meeting, the Commission shall adopt by consensus its budget and the budget of the Scientific Committee.
2. A draft budget for the Commission and the Scientific Committee and any subsidiary bodies shall be prepared by the Executive Secretary and submitted to the Members of the Commission at least sixty days before the annual meeting of the Commission.
3. Each Member of the Commission shall contribute to the budget. Until the expiration of five years after the entry into force of this Convention, the contribution of each Member of the Commission shall be equal. Thereafter the contribution shall be determined in accordance with two criteria: the amount harvested and an equal sharing among all Members of the Commission. The Commission shall determine by consensus the proportion in which these two criteria shall apply.
4. The financial activities of the Commission and Scientific Committee shall be conducted in accordance with financial regulations adopted by the Commission and shall be subject to an annual audit by external auditors selected by the Commission.
5. Each Member of the Commission shall meet its own expenses arising from attendance at meetings of the Commission and of the Scientific Committee.
6. A Member of the Commission that fails to pay its contributions for two consecutive years shall not, during the period of its default, have the right to participate in the taking of decisions in the Commission.

Article XX

1. The Members of the Commission shall, to the greatest extent possible, provide annually to the Commission and to the Scientific Committee such statistical, biological and other data and information as the Commission and Scientific Committee may require in the exercise of their functions.
2. The Members of the Commission shall provide, in the manner and at such intervals as may be prescribed,

information about their harvesting activities, including fishing areas and vessels, so as to enable reliable catch and effort statistics to be compiled.
3. The Members of the Commission shall provide to the Commission at such intervals as may be prescribed information on steps taken to implement the conservation measures adopted by the Commission.
4. The Members of the Commission agree that in any of their harvesting activities, advantage shall be taken of opportunities to collect data needed to assess the impact of harvesting.

Article XXI

1. Each Contracting Party shall take appropriate measures within its competence to ensure compliance with the provisions of this Convention and with conservation measures adopted by the Commission to which the Party is bound in accordance with Article IX of this Convention.
2. Each Contracting Party shall transmit to the Commission information on measures taken pursuant to paragraph 1 above, including the imposition of sanctions for any violation.

Article XXII

1. Each Contracting Party undertakes to exert appropriate efforts, consistent with the Charter of the United Nations, to the end that no one engages in any activity contrary to the objective of this Convention.
2. Each Contracting Party shall notify the Commission of any such activity which comes to its attention.

Article XXIII

1. The Commission and the Scientific Committee shall co-operate with the Antarctic Treaty Consultative Parties on matters falling within the competence of the latter.
2. The Commission and the Scientific Committee shall co-operate, as appropriate, with the Food and Agriculture Organisation of the United Nations and with other Specialised Agencies.
3. The Commission and the Scientific Committee shall seek to develop co-operative working relationships, as appropriate, with inter-governmental and non-governmental organizations which could contribute to their work, including the Scientific Committee on Antarctic Research, the Scientific Committee on Oceanic Research and the International Whaling Commission.
4. The Commission may enter into agreements with the organizations referred to in this Article and with other organizations as may be appropriate. The Commission and the Scientific Committee may invite such organizations to send observers to their meetings and to meetings of their subsidiary bodies.

Article XXIV

1. In order to promote the objective and ensure observance of the provisions of this Convention, the Contracting Parties agree that a system of observation and inspection shall be established.
2. The system of observation and inspection shall be elaborated by the Commission on the basis of the following principles:
    a. Contracting Parties shall co-operate with each other to ensure the effective implementation of the system of observation and inspection, taking account of the existing international practice. This system shall include, inter alia, procedures for boarding and inspection by observers and inspectors designated by the Members of the Commission and procedures for flag state prosecution and sanctions on the basis of evidence resulting from such boarding and inspections. A report of such prosecutions and sanctions imposed shall be included in the information referred to in Article XXI of this Convention;
    b. in order to verify compliance with measures adopted under this Convention, observation and inspection shall be carried out on board vessels engaged in scientific research or harvesting of marine living resources in the area to which this Convention applies, through observers and inspectors designated by the Members of the Commission and operating under terms and conditions to be established by the Commission.
    c. designated observers and inspectors shall remain subject to the jurisdiction of the Contracting Party of which they are nationals. They shall report to the Member of the Commission by which they have been designated which in turn shall report to the Commission.
3. Pending the establishment of the system of observation and inspection, the Members of the Commission shall seek to establish interim arrangements to designate observers and inspectors and such designated observers and inspectors shall be entitled to carry out inspections in accordance with the principles set out in paragraph 2 above.

Article XXV

1. If any dispute arises between two or more of the Contracting Parties concerning the interpretation or application of this Convention, those Contracting Parties shall consult among themselves with a view to having the dispute resolved by negotiation, inquiry, mediation, conciliation, arbitration, judicial settlement or other peaceful means of their own choice.
2. Any dispute of this character not so resolved shall, with the consent in each case of all Parties to the

dispute, be referred for settlement to the International Court of Justice or to arbitration; but failure to reach agreement on reference to the International Court or to arbitration shall not absolve Parties to the dispute from the responsibility of continuing to seek to resolve it by any of the various peaceful means referred to in paragraph 1 above.
3. In cases where the dispute is referred to arbitration, the arbitral tribunal shall be constituted as provided in the Annex to this Convention.

Article XXVI

1. This Convention shall be open for signature at Canberra from 1 August to 31 December 1980 by the States participating in the Conference on the Conservation of Antarctic Marine Living Resources held at Canberra from 7 to 20 May 1980.
2. The States which so sign will be the original signatory States of the Convention.

Article XXVII

1. This Convention is subject to ratification, acceptance or approval by signatory States.
2. Instruments of ratification, acceptance or approval shall be deposited with the Government of Australia, hereby designated as the Depositary.

Article XXVIII

1. This Convention shall enter into force on the thirtieth day following the date of deposit of the eighth instrument of ratification, acceptance or approval by States referred to in paragraph 1 of Article XXVI of this Convention.
2. With respect to each State or regional economic integration organization which subsequent to the date of entry into force of this Convention deposits an instrument of ratification, acceptance, approval or accession, the Convention shall enter into force on the thirtieth day following such deposit.

Article XXIX

1. This Convention shall be open for accession by any State interested in research or harvesting activities in relation to the marine living resources to which this Convention applies.
2. This Convention shall be open for accession by regional economic integration organizations constituted by sovereign States which include among their members one or more State's Members of the Commission and to which the State's members of the organization have transferred, in whole or in part, competences with regard to the matters covered by this Convention. The accession of such regional economic integration organizations shall be the subject of consultations among Members of the Commission.

Article XXX

1. This Convention may be amended at any time.
2. If one-third of the Members of the Commission request a meeting to discuss a proposed amendment the Depositary shall call such a meeting.
3. An amendment shall enter into force when the Depositary has received instruments of ratification, acceptance or approval thereof from all the Members of the Commission.
4. Such amendment shall thereafter enter into force as to any other Contracting Party when notice of ratification, acceptance or approval has been received by the Depositary. Any such Contracting Party from which no such notice has been received within a period of one year from the date of entry into force of the amendment in accordance with paragraph 3 above shall be deemed to have withdrawn from this Convention.

Article XXXI

1. Any Contracting Party may withdraw from this Convention on 30 June of any year, by giving written notice not later than 1 January of the same year to the Depositary, which, upon receipt of such a notice, shall communicate it forthwith to the other Contracting Parties.
2. Any other Contracting Party may, within sixty days of the receipt of a copy of such a notice from the Depositary, give written notice of withdrawal to the Depositary in which case the Convention shall cease to be in force on 30 June of the same year with respect to the Contracting Party giving such notice.
3. Withdrawal from this Convention by any Member of the Commission shall not affect its financial obligations under this Convention.

Article XXXII

The Depositary shall notify all Contracting Parties of the following:

a. signatures of this Convention and the deposit of instruments of ratification, acceptance, approval or accession;
b. the date of entry into force of this Convention and of any amendment thereto.

Article XXXIII

1. This Convention, of which the English, French, Russian and Spanish texts are equally authentic, shall be deposited with the Government of Australia which shall transmit duly certified copies thereof to all signatory and acceding Parties.
2. This Convention shall be registered by the Depositary pursuant to Article 102 of the Charter of the United Nations. Drawn up at Canberra this twentieth day of May 1980.

IN WITNESS THEREOF the undersigned, being duly authorized, have signed this Convention.

Annex for an Arbitral Tribunal

The arbitral tribunal referred to in paragraph 3 of Article XXV shall be composed of three arbitrators who shall be appointed as follows:

The Party commencing proceedings shall communicate the name of an arbitrator to the other Party which, in turn, within a period of forty days following such notification, shall communicate the name of the second arbitrator. The Parties shall, within a period of sixty days following the appointment of the second arbitrator, appoint the third arbitrator, who shall not be a national of either Party and shall not be of the same nationality as either of the first two arbitrators. The third arbitrator shall preside over the tribunal.

If the second arbitrator has not been appointed within the prescribed period, or if the Parties have not reached agreement within the prescribed period on the appointment of the third arbitrator, that arbitrator shall be appointed, as the request of either Party, by the Secretary-General of the Permanent Court of Arbitration, from among persons of international standing not having the nationality of a State which is a Party to this Convention.

The arbitral tribunal shall decide where its headquarters will be located and shall adopt its own rules of procedure.

The award of the arbitral tribunal shall be made by a majority of its members, who may not abstain from voting.

Any Contracting Party which is not a Party to the dispute may intervene in the proceedings with the consent of the arbitral tribunal.

The award of the arbitral tribunal shall be final and binding on all Parties to the dispute and on any Party which intervenes in the proceedings and shall be complied with without delay. The arbitral tribunal shall interpret the award at the request of one of the Parties to the dispute or of any intervening Party.

Unless the arbitral tribunal determines otherwise because of the particular circumstances of the case, the expenses of the tribunal, including the remuneration of its members, shall be borne by the Parties to the dispute in equal shares.

Appendix D

Protocol on Environmental Protection to the Antarctic Treaty

Preamble

The States Parties to this Protocol to the Antarctic Treaty, hereinafter referred to as the Parties,

Convinced of the need to enhance the protection of the Antarctic environment and dependent and associated ecosystems;

Convinced of the need to strengthen the Antarctic Treaty system so as to ensure that Antarctica shall continue forever to be used exclusively for peaceful purposes and shall not become the scene or object of international discord;

Bearing in mind the special legal and political status of Antarctica and the special responsibility of the Antarctic Treaty Consultative Parties to ensure that all activities in Antarctica are consistent with the purposes and principles of the Antarctic Treaty;

Recalling the designation of Antarctica as a Special Conservation Area and other measures adopted under the Antarctic Treaty system to protect the Antarctic environment and dependent and associated ecosystems;

Acknowledging further the unique opportunities Antarctica offers for scientific monitoring of and research on processes of global as well as regional importance;

Reaffirming the conservation principles of the Convention on the Conservation of Antarctic Marine Living Resources;

Convinced that the development of a comprehensive regime for the protection of the Antarctic environment and dependent and associated ecosystems is in the interest of mankind as a whole;

Desiring to supplement the Antarctic Treaty to this end;
Have agreed as follows:

Article 1: Definitions

For the purposes of this Protocol:

a. 'The Antarctic Treaty' means the Antarctic Treaty done at Washington on 1 December 1959;
b. 'Antarctic Treaty area' means the area to which the provisions of the Antarctic Treaty apply in accordance with Article VI of that Treaty;
c. 'Antarctic Treaty Consultative Meetings' means the meetings referred to in Article IX of the Antarctic Treaty;
d. 'Antarctic Treaty Consultative Parties' means the Contracting Parties to the Antarctic Treaty entitled to appoint representatives to participate in the meetings referred to in Article IX of that Treaty;
e. 'Antarctic Treaty system' means the Antarctic Treaty, the measures in effect under that Treaty, its associated separate international instruments in force and the measures in effect under those instruments;
f. 'Arbitral Tribunal' means the Arbitral Tribunal established in accordance with the Schedule to this Protocol, which forms an integral part thereof;
g. 'Committee' means the Committee for Environmental Protection established in accordance with Article 11.

Article 2: Objective and designation

The Parties commit themselves to the comprehensive protection of the Antarctic environment and dependent and associated ecosystems and hereby designate Antarctica as a natural reserve, devoted to peace and science.

Article 3: Environmental principles

1. The protection of the Antarctic environment and dependent and associated ecosystems and the intrinsic value of Antarctica, including its wilderness and aesthetic values and its value as an area for the conduct of scientific research, in particular research essential to understanding the global environment, shall be fundamental considerations in the planning and conduct of all activities in the Antarctic Treaty area.
2. To this end:
   a. activities in the Antarctic Treaty area shall be planned and conducted so as to limit adverse impacts on the Antarctic environment and dependent and associated ecosystems;
   b. activities in the Antarctic Treaty area shall be planned and conducted so as to avoid:
      i. adverse effects on climate or weather patterns;
      ii. significant adverse effects on air or water quality;
      iii. significant changes in the atmospheric, terrestrial (including aquatic), glacial or marine environments;
      iv. detrimental changes in the distribution, abundance or productivity of species or populations of species of fauna and flora;

v. further jeopardy to endangered or threatened species or populations of such species; or
vi. degradation of, or substantial risk to, areas of biological, scientific, historic, aesthetic or wilderness significance;
c. activities in the Antarctic Treaty area shall be planned and conducted on the basis of information sufficient to allow prior assessments of, and informed judgments about, their possible impacts on the Antarctic environment and dependent and associated ecosystems and on the value of Antarctica for the conduct of scientific research; such judgments shall take account of:
i. the scope of the activity, including its area, duration and intensity;
ii. the cumulative impacts of the activity, both by itself and in combination with other activities in the Antarctic Treaty area;
iii. whether the activity will detrimentally affect any other activity in the Antarctic Treaty area;
iv. whether technology and procedures are available to provide for environmentally safe operations;
v. whether there exists the capacity to monitor key environmental parameters and ecosystem components so as to identify and provide early warning of any adverse effects of the activity and to provide for such modification of operating procedures as may be necessary in the light of the results of monitoring or increased knowledge of the Antarctic environment and dependent and associated ecosystems; and
vi. whether there exists the capacity to respond promptly and effectively to accidents, particularly those with potential environmental effects;
d. regular and effective monitoring shall take place to allow assessment of the impacts of ongoing activities, including the verification of predicted impacts;
e. regular and effective monitoring shall take place to facilitate early detection of the possible unforeseen effects of activities carried on both within and outside the Antarctic Treaty area on the Antarctic environment and dependent and associated ecosystems.
3. Activities shall be planned and conducted in the Antarctic Treaty area so as to accord priority to scientific research and to preserve the value of Antarctica as an area for the conduct of such research, including research essential to understanding the global environment.
4. Activities undertaken in the Antarctic Treaty area pursuant to scientific research programmes, tourism and all other governmental and non-governmental activities in the Antarctic Treaty area for which advance notice is required in accordance with Article VII (5) of the Antarctic Treaty, including associated logistic support activities, shall:
a. take place in a manner consistent with the principles in this Article; and
b. be modified, suspended or cancelled if they result in or threaten to result in impacts upon the Antarctic environment or dependent or associated ecosystems inconsistent with those principles.

Article 4: Relationship with the other components of the Antarctic Treaty System

1. This Protocol shall supplement the Antarctic Treaty and shall neither modify nor amend that Treaty.
2. Nothing in this Protocol shall derogate from the rights and obligations of the Parties to this Protocol under the other international instruments in force within the Antarctic Treaty system.

Article 5: Consistency with the other components of the Antarctic Treaty System

The Parties shall consult and co-operate with the Contracting Parties to the other international instruments in force within the Antarctic Treaty system and their respective institutions with a view to ensuring the achievement of the objectives and principles of this Protocol and avoiding any interference with the achievement of the objectives and principles of those instruments or any inconsistency between the implementation of those instruments and of this Protocol.

Article 6: Co-operation

1. The Parties shall co-operate in the planning and conduct of activities in the Antarctic Treaty area. To this end, each Party shall endeavour to:
a. promote co-operative programmes of scientific, technical and educational value, concerning the protection of the Antarctic environment and dependent and associated ecosystems;
b. provide appropriate assistance to other Parties in the preparation of environmental impact assessments;
c. provide to other Parties upon request information relevant to any potential environmental risk and assistance to minimize the effects of accidents which may damage the Antarctic environment or dependent and associated ecosystems;
d. consult with other Parties with regard to the choice of sites for prospective stations and other facilities so as to avoid the cumulative impacts caused by their excessive concentration in any location;
e. where appropriate, undertake joint expeditions and share the use of stations and other facilities; and

f. carry out such steps as may be agreed upon at Antarctic Treaty Consultative Meetings.
2. Each Party undertakes, to the extent possible, to share information that may be helpful to other Parties in planning and conducting their activities in the Antarctic Treaty area, with a view to the protection of the Antarctic environment and dependent and associated ecosystems.
3. The Parties shall co-operate with those Parties which may exercise jurisdiction in areas adjacent to the Antarctic Treaty area with a view to ensuring that activities in the Antarctic Treaty area do not have adverse environmental impacts on those areas.

Article 7: Prohibition of mineral resource activities

Any activity relating to mineral resources, other than scientific research, shall be prohibited.

Article 8: Environmental impact assessment

1. Proposed activities referred to in paragraph 2 below shall be subject to the procedures set out in Annex I for prior assessment of the impacts of those activities on the Antarctic environment or on dependent or associated ecosystems according to whether those activities are identified as having:
   a. less than a minor or transitory impact;
   b. a minor or transitory impact; or
   c. more than a minor or transitory impact.
2. Each Party shall ensure that the assessment procedures set out in Annex I are applied in the planning processes leading to decisions about any activities undertaken in the Antarctic Treaty area pursuant to scientific research programmes, tourism and all other governmental and non-governmental activities in the Antarctic Treaty area for which advance notice is required under Article VII (5) of the Antarctic Treaty, including associated logistic support activities.
3. The assessment procedures set out in Annex I shall apply to any change in an activity whether the change arises from an increase or decrease in the intensity of an existing activity, from the addition of an activity, the decommissioning of a facility, or otherwise.
4. Where activities are planned jointly by more than one Party, the Parties involved shall nominate one of their number to co-ordinate the implementation of the environmental impact assessment procedures set out in Annex I.

Article 9: Annexes

1. The Annexes to this Protocol shall form an integral part thereof.
2. Annexes, additional to Annexes I-IV, may be adopted and become effective in accordance with Article IX of the Antarctic Treaty.
3. Amendments and modifications to Annexes may be adopted and become effective in accordance with Article IX of the Antarctic Treaty, provided that any Annex may itself make provision for amendments and modifications to become effective on an accelerated basis.
4. Annexes and any amendments and modifications thereto which have become effective in accordance with paragraphs 2 and 3 above shall, unless an Annex itself provides otherwise in respect of the entry into effect of any amendment or modification thereto, become effective for a Contracting Party to the Antarctic Treaty which is not an Antarctic Treaty Consultative Party, or which was not an Antarctic Treaty Consultative Party at the time of the adoption, when notice of approval of that Contracting Party has been received by the Depositary.
5. Annexes shall, except to the extent that an Annex provides otherwise, be subject to the procedures for dispute settlement set out in Articles 18 to 20.

Article 10: Antarctic Treaty Consultative Meetings

1. Antarctic Treaty Consultative Meetings shall, drawing upon the best scientific and technical advice available:
   a. define, in accordance with the provisions of this Protocol, the general policy for the comprehensive protection of the Antarctic environment and dependent and associated ecosystems; and
   b. adopt measures under Article IX of the Antarctic Treaty for the implementation of this Protocol.
2. Antarctic Treaty Consultative Meetings shall review the work of the Committee and shall draw fully upon its advice and recommendations in carrying out the tasks referred to in paragraph 1 above, as well as upon the advice of the Scientific Committee on Antarctic Research.

Article 11: Committee for Environmental Protection

1. There is hereby established the Committee for Environmental Protection.
2. Each Party shall be entitled to be a member of the Committee and to appoint a representative who may be accompanied by experts and advisers.
3. Observer status in the Committee shall be open to any Contracting Party to the Antarctic Treaty which is not a Party to this Protocol.
4. The Committee shall invite the President of the Scientific Committee on Antarctic Research and the Chairman of the Scientific Committee for the Conservation of Antarctic Marine Living Resources to participate as observers at its sessions. The Committee may also, with the approval of the Antarctic Treaty Consultative Meeting, invite such other relevant scientific, environmental

and technical organisations which can contribute to its work to participate as observers at its sessions.
5. The Committee shall present a report on each of its sessions to the Antarctic Treaty Consultative Meeting. The report shall cover all matters considered at the session and shall reflect the views expressed. The report shall be circulated to the Parties and to observers attending the session, and shall thereupon be made publicly available.
6. The Committee shall adopt its rules of procedure which shall be subject to approval by the Antarctic Treaty Consultative Meeting.

Article 12: Functions of the Committee

1. The functions of the Committee shall be to provide advice and formulate recommendations to the Parties in connection with the implementation of this Protocol, including the operation of its Annexes, for consideration at Antarctic Treaty Consultative Meetings, and to perform such other functions as may be referred to it by the Antarctic Treaty Consultative Meetings. In particular, it shall provide advice on:
    a. the effectiveness of measures taken pursuant to this Protocol;
    b. the need to update, strengthen or otherwise improve such measures;
    c. the need for additional measures, including the need for additional Annexes, where appropriate;
    d. the application and implementation of the environmental impact assessment procedures set out in Article 8 and Annex I;
    e. means of minimising or mitigating environmental impacts of activities in the Antarctic Treaty area;
    f. procedures for situations requiring urgent action, including response action in environmental emergencies;
    g. the operation and further elaboration of the Antarctic Protected Area system;
    h. inspection procedures, including formats for inspection reports and checklists for the conduct of inspections;
    i. the collection, archiving, exchange and evaluation of information related to environmental protection;
    j. the state of the Antarctic environment; and
    k. the need for scientific research, including environmental monitoring, related to the implementation of this Protocol.
2. In carrying out its functions, the Committee shall, as appropriate, consult with the Scientific Committee on Antarctic Research, the Scientific Committee for the Conservation of Antarctic Marine Living Resources and other relevant scientific, environmental and technical organizations.

Article 13: Compliance with this Protocol

1. Each Party shall take appropriate measures within its competence, including the adoption of laws and regulations, administrative actions and enforcement measures, to ensure compliance with this Protocol.
2. Each Party shall exert appropriate efforts, consistent with the Charter of the United Nations, to the end that no one engages in any activity contrary to this Protocol.
3. Each Party shall notify all other Parties of the measures it takes pursuant to paragraphs 1 and 2 above.
4. Each Party shall draw the attention of all other Parties to any activity which in its opinion affects the implementation of the objectives and principles of this Protocol.
5. The Antarctic Treaty Consultative Meetings shall draw the attention of any State which is not a Party to this Protocol to any activity undertaken by that State, its agencies, instrumentalities, natural or juridical persons, ships, aircraft or other means of transport which affects the implementation of the objectives and principles of this Protocol.

Article 14: Inspection

1. In order to promote the protection of the Antarctic environment and dependent and associated ecosystems, and to ensure compliance with this Protocol, the Antarctic Treaty Consultative Parties shall arrange, individually or collectively, for inspections by observers to be made in accordance with Article VII of the Antarctic Treaty.
2. Observers are:
    a. observers designated by any Antarctic Treaty Consultative Party who shall be nationals of that Party; and
    b. any observers designated at Antarctic Treaty Consultative Meetings to carry out inspections under procedures to be established by an Antarctic Treaty Consultative Meeting.
3. Parties shall co-operate fully with observers undertaking inspections, and shall ensure that during inspections, observers are given access to all parts of stations, installations, equipment, ships and aircraft open to inspection under Article VII (3) of the Antarctic Treaty, as well as to all records maintained thereon which are called for pursuant to this Protocol.
4. Reports of inspections shall be sent to the Parties whose stations, installations, equipment, ships or aircraft are covered by the reports. After those Parties have been given the opportunity to comment, the reports and any comments thereon shall be circulated to all the Parties and to the Committee, considered at the next Antarctic Treaty Consultative Meeting, and thereafter made publicly available.

Article 15: Emergency response action

1. In order to respond to environmental emergencies in the Antarctic Treaty area, each Party agrees to:
   a. provide for prompt and effective response action to such emergencies which might arise in the performance of scientific research programmes, tourism and all other governmental and non-governmental activities in the Antarctic Treaty area for which advance notice is required under Article VII (5) of the Antarctic Treaty, including associated logistic support activities; and
   b. establish contingency plans for response to incidents with potential adverse effects on the Antarctic environment or dependent and associated ecosystems.
2. To this end, the Parties shall:
   a. co-operate in the formulation and implementation of such contingency plans; and
   b. establish procedures for immediate notification of, and co-operative response to, environmental emergencies.
3. In the implementation of this Article, the Parties shall draw upon the advice of the appropriate international organisations.

Article 16: Liability

Consistent with the objectives of this Protocol for the comprehensive protection of the Antarctic environment and dependent and associated ecosystems, the Parties undertake to elaborate rules and procedures relating to liability for damage arising from activities taking place in the Antarctic Treaty area and covered by this Protocol. Those rules and procedures shall be included in one or more Annexes to be adopted in accordance with Article 9 (2).

Article 17: Annual report by Parties

1. Each Party shall report annually on the steps taken to implement this Protocol. Such reports shall include notifications made in accordance with Article 13 (3), contingency plans established in accordance with Article 15 and any other notifications and information called for pursuant to this Protocol for which there is no other provision concerning the circulation and exchange of information.
2. Reports made in accordance with paragraph 1 above shall be circulated to all Parties and to the Committee, considered at the next Antarctic Treaty Consultative Meeting, and made publicly available.

Article 18: Dispute settlement

If a dispute arises concerning the interpretation or application of this Protocol, the parties to the dispute shall, at the request of any one of them, consult among themselves as soon as possible with a view to having the dispute resolved by negotiation, inquiry, mediation, conciliation, arbitration, judicial settlement or other peaceful means to which the parties to the dispute agree.

Article 19: Choice of dispute settlement procedure

1. Each Party, when signing, ratifying, accepting, approving or acceding to this Protocol, or at any time thereafter, may choose, by written declaration, one or both of the following means for the settlement of disputes concerning the interpretation or application of Articles 7, 8 and 15 and, except to the extent that an Annex provides otherwise, the provisions of any Annex and, insofar as it relates to these Articles and provisions, Article 13:
   a. the International Court of Justice;
   b. the Arbitral Tribunal.
2. A declaration made under paragraph 1 above shall not affect the operation of Article 18 and Article 20 (2).
3. A Party which has not made a declaration under paragraph 1 above or in respect of which a declaration is no longer in force shall be deemed to have accepted the competence of the Arbitral Tribunal.
4. If the parties to a dispute have accepted the same means for the settlement of a dispute, the dispute may be submitted only to that procedure, unless the parties otherwise agree.
5. If the parties to a dispute have not accepted the same means for the settlement of a dispute, or if they have both accepted both means, the dispute may be submitted only to the Arbitral Tribunal, unless the parties otherwise agree.
6. A declaration made under paragraph 1 above shall remain in force until it expires in accordance with its terms or until three months after written notice of revocation has been deposited with the Depositary.
7. A new declaration, a notice of revocation or the expiry of a declaration shall not in any way affect proceedings pending before the International Court of Justice or the Arbitral Tribunal, unless the parties to the dispute otherwise agree.
8. Declarations and notices referred to in this Article shall be deposited with the Depositary who shall transmit copies thereof to all Parties.

Article 20: Dispute settlement procedure

1. If the parties to a dispute concerning the interpretation or application of Articles 7, 8 or 15 or, except to the extent that an Annex provides otherwise, the provisions of any Annex or, insofar as it relates to these Articles and provisions, Article 13, have not agreed on a means for resolving it within 12 months of the request for consultation pursuant to Article 18, the dispute shall be referred, at the request of any party to the dispute, for

settlement in accordance with the procedure determined by Article 19 (4) and (5).

2. The Arbitral Tribunal shall not be competent to decide or rule upon any matter within the scope of Article IV of the Antarctic Treaty. In addition, nothing in this Protocol shall be interpreted as conferring competence or jurisdiction on the International Court of Justice or any other tribunal established for the purpose of settling disputes between Parties to decide or otherwise rule upon any matter within the scope of Article IV of the Antarctic Treaty.

Article 21: Signature

This Protocol shall be open for signature at Madrid on the 4th of October 1991 and thereafter at Washington until the 3rd of October 1992 by any State which is a Contracting Party to the Antarctic Treaty.

Article 22: Ratification, acceptance, approval or accession

1. This Protocol is subject to ratification, acceptance or approval by signatory States.
2. After the 3rd of October 1992 this Protocol shall be open for accession by any State which is a Contracting Party to the Antarctic Treaty.
3. Instruments of ratification, acceptance, approval or accession shall be deposited with the Government of the United States of America, hereby designated as the Depositary.
4. After the date on which this Protocol has entered into force, the Antarctic Treaty Consultative Parties shall not act upon a notification regarding the entitlement of a Contracting Party to the Antarctic Treaty to appoint representatives to participate in Antarctic Treaty Consultative Meetings in accordance with Article IX (2) of the Antarctic Treaty unless that Contracting Party has first ratified, accepted, approved or acceded to this Protocol.

Article 23: Entry into force

1. This Protocol shall enter into force on the thirtieth day following the date of deposit of instruments of ratification, acceptance, approval or accession by all States which are Antarctic Treaty Consultative Parties at the date on which this Protocol is adopted.
2. For each Contracting Party to the Antarctic Treaty which, subsequent to the date of entry into force of this Protocol, deposits an instrument of ratification, acceptance, approval or accession, this Protocol shall enter into force on the thirtieth day following such deposit.

Article 24: Reservations

Reservations to this Protocol shall not be permitted.

Article 25: Modification or amendment

1. Without prejudice to the provisions of Article 9, this Protocol may be modified or amended at any time in accordance with the procedures set forth in Article XII (1) (a) and (b) of the Antarctic Treaty.
2. If, after the expiration of 50 years from the date of entry into force of this Protocol, any of the Antarctic Treaty Consultative Parties so requests by a communication addressed to the Depositary, a conference shall be held as soon as practicable to review the operation of this Protocol.
3. A modification or amendment proposed at any Review Conference called pursuant to paragraph 2 above shall be adopted by a majority of the Parties, including 3/4 of the States which are Antarctic Treaty Consultative Parties at the time of adoption of this Protocol.
4. A modification or amendment adopted pursuant to paragraph 3 above shall enter into force upon ratification, acceptance, approval or accession by 3/4 of the Antarctic Treaty Consultative Parties, including ratification, acceptance, approval or accession by all States which are Antarctic Treaty Consultative Parties at the time of adoption of this Protocol.
5.
    a. With respect to Article 7, the prohibition on Antarctic mineral resource activities contained therein shall continue unless there is in force a binding legal regime on Antarctic mineral resource activities that includes an agreed means for determining whether, and, if so, under which conditions, any such activities would be acceptable. This regime shall fully safeguard the interests of all States referred to in Article IV of the Antarctic Treaty and apply the principles thereof. Therefore, if a modification or amendment to Article 7 is proposed at a Review Conference referred to in paragraph 2 above, it shall include such a binding legal regime.
    b. If any such modification or amendment has not entered into force within 3 years of the date of its adoption, any Party may at any time thereafter notify to the Depositary of its withdrawal from this Protocol, and such withdrawal shall take effect 2 years after receipt of the notification by the Depositary.

Article 26: Notifications by the Depositary

The Depositary shall notify all Contracting Parties to the Antarctic Treaty of the following:

a. signatures of this Protocol and the deposit of instruments of ratification, acceptance, approval or accession;
b. the date of entry into force of this Protocol and any additional Annex thereto;

c. the date of entry into force of any amendment or modification to this Protocol;
d. the deposit of declarations and notices pursuant to Article 19; and
e. any notification received pursuant to Article 25 (5) (b).

Article 27: Authentic texts and registration with the United Nations

1. This Protocol, done in the English, French, Russian and Spanish languages, each version being equally authentic, shall be deposited in the archives of the Government of the United States of America, which shall transmit duly certified copies thereof to all Contracting Parties to the Antarctic Treaty.
2. This Protocol shall be registered by the Depositary pursuant to Article 102 of the Charter of the United Nations.

Schedule to the Protocol: Arbitration

Article 1

1. The Arbitral Tribunal shall be constituted and shall function in accordance with the Protocol, including this Schedule.
2. The Secretary referred to in this Schedule is the Secretary General of the Permanent Court of Arbitration.

Article 2

1. Each Party shall be entitled to designate up to three Arbitrators, at least one of whom shall be designated within three months of the entry into force of the Protocol for that Party. Each Arbitrator shall be experienced in Antarctic affairs, have thorough knowledge of international law and enjoy the highest reputation for fairness, competence and integrity. The names of the persons so designated shall constitute the list of Arbitrators. Each Party shall at all times maintain the name of at least one Arbitrator on the list.
2. Subject to paragraph 3 below, an Arbitrator designated by a Party shall remain on the list for a period of five years and shall be eligible for redesignation by that Party for additional five year periods.
3. A Party which designated an Arbitrator may withdraw the name of that Arbitrator from the list. If an Arbitrator dies or if a Party for any reason withdraws from the list the name of an Arbitrator designated by it, the Party which designated the Arbitrator in question shall notify the Secretary promptly. An Arbitrator whose name is withdrawn from the list shall continue to serve on any Arbitral Tribunal to which that Arbitrator has been appointed until the completion of proceedings before the Arbitral Tribunal.
4. The Secretary shall ensure that an up-to-date list is maintained of the Arbitrators designated pursuant to this Article.

Article 3

1. The Arbitral Tribunal shall be composed of three Arbitrators who shall be appointed as follows:
    a. The party to the dispute commencing the proceedings shall appoint one Arbitrator, who may be its national, from the list referred to in Article 2. This appointment shall be included in the notification referred to in Article 4.
    b. Within 40 days of the receipt of that notification, the other party to the dispute shall appoint the second Arbitrator, who may be its national, from the list referred to in Article 2.
    c. Within 60 days of the appointment of the second Arbitrator, the parties to the dispute shall appoint by agreement the third Arbitrator from the list referred to in Article 2. The third Arbitrator shall not be either a national of a party to the dispute, or a person designated for the list referred to in Article 2 by a party to the dispute, or of the same nationality as either of the first two Arbitrators. The third Arbitrator shall be the Chairperson of the Arbitral Tribunal.
    d. If the second Arbitrator has not been appointed within the prescribed period, or if the parties to the dispute have not reached agreement within the prescribed period on the appointment of the third Arbitrator, the Arbitrator or Arbitrators shall be appointed, at the request of any party to the dispute and within 30 days of the receipt of such request, by the President of the International Court of Justice from the list referred to in Article 2 and subject to the conditions prescribed in subparagraphs (b) and (c) above. In performing the functions accorded him or her in this subparagraph, the President of the Court shall consult the parties to the dispute.
    e. If the President of the International Court of Justice is unable to perform the functions accorded him or her in subparagraph (d) above or is a national of a party to the dispute, the functions shall be performed by the Vice-President of the Court, except that if the Vice-President is unable to perform the functions or is a national of a party to the dispute the functions shall be performed by the next most senior member of the Court who is available and is not a national of a party to the dispute.
2. Any vacancy shall be filled in the manner prescribed for the initial appointment.
3. In any dispute involving more than two Parties, those Parties having the same interest shall appoint one

Arbitrator by agreement within the period specified in paragraph 1 (b) above.

Article 4

The party to the dispute commencing proceedings shall so notify the other party or parties to the dispute and the Secretary in writing. Such notification shall include a statement of the claim and the grounds on which it is based. The notification shall be transmitted by the Secretary to all Parties.

Article 5

1. Unless the parties to the dispute agree otherwise, arbitration shall take place at The Hague, where the records of the Arbitral Tribunal shall be kept. The Arbitral Tribunal shall adopt its own rules of procedure. Such rules shall ensure that each party to the dispute has a full opportunity to be heard and to present its case and shall also ensure that the proceedings are conducted expeditiously.
2. The Arbitral Tribunal may hear and decide counter-claims arising out of the dispute.

Article 6

1. The Arbitral Tribunal, where it considers that *prima facie* it has jurisdiction under the Protocol, may:
   a. at the request of any party to a dispute, indicate such provisional measures as it considers necessary to preserve the respective rights of the parties to the dispute;
   b. prescribe any provisional measures which it considers appropriate under the circumstances to prevent serious harm to the Antarctic environment or dependent or associated ecosystems.
2. The parties to the dispute shall comply promptly with any provisional measures prescribed under paragraph 1 (b) above pending an award under Article 10.
3. Notwithstanding the time period in Article 20 of the Protocol, a party to a dispute may at any time, by notification to the other party or parties to the dispute and to the Secretary in accordance with Article 4, request that the Arbitral Tribunal be constituted as a matter of exceptional urgency to indicate or prescribe emergency provisional measures in accordance with this Article. In such case, the Arbitral Tribunal shall be constituted as soon as possible in accordance with Article 3, except that the time periods in Article 3 (1) (b), (c) and (d) shall be reduced to 14 days in each case. The Arbitral Tribunal shall decide upon the request for emergency provisional measures within two months of the appointment of its Chairperson.
4. Following a decision by the Arbitral Tribunal upon a request for emergency provisional measures in accordance with paragraph 3 above, settlement of the dispute shall proceed in accordance with Articles 18, 19 and 20 of the Protocol.

Article 7

Any Party which believes it has a legal interest, whether general or individual, which may be substantially affected by the award of an Arbitral Tribunal, may, unless the Arbitral Tribunal decides otherwise, intervene in the proceedings.

Article 8

The parties to the dispute shall facilitate the work of the Arbitral Tribunal and, in particular, in accordance with their law and using all means at their disposal, shall provide it with all relevant documents and information, and enable it, when necessary, to call witnesses or experts and receive their evidence.

Article 9

If one of the parties to the dispute does not appear before the Arbitral Tribunal or fails to defend its case, any other party to the dispute may request the Arbitral Tribunal to continue the proceedings and make its award.

Article 10

1. The Arbitral Tribunal shall, on the basis of the provisions of the Protocol and other applicable rules and principles of international law that are not incompatible with such provisions, decide such disputes as are submitted to it.
2. The Arbitral Tribunal may decide, *ex aequo et bono*, a dispute submitted to it, if the parties to the dispute so agree.

Article 11

1. Before making its award, the Arbitral Tribunal shall satisfy itself that it has competence in respect of the dispute and that the claim or counterclaim is well founded in fact and law.
2. The award shall be accompanied by a statement of reasons for the decision and shall be communicated to the Secretary who shall transmit it to all Parties.
3. The award shall be final and binding on the parties to the dispute and on any Party which intervened in the proceedings and shall be complied with without delay. The Arbitral Tribunal shall interpret the award at the request of a party to the dispute or of any intervening Party.
4. The award shall have no binding force except in respect of that particular case.
5. Unless the Arbitral Tribunal decides otherwise, the expenses of the Arbitral Tribunal, including the remuneration of the Arbitrators, shall be borne by the parties to the dispute in equal shares.

Article 12

All decisions of the Arbitral Tribunal, including those referred to in Articles 5, 6 and 11, shall be made by a majority of the Arbitrators who may not abstain from voting.

Article 13

1. This Schedule may be amended or modified by a measure adopted in accordance with Article IX (1) of the Antarctic Treaty. Unless the measure specifies otherwise, the amendment or modification shall be deemed to have been approved, and shall become effective, one year after the close of the Antarctic Treaty Consultative Meeting at which it was adopted, unless one or more of the Antarctic Treaty Consultative Parties notifies the Depositary, within that time period, that it wishes an extension of that period or that it is unable to approve the measure.
2. Any amendment or modification of this Schedule which becomes effective in accordance with paragraph 1 above shall thereafter become effective as to any other Party when notice of approval by it has been received by the Depositary.

Annex I. Environmental impact assessment

Article 1: Preliminary stage

1. The environmental impacts of proposed activities referred to in Article 8 of the Protocol shall, before their commencement, be considered in accordance with appropriate national procedures.
2. If an activity is determined as having less than a minor or transitory impact, the activity may proceed forthwith.

Article 2: Initial environmental evaluation

1. Unless it has been determined that an activity will have less than a minor or transitory impact, or unless a Comprehensive Environmental Evaluation is being prepared in accordance with Article 3, an Initial Environmental Evaluation shall be prepared. It shall contain sufficient detail to assess whether a proposed activity may have more than a minor or transitory impact and shall include:
   a. a description of the proposed activity, including its purpose, location, duration and intensity; and
   b. consideration of alternatives to the proposed activity and any impacts that the activity may have, including consideration of cumulative impacts in the light of existing and known planned activities.
2. If an Initial Environmental Evaluation indicates that a proposed activity is likely to have no more than a minor or transitory impact, the activity may proceed, provided that appropriate procedures, which may include monitoring, are put in place to assess and verify the impact of the activity.

Article 3: Comprehensive Environmental Evaluation

1. If an Initial Environmental Evaluation indicates or if it is otherwise determined that a proposed activity is likely to have more than a minor or transitory impact, a Comprehensive Environmental Evaluation shall be prepared.
2. A Comprehensive Environmental Evaluation shall include:
   a. a description of the proposed activity including its purpose, location, duration and intensity, and possible alternatives to the activity, including the alternative of not proceeding, and the consequences of those alternatives;
   b. a description of the initial environmental reference state with which predicted changes are to be compared and a prediction of the future environmental reference state in the absence of the proposed activity;
   c. a description of the methods and data used to forecast the impacts of the proposed activity;
   d. estimation of the nature, extent, duration, and intensity of the likely direct impacts of the proposed activity;
   e. consideration of possible indirect or second order impacts of the proposed activity;
   f. consideration of cumulative impacts of the proposed activity in the light of existing activities and other known planned activities;
   g. identification of measures, including monitoring programmes, that could be taken to minimise or mitigate impacts of the proposed activity and to detect unforeseen impacts and that could provide early warning of any adverse effects of the activity as well as to deal promptly and effectively with accidents;
   h. identification of unavoidable impacts of the proposed activity;
   i. consideration of the effects of the proposed activity on the conduct of scientific research and on other existing uses and values;
   j. an identification of gaps in knowledge and uncertainties encountered in compiling the information required under this paragraph;
   k. a non-technical summary of the information provided under this paragraph; and
   l. the name and address of the person or organization which prepared the Comprehensive Environmental Evaluation and the address to which comments thereon should be directed.

3. The draft Comprehensive Environmental Evaluation shall be made publicly available and shall be circulated to all Parties, which shall also make it publicly available, for comment. A period of 90 days shall be allowed for the receipt of comments.
4. The draft Comprehensive Environmental Evaluation shall be forwarded to the Committee at the same time as it is circulated to the Parties, and at least 120 days before the next Antarctic Treaty Consultative Meeting, for consideration as appropriate.
5. No final decision shall be taken to proceed with the proposed activity in the Antarctic Treaty area unless there has been an opportunity for consideration of the draft Comprehensive Environmental Evaluation by the Antarctic Treaty Consultative Meeting on the advice of the Committee, provided that no decision to proceed with a proposed activity shall be delayed through the operation of this paragraph for longer than 15 months from the date of circulation of the draft Comprehensive Environmental Evaluation.
6. A final Comprehensive Environmental Evaluation shall address and shall include or summarise comments received on the draft Comprehensive Environmental Evaluation. The final Comprehensive Environmental Evaluation, notice of any decisions relating thereto, and any evaluation of the significance of the predicted impacts in relation to the advantages of the proposed activity, shall be circulated to all Parties, which shall also make them publicly available, at least 60 days before the commencement of the proposed activity in the Antarctic Treaty area.

Article 4: Decisions to be based on Comprehensive Environmental Evaluations

Any decision on whether a proposed activity, to which Article 3 applies, should proceed, and, if so, whether in its original or in a modified form, shall be based on the Comprehensive Environmental Evaluation as well as other relevant considerations.

Article 5: Monitoring

1. Procedures shall be put in place, including appropriate monitoring of key environmental indicators, to assess and verify the impact of any activity that proceeds following the completion of a Comprehensive Environmental Evaluation.
2. The procedures referred to in paragraph 1 above and in Article 2 (2) shall be designed to provide a regular and verifiable record of the impacts of the activity in order, *inter alia*, to:
   a. enable assessments to be made of the extent to which such impacts are consistent with the Protocol; and
   b. provide information useful for minimising or mitigating impacts, and, where appropriate, information on the need for suspension, cancellation or modification of the activity.

Article 6: Circulation of information

1. The following information shall be circulated to the Parties, forwarded to the Committee and made publicly available:
   a. a description of the procedures referred to in Article 1;
   b. an annual list of any Initial Environmental Evaluations prepared in accordance with Article 2 and any decisions taken in consequence thereof;
   c. significant information obtained, and any action taken in consequence thereof, from procedures put in place in accordance with Articles 2 (2) and 5; and
   d. information referred to in Article 3 (6).
2. Any Initial Environmental Evaluation prepared in accordance with Article 2 shall be made available on request.

Article 7: Cases of emergency

1. This Annex shall not apply in cases of emergency relating to the safety of human life or of ships, aircraft or equipment and facilities of high value, or the protection of the environment, which require an activity to be undertaken without completion of the procedures set out in this Annex.
2. Notice of activities undertaken in cases of emergency, which would otherwise have required preparation of a Comprehensive Environmental Evaluation, shall be circulated immediately to all Parties and to the Committee and a full explanation of the activities carried out shall be provided within 90 days of those activities.

Article 8: Amendment or modification

1. This Annex may be amended or modified by a measure adopted in accordance with Article IX (1) of the Antarctic Treaty. Unless the measure specifies otherwise, the amendment or modification shall be deemed to have been approved, and shall become effective, one year after the close of the Antarctic Treaty Consultative Meeting at which it was adopted, unless one or more of the Antarctic Treaty Consultative Parties notifies the Depositary, within that period, that it wishes an extension of that period or that it is unable to approve the measure.
2. Any amendment or modification of this Annex which becomes effective in accordance with paragraph 1 above shall thereafter become effective as to any other Party when notice of approval by it has been received by the Depositary.

Annex II. Conservation of Antarctic Fauna and Flora

Article 1: Definitions

For the purposes of this Annex:

a. 'native mammal' means any member of any species belonging to the Class Mammalia, indigenous to the Antarctic Treaty area or occurring there seasonally through natural migrations;
b. 'native bird' means any member, at any stage of its life cycle (including eggs), of any species of the Class Aves indigenous to the Antarctic Treaty area or occurring there seasonally through natural migrations;
c. 'native plant' means any terrestrial or freshwater vegetation, including bryophytes, lichens, fungi and algae, at any stage of its life cycle (including seeds, and other propagules), indigenous to the Antarctic Treaty area;
d. 'native invertebrate' means any terrestrial or freshwater invertebrate, at any stage of its life cycle, indigenous to the Antarctic Treaty area;
e. 'appropriate authority' means any person or agency authorized by a Party to issue permits under this Annex;
f. 'permit' means a formal permission in writing issued by an appropriate authority;
g. 'take' or 'taking' means to kill, injure, capture, handle or molest, a native mammal or bird, or to remove or damage such quantities of native plants that their local distribution or abundance would be significantly affected;
h. 'harmful interference' means:
   i. flying or landing helicopters or other aircraft in a manner that disturbs concentrations of birds and seals;
   ii. using vehicles or vessels, including hovercraft and small boats, in a manner that disturbs concentrations of birds and seals;
   iii. using explosives or firearms in a manner that disturbs concentrations of birds and seals;
   iv. wilfully disturbing breeding or moulting birds or concentrations of birds and seals by persons on foot;
   v. significantly damaging concentrations of native terrestrial plants by landing aircraft, driving vehicles, or walking on them, or by other means; and
   vi. any activity that results in the significant adverse modification of habitats of any species or population of native mammal, bird, plant or invertebrate.
i. 'International Convention for the Regulation of Whaling' means the Convention done at Washington on 2 December 1946.

Article 2: Cases of emergency

1. This Annex shall not apply in cases of emergency relating to the safety of human life or of ships, aircraft, or equipment and facilities of high value, or the protection of the environment.
2. Notice of activities undertaken in cases of emergency shall be circulated immediately to all Parties and to the Committee.

Article 3: Protection of native fauna and flora

1. Taking or harmful interference shall be prohibited, except in accordance with a permit.
2. Such permits shall specify the authorized activity, including when, where and by whom it is to be conducted and shall be issued only in the following circumstances:
   a. to provide specimens for scientific study or scientific information;
   b. to provide specimens for museums, herbaria, zoological and botanical gardens, or other educational or cultural institutions or uses; and
   c. to provide for unavoidable consequences of scientific activities not otherwise authorized under sub-paragraphs (a) or (b) above, or of the construction and operation of scientific support facilities.
3. The issue of such permits shall be limited so as to ensure that:
   a. no more native mammals, birds, or plants are taken than are strictly necessary to meet the purposes set forth in paragraph 2 above;
   b. only small numbers of native mammals or birds are killed and in no case more native mammals or birds are killed from local populations than can, in combination with other permitted takings, normally be replaced by natural reproduction in the following season; and
   c. the diversity of species, as well as the habitats essential to their existence, and the balance of the ecological systems existing within the Antarctic Treaty are maintained.
4. Any species of native mammals, birds and plants listed in Appendix A to this Annex shall be designated 'Specially Protected Species', and shall be accorded special protection by the Parties.
5. A permit shall not be issued to take a Specially Protected Species unless the taking:
   a. is for a compelling scientific purpose;
   b. will not jeopardize the survival or recovery of that species or local population; and
   c. uses non-lethal techniques where appropriate.
6. All taking of native mammals and birds shall be done in the manner that involves the least degree of pain and suffering practicable.

Article 4: Introduction of non-native species, parasites and diseases

1. No species of animal or plant not native to the Antarctic Treaty area shall be introduced onto land or ice shelves,

or into water in the Antarctic Treaty area except in accordance with a permit.
2. Dogs shall not be introduced onto land or ice shelves and dogs currently in those areas shall be removed by April 1, 1994.
3. Permits under paragraph 1 above shall be issued to allow the importation only of the animals and plants listed in Appendix B to this Annex and shall specify the species, numbers and, if appropriate, age and sex and precautions to be taken to prevent escape or contact with native fauna and flora.
4. Any plant or animal for which a permit has been issued in accordance with paragraphs 1 and 3 above, shall, prior to expiration of the permit, be removed from the Antarctic Treaty area or be disposed of by incineration or equally effective means that eliminates risk to native fauna or flora. The permit shall specify this obligation. Any other plant or animal introduced into the Antarctic Treaty area not native to that area, including any progeny, shall be removed or disposed of, by incineration or by equally effective means, so as to be rendered sterile, unless it is determined that they pose no risk to native flora or fauna.
5. Nothing in this Article shall apply to the importation of food into the Antarctic Treaty area provided that no live animals are imported for this purpose and all plants and animal parts and products are kept under carefully controlled conditions and disposed of in accordance with Annex III to the Protocol and Appendix C to this Annex.
6. Each Party shall require that precautions, including those listed in Appendix C to this Annex, be taken to prevent the introduction of micro-organisms (e.g., viruses, bacteria, parasites, yeasts, fungi) not present in the native fauna and flora.

Article 5: Information

Each Party shall prepare and make available information setting forth, in particular, prohibited activities and providing lists of Specially Protected Species and relevant Protected Areas to all those persons present in or intending to enter the Antarctic Treaty area with a view to ensuring that such persons understand and observe the provisions of this Annex.

Article 6: Exchange of information

1. The Parties shall make arrangements for:
   a. collecting and exchanging records (including records of permits) and statistics concerning the numbers or quantities of each species of native mammal, bird or plant taken annually in the Antarctic Treaty area;
   b. obtaining and exchanging information as to the status of native mammals, birds, plants, and invertebrates in the Antarctic Treaty area, and the extent to which any species or population needs protection;
   c. establishing a common form in which this information shall be submitted by Parties in accordance with paragraph 2 below.
2. Each Party shall inform the other Parties as well as the Committee before the end of November of each year of any step taken pursuant to paragraph 1 above and of the number and nature of permits issued under this Annex in the preceding period of 1st July to 30th June.

Article 7: Relationship with other agreements outside the Antarctic Treaty System

Nothing in this Annex shall derogate from the rights and obligations of Parties under the International Convention for the Regulation of Whaling.

Article 8: Review

The Parties shall keep under continuing review measures for the conservation of Antarctic fauna and flora, taking into account any recommendations from the Committee.

Article 9: Amendment or modification

1. This Annex may be amended or modified by a measure adopted in accordance with Article IX (1) of the Antarctic Treaty. Unless the measure specifies otherwise, the amendment or modification shall be deemed to have been approved, and shall become effective, one year after the close of the Antarctic Treaty Consultative Meeting at which it was adopted, unless one or more of the Antarctic Treaty Consultative Parties notifies the Depositary, within that time period, that it wishes an extension of that period or that it is unable to approve the measure.
2. Any amendment or modification of this Annex which becomes effective in accordance with paragraph 1 above shall thereafter become effective as to any other Party when notice of approval by it has been received by the Depositary.

Appendix A: Specially Protected Species

All species of the genus *Arctocephalus*, Fur Seals. *Ommatophoca rossii*, Ross Seal.

Appendix B: Importation of animals and plants

The following animals and plants may be imported into the Antarctic Treaty area in accordance with permits issued under Article 4 of this Annex:

a. domestic plants; and
b. laboratory animals and plants including viruses, bacteria, yeasts and fungi.

Appendix C: Precautions to prevent introductions of micro-organisms

1. Poultry. No live poultry or other living birds shall be brought into the Antarctic Treaty area. Before dressed

poultry is packaged for shipment to the Antarctic Treaty area, it shall be inspected for evidence of disease, such as Newcastle's Disease, tuberculosis, and yeast infection. Any poultry or parts not consumed shall be removed from the Antarctic Treaty area or disposed of by incineration or equivalent means that eliminates risks to native flora and fauna.

2. The importation of non-sterile soil shall be avoided to the maximum extent practicable.

Annex III. Waste disposal and waste management

Article 1: General obligations

1. This Annex shall apply to activities undertaken in the Antarctic Treaty area pursuant to scientific research programmes, tourism and all other governmental and non-governmental activities in the Antarctic Treaty area for which advance notice is required under Article VII (5) of the Antarctic Treaty, including associated logistic support activities.
2. The amount of wastes produced or disposed of in the Antarctic Treaty area shall be reduced as far as practicable so as to minimise impact on the Antarctic environment and to minimise interference with the natural values of Antarctica, with scientific research and with other uses of Antarctica which are consistent with the Antarctic Treaty.
3. Waste storage, disposal and removal from the Antarctic Treaty area, as well as recycling and source reduction, shall be essential considerations in the planning and conduct of activities in the Antarctic Treaty area.
4. Wastes removed from the Antarctic Treaty area shall, to the maximum extent practicable, be returned to the country from which the activities generating the waste were organized or to any other country in which arrangements have been made for the disposal of such wastes in accordance with relevant international agreements.
5. Past and present waste disposal sites on land and abandoned work sites of Antarctic activities shall be cleaned up by the generator of such wastes and the user of such sites. This obligation shall not be interpreted as requiring:
   a. the removal of any structure designated as a historic site or monument; or
   b. the removal of any structure or waste material in circumstances where the removal by any practical option would result in greater adverse environmental impact than leaving the structure or waste material in its existing location.

Article 2: Waste disposal by removal from the Antarctic Treaty area

1. The following wastes, if generated after entry into force of this Annex, shall be removed from the Antarctic Treaty area by the generator of such wastes:
   a. radioactive materials;
   b. electrical batteries;
   c. fuel, both liquid and solid;
   d. wastes containing harmful levels of heavy metals or acutely toxic or harmful persistent compounds;
   e. polyvinyl chloride (PVC), polyurethane foam, polystyrene foam, rubber and lubricating oils, treated timbers and other products which contain additives that could produce harmful emissions if incinerated;
   f. all other plastic wastes, except low density polyethylene containers (such as bags for storing wastes), provided that such containers shall be incinerated in accordance with Article 3 (1);
   g. fuel drums; and
   h. other solid, non-combustible wastes;
   provided that the obligation to remove drums and solid non-combustible wastes contained in subparagraphs (g) and (h) above shall not apply in circumstances where the removal of such wastes by any practical option would result in greater adverse environmental impact than leaving them in their existing locations.
2. Liquid wastes which are not covered by paragraph 1 above and sewage and domestic liquid wastes, shall, to the maximum extent practicable, be removed from the Antarctic Treaty area by the generator of such wastes.
3. The following wastes shall be removed from the Antarctic Treaty area by the generator of such wastes, unless incinerated, autoclaved or otherwise treated to be made sterile:
   a. residues of carcasses of imported animals;
   b. laboratory culture of micro-organisms and plant pathogens; and
   c. introduced avian products.

Article 3: Waste disposal by incineration

1. Subject to paragraph 2 below, combustible wastes, other than those referred to in Article 2 (1), which are not removed from the Antarctic Treaty area shall be burnt in incinerators which to the maximum extent practicable reduce harmful emissions. Any emission standards and equipment guidelines which may be recommended by, *inter alia*, the Committee and the Scientific Committee on Antarctic Research shall be taken into account. The solid residue of such incineration shall be removed from the Antarctic Treaty area.
2. All open burning of wastes shall be phased out as soon as practicable, but no later than the end of the 1998/1999 season. Pending the completion of such

phase-out, when it is necessary to dispose of wastes by open burning, allowance shall be made for the wind direction and speed and the type of wastes to be burnt to limit particulate deposition and to avoid such deposition over areas of special biological, scientific, historic, aesthetic or wilderness significance including, in particular, areas accorded protection under the Antarctic Treaty.

Article 4: Other waste disposal on land

1. Wastes not removed or disposed of in accordance with Articles 2 and 3 shall not be disposed of onto ice-free areas or into fresh water systems.
2. Sewage, domestic liquid wastes and other liquid wastes not removed from the Antarctic Treaty area in accordance with Article 2, shall, to the maximum extent practicable, not be disposed of onto sea ice, ice shelves or the grounded ice-sheet, provided that such wastes which are generated by stations located inland on ice shelves or on the grounded ice-sheet may be disposed of in deep ice pits where such disposal is the only practicable option. Such pits shall not be located on known ice-flow lines which terminate at ice-free areas or in areas of high ablation.
3. Wastes generated at field camps shall, to the maximum extent practicable, be removed by the generator of such wastes to supporting stations or ships for disposal in accordance with this Annex.

Article 5: Disposal of waste in the sea

1. Sewage and domestic liquid wastes may be discharged directly into the sea, taking into account the assimilative capacity of the receiving marine environment and provided that:
   a. such discharge is located, wherever practicable, where conditions exist for initial dilution and rapid dispersal; and
   b. large quantities of such wastes (generated in a station where the average weekly occupancy over the austral summer is approximately 30 individuals or more) shall be treated at least by maceration.
2. The by-product of sewage treatment by the Rotary Biological Contacter process or similar processes may be disposed of into the sea provided that such disposal does not adversely affect the local environment, and provided also that any such disposal at sea shall be in accordance with Annex IV to the Protocol.

Article 6: Storage of waste

All wastes to be removed from the Antarctic Treaty area, or otherwise disposed of, shall be stored in such a way as to prevent their dispersal into the environment.

Article 7: Prohibited products

No polychlorinated biphenyls (PCBs), non-sterile soil, polystyrene beads, chips or similar forms of packaging, or pesticides (other than those required for scientific, medical or hygiene purposes) shall be introduced onto land or ice shelves or into water in the Antarctic Treaty area.

Article 8: Waste management planning

1. Each Party which itself conducts activities in the Antarctic Treaty area shall, in respect of those activities, establish a waste disposal classification system as a basis for recording wastes and to facilitate studies aimed at evaluating the environmental impacts of scientific activity and associated logistic support. To that end, wastes produced shall be classified as:
   a. sewage and domestic liquid wastes (Group 1);
   b. other liquid wastes and chemicals, including fuels and lubricants (Group 2);
   c. solids to be combusted (Group 3);
   d. other solid wastes (Group 4); and
   e. radioactive material (Group 5).
2. In order to reduce further the impact of waste on the Antarctic environment, each such Party shall prepare and annually review and update its waste management plans (including waste reduction, storage and disposal), specifying for each fixed site, for field camps generally, and for each ship (other than small boats that are part of the operations of fixed sites or of ships and taking into account existing management plans for ships):
   a. programmes for cleaning up existing waste disposal sites and abandoned work sites;
   b. current and planned waste management arrangements, including final disposal;
   c. current and planned arrangements for analysing the environmental effects of waste and waste management; and
   d. other efforts to minimise any environmental effects of wastes and waste management.
3. Each such Party shall, as far as is practicable, also prepare an inventory of locations of past activities (such as traverses, field depots, field bases, crashed aircraft) before the information is lost, so that such locations can be taken into account in planning future scientific programmes (such as snow chemistry, pollutants in lichens or ice core drilling).

Article 9: Circulation and review of waste management plans

1. The waste management plans prepared in accordance with Article 8, reports on their implementation, and the inventories referred to in Article 8 (3), shall be included in the annual exchanges of information in accordance with Articles III and VII of the Antarctic Treaty and related Recommendations under Article IX of the Antarctic Treaty.
2. Each Party shall send copies of its waste management plans, and reports on their implementation and review, to the Committee.

3. The Committee may review waste management plans and reports thereon and may offer comments, including suggestions for minimising impacts and modifications and improvement to the plans, for the consideration of the Parties.
4. The Parties may exchange information and provide advice on, *inter alia*, available low waste technologies, reconversion of existing installations, special requirements for effluents, and appropriate disposal and discharge methods.

Article 10: Management plans

Each Party shall:

a. designate a waste management official to develop and monitor waste management plans; in the field, this responsibility shall be delegated to an appropriate person at each site;
b. ensure that members of its expeditions receive training designed to limit the impact of its operations on the Antarctic environment and to inform them of requirements of this Annex; and
c. discourage the use of polyvinyl chloride (PVC) products and ensure that its expeditions to the Antarctic Treaty area are advised of any PVC products they may introduce into that area in order that these products may be removed subsequently in accordance with this Annex.

Article 11: Review

This Annex shall be subject to regular review in order to ensure that it is updated to reflect improvement in waste disposal technology and procedures and to ensure thereby maximum protection of the Antarctic environment.

Article 12: Cases of emergency

1. This Annex shall not apply in cases of emergency relating to the safety of human life or of ships, aircraft or equipment and facilities of high value or the protection of the environment.
2. Notice of activities undertaken in cases of emergency shall be circulated immediately to all Parties and to the Committee.

Article 13: Amendment or modification

1. This Annex may be amended or modified by a measure adopted in accordance with Article IX (1) of the Antarctic Treaty. Unless the measure specifies otherwise, the amendment or modification shall be deemed to have been approved, and shall become effective, one year after the close of the Antarctic Treaty Consultative Meeting at which it was adopted, unless one or more of the Antarctic Treaty Consultative Parties notifies the Depositary, within that time period, that it wishes an extension of that period or that it is unable to approve the amendment.
2. Any amendment or modification of this Annex which becomes effective in accordance with paragraph 1 above shall thereafter become effective as to any other Party when notice of approval by it has been received by the Depositary.

Annex IV. Prevention of marine pollution

Article 1: Definitions

For the purpose of this Annex:

a. 'discharge' means any release howsoever caused from a ship and includes any escape, disposal, spilling, leaking, pumping, emitting or emptying;
b. 'garbage' means all kinds of victual, domestic and operational waste excluding fresh fish and parts thereof, generated during the normal operation of the ship, except those substances which are covered by Articles 3 and 4;
c. 'MARPOL 73/78' means the International Convention for the Prevention of Pollution from Ships, 1973, as amended by the Protocol of 1978 relating thereto and by any other amendment in force thereafter;
d. 'noxious liquid substance' means any noxious liquid substance as defined in Annex II of MARPOL 73/78;
e. 'oil' means petroleum in any form including crude oil, fuel oil, sludge, oil refuse and refined oil products (other than petrochemicals which are subject to the provisions of Article 4);
f. 'oily mixture' means a mixture with any oil content; and
g. 'ship' means a vessel of any type whatsoever operating in the marine environment and includes hydrofoil boats, air-cushion vehicles, submersibles, floating craft and fixed or floating platforms.

Article 2: Application

This Annex applies, with respect to each Party, to ships entitled to fly its flag and to any other ship engaged in or supporting its Antarctic operations, while operating in the Antarctic Treaty area.

Article 3: Discharge of oil

1. Any discharge into the sea of oil or oily mixture shall be prohibited, except in cases permitted under Annex I of MARPOL 73/78. While operating in the Antarctic Treaty area, ships shall retain on board all sludge, dirty ballast, tank washing waters and other oily residues and mixtures which may not be discharged into the sea. Ships shall discharge these residues only outside

the Antarctic Treaty area, at reception facilities or as otherwise permitted under Annex I of MARPOL 73/78.
2. This Article shall not apply to:
   a. the discharge into the sea of oil or oily mixture resulting from damage to a ship or its equipment:
      i. provided that all reasonable precautions have been taken after the occurrence of the damage or discovery of the discharge for the purpose of preventing or minimising the discharge; and
      ii. except if the owner or the Master acted either with intent to cause damage, or recklessly and with the knowledge that damage would probably result; or
   b. the discharge into the sea of substances containing oil which are being used for the purpose of combating specific pollution incidents in order to minimise the damage from pollution.

Article 4: Discharge of noxious liquid substances

The discharge into the sea of any noxious liquid substance, and any other chemical or other substances, in quantities or concentrations that are harmful to the marine environment, shall be prohibited.

Article 5: Disposal of garbage

1. The disposal into the sea of all plastics, including but not limited to synthetic ropes, synthetic fishing nets, and plastic garbage bags, shall be prohibited.
2. The disposal into the sea of all other garbage, including paper products, rags, glass, metal, bottles, crockery, incineration ash, dunnage, lining and packing materials, shall be prohibited.
3. The disposal into the sea of food wastes may be permitted when they have been passed through a comminuter or grinder, provided that such disposal shall, except in cases permitted under Annex V of MARPOL 73/78, be made as far as practicable from land and ice shelves but in any case not less than 12 nautical miles from the nearest land or ice shelf. Such comminuted or ground food wastes shall be capable of passing through a screen with openings no greater than 25 millimeters.
4. When a substance or material covered by this article is mixed with other such substance or material for discharge or disposal, having different disposal or discharge requirements, the most stringent disposal or discharge requirements shall apply.
5. The provisions of paragraphs 1 and 2 above shall not apply to:
   a. the escape of garbage resulting from damage to a ship or its equipment provided all reasonable precautions have been taken, before and after the occurrence of the damage, for the purpose of preventing or minimising the escape; or
   b. the accidental loss of synthetic fishing nets, provided all reasonable precautions have been taken to prevent such loss.
6. The Parties shall, where appropriate, require the use of garbage record books.

Article 6: Discharge of sewage

1. Except where it would unduly impair Antarctic operations:
   a. each Party shall eliminate all discharge into the sea of untreated sewage ("sewage" being defined in Annex IV of MARPOL 73/78) within 12 nautical miles of land or ice shelves;
   b. beyond such distance, sewage stored in a holding tank shall not be discharged instantaneously but at a moderate rate and, where practicable, while the ship is *en route* at a speed of no less than 4 knots. This paragraph does not apply to ships certified to carry not more than 10 persons.
2. The Parties shall, where appropriate, require the use of sewage record books.

Article 7: Cases of emergency

1. Articles 3, 4, 5 and 6 of this Annex shall not apply in cases of emergency relating to the safety of a ship and those on board or saving life at sea.
2. Notice of activities undertaken in cases of emergency shall be circulated immediately to all Parties and to the Committee.

Article 8: Effect on dependent and associated ecosystems

In implementing the provisions of this Annex, due consideration shall be given to the need to avoid detrimental effects on dependent and associated ecosystems, outside the Antarctic Treaty area.

Article 9: Ship retention capacity and reception facilities

1. Each Party shall undertake to ensure that all ships entitled to fly its flag and any other ship engaged in or supporting its Antarctic operations, before entering the Antarctic Treaty area, are fitted with a tank or tanks of sufficient capacity on board for the retention of all sludge, dirty ballast, tank washing water and other oil residues and mixtures, and have sufficient capacity on board for the retention of garbage, while operating in the Antarctic Treaty area and have concluded arrangements to discharge such oily residues and garbage at a reception facility after leaving that area. Ships shall also have sufficient capacity on board for the retention of noxious liquid substances.
2. Each Party at whose ports ships depart *en route* to or arrive from the Antarctic Treaty area undertakes to

ensure that as soon as practicable adequate facilities are provided for the reception of all sludge, dirty ballast, tank washing water, other oily residues and mixtures, and garbage from ships, without causing undue delay, and according to the needs of the ships using them.

3. Parties operating ships which depart to or arrive from the Antarctic Treaty area at ports of other Parties shall consult with those Parties with a view to ensuring that the establishment of port reception facilities does not place an inequitable burden on Parties adjacent to the Antarctic Treaty area.

Article 10: Design, construction, manning and equipment of ships

In the design, construction, manning and equipment of ships engaged in or supporting Antarctic operations, each Party shall take into account the objectives of this Annex.

Article 11: Sovereign immunity

1. This Annex shall not apply to any warship, naval auxiliary or other ship owned or operated by a State and used, for the time being, only on government non-commercial service. However, each Party shall ensure by the adoption of appropriate measures not impairing the operations or operational capabilities of such ships owned or operated by it, that such ships act in a manner consistent, so far as is reasonable and practicable, with this Annex.
2. In applying paragraph 1 above, each Party shall take into account the importance of protecting the Antarctic environment.
3. Each Party shall inform the other Parties of how it implements this provision.
4. The dispute settlement procedure set out in Articles 18 to 20 of the Protocol shall not apply to this Article.

Article 12: Preventive measures and emergency preparedness and response

1. In order to respond more effectively to marine pollution emergencies or the threat thereof in the Antarctic Treaty area, the Parties, in accordance with Article 15 of the Protocol, shall develop contingency plans for marine pollution response in the Antarctic Treaty area, including contingency plans for ships (other than small boats that are part of the operations of fixed sites or of ships) operating in the Antarctic Treaty area, particularly ships carrying oil as cargo, and for oil spills, originating from coastal installations, which enter into the marine environment. To this end they shall:
    a. co-operate in the formulation and implementation of such plans; and
    b. draw on the advice of the Committee, the International Maritime Organization and other international organizations.

2. The Parties shall also establish procedures for co-operative response to pollution emergencies and shall take appropriate response actions in accordance with such procedures.

Article 13: Review

The Parties shall keep under continuous review the provisions of this Annex and other measures to prevent, reduce and respond to pollution of the Antarctic marine environment, including any amendments and new regulations adopted under MARPOL 73/78, with a view to achieving the objectives of this Annex.

Article 14: Relationship with MARPOL 73/78

With respect to those Parties which are also Parties to MARPOL 73/78, nothing in this Annex shall derogate from the specific rights and obligations thereunder.

Article 15: Amendment or modification

1. This Annex may be amended or modified by a measure adopted in accordance with Article IX (1) of the Antarctic Treaty. Unless the measure specifies otherwise, the amendment or modification shall be deemed to have been approved, and shall become effective, one year after the close of the Antarctic Treaty Consultative Meeting at which it was adopted, unless one or more of the Antarctic Treaty Consultative Parties notifies the Depositary, within that time period, that it wishes an extension of that period or that it is unable to approve the measure.
2. Any amendment or modification of this Annex which becomes effective in accordance with paragraph 1 above shall thereafter become effective as to any other Party when notice of approval by it has been received by the Depositary.

Annex V. Area Protection and Management

Article 1: Definitions

For the purposes of this Annex:

a. 'appropriate authority' means any person or agency authorized by a Party to issue permits under this Annex;
b. 'permit' means a formal permission in writing issued by an appropriate authority;
c. 'Management Plan' means a plan to manage the activities and protect the special value or values in an Antarctic Specially Protected Area or an Antarctic Specially Managed Area.

Article 2: Objectives

For the purposes set out in this Annex, any area, including any marine area, may be designated as an Antarctic Specially Protected Area or an Antarctic Specially Managed Area. Activities in those Areas shall be prohibited, restricted or managed in accordance with Management Plans adopted under the provisions of this Annex.

Article 3: Antarctic Specially Protected Areas

1. Any area, including any marine area, may be designated as an Antarctic Specially Protected Area to protect outstanding environmental, scientific, historic, aesthetic or wilderness values, any combination of those values, or ongoing or planned scientific research.
2. Parties shall seek to identify, within a systematic environmental-geographical framework, and to include in the series of Antarctic Specially Protected Areas:
   a. areas kept inviolate from human interference so that future comparisons may be possible with localities that have been affected by human activities;
   b. representative examples of major terrestrial, including glacial and aquatic, ecosystems and marine ecosystems;
   c. areas with important or unusual assemblages of species, including major colonies of breeding native birds or mammals;
   d. the type locality or only known habitat of any species;
   e. areas of particular interest to on-going or planned scientific research;
   f. examples of outstanding geological, glaciological or geomorphological features;
   g. areas of outstanding aesthetic and wilderness value;
   h. sites or monuments of recognized historic value; and
   i. such other areas as may be appropriate to protect the values set out in paragraph 1 above.
3. Specially Protected Areas and Sites of Special Scientific Interest designated as such by past Antarctic Treaty Consultative Meetings are hereby designated as Antarctic Specially Protected Areas and shall be renamed and renumbered accordingly.
4. Entry into an Antarctic Specially Protected Area shall be prohibited except in accordance with a permit issued under Article 7.

Article 4: Antarctic Specially Managed Areas

1. Any area, including any marine area, where activities are being conducted or may in the future be conducted, may be designated as an Antarctic Specially Managed Area to assist in the planning and co-ordination of activities, avoid possible conflicts, improve co-operation between Parties or minimize environmental impacts.
2. Antarctic Specially Managed Areas may include:
   a. areas where activities pose risks of mutual interference or cumulative environmental impacts; and
   b. sites or monuments of recognized historic value.
3. Entry into an Antarctic Specially Managed Area shall not require a permit.
4. Notwithstanding paragraph 3 above, an Antarctic Specially Managed Area may contain one or more Antarctic Specially Protected Areas, entry into which shall be prohibited except in accordance with a permit issued under Article 7.

Article 5: Management plans

1. Any Party, the Committee, the Scientific Committee for Antarctic Research or the Commission for the Conservation of Antarctic Marine Living Resources may propose an area for designation as an Antarctic Specially Protected Area or an Antarctic Specially Managed Area by submitting a proposed Management Plan to the Antarctic Treaty Consultative Meeting.
2. The area proposed for designation shall be of sufficient size to protect the values for which the special protection or management is required.
3. Proposed Management Plans shall include, as appropriate:
   a. a description of the value or values for which special protection or management is required;
   b. a statement of the aims and objectives of the Management Plan for the protection or management of those values;
   c. management activities which are to be undertaken to protect the values for which special protection or management is required;
   d. a period of designation, if any;
   e. a description of the area, including:
      i. the geographical co-ordinates, boundary markers and natural features that delineate the area;
      ii. access to the area by land, sea or air including marine approaches and anchorages, pedestrian and vehicular routes within the area, and aircraft routes and landing areas;
      iii. the location of structures, including scientific stations, research or refuge facilities, both within the area and near to it; and
      iv. the location in or near the area of other Antarctic Specially Protected Areas or Antarctic Specially Managed Areas designated under this Annex, or other protected areas designated in accordance with measures adopted under other components of the Antarctic Treaty System;
   f. the identification of zones within the area, in which activities are to be prohibited, restricted or managed for the purpose of achieving the aims and objectives referred to in subparagraph b. above;
   g. maps and photographs that show clearly the boundary of the area in relation to surrounding features and key features within the area;
   h. supporting documentation;

i. in respect of an area proposed for designation as an Antarctic Specially Protected Area, a clear description of the conditions under which permits may be granted by the appropriate authority regarding:
   i. access to and movement within or over the area;
   ii. activities which are or may be conducted within the area, including restrictions on time and place;
   iii. the installation, modification, or removal of structures;
   iv. the location of field camps;
   v. restrictions on materials and organisms which may be brought into the area;
   vi. the taking of or harmful interference with native flora and fauna;
   vii. the collection or removal of anything not brought into the area by the permit holder;
   viii. the disposal of waste;
   ix. measures that may be necessary to ensure that the aims and objectives of the Management Plan can continue to be met; and
   x. requirements for reports to be made to the appropriate authority regarding visits to the area;
j. in respect of an area proposed for designation as an Antarctic Specially Managed Area, a code of conduct regarding:
   i. access to and movement within or over the area;
   ii. activities which are or may be conducted within the area, including restrictions on time and place;
   iii. the installation, modification, or removal of structures;
   iv. the location of field camps;
   v. the taking of or harmful interference with native flora and fauna;
   vi. the collection or removal of anything not brought into the area by the visitor;
   vii. the disposal of waste; and
   viii. any requirements for reports to be made to the appropriate authority regarding visits to the area; and
k. provisions relating to the circumstances in which Parties should seek to exchange information in advance of activities which they propose to conduct.

Article 6: Designation procedures

1. Proposed Management Plans shall be forwarded to the Committee, the Scientific Committee on Antarctic Research and, as appropriate, to the Commission for the Conservation of Antarctic Marine Living Resources. In formulating its advice to the Antarctic Treaty Consultative Meeting, the Committee shall take into account any comments provided by the Scientific Committee on Antarctic Research and, as appropriate, by the Commission for the Conservation of Antarctic Marine Living Resources. Thereafter, Management Plans may be approved by the Antarctic Treaty Consultative Parties by a measure adopted at an Antarctic Treaty Consultative Meeting in accordance with Article IX(1) of the Antarctic Treaty. Unless the measure specifies otherwise, the Plan shall be deemed to have been approved 90 days after the close of the Antarctic Treaty Consultative Meeting at which it was adopted, unless one or more of the Consultative Parties notifies the Depositary, within that time period, that it wishes an extension of that period or is unable to approve the measure.
2. Having regard to the provisions of Articles 4 and 5 of the Protocol, no marine area shall be designated as an Antarctic Specially Protected Area or an Antarctic Specially Managed Area without the prior approval of the Commission for the Conservation of Antarctic Marine Living Resources.
3. Designation of an Antarctic Specially Protected Area or an Antarctic Specially Managed Area shall be for an indefinite period unless the Management Plan provides otherwise. A review of a Management Plan shall be initiated at least every five years. The Plan shall be updated as necessary.
4. Management Plans may be amended or revoked in accordance with paragraph 1 above.
5. Upon approval Management Plans shall be circulated promptly by the Depositary to all Parties. The Depositary shall maintain a record of all currently approved Management Plans.

Article 7: Permits

1. Each Party shall appoint an appropriate authority to issue permits to enter and engage in activities within an Antarctic Specially Protected Area in accordance with the requirements of the Management Plan relating to that Area. The permit shall be accompanied by the relevant sections of the Management Plan and shall specify the extent and location of the Area, the authorized activities and when, where and by whom the activities are authorized and any other conditions imposed by the Management Plan.
2. In the case of a Specially Protected Area designated as such by a past Antarctic Treaty Consultative Meeting which does not have a Management Plan, the appropriate authority may issue a permit for a compelling scientific purpose which cannot be served elsewhere and which will not jeopardize the natural ecological system in that Area.

3. Each Party shall require a permit-holder to carry a copy of the permit while in the Antarctic Specially Protected Area concerned.

Article 8: Historic Sites and Monuments

1. Sites or monuments of recognized historic value which have been designated as Antarctic Specially Protected Areas or Antarctic Specially Managed Areas, or which are located within such Areas, shall be listed as Historic Sites and Monuments.
2. Any Party may propose a site or monument of recognized historic value which has not been designated as an Antarctic Specially Protected Area or an Antarctic Specially Managed Area, or which is not located within such an Area, for listing as a Historic Site or Monument. The proposal for listing may be approved by the Antarctic Treaty Consultative Parties by a measure adopted at an Antarctic Treaty Consultative Meeting in accordance with Article IX(1) of the Antarctic Treaty. Unless the measure specifies otherwise, the proposal shall be deemed to have been approved 90 days after the close of the Antarctic Treaty Consultative Meeting at which it was adopted, unless one or more of the Consultative Parties notifies the Depositary, within that time period, that it wishes an extension of that period or is unable to approve the measure.
3. Existing Historic Sites and Monuments which have been listed as such by previous Antarctic Treaty Consultative Meetings shall be included in the list of Historic Sites and Monuments under this Article.
4. Listed Historic Sites and Monuments shall not be damaged, removed or destroyed.
5. The list of Historic Sites and Monuments may be amended in accordance with paragraph 2 above. The Depositary shall maintain a list of current Historic Sites and Monuments.

Article 9: Information and publicity

1. With a view to ensuring that all persons visiting or proposing to visit Antarctica understand and observe the provisions of this Annex, each Party shall make available information setting forth, in particular:
   a. the location of Antarctic Specially Protected Areas and Antarctic Specially Managed Areas;
   b. listing and maps of those Areas;
   c. the Management Plans, including listings of prohibitions relevant to each Area;
   d. the location of Historic Sites and Monuments and any relevant prohibition or restriction.
2. Each Party shall ensure that the location and, if possible, the limits of Antarctic Specially Protected Areas, Antarctic Specially Managed Areas and Historic Sites and Monuments are shown on its topographic maps, hydrographic charts and in other relevant publications.
3. Parties shall co-operate to ensure that, where appropriate, the boundaries of Antarctic Specially Protected Areas, Antarctic Specially Managed Areas and Historic Sites and Monuments are suitably marked on the site.

Article 10: Exchange of information

1. The Parties shall make arrangements for:
   a. collecting and exchanging records, including records of permits and reports of visits, including inspection visits, to Antarctic Specially Protected Areas and reports of inspection visits to Antarctic Specially Managed Areas;
   b. obtaining and exchanging information on any significant change or damage to any Antarctic Specially Managed Area, Antarctic Specially Protected Area or Historic Site or Monument; and
   c. establishing common forms in which records and information shall be submitted by Parties in accordance with paragraph 2 below.
2. Each Party shall inform the other Parties and the Committee before the end of November of each year of the number and nature of permits issued under this Annex in the preceding period of 1st July to 30th June.
3. Each Party conducting, funding or authorizing research or other activities in Antarctic Specially Protected Areas or Antarctic Specially Managed Areas shall maintain a record of such activities and in the annual exchange of information in accordance with the Antarctic Treaty shall provide summary descriptions of the activities conducted by persons subject to its jurisdiction in such areas in the preceding year.
4. Each Party shall inform the other Parties and the Committee before the end of November each year of measures it has taken to implement this Annex, including any site inspections and any steps it has taken to address instances of activities in contravention of the provisions of the approved Management Plan for an Antarctic Specially Protected Area or Antarctic Specially Managed Area.

Article 11: Cases of emergency

1. The restrictions laid down and authorized by this Annex shall not apply in cases of emergency involving safety of human life or of ships, aircraft, or equipment and facilities of high value or the protection of the environment.
2. Notice of activities undertaken in cases of emergency shall be circulated immediately to all Parties and to the Committee.

Article 12: Amendment or modification

1. This Annex may be amended or modified by a measure adopted in accordance with Article IX(1) of

the Antarctic Treaty. Unless the measure specifies otherwise, the amendment or modification shall be deemed to have been approved, and shall become effective, one year after the close of the Antarctic Treaty Consultative Meeting at which it was adopted, unless one or more of the Antarctic Treaty Consultative Parties notifies the Depositary, within that time period, that it wishes an extension of that period or that it is unable to approve the measure.

2. Any amendment or modification of this Annex which becomes effective in accordance with paragraph 1 above shall thereafter become effective as to any other Party when notice of approval by it has been received by the Depositary.

Appendix E

Text of the Antarctic Treaty

The Governments of Argentina, Australia, Belgium, Chile, the French Republic, Japan, New Zealand, Norway, the Union of South Africa, the Union of Soviet Socialist Republics, the United Kingdom of Great Britain and Northern Ireland, and the United States of America,

Recognizing that it is in the interest of all mankind that Antarctica shall continue for ever to be used exclusively for peaceful purposes and shall not become the scene or object of international discord;

Acknowledging the substantial contributions to scientific knowledge resulting from international cooperation in scientific investigation in Antarctica;

Convinced that the establishment of a firm foundation for the continuation and development of such cooperation on the basis of freedom of scientific investigation in Antarctica as applied during the International Geophysical Year accords with the interests of science and the progress of all mankind;

Convinced also that a treaty ensuring the use of Antarctica for peaceful purposes only and the continuance of international harmony in Antarctica will further the purposes and principles embodied in the Charter of the United Nations;

Have agreed as follows:

Article I

1. Antarctica shall be used for peaceful purposes only. There shall be prohibited, *inter alia*, any measure of a military nature, such as the establishment of military bases and fortifications, the carrying out of military manoeuvres, as well as the testing of any type of weapon.
2. The present Treaty shall not prevent the use of military personnel or equipment for scientific research or for any other peaceful purpose.

Article II

Freedom of scientific investigation in Antarctica and cooperation toward that end, as applied during the International Geophysical Year, shall continue, subject to the provisions of the present Treaty.

Article III

1. In order to promote international cooperation in scientific investigation in Antarctica, as provided for in Article II of the present Treaty, the Contracting Parties agree that, to the greatest extent feasible and practicable:
   a. information regarding plans for scientific programs in Antarctica shall be exchanged to permit maximum economy of and efficiency of operations;
   b. scientific personnel shall be exchanged in Antarctica between expeditions and stations;
   c. scientific observations and results from Antarctica shall be exchanged and made freely available.

Article IV

1. Nothing contained in the present Treaty shall be interpreted as:
   a. a renunciation by any Contracting Party of previously asserted rights of or claims to territorial sovereignty in Antarctica;
   b. a renunciation or diminution by any Contracting Party of any basis of claim to territorial sovereignty in Antarctica which it may have whether as a result of its activities or those of its nationals in Antarctica, or otherwise;
   c. prejudicing the position of any Contracting Party as regards its recognition or non-recognition of any other State's rights of or claim or basis of claim to territorial sovereignty in Antarctica.
2. No acts or activities taking place while the present Treaty is in force shall constitute a basis for asserting, supporting or denying a claim to territorial sovereignty in Antarctica or create any rights of sovereignty in Antarctica. No new claim, or enlargement of an existing claim, to territorial sovereignty in Antarctica shall be asserted while the present Treaty is in force.

Article V

1. Any nuclear explosions in Antarctica and the disposal there of radioactive waste material shall be prohibited.
2. In the event of the conclusion of international agreements concerning the use of nuclear energy, including nuclear explosions and the disposal of radioactive waste material, to which all of the Contracting Parties whose representatives are entitled to participate in the meetings provided for under Article IX are parties, the rules established under such agreements shall apply in Antarctica.

Article VI

The provisions of the present Treaty shall apply to the area south of 60° South Latitude, including all ice shelves, but nothing in the present Treaty shall prejudice or in any way affect the rights, or the exercise of the rights, of any State under international law with regard to the high seas within that area.

Article VII

1. In order to promote the objectives and ensure the observance of the provisions of the present Treaty, each Contracting Party whose representatives are entitled to participate in the meetings referred to in Article IX of the Treaty shall have the right to designate observers to carry out any inspection provided for by the present Article. Observers shall be nationals of the Contracting Parties which designate them. The names of observers shall be communicated to every other Contracting Party having the right to designate observers, and like notice shall be given of the termination of their appointment.
2. Each observer designated in accordance with the provisions of paragraph 1 of this Article shall have complete freedom of access at any time to any or all areas of Antarctica.
3. All areas of Antarctica, including all stations, installations and equipment within those areas, and all ships and aircraft at points of discharging or embarking cargoes or personnel in Antarctica, shall be open at all times to inspection by any observers designated in accordance with paragraph 1 of this Article.
4. Aerial observation may be carried out at any time over any or all areas of Antarctica by any of the Contracting Parties having the right to designate observers.
5. Each Contracting Party shall, at the time when the present Treaty enters into force for it, inform the other Contracting Parties, and thereafter shall give them notice in advance, of
   a. all expeditions to and within Antarctica, on the part of its ships or nationals, and all expeditions to Antarctica organized in or proceeding from its territory;
   b. all stations in Antarctica occupied by its nationals; and
   c. any military personnel or equipment intended to be introduced by it into Antarctica subject to the conditions prescribed in paragraph 2 of Article I of the present Treaty.

Article VIII

1. In order to facilitate the exercise of their functions under the present Treaty, and without prejudice to the respective positions of the Contracting Parties relating to jurisdiction over all other persons in Antarctica, observers designated under paragraph 1 of Article VII and scientific personnel exchanged under sub-paragraph 1(b) of Article III of the Treaty, and members of the staffs accompanying any such persons, shall be subject only to the jurisdiction of the Contracting Party of which they are nationals in respect of all acts or omissions occurring while they are in Antarctica for the purpose of exercising their functions.
2. Without prejudice to the provisions of paragraph 1 of this Article, and pending the adoption of measures in pursuance of sub-paragraph 1(e) of Article IX, the Contracting Parties concerned in any case of dispute with regard to the exercise of jurisdiction in Antarctica shall immediately consult together with a view to reaching a mutually acceptable solution.

Article IX

1. Representatives of the Contracting Parties named in the preamble to the present Treaty shall meet at the City of Canberra within two months after the date of entry into force of the Treaty, and thereafter at suitable intervals and places, for the purpose of exchanging information, consulting together on matters of common interest pertaining to Antarctica, and formulating and considering, and recommending to their Governments, measures in furtherance of the principles and objectives of the Treaty, including measures regarding:
   a. use of Antarctica for peaceful purposes only;
   b. facilitation of scientific research in Antarctica;
   c. facilitation of international scientific cooperation in Antarctica;
   d. facilitation of the exercise of the rights of inspection provided for in Article VII of the Treaty;
   e. questions relating to the exercise of jurisdiction in Antarctica;
   f. preservation and conservation of living resources in Antarctica.
2. Each Contracting Party which has become a party to the present Treaty by accession under Article XIII shall be entitled to appoint representatives to participate in the meetings referred to in paragraph 1 of the present Article, during such times as that Contracting Party demonstrates its interest in Antarctica by conducting substantial research activity there, such as the establishment of a scientific station or the despatch of a scientific expedition.
3. Reports from the observers referred to in Article VII of the present Treaty shall be transmitted to the representatives of the Contracting Parties participating in the meetings referred to in paragraph 1 of the present Article.
4. The measures referred to in paragraph 1 of this Article shall become effective when approved by all the

Contracting Parties whose representatives were entitled to participate in the meetings held to consider those measures.

5. Any or all of the rights established in the present Treaty may be exercised as from the date of entry into force of the Treaty whether or not any measures facilitating the exercise of such rights have been proposed, considered or approved as provided in this Article.

Article X

Each of the Contracting Parties undertakes to exert appropriate efforts, consistent with the Charter of the United Nations, to the end that no one engages in any activity in Antarctica contrary to the principles or purposes of the present Treaty.

Article XI

1. If any dispute arises between two or more of the Contracting Parties concerning the interpretation or application of the present Treaty, those Contracting Parties shall consult among themselves with a view to having the dispute resolved by negotiation, inquiry, mediation, conciliation, arbitration, judicial settlement or other peaceful means of their own choice.
2. Any dispute of this character not so resolved shall, with the consent, in each case, of all parties to the dispute, be referred to the International Court of Justice for settlement; but failure to reach agreement on reference to the International Court shall not absolve parties to the dispute from the responsibility of continuing to seek to resolve it by any of the various peaceful means referred to in paragraph 1 of this Article.

Article XII

1.
   a. The present Treaty may be modified or amended at any time by unanimous agreement of the Contracting Parties whose representatives are entitled to participate in the meetings provided for under Article IX. Any such modification or amendment shall enter into force when the depositary Government has received notice from all such Contracting Parties that they have ratified it.
   b. Such modification or amendment shall thereafter enter into force as to any other Contracting Party when notice of ratification by it has been received by the depositary Government. Any such Contracting Party from which no notice of ratification is received within a period of two years from the date of entry into force of the modification or amendment in accordance with the provision of sub-paragraph 1(a) of this Article shall be deemed to have withdrawn from the present Treaty on the date of the expiration of such period.

2.
   a. If after the expiration of thirty years from the date of entry into force of the present Treaty, any of the Contracting Parties whose representatives are entitled to participate in the meetings provided for under Article IX so requests by a communication addressed to the depositary Government, a Conference of all the Contracting Parties shall be held as soon as practicable to review the operation of the Treaty.
   b. Any modification or amendment to the present Treaty which is approved at such a Conference by a majority of the Contracting Parties there represented, including a majority of those whose representatives are entitled to participate in the meetings provided for under Article IX, shall be communicated by the depositary Government to all Contracting Parties immediately after the termination of the Conference and shall enter into force in accordance with the provisions of paragraph 1 of the present Article.
   c. If any such modification or amendment has not entered into force in accordance with the provisions of sub-paragraph 1(a) of this Article within a period of two years after the date of its communication to all the Contracting Parties, any Contracting Party may at any time after the expiration of that period give notice to the depositary Government of its withdrawal from the present Treaty; and such withdrawal shall take effect two years after the receipt of the notice by the depositary Government.

Article XIII

1. The present Treaty shall be subject to ratification by the signatory States. It shall be open for accession by any State which is a Member of the United Nations, or by any other State which may be invited to accede to the Treaty with the consent of all the Contracting Parties whose representatives are entitled to participate in the meetings provided for under Article IX of the Treaty.
2. Ratification of or accession to the present Treaty shall be effected by each State in accordance with its constitutional processes.
3. Instruments of ratification and instruments of accession shall be deposited with the Government of the United States of America, hereby designated as the depositary Government.

4. The depositary Government shall inform all signatory and acceding States of the date of each deposit of an instrument of ratification or accession, and the date of entry into force of the Treaty and of any modification or amendment thereto.

5. Upon the deposit of instruments of ratification by all the signatory States, the present Treaty shall enter into force for those States and for States which have deposited instruments of accession. Thereafter the Treaty shall enter into force for any acceding State upon the deposit of its instruments of accession.

6. The present Treaty shall be registered by the depositary Government pursuant to Article 102 of the Charter of the United Nations.

Article XIV

The present Treaty, done in the English, French, Russian and Spanish languages, each version being equally authentic, shall be deposited in the archives of the Government of the United States of America, which shall transmit duly certified copies thereof to the Governments of the signatory and acceding States.

Further reading

Alberts, F. G. 1995. *Geographic names of the Antarctic*. 2nd edition. Washington, National Science Foundation (NSF 95–157).

Amundsen, R. 1912. *The South Pole: An account of the Norwegian Antarctic expedition in the 'Fram', 1910–1912*. 2 vols. London, John Murray.

Armitage, A. B. 1905. *Two years in the Antarctic*. London, Arnold

Auburn, F. M. 1982. *Antarctic law and politics*. London, Hurst.

Bagshawe, T. W. 1939. *Two men in the Antarctic. An expedition to Graham Land 1920–1922*. Cambridge, Cambridge University Press.

Bailey, A. M. and Sorensen, J. H. 1962. *Sub-Antarctic Campbell Island*. Wellington, A. H. and A. W. Reed.

Bailey, H. P. 1964. Toward a unified concept of the temperate climate. *Geographical review* **54**: 506–545.

Bakayev, V. G. 1966. *Atlas of Antarctica* Vol. 1. Moscow and Leningrad, Ministry of Geology.

Barnes, J. 1984. Antarctica: The politics of protection. *Report of the 16th Technical meeting, IUCN. Symposium B: Protecting Antarctica and the Southern Ocean. Madrid 5–14 November 1984.*

Barnes, J. N. 1982. *Let's save Antarctica!*. Victoria, Greenhouse Publications.

Bauer, T. 2001. *Tourism in the Antarctic: Opportunities, constraints and future prospects*. New York, Haworth Press.

Baughman, T. 1994. *Before the heroes came: Antarctica in the 1890s*. Lincoln and London: University of Nebraska Press.

Beaglehole, J. C. (editor). 1955. *The journals of Captain James Cook on his voyages of discovery. I. The voyage of the Endeavour 1768–71*. London, Hakluyt Society.

Beaglehole, J. C. (editor). 1961. *The journals of Captain James Cook on his voyages of discovery. II. The voyage of the Resolution and Adventure 1772–75*. London, Hakluyt Society.

Beaglehole, J. C. (editor). 1962. *The Endeavour journals of Joseph Banks*. 2 vols. Sydney: Angus and Robertson.

Beaglehole, J. C. (editor). 1967. *The journals of Captain James Cook on his voyages of discovery. III. The voyage of the Resolution and Discovery 1776–80*. London, Hakluyt Society.

Beck, P. 1986. *The international politics of Antarctica*. London, Croom Helm.

Behrendt, J. C. 1991. Scientific studies relevant to the question of Antarctica's petroleum resource potential. In: Tingey, R. J. (editor). *The geology of Antarctica*. Oxford, Clarendon Press: 588–616.

Beltramino, J. C. M. 1993. *The structure and dynamics of Antarctic population*. New York, Vantage Press.

Bernacchi, L. 1901. *To the south polar regions: Expedition of 1898–1900*. London, Hurst and Blackett.

Beyer, L. and Bölter, M. 2001. *Geoecology of terrestrial Antarctic oases*. Berlin, Springer-Verlag.

Bonner, W. N. 1982. *Seals and man: A study of interactions*. Seattle, University of Washington Press.

Bonner, W. N. 1989. *Whales of the world*. London, Blandford Press.

Borchgrevink, C. E. 1901. *First on the Antarctic continent, being an account of the British Antarctic Expedition, 1898–1900*. London, George Newnes.

Borman, P. and Fritzsch, D. (editors). 1996. *The Schirmacher Oasis, Queen Maud Land, East Antarctica, and its surroundings*. Gotha, Justus Perthes Verlag.

Brent, P. *Captain Scott and the Antarctic tragedy*. London, Weidenfeld and Nicolson.

Brownell, R. L., Best, P. B. and Prescott, J. H. (editors). 1986. *Right whales: Past and present status*. Cambridge, International Whaling Commission (Reports of the IWC, 10).

Bruce, W. S. 1896. Cruise of the *Balaena* and the *Active* in the Antarctic seas, 1892–93. Part 1. The *Balaena*. *Geographical Journal* **7**:502–521.

Bruce, W. S. (no date). *Report on the work of the Scottish National Antarctic Expedition*. Edinburgh, Scottish Oceanographic Laboratory.

Bruce, W. S. 1911. *Polar Exploration*. London, Williams and Norgate.

Bull, H. J. 1896. *The cruise of the 'Antarctic' to the south polar regions*. London, Edward Arnold.

Burton, A. C. and Edholm, O. G. 1955. *Man in a cold environment*. London, Edward Arnold.

Bushnell, V., (editor). 1964–75. *Antarctic map folio series*. New York: American Geographical Society.

Byrd, R. E. 1931. *Little America: Aerial exploration in the Antarctic*. New York, G. P. Putnam's sons.

Byrd, R. E. 1935. *Discovery: The story of the second Byrd Antarctic expedition*. New York, G. P. Putnam's sons.

Campbell, Lord George. 1876. *Log letters from the 'Challenger'*. London, Macmillan.

Campbell, L. B. and Claridge, G. G. C. 1987. *Antarctica: Soils, weathering processes and environments*. Amsterdam, Elsevier.

Carwardine, M. 1995. *Whales, dolphins and porpoises: The visual guide to all the world's cetaceans*. London, Dorling Kindersley.

Cassidy, W. A. 1991. Meteorites from Antarctica. In: Tingey, R. J. (editor). *The geology of Antarctica*. Oxford, Clarendon Press: 652–666.

Central Intelligence Agency. 1978. *Polar regions atlas*. Washington, DC: Government Printing Office.

Charcot, J. -B. 1906. *Le 'Francais' au Pôle Sud. Journal de l'Expédition Antarctique Français, 1903–1905*. Paris, Flammarion.

Charcot, J. -B. 1911. *The voyage of the 'Why not?' in the Antarctic. The journal of the second French South Polar Expedition, 1908–1910*. London, Hodder and Stoughton.

Chaturvedi, S. 1996. *The polar regions: A political geography*. London, John Wiley.

Christensen, Lars. 1935. *Such is the Antarctic*. London, Hodder and Stoughton.
Christie, E. W. H. 1951. *The Antarctic problem*. London, George Allen and Unwin.
Clark, M. R. and Dingwall, P. R. 1985. *Conservation of islands in the Southern Ocean*. Gland, IUCN.
Codling, R. J. 1995. The precursors of tourism in the Antarctic. In: Hall, C. M. and Johnston, M. E. (editors). *Polar tourism: Tourism in the Arctic and Antarctic regions*: 167–177.
Colbert, E. H. 1991. Mesozoic and Cainozoic fossils from Antarctica. In: Tingey, R. J. (editor). *The geology of Antarctica*. Oxford, Clarendon Press: 568–587.
Conrad, L. J., comp. 1999. *Bibliography of Antarctic exploration: Expedition accounts from 1768 to 1960*. Washougal, WA: The Author.
Cook, F. A. 1900. *Through the first Antarctic night, 1898–99*. London, Heinemann.
Cooper, R. A. and Shergold, J. H. 1991. Palaeozoic invertebrates of Antarctica. In: Tingey, R. J. (editor). *The geology of Antarctica*. Oxford, Clarendon Press: 455–486.
Crawford, A. 1982. *Tristan da Cunha and the Roaring Forties*. Edinburgh, Charles Skilton.
Croxall, J. P., Evans, P. G. H. and Schreiber, R. W. (editors). 1985. The status and conservation of the world's seabirds. *ICBP Technical publication 2*.
Dalrymple, P. C. and Frostman, T. O. 1971. Some aspects of the climate of interior Antarctica. In: Quam, L. O. and Porter, H. D. (editors). *Research in the Antarctic*. Washington DC, American Association for the Advancement of Science: 429–442.
Darlington, J. 1949. *My Antarctic honeymoon: A year at the bottom of the world*. London, Frederick Mueller.
Davis, B. W. and Herr, R. A. 1993. The legitimacy of the Convention on the Conservation of Antarctic Marine Living Resources. *International Antarctic Regime Project Publications Series 6*.
De Wit, M. J. 1985. *Minerals and mining in Antarctica: Science and technology, economics and politics*. Oxford, Clarendon Press.
Deacon, G. 1984. *The Antarctic circumpolar ocean*. Cambridge, Cambridge University Press.
Debenham, F. (editor). 1945. *The voyage of Captain Bellingshausen in the Antarctic Seas, 1819–21. 2 vols*. London, Hakluyt Society.
Dodds, K. 1997. *Geopolitics in Antarctica*. Chichester, John Wiley.
Donald, C. W. 1896. Cruise of the *Balaena* and the *Active* in the Antarctic seas, 1892–93. Part 2. The *Active*. *Geographical Journal* 7: 625–643.
Drewry, D. J. (editor). 1983. *Antarctica: Glaciological and geophysical folio*. Cambridge, Scott Polar Research Institute.
Eastman, J. T. 1992. *Antarctic Fish Biology*. New York, Academic Press.
Eden, A. W. 1955. *Islands of despair*. London, Melrose.
Edholm, O. G. 1978. *Man – hot and cold*. London, Edward Arnold.
Edwards, D. and Heap, J. 1981. Convention on the Conservation of Antarctic Marine Living Resources: A commentary. *Polar Record* **20**(127): 353–362.

Ellsworth, L. 1938. *Beyond horizons*. London, William Heinemann.
Enzenbacher, D. J. 1995. The regulation of Antarctic tourism. In: Hall, C. M. and Johnston, M. E. (editors). *Polar tourism: Tourism in the Arctic and Antarctic regions*: 179–215.
Evans, E. R. G. R. 1921. *South with Scott*. London, Collins.
Evans, P. G. H. 1987. *The natural history of whales and dolphins*. London, Christopher Helm.
Everson, I. (editor). 2000. *Krill Ecology and Fisheries*. Oxford, Blackwell Science.
Farman, J. C., Gardiner, B. G. and Shanklin, J. D. 1995. Large losses of total ozone in Antarctica reveal seasonal ClOx/NOx interaction. *Nature* **315**: 207–210.
Fifield, R. 1987. *International Research in the Antarctic*. Cambridge, SCAR/ICSU and Oxford University Press.
Fisher, M. and Fisher, J. 1957. *Shackleton*. London, Barrie.
Fraser, C. 1986. *Beyond the roaring forties: New Zealand's sub-Antarctic islands*. Wellington, Government Printing Office.
Fuchs, V. E. 1982. *Of ice and men*. Oswestry, Anthony Nelson.
Fuchs, V. E. 1990. *A time to speak*. Oswestry, Anthony Nelson.
Fuchs, V. E. and Hillary, E. 1958. *The crossing of Antarctica*. London, Cassell.
Gaskin, D. E. 1982. *The ecology of whales and dolphins*. London, Heinemann.
Gentry, R. L. & Kooyman, G. L. (editors). 1986. *Fur seals: Maternal strategies on land and at sea*. Princeton, N.J., Princeton University Press.
Giæver, J. 1954. *The white desert: The official account of the Norwegian-British-Swedish Antarctic Expedition*. London, Chatto and Windus.
Gildea, D. 1998. *The Antarctic mountaineering chronology*. Fyshwick, ACT: The Author.
Gloerson, P. et al. 1992. *Arctic and Antarctic sea ice, 1978–1987. Satellite passive-microwave observations and analysis (NASA SP-511/Atlas)*. Washington DC, NASA.
Gould, L. M. 1931. *Cold. The record of an Antarctic sledge journey*. New York, Brewer, Warren and Putnam.
Grierson, J. 1949. *Air whaler*. London, Sampson Low.
Grierson, J. 1960. *Sir Hubert Wilkins. Enigma of exploration*. London, Robert Hale.
Hamre, I. 1933. The Japanese South Polar Expedition of 1911–12: A little-known episode in Antarctic exploration. *Geographical Journal* **82** (5): 411–423.
Hardy, A. C. 1956. *The open sea: The world of plankton*. London, Collins.
Hardy, A. C. 1959. *The open sea II: Fish and fisheries*. London, Collins.
Hardy, A. C. 1967. *Great waters*. London, Collins.
Hart, I. B. 2001. *Pesca: A history of the pioneer modern whaling company in the Antarctic*. Salcombe, Aiden Ellis.
Harvey, J. G. 1976. *Atmosphere and ocean: Our fluid environments*. Sussex, Artemis.
Hattersley-Smith, G. 1980. The history of place-names in the Falkland Islands Dependencies (South Georgia and the South Sandwich Islands). *British Antarctic Survey Scientific Reports* No. 101.
Hattersley-Smith, G. 1991. The history of place-names in the British Antarctic Territory. *British Antarctic Survey Scientific Reports* No. 113.

Hattersley-Smith, G. and Roberts, A. 1993. *Gazetteer of the British Antarctic Territory*. 2nd edition. London: HMSO.

Headland, R. K. 1984. *The island of South Georgia*. Cambridge, Cambridge University Press.

Headland, R. K. 1989. *Chronological list of Antarctic expeditions and related historical events*. Cambridge, Cambridge University Press.

Heap, J. 1994. *Antarctic Treaty Handbook*. Eighth Edition. Washington DC, US Department of State.

Hedgepeth, J. W. 1971. James Eights of the Antarctic (1798–1882). In: Quam, L. O. and Porter, H. D. (editors). *Research in the Antarctic*. Washington, American Association for the Advancement of Science: 3–45.

Holdgate, M. 1977. Terrestrial ecosystems in the Antarctic. *Philosophical Transactions of the Royal Society of London*, Series B: **279**: 5–25.

Holdgate, M. W. and Tinker, J. 1978. *Oil and other minerals in the Antarctic*. Cambridge, SCAR.

Holdgate, M. W. 1958 *Mountains in the sea*. London, Macmillan.

Huntford, R. 1979. *Scott and Amundsen*. London, Hodder and Stoughton.

Huntford, R. 1985. *Shackleton*. London, Hodder and Stoughton.

Huxley, L. 1914. *Scott's last expedition. 2 vols*. London, Smith Elder.

Jacka, T. H., Christou, L. and Cook, B. J. 1984. A data bank of mean monthly and annual surface temperatures for Antarctica, the Southern Ocean and South Pacific Ocean. *ANARE Research Notes 22*.

Jeannel, R. 1941. *Au seuil de l'Antarctique*. Paris, Presses Universitaires de France.

Jones, A. and Shanklin, J. D. 1995. Continued decline of total ozone over Halley, Antarctica, since 1985. *Nature* **376**: 409–411.

Jones, A. G. E. 1992. *Polar portraits: Collected papers*. Whitby, Caedmon.

Jørgensen-Dahl, A. and Østreng, W. 1991. *The Antarctic Treaty System in world politics*. Basingstoke, Macmillan.

Joyce, E. E. M. 1929. *South polar trail: The log of the Imperial Trans-Antarctic Expedition*. London, Duckworth.

King, J. C. and Turner, J. 1997. *Antarctic meteorology and climatology*. Cambridge, Cambridge University Press.

King, J. E. 1983. *Seals of the world*. Oxford, Oxford University Press.

Knox, G. A. 1994. *The biology of the Southern Ocean*. Cambridge, Cambridge University Press.

Kock, K. H. 1992. *Antarctic Fish and Fisheries*, Cambridge, Cambridge University Press.

Kohl-Larsen, L. 1930. *An den Toren der Antarktis*. Stuttgart, Strecker und Schroder.

Kooyman, G. L. 1981. *Weddell seal, consummate diver*. Cambridge, Cambridge University Press.

Kooyman, G. L. 1989. *Diverse divers: Physiology and behaviour*. Berlin, Springer-Verlag.

Laws, R. M. (editor). 1993. *Antarctic seals: Research methods and techniques*. Cambridge, Cambridge University Press.

Leader-Williams, N. 1988. *Reindeer on South Georgia*. Cambridge, Cambridge University Press.

Le Boeuf, B. J. and Laws, R. M. (editors). 1994. *Elephant seals: Population ecology, behavior, and physiology*. Berkeley, University of California Press.

Le Masurier, W. E. and Thomson, J. W. (editors). 1990. Volcanoes of the Antarctic plate and Southern Oceans. *American Geophysical Union Antarctic Research Series 48*.

Levick, G. M. 1914. *Antarctic penguins: A study of their social habits*. London, Heinemann.

Linklater, Eric. 1972. *The voyage of the Challenger*. London, John Murray.

Lockwood, J. G. 1974. *World climatology: An environmental approach*. London, Edward Arnold.

Longton, R. E. 1988. *Biology of polar bryophytes and lichens*. Cambridge, Cambridge University Press.

Lysaght, A. (editor). 1971. *Joseph Banks in Newfoundland and Labrador 1766.; his diary manuscripts and collections*. London, British Museum.

Lyster, S. 1985. *International wildlife law*. Cambridge, Grotius Publications.

Madigan, C. T. 1929. Meteorology of the Cape Denison station. *Australasian Antarctic Expedition 1911–14, Scientific Reports*, **B** (4), 1–286.

Malaurie, J. 1989. J.-B. Charcot; father of French polar research. *Polar Record* **25**(154): 191–196.

Marchal, A. 1989. Convention for the Conservation of Antarctic Seals: 1988 review of operations. *Polar Record* **25** (153): 142–143.

Marr, J. W. S. 1923. *Into the frozen south*. London, Cassell.

Marsh, John H. 1948. *No pathway here*. London, Hodder and Stoughton.

Massom, R. 1991. *Satellite remote sensing of polar regions*. London, Belhaven Press.

Mawson, D. 1915. *The home of the blizzard. 2 vols*. London, Heinemann.

Mawson, D. 1932. The B.A.N.Z. Antarctic Research Expedition 1929–31. *Geographical Journal*. **80**: 101–131.

Meadows, J., Mills, W. J. and King, H. G. R. comps. 1994. *The Antarctic*. Oxford: Clio Press.

Mills, H. R. 1905. *The siege of the South Pole*. London, Alston, Rivers.

Mills, W. J. (forthcoming). *The encyclopedia of polar exploration*. Santa Barbara, CA: ABC-Clio.

Mills, W. J. and P. Speak. 1998. *Keyguide to information sources on the polar and cold regions*. London: Mansell.

Mott, P. G. 1986. *Wings over ice: An account of the Falkland Islands and Dependencies Aerial Survey Expedition*. Long Sutton, Peter Mott.

Murdoch, W. G. B. 1894. *From Edinburgh to the Antarctic*. London and New York, Longmans, Green and Co.

Murphy, R. C. 1936. *Oceanic birds of South America*. New York, Macmillan and the American Museum of Natural History.

Murphy, R. C. 1947. *Logbook for Grace*. London, Robert Hale.

Nicol, S. and Allison, I. 1997. The frozen skin of the Southern Ocean. *American Scientist* **85** (Sept–Oct): 426–439.

Nordenskjöld, N. O. G. and others: 1905. *Antarctica: Or two years amongst the ice of the South Pole*. London, Hurst and Blackett.

O'Brian, P. 1987. *Joseph Banks*. London, Collins Harvill.

Ommanney, F. D. 1938. *South latitude*. London, Longmans, Green.
Ommanney, F. D. 1971. *Lost leviathan, whales and whaling*. London, Hutchinson.
Orego Vicuña, F. 1983. *Antarctic resources policy: Scientific, legal and political issues*. Cambridge, Cambridge University Press.
Orego Vicuña, F. 1988. *Antarctic mineral exploitation: The emerging legal framework*. Cambridge, Cambridge University Press.
Parry, R. B. and Perkins, C. R. editors. 2000. *World mapping today*. 2nd ed. East Grinstead, England: Bowker-Saur.
Pennycuik, C. J. 1987. Flight of seabirds. In *Seabirds: Feeding ecology and role in marine ecosystems* (ed. J. P. Croxall), pp. 43–62. Cambridge, Cambridge University Press.
Pickard, J. (editor). 1986. *Antarctic oasis: Terrestrial environments and history of the Vestfold Hills*. Sydney, Academic Press.
Polar Record. 1941. A summary of the activities of the US Antarctic Expedition 1939–41. *Polar Record*, **3**(22), 427–449.
Priestley, R. E. 1914. *Antarctic adventure, Scott's northern party*. London, T. Fisher Unwin.
Puissochet, J.-P. 1993. CCAMLR – a critical assessment. In: Jørgensen-Dahl, A. and Østreng, W. 1991. *The Antarctic Treaty System in world politics*. Basingstoke, Macmillan: 70–76.
Rankin, A. N. 1951. *Antarctic isle. Wildlife in South Georgia*. London, Collins.
Reader's Digest. 1985. *Antarctica: Great stories from the frozen continent*. Sydney, Reader's Digest.
Reader's Digest. 1990. *Antarctica: The extraordinary story of man's conquest of the frozen continent*. 2nd edition. Sydney: Reader's Digest.
Rees, W. G. 1993. A new wind-chill nomogram. *Polar Record* **29**(170): 229–234.
Riedman, M. 1990. *The pinnipeds: Seals, sea lions, and walruses*. Berkeley, University of California Press.
Ronne, F. 1949. *Antarctic conquest*. New York, Van Rees Press.
Rose, L. A. 1980. *Assault on eternity: Richard E. Byrd and the exploration of Antarctica, 1946–47*. Annapolis, MD, Naval Institute Press.
Rosove, M. H., comp. 2001. *Antarctica, 1772–1922: Freestanding Publications through 1999*. Santa Monica, CA: Adélie Books.
Ross, J. C. 1847. *A voyage of discovery and research in the southern and Antarctic regions during the years 1839–43*. London, John Murray.
Rowley, P. D., Ford, A. B., Williams, P. L. and Pride, D. E. 1991. Metallic and non-metallic mineral resources of Antarctica. In: Tingey, R. J. (editor). *The geology of Antarctica*. Oxford, Clarendon Press: 617–651.
Rubin, M. J. 1982. Thaddeus Bellingshausen's scientific programme in the Southern Ocean, 1818–21. *Polar Record* **21**(132): 215–229.
Rusin, N. P. 1964. *Meteorological and radiational regimes of Antarctica*. Jerusalem, Program for Scientific Translations.
Rycroft, M. J. 1990. The Antarctic atmosphere: A hot topic in a cold cauldron. *Geographical Journal* **156**(1): 1–11.

Rymill, J. R. 1938. *Southern lights: The official account of the British Graham Land Expedition, 1934–1937*. London, Chatto and Windus.
Savours, A. (editor) 1966. *Edward Wilson: Diary of the Discovery Expedition*. London, Blandford Press.
SCAR. 1993. *Antarctic Digital Database*. Cambridge: SCAR.
SCAR Bulletin. 2000. Life under the ice and the climatic evolution of Antarctica. SCAR Bulletin **138**: 272.
Schmidt-Nielsen, K. 1997. *Animal physiology: Adaptation and environment*. Fifth edition. Cambridge, Cambridge University Press.
Schwerdtfeger, W. 1984. *Weather and climate of the Antarctic*. Amsterdam, Elsevier.
Scientific Committee on Antarctic Research. 1998 *Composite Gazetteer of Antarctica* (Second printing: 2 vols plus supplement). Cambridge, SCAR.
Scott, R. F. 1905. *The Voyage of the 'Discovery'*. London, Smith Elder and Co.
Selkirk, P. M., Seppelt, R. D. and Selkirk, D. R. 1990. *Sub-Antarctic Macquarie Island*. Cambridge, Cambridge University Press.
Shackleton, E. H. 1909. *The heart of the Antarctic. 2 vols*. London, Heinemann.
Shackleton, Lord. 1982. *Falkland Islands. Economic Report*. London, HMSO.
Simmonds, M. P. and Hutchinson, J. D. 1996. *The conservation of whales and dolphins*. Chichester, Wiley.
Siple, P. A. 1931. *A boy scout with Byrd*. New York, Putnam.
Siple, P. A. 1936. *Scout to explorer. Back with Byrd to the Antarctic*. New York, Putnam.
Siple, P. A. 1959. *90° South: the story of the American South Pole conquest*. New York, Putnam.
Small, G. L. 1971. *The blue whale*. New York, Columbia University Press.
Solomon, S. 2001. *The coldest march. Scott's fateful Antarctic expedition*. New Haven, Yale University Press.
Speak, P. 1992. (editor). *The log of the Scotia by William Spiers Bruce*. Edinburgh, University of Edinburgh Press.
Spence, S. A., comp. 1980. *Antarctic miscellany; books, periodicals and maps relating to the discovery and exploration of Antarctica*. London: J.J.H. and J.I. Simper.
Stamp, T. and Stamp, C. 1978. *James Cook, maritime scientist*. Whitby, Caedmon.
Stokke, O. S. 1993. The effectiveness of the Convention on the Conservation of Antarctic Marine Living Resources. *International Antarctic Regime Project Publications* Series 4.
Stonehouse, B. 1982. La zonation écologique sous les hautes latitudes australes. In: Jouventin, P., Massé, L. and Tréhen, P. (editors). *Colloque sur les ecosystémes subantarctiques*. Paris, Comité national français des récherches antarctiques: 531–536.
Stonehouse, B. 1989. *Polar ecology*. New York, Chapman and Hall.
Stonehouse, B. 2000. *The last continent: Discovering Antarctica*. Burgh Castle: SCP Books.
Suter, K. 1991. *Antarctica: Private property or public heritage?* London, Zed Books.
Tauber, G. M. 1960. Characteristics of Antarctic katabatic winds. In: [Australian Academy of Science]. *Antarctic meteorology: Proceedings of the symposium held in Melbourne 1959*. Oxford, Pergamon Press.

Thomson, J. 1999. *Shackleton's captain: A biography of Frank Worsley*. Christchurch, Hazard Press.

Thomson, M. R. A. 1991. Antarctic invertebrate fossils: The Mezozoic–Cainozoic record. In: Tingey, R. J. (editor). *The geology of Antarctica*. Oxford, Clarendon Press.

Thomson, W. 1877. *The voyage of the 'Challenger'*. 2 vols. London, Macmillan and Co.

Times Books. 1999. *Times comprehensive atlas of the world*. 10th edition. London: Times Books.

Tingey R. J. 1991. *The geology of Antarctica*. Oxford, Clarendon Press, 680 pp.

Tolstikov, Ye. I., (editor). 1966, 1969. *Atlas Antarktiki*. Moscow and Leningrad: Glavnoye Upravleniye Geodezii I Kartografii MG SSR.

Tønnesen, J. N. 1970. Norwegian Antarctic whaling, 1905–68: An historical appraisal. *Polar Record* **15** (96): 283–290.

Tønnesen, J. N. and Johnsen, A. O. 1982. *The history of modern whaling*. London, Hurst.

Triggs, G. D. 1986. *International law and Australian sovereignty in Antarctica*. Sydney, Legal Books Pty Ltd.

Triggs, G. D. 1987. *The Antarctic Treaty Regime*. Cambridge, Cambridge University Press.

Trusswell, E. M. 1991. Antarctica: A history of terrestrial vegetation. In: Tingey, R. J. (editor). *The geology of Antarctica*. Oxford, Clarendon Press: 499–537.

Udvardy, M. D. F. 1975. *A Classification of the Biogeographical Provinces of the World*. IUCN Occasional paper 18.

Vamplew, W. 1975. *Salvesen of Leith*. Edinburgh, Scottish Academic Press.

Van Zinderen Bakker, E. M., Winterbottom, J. M. and Dyer, R. A. 1971. *Marion and Prince Edward Islands*. Cape Town, A. A. Balkema.

Venter, R. J. 1957. Sources of meteorological data for the Antarctic. In: van Rooy, M. P. (editor). *Meteorology of the Antarctic*. Pretoria, Weather Bureau, South Africa: 17–38.

Vincent, W. V. 1988. *Microbial ecosystems of Antarctica*. Cambridge, Cambridge University Press.

Wadhams, P. 2000. *Ice in the ocean*. Amsterdam, Gordon and Breach.

Waldman, C. and Wexler, A. 1992. *Who was who in world exploration*. New York, Facts on File.

Walton, K. and Atkinson, R. 1996. *Of dogs and men: Fifty years in the Antarctic*. Malvern, Images.

Warham, J. 1990. *The petrels. Their ecology and breeding systems*. London, Academic Press.

Warham, J. 1996. *The behaviour, population biology and physiology of the petrels*. London, Academic Press.

Watson, L. 1981. *Sea guide to whales of the world*. London, Hutchinson.

Webster, W. H. B. 1834. *Narrative of a voyage to the southern Atlantic Ocean, in the years 1828–1830, performed in Her Majesty's Sloop 'Chanticleer', etc*. 2 vols. London, privately published.

Weddell, J. 1825. *Voyage towards the South Pole*.

West, J. and Credland, A. G. 1995. *Scrimshaw: The art of the whaler*. Cherry Burton, Hutton Press.

Wild, F. 1923. *Shackleton's last voyage: The story of the Quest*. London, Cassell.

Wilkes, C. 1845. *Narrative of the United States Exploring Expedition, etc*. 5 vols. Philadelphia, Lea and Blanchard.

Wilkins, Sir G. H. 1929. The Wilkins-Hurst Antarctic Expedition 1928–29. *Geographical Review* **19**(3): 353–376.

Wilkins, Sir G. H. 1930. Further Antarctic explorations. *Geographical Review* **20**(3): 357–388.

Williams, P. J. and Smith, M. W. 1989. *The frozen earth. Fundamentals of geocryology*. Cambridge, Cambridge University Press.

Williams, T. D. 1995. *The penguins: Spheniscidae*. Oxford, Oxford University Press.

Winn, L. K. and Winn, H. E. 1985. *Wings in the sea: The humpback whale*. Hanover, University Press of New England.

Worsley, F. A. 1931. *Endurance. An epic of polar adventure*. London, Philip Allan.

Young, G. C. 1991. Fossil fishes from Antarctica. In: Tingey, R. J. (editor). *The geology of Antarctica*. Oxford, Clarendon Press: 538–567.

Zumberge, J. H. 1979. *Possible environmental effects of mineral exploration and exploitation in Antarctica*. Cambridge, Scientific Committee on Antarctic Research.

Study Guide: Climate and Life

Earth, ocean and atmosphere, a dynamically interactive system, provide the range of environments in which living processes occur. Without its atmosphere, Earth would experience diurnal and annual shifts of temperature that alone would make life impossible. The atmosphere filters, entraps and modifies solar radiation, and together with the ocean, circulates and redistributes solar energy in ways that facilitate life in both equatorial and polar regions. For insights into the solar inputs, radiation budgets and balances underlying climate formation see entries on **solar radiation**, **albedo**, **atmosphere**, **atmospheric circulation** and **atmospheric heat transport**. Our knowledge of southern polar and subpolar climates is based on long-term weather records: for details of wind, cloud and precipitation see **weather systems**. How the data are gathered appears in **weather data** and **weather stations**.

Antarctica and its surrounding regions, notoriously cold, snow-covered and windy, tend to be written off as uniformly insufferable. However, like any other large region of the globe, the southern polar region demonstrates a wide range of climates, from the continuous unyielding cold of the high polar plateau to the relative warmth of the lower plateau, to the far warmer but also much windier coastal slopes, to the milder maritime conditions of Antarctic Peninsula and the southern islands of the Scotia Arc. Precipitation falls rarely, and then mainly as snow or ice spicules, on the high plateau. Around the coasts and islands successive depressions bring heavy snow, which packs down with wind and its own weight to form ice. The wide range of climates experienced in Antarctica and on the oceanic islands is dealt with in **climatic zones**, **climatic data: sources** and **climatic records**. For some of the most difficult climatic conditions experienced by man in Antarctica see **Home of the blizzard** and **wind chill**. Polar climates, like those elsewhere in the world, undergo secular shifts due to a variety of causes. From the Antarctic region, with its patchy and transient settlements, we have few records long enough to show long-term trends: see **climate change**. Additional evidence comes from the ice covering Antarctica (**ice sheet, Antarctica**): see also **ice cores** and **ice core analysis**.

LAND LIFE

The geographical range of this encyclopedia spans **climatic zones** from extreme polar in the south to cool temperate in the north, traversing a range of biological zones from high polar plateau to temperate islands of the southern oceans. At the southern end, the 98 per cent or more of continental surface covered by permanent ice (**ice sheet, Antarctica**) represents one of the world's coldest, driest and least hospitable environments, and is virtually sterile. At the northern periphery, the **Southern temperate islands** are home to a variety of plants and soil fauna, and breeding platforms for myriads of sea-feeding birds and mammals.

Though most of Antarctica's snows and ice show little evidence of supporting life, patches close to the coasts are often tinted brilliant red or green by **snow algae**, growing usually where seabird droppings and sea spray provide nutrients. Where the ice has retreated and the snows melt seasonally, the bare rocks of the underlying continent, exposed as cliffs and nunataks, may support crustose and dendritic **lichens**, capable of growing without soil. Fierce **weathering** processes may erode the rocks faster than the plants can settle, ultimately yielding **soils**, in which a wider range of vegetation may seek footing. However, continental Antarctica and the Antarctic islands provide very little land where soils can advance beyond the most primitive stage, with little organic content or water-holding capacity. **Permafrost** and **patterned ground** inhibit soil penetration and settlement. Remoteness from sources of propagules (spores, seeds, fragments of plants capable of growth), and chronically low temperatures, searing winds and aridity, discourage all but the hardiest plants with adequate **survival strategies**, that are fortunate enough to find a favourable **microhabitat**. The **colonization of Antarctic environments** is therefore a hazardous and difficult process. Antarctic **oases** – expanses of ice-free ground, usually but not always coastal – are disappointingly misnamed: they are ice-free because precipitation is low, so tend to be arid deserts. Melt-water from peripheral glaciers may provide streams and **freshwater lakes**, some with unusual saline stratification or patterns of annual melting.

Antarctica in consequence has a meagre terrestrial flora, restricted almost entirely to protists, algae, lichens and mosses, that may be capable of active growth for only a few days per year following **spring warming**. The continent's two species of **flowering plants** grow only on Antarctic Peninsula and islands of the Scotia Arc. Antarctica has no equivalent of Arctic tundra: the continent's richest vegetation compares with the northernmost and most meagre Arctic polar desert vegetation. However, fragments of vegetation grow on all known rocky outcrops, to within 330 km

(180 mi) of the South Pole, and on favourable coastal situations on the Antarctic islands, **moss peat** and **moss mires**, can be surprisingly prolific (see also **peats and peat formation, and vegetation, sub-fossil**). Sparse vegetation provides food and shelter for an equally meagre fauna of protozoans, nematodes, tardigrades, collembolae (primitive insects that browse on fungi and detritus) and carnivorous mites.

The **southern islands** present a graded series of increasingly rich soils, soil flora and fauna, and larger plants (See **Study Guide: Southern Oceans and Islands**). While the heavily glaciated **Antarctic islands** show little advance on the continent itself, the **sub-Antarctic islands** beyond the mean northern limit of pack ice, with markedly warmer winters, have accumulated a wider range of soils, soil fauna and flowering plants. These are particularly well developed on the coastal flats, where dense stands of tussock grasses may line the shore, but extend to upland fellfields that approach the permanent snowline. The **cool temperate islands** and **warm temperate islands** north of the **Antarctic Convergence**, with much milder climates year-round, show lush coastal vegetation (again dominated by tussock grasses), and rolling moorlands with grasses and shrubs. The northernmost islands support stands of trees, particularly in areas sheltered from the prevailing westerly winds. For details see accounts of individual islands and groups.

Isolation has kept all the southern islands free from indigenous land mammals, but almost all of the sub-Antarctic and temperate islands support stocks of mammals introduced purposely or accidentally by man, and some of the temperate islands support cattle, sheep and other domestic stock. Several of the islands have indigenous species of birds, mostly showing affinities with South American or Australasian species from which they are probably derived.

MARINE LIFE

Seaweeds, including long-stranded kelps, proliferate on sub-Antarctic islands, but are inhibited on Antarctic islands where floating sea ice scrapes the shores and shallow waters. The cold waters of the southern oceans, though relatively unproductive in winter, in summer support masses of phytoplankton (see **seasonality, Southern Ocean**), especially in areas of up-welling or stirring where nutrient-rich waters are brought to the surface. These form the basis of the **food chains** that provide a rich source of food for marine animals. Phytoplankton is browsed by zooplankton, including shrimp-like crustaceans (notably **krill**), copepods, arrow-worms, jelly-fish, fish larvae, and larval forms of bottom-living starfish, bristle-worms, anemones and molluscs. Zooplankton provides food for fishes and squid, both of which are abundant in Antarctic waters (see **Fishes, Southern Ocean**). Concentrated swarms of zooplankton

Adélie penguins in strong drift. Photo: BS

(especially of krill and young fish), together with larger fish and squid, provide food for the seals, whales and seabirds that are the major predators of southern waters.

Seven species of seal breed within the area, some on southern island shores, others exclusively within the pack ice. For a general account see **seals, Southern Ocean**; see also accounts of individual species. Seven species of whale and eight species of dolphin feed in the area, several penetrating far south into the pack ice in summer. For a general account see **whales, Southern Ocean**; see also accounts of individual species.

Over 40 species of seabird, including nine species of **penguin**, 38 species of **albatross** and lesser **petrel**, and inshore-feeding **cormorants**, **gulls**, **skuas**, **terns** and **sheathbills** breed within the region, mainly on islands and continental coasts. By contrast only five species of land-feeding bird nest in the same area. **Seabird flight** is particularly well developed in some of the southern species that travel long distances between nests and feeding grounds. **Diving in birds and mammals** too is a conspicuous feature of polar and sub-polar species, allowing them to partition the food resources of the Southern Ocean vertically as well as horizontally. **Thermoregulation** is an issue affecting all homeotherms (warm-blooded animals), particularly those of polar regions. Other **survival strategies**, as vital to animals as to plants, include **migration**, which takes many birds and mammals away from the Antarctica region before the onset of winter, spreading them as far as the north Pacific and Atlantic oceans. There is no massive spring influx of shorebirds and waterfowl, matching that of the Arctic: there is no summer-rich tundra to attract them in the far south.

EXPLOITATION AND CONSERVATION

Both seals and whales have in the past been subject to heavy predation by man. Fur seals and southern elephant seals of Antarctic fringe islands were hunted throughout the 19th century for skins and oil: for a general account see **Sealing** in the Southern Ocean. They are currently protected by an instrument of the Antarctic Treaty System, the **Convention for the Conservation of Antarctic Seals**. Whales, especially humpbacks and rorquals, formed the basis of a major industry that began in 1904, reached its zenith during the years following World War II, then declined, despite regulatory efforts of the International Whaling Commission. For a general account of the industry see **Whaling in the Southern Ocean**. For early attempts to start southern whaling, see **Enderby settlement** (a colony established on the inhospitable shores of the **Auckland Islands** by the Enderby Brothers (see **Enderby, Messrs.**) an enterprising whaling firm that encouraged its captains to explore southern waters). For later attempts see accounts of the **German Whaling and Sealing Expedition 1873–74**, the **Norwegian (Sandefjord) Whaling Expedition 1892–94**, and the **Norwegian (Tønsberg) Whaling Expedition 1893–95**. For accounts of more recent and successful whaling enterprises see **Salvesens of Leith** and **Compania Argentina de Pesca**, whose land-based whaling operations on **South Georgia** ended in the 1960s with the closure of their whaling stations. Pelagic whaling in the Southern Ocean continues to the present, though conducted by only one nation (Japan), on a much reduced scale, in an area that other interested nations have declared a **Southern Ocean Sanctuary** for whales.

Both sealing and whaling were important in the development of southern exploration and politics. Sealers were among the first to discover and chart many of the southern islands, and may have been the first to see Antarctica itself. Whalers too explored, initially to examine the feasibility of Antarctic whaling, then to find new anchorages and unclaimed islands for whaling stations. Later they explored for new hunting grounds, and ultimately for the sake of exploration (see **Norwegian (Christensen) Whaling Expeditions 1926–37**): to follow this development see **Study guide: Exploration**). Sealing alerted the commercial world to the presence of islands and a mysterious, ice-covered continent in the Southern Ocean. Whaling triggered the round of sovereignty claims that ultimately divided Antarctica between Britain and six other claimant nations. Revenues from commercial whaling supported **Discovery Investigations 1925–51**, a long-term series of integrated oceanographic expeditions that set a world standard for biological oceanography.

The biological wealth of the Southern Ocean continues to attract industry – currently commercial fisheries interested in its enormous stocks of krill, squid and fin fish. Fish stocks, exploited almost entirely since World War II, are now heavily dented (see for example **toothfish**). The **Convention on the Conservation of Antarctic Marine Living Resources** (CCAMLR), an instrument of the Antarctic Treaty System, points the need for a firm foundation of data and understanding of ecological processes in the Southern Ocean, on which may be based a sensible regulatory regime for exploitation of these resources. CCAMLR has stimulated a wealth of coordinated research on Southern Ocean ecosystems, details of which can be found in its annual reports and website.

ANTARCTIC PROTECTED AREAS

Some 29 Specially Protected Areas (SPAs) and 36 Sites of Special Scientific Interest (SSSIs) have been established under the Antarctic Treaty, almost all with the objective of protecting concentrations of plants or animals, terrestrial or marine. For details see **Study Guide: Protected Areas under the Antarctic Treaty**, and individual entries.

Study Guide: Exploration

INTRODUCTION

This study guide summarizes the history of southern exploration. Further historical details appear in individual entries for explorers (e.g. **Cook**, James), explorations, expeditions (e.g. **Scottish National Antarctic Expedition 1902–4**), place names (e.g. **McMurdo Sound**), and industries and enterprises (e.g. **Sealing in the Southern Ocean, Tourism, Antarctic, Whaling in the Southern Ocean**).

DISCOVERING THE SOUTHERN HEMISPHERE

European awareness of the southern hemisphere and its geography remained largely speculative until the mid to late fifteenth century. Over 2000 years earlier, geographers of the Alexandrian School had deduced Earth's spherical shape and estimated its diameter to within about 15 per cent (**Eratosthenes of Alexandria**). They and their successors speculated on the existence of a southern hemisphere with torrid, temperate and polar zones, hypothesizing the existence of a huge circumpolar landmass, **Terra australis incognita**, of continental size, extending from pole to tropics. Though Phoenician traders may have navigated far south along both coasts of Africa by the eighth century BC, European exploration into the Atlantic Ocean began with voyages sponsored and supported by Portuguese Prince Henry the Navigator (1394–1460) during the early fifteenth century. These started a tradition that ended with the eighteenth-century discovery of the Southern Ocean, and the early nineteenth-century discovery of Antarctica.

DISPELLING A MYTH: 1487–1780

Following early fifteenth century Portuguese discoveries of the Madeira, Azores and Cape Verde archipelagos, European exploration continued southward in efforts to establish sea routes to the wealth of India and the Far East. Portuguese navigators rounding the Cape of Good Hope and crossing the Indian Ocean distanced both Africa and southern Asia from **Terra australis incognita**. Discovery of central and southern America, the **Strait of Magellan** and **Drake Passage** similarly distanced South America, incidentally confirming the possibility of world circumnavigation via the southern hemisphere. Dutch interest in the East Indies stimulated seventeenth-century exploration eastward

Table 1 Expeditions. 1487–1780

Date	Expedition
1487–88	Bartholomeu Diaz de Novães and João Infante discovered and rounded the Cape of Good Hope and Cape Agulhas, opening a sea route to the Indian Ocean.
1497–99	Vasco da Gama led an expedition from Portugal to Calicut, India.
1501–2	Amerigo Vespucci reported land in 52°S, possibly Patagonia or South Georgia.
1506–7	Tristão da Cunha discovered the Tristan da Cunha group.
1519–22	Fernão de Magalhães discovered the Strait of Magellan.
1577–80	Francis Drake discovered Drake Passage, south of Tierra del Fuego.
1586–88	John Davis, of Thomas Cavendish's expedition, discovered the Falkland Islands.
1615–16	Jakob le Maire discovered Cape Horn and Le Maire Strait.
1617–18	Haevik Klasszoon van Hillegom and Adriaen de Wale discovered Ile Amsterdam and Ile St Paul.
1618–19	Bartolomé de Nodal circumnavigated Tierra del Fuego and discovered Islas Diego Ramirez.
1642–43	Abel Janszoon Tasman surveyed the coasts of Staten Landt (eastern New Zealand) and Van Diemen's Landt (Tasmania).
1644–45	Abel Janszoon Tasman surveyed the coast of New Holland (Australia).
1663	Barend Barendszoon Lam discovered Prince Edward Islands.
1738–39	J.-B. C. Bouvet de Lozier discovered Bouvetøya.
1768–71	J. Cook's first expedition explored Tierra del Fuego and Drake Passage, and charted coasts of Australia and New Zealand.
1771–72	Y. J. de Kerguelen-Tremarec discovered Iles Kerguelen.
1771–73	M. M. Marion du Fresne visited Prince Edward Islands, discovered Iles Crozet.
1772–75	J. Cook's second expedition identified and circumnavigated the Southern Ocean, crossed the Antarctic circle, penetrated south to 71°10'S, charted the north coast of South Georgia and the southernmost South Sandwich Islands.
1773–74	Y. J. de Kerguelen-Tremarec revisited Iles Kerguelen.
1776–80	J. Cook's third expedition visited and named Prince Edward Islands: visited Iles Crozet and Kerguelen.

beyond the Indian Ocean, identifying and isolating Australia from the southern continent. French interest led to the discovery of **Iles Kerguelen** and other isolated islands in the southern Indian Ocean. The three voyages of British navigator James Cook identified the **Southern Ocean**, discovering within it **South Georgia** and the **South Sandwich Islands**. Though Cook may have seen the ice-bound shores of the southern continent, he did not claim to have discovered it. However, his voyages reduced considerably the expanse of ocean in which it might still be found.

DISCOVERING ANTARCTICA (1780–1873)

Though fur seals had for several years been taken from the Falkland Islands, Cook's published observations that they were abundant also on South Georgia brought a rush of sealing vessels to southern waters. From the late 1780s, many hundreds of voyages are logged. The sealers' search for new, unexploited stocks of seals led to the discovery of many other southern islands. Others again were sighted by naval or merchant ships using eastern or western routes across the Southern Ocean. A major exploring expedition of this period, the **Russian Naval Expedition** led by Fabian von Bellingshausen, was routed specifically to fill the gaps that Cook had left. In 1829 veteran sealer Benjamin Pendleton led a US government-sponsored scientific expedition to Staten Island and the South Shetland Islands. Between 1837 and 1843 three major scientific expeditions, French, American and British, explored high southern latitudes to record variation in Earth's magnetic field. Their combined results established the presence of a major continent close to the latitude of the polar circle. US sealer Mercator Cooper, exploring for seals off Victoria Land, may have made the first ever landing on the new continent.

WHALERS AND SCIENTIFIC EXPEDITIONS (1873–94)

Whales were by this time known to be plentiful in southern waters, and initiatives for further exploration passed from sealers to whalers. Dismayed by falling catches in the north, German, Scottish and Norwegian whalers explored first the area of Antarctic Peninsula and South Georgia, later the coast of Victoria Land, in search of slow-moving right whales. During the same period scientific expeditions recorded astronomical and other observations from temporary stations on cold temperate and periantarctic islands. The German International Polar Year expedition to Royal Bay, South Georgia, became the first scientific party to winter south of the Antarctic Convergence. H. J. Bull's Norwegian whaling expedition made the first authenticated landing on Antarctica on 24 January 1895.

Table 2 Expeditions. 1786–1876

Date	Expedition
1786	T. Delano: first recorded sealing voyage to South Georgia.
1800	H. Waterhouse discovered Antipodes Islands.
1805–6	A. Bristow discovered Auckland Islands.
1807	W. Moody discovered Bounty Islands.
1809–10	F. Hasselburg discovered Campbell Island and Macquarie Island.
1819	W. Smith discovered South Shetland Islands 19 February, revisited 16 October.
1819–20	W. Smith revisited South Shetland Islands: sighted northwest coast of Antarctic Peninsula 30 Jan 1820.
1819–21	F. von Bellingshausen circumnavigated the Southern Ocean, surveying parts of South Georgia and South Sandwich Islands: probably sighted mainland Antarctica (coast of Dronning Maud Land) 27 January 1820: discovered Peter I Øy and Alexander Island.
1820–21	N. Palmer sighted northern tip of Antarctic Peninsula 16 November 1820.
1821–22	G. Powell, N. Palmer discovered South Orkney Islands.
1829–30	B. Pendleton led a US government-sponsored expedition to Staten Island and South Shetland Islands.
1830–33	J. Biscoe and J. Avery circumnavigated the Southern Ocean, discovered Enderby Land (1831) and Adelaide Island and Graham Land (1832).
1833	P. Kemp probably first sighted Heard Island, discovered Kemp Land.
1837–40	Dumont d'Urville's French Antarctic Expedition visited South Shetland Islands, South Orkney Islands, Antarctic Peninsula, and discovered Terre Adélie.
1838–39	J. Balleny and T. Freeman discovered and charted Balleny Islands, and discovered Sabrina Coast.
1838–42	US Exploring Expedition: six ships visited the South Shetland Islands and Antarctic Peninsula, and discovered the Wilkes Land coast.
1839–43	J. C. Ross visited Antarctic Peninsula, discovered the Ross Sea and Victoria Land, penetrating south to 78°S.
1853	M. Cooper penetrated pack ice to land on Oates Coast of East Antarctica.
1853–54	J. J. Heard discovered Heard Island.
1853–54	W. McDonald discovered McDonald Islands.
1872–76	*Challenger* Expedition visited many southern islands; the first steamship to cross the Antarctic Circle

GEOGRAPHICAL EXPLORATION: THE 'HEROIC AGE' (1895–1922)

The Sixth International Geographical Congress, held in London in 1895, recorded that 'exploration of the Antarctic

Table 3 Expeditions. 1873–94

Date	Expedition
1873–74	German Whaling and Sealing Expedition visited South Shetland Islands, South Orkney Islands and Antarctic Peninsula coast, investigating possibilities for whaling.
1874–76	Transit of Venus expeditions: temporary German, French, US and British observatories established on Aukland Islands, Campbell Island, Iles Kerguelen and Ile St Paul.
1882–83	German International Polar Year Expedition over-wintered in Royal Bay, South Georgia, and on the Falkland Islands.
1892–93	Dundee Whaling Expedition: four ships explored the Falkland Islands and northern Antarctic Peninsula, investigating possibilities for whaling.
1892–93	Norwegian (Sandefjord) Whaling Expedition explored Antarctic Peninsula and western Weddell Sea, investigating possibilities for whaling.
1893–94	Second Norwegian (Sandefjord) Whaling Expedition: three ships explored Antarctic Peninsula, South Orkney Islands and South Georgia, investigating possibilities for whaling.
1893–94	Norwegian (Tønsberg) Whaling Expedition visited Tristan da Cunha, Prince Edward Islands, Iles Crozet and Kerguelen, Auckland Islands, Campbell Island and Antarctica, landing at Cape Adare.

Table 4 Expeditions. 1897–1922

Date	Expedition
1897–99	Belgian Antarctic Expedition explored the South Shetland Islands and Antarctic Peninsula, wintering in Bellingshausen Sea.
1898–99	German Deep Sea Expedition: oceanographic exploration of the Southern Ocean, visiting Bouvetøya and Iles Kerguelen, St Paul and Amsterdam.
1898–1900	British Antarctic (*Southern Cross*) Expedition wintered on land at Cape Adare, completing a comprehensive programme of scientific research
1901–3	German South Polar Expedition wintered in pack ice off Gaussberg, East Antarctica, and on Iles Kerguelen.
1901–4	Swedish South Polar Expedition explored the tip of Antarctic Peninsula: parties wintered at Snow Hill Island, Hope Bay and Paulet Island.
1901–4	British National Antarctic Expedition wintered in McMurdo Sound, exploring extensively overland to 82°S, and across the Ross Ice Shelf toward the South Pole.
1902–4	Scottish National Antarctic Expedition wintered off Laurie Island, South Orkney Islands, explored the Scotia Sea and eastern Weddell Sea coast.

Table 4 (*Continued*)

Date	Expedition
1903–5	First French Antarctic Expedition wintered off Booth Island, Palmer Archipelago, exploring south to Adelaide Island.
1904–5	C. A. Larsen established the first Antarctic whaling station at Grytviken, South Georgia.
1907–9	British Antarctic (*Nimrod*) Expedition wintered at Cape Royds: parties climbed Mount Erebus, reached the South Magnetic Pole, crossed the polar plateau to 88°23′S.
1908–10	Charcot's Second French Antarctic Expedition wintered at Petermann Island; explored south into Bellingshausen Sea, discovering Marguerite Bay and Charcot Land.
1910–12	Norwegian South Polar Expedition wintered at Framheim, on the Ross Ice Shelf: a dog-sledging party reached the South Pole.
1910–12	Japanese National Antarctic Expedition sledged inland from Bay of Whales toward King Edward VII Land.
1910–13	British Antarctic (*Terra Nova*) Expedition wintered in McMurdo Sound: a man-hauling party reached the South Pole, but all died during the return journey.
1911–12	German South Polar Expedition attempted to cross Antarctica: wintered in Weddell Sea pack ice.
1911–14	Australasian Antarctic Expedition wintered at Cape Denison, Shackleton Ice Shelf and Macquarie Island, explored King George V Land and Queen Mary Land.
1914–16	Imperial Trans-Antarctic Expedition wintered in Weddell Sea and McMurdo Sound, attempted to cross Antarctica.
1920–22	British Imperial Antarctic Expedition wintered at Waterboat Point, Antarctic Peninsula.
1921–22	Shackleton–Rowett Antarctic Expedition visited South Georgia, South Sandwich Islands and South Shetland Islands.

Regions is the greatest piece of geographical exploration still to be undertaken', and recommended scientific societies throughout the world urge that the work be undertaken before the end of the century. There resulted a series of scientific and exploratory expeditions: Belgian, German, British, French, Swedish, Norwegian and Japanese. The Belgian expedition, beset in pack ice, was the first to winter south of the Antarctic Circle. Simultaneously a German maritime expedition explored many scientific aspects of the Southern Ocean. Borchgrevink's privately funded British expedition was the first to winter on land in Antarctica. Scott's first expedition explored extensively inland from McMurdo Sound. Shackleton's first expedition pioneered a route to the South Pole, later taken by Scott. Amundsen's expedition was the first to reach the South Pole on 7 December 1911. German and British expeditions tried but failed to cross Antarctica: Mawson's Australian-based expedition explored new ground in East Antarctica. Shackleton's final expedition and death on South Georgia brought the 'heroic age' to a close.

WHALING AND AVIATION (1927–39)

The commercial whaling that began in South Georgia in 1904–5 spread rapidly to other southern islands and Antarctic Peninsula, based mainly on moored factory ships. In 1912–13 factory ships and catchers operated for the first time pelagically off the edge of the pack ice, starting a trend that liberated whaling from both land stations and moorings. In 1923–24 pelagic whaling spread to the Ross Sea and 1925 marked the start of long-term whaling research sponsored by the British 'Discovery' Committee, based on a permanent laboratory at King Edward Point, South Georgia, and cruises of RRS *Discovery* (later replaced by RRSs *Discovery II* and *William Scoresby*). All three ships became extensively involved in hydrographic survey and exploration, especially though not exclusively in the Scotia Arc and Antarctic Peninsula areas. In 1927 Norwegian whaling entrepreneur Lars Christensen began using his factory ships and catchers in a series of seven annual research expeditions to explore Norway's recently claimed Antarctic possessions.

Scott and Drygalski had used hydrogen balloons for observation in early 1902. In 1911 Mawson's Australasian Antarctic Expedition took south a fixed-wing aircraft that never flew. The first successful use of aircraft in Antarctic exploration was a flight in December 1928 by Australian aviator Hubert Wilkins over Antarctic Peninsula. In January 1929 Richard Byrd, in the first of a series of expeditions based at the Bay of Whales and involving aerial survey, flew eastward over Edward VII Peninsula. In November of the same year he flew over the Ross Ice Shelf and polar plateau towards the South Pole. Wilkins made several flights during a second expedition in 1929–30. From this date onward most Antarctic expeditions made use of one or more aircraft. Byrd's second expedition concentrated on extensive aerial photography, with sledging parties providing essential ground control. The later Christensen whaling expeditions, Mawson's BANZARE and Rymill's BGLE enhanced their observations with aerial reconnaissance. Ritscher's *Schwabenland* expedition flew extensively over Dronning Maud Land, though without adequate ground control.

Table 5   Expeditions. 1927–39

Date	Expedition
1927–35	Christensen's Norwegian whaling expeditions: seven summer expeditions using ships and aircraft to explore coasts of the Norwegian sector of East Antarctica, Bouvetøya and Peter I Øy.
1928–29	First Wilkins Antarctic Expedition: first flight by fixed-wing aircraft over Antarctica, from Deception Island to Bellingshausen Sea.
1928–30	First Byrd Antarctic Expedition: wintered at Bay of Whales. Flights over Edward VII Peninsula and toward the South Pole: geological survey in Queen Maud Mts.
1929–30	Second Wilkins Antarctic Expedition: flights over Charcot Island and Antarctic Peninsula
1929–31	British, Australian and New Zealand Antarctic Research Expedition: in two summers, explored coast of the Australian sector of East Antarctica by ship and aircraft.
1933–35	Second Byrd Antarctic Expedition: wintered at Bay of Whales. Flights over and sledging journeys to Marie Byrd Land
1934–36	First and Second Ellsworth Expeditions: in two summer seasons achieved a flight across West Antarctica from Dundee Island to Bay of Whales.
1934–37	British Graham Land Expedition: wintered in Graham Land, exploring south to southern Alexander Island by ship, aircraft and sledges.
1938–39	German Antarctic (*Schwabenland*) Expedition: aerial survey flights over the western sector of Dronning Maud Land.

Table 6   Expeditions. 1939–59

Date	Expedition
1939–41	US Antarctic Service Expedition: established stations at Bay of Whales and Stonington Island: explored Marie Byrd Land and southern Antarctic Peninsula.
1943–44	Operation Tabarin: established long-term stations at Port Lockroy and Deception Island.
1946–47	US Naval Operation Highjump: naval task forces undertook aerial surveys of much of the Antarctic coastline.
1947–48	US Naval Operation Windmill: naval task force continued aerial surveys of the Antarctic coastline.
1955–57	Falkland Islands Dependencies Aerial Survey Expedition: long-range aircraft made aerial surveys of Antarctic Peninsula and neighbouring islands to 68°S.
1955–58	Commonwealth Trans-Antarctic Expedition: crossed Antarctica by tractor train and sledges, with aircraft support, from Filchner Ice Shelf to McMurdo Sound.
1957–58	International Geophysical Year expeditions: over 50 stations operated by 12 nations undertook coordinated scientific research and exploration.

PERMANENT STATIONS: OCCUPATION (1939–PRESENT)

The two research stations of the US Antarctic Service Expedition were the first to be built with the intention of long-term occupation, an objective defeated by entry of the USA into World War II. The British naval expedition, Operation Tabarin, established the first two stations of a long-term operation that became successively the Falkland Islands Dependencies Survey and the British Antarctic Survey. This initiative was followed immediately by Argentina and Chile, later by France, Australia and New Zealand, and subsequently by other states that, for one reason or another, developed long-term interests in Antarctica. The International Geophysical Year (1957–58) stimulated a further wave of interest in long-term programmes. Several of these government-run operations undertook extensive sledging and aerial surveys, adding substantial detail to topographic and scientific knowledge of their areas: many continue today.

Immediately after World War II the Unites States Navy returned south with two ship-based task forces, code-named Operations Highjump and Windmill. These achieved aerial photography of enormous tracts of Antarctica, both coastal and inland, that earlier expeditions had reconnoitred or missed. The Falkland Islands Dependencies Aerial Survey Expedition undertook comprehensive photographic coverage of Antarctic Peninsula and neighbouring islands, with unusually thorough ground control. The Commonwealth Trans-Antarctic Expedition was perhaps Antarctica's last major three-season exploration involving tractors, dog-teams and sledges with aircraft support.

The 1957–58 International Geophysical Year, in which 12 nations operated research stations and exploratory expeditions in Antarctica and the sub-Antarctic, marked a final turning point in Antarctic exploration. Since then, geographical discovery has largely been replaced by long-term scientific research programmes undertaken by national expeditions.

STUDY GUIDE: EXPLORATION 353

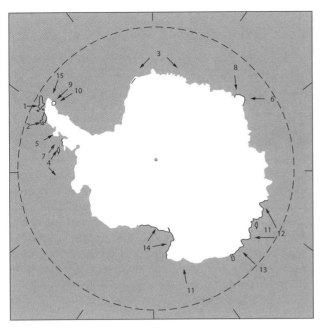

(a)
1. Smith 1819
2. Bransfield 1820
3. Bellingshausen 1820
4. Bellingshausen 1821
5. Palmer 1821-22
6. Biscoe 1831
7. Biscoe 1832
8. Kemp 1833
9. Dumont d'Urville 1839
10. Wilkes 1839
11. Balleny 1839
12. Wilkes 1840
13. Dumont d'Urville 1840
14. Ross 1841-42
15. Ross 1843

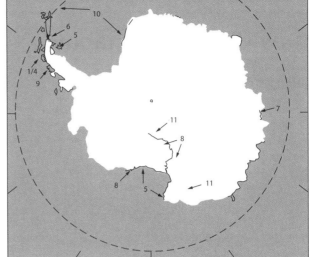

(b)
1. Dallman 1873-74
2. Dundee Whalers 1892-93
3. Larsen 1893-94
4. Gerlache 1897-99
5. Borchgrevink 1898-1900
6. Nordenskjold 1901-04
7. Drygalski 1901-03
8. Scott 1901-04
9. Charcot 1903-05
10. Bruce 1902-04
11. Shackleton 1908-09

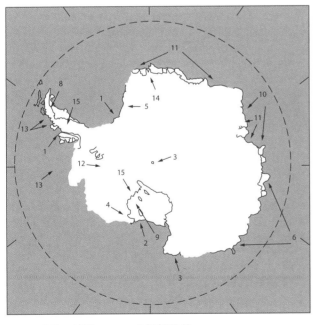

(c)
1. Charcot 1909
2. Admundsen 1910-12
3. Scott 1910-12
4. Shirase 1912
5. Filchner 1911-12
6. Mawson 1912-13
7. Shackleton 1914-16
8. Wilkins 1928-29
9. Byrd 1928-35
10. Mawson BANZARE 1929-31
11. Christensen 1926-37
12. Ellsworth 1933-39
13. Rymill 1935-37
14. Ritscher 1938-39
15. USASE 1939-41

(d)
1. Operation Highjump 1946-47
2. Ronne 1946-48
3. Giaever 1950-52
4. Fuchs 1955-58

Localities of the major expeditions. (a) 1819–43; (b) 1873–1909; (c) 1909–41; (d) 1946–58.

Study Guide: Geography

INTRODUCTION

The entries **Antarctic** and **Antarctic region** explain the derivation of the word and the region occupying the southernmost 11 per cent of Earth's surface. The entry **Antarctica** describes the continent – the southernmost, highest, coldest, windiest and fourth-largest of the continents, unique for its almost total ice cover, and for lacking both indigenous population and permanent settlers. It is also the continent most recently discovered and exploited by man (see **Study Guide: Exploration**). **East Antarctica** and **West Antarctica** define and describe the continent's two major provinces, the latter including **Antarctic Peninsula**. **Antarctic fringe islands** describe the islands of the continental shelf. Islands more remote from the continent are featured in **Southern islands**.

LOCATION, GEOLOGICAL HISTORY

Antarctica lies in isolation 1000 km (560 miles) from its nearest neighbour South America, 4350 km (2420 miles) from South Africa and 3000 km (1670 miles) from Australia. The subcontinental bloc of East Antarctica, formerly part of the Gondwana supercontinent, broke free from Australia some 180 million years ago and drifted south into a polar position close to the southern extremity of the Andean chain, now West Antarctica. Strong latitudinal atmospheric and oceanic circulation (**Study Guide: Southern Oceans and Islands**) isolated the continent from warming tropical influences, and orogeny elevated its surface, allowing the development of the permanent ice cap.

TOPOGRAPHY

In satellite images and maps based on polar projections, continental Antarctica appears comma-shaped, with the round body centred some distance from the geographic South Pole, and the long peninsular tail curving toward South America. By convention, polar projections are usually oriented with the Greenwich meridian running from top to bottom. Thus the bulk of East Antarctica lies in the eastern hemisphere, West Antarctica entirely in the western hemisphere.

The continent's mean elevation of over 2000 m (6500 ft) is due largely to the immense ice cap, of mean thickness 2160 m (7087 ft). The underlying continent, much smaller and generally lower, shows as rocky outcrops amounting to only 2.4 per cent of the total area. These include several spectacular ranges, notably the **Transantarctic Mountains**, which cross the continent, and the **Antarctic Peninsula**. Vinson Massif (5140 m, 16 859 ft), Antarctica's highest peak, forms the summit of the **Sentinel Range**, part of the **Ellsworth Mountains** of West Antarctica. Only about 2 per cent of the coast appears as rock cliffs or beaches, much of it along Antarctic Peninsula. The rest is made up of ice cliffs, backed by wide, sloping ice shelves that over-ride the true shoreline.

CLIMATES

Antarctica has many climates, all cold but differing considerably in severity (**Study Guide: Climate and Life**). Latitude and elevation combine to make it overall by far the coldest continent. Air temperatures on the high plateau fall below $-80\,°C$ in winter, and rise only to about $-30\,°C$ in summer. The Russian research station **Vostok**, close to the highest point of the polar plateau, records the world's lowest mean annual temperature ($-55.4\,°C$) and lowest minimum temperatures (usually below $-85\,°C$), in late winter. Points lower on the plateau (represented by **Amundsen-Scott**, **Byrd**) are warmer, depending more on altitude than latitude. Coastal sectors are warmer still, with latitude more critical: **McMurdo** in 78°S is much colder than **Mawson** in 67°S, especially in winter. Strong winds are rare on the plateau but prevail elsewhere, especially katabatic (downslope) winds descending the coastal ice plains. Depressions skirting the continent bring plentiful snow to the coasts, constantly renewing the ice shelves. Precipitation is lower along the rocky coasts, and lowest of all on the high continental plateau.

EXPLORATION

First sighted in the early nineteenth century, long after exploration had revealed Africa, South America and Australasia, Antarctica excited little interest: it was perceived as a cold land, with no native peoples, forests, land animals or obvious natural resources (**Study Guide: Exploration**). By the end of the nineteenth century, sealers (who were first to exploit the region's natural resources: **Sealing in the Southern Ocean**) and scientific expeditions had explored less than one-third of its coasts, while the interior remained unknown. Explorers first wintered on the continent in 1901, and first reached the South Pole ten years later. Whalers (**Whaling in the Southern Ocean**) hunted Antarctic waters from 1904 onward, and explorers charted more of its coasts. The first long-term settlements – scientific stations manned for several consecutive years – were established from 1943 (**Operation Tabarin 1943–45**). Since then the pace of exploration has accelerated. Today Antarctica is valued mainly by scientists for research, and by tourists (**Tourism, Antarctic**) for majestic scenery and wildlife.

SOVEREIGNTY

Seven nations (Argentina, Australia, Chile, France, New Zealand, Norway and the United Kingdom) claim sovereignty over sectors of Antarctica, including all but the sector 90°–150°W (see **Sovereignty in the Antarctic region**). In the Peninsula area, claims of three nations overlap. Other nations, including the USA and Russia, do not acknowledge these claims and make no claims of their own, but reserve rights to claim sovereignty in the future. Since 1961 the continent has been administered under the international **Antarctic Treaty System**.

PLACE NAMES

Geographical features were at first named mainly by explorers who, in recording their discoveries, exercised the privilege of naming them descriptively or reminiscently, or dedicating them to patrons and sponsors, royalty, family members, acquaintances, fellow explorers or themselves. Later the responsibility fell on map-makers, identifying features from field records, sketches and photographs (especially aerial photographs). Most countries with interests in Antarctica now have place-names committees with responsibilities for adjudicating on use of old names and new. Over 21000 major features – seas, coasts, mountain ranges, peaks, glaciers, ice streams and coastal features – in the Antarctic Treaty area (i.e. south of 60°S) have now been assigned names (Scientific Committee on Antarctic Research, 1998), and many more on southern islands. However, no single international authority has overall control, and confusion remains, especially in the South American sector where the sovereignty claims of three nations overlap, and naming assumes political significance.

Most names in Antarctica have been assigned by British or US authorities and are rendered in English (see Alberts, 1995 and Hattersley-Smith, 1991, both invaluable reference works that explain the principles on which they were compiled). Complications arise over names assigned in other languages. Norwegian, French, and Spanish-speaking Chileans and Argentines assign names in their own languages to features within the territories that they claim.

Britons accept French and Norwegian rights, but favour British names over Spanish in the Peninsula area where Argentine, British and Chilean sovereignty claims overlap. American authorities anglicize all names as far as possible, claiming interests of clarity rather than of politics. Thus the Norwegian Kronprinsesse Märtha Kyst and Heimfrontfjella are rendered thus in this encyclopedia, but in the official American listing (Alberts, 1995) become Princess Martha Coast and Heimfront Range. Britons accept Terre Adélie as the French name for an Antarctic territory claimed by France, while the American publication admits only Adélie Coast, a portion of the coast of a larger land named for the US explorer Charles Wilkes. The *Composite Gazetteer of Antarctica* (Scientific Committee on Antarctic Research, 1998) has valiantly coordinated 32955 Antarctic place names from 20 national gazetteers, assigning them to the 21,552 places actually named, an invaluable service to all who try to use Antarctic names accurately.

LOCATIONS

The names East and West Antarctica are now generally agreed for the major provinces, though some geographers continue to use Greater and Lesser Antarctica. The two provinces are subdivided into 17 lands: for a listing see **Lands, Antarctic**. Not all lands so designated are universally accepted (see above), and while some are bounded by coasts and meridians, others have less clearly defined limits. Most of the coastline of Antarctica has been divided into 55 lengths of coast: for a listing see **Coasts, Antarctic**. Both lands and coasts provide useful location addresses, e.g. Mt Faraway, Theron Mountains, Coats Land; Cuverville Island, Danco Coast.

HUMAN OCCUPATION

During the exploratory period of Antarctic history, scientific research was less important than discovery, and expedition bases were seldom manned for more than one or two years. Richard Byrd's **US Antarctic Service Expedition 1939–41** introduced a new concept of permanent stations, with science the main objective. The expedition's two stations, at Bay of Whales and Stonington Island, opened in 1941, but closed after a year shortly before the USA entered World War II. In 1943 Britain set up permanent stations in the maritime Antarctic, manned by scientists, mainly to assert sovereignty against Argentine, Chilean and possible American claims (**Operation Tabarin 1943–45**). So began the era of scientific research in Antarctica. Soon afterwards Argentina, Australia, Chile and France established permanent national expeditions, while the USA reverted to massive pan-Antarctic exploration using ships, aircraft and temporary land stations.

The **International Geophysical Year 1957–58** (IGY), an 18-month period of world-wide coordinated geophysical research, involved 49 stations in Antarctica, manned by over 5000 scientists and support staff from 12 nations, with programmes covering a wide range of geophysical topics from upper atmosphere physics to meteorology, oceanography, glaciology, seismology and geology. Research was coordinated by a group established in 1958, the Special Committee, later **Scientific Committee on Antarctic Research** (SCAR), which continues in that role today. The international accord and overall success of the IGY made continuing cooperation desirable. The governments of the 12 nations developed the **Antarctic Treaty,** an agreement to extend cooperation in Antarctica after the IGY ended. Despite conflicting claims over **Sovereignty in the Antarctic region** and the prevailing cold war, the Treaty provided for continuing scientific cooperation, paving the way for a continuing wave of research from permanent stations.

SCIENTIFIC RESEARCH

Geologists and solid-earth geophysicists found in Antarctica the key to plate tectonics: though much of the basic geological mapping is now completed, it is still a valued source of geological information and inspiration. Glaciologists measure the ice cap, model its dynamics, and deduce climate history from its layers. Climatologists find Antarctica important in atmospheric circulation modelling; upper atmosphere scientists use it as a platform for studying geospace phenomena. Terrestrial biologists model the continent's relatively simple ecosystems, study responses of flora and fauna to hostile environments, and measure the impacts of man on the polar environment. Marine biologists budget energy flow through the more complex marine food webs.

LIVING IN ANTARCTICA

Currently (2001) 17 nations, at 36 stations, operate year-round in the Antarctic region. Most stations are coastal, on rocky shores or ice slopes, and are relieved annually by ship. A few are far inland on the ice cap, cut off from the outside world except by radio. Small stations, for example the Polish station **Henryk Arctowski**, accommodate up to a dozen scientists and support staff over winter. Large stations take two to three times as many: the largest, for example **McMurdo Station**, may accommodate several thousand visitors in summer, mostly transients on their way to inland stations or field camps. The largest are small towns, with stores, a cinema, a chapel, possibly a bank, offices, laboratory blocks, garages, powerhouses, an airstrip, and hostels for residents and visitors.

Antarctica's resident **human population** seldom exceeds two or three hundred, with men far outnumbering women. A minority spend one or two years there at a time; more visit just for the summer months, when good weather facilitates fieldwork. Winters are long and dark: stations south of the **Antarctic Circle** lose the sun altogether for days or weeks, and the weather is cold and windy. Summers are short, but often bright, with clear skies and warm sunshine, making it possible to work outside in shirt-sleeves, though with shelter and windproof clothing close at hand. Scientists who work indoors, in laboratories or offices, servicing self-recording instruments or collecting data by radio from remote instruments, may hardly be aware of the cold world outside. Field scientists travelling in small parties by tractor or skidoo, surveying or collecting specimens, and camping for weeks on end, live closer to the real Antarctica, appreciating more fully its challenges and uniqueness.

Formerly exclusive to explorers, until recently the preserve of scientists, Antarctica has currently slipped into public awareness, both as a wilderness in need of conservation, and as a venue for **tourism**, which began in 1958 and grows slowly. The two trends occurred together: early tourists in the 1960s drew attention to accumulating rubbish and derelict buildings littering Antarctica – relics of installations used and discarded by scientists – against which Greenpeace, WWF and other environmental organizations campaigned vigorously. The environmentalists marshalled public opinion effectively, not only against rubbish, but also against mining in Antarctica. Currently they continue to press for high standards of environmental protection. Some seek to remove Antarctica altogether from the control of the Treaty System, and administer it on sounder environmental principles as a World Park.

Currently about 14000 tourists visit Antarctica, the **Scotia Arc** and **South Georgia** annually between late October and early April. Most are shipborne, as yet requiring few facilities ashore. Several thousands more each year join sight-seeing flights over the continent from Australia and New Zealand. On their current scale these forms of tourism make few demands on the environment, and interfere little with science. In introducing non-scientists to the scenery, wildlife and mystery of Antarctica, they may well be helping to broaden public interest in Antarctica, and ensure it a safer future.

POLITICAL MANAGEMENT

Starting as a tentative exercise in scientific cooperation, the **Antarctic Treaty System** has gradually acquired a management role in Antarctica, addressing wide-ranging issues from commemorative postage stamps to conservation. Its most significant instruments are the **Agreed Measures for the Conservation of Antarctic Fauna and Flora** (1964, now largely redundant, Appendix A in this encyclopedia), the **Convention on the Conservation of**

Antarctic Seals (CCAS, 1972, Appendix B), the **Convention on the Conservation of Antarctic Marine Living Resources** (CCAMLR, 1980, Appendix C), and the **Protocol on Environmental Protection to the Antarctic Treaty** ('Madrid Protocol', 1991, Appendix D). A **Convention on the Regulation of Antarctic Mineral Resource Activities** (CRAMRA), prepared in 1988, was opposed by environmentalists and scientists, and was never adopted.

Study Guide: Geology and Glaciology

INTRODUCTION

Viewed from space, Antarctica presents the world's largest sheet of continuous land ice, with less than 2 per cent of its surface exposed as rock outcrops. The presence of a solid continent underlying the ice sheet was deduced by some of the earliest explorers. Despite the relative dearth of exposed rocks, every land-based expedition from the late nineteenth and early twentieth centuries included at least one geologist, and geological research became a major feature of expeditions during and after the **International Geophysical Year 1957–58**. Information from the very limited sample of rocks available to classical geologists has more recently been enhanced by a wide range of modern geophysical techniques. A very substantial literature, including many hundreds of expedition reports and reviews, provides a sound basis for understanding the geological history and structure of the underlying continent. Even its shape under the ice has largely been elucidated (**Bedrock surface of Antarctica**).

The ice that hides so much of the continent from geologists has provided unparalleled opportunities for glaciologists, demanding new techniques and generating whole new areas of glaciological study. Thanks largely to aerial photography and airborne glaciological techniques, the ice sheet itself is now well mapped and studied in depth, and some understanding of its dynamics has been achieved (**Ice sheet: Antarctic**). The role of the massive ice cap in determining southern hemisphere climates is becoming apparent, and current changes in its structure, including massive losses from peripheral ice shelves, are being monitored with interest. Its further role in holding millennia-long records of past climates and atmospheres has only recently been demonstrated. Drilling and coring for samples from deep in the ice sheet currently provides the major focus of glaciological research in Antarctica.

EAST ANTARCTICA

The huge, complex ice dome of East Antarctica rises from coastal plains to a high plateau exceeding 4000 m (13 100 ft) at its highest point. Along its inland flank, where the ice sheet abuts onto West Antarctica, the Transantarctic Mountains present the uplifted edge of an underlying continental bloc, forming a massive dam that holds back the plateau ice. The dam is breached and penetrated by glaciers that flow into the Ross and Ronne–Filchner ice shelves. Behind the mountains, the ice cap slopes downward in the direction of the Indian and Pacific oceans. Lesser mountain ranges and nunataks appear toward the periphery, in a wide arc from Dronning Maud Land to Queen Mary Land. A deep channel between Lars Christensen and Ingrid Christensen coasts drains ice, notably the massive Lambert Glacier, into the Amery Ice Shelf. Ice domes high on the plateau (e.g. Beacon Dome, Dome C) form in relation to underlying topography.

Beneath the ice East Antarctica consists of a basement complex of ancient gneisses, schists and other metamorphic rocks, with massive intrusions, overlain by sediments of Lower Cambrian to Permian age (**Geology: stratigraphy and structure**). These reflect a complex climatic history from glacial conditions to tropical forests and deserts. Structurally East Antarctica shows common origins with South America, South Africa, India and Australia, with all of which it was for long united in the Gondwana supercontinent (**Plate tectonics**). Evidence from mid-oceanic ridges surrounding Antarctica indicates initial fragmentation some 150 million years ago, followed by the continent's gradual drift to its present polar position, and acquisition of its ice cap. Though the oceanic ridges remain tectonically active, East Antarctica itself shows little volcanic activity.

Radio echo sounding has revealed the presence of at least one huge mountain range, the Gamburtsev Subglacial Mountains, underlying the ice cap of East Antarctica. Also revealed are several freshwater lakes in deep subglacial basins, notably Lake Vostok, one of the largest, close to Vostok Station.

WEST ANTARCTICA

West Antarctica consists of a much lower, undulating ice sheet, covering a complex of mountainous islands that are separated by deep subglacial trenches. Its northern extension is the heavily-glaciated but more accessible Antarctic Peninsula. The emergent mountains, including those of the Peninsula, stand on a Precambrian basement complex. They consist mainly of Carboniferous, Mesozoic and Tertiary volcanic sediments, including greywackes, siltstones, mudstones and shales, folded and metamorphosed by later tectonic movements. Overlying plant-bearing sediments of mid-Jurassic age, and later tertiary volcanic sediments, indicate a geological history quite distinct from that of East

Antarctica, and generally more disturbed by earth movements (**Geology: geological history and palaeontology**). The geology of all the major ranges of West Antarctica has now been examined by field parties. No current volcanic activity is reported.

The ice cap covering the province appears to be chronically less stable than that of East Antarctica. Unusual amounts of ice have recently been shed from the extensive shelves fringing the eastern coast of Antarctic Peninsula, narrowing the Larsen Ice Shelf and creating new channels among the northern islands, for example around **James Ross Island**.

THE SOUTHERN ISLANDS

The geology and glaciology of the southern islands are outlined in accounts of individual islands and groups, and in the **Study Guide: Southern Oceans and Islands**. The Antarctic fringe islands, geologically similar to their nearest neighbouring points on the continent, are all heavily glaciated. Among the Antarctic islands, those of the **Scotia Arc** are structurally similar both to the Andes and to Antarctic Peninsula, but include several that show evidence of recent vulcanism, and some that are currently active, for example Deception Island in the South Shetland Islands, and several in the South Sandwich Islands. All are glaciated, though many show evidence of relatively recent loss of ice, including isostatic uplift. The more remote islands of the sub-Antarctic and cool temperate zones are mostly of relatively recent volcanic origin, showing varying evidence of current or past glaciation.

MINERAL RESOURCES

Antarctica is well endowed with minerals, but the ice cap, environmental conditions, and remoteness from possible markets severely restrict possibilities of mining (**Minerals and mining in Antarctica**). The only mineral resources currently used are sand, gravel and crushed rocks, for levelling airstrips and building foundations. The Transantarctic Mountains include huge deposits of coal, also quantities of copper, lead, zinc, silver, tin and gold. Antarctic Peninsula has copper and molybdenum ores; Dufek Massif includes ores of chromium, platinum, copper and nickel, and there is a strong likelihood that petroleum and natural gas will be found in the continental shelves, for example underlying the Ross Sea. Under Article 7 of the Protocol on Environmental Protection to the Antarctic Treaty, members have agreed not to engage in activities relating to mineral resources other than scientific research, a moratorium effective for 50 years from the date of entry into force of the Protocol.

THE ANTARCTIC ICE SHEET

Ice is solid water, formed mainly from water vapour, initially as snow or hoar frost. Snow is redistributed by wind, carved into **sastrugi**, and accumulations on steep slopes may be fluid enough to fall downhill due to gravity (**avalanche**). Accumulations of crystals, packing down and recrystallizing under their own weight, consolidate into **firn** or névé, a granular, air-filled intermediate stage. As more snow piles on top, further compaction and recrystallization eliminate much of the air, and the ice becomes clear and blue. Though hard and brittle, ice under pressure flows under its own weight, slumping downhill, filling valleys and hollows, surging in glaciers and ice streams at rates of a few metres to a few kilometres per year. Their bulk provides enormous power, and capacity to carve and transport rock, erode mountains, reduce high ranges to foothills, and create the faceted pinnacles, U-shaped valleys and moraines characteristic of glacial scenery.

Antarctica's ice sheet grew originally from many separate ice sheets, each with its own geography and dynamics (see **Ice sheet, Antarctica: history and development**). The bulk of the sheet changes constantly, some parts rapidly, others very slowly, reshaped by slumping and downhill flow, eroded and wasted by melting, vaporization, ablation and calving (**Accumulation of snow and ice**). On the coast of Dronning Maud Land, where several metres of snow fall per year, huts built on the surface disappear in two or three seasons, and cliffs of soft, air-filled ice calve constantly into the sea. In the near-desert conditions of the high polar plateau, precipitation is low, lateral movement slight, and buildings remain on the surface for decades (see **Climatic zones**). Where near-desert conditions prevail along the coast, the ice cap is locally starved and disappears, creating oases or anomalous areas of bare ground.

Ultimately the ice slumps seaward (**Ice sheet movement**), creating the ice cliffs, ice shelves, glacier fronts and extended glacier tongues that combine to form over 95 per cent of Antarctica's coastline. By constant calving, enormous quantities of ice islands, icebergs, growlers and lesser fragments are shed annually into the Southern Ocean. The slow surging movements of the ice cap can be monitored from space (see **Ice monitoring by remote sensing**).

The accumulated ice contains within its layers an invaluable record of past climates, obtained by ice coring and interpreted by subsequent ice core analysis, including radiometric dating. The lateral moraines lining Antarctic glaciers have been found to contain unusual concentrations of meteorites, well preserved and relatively free of contaminants.

Study Guide: Information Sources

INTRODUCTION

Information on Antarctica and the Southern Ocean, once restricted to specialist libraries, is now available from a wide range of printed and electronic sources. Matching a growing public awareness of Antarctica, information has simultaneously become more plentiful and more accessible, though often originating from more ephemeral sources. A recently published guide to information sources on polar regions (Mills and Speak, 1998), already has a proportion of seemingly dated advice. This survey seeks only to give guidance on generally applicable topics. Since the addresses of websites are in constant flux, addresses of parent organizations where appropriate are provided also, in the expectation that they may prove more constant.

BIBLIOGRAPHIES AND BIBLIOGRAPHIC DATABASES

Antarctica's unique political status places it outside the jurisdiction of any nation-state: its literature is therefore not collected and listed within normally applicable systems of national bibliographies. Three major libraries provide remedies. The Scott Polar Research Institute (SPRI), Cambridge, UK, has aimed at a comprehensive collection of Antarctic publications since its foundation in 1920. The US Library of Congress between 1965 and 1998, and since then the American Geological Institute (AGI), have compiled *Antarctic Bibliography,* in both cases with assistance from SPRI. *SPRILIB Antarctica,* which lists SPRI's own collections, may be accessed on the SPRI website (www.spri.cam.ac.uk). The *Antarctic Bibliography* may be searched on the AGI website (www.agiweb.org); alternatively it can be searched, together with the SPRI and other relevant databases, as *Arctic and Antarctic Regions* (AAR) on the NISC website (www.nisc.com). AAR is available also on CD-ROM.

The *Antarctic Bibliography* aims to be comprehensive and to cover all types of publications, including articles and conference papers, on all subjects and in all languages. More selective bibliographies include the Antarctic volume in the *World Bibliographical Series* (Meadows *et al.*, 1994), which presents broad coverage of the more significant monographic literature. For collectors of historic books, Spence (1980) has yet to be supplanted. Conrad (1999) too is excellent for the older literature and, in contrast to Spence's alphabetical treatment, lists publications under the expeditions to which they relate, preceded by useful expedition summaries. Both are usefully complemented by Rosove (2001), an exhaustive physical bibliography which is likely to be the last word on its subject.

Although they should contain nothing that is not listed in the *Antarctic Bibliography,* subject-centred databases and bibliographies may also include many relevant publications. They may also be more up-to-date and, because of differences in indexing, may in certain cases yield superior recall. For current awareness, the Institute of Scientific Information (ISI) *Current Contents* service on CD-ROM and via the ISI website (www.isinet.com) remains the best source for listings of the most recent English-language journal issues.

ATLASES, MAPS AND GAZETTEERS

Antarctica is poorly represented in standard world atlases. Among the best is a recent edition of the *Times Comprehensive Atlas of the World* (1999), which shows the continent at 1:15 m in a double-page spread. This is more generous than the 1:17.5 m map included in the more specialist *Polar Regions Atlas* (CIA, 1978). Nevertheless, when the much smaller continent of Europe is covered by 39 maps, a single map for all Antarctica seems barely adequate. The tendency to regard it as an undifferentiated whole, without regional coverage, is symptomatic of much Antarctic literature – equivalent to treating British Columbia, Newfoundland, and Nunavut indiscriminately as 'Canada'.

Antarctica's regions are given generous treatment in Tolstikov's *Atlas Antarktiki* (1966, 1969), an atlas that one reviewer described as 'without doubt the most exciting cartographic document which has ever been produced in the Soviet Union' (*Cartographic Journal,* 1967, p. 54). Unfortunately it is rare and difficult to find. In addition to topographic maps showing the continent at 1:10 m and the regions at 1:5 m, *Atlas Antarktiki* includes an excellent selection of thematic maps. Texts and legends are in Russian, but English translations are provided in special issues of *Soviet Geography* (vol. 8, nos 5–6, 1967). Useful thematic maps on a wide range of subjects may also be found in the 19 folios of the *Antarctic Map Folio Series* (Bushnell, 1964–75). More specialized is *Antarctica: A Glaciological and Geophysical Folio* (Drewry, 1983) presenting the fascinating results of a research programme, which used radio-echo sounding to measure the varying thickness of the ice sheet. This essentially laid bare the topography of the continent under the ice.

Prior to the International Geophysical Year (1957–58), most Antarctic mapping was conducted on individual expeditions. The resulting maps provided detailed but patchy coverage for areas visited. Following the establishment of the Scientific Committee on Antarctic Research (SCAR) in 1958, mapping has been coordinated by the SCAR Working Group on Geodesy and Geographic Information. In addition to setting standards to which maps should conform, this working group publishes a catalogue of maps, posted on the web and available through the SCAR website (www.scar.org). To ensure widespread distribution of maps, SCAR members are asked to designate an Antarctic Mapping Centre, to which participating countries are expected to send copies of all maps and gazetteers. This appears to have worked better in the past than currently: thematic maps particularly are poorly distributed and some countries are more diligent than others in the distribution of topographic and satellite image series. Nevertheless, each country that is active in Antarctic research should in time possess at least one good collection of Antarctic maps, most likely to be located in whichever organization has been designated the national Antarctic operating agency.

In a major project, the British Antarctic Survey, Scott Polar Research Institute and World Monitoring Centre collaborated with ten national Antarctic mapping agencies to produce the *SCAR Antarctic Digital Database* (ADD), currently available via the SCAR website. This incorporates map and satellite data from 11 countries and may be used to generate maps for educational and research purposes at a variety of scales. For large-scale maps, one need look no further. For the southern oceanic islands north of 60°S, outside the area covered by the Antarctic Treaty and SCAR, the best source for information on published maps is Parry and Perkins (2000).

Place-names are political. Though territorial claims are suspended under the Antarctic Treaty, none has been abrogated, and each claimant country maintains organizations responsible for nomenclature in the territories claimed. Most have published gazetteers listing approved names, and these may conflict where territorial claims overlap, as in the Antarctic Peninsula region. The single gazetteer covering the entire continent is Alberts (1995), an authoritative compilation of names approved by the US Board of Geographic Names. However, readers should be aware that not all of the names there listed are necessarily approved by other countries. Hattersley-Smith (1991) and Hattersley-Smith and Roberts (1993) list British-approved names for the sector claimed by Great Britain. Alberts (1995) and Hattersley-Smith (1991) provide a mass of otherwise hard-to-discover historical information on origins of place names.

JOURNALS

There are no popular magazines for the Antarctic enthusiast. The market might well now exist to support such a publication, but no one has yet attempted to supply it. *Antarctic,* the journal of the New Zealand Antarctic Society, offers a useful overview of recent work on the continent and is particularly informative on the New Zealand, Italian, and United States programmes. Some historical pieces are also included. The Scott Polar Research Institute has published *Polar Record* since 1931. Now distributed by Cambridge University Press, this is an academic journal covering all subjects for both polar regions, but is especially strong on history, the social sciences (e.g. law and psychology), and the environmental implications of tourism. It includes excellent book reviews. Reflecting general trends in publication practices, substantial scientific research is now generally reported in the major disciplinary journals. In consequence, institutional and more regionally oriented journals have become less important. The major exceptions are *Antarctic Science* (Cambridge University Press), *Polar Biology* (Springer), and *Terra Antarctica* (Department of Earth Science, University of Siena).

OTHER REFERENCE SOURCES

Key works on particular topics are recorded in the references given for those entries. In addition to these, there are a number of more general publications to note, both print and electronic. All historical enquiries should begin by consulting Headland (1989), which includes much information not readily traceable elsewhere. The brief expedition summaries are especially useful. More detailed accounts of the major expeditions may be found in the well-illustrated *Reader's Digest* (1990) compilation. *The Encyclopedia of Polar Exploration* (Mills, forthcoming) will provide more comprehensive coverage of all expeditions making a significant contribution to Antarctic exploration. For the first time it will also provide summaries of the process whereby Antarctica's individual regions and islands were discovered and explored. This work will be made available both as a conventional book and via the Internet as an e-book on the ABC-Clio site (www.abc-clio.com). For mountaineers, Gildea's (1998) chronology is the essential starting point. *The Index to Antarctic Expeditions* on the SPRI website (see above) provides near-comprehensive listing of all publications written about individual expeditions as well as links to relevant other websites and photographs. A more selective list is available in Conrad (1999).

Other subjects are less well covered by reference works, though Heap (1994) deserves mention as an indispensable guide to the system of governance under which Antarctica is administered. A more complete list of reference works appears in Mills and Speak (1998).

LIBRARIES AND ARCHIVES

The development of the Internet has in many ways made libraries and archives more rather than less important as sources of information about the Antarctic, though now as suppliers of information and not simply as repositories. For a list of libraries and archives likely to possess relevant collections, see *Polar and Cold Regions Library Resources: A Directory* on the website of the Polar Libraries Colloquy (www.urova.fi:80/home/arktinen/polarweb/polarweb.htm). In addition to organizations listed here, most large academic and public libraries will include at least some publications about Antarctica, though truly comprehensive collections are unlikely to be found outside agencies actively involved in Antarctic research. These are listed in the *Directory of Polar and Cold Regions Organizations* maintained on the SPRI website (see above). Since some of these collections may not be open to the public or offer at best restrictive viewing arrangements, prospective readers are strongly advised to apply in advance for permission to visit.

Prior application is also strongly recommended for archives, which are likely to insist on appointments being booked. Archives relevant to Antarctica's history are exceptionally scattered. In general, those most likely to possess pertinent material are national archives and libraries, organizations conducting Antarctic research, and whaling museums, in all countries active at any time in the region. There is no comprehensive guide to what is available, a situation now partially but not entirely relieved by the increasing availability of Internet tools by means of which subject searches can be run across all national repositories whose collections are represented in these services. Needless to say, this does not apply to many of the small and poorly funded repositories in which much Antarctic material is held.

ANTARCTIC ORGANIZATIONS

Many agencies engaged in Antarctic research would not welcome the suggestion that in certain circumstances it may be appropriate for you to apply to them directly for further information. Scientists working in Antarctica are busy people, and while it is generally inadvisable to apply directly to a named individual, the agencies in which they work may have librarians and information officers. Information seekers should first read all they can, then check to see if the agency has a website (most have: see below), then, if still necessary, write (rather than e-mail, if they want anything other than the briefest of replies), keeping their enquiry as specific as possible. Information officers too are busy, but may be prepared to answer questions that do not presume too much on their time. Each year, the library of the Scott Polar Research Institute receives hundreds of enquiries. Most are answered, but enquirers who seek to know everything about Antarctica will in future be advised first to read this encyclopedia. (WJM)

ADDITIONAL WEBSITES

Many websites provide information on different aspects of Antarctic life and research. A useful introductory site is that of the Council of Managers of National Antarctic Programs (COMNAP), which gives access to summaries of all the national programmes, plus the Antarctic Treaty Searchable Database (see below). *National programmes* sites cover each state's current and future activities and research programmes, with information on stations, national organizations, logistics, etc. *Research institutions* give information on their own activities and current research, some of which is Antarctic-oriented. *Sites relating to the Antarctic Treaty* give texts of Treaty documents and other Treaty matters, including reports of consultative meetings, etc. *Scientific groups* include websites that from time to time cover Antarctic issues. *Non-governmental organizations* include conservation groups, and a group representing the Antarctic tourist industry. *News and general information* includes websites with up-to-date information on Antarctic affairs: the ANAN twice-monthly newsletter and archive of back issues is particularly informative. The following websites are currently (2001) available:

NATIONAL PROGRAMMES

Argentina. Argentine Antarctic Institute: www.dna.gov.ar
Australia: Australian Antarctic Division:
    www.antdiv.gov.au
Belgium: Belgian Antarctic Programme:
    www.belspo.be/antar
Brazil: Brazilian Antarctic Program (PROANTAR):
    www.mar.mil.br/~secirm/proantar.htm
Canada: Canadian Polar Commission: www.polar.gc.ca
Chile: Chilean Antarctic Institute: www.inach.cl
France: French Polar Institute (IFRTP):
    www.ifremer.fr/ifrtp
Germany: Alfred Wegner Institute:
    www.awi-bremerhaven.de
Italy: Italian National Antarctic Research Program
    (PNRA): www.pnra.it
Japan: National Institute of Polar Research:
    www.nipr.ac.jp
Korea: Korean Antarctic Research Program (KARP):
    www.kordi.re.kr
Netherlands: Council for Earth and Life Sciences:
    www.now.nl/english/alw/programmes/antarctica
New Zealand: Antarctica New Zealand:
    www.antarcticanz.govt.nz

Norway: Norwegian Polar Institute:
www.npolar.no/
Peru: Commission of National Antarctic Affairs:
www.rree.gob.pe/conaan
Russia: Arctic and Antarctic Research Institute:
www.aari.nw.ru
South Africa: South African National Antarctic Program (SANAP): home.intekom.com/sanae/
Spain: Spanish Polar Committee (CPE):
www. Myct.es/sepct/ACT REGISTROS/comitepolar...
Sweden: Swedish Polar Research Secretariat:
www.polar.kva.se
United Kingdom: British Antarctic Survey:
www.antarctica.ac.uk
United States of America. US Antarctic Program:
www.nsf.gov/od/opp/antarct
Uruguay: Uruguayan Antarctic Institute:
www.iau.gub.uy

RESEARCH INSTITUTIONS

Byrd Polar Research Centre: www-bprc.mps.ohio-state.edu
Norsk Polarinstitutt (Norway): www.npolar.no
Scott Polar Research Institute (SPRI):
www.spri.cam.ac.uk
Canadian Polar Commission: www.polarcom.gc.ca
Gateway Antarctica, (New Zealand):
www.anta.canterbury.ac.nz

SITES RELATING TO THE ANTARCTIC TREATY

Antarctic Treaty Searchable Database (ATSD):
http://webhost.nvi.net/aspire
XXIV Antarctic Treaty Consultative Party Meeting:
www.24atcm.mid.ru

Convention on the Conservation of Antarctic Marine Living Resources (CCAMLR): www.ccamlr.org
Council for Environmental Protection (CEP):
www.npolar.no/cep
Council of Managers of National Antarctic Programs (COMNAP): www.comnap.aq

SCIENTIFIC GROUPS

International Council for Science (ICSU): www.icsu.org
International Biosphere-Hemisphere Programme (IGBP):
www.igbp.kva.se
World Climate Research Programme (WRCP):
www.wmo.ch/web/wcrp/wcrp-home.html
Scientific Committee on Antarctic Research (SCAR):
www.scar.org

NON-GOVERNMENTAL ORGANIZATIONS

Antarctic and Southern Ocean Coalition (ASOC):
www.asoc.org
Greenpeace: www.greenpeace.org
International Association of Antarctica Tour Operators (IAATO): www.iaato.org
Tourism Board of Tierra del Fuego:
www.tierradelfuego.org.ar

NEWS AND GENERAL INFORMATION

Antarctican – latest news and comment:
www.antarctican.com
Cool Antarctica: links to scientific and educational sites:
www.coolantarctica.com
Antarctic Non-governmental Activities News (ANAN):
www.antdiv.gov.au/goingsouth
(EKRB)

Study Guide: National Interests in Antarctica

Information on the interests of the many nations that are involved in Antarctica is scattered widely throughout the encyclopedia, for example, in accounts of expeditions, stations, sovereignty and Antarctic Treaty issues. This guide summarizes some of the interests of states that are signatory to the Antarctic Treaty. Most national operations have informative, up-to-date websites (see **Study Guide: Information Sources**), for which the COMNAP website is a good starting point.

ARGENTINA

Argentina claims sovereignty over **Antártida Argentina** (see **Sovereignty in the Antarctic region),** administering it as part of the province of Tierra del Fuego, Antarctica and the South Atlantic islands. The weather station **Orcadas**, on the South Orkney Islands, has operated continuously since 1905. In January 1942 and 1943 Argentine naval expeditions undertook hydrographic surveys in the South Shetland, Melchior and Argentine Islands and Marguerite Bay, and from 1947 to 1955 established seven permanent year-round stations: **Primero de Mayo** (later **Decepción**), **Melchior, General San Martin, Almirante Brown, Esperanza, Teniente Cámara** and **General Belgrano**, plus a scattering of refuge huts, some of which later became stations. These operated throughout the International Geophysical Year 1957–1958 into the 1960s. From 1959 Argentina took over and operated **Ellsworth Station** for three seasons, closing it in December 1962. In 1961 Argentina became one of the first 12 nations to sign the Antarctic Treaty.

Further stations followed, in 1962 **Teniente Matienzo**, in 1965 **Sobral** and in 1967 **Petrel**. In 1969 came **Vicecomodoro Marambio**, in March 1977 **Primavera,** and in 1979 General Belgrano was replaced by General Belgrano II, followed by General Belgrano III in 1980 (see **General Belgrano Stations**). In 1982 a refuge in Potter Cove, King George Island, became the permanent station **Teniente Jubany**. Argentina currently (2001) operates six year-round stations (Esperanza, General Belgrano III, Orcadas, San Martin, Teniente Jubany and Vicecomodoro Marambio), and up to six additional summer stations. The Argentine navy continues to be responsible for logistics, using among other ships the icebreaker *Almirante Irizar*. Vicecomodoro Marambio Station has become the centre for air operations.

Scientific research is coordinated through the Instituto Antártico Argentino, Buenos Aires.

AUSTRALIA

Geographical proximity has given Australia and Australian scientists a long-standing interest in Antarctica. Though the **British, Australian and New Zealand Antarctic Research Expedition 1929–31**, followed by the designation of Australian Antarctic Territory, had little direct effect on Australian policies toward the south, since 1947 Australian National Antarctic Research Expeditions (ANARE) has maintained a continuous Australian presence on the continent and Macquarie Island, and intermittent presence on Heard Islands. Australian scientists took part in the International Geophysical Year 1957–58, and Australia was among the first to sign the Antarctic Treaty (1961). Administered by the Australian Antarctic Division ANARE operates permanent stations **Casey, Davis** and **Mawson** on the continent, plus a number of smaller field stations, and a permanent station on **Macquarie Island**. The icebreaker *Aurora Australis* provides logistic support and is a platform for oceanographic research.

BELGIUM

Belgian interest in Antarctica began with the **Belgian Antarctic Expedition 1897–99**. Belgium contributed to the International Geophysical Year 1957–58, operating **Roi Baudouin Station** on Prinsesse Ragnhild Kyst from December 1957, and was among the first dozen states to sign the Antarctic Treaty (1961). The station was closed in 1961, but a new station built nearby in 1963 was occupied by a Belgian/Netherlands expedition, involving year-round sledging, survey and scientific research in the Sør Rondane to early 1967. Interest was again revived in 1985 with the start of the Belgian Scientific Research Programme, funded, managed and coordinated by the OSTC (Federal Office for Scientific, Technical, and Cultural Affairs). Participating Belgian scientists work by invitation from the stations and ships of other nations.

BRAZIL

Brazil acceded to the Antarctic Treaty in May 1975. Following the establishment in 1982 of PROANTAR, its national

Antarctic programme, Brazil achieved consultative status in 1983, and since February 1983 has operated the permanent year-round research **Commandante Ferraz Station** on Keller Peninsula, King George Island. Research centres on glaciology, geology and atmospheric studies, and shipborne studies of coastal and marine biology.

BULGARIA

Bulgarian scientists worked with Soviet and other expeditions from the 1960s onwards. Bulgaria acceded to the Antarctic Treaty in 1978. In 1993–95 the Bulgarian Antarctic Institute and Atlantic Club of Bulgaria, with logistic help from the Spanish Antarctic programme, established the summer station **St Kliment Ochridski** in South Bay, Livingston Island, South Shetland Islands. Bulgaria achieved Treaty consultative status in May 1998. The Bulgarian Nation Programme for Antarctic Research coordinates research, guided by the Bulgarian Antarctic Institute.

CANADA

Canada acceded to the Antarctic Treaty in May 1988. The Canadian Antarctic Research Program/Program de récherche de l'Antartique de Canada (CARP/PRAC), sponsored by the Canadian Polar Commission in 1993, expresses interest in Antarctic research and stresses possibilities offered by Canada's Arctic stations for bipolar research. Canada has no Antarctic station of its own, but Canadian scientists from time to time work on Antarctic programmes of other nations.

CHILE

Claiming sovereignty to the maritime sector of Antarctica based on rights inherited from the Spanish Crown, Chile in 1940 defined Territorio Chileno Antártico, an area deemed to be contiguous with its mainland. From 1947 Chilean naval expeditions have visited the territory. The station Soberania (later renamed **Capitán Arturo Prat**, a naval station, was established in 1947 on Greenwich Island, South Shetland Islands, and in 1948 **General Bernardo O'Higgins**, an army station on Trinity Peninsula. In 1951 came **Presidente Gonzala Videla**, an air force station in Paradise Harbour, and in 1955 Presidente Pedro Aguirre Cerda Station on Deception Island, South Shetland Islands. These stations operated during the International Geophysical Year 1957–58, and in 1961 Chile was among the first dozen states to sign the Antarctic Treaty. Presidente Pedro Aguirre Cerda Station was destroyed by fire during the volcanic eruptions of 1969: in the same year **Presidente Eduardo Frei Station** was established on King George Island, later to become part of **Teniente Rodolfo Marsh Martin**, an air station providing a direct link with Tierra del Fuego. Several refuges have also been erected. Logistics of the annual expeditions are managed by the Chilean armed forces: research in a wide range of sciences is coordinated through the Instituto Antárctico Chileno (INACH).

CHINA

The People's Republic of China acceded to the Antarctic Treaty in May 1983. Great Wall Station was established in 1985 on King George Island, South Shetland Islands, and China attained consultative status in the same year. In 1989 a second station, **Zhongshan**, was established in the Larsemann Hills, Prydz Bay. Both are permanent year-round stations. Research is coordinated by the Polar Research Institute of China under the Chinese Antarctic Administration, which operates the research and re-supply ship *Xuelong*.

ECUADOR

Ecuador acceded to the Antarctic treaty in September 1987, and in 1989–90 installed the summer-only scientific station **Pedro Vicente Maldonado** at Spark Point (incorrectly called 'Fort William'), on the southern coast of Greenwich Island, South Shetland Islands. Consultative status was granted in November 1990. National operations are conducted by Programa Antártico Equatoriano, Guayaquil.

FINLAND

Acceding to the Antarctic Treaty in 1984, Finland established the small summer-only research station **Aboa** in western Dronning Maud Land in 1988, and achieved consultative status in the following year. Expeditions since 1989 have operated in cooperation with Sweden and Norway, at times using the research ship RV *Aranda* of the Finnish Institute of Marine Research.

FRANCE

French interest in Antarctica dates back to the **French Naval Expedition 1837–40**, in which Terre Adélie was discovered and claimed. Interest was revived by two expeditions of the early twentieth century, the **French Antarctic (*Français*) Expedition 1903–5** and the **French Antarctic (*Pourquoi Pas?*) Expedition 1907–9**, both to the Peninsula region. In 1950 Expeditions Polaires Françaises established Port-Martin Station, later Pointe Géologie and Dumont d'Urville stations, on the coast of Terre Adélie. France participated in the International Geophysical Year 1957–58, was one of the first dozen to accede to the Antarctic Treaty (1961), and has since maintained a continuous presence in Terre Adélie. From 1992 research has been organized through the Institut Français pour la Récherche et la Technologie Polaires (IFRTP). The European Deep Ice Coring Project, which began in 1986, is a joint French/Italian project involving the construction and maintenance of **Concordia**, an inland ice station. France maintains permanent year-round stations also on **Iles Crozet**, **Iles Kerguelen** and **Ile Amsterdam** in the southern Indian Ocean.

GERMANY

Early German interest in Antarctica may be traced from the **German Whaling and Sealing Expedition 1873–74**, through five further expeditions (including one oceanographic and two to South Georgia), to the **German Antarctic (*Schwabenland*) Expedition 1938–39**. Pre-World War II whaling and post-war fishing maintained the maritime interest. The German Federal Republic acceded to the Antarctic Treaty in February 1979, achieving consultative status in March 1981. **Georg von Neumayer Station**, a permanent year-round research facility, was established in that year on Ekströmeisen, Dronning Maud Land, and has operated continuously since. The icebreaker *Polarstern* makes annual cruises in southern waters. A further research facility, the **Dallmann Laboratory**, was opened in conjunction with Argentina at Jubany Station, King George Island. The German Democratic Republic acceded to the Treaty in November 1974 and became a consultative member in October 1987. Since unification in 1990 all German polar research is coordinated through the government-sponsored Alfred Wegener Institute (AWI), Bremerhaven.

INDIA

Indian involvement in Antarctica began with summer expeditions in 1981–82 and 1982–83 to Prinsesse Astrid Kyst, Dronning Maud Land. The year-round research station **Dakshin Gangotri** was established in 1983 on shelf ice. In August of the same year India acceded to the Antarctic Treaty, and in the following month was accorded consultative status. Dakshin Gangotri was in 1989 replaced by **Maitri**, a permanent year-round station in Schirmacheroasen. Current Antarctic programmes are coordinated by the National Centre for Antarctic and Ocean Research (NCAOR), an autonomous institution within the governmental Department of Ocean Development.

ITALY

Acceding to the Antarctic Treaty in March 1981, Italy established a national Antarctic research programme four years later. The first expedition in 1985–86 set up **Terra Nova Bay (TNB) Station** on Scott Coast, Victoria Land. Italy became a consultative party to the Treaty in 1987. Apart from research at TNB, Italian scientists are involved with France in the **Concordia** project, and with other countries in drilling shallower ice cores and the ocean floor, to elucidate climate changes.

JAPAN

Japanese interests in Antarctica began with the **Japanese Antarctic Expedition 1910–12**, continuing with whaling and deep-water fishing. The first post-war Japanese Antarctic Research Expedition in 1956 established **Syowa Station** on East Ongul Island, Prins Olav Kyst, for the International Geophysical Year 1957–58. Japan was among the first to sign the Antarctic Treaty in August 1960, and has maintained a continuous presence in Syowa since 1959. Summer inland stations include **Asuka**, **Dome Fuji** (occasionally used for wintering) and **Mizuho**. Syowa is replenished annually by the icebreaker and oceanic research vessel *Shirase*. The National Institute of Polar Research (NIPR) coordinates Japanese Antarctic research programmes.

KOREA

The Republic of Korea acceded to the Antarctic Treaty in November 1986. The Korea Antarctic Research Program (KARP), inaugurated in 1987, has operated **King Sejong Station**, a permanent year-round research facility in Marian Cove, King George Island continuously since 1988. Korea was accorded Treaty consultative status in October 1989. Research programmes are coordinated by KARP and Korea Ocean Research and Development Institute (KORDI).

THE NETHERLANDS

Dutch interest in southern waters dates back to trading voyages of the sixteenth century, more recently to post-World War II Southern Ocean pelagic whaling. The Netherlands joined with Belgium in three annual survey and sledging expeditions based on **Roi Baudouin Station** in 1965–68, and was accorded consultative status in November 1990. Dutch scientists worked for a year at Henryk Arctowski Station 1990–91. The Council for Earth and Life Sciences, the coordinating agency, continues to arrange full annual research programmes, using the stations and research ships of other nations.

NEW ZEALAND

New Zealand in 1923 undertook responsibility for the newly-defined Ross Dependency, and participated in the **British, Australian and New Zealand Antarctic Research Expedition 1929–31**. Spurred by a lively and compelling Antarctic Society, the government in 1956 sponsored participation in the **International Geophysical Year 1957–58** and the **Commonwealth Trans-Antarctic Expedition 1955–58**, jointly housed in Scott Base. One of the dozen original signatories of the Antarctic Treaty, New Zealand has since maintained a year-round presence at **Scott Base**, and encouraged participation of universities, private research organizations, recreational clubs and other non-scientific bodies in annual expeditions. The coordinating agency is Antarctica New Zealand.

NORWAY

Norwegian interest in Antarctica began with two late nineteenth-century exploratory whaling expeditions. They flourished through the first half of the twentieth century, when Amundsen's **Norwegian South Polar Expedition 1910–12** successfully reached the South Pole, and Norway came to dominate the southern whaling industry. The **Norwegian (*Christensen*) Whaling Expeditions 1926–37** secured Norwegian claims to Bouvetøya, Peter I Øy and Dronning Maud Land. **Norway Station** operated during the International Geophysical Year 1957–58, and Norway was one of the first dozen signatories to the Antarctic Treaty. Research interests revived in 1989–90 with the establishment of **Troll Station**, a permanent year-round research facility 200 km (78 miles) inland in Gjelsvikfjella, and smaller **Tor Station**, built in 1993 in Svarthmaren. Since 1991 Norway, Sweden and Finland have cooperated in annual expeditions. Norwegian polar research is coordinated through the Norsk Polarinstitut, Tromsö.

PAPUA NEW GUINEA

Papua New Guinea was deemed to have acceded to the Treaty on gaining independence from Australia in March 1981. It has played no active role in Antarctic or Treaty affairs.

PERU

Peru acceded to the Antarctic Treaty in April 1982, and was accorded consultative status in October 1989, the year in which a permanent station, **Macchu Picchu**, was established on Crepin Point, King George Island. The station has been occupied intermittently by summer research parties. Peruvian polar research is coordinated by the Comisión Nacional de Assuntos Abtárticos, Ministerio de Relaciones Exteriores, Lima.

POLAND

Polish scientists **Henryk Arctowski** and **Anton Dobrowolski**, participating in the **Belgian Antarctic Expedition 1897–99**, initiated Polish interest in polar research. Deep-water fishing research brought Polish scientists back to Antarctica after World War II. Poland acceded to the Treaty in June 1961, and was accorded consultative status in July 1977, soon after **Henryk Arctowski Station**, a permanent year-round research facility, was established in Admiralty Bay. The station is maintained by the Department of Antarctic Biology, Polish Academy of Sciences.

RUSSIA

Russian interest in Antarctica dates from the **Russian Naval Expedition 1819–21**. The Soviet Union re-entered the field in 1956, opening six stations in the International Geophysical Year 1957–58: **Mirnyy, Pionerskaya, Oazis, Komsomol'skaya, Vostok** and **Sovetskaya**. Pionerskaya, Komsomol'skaya and Sovetskaya closed at the end of the IGY and Oazis was passed to Polish scientists. Among the first dozen signatories to the Antarctic Treaty, the Soviet Union continued to sponsor annual expeditions, opening successively **Lazarev** (1959), **Novolazarevskaya** (1961), **Molodezhnaya** (1962), **Bellingshausen** (1968), Leningradskaya (1970), **Druzhnaya** (1976), **Russkaya** (1979), **Soyuz** (1982) and **Progress** (1989). On dissolution of the Union in 1992, the Russian Federation inherited responsibility for the programme. The current (2001) Russian Antarctic Expedition (RAE) is led and coordinated by the Federal Service on Hydrometeorology and Environmental Monitoring (RosHydroMet). Programmes are agreed with the Council of Research of the Russian Academy of Science. Four to six stations operate in most winters, with two or three additional ones for summer parties. Two ice-strengthened ships, *Mikhail Somov* and *Akademik Fedorov*, are deployed for station resupply.

SOUTH AFRICA

South Africa established **SANAE Station** in January 1960, in the same year becoming one of the first dozen states to accede to the Antarctic Treaty. A continuous presence has been maintained since then in successive stations of the same name, currently SANAE IV. Research is coordinated by the South African National Antarctic Program (SANAP). South Africa maintains stations also on **Gough Island** and **Marion Island**. The research and transport vessel SA *Agulhas* resupplies the stations and provides scientific support.

SPAIN

Spain acceded to the Antarctic Treaty in 1988 as a consultative party. The National Program for Research in Antarctica, established in that year, is supported by the Interministerial Commission for Science and Technology, which acts also as the state's National Antarctic Committee. Two summer-only stations, **Juan Carlos Island** on Livingston Island and **Gabriele de Castilla** on Deception Island operate in summers only. The oceanographic research ship RV *Hesperides* operates annually in Antarctic waters.

SWEDEN

Swedish interest in Antarctica began with the **Swedish South Polar Expedition 1901–4**. In 1984 Sweden established a Polar Research Secretariat, and in April of that year acceded to the Antarctic Treaty. In September 1988 consultative status was accorded following the establishment of **Wasa** and **Svea** stations in Dronning Maud Land. Finland, Norway, and Sweden share responsibility for joint operations, each country in turn arranging an expedition on a three-year rotation.

SWITZERLAND

Switzerland acceded to the Antarctic Treaty in November 1990, but has played no active role in operations or research.

UKRAINE

Ukraine acceded to the Antarctic Treaty in October 1992, following the dissolution of USSR, and in February 1996 took over responsibility for the British station Faraday, Argentine Islands, renaming it **Akademik Vernadsky Station** and maintaining continuous year-round operation.

UNITED KINGDOM

Britain's Antarctic interests began with the voyages of Capt. James **Cook**, developed with the sealing industry during the early nineteenth century, and were maintained by a succession of Royal Navy or navy-dominated expeditions into the early twentieth century. The small, private **British Graham Land Expedition 1934–37** was remarkable for its cheapness and efficiency. For British involvement during and after World War II see **Operation Tabarin 1943–45**. From this developed successively the **Falkland Islands Dependencies Survey 1946–61** and **British Antarctic Survey** (BAS), which coordinates British Antarctic programmes. Five stations are currently (2001) operational: **Bird Island** and King Edward Point on South Georgia, and continental **Halley** and **Rothera** operate year-round, while **Signy**, on Signy Island, South Orkney Islands, is a summer-only station. The icebreaker RRS *James Clark Ross* and ice-strengthened transport RRS *Bransfield* resupply the stations and undertake oceanographic research, helped by the naval hydrographic ship HMS *Endurance*. The air fleet includes a four-engined Dash-7 and four Twin Otter aircraft.

UNITED STATES

United States sealers were prominent in southern waters during the early nineteenth century, and ships of the **United States Exploring Expedition 1838–42** explored a wide sector of the East Antarctica coast. **Byrd's First Antarctic Expedition 1928–30** and **Byrd's Second Antarctic Expedition 1933–35**, followed by the **United States Antarctic Service Expedition 1939–41** established a pattern of sledging and aerial survey that the **Ronne Antarctic Research Expedition 1946–48** continued after World War II. Large-scale aerial survey was the keynote of the **US Navy Development Project 1946–47 (Operation Highjump)**, backed by the **US Navy Second Development Project 1947–48 (Operation Windmill)**.

From 1954 the US Navy sought sites for stations for the forthcoming International Geophysical Year 1957–58, starting a series of annual expeditions code-named **Operation Deep-Freeze**. During the IGY, US scientists worked at and ran extensive overland traverses from 'Little America V' on Ross Ice Shelf, and **Byrd, Wilkes, Ellsworth, Hallett** (with New Zealand) and **Amundsen-Scott Station**: the latter, at the South Pole, was serviced from a naval air facility in McMurdo Sound that later became **McMurdo Station**. The United States was one of the first dozen nations to sign the Antarctic Treaty. Annual expeditions continued, with the naval logistics (both sea and air) continuing as Operation Deep-Freeze, and scientific research coordinated by the United States Antarctic Research Program (USARP), a civilian agency of the National Science Foundation. In later years the USA opened **Palmer Station** (1965) and **Siple** (1969). The USA currently operates three year-round stations, McMurdo, Amundsen-Scott and Palmer, serviced by the research icebreaker RV *Nathaniel B. Palmer* and ice-strengthened research ship *Laurence M. Gould*. Research is coordinated by the Office of Polar Programs, National Science Foundation.

URUGUAY

The Uruguayan Antarctic Institute, formed in 1975 within the Ministry of National Defence, stimulated Uruguay to accede to the Antarctic Treaty in January 1980. In 1984 it opened **Artigas Station** on Fildes Peninsula, King George Island, and consultative status was accorded in 1985. Artigas has since operated year-round, providing facilities mainly for marine and terrestrial biology and atmospheric research, with logistic support provided by the Uruguayan armed services.

VENEZUELA

Venezuela acceded to the Antarctic Treaty in March 1990, but has played no active role in operations or research.

Study Guide: Protected Areas under the Antarctic Treaty

INTRODUCTION

The **Antarctic Treaty** (see Appendix E) and **Antarctic Treaty System** apply to continental Antarctica and all islands and ice shelves south of 60°S. Under Article IX, Section 1(f) of the Treaty, signatories undertook to consider measures regarding 'preservation and conservation of living resources in Antarctica'. From the first Antarctic Treaty Consultative Meeting (ATCM I) in 1961, conservation of flora and fauna has been a regular topic for deliberation and recommendation. From the third meeting (ATCM III, 1964) came the **Agreed Measures for the Conservation of Antarctic Flora and Fauna** (see Appendix A), of which Article VIII provided for the establishment of **Specially Protected Areas** (SPAs), essentially to protect ongoing scientific research from interference. The seventh meeting (ATCM VII, 1972) assigned a more rigorous role to SPAs: under Recommendation VII-2 they became reserves of particular scientific interest, to be kept inviolate and entered only under permit. At the same meeting Recommendation VII-3 introduced a new category, **Sites of Special Scientific Interest** (SSSIs), as areas set aside for scientific research.

Recommendation V-4 of ATCM V (1968) provided for the recognition and protection of Historic Sites and Monuments (HSMs). Later came Marine Sites of Special Scientific Interest (MSSSIs, XIV-6, 1987), Specially Reserved Areas (SRAs, XV-10, 1989), to protect examples of geological, glaciological and geomorphological features and areas of outstanding aesthetic, scenic, and wilderness value, and Multiple-use Planning Areas (MPAs, XV-11, 1989). In addition, Seal Reserves were established under the **Convention for the Conservation of Antarctic Seals** (1972), and Environmental Monitoring Programme sites under the **Convention on the Conservation of Antarctic Marine Living Resources** (1980). A single '**Tomb**' was designated in 1981 in Lewis Bay, on the northern flank of Mt Erebus, Ross Island, at the site of a 1979 air accident.

CHANGES UNDER THE PROTOCOL

The 1991 **Protocol on Environmental Protection to the Antarctic Treaty** (Appendix D), which overtook and to some degree replaced the Agreed Measures of 1964, recognized the complexity of the system and provided, in Annex V, for two new categories of protected area that would replace existing designations. Antarctic Specially Protected Areas (ASPAs) are conceived of as protecting environmental, scientific, historic, aesthetic or wilderness values, and intended to replace existing SPAs, SSSIs, MSSSIs and SRAs (see below). Antarctic Specially Managed Areas (ASMAs) include such areas as scientific stations where active day-to-day management is possible, and may replace MPAs. Both require management plans, the form of which is prescribed in Article 5 of the Annex. ASPAs can be entered only under permit: ASMAs may be entered without permit. Historic Monuments may receive either designation, depending on whether or not they are under supervision.

New sites have been proposed for SPAs, SSSIs, MSSSIs and HSMs at almost every Antarctic Treaty Consultative Meeting: several sites have been reconsidered or switched from one designation to another.

Ten years after its introduction, Annex V of the Protocol remains unratified by one signatory (India), so its provisions cannot yet (2002) come into force. Meanwhile protected areas remain under their old designations arising from the Agreed Measures and subsequent recommendations. Many SPAs and SSSIs have been provided with management plans in readiness for their conversion to ASPAs, when they will also be renumbered. Only one draft ASMA has so far been submitted, covering neighbouring Polish, US, Brazilian and Ecuadorean stations in Admiralty Bay.

Conservation of southern oceanic islands north of 60°S resides with the national authorities that claim responsibilities for them. For details, see individual islands and island groups.

SPECIALLY PROTECTED AREAS (SPAs)

The purpose of SPAs is to preserve unique and representative examples of the natural ecological systems of areas which are of outstanding scientific interest. Entry is prohibited without a permit, and permits may be issued only for a compelling scientific purpose which cannot be served elsewhere. The collection of any native plant or animal requires a further and more specific permit. Vehicles of any kind are forbidden. SPAs are numbered in the order in which they were nominated and identified by localities. For details of individual SPAs, see individual entries under locality names.

Archipel de Pointe Géologie (SPA No. 24)
Ardery Island and Odbert Island (SPA No. 3)

Avian Island (SPA No. 21)
Beaufort Island (SPA No. 5)
Cape Adare (SPA No. 29)
Cape Crozier (SPA No. 6): redesignated SSSI No. 4
Cape Evans (SPA No. 25)
Cape Hallett (SPA No. 7)
Cape Royds (SPA No. 27)
Cape Shirreff (SPA No. 11): redesignated SSSI No. 32
Caughley Beach (SPA No. 10): redesignated SSSI No. 10
Coppermine Peninsula (SPA No. 16)
Cryptogam Ridge (SPA No. 22)
Dion Islands (SPA No. 8)
Fildes Peninsula (SPA No. 12): redesignated SSSI No. 5
Forlidas Pond (SPA No. 23)
Green Island (SPA No. 9)
Hut Point (SPA No. 28)
Lagotellerie Island (SPA No. 19)
Lewis Bay (SPA No. 26)
Litchfield Island (SPA No. 17)
Lynch Island (SPA No.14)
Moe Island (SPA No. 13)
New College Valley (SPA No. 20)
North Coronation Island (SPA No. 18)
Rookery Islands (SPA No. 2)
Sabrina Island (SPA No. 4)
Southern Powell Island (SPA No. 15)
Taylor Rookery (SPA No. 1)

Canada Glacier (SSSI No. 12)
Cape Crozier (SSSI No. 4)
Cape Royds (SSSI No. 1)
Cape Shirreff (SSSI No. 32)
Caughley Beach (SSSI No. 10)
Chile Bay (SSSI No. 26)
Cierva Point (SSSI No. 15)
Clark Peninsula (SSSI No. 17)
Dallman Bay (SSSI No. 36)
Deception Island, Parts of (SSSI No. 21)
Fildes Peninsula (SSSI No. 5)
Harmony Point (SSSI No. 14)
Haswell Island (SSSI No. 7)
Linnaeus Terrace (SSSI No. 19)
Lions Rump (SSSI No. 34)
Marine Plain (SSSI No. 25)
Mt Flora (SSSI No. 31)
Mt Melbourne Summit of (SSSI No. 24)
Port Foster (SSSI No. 27)
Potter Peninsula (SSSI No. 13)
Rothera Point (SSSI No. 9)
South Bay (SSSI No. 28)
Svarthamaren (SSSI No. 23)
Tramway Ridge (SSSI No. 11)
White Island, Northwest (SSSI No. 18)
Yukidori Valley (SSSI No. 22)

SITES OF SPECIAL SCIENTIFIC INTEREST (SSSIs)

The purpose of SSSIs is to protect ongoing scientific investigation or to set aside undisturbed reference areas until needed. They can be designated only where there is a demonstrable risk of harmful interference, and are subject to time limits that can be reviewed or extended. Each requires a management plan, specifying the work to be done, possible forms of interference, and suggesting kinds of research that would not interfere with the main purposes of designation. SSSIs are numbered in the order in which they were nominated and identified by localities. For details of individual SSSIs, see individual entries under locality names.

Ablation Point (SSSI No. 29)
Admiralty Bay, Western shore (SSSI No. 8)
Ardley Island (SSSI No. 33)
Arrival Heights (SSSI No. 2)
Avian Island (SSSI No. 30): redesignated SPA No. 21
Bailey Peninsula, Northeast (SSSI No. 16)
Barwick Valley (SSSI No. 3)
Biscoe Point (SSSI No. 20)
Bransfield Strait, Western (SSSI No. 35)
Byers Peninsula (SSSI No. 6)

MARINE SITES OF SPECIAL SCIENTIFIC INTEREST (MSSSIs)

Such sites as Bransfield Strait, Western, and Dallmann Bay, though distinguished originally under this heading, were numbered consecutively with land-based SSSIs and have received similar treatment throughout.

HISTORIC SITES AND MONUMENTS (HSMs)

The register of HSMs puts on record places and human artefacts that governments regard as being worthy of preservation and protection from damage. To achieve on-the-spot recognition, a site or monument should be marked with notices in English, French, Spanish and Russian, indicating that it is scheduled for preservation in accordance with provisions of the Antarctic Treaty. HSMs are numbered in the order in which they were nominated, and identified by localities.

Admiralty Bay. 62°09′S, 58°28′W. Grave near Arctowski Station of Polish artist and film producer Wlodzimierz Puchalski, who died there in January 1979. (HSM No. 51).
Argentine Islands. 65°15′S, 64°16′W. Expedition hut Wordie House on Winter Island formerly Base F of the Falkland Islands Dependencies Survey. (HSM No. 62).

Arturo Prat Station, Greenwich Island. 62°29′S, 59°41′W. Shelter, cross and plaque commemorating the death of station leader Gonzales Pacheco. (HSM No. 33).

Arturo Prat Station, Greenwich Island. 62°30′S, 59°41′W. Bust of Chilean naval hero Capt. Arturo Prat, erected in 1947. (HSM No. 34).

Arturo Prat Station, Greenwich Island. 62°30′S, 59°41′W. Wooden cross and statue of the Virgin of Carmen. (HSM No. 35).

Barry Island, Debenham Islands. 68°08′S, 67°08′W. Cross, flag, mast, and monolith at the Argentine station General San Martin. (HSM No. 26).

Betty, Mt. 85°11′S, 163°45′W. Cairn erected by Amundsen's party during their return from the South Pole. (HSM No. 24).

Bunger Hills. 66°16′S, 100°45′E. Pillar at Dobrowolski Station used for gravity measurements by the First Polish Antarctic Expedition in January 1959. (HSM No. 49).

Buromskiy Island, Haswell Islands. 66°32′S, 93°01′E. Cemetery with graves of Russian and other scientists. (HSM No. 9).

Cape Adare. 71°11′S, 170°15′E. Two expedition huts of the British Antarctic (*Southern Cross*) Expedition 1898–1900. (HSM No. 22).

Cape Adare. 71°17′S, 170°15′E. Grave of Norwegian biologist Nicolai Hanson, of the British Antarctic (*Southern Cross*) Expedition 1898–1900.(HSM No. 23).

Cape Bruce. 67°25′S, 60°47′E. Cairn and plaque commemorating landing by the British, Australian and New Zealand Antarctic Research Expedition 1929–31. (HSM No. 5).

Cape Crozier. 77°32′S, 169°18′E. Remains of stone hut built in July 1911 by a party of the British Antarctic (*Terra Nova*) Expedition 1910–13. (HSM No. 21).

Cape Crozier. 77°21′S, 169°16′E. Message post left by the British National Antarctic Expedition 1901–4. (HSM No. 69).

Cape Denison. 67°00′S, 142°42′E. Cross and plaque commemorating expedition members B. E. S. Ninnis and X. Mertz, who died in 1913. (HSM No. 12).

Cape Denison. 67°00′S, 142°42′E. Expedition hut of the Australasian Antarctic Expedition 1911–14. (HSM No. 13).

Cape Evans. 77°38′S, 166°24′E. Cross on Wind Vane Hill commemorating three members of the Imperial Trans-Antarctic Expedition 1914–16, who died in 1916. (HSM No. 17).

Cape Evans. 77°38′S, 166°24′E. Expedition hut of the British Antarctic (*Terra Nova*) Expedition 1910–13. (HSM No. 16).

Cape Royds. [See also SPA 27.] 77°38′S, 166°01′E. Expedition hut of the British Antarctic (*Nimrod*) Expedition 1907–9. (HSM No. 15).

Charcot, Port, Booth Island. 65°03′S, 64°01′W. Cairn and plaque commemorating the French Antarctic (*Français*) Expedition 1903–5. (HSM No. 28).

Coulman Island. 73°19′S, 169°41′E. Message post at Cape Wadworth, left by the British National Antarctic Expedition 1901–4. (HSM No. 70).

Elephant Island. 61°03′S, 54°50′W. Bust of Piloto Pardo and plaque commemorating the rescue of survivors of *Endurance* by the Chilean naval tug *Yelcho* in August 1916. (HSM No. 53).

Elephant Island. 61°10′S, 55°24′W to 61°17′S, 55°13′W. Wreckage of a wooden ship in a bay on the western end of the island. (HSM No. 74).

Filchner Ice Shelf. 77°49′S, 38°02′W. Cross erected in 1955 1300 m north-east of General Belgrano Station. (HSM No. 43).

Fildes Peninsula, King George Island. 62°12′S, 58°54′W. Plaque in memory of Polish expedition members Siedlecki and Tazar, who landed in February 1976. (HSM No. 50).

Fildes Peninsula. 62°13′S, 58°58′W. Monolith commemorating the establishment of Chinese Great Wall Station in February1985. (HSM No. 52).

Framnesodden, Peter I Øy. 68°41′S, 90°42′W. Hut and plaque inscribed 'Norvegia-ekspedisjonen 2/2 1929', left by Norwegian Capt. Nils Larsen in February 1929. Reported missing, May 1998. (HSM No. 25).

General Bernardo O'Higgins Station, Trinity Peninsula. 63°19′S, 57°54′W. Statue of Chilean liberator Bernard O'Higgins. (HSM No. 37).

Granite Harbour, Scott Coast. 77°00′S, 162°32′E. Rock shelter 'Granite House', Cape Geology, built by members of the British Antarctic (*Terra Nova*) Expedition 1910–13. (HSM No. 67).

Greenwich Island. 62°29′S, 59°40′W. Monolith near Arturo Prat station used as a reference point for Chilean hydrographic surveys. (HSM No. 32).

Half Moon Beach, Livingston Island. 69°29′S, 60°41′W. Cairn commemorating those lost in the Spanish ship *San Telmo,* which sank nearby in September 1819. (HSM No. 59).

Hell's Gate Moraine, Terra Nova Bay. 74°56′S, 168°48′E. Emergency depot on Inexpressible Island, left by the British Antarctic (*Terra Nova*) Expedition 1910–13. (HSM No. 68).

Hope Bay. 63°24′S, 56°59′W. Bust of General San Martin, grotto, cemetery and flag mast at Esperanza Station. (HSM No. 40).

Hope Bay. 63°24′S, 56°59′W. Stone shelter built by a party of the Swedish South Polar Expedition 1901–4. (HSM No. 39).

Horseshoe Island, Marguerite Bay. 67°49′S, 67°18′W. Expedition hut Base Y of the Falkland Islands Dependencies Survey, and refuge hut on nearby Blaiklock Island. (HSM No. 63).

Hunger Hills, Queen Mary Land. 66°16′S, 100°45′E. Plaque at Dobrowolski Station, commemorating its opening as Oazis Station in 1956. (HSM No. 10).

Hut Point, Ross Island. 77°51'S, 166°31'E. Cross commemorating the death of George T. Vince, a member of the British National Antarctic Expedition 1901–4. (HSM No. 19).

Hut Point, Ross Island. 77°51'S, 166°31'E. Expedition store hut of the British National Antarctic Expedition 1901–4. (HSM No. 18).

Ile de Pétrels, Terre Adélie. 66°40'S, 140°01'E. 'Base Marret' where members of a French expedition wintered in 1952. (HSM No. 47).

Ile de Pétrels, Terre Adélie. 66°40'S, 140°01'E. Cross on the northeast headland of the island commemorating André Prudhomme, expedition meteorologist who disappeared during a storm in January 1959. (HSM No. 48).

Inexpressible Island. 74°54'S, 163°43'E. Site of cave shelter where the northern party of the British Antarctic (*Terra Nova*) Expedition 1910–13 wintered in 1912. (HSM No. 14).

Lambda I. Melchior Islands. 64°18'S, 62°59'W. Argentine lighthouse Primero de Mayo built in 1942. (HSM No. 29).

Lewis Bay, Ross Island. 77°21'33"S, 167°33'27"E. Memorial cross on a rocky promontory 3 km from the site of the 1979 Mount Erebus air disaster. (HSM No. 73).

Mabus Point. 66°33'S, 93°01'E. Metal sledge with plaque at Mirnyy Station commemorating driver-mechanic Anatoliy Shcheglov who died in 1960. (HSM No. 8).

Mabus Point. 66°33'S, 93°01'E. Stone with plaque at Mirnyy Station commemorating driver-mechanic Ivan Kharma, who died in 1956. (HSM No. 7).

McMurdo Station, Ross Island. 77°51'S, 166°40'E. Bust on plinth commemorating the polar achievements of Rear Admiral Richard E. Byrd. (HSM No. 54).

Megalestris Hill, Petermann Island. 65°10'S, 64°10'W. Cairn and plaque commemorating the French Antarctic (*Pourquoi Pas?*) Expedition 1908–10. The plaque has been removed for safe keeping. (HSM No. 27).

Metchnikoff Point, Brabant Island. 64°02'S, 62°34'W. Plaque erected by the Joint Services Expedition 1983–85 to commemorate the first landing on Brabant Island by the Belgian Antarctic Expedition 1897–99. (HSM No. 45).

Nivlisen, Prinsesse Astrid Kyst. 70°45'S, 11°38'E. Plaque at the Indian station Dakshin Gangotri commemorating the First Indian Antarctic Expedition, January 1982. (HSM No. 44).

Observation Hill. 77°51'S, 166°40'E. Cross erected by the British Antarctic (*Terra Nova*) Expedition 1910–13, commemorating the five members of the polar party who died on returning from the South Pole. (HSM No. 20).

Ongul Island. 69°00'S, 39°35'E. Cairn and plaques at Syowa Station, commemorating Japanese expedition member Shin Fukushima, who died in 1960. (HSM No. 2).

Paradise Harbour. 64°49'S, 62°51'W. Shelter near the Chilean station Gabriel Gonzales Videla commemorating President Gonzalez Videla. (HSM No. 30).

Paulet Island. 63°35'S, 55°41'W. Stone hut of the Swedish South Polar Expedition 1901–4, with nearby grave, and rock cairn on the island's peak. (HSM No. 41).

Penguin Bay, Seymour Island. 64°11'S, 56°38'W. Plaque and cairn commemorating help given by the Argentine corvette *Uruguay* to the Swedish Antarctic Expedition 1901–4. (HSM No. 60).

Pole of Inaccessibility. 83°06'S, 54°58'E. Station building with a plaque commemorating Soviet explorers who reached this pole in 1958. (HSM No. 4).

Port Lockroy, Wienke Island. 64°49'S, 63°31'W. Expedition hut and installations of Base A, built by Operation Tabarin in 1944. (HSM No. 61).

Port-Martin, Terre Adélie. 66°49'S, 141°24'E. Buildings of Port-Martin station, which were partly destroyed by fire in January 1952. (HSM No. 46).

Potter Cove, King George Island. 62°13'S, 58°42'W. Metal plaque dated 1 March 1874 commemorating the visit by Capt. Eduard Dallmann. [Reported missing.] (HSM No. 36).

Proclamation Island. 65°51'S, 53°41'E. Cairn and plaque commemorating landing by the British, Australian and New Zealand Antarctic Research Expedition 1929–31. (HSM No. 3).

Scotia Bay, Laurie Island. 60°46'S, 44°40'W. Stone hut of the Scottish National Antarctic Expedition 1902–4, a nearby meteorological hut and magnetic observatory, and cemetery dating from 1903. (HSM No. 42).

Scott Nunataks, Queen Alexandra Mountains. 77°12'S, 154°30'W. Cairn left by K. Prestrud during the Norwegian Antarctic Expedition 1910–12. (HSM No. 66).

Snow Hill Island, Antarctic Peninsula. 64°24'S, 57°00'W. Expedition hut of the Swedish South Polar Expedition 1901–4. (HSM No. 38).

South Pole. 90°S. Flag mast erected in December 1965 by the first Argentine overland traverse. (HSM No. 1).

Stonington Island. 68°11'S, 67°00'W. Expedition hut Base E of the Falkland Islands Dependencies Survey. (HSM No. 64).

Stonington Island. 68°11'S, 67°00'W: Huts of East Base of the United States Antarctic Service Expedition 1940–41. (HSM No. 55).

Svend Foyn Island, Ross Sea. 71°52'S, 171°10'E. Message post and box left by the Norwegian (*Tønsberg*) Whaling Expedition 1894–95. (HSM No. 65).

Tryne Islands, Ingrid Christensen Coast. 68°22'34"S, 78°24'33"E. Cairn and flag staff erected by Klarius Mikkelsen in February 1935, commemorating the landing

of Caroline Mikkelsen, first woman to land on Antarctica. (HSM No. 72).
Vostok Station. 78°28′S, 106°48′E. Tractor with plaque commemorating the station's opening in 1957. (HSM No. 11).
Walkabout Rocks, Vestfold Hills. 68°22′S, 78°33′E. Cairn with canister commemorating a visit in 1939 by Sir Hubert Wilkins. (HSM No. 6).
Waterboat Point, Paradise Harbour. 64°49′S, 62°52′W. Remains of the waterboat occupied by two members of the British Imperial Antarctic Expedition 1920–21. (HSM No. 56).
Whalers Bay, Deception Island. 62°59′S, 60°34′W. Plaque honouring Capt. A. A. Andresen, who established whaling on the island in 1906. (HSM No. 58).
Whalers Bay, Deception Island. 62°59′S, 60°34′W. Plaque marking the position of a Norwegian whalers' cemetery (HSM No. 31).
Whalers Bay, Deception Island. 62°59′S, 60°34′W. Remnants of the Norwegian whaling station established in 1912. (HSM No. 71).
Yankee Harbour, Greenwich Island. 62°32′S, 59°45′W. Plaque commemorating Capt. R. MacFarlane, explored in the brigantine *Dragon* in 1820. (HSM No. 57).

SPECIALLY RESERVED AREAS (SRAs)

SRAs are designated to protect representative examples of (a) Antarctica's major geological, glaciological and geomorphological features, and (b) areas of outstanding aesthetic, scenic, and wilderness value. A single plan has been put forward for North Dufek Massif, Pensacola Mountains.

MULTIPLE-USE PLANNING AREAS (MPAs)

MPAs were proposed to coordinate human activities in areas where mutual interference or cumulative environmental effects might prove harmful. A single plan has been put forward for Southwest Anvers Island and vicinity, the area close to and including Palmer Station.

ANTARCTIC SPECIALLY PROTECTED AREAS (ASPAs): ANTARCTIC SPECIALLY MANAGED AREAS (ASMAs)

No ASPA has so far (2001) been designated. A single ASMA management plan has been drafted and adopted for **Admiralty Bay, King George Island, South Shetland Islands,** designed to reduce risks of mutual interference among stations of four nations, and minimize adverse environmental effects. The site includes SSSI 28 and Historic Site 51. The plan is for voluntary compliance only, to be reviewed when Annex V of the Protocol becomes effective.

SEALING ZONES AND RESERVES

Under the **Convention for the Conservation of Antarctic Seals**, sealing is prohibited in each of six nominated zones of the Antarctic region in rotation between 1 September and the last day of February inclusive. Killing or capturing seals is forbidden in three nominated reserves, which are recognized seal breeding areas or sites of long-term research.

ECOSYSTEM MONITORING SITES (EMSs)

Scheduled under the **Convention on the Conservation of Antarctic Marine Living Resources**, their purpose is to preserve representative samples of marine ecosystems to meet the objectives of the Convention. Only two have been nominated. For details see individual entries.

Seal Islands, South Shetland Islands (EMS No. 1)
Cape Shirreff and Telmo Island, South Shetland Islands (EMS No. 2)

Study Guide: Southern Oceans and Islands

INTRODUCTION

The **Southern Ocean** surrounding Antarctica is the world's only annular ocean. Alternatively named the Antarctic Ocean, it is one of the smaller oceans, though it includes some of the world's deepest waters. With an area of about 28 million km^2 (10.94 million sq miles), the ocean is bounded to the south by the Antarctic continent, and to the north by the Antarctic **Convergence** or Polar Front, a zone a few km wide winding between latitudes 47° and 62°S, where cold Antarctic surface waters are over-ridden by warmer southern waters of the Atlantic, Indian and Pacific oceans. The general entry **Southern Ocean** summarizes the ocean's dimensions and main characteristics under four subheadings: *Ocean floor*, *Winds and currents*, *Sea ice*, *Study techniques*. The encyclopedia deals also with southern extremities of the Atlantic, Indian and Pacific oceans (see below), and with groups of islands scattered from the southern continent, across the Southern Ocean, to the warmer waters beyond the Antarctic Convergence.

SHELF SEAS AND COASTS

Hydrographers of different nations have divided the southern waters of the Southern Ocean, i.e. those nearest to the continent, into 13 individual shelf seas. However, no single international authority controls nomenclature, and not all the names and identifications are universally accepted. Boundaries of individual seas are illustrated in the map accompanying **Seas, Antarctic**, together with a listing in clockwise order. Each sea receives an individual entry.

Similarly, Antarctica's coastline is divided into over 50 named lengths of coast. Most border the Southern Ocean directly: the remainder border shelf ice, so the actual boundary between sea and land may underlie several hundred metres of solid ice. Each coast has an individual entry and map in this encyclopedia.

DISCOVERY AND EXPLORATION

Though first visited in the sixteenth century and possibly earlier, the Southern Ocean was first explored, defined and named by Capt. James **Cook** in the mid-eighteenth century. Since then it has been crossed and recrossed by almost every Antarctic expedition. Of these, almost all kept logs recording weather, sea temperature and other basic data, and many included sounding, trawling, dredging and other marine studies in their itineraries. Most drew attention to the rough weather, near-freezing temperatures and **sea ice** that characterize the ocean. The latter in particular proved hazardous to many expedition ships, especially to sailing ships and early steamships with low power and vulnerable propellers. Sea ice effectively limited exploration to a few weeks of the year in favourable areas, and excluded some areas (for example, the Amundsen Sea) altogether from seaborne investigation until the advent of icebreakers. Expeditions, including many of the early sealing voyages, are listed in **Study Guide: Exploration**: individual entries provide accounts of particular voyages. From the late eighteenth century onward **sealers** explored widely and discovered many of the southern islands. Voyages investigating possibilities for **whaling** contributed to Southern Ocean exploration from the 1870s, and whalers made further important contributions during the early decades of the industry up to World War II. Hydrological and biological studies of the Southern Ocean were the main purpose of the **Discovery Investigations**, which involved annual cruises between 1925 and 1939, and produced a comprehensive series of reports.

GEOLOGICAL EVOLUTION

The geological development of the Southern Ocean basin is mentioned in several entries (see **Geology**). In geological history, the annular ocean formed some 180 million years ago, when East Antarctica began to split from Southern Australia and drifted poleward. The resulting circumpolar ring of atmospheric and oceanic surface currents isolated Antarctica from warming influences, starting the chain of events that led to the development of the ice cap. The submarine platforms surrounding some of the southern islands (e.g. **Balleny Islands**, **Iles Kerguelen**) may represent remnants of larger landmasses or archipelagos that, like the present-day Scotia Arc, would to some degree have maintained continuity between the dispersing southern continents. The mid-ocean ridge remains tectonically active: the floor of the Southern Ocean continues to spread (see **Plate tectonics**). Most of the southern islands are of volcanic origin: details will be found under individual entries.

BIOLOGY AND BIOLOGICAL EXPLOITATION

The general entry **Southern Ocean: biology** outlines basic biological processes in the Southern Ocean, under the subheadings *Sea ice and annual productivity, Biological communities, Food chains and webs, Phytoplankton, Zooplankton and other consumers*. The Southern Ocean is biologically rich, well endowed with plankton in summer, with a rich coastal fauna and deep-water benthos. **Krill, fishes, birds, seals** and **whales** are covered by separate entries, as are many individual species (e.g. **Weddell seal**). Oceanic resources, first reported by James Cook, attracted a succession of exploiters from the late eighteenth century onward. See historical accounts of **sealing** and **whaling**: penguins also were exploited for oil. Conservation of living resources (other than whales) in the open ocean is currently effected under the **Antarctic Treaty System**: see **Convention on the Conservation of Antarctic Marine Living Resources** and **Convention for the Conservation of Antarctic Seals**. Full texts of both appear in Appendix B and Appendix C. Whales are protected under regulations promulgated by the International Whaling Commission, which has closed the Southern Ocean to commercial whaling: see **Southern Ocean Sanctuary**. Current exploitation includes fishing for krill, squid and fin-fish. Conservation in waters surrounding some southern islands (e.g. **South Georgia, Iles Kerguelen**) is effected by claimant governments: for details, see entries on individual island groups.

SEA ICE

The southern half to two-thirds of the Southern Ocean freezes over each winter, to a maximum area of some 20 million km^2 (7.8 million sq miles) by September. **Fast ice** is attached to the land; **pack ice** drifts with wind and currents. See the general entry on **Sea ice**, and individual entries covering stages in ice formation, e.g. **frost smoke, nilas, frazil ice, grease ice, pancake ice, first-year ice, brash ice, anchor ice**. Sea ice is subject to constant stresses from wind and currents: see also **lead, pressure ridge, polynya**. From September onward the northern edges of the ice sheet start to melt. The area of ocean covered reduces, slowly at first, then more rapidly, to a summer minimum of about 4 million km^2 (1.56 million sq miles) in summer. This remnant forms the basis of persistent **second-year ice**. The ocean also contains drifting **ice bergs** and **ice islands**, which break from continental glaciers and ice cliffs. Sea ice has many implications for navigation: during the days of sailing ships it limited exploration virtually to less than half the year, and heavy, persistent ice still restricts access to all but icebreakers along several lengths of Antarctica's coastline. It has many implications too for biology, both of the sea (**Southern Ocean: biology**) and of the land. Islands and coasts invested by sea ice are isolated from warming influences of the sea, and are much colder in winter than ice-free coasts in similar latitudes.

THE SOUTHERN ISLANDS

Some 26 groups of islands dot the Southern Ocean and neighbouring southern areas of the Atlantic, Indian and Pacific oceans. Most are widely separated from their neighbours. Nearly all are remarkable for their wealth of wild life, notably breeding colonies of seabirds and seals. Two of the groups have resident human populations that live mainly by farming and fishing: several others have long-standing research stations. Almost all have been modified, some severely, by past human activities, including burning, grazing and the purposeful or accidental introduction of such alien species as fodder grasses, weeds, pigs, goats, cats, dogs, rabbits, rats and mice. All are now to some degree protected by conservation legislation. Several are currently becoming popular for visits by shipborne tourists.

The islands have been described under a bewildering range of regional group names, from 'Antarctic' to 'sub-Antarctic' (a term that, with various forms of spelling, has been used indiscriminately for islands standing anywhere in a broad zone from southern New Zealand to Antarctica) and even 'subtropical'. This account follows a simple, ecologically-based system of classification based on Stonehouse (1982, 1989). The islands considered fall into two major categories:

1. **Southern Ocean islands**: islands that stand south of the Antarctic Convergence, i.e. in the Southern Ocean itself;

2. **Southern temperate islands**: islands standing north of the Convergence, i.e. in cool southern waters of the Atlantic, Indian and Pacific oceans.

Southern temperate islands are so-called because they occupy a zone that, latitudinally and climatically, fits into climatologists' and geographers' classic diagnosis of 'temperate' (e.g. Bailey, 1964). Two of the group (**Macquarie Island** and **Iles Kerguelen**), stand very close to the Convergence, close enough for them from time to time to be washed by Southern Ocean waters as that boundary shifts north and south. Not surprisingly, they share many characteristics of Southern Ocean islands. Practically all the southern temperate islands have close historical, political and biogeographic links with Southern Ocean islands. Some (e.g. **Iles Kerguelen**) are currently glaciated: others show evidence of recent glaciation, and share periglacial features with islands south of the Convergence. Several share species, including plants, insects and marine birds and mammals.

Table 7 includes all the islands and island groups that are given individual entries. Excluded from the table are

Table 7 This table lists all of the southern islands within their two main ecological zones, Southern Ocean and Southern temperate, each of which is divided into two sub-zones. The table shows each group of islands in both its climatic zone and its longitudinal setting, with some characteristics of the vegetation of each category of islands.

	Atlantic sector	Indian sector	Pacific sector	Characteristic vegetation
SOUTHERN TEMPERATE ISLANDS				
Warm Temperate Islands	Tristan da Cunha Nightingale Island Inaccessible Island Gough Island	Island St. Paul Island Amsterdam	Auckland Islands Bounty Islands	Trees, shrubs, bushes and tussock meadows on coasts, shrubs and fellfields on uplands. Soils are mature and fertile, with thick peat deposits
Boundary: 10°C isotherm for the warmest month				
Cool Temperate Islands	Falkland Islands Islands Diego Ramirez	Marion Island Prince Edward Islands Island Crozet Island Kerguelen	Macquarie Island Campbell Island Antipodes Islands	Grasses and herbs form coastal and upland meadows. Small shrubs predominate: soils are mature and fertile, with thin peat deposits.
Boundary: Antarctic convergence				
SOUTHERN OCEAN ISLANDS				
Sub-Antarctic Islands	South Georgia Bouvetøya	Heard Island McDonald Islands		Several species of flowering plants are prominent at sea level. Tussock grass and fellfields predominate, with mature soils patchy.
Boundary: Northern limit of pack ice				
Antarctic Islands	South Shetland Islands South Orkney Islands South Sandwich Islands		Balleny Islands Peter I Øy Scott Island	Ground mostly bare. Only two species of flowering plants in the Atlantic sector only. Algae, lichens and mosses predominate: little or no mature soil.

Antarctic fringe islands, i.e. islands and archipelagos that stand on the continental shelf, physically close enough to Antarctica to come under strong continental influences, and to be included with it biogeographically. Examples are Alexander Island, Charcot Island, Palmer Archipelago and the Windmill Islands. For details of these and other fringe islands, see individual entries.

Southern Ocean islands fall into two categories, (a) Antarctic islands, that are surrounded by sea ice in winter, and (b) Sub-Antarctic islands, that stand north of the northern limit of pack ice, and are relatively ice-free. (This is the only sense in which the term 'sub-Antarctic' is used throughout the encyclopedia.) Chilled by the presence of sea ice for up to seven or eight months of the year, Antarctic islands have longer and colder winters than sub-Antarctic islands, and a correspondingly shorter growing season. They have a correspondingly restricted flora, poorer and less mature **soils**, and a smaller soil fauna and microfauna. A most striking difference is the presence of tussock grasses on the coastal fringes of sub-Antarctic islands (Bouvetøya, with its steep, wave-washed cliffs, is an exception): Antarctic islands are comparatively bare.

The Southern Ocean islands are:

Antarctic islands
South Shetland Islands
South Orkney Islands
South Sandwich Islands
Peter I Øy
Balleny Islands
Scott Island

Sub-Antarctic islands
> South Georgia
> Bouvetøya
> Heard Island
> McDonald Islands

Southern temperate islands are also divided into two categories, (a) Cool temperate islands, that occupy the zone immediately north of the Antarctic Convergence, south of the 10 °C isotherm for the warmest month: and (b) Warm temperate islands, that occupy the zone north of the 10 °C summer isotherm, to an arbitrary boundary of latitude 38°S. With mild climates, relatively free of snow and ice for much of the year, the cool temperate islands support rich lowland meadows of grasses and shrubs. The **Falkland Islands** support farming communities, and farming on others has been inhibited more by isolation, particularly from markets, than by climate. Warm temperate islands are positively mild, and well vegetated (except for the almost-bare Bounty Islands), some with trees and shrubs. **Tristan da Cunha** is the only warm temperate island with a resident human population: farming, fishing and whaling have been tried on several others.

The Southern temperate islands are:

Cool temperate islands
> Falkland Islands
> Islas Diego Ramirez
> Prince Edward Islands
> Iles Crozet
> Iles Kerguelen
> Macquarie Island
> Campbell Island
> Antipodes Islands

Warm temperate islands
> Tristan da Cunha
> Inaccessible Island
> Nightingale Island
> Gough Island
> Ile Amsterdam
> Ile St Paul
> Auckland Islands
> Bounty Islands

A-Z Listing of Encyclopedia Entries

Aagaard, Bjarne	1	Antarctic and Southern Ocean Coalition (ASOC)	7
Abbot Ice Shelf	1	Antarctic Circle	7
Abbott, George P	1	Antarctic fringe islands	8
Ablation	1	Antarctic islands	8
Ablation Point	1	Antarctic Peninsula	8
Ablation Valley	1	Antarctic region	8
Aboa Station	1	Antarctic Sound	9
Accumulation of snow and ice	1	Antarctic Treaty System	9
Adare, Cape	1	Antarctica	12
Adelaide Island	1	Antarctica, East	14
Adelaide (Station T)	1	Antarctica, Greater	14
Adélie Coast	2	Antarctica, Lesser	14
Adélie Land	2	Antarctica Project	14
Adélie penguin	2	Antarctica, West	14
Admiralty Bay	2	Antártida Argentina	14
Admiralty Bay (Base G)	2	Antipodes Islands	14
Admiralty Mountains	3	Anvers Island	14
Agreed Measures for the Conservation of Antarctic Fauna and Flora	3	Anvers Island (Base N)	15
Ahlmannryggen	3	Aramis Range	15
Aiken, Alexander	3	Archer, Colin	15
Aitcho Islands	3	Archer, Walter William	15
Akademik Vernadsky Station	3	Arctowski, Henryk	15
Albatross Island	4	Arctowski Peninsula	15
Albatrosses, southern oceanic	4	Arctowski Station	15
Albedo	4	Ardery Island	15
Alexander Island	4	Ardley, Richard Arthur Blyth	15
Alexandra, Cape	5	Ardley Island	15
Alexandra Mountains	5	Arena Valley	15
Alfred-Faure Station	5	Argentine Islands	15
Allardyce Range	5	Argus, Dome	16
Almirante Brown Station	5	Armitage, Albert Borlase	16
Amery Ice Shelf	5	Armytage, Bertram	16
Ames Range	5	Arrival Heights	16
Amphibolite Point	5	Arrowsmith Peninsula	16
Amsterdam, Ile	5	Arthur Harbour	16
Amundsen, Roald Engelbregt Gravning	6	Artigas Station	16
Amundsen Coast	6	Asgard Range	16
Amundsen-Scott Station	6	ASOC	16
Amundsen Sea	7	Astrolabe Island	16
Anare Mountains	7	Asuka Station	16
Anchor ice	7	Athos Range	16
Andersen Harbour	7	Atkinson, Edward Leicester	16
Anderson, William Ellery	7	Atmosphere	17
Anderson Massif	7	Atmospheric circulation	17
Andersson, Johan Gunnar	7	Atmospheric Heat Transport	18
Andvord Bay	7	Auckland Islands	18
Annenkov Island	7	Aurora australis	19
Antarctic	7	Australasian Antarctic Expedition 1911–14	19
		Australian Antarctic Territory	21

Automatic weather stations	21
Avalanche	21
Avery Plateau	21
Avian Island	21
Bach Ice Shelf	22
Bacharach, Alfred Louis	22
Bage, Robert	22
Bagshawe, Thomas Wyatt	22
Bailey Ice Stream	22
Bailey Peninsula	22
Bakewell, William L	22
Bakutis Coast	22
Balleny, John	23
Balleny Islands	23
Banana belt	24
Banks, Joseph	24
Banzare Coast	24
Barff Peninsula	24
Barlas, William	24
Barne, Michael	24
Barrier	24
Barrier winds	24
Barwick Valley	24
Bastin, François	25
Beacon Dome	25
Beaked whale, southern	25
Bear Peninsula	25
Beaufort Island	25
Béchervaise, John Mayston	25
Beckmann Fjord	25
Bedrock surface of Antarctica	25
Belgian Antarctic Expedition 1897–99	25
Belgicafjella	26
Belgrano II Station	26
Bellingshausen, Fabian Gottlieb Benjamin von	26
Bellingshausen Island	27
Bellingshausen Sea	27
Bellingshausen Station	27
Belt of pack ice	27
Bennett, Messrs. Daniel, and Son	27
Berg	27
Berg, Thomas Erik	27
Berg Ice Stream	27
Bergschrund	27
Bergy bit	27
Berkner, Lloyd Viel	27
Berkner Island	27
Bernacci, Louis Charles	28
Bertram, George Colin Lawder	28
Bertrand, Kenneth J	28
Bickerton, Francis H	28
Bigourdan Fjord	28
Bingham, Edward William	28
Bird Island	28
Bird Island Station (Base BI)	28
Birds, southern oceanic	28
Biscoe, John	32
Biscoe Islands	33
Biscoe Point	33
Bismarck Strait	33
Bjåland, Olav	33
Black Coast	33
Black Island	33
Blaiklock Island	33
Blissett, Harry Arthur	33
Blowing snow	33
Blue whale	33
Blue Whale Harbour	34
Bonner, William Nigel	34
Booth Island	34
Borchgrevink, Carsten Egeberg	34
Borchgrevink Coast	34
Borga Station	35
Borgmassivet	35
Borradaile Island	35
Bottlenose whale, southern	35
Boucot Plateau	35
Bounty Islands	35
Bourgeois Fjord	36
Bouvet de Lozier, Jean-Baptiste Charles	36
Bouvetøya	36
Bowers, Henry Robertson (Birdie)	37
Bowers Mountains	37
Bowling Green Plateau	37
Bowman Coast	37
Boyd Strait	37
Boyd, Vernon Davis	37
Brabant Island	37
Bransfield, Edward	37
Bransfield Island	38
Bransfield Strait	38
Brash ice	38
Breakbones Plateau	38
Bridgeman Island	38
Brisbane, Matthew	38
Bristol Island	38
Bristow, Abraham	38
Britannia Range	38
British Antarctic (*Nimrod*) Expedition 1907–9	38
British Antarctic (*Southern Cross*) Expedition 1898–1900	39
British Antarctic (*Terra Nova*) Expedition 1910–13	40
British Antarctic Survey (BAS)	42
British Antarctic Territory	43
British, Australian and New Zealand Antarctic Research Expedition 1929–31	43
British Graham Land Expedition 1934–37	44

British Imperial Antarctic Expedition 1920–22	45
British National Antarctic Expedition 1901–4	45
British Naval Expedition 1839–43	46
Brocklehurst, Sir Philip Lee, Bart	47
Bruce, Wilfred Montague	48
Bruce, William Spiers	48
Bruce Plateau	48
Brunt Ice Shelf	48
Bryan Coast	48
Buckle Island	49
Bucknell, Ernest Selwyn	49
Budd Coast	49
Bunger Hills	49
Burdwood Bank	49
Bursey, Jacob	49
Byers Peninsula	49
Byrd, Richard Evelyn, Jr	49
Byrd, Richard E. III	50
Byrd's First Antarctic Expedition 1928–30	50
Byrd's Second Antarctic Expedition 1933–35	51
Byrd Station	52
Caird Coast	53
California Plateau	53
Calving	53
Campbell, Victor Lindsey Arbuthnot	53
Campbell Island	53
Campbell Island Station	54
Canada Glacier	54
Canadian sealing expedition 1901–2	54
Candlemas Islands	54
Cape Geddes (Base C)	54
Capitán Arturo Prat Station	54
Carvajal Station	54
Casey Station	54
Caughley Beach	54
Cavendish, Thomas	55
CCAMLR	55
CCAS	55
Challenger Expedition 1872–76	55
Chanticleer Expedition 1828–29	55
Charcot, Jean-Baptiste Etienne Auguste	55
Charcot Island	56
Charcot, Port	56
Charlie, Dome	56
Cheeseman, Al	56
Cheetham, Alfred	56
Cherry-Garrard, Apsley	56
Chinstrap penguin	57
Christensen, Lars	57
Churchill Mountains	58
Cierva Point	58
Circumcision, Port	58
Cirque Fjord	58
Cirque glacier	58
Clarence Island	58
Clarie Coast	58
Clark, Robert Selbie	58
Clark Peninsula	58
Clerke Rocks	58
Climate change	58
Climatic data: sources	59
Climatic records	59
Climatic zones	60
Clissold, Thomas Charles	65
Clo unit	65
Clothier Harbour	65
Clowes, Archibald John	65
Coal Harbour	65
Coasts, Antarctic	65
Coats Land	66
Colbeck, William	66
Colbeck, William Robinson	66
Colbeck Archipelago	66
Cold climate survival strategies	66
Cold, human responses	67
Colonization of Antarctic environments	68
Coman, Francis Dana	68
Commandante Ferraz Station	68
Commerson's dolphin	68
Commonwealth Meteorological Expedition 1913–15	68
Commonwealth Range	68
Commonwealth Trans-Antarctic Expedition 1955–58	68
Community development	70
COMNAP	70
Compañia Argentina de Pesca	70
Concordia Station	71
Convention for the Conservation of Antarctic Seals	71
Convention on the Conservation of Antarctic Marine Living Resources (CCAMLR)	72
Convention on the Regulation of Antarctic Mineral Resource Activities (CRAMRA)	74
Convergence, Antarctic	74
Cook, James	75
Cook Ice Shelf	76
Cook Island	76
Cool temperate islands	77
Cooper Sound	77
Cooperation Sea	77
Coppermine Peninsula	77
Cormorants, southern oceanic	77
Cornice	77
Coronation Island	77
Cosgrove Ice Shelf	77
Cosmonaut Sea	77
Cotton Plateau	78
Coughtrey Peninsula	78

Entry	Page
Council of Managers of National Antarctic Programmes (COMNAP)	78
Couzens, Thomas	78
Covadonga Harbour	78
Crabeater seal	78
CRAMRA	78
Crean, Thomas	78
Crevasse	78
Cross, Jacob	79
Crosson Ice Shelf	79
Crozet, Iles	79
Crozier, Cape	79
Crust	79
Cruzon, Richard L	79
Cryptogam Ridge	80
Crystal Sound	80
Currents, convergences and divergences	80
Czegka, Victor H	80
Dakshin Gangotri	81
Dallmann Bay	81
Dallmann Laboratory	81
Damoy Point	81
Dana, James Dwight	81
Danco Coast	81
Danco Island (Base O)	81
Darwin Mountains	81
Dater, Henry Murray	82
David, Tannat William Edgeworth	82
Davis, John	82
Davis, John King	82
Davis Coast	82
Davis Sea	82
Davis Station	83
Davis Valley Ponds	83
Davys, John	83
Day length	83
Deacon, George Edward Raven	84
Debenham, Frank	84
De Bougainville, Comte Louis-Antoine	84
De Brosses, Charles	84
Decepción Station	84
Deception Island	84
Deception Island (Base B)	85
De Gerlache de Gomery, Baron Adrian Victor	85
De Haven, Edwin Jesse	85
Dell, James William	85
Detaille Island (Base W)	86
Detroit Plateau	86
Dickason, Harry	86
Diego Ramirez, Islas	86
Dion Islands	86
Discovery Bay	86
Discovery Investigations (1925–51)	86
Discovery Sound	87
Diving in birds and mammals	87
Diving petrels	88
Dobrowolski, Antoni Bolesaw	88
Dobson unit	88
Dome Fuji Station	88
Dominion Range	88
Doorly, Gerald Stokely	88
Dotson Ice Shelf	88
Douglas, Eric	89
Douglas, George Vibert	89
Doumer Island	89
Dovers, George Harris Sargeant	89
Drake, Francis	89
Drake Passage	89
Dronning Fabiolafjella	89
Dronning Maud Land	89
Druzhnaya stations	90
Drygalski, Erich von	90
Drygalski Fjord	90
Ducks and geese, southern oceanic	90
Dufek, George J	90
Dufek Coast	90
Dufek Massif	91
Dumont d'Urville, Jules Sébastien César	91
Dumont d'Urville Sea	91
Dumont d'Urville Station	91
Dundee Whaling Expedition 1892–93	91
Dusky dolphin	92
Dyer Plateau	92
Earthquakes and earth movements	93
East Antarctica	93
Edholm, O.G	93
Edward VII Peninsula	93
Edward VIII Plateau	93
Eielson Peninsula	94
Eights, James	94
Eights Coast	94
Eights Station	94
Eklund, Carl Robert	94
Eklund Islands	94
Ekström, Bertil. A. W	94
Ekströmisen	95
Elephant Island	95
Elephant seal	95
Ellefsen Harbour	95
Ellsworth, Lincoln	95
Ellsworth Land	95
Ellsworth Mountains	95
Ellsworth Station	95
Ellsworth's Antarctic Expeditions 1933–39	95
Emperor Island	96
Emperor penguin	96

Enderby Land	97
Enderby, Messrs	97
Enderby Settlement	97
England, Rupert G	97
English Coast	97
English Strait	98
Enterprise Island	98
Equilibrium line	98
Eratosthenes of Alexandria	98
Erebus, Mt	98
Escudero Station	98
Esperanza Station	98
Esther Harbour	98
European Project for Ice Coring in Antarctica (EPICA)	98
Evans, Edgar	98
Evans, Edward Ratcliffe Garth Russell	99
Evans, Hugh Blackwall	99
Evans, Cape	99
Evans Ice Stream	99
Falkland Harbour	100
Falkland Islands	100
Falkland Islands and Dependencies Aerial Survey Expeditions 1955–57 (FIDASE)	101
Falkland Islands Dependencies	101
Falkland Islands Dependencies Survey (FIDS) 1945–61	102
Falla, Robert	102
Fallières Coast	102
Faraday Station (Base F)	103
Fast Ice	103
Faure Islands	103
Ferrar, Hartley Travers	103
Filchner, Wilhelm	103
Filchnerfjella	103
Filchner Ice Shelf	103
Fildes Peninsula	103
Fildes Strait	104
Fimbulisen	104
Fin whale	104
Firn	104
First-year ice	104
Fishes, Southern Ocean	104
Fitzsimmons, Roy	104
Fleming, William Launcelot Scott	105
Flight in Antarctic seabirds	105
Floe	105
Flood Range	105
Flora, Mt	105
Flowering plants, Antarctic	105
Forbidden Plateau	106
Ford, Charles Reginald	106
Ford Massif	106
Ford Ranges	106
Forde, Robert	106
Forlidas Pond	106
Forrestal Range	106
Forster, Johann Georg Adam	106
Forster, Johann Reinhold	106
Fosdick Mountains	107
Fossil Bluff Station (Base KG)	107
Foster Plateau	107
Foster, Port	107
Foundation Ice Stream	107
Foyn Coast	107
Foyn Harbour	107
Framnes Mountains	107
Francis, Samuel John	107
Franklin Island	107
Frazier, Russell G	107
Frazil ice	108
Frei Station	108
French Antarctic (*Français*) Expedition 1903–5	108
French Antarctic (*Pourquoi Pas?*) Expedition 1908–10	108
French Naval Expedition 1837–40	109
Fridtjof Sound	110
Friis, Herman R	110
Frolov, Vyacheslav Vasil'yevich	110
Frost smoke	110
Fuchs, Vivian Ernest	110
Fuchs Dome	110
Fuchs Ice Piedmont	110
Fulmars, southern oceanic	110
Furneaux, Tobias	111
Fur seals	111
Gabriele de Castilla	112
Gadfly petrels	112
Gaimard, Joseph-Paul	112
Gamburtsev Subglacial Mountains	112
Gaussberg	112
General Belgrano stations	112
General Bernardo O'Higgins Station	112
General Ramon Cañas Montalva Station	113
General San Martín Station	113
Gentoo penguin	113
Geographic Pole, South	113
Geology: geological history and palaeontology	113
Geology: stratigraphy and structure	115
Geomagnetic Pole, South	117
Geophysical techniques	118
Georg Forster Station	119
Georg von Neumayer Station	119
George V Coast	119
George V Land	119
George VI Ice Shelf	119
George VI Sound	119
Gerlache Strait	119

German Antarctic (*Deutschland*) Expedition 1911–13	119
German Antarctic (*Schwabenland*) Expedition 1938–39	120
German Deep Sea Expedition 1898–99	120
German International Polar Year Expedition 1882–83	120
German South Georgia Expedition 1928–29	121
German South Polar Expedition 1901–3	121
German Whaling and Sealing Expedition 1873–74	121
Getz Ice Shelf	122
Giæver, John Schelderup	122
Giant petrels	122
Gibbs Island	122
Gibson-Hill, Carl Alexander	122
Gilbert Strait	122
Girev, Dmitrii Semenovich	122
Gjelsvikfjella	122
Glaciers and ice streams	122
Glacier Strait	123
Glacier tongue	123
Glaze ice	123
Glossopteris	123
Gold Harbour	124
Goodale, Edward E	124
Goudier Island	124
Gough Island	124
Gough Island Scientific Survey 1955–56	124
Gough Island Station	124
Gould, Laurence McKinley	125
Gould, Rupert Thomas	125
Gould Coast	125
Gouvernøren Harbour	125
Graham Coast	125
Graham Land	126
Gran, Tryggve	126
Granite Harbour	126
Grearson Oasis	126
Grease ice	126
Great Wall Station	126
Green, Charles John	126
Green Island	126
Greene, Stanley Wilson	126
Greenhouse effect	127
Greenwich Island	127
Gressitt, Linsley	127
Grierson, John	127
Grindley Plateau	127
Grove Mountains	127
Growler	127
Grytviken	127
Grytviken (Base M)	127
Guesalaga Peninsula	128
Gulls, southern oceanic	128
Hail	129
Half Moon Island	129
Hallett, Cape	129
Hallett Station	129
Halley Station (Base Z)	129
Hanson, Malcolm P	129
Hanssen, Helmer Julius	129
Harbord, Arthur Edward	129
Hardy, Alister Clavering	130
Hare, Clarence H	130
Harmony Point	130
Haslop, Gordon Murray	130
Hassel, Helge Sverre	130
Haswell Islands	130
Hatherton, Trevor	130
Heard Island	130
Heard Island Station	131
Hearst Island	131
Heimefrontfjella	131
Henkes Islands	131
Henryk Arctowski Station	131
Herbert Plateau	131
Herdman, Henry Franceys Porter	131
Heritage Range	131
Highjump Archipelago	132
Hillary Coast	132
Hjort, Johan	132
Hoar frost	132
Hobbs, William Herbert	132
Hobbs Coast	132
Hoelfjella	132
Holland Range	133
Hollick-Kenyon Peninsula	133
Hollick-Kenyon Plateau	133
Home of the blizzard	133
Hooker, Joseph Dalton	133
Hooker's (New Zealand) sea lion	133
Hooper, Frederick J	133
Hope Bay (Base D)	133
Horlick Ice Stream	133
Horlick Mountains	133
Horseshoe Harbour	134
Horseshoe Island	134
Horseshoe Island (Base Y)	134
Hoseason Island	134
Hourglass dolphin	134
Hovgaard Island	134
Hudson, Hubert T	134
Hughes Range	134
Human occupation	134
Hummock	135
Humpback whale	135
Hunter, John George	136
Hurley, James Francis	136

Hussey, Leonard Duncan Albert	136
Husvik Harbour	136
Hut Point	136
Hutton Mountains	136
IAATO	137
Ice	137
Iceberg	137
Ice blink	138
Ice cap	138
Ice core	138
Ice core analysis	138
Ice edge	139
Ice fall	139
Ice field	139
Ice fog	139
Ice foot	139
Ice front	139
Ice Island	139
Ice limit	139
Ice monitoring by remote sensing	139
Ice piedmont	140
Ice port	140
Ice prisms	140
Ice rind	140
Ice sheet	140
Ice sheet, Antarctic	140
Ice sheet, Antarctic: history and development	142
Ice sheet movement	144
Ice shelf	144
Ice stream	144
Ice tongue	144
Imperial Trans-Antarctic Expedition 1914–17	144
Inaccessible Island	146
Inexpressible Island	147
Ingrid Christensen Coast	147
Institute Ice Stream	147
International Association of Antarctica Tour Operators	147
International Geographical Congress, Sixth	147
International Geophysical Year 1957–58	147
Inverleith Harbour	148
Iroquois Plateau	148
Islands, non-existent	148
Isostasy	149
Isotope	149
Jack, Andrew Keith	150
James, Reginald William	150
James Ross Island	150
Japanese Antarctic Expedition 1910–12	150
Jeannel, René	150
Jelbart, John Ellis	151
Jelbartisen	151
Jenny Island	151
Johnson, Charles Ocean	151
Johnston, T. Harvey	151
Johnston, William	151
Joint Services Expedition to Brabant Island 1983–85	151
Joinville Island	151
Jones, Sydney Evan	151
Jones Mountains	151
Joyce, Ernest Edward Mills	152
Juan Carlos Island	152
Jubany Station	152
June, Harold I	152
Katabatic wind	153
Keller Peninsula	153
Kelly Plateau	153
Kemp, Peter	153
Kemp Coast	153
Kemp Land	154
Keohane, Patrick	154
Kerguelen, Iles	154
Kernlos effect	155
Kershaw, John Edward Giles	155
Killer whale	155
King, Philip Parker	155
King George Island	155
King Håkon VII Sea	155
King Leopold and Queen Astrid Coast	156
King penguin	156
King Sejong Station	156
Kirkwood, Harold	156
Knox Coast	156
Kohl Plateau	157
Kohler Range	157
Komsomol'skaya Station	157
Kraulberga	157
Krebs, Manson	157
Krill, Antarctic	157
Kronprins Olav Kyst	157
Kronprinsesse Märtha Kyst	158
Laclavère Plateau	159
Lagotellerie Island	159
Lakes, Antarctic	159
Lakes beneath the ice sheet	159
Lallemand Fjord	160
Lamb, Ivan Mackenzie	160
Land birds, southern oceanic island	160
Lands, Antarctic	161
Langhovde	161
Larkman, Alfred Herbert	161
Lars Christensen Coast	161
Lars Christensentoppen Peak	161
Larsemann Hills	161

Larsen Harbour	162	Mac.Robertson Land	168
Larsen Ice Shelf	162	Madigan, Cecil Thomas	168
Laseron, Charles Francis	162	Magellan, Strait of	169
Lashly, William	162	Magellanic penguin	169
Lassiter Coast	162	Magnetic anomalies	169
Latady Island	162	Magnetic Pole, South	169
Latitude and longitude distances	162	Magnetism, terrestrial	170
Laubeuf Fjord	162	Maitri Station	170
Laurie Island	162	Marie Byrd Land	170
Law Base	162	Marion du Fresne, Marc Macé	170
Law Dome	163	Marion Island	170
Lazarev Ice Shelf	163	Marion Island Station	170
Lazarev Sea	163	Maritime Antarctic	171
Lazarev Station	163	Markham Plateau	171
Lead	163	Marr, James William Slesser	171
Lecointe Island	163	Marsh Station	171
Leith Harbour	163	Marshall, Edward Hillis	171
Lemaire Channel	163	Marshall, Eric Stewart	171
Léonie Islands	163	Marshall Archipelago	171
Leopard seal	163	Marston, George Edward	171
Leskov Island	163	Martin, Port	171
Lester, Maxime Charles	163	Martin-de-Viviès Station	171
Levick, George Murray	164	Mason, Douglas P	171
Lewis Bay	164	Matha Strait	172
Lewthwaite Strait	164	Mather, John Hugh	172
Lichens	164	Matthews, Leslie Harrison	172
Lidke Ice Stream	164	Mauger, Clarence Charles	172
Liège Island	164	Mawson, Douglas	172
Linnaeus Terrace	164	Mawson Coast	172
Lion Island	165	Mawson Escarpment	173
Lion Sound	165	Mawson Sea	173
Lions Rump	165	Mawson Station	173
Litchfield Island	165	McCarthy, Mortimer	173
Little America stations	165	McCarthy, Timothy	173
Livingston Island	165	McDonald Islands	173
Lockroy, Port	165	McFarlane Strait	173
Longhurst Plateau	165	McIlroy, James Archibald	174
Loranchet, Jean	165	McKenzie, Edward A	174
Loubet Coast	165	McKinley, Ashley C	174
Louis Philippe Plateau	165	McMurdo Ice Shelf	174
Luitpold Coast	166	McMurdo Oasis	174
Lynch Island	166	McMurdo Sound	174
Lystad, Isak	166	McMurdo Station	174
		Meinardus, William	174
Mabus Point	167	Melba Peninsula	174
Macaroni penguin	167	Melbourne, Mt	174
Macfie Sound	167	Melchior Islands	175
Machu Picchu Station	167	Melchior Station	175
Mackellar Islands	167	Melsom, Henrik Govenius	175
Mackenzie, Kenneth Norman	167	Meteorites	175
Mackintosh, Neil Alison	167	Michigan Plateau	175
Macklin, Alexander Hepburne	167	Microclimate	175
Macquarie Island	168	Microhabitat	175
Macquarie Island Station	168	Migration, seasonal	175

Mikkelsen Harbour	176
Miller, Joseph Holmes	176
Miller Range	176
Minerals and mining in Antarctica	176
Minke whale	177
Mirnyy Station	178
Mizuho Plateau	178
Moe Island, South Orkney Islands	178
Mollymawks	178
Molodezhnaya Station	178
Moltke Harbour	178
Montagu Island	178
Moraine	178
Morgan, Charles Gill	178
Morton Strait	178
Moss mires	178
Moss peat	179
Moyes, Morton Henry	179
Mühlig-Hofmannfjella	180
Mule Peninsula	180
Mulock, George Francis Arthur	180
Murphy, Robert Cushman	180
Murray Harbour	180
Mushroom Island	180
Nansen, Fridtjof	181
Nansen Island	181
Napier Mountains	181
Nares, George Strong	181
Nash Range	181
National expeditions	181
Neko Harbour	182
Nelson, Andrew Laidlaw	182
Nelson Island	182
Neptune Range	182
Neumayer Channel	182
Névé	182
New College Valley	182
New Harbour	182
New ice	182
New Zealand Cape Expeditions 1941–45	182
Nickerson Ice Shelf	183
Nightingale Island	183
Nilas	183
Nilsen Plateau	183
Nordenskjöld Coast	183
Normanna Strait	183
Norsk Polarinstitutt	183
Norway Station	183
Norwegian Antarctic (*Brategg*) Expedition 1947–48	183
Norwegian-British-Swedish Antarctic Expedition 1949–52	184
Norwegian (Christensen) Whaling Expeditions 1927–37	184
Norwegian (Sandefjord) Whaling Expeditions 1892–94	186
Norwegian South Polar Expedition 1910–12	186
Norwegian (Tønsberg) Whaling Expedition 1893–95	187
Novolazarevskaya Station	188
Nunatak	188
Oases	189
Oates, Lawrence Edward Grace	189
Oates Coast	189
Oates Land	190
Oazis Station	190
Ocean Harbour	190
Odbert Island	190
Oddera, Alberto J	190
O'Higgins Station	191
Ohio Range	191
Ohridiskii Station	191
Olympus Range	191
Omelchenko, Anton	191
Ommanney, Francis Downes	191
Ongul Island	191
Operation Deep-Freeze	191
Operation Highjump	191
Operation Tabarin 1943–45	191
Operation Windmill	192
Orca	192
Orcadas Station	192
Orde-Lees, Thomas Hans	192
Orléans Strait	192
Orne Harbour	192
Orville Coast	192
Orvinfjella	192
Oscar II Coast	192
Outlet glacier	193
Oxygen isotope ratios	193
Ozawa, Keijiro	193
Ozone hole, Antarctic	193
Pack ice	195
Palmer, Nathaniel Brown	195
Palmer Archipelago	195
Palmer Coast	195
Palmer Land	195
Palmer Station	195
Pancake ice	195
Paradise Harbour	196
Parker, Alton N	196
Patterned ground	196
Patuxent Ice Stream	196
Patuxent Range	196
Paulet Island	196
Peacock Sound	196
Peale's (blackchin) dolphin	196

Peats and peat formation	196	Prince Charles Strait	203
Pedro Vicente Maldonado	197	Prince Edward Islands	204
Pendleton Strait	197	Prince Olav Harbour	204
Penfold, David	197	Prince Olav Mountains	204
Penguin Island	197	Princess Elizabeth Land	204
Penguins, southern oceanic	197	Prins Harald Kyst	204
Pennell Coast	197	Prinsesse Astrid Kyst	205
Penola Strait	198	Prinsesse Ragnhild Kyst	205
Pensacola Mountains	198	Prions, southern oceanic	205
Periantarctic islands	198	Proclamation Island	206
Permafrost	198	Progress Station	206
Peter I Øy	198	Prospect Point (Base J)	206
Petermann Island	198	Protocol on Environmental Protection to the	
Peters, Nikolaus	199	Antarctic Treaty	206
Petersen, Carl, O	199	Prudhomme, André	207
Petrel Station	199	Publications Ice Shelf	207
Petrels, southern oceanic	199		
Phleger Dome	199	Quar, Leslie Arthur	208
Pickersgill Islands	199	Quarisen	208
Piedmont glacier	199	Quartermain, Leslie Bowden	208
Pieter J. Lenie Station	199	Queen Alexandra Range	208
Piked whale	199	Queen Elizabeth Range	208
Pionerskaya Station	199	Queen Mary Coast	208
Pirie, James Hunter Harvey	199	Queen Mary Land	208
Plate tectonics	200	Queen Maud Land	209
Plateau Station	200	Queen Maud Mountains	209
Pléneau Island	200	Quoy, Jean-René Constant	209
Plumley, Frank	200		
Pobeda Ice Island	200	Radio echo sounding	210
Pointe Géologie, Archipel de	200	Radiometric dating	210
Pole of Inaccessibility	201	Rafting	211
Polynya	201	Rankin, Arthur Niall	211
Pomona Plateau	201	Rasmussen, Johan	211
Port aux Français	201	Rawson Plateau	211
Port Lockroy (Base A)	201	Raymond, John East	211
Portal Point Refuge	202	Reece, Alan	211
Porthos Range	202	Richards, Richard Walter	211
Possession Islands	202	Right whale, southern	211
Potter Peninsula	202	Riiser-Larsen, Hjalmar	212
Pourquoi Pas Island	202	Riiser-Larsen Halvøya	212
Powder snow	202	Riiser-Larsen Sea	212
Powell, George	202	Riiser-Larsenisen	212
Powell Island	202	Rime	212
Prebble, Michael	203	Ripamonti (Luis Ripamonti)	212
Presidente Eduardo Frei Station	203	Risopatron	212
Presidente González Videla Station	203	Ritscher, Alfred	212
Pressure ice	203	Robert Island	212
Pressure ridge	203	Roberts, Brian Birley	212
Priestley, Raymond Edward	203	Roberts, Cape	213
Primavera Station	203	Robertson, MacPherson	213
Primero de Mayo Station	203	Rockefeller Mountains	213
Prince Albert Mountains	203	Rockefeller Plateau	213
Prince Andrew Plateau	203	Rockhopper penguin	213
Prince Charles Mountains	203	Rogers, Allan Frederick	213

Roi Baudouin Station	214
Rongé Island	214
Ronne, Finn	214
Ronne Antarctic Research Expedition 1947–48	214
Ronne Ice Shelf	214
Rookery Islands	214
Rooney, Felix	214
Roosevelt Island	214
Rosamel Island	214
Rosita Harbour	215
Ross, James Clark	215
Ross Archipelago	216
Ross Dependency	216
Ross Island	216
Ross Ice Shelf	216
Ross Sea	216
Ross seal	216
Rothera Point	216
Rothera Station	216
Rothschild Island	216
Royal penguin	216
Royal Society International Geophysical Year Antarctic Expedition 1955–57	216
Royal Society Range	217
Royds, Charles William Rawson	217
Royds, Cape	217
Rudmose-Brown, Robert Neal	217
Ruppert Coast	218
Russkaya Station	218
Russian Naval Expedition 1819–21	218
Rutford Ice Stream	219
Rymill, John Riddoch	219
Rymill Coast	219
Sabrina Coast	220
Sabrina Island	220
Saffery, John Hugh	220
Salvesen, Harold Keith	220
Salvesen Range	220
Salvesens of Leith	220
Sandefjord Bay (Base C, later Base P)	221
SANAE Stations	221
San Martin Station	221
San Telmo Island	221
Sarie Marais Station	221
Sastrugi	221
Satellite imagery	221
Saunders, Harold E	222
Saunders Coast	222
Saunders Island	222
SCALOP	223
SCAR	223
Schirmacheroasen	223
Schouten, Willem Korneliszoon	223
Scientific Committee on Antarctic Research (SCAR)	224
Scotia Arc	224
Scotia Sea	224
Scott, Robert Falcon	224
Scott Base	225
Scott Coast	225
Scott Island	225
Scott Mountains	226
Scott Polar Research Institute	226
Scottish National Antarctic Expedition 1902–4	226
Scrimshaw, Antarctic and southern	227
Sea ice: Southern Ocean	227
Seal Islands	228
Sealing in the Southern Ocean	229
Sealing zones and reserves	229
Seals, southern oceanic	229
Seas, Antarctic	230
Seasonality	230
Seasons	230
Second-year ice	231
Sei whale	231
Sentinel Range	232
Seymour Island	232
Shackleton, Ernest Henry	232
Shackleton Coast	233
Shackleton Ice Shelf	233
Shackleton Range	233
Shackleton–Rowett Antarctic Expedition 1921–22	234
Shackleton Station	234
Shag Rocks	234
Shcherbakov, Dimitriy Ivanovich	234
Shearwaters	234
Sheathbills	234
Shimizu Ice Stream	235
Shirase, Choku	235
Shirase Coast	235
Shirreff, Cape	235
Shore Lead	235
Shuga	235
Signy Island	235
Signy Station (Base H)	235
Simpson, George Clarke	235
Siple, Paul Allman	236
Siple Coast	236
Siple Island	236
Siple Station	236
Sites of Special Scientific Interest (SSSIs)	236
Skelton, Reginald	236
Skottsberg, Carl Johan Fredrik	237
Skuas, southern oceanic	237

Entry	Page
Slava Ice Shelf	237
Sledge dogs	237
Sleet	238
Smith Island	238
Snow	238
Snow algae	239
Snow bridge	239
Snow drift	239
Snow field	239
Snow Hill Island	239
Snow Island	239
Snow line	239
Snow petrels	239
Sobral, José Maria	239
Sobral Station	239
Soils	239
Solander, Daniel Carl	240
Solar radiation	240
Somov, Mikhail Mikhaylovich	241
Sooty albatrosses	242
Sør Rondane	242
South American sea lion	242
South Georgia	242
South Georgia Survey Expeditions 1951–57	243
South Orkney Islands	243
South Polar Times	244
South Sandwich Islands	244
South Shetland Islands	245
Southern elephant seal	246
Southern islands	246
Southern lights	247
Southern Ocean	247
Southern Ocean: biology	249
Southern Ocean islands	251
Southern Ocean Sanctuary	252
Southern temperate islands	252
Southern Thule	252
Southwest Anvers Island and vicinity	252
Sovereignty in the Antarctic region	253
Sovetskaya Station	255
Soyuz Station	255
Spaatz Island	255
Specially Protected Areas (SPAs)	255
Sperm whale	255
Spring warming	255
Standing Committee on Antarctic Logistics and Operations (SCALOP)	255
Stanford Plateau	255
Stange Ice Shelf	256
Stefansson Sound	256
Stenhouse, Joseph Russel	256
Stephenson, Alfred	256
Stevens, Alexander	256
Stewart VII, Duncan	256
Stillwell, Frank Leslie	256
St Kliment Ohridiski Station	257
Stonington Island	257
Stonington Island Station (Base E)	257
Storm-petrels	257
St Paul, Ile	257
Strand crack	258
Stratton, David George	258
Stromness Harbour	258
Sturge Island	258
Sub-Antarctic	258
Sub-Antarctic islands	258
Sulzberger Ice Shelf	258
Svarthamaren	258
Svea Station	258
Sverdrup, Harald Ulrik	258
Sverdrupfjella	258
Swedish South Polar Expedition 1901–4	259
Syowa Station	260
Tabarin Peninsula	261
Tasman, Abel Janszoon	261
Taylor, Thomas Griffith	261
Taylor Rookery, Mac.Robertson Land	261
Taylor Valley	261
Temperature inversion	261
Teniente Cámara Station	261
Teniente Carvajal Station	261
Teniente Jubany Station	262
Teniente Matienzo Station	262
Teniente Rodolfo Marsh (Escudero) Station	262
Terns, southern oceanic	262
Terra australis incognita	262
Terra Nova Bay (TNB) Station	262
Terre Adélie	263
Terres australes et antarctiques françaises (TAAF)	263
Thala Hills	263
Thermoregulation	263
Theron Mountains	264
Thomas, Charles W	264
Thompson, Andrew A	264
Thomson, Charles Wyville	264
Thule Island	264
Thule Islands	264
Thurston Island	264
Tide crack	265
Tierra de O'Higgins	265
Tierra San Martín	265
Till	265
Tomb	265

Tonkin, John Eliot	265
Toothfish	265
Tor Station	265
Tordesillas, Treaty of	265
Torgersen Island	265
Tourism, Antarctic	265
Tourmaline Plateau	267
Tower Island	267
Tractors, early	267
Tramway Ridge	268
Transantarctic Mountains	268
Traversay Islands	268
Trinity Peninsula	268
Tristan da Cunha group	269
Troll Station	269
Tula Mountains	269
Undine Harbour	270
United States Antarctic Service Expedition 1939–41	270
United States Exploring Expedition 1838–42	270
United States Navy Antarctic Development Project 1946–47	271
United States Navy Antarctic Expeditions (Operation Deep-Freeze) 1955–98	272
United States Navy Second Antarctic Development Project 1947–48	273
Usarp Mountains	273
Valkyrjedomen	274
Van Oordt, Gregorius Johannes	274
Vanda, Lake	274
Vanda Station	274
Vega Island	274
Vegetation, subfossil	274
Venable Ice Shelf	274
Vernadsky Station	275
Vestfjella	275
Vestfold Hills	275
Vicecomodoro Marambio Station	275
Victoria Land	275
Victoria Land Dry Valleys	275
Victory Mountains	275
View Point (Base V)	276
Vindication Island	276
Vinson Massif	276
Vishniac, Wolf V	276
Visokoi Island	276
Volcanic activity	276
Vostok, Lake	276
Vostok Station	277
Voyeykov Ice Shelf	277

Waite, Amory H. (Bud)	278
Walgreen Coast	278
Walker, Paul	278
Warburton, Keith	278
Warm temperate islands	278
Wasa Station	278
Washington Strait	278
Watson, Andrew Dougal	279
Wauwermans Islands	279
Way Archipelago	279
Weather, Antarctic	279
Weather buoys	280
Weather data	280
Weather stations	280
Weather systems	281
Weathering	281
Webb, Eric N	281
Weddell, James	282
Weddell Sea	283
Weddell seal	283
Welcome Islands	283
West Antarctica	283
West Ice Shelf	283
Wexler, Harry	284
Whales, Southern Ocean	284
Whaling in the Southern Ocean	285
White Island	288
Whiteout	288
Whitmore Mountains	288
Wiencke Island	288
Wild, Harry Ernest	288
Wild, John Robert Francis (Frank)	288
Wilhelm Archipelago	288
Wilhelm II Coast	288
Wilhelm II Land	289
Wilkes, Charles	289
Wilkes Coast	290
Wilkes Land	290
Wilkes Station	290
Wilkins, George Hubert	290
Wilkins Coast	291
Wilkins-Hearst Antarctic Expeditions 1928–30	291
Wilkins Ice Shelf	292
Wilkins Sound	292
Wilkinson, John Valentine	292
William Scoresby Archipelago	292
Willis Islands	292
Wilson, Edward Adrian	292
Wind chill	293
Windmill Islands	294
Wisconsin Plateau	294
Wisconsin Range	294

Wisting, Oscar	294	Yalour Sound	296
Wohlthat Massivet	294	Yankee Harbour	296
Worcester Range	294	Yelcho	296
Wordie, James Mann	294	Young Island	296
Wordie Ice Shelf	294	Yukidori Valley	296
World War II in the southern islands	294		
Worsley, Frank Arthur	295	Zavodovski Island	297
Wright, Charles Seymour	295	Zhongshan Station	297
Wright Valley	295	Zumberge Coast	297